T0223499

Lecture Notes in Computer Science 9492

Commenced Publication in 1973
Founding and Former Series Editors:
Gerhard Goos, Juris Hartmanis, and Jan van Leeuwen

Editorial Board

David Hutchison
Lancaster University, Lancaster, UK
Takeo Kanade
Carnegie Mellon University, Pittsburgh, PA, USA
Josef Kittler
University of Surrey, Guildford, UK
Jon M. Kleinberg
Cornell University, Ithaca, NY, USA
Friedemann Mattern
ETH Zurich, Zürich, Switzerland
John C. Mitchell
Stanford University, Stanford, CA, USA
Moni Naor
Weizmann Institute of Science, Rehovot, Israel
C. Pandu Rangan
Indian Institute of Technology, Madras, India
Bernhard Steffen
TU Dortmund University, Dortmund, Germany
Demetri Terzopoulos
University of California, Los Angeles, CA, USA
Doug Tygar
University of California, Berkeley, CA, USA
Gerhard Weikum
Max Planck Institute for Informatics, Saarbrücken, Germany

More information about this series at http://www.springer.com/series/7407

Sabri Arik · Tingwen Huang
Weng Kin Lai · Qingshan Liu (Eds.)

Neural
Information Processing

22nd International Conference, ICONIP 2015
Istanbul, Turkey, November 9–12, 2015
Proceedings, Part IV

 Springer

Editors

Sabri Arik
University of Istanbul
Istanbul
Turkey

Tingwen Huang
University at Qatar
Doha
Qatar

Weng Kin Lai
Tunku Abdul Rahman University College
Kuala Lumpur
Malaysia

Qingshan Liu
University of Science Technology
Wuhan
China

ISSN 0302-9743   ISSN 1611-3349   (electronic)
Lecture Notes in Computer Science
ISBN 978-3-319-26560-5   ISBN 978-3-319-26561-2   (eBook)
DOI 10.1007/978-3-319-26561-2

Library of Congress Control Number: 2015954339

LNCS Sublibrary: SL1 – Theoretical Computer Science and General Issues

Springer Cham Heidelberg New York Dordrecht London

Printed on acid-free paper

Springer International Publishing AG Switzerland is part of Springer Science+Business Media
(www.springer.com)

Preface

This volume is part of the four-volume proceedings of the 22nd International Conference on Neural Information Processing (ICONIP 2015), which was held in Istanbul, Turkey, during November 9–12, 2015. The ICONIP is an annual conference of the Asia Pacific Neural Network Assembly (APNNA; which was reformed in 2015 as the Asia Pacific Neural Network Society, APNNS). This series of ICONIP conferences has been held annually since 1994 in Seoul and has become one of the leading international conferences in the areas of artificial intelligence and neural networks.

ICONIP 2015 received a total of 432 submissions by scholars coming from 42 countries/regions across six continents. Based on a rigorous peer-review process where each submission was evaluated by an average of two qualified reviewers, a total of 301 high-quality papers were selected for publication in the reputable series of *Lecture Notes in Computer Science* (LNCS). The selected papers cover major topics of theoretical research, empirical study, and applications of neural information processing research. ICONIP 2015 also featured the Cybersecurity Data Mining Competition and Workshop (CDMC 2015), which was jointly held with ICONIP 2015. Nine papers from CDMC 2015 were selected for the conference proceedings.

In addition to the contributed papers, the ICONIP 2015 technical program also featured four invited speakers, Nik Kasabov (Auckland University of Technology, New Zealand), Jun Wang (The Chinese University of Hong Kong), Tom Heskes (Radboud University, Nijmegen, The Netherlands), and Michel Verleysen (Université catholique de Louvain, Belgium).

We would like to sincerely thank to the members of the Advisory Committee and Program Committee, the APNNS Governing Board for their guidance, and the members of the Organizing Committee for all their great efforts and time in organizing such an event. We would also like to take this opportunity to express our deepest gratitude to all the reviewers for their professional review that guaranteed high-quality papers.

We would like to thank Springer for publishing the proceedings in the prestigious series of *Lecture Notes in Computer Science*. Finally, we would like to thank all the speakers, authors, and participants for their contribution and support in making ICONIP 2015 a successful event.

November 2015

Sabri Arik
Tingwen Huang
Weng Kin Lai
Qingshan Liu

Organization

General Chair

Sabri Arik — Istanbul University, Turkey

Honorary Chair

Shun-ichi Amari — Brain Science Institute, RIKEN, Japan

Program Chairs

Tingwen Huang — Texas A&M University at Qatar, Qatar
Weng Kin Lai — School of Technology, Tunku Abdul Rahman College (TARC), Malaysia
Qingshan Liu — Huazhong University of Science Technology, China

Advisory Committee

P. Balasubramaniam — Deemed University, India
Jinde Cao — Southeast University, China
Jonathan Chan — King Mongkut's University of Technology, Thailand
Sung-Bae Cho — Yonsei University, Korea
Tom Gedeon — Australian National University, Australia
Akira Hirose — University of Tokyo, Japan
Tingwen Huang — Texas A&M University at Qatar, Qatar
Nik Kasabov — Auckland University of Technology, New Zealand
Rhee Man Kil — Korea Advanced Institute of Science and Technology (KAIST), Korea
Irwin King — Chinese University of Hong Kong, SAR China
James Kwok — Hong Kong University of Science and Technology, SAR China
Weng Kin Lai — School of Technology, Tunku Abdul Rahman College (TARC), Malaysia
James Lam — The University of Hong Kong, Hong Kong, SAR China
Kittichai Lavangnananda — King Mongkut's University of Technology, Thailand
Minho Lee — Kyungpook National University, Korea
Andrew Chi-Sing Leung — City University of Hong Kong, SAR China
Chee Peng Lim — University Sains Malaysia, Malaysia
Derong Liu — The Institute of Automation of the Chinese Academy of Sciences (CASIA), China

Chu Kiong Loo	University of Malaya, Malaysia
Bao-Liang Lu	Shanghai Jiao Tong University, China
Aamir Saeed Malik	Petronas University of Technology, Malaysia
Seichi Ozawa	Kobe University, Japan
Hyeyoung Park	Kyungpook National University, Korea
Ju. H. Park	Yeungnam University, Republic of Korea
Ko Sakai	University of Tsukuba, Japan
John Sum	National Chung Hsing University, Taiwan
DeLiang Wang	Ohio State University, USA
Jun Wang	Chinese University of Hong Kong, SAR China
Lipo Wang	Nanyang Technological University, Singapore
Zidong Wang	Brunel University, UK
Kevin Wong	Murdoch University, Australia

Program Committee Members

Syed Ali, India
R. Balasubramaniam, India
Tao Ban, Japan
Asim Bhatti, Australia
Jinde Cao, China
Jonathan Chan, Thailand
Tom Godeon, Australia
Denise Gorse, UK
Akira Hirose, Japan
Lu Hongtao, China
Mir Md Jahangir Kabir, Australia
Yonggui Kao, China
Hamid Reza Karimi, Norway
Nik Kasabov, New Zealand
Weng Kin Lai, Malaysia
S. Lakshmanan, India
Minho Lee, Korea
Chi Sing Leung, Hong Kong, SAR China
Cd Li, China

Ke Liao, China
Derong Liu, USA
Yurong Liu, China
Chu Kiong Loo, Malaysia
Seiichi Ozawa, Japan
Serdar Ozoguz, Turkey
Hyeyoung Park, South Korea
Ju Park, North Korea
Ko Sakai, Japan
Sibel Senan, Turkey
Qianqun Song, China
John Sum, Taiwan
Ying Tan, China
Jun Wang, Hong Kong, SAR China
Zidong Wang, UK
Kevin Wong, Australia
Mustak Yalcin, Turkey
Enes Yilmaz, Turkey

Special Sessions Chairs

Zeynep Orman	Istanbul University, Turkey
Neyir Ozcan	Uludag University, Turkey
Ruya Samli	Istanbul University, Turkey

Publication Chair

Selcuk Sevgen   Istanbul University, Turkey

Organizing Committee

Emel Arslan   Istanbul University, Turkey
Muhammed Ali Aydin   Istanbul University, Turkey
Eylem Yucel Demirel   Istanbul University, Turkey
Tolga Ensari   Istanbul University, Turkey
Ozlem Faydasicok   Istanbul University, Turkey
Safak Durukan Odabasi   Istanbul University, Turkey
Sibel Senan   Istanbul University, Turkey
Ozgur Can Turna   Istanbul University, Turkey

Contents – Part IV

Image and Signal Processing

Intelligent Social Networks

Deep Feature-Action Processing with Mixture of Updates

Abdulrahman Altahhan[✉]

Department of Computing, Coventry University, Coventry, CV1 5FB, UK
abdulrahman.altahhan@coventry.ac.uk

Abstract. This paper explores the possibility of combining an actor and critic in one architecture and uses a mixture of updates to train them. It describes a model for robot navigation that uses architecture similar to an actor-critic reinforcement learning architecture. It sets up the actor as a layer seconded by another layer which deduce the value function. Therefore, the effect is to have similar to a critic outcome combined with the actor in one network. The model hence can be used as the base for a truly deep reinforcement learning architecture that can be explored in the future. More importantly this work explores the results of mixing conjugate gradient update with gradient update for the mentioned architecture. The reward signal is back propagated from the critic to the actor through conjugate gradient eligibility trace for the second layer combined with gradient eligibility trace for the first layer. We show that this mixture of updates seems to work well for this model. The features layer have been deeply trained by applying a simple PCA on the whole set of images histograms acquired during the first running episode. The model is also able to adapt to a reduced features dimension autonomously. Initial experimental result on real robot shows that the agent accomplished good success rate in reaching a goal location.

Keywords: Reinforcement learning · Deep learning · Actor-critic · Neural networks · Robot navigation

1 Introduction

Reinforcement Learning, or RL, is considered ideal for learning novel scenario where it is impractical to obtain a model of the environment [4]. This is the norms for animals once they face a new situation. In that moment the actual learning begins, and we are interested to know what could be going on at that moment. RL does not require planning, but planning-like behavior comes as a natural consequence of it [4]. Actor-critic architecture have been proposed and studied in plenty of scenarios [5–7] that ranges from simulation to real problems and recently an interesting development has been proposed [8].

1.1 Actor-Critic and Neural Networks

The actor-critic architecture is interesting due to the fact that it allows for explicit natural separation of concern between a performer, that tries to learn the best set of actions in

© Springer International Publishing Switzerland 2015
S. Arik et al. (Eds.): ICONIP 2015, Part IV, LNCS 9492, pp. 1–10, 2015.
DOI: 10.1007/978-3-319-26561-2_1

certain situations, and a critic that tries to maximise overall future gain strategically. For the learner everything can be summed up in terms of a reward signal (food, shelter etc.) and the sensory data that feedback to it, as well as the actuators that the animal needs to use in order to reach its target and maximize the long and short term rewards. Shelter is considered a complex but important urge in animals. It comes after food as it is less urgent in the short term however it could be very important in the long term and could be vital in some dangerous situations. Therefore, animals have developed a complex behavior around the shelter need and through evolution it becomes intrinsic in the brains of higher animals. To trace all that back one needs to understand how primitive rewards form skills that governs complex behavior.

1.2 Deep Learning and RL

Deep learning has been shown to overcome the bottleneck representation problem that has long set back the success of machine learning applications [9, 10]. It is especially important for RL since RL normally takes a long time to converge due to the fact that there is no direct answer to the input, the answer is just a signal that indicates how good or bad the current action is (in the long run), and hence how good or bad the overall behavior is [14]. Deep learning showed better results when combined with supervised learning [11, 12]. When combined with RL it is also believed to have a good potential. The aim of this research is to set up a suitable architecture that permits deep learning as a first step towards realizing a deep reinforcement learning framework, we show that the architecture works as expected when operating under normal training conditions and we show some of the theoretical aspects and rules that govern it, while we retain to show how to train this architecture in a layer by layer fashion for the future.

1.3 Robot Homing

Robot homing is considered one of the important special cases of navigation. It pertains to all animals naturally and is a must for most of the commercial and entertaining robotics application. Central to this ability is the skill of orienting towards the home and recognizing it once the agent is around it. Animals do that by wiring the surrounding visual memory somehow to their neural map of the environment. How they do that is yet to be discovered. Traditionally it has been linked to distinctive position or places in the environment i.e. landmarks. However, the way animals do its navigation and find their home suggests something more subtle than only landmarks [1–3]. In this paper it is argued that a plausible proposition is that the brain hard wire the scenes to itself and compare it with the look of the home. It forms a frame of reference which is used to compare all information passed through to obtain internal map of the home as opposed to the different paths to its home. It is argued that the animal cannot remember all paths to its home neither it can remember the links between the landmarks consciously. It simply react to what it sees once it set back home and it does not normally plan it unless something novel happened. It uses this internal representation to guide itself along with other visual aids such as landmarks for the long distance. However for the short distance it uses this internal map solely.

2   The Model

Our model starts by learning a concise and reduced feature representation. The model obtains a reduced representation in the first episode by applying a simple PCA on the whole set of images acquired during this first episode(s). This explorative episode aligns with how animals normally explore a place by looking around. Then the model shrinks the whole architecture to fit the new reduced number of features, and this concludes the deep learning phase for the feature representation layer. It should be noted that other methods could be better for object detection or recognizing a certain pattern such as a hand-written digits [13]. However, for the purposes of our model we need something fast and low-level and we do not need to obtain features that could be used for recognizing an object. We just need a set of features that is good enough to distinguish the goal from other locations.

2.1   Model Architecture and Components

Next is the turn of actor-critic architecture [8]. After the actor layer takes its input form the PCA layer, it then decides to do a certain action, accordingly the critic layer punishes or rewards the Actor depending on the reward it receives form the environment.

Formally, the presented model uses the following components/stages shown in Fig. 1:

- Goal representation: As opposed to many models the goal or the home is represented by just few snapshots taken for that location with the desired orientation of the robot. In fact the method used to identify the goal is transitionally invariant. Hence the goal location could in deed by identified by the agent from an angle different to the one it has originally taken from as we shall emphasize later by our stopping condition. Feature vector is calculated in two stages
- The preliminary stage is to learn a reduced feature vector by deducing a representative Eigen vectors that gives results similar to a simple autoencoder. This is done by running the robot in an explorative episode(s) and analyzing the different scenes images in each step in an online fashion where the mean of the entire Eigen vector set is calculated at the end of the episode. And the max dimension is taken in case of more than one explorative episode is performed by the agent. This stage can be extended to span more than one episode to give a stable acceptable dense sample of the environment. However almost always the agent had to spend relatively long time in the first episode due to the fact that the weights are initialized to completely random small values (close to zero), hence they encouraged disoriented behavior that is explorative by nature. Then the model shrinks the whole architecture to fit the new reduced number of features (by picking Eigen vectors that corresponds to Eigen values of certain significance). This concludes the deep learning stage for the feature representation layer. This stage is done once and will not be repeated.
- The primary step in which the Eigen vectors deduced in the preliminary stage is used to calculate a reduced features vector. This step helped in focusing the policy learning step which is inevitably much longer. In both stages the initial

features used are differential Radial Basis features [14] that make the goal image its referential point and make all the views relative to that goal. This is consistent with home-aware localization and allows the agent to view the world from the perspective of its current homing task. The initial features are given by:

$$\varphi_f\left(s_t(c,j)\right) = \exp\left(-\frac{\left(h_f(s_t(c))-h_f(v(c,j))\right)^2}{2\hat{\sigma}^2}\right).$$ Where $h_f\left(v(c,j)\right)$ is histogram, bin f of

channel $v(c,j)$, and $h_i\left(s_t(c)\right)$ is histogram bin f of channel c of current image. Using the mean Eigen vector that has been calculated in the explorative episode, the new reduced features are given by $F_k = \Omega_{if} \times \varphi_f\left(s_t(c,j)\right)$. The dimension of the features φ_f is $d_1 \times d_2 \times 3$ where d_1 and d_2 are the dimensions of the images and 3 coming from having three channels. The dimension of the reduced features F_k is $n < <d_1 \times d_2 \times 3$.

- Another component of the model is a similarity measure that specifies the termination of the episode and is given by $NRB(s_t) = \sum_{k=1}^{n} \vec{F}_k\left(s_t\right)\Big/ n$. This measure has been used along with two thresholds to set the stopping and approaching conditions for the agent. The reward signal is calculated using the weights of the Critic which is constituted of three parameters in accordance with the number of actions allowed in this model. These actions are forward, left and right respectively, where left and right have been set to equal speeds. The reward function is given as a combination of step cost in addition to a reward for going towards (approaching) the goal as well as a reward for reaching the goal itself (which is proportional to how fast the agent reached the goal in terms of number of steps) as in [14]. Further a higher cost has been associated with turning actions. I.e. when the agent turns it will acquire higher costs than when it goes strait. This had the desired consequence of suppressing unnecessary turns and emphasizing going strait. Also a punishment for taking any action that leads to a reactive behavior has been set. This is also to help reduce the costal behavior and to encourage going directly towards the goal.

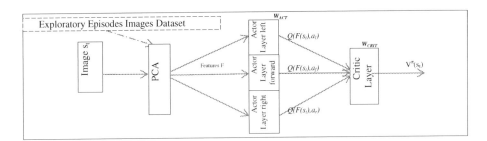

Fig. 1. The Model's Components and Stages.

2.2 Actor-Critic Combined Network with Double Eligibility Traces

In this section we show the derivation of the learning formulae for the layered actor-critic architecture. When function approximation techniques are used to learn a parametric estimate of the value function $V^\pi(s)$, $V_t(s_t)$ is expressed in terms of a set of parameters Θ_t. The mean squared error performance function [4] can be used to drive the learning process:

$$F_t = MSE(\Theta_t) = \sum_{s \in S} pr(s) \left[V^\pi(s) - V_t(s) \right]^2 \tag{1}$$

pr is a probability distribution weighting the errors $Err_t^2(s) = \left[V^\pi(s) - V_t(s) \right]^2$ of each state, and expresses the fact that better estimates should be obtained for more frequent states. The function F_t needs to be minimized in order to find an optimal solution Θ_t^* that best approximates the value function. For on-policy learning if the sample trajectories are being drawn according to pr through real or simulated experience, then one can concentrate on minimizing the minimizes the error function $Err_t^2(s)$. By using two layered neural network (one hidden layer and an output layer) and two sigmoid activations the reinforcement learning problem of learning an approximation of the value function and the action-value function can be written as:

$$V_t(s_t) = \cfrac{1}{1 + e^{\sum\limits_{i=1}^{I} Q_t(i)w_t(i)}}, \qquad Q_t(i) = \cfrac{1}{1 + e^{\sum\limits_{k=1}^{K} F_t(k)\theta_t(i,k)}} \tag{2}$$

The update rule can be written as $\Delta\Theta_t = \frac{1}{2}\alpha_t\vec{d_t}$, $\vec{d_t}$ is a vector that drives the search for Θ_t^* in the direction that minimizes the error function $Er_t^2(s)$, and $0 < \alpha_t \leq 1$ is a step size. This direction can be chosen in several ways. For example the update rules for weights that go opposite to the gradient direction are:

$$\Delta w_i(i) = \alpha_t \left(V^\pi(s_t) - V_t(s_t) \right) \frac{\partial V_t(s_t)}{\partial w_t(i)}, \quad \Delta\theta_i(i,k) = \alpha_t \left(V^\pi(s_t) - V_t(s_t) \right) \frac{\partial V_t(s_t)}{\partial w_t(i,k)} \tag{3}$$

By bootstrapping and using the $r_{t+1} + \gamma V_t(s_{t+1})$ as an approximation for $V^\pi(s_t)$ we have:

$$\Delta w_t(i) = \alpha_t\delta_t \frac{\partial V_t(s_t)}{\partial w_t(i)}, \quad \Delta\theta_t(i,k) = \alpha_t\delta_t \frac{\partial V_t(s_t)}{\partial\theta_t(i,k)} \tag{4}$$

$$\delta_t = r_{t+1} + \gamma V_t(s_{t+1}) - V_t(s_t) \tag{5}$$

It should be noted that these rules are approximate gradient descend.

2.3   Mixing Gradient with Conjugate Gradient Updates

Now by following the same analogy of [14] we update the error in opposite to the direction of the conjugate gradient for the output layer and leaving the action layer as it is we have instead of (6).

$$\Delta w_t(i) = \alpha_t Err_t \frac{\partial V_t(s_t)}{\partial w_t(i)} + \beta_t p_{t-1}(i), \quad \Delta\theta_t(i,k) = \alpha_t Err_t \frac{\partial V_t(s_t)}{\partial \theta_t(i,k)} \tag{6}$$

Where beta factor can be specified in several ways, for example:

$$\beta_t^{(HS)} = \frac{\Delta \vec{g}_{t-1}^T \vec{g}_t}{\Delta \vec{g}_{t-1}^T \vec{p}_{t-1}}, \quad \beta_t^{(FR)} = \frac{\vec{g}_t^T \vec{g}_t}{\vec{g}_{t-1}^T \vec{g}_{t-1}}, \quad \beta_t^{(PR)} = \frac{\Delta \vec{g}_{t-1}^T \vec{g}_t}{\vec{g}_{t-1}^T \vec{g}_{t-1}} \tag{7}$$

Conjugate gradient direction for the critic

$$\vec{p}_t = -\vec{g}_{\vec{w}_t} + \beta_t \vec{p}_{t-1}, \quad \vec{p}_0 = -\nabla_{\vec{w}} Err_t \tag{8}$$

$$\vec{g}_{\vec{\theta}_t(k)} = \nabla_{\vec{\theta}_t} Err_t^2(s_t) = 2\left[V^\pi(s_t) - V_t(s_t)\right] \cdot \frac{\partial V_t(s_t)}{\partial \vec{\theta}_t(k)},$$
$$\vec{g}_{\vec{w}_t} = \nabla_{\vec{w}_t} Err_t^2(s_t) = 2\left[V^\pi(s_t) - V_t(s_t)\right] \cdot \frac{\partial V_t(s_t)}{\partial \vec{w}_t} \tag{9}$$

Now we define $\vec{e}_{t-1}^{(conj)}$ as follows:

$$\gamma_t \lambda_t \vec{e}_{t-1}^{(conj)} = \frac{\beta_t}{Err_t} \vec{p}_{t-1} \tag{10}$$

We can evaluate this in several ways; for example:

$$\gamma_t \lambda_t = \frac{\beta_t}{Err_t} \tag{11}$$

The updates can be rewritten as:

$$\Delta \vec{w}_t = \alpha_t \left(V^\pi(s_t) - V_t(s_t)\right) \vec{e}_t^{(conj)}, \quad \Delta\vec{\theta}_t(k) = \alpha_t \left(V^\pi(s_t) - V_t(s_t)\right) \vec{e}_t'(k) \tag{12}$$

where

$$\vec{e}_t^{(conj)} = \frac{\partial V_t(s_t)}{\partial \vec{w}_t} + \gamma_t \lambda_t \vec{e}_{t-1}^{(conj)}, \quad \vec{e}_t'(k) = \frac{\partial V_t(s_t)}{\partial \vec{\theta}_t(k)} + \gamma_t' \lambda_t \vec{e}_{t-1}'(k) \tag{13}$$

This proves that eligibility traces are in fact independent of the error that we use whether it is approximation or exact. In fact it distinctively says that for any error or view, whether it is forward or backward there is an eligibility trace that coincides with the conjugate gradient and it is in fact variant with the reverse of the error, so the more error we have the less blame we should go deeper and the less error we have we can blame deeper [14]. It should be noted that λ_t can be chosen in many ways the presented is one of them. Also, this states that the discount should be varied according to the beta discount that set the relationship between consecutive conjugate directions, in general best is $\beta_t^{(PR)}$ due to its stable properties for nonlinear error functions [17] as this is the case for the sigmoid functions that we have chosen. This is in alignment with recent findings of [18] which is based on complex definition for the reward function. Our results shows that indeed we can find canonical eligibility traces that varies with the error no matter what type of error we are using, and that the reward discount should be varied according to the direction of the conjugate. For example if we approximate the error using. Two layers with and bootstrapping and using the $r_{t+1} + \gamma V_t(s_{t+1})$ as an approximation for $V^\pi(s_t)$ we have:

$$\Delta \vec{w}_t = \alpha_t \delta_t \gamma \lambda \vec{e}_{t-1}^{(conj)}, \quad \Delta \vec{\theta}_t(k) = \alpha_t \delta_t \gamma' \lambda \vec{e}_{t-1}'(k) \tag{14}$$

$$\vec{e}_t^{(conj)} = \frac{\partial V_t(s_t)}{\partial \vec{w}_i} + \gamma_t \lambda_t \vec{e}_{t-1}^{(conj)}, \quad \vec{e}_t'(k) = \frac{\partial V_t(s_t)}{\partial \vec{\theta}_t(k)} + \gamma' \lambda \vec{e}_{t-1}'(k) \tag{15}$$

It should be noted that eligibility traces in reinforcement learning framework is similar to the momentum for supervised learning. It constitutes a way to accommodate previous updates into current updates to guide the search for the local optima. In RL it traces blame of current decision back to older decisions that lead to the current situation. In addition, two regularizers have been multiplied by the two parameter sets to discount the old values of the parameters (hence prevent ovefitting).

2.4 Deep Blended Actor-Critic Architecture

By setting the second layer to three parameters; one for each action and calculating the error signal for the actor layer (which is responsible for taking the actions; its decision is the one that is being carried out). And by allowing this second layer to act as a critic that contemplates the consequences of the actions of the actor layer and sends a conjugate gradient signal to it to indicate how well its current policy is. And by making the two layers to work as hidden and output layers of a one neural network, we are creating deep blended actor-critic architecture in one sound system that depends on two eligibility traces. The value function layer (we are calling it critic layer but it is not in the strict conventional sense as it does not take input directly from the features, so one can call it an evaluator layer as well) itself is taking its feedback form the reward function. The action layer can learn independently form the critic layer by utilizing an action-value function approach while the critic layer cannot act independently, it still needs the actor layer to calculate the estimated value function. In that sense the learning process can be thought of as a layer by layer learning or deep learning enabled. In the future we will explore training in each layer independently by freezing learning in each layer and then fine-tune by utilizing the presented approach. So this model is deep in terms of its feature representation and has the potential to be deep in terms of its action representation.

3 Experimental Results

Figure 2 shows the used robot and its environment. It is basically an updated version of Lego Mindstorm that has been used with additional camera module and processing unit that was mounted and attached on top of it. This robot has relatively a low level of sophistication in terms of the motor commands, balance, senor reading as well as its shape. Matlab have been used throughout the model in the form of a set of library functions that have been written specifically for this model. In addition the RWTH- Mindstorms NXT Toolbox for MATLAB has been used to provide the sensory reading and the actuator commands form the NXT robot to Matlab.

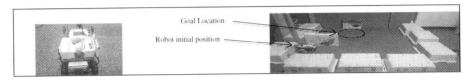

Fig. 2. Left: A snapshot of the built robot with its sensors, actuators, and camera module. Right: The training environment.

The robot was let to run for 30 episodes. Each episode starts by going from behind the barrier location in the environment to the goal/home location. It was allowed to run for a 500 steps before the episode is considered a failure. Images with resolution of 160×120 were sent form the Raspberry PI via wireless network adaptor to an off-board computer for processing where learning is taking place, then the required commands is sent to the actuators of the robot via Bluetooth. Threshold that specifies reaching the goal was set to 0.91.

3.1   Agent Learning Behavior and Convergence

First the agent developed a primitive behavior of turning only and occasionally going forward as because of the confined space this behavior pays best at start, then the agent started to develop inclination of going forwards more. In fact turning in one direction was preferred and enforced over the other. Next all actions received same enforcement. Next stage of behavior development was a bit of turning and some costal behavior which was later developed into direct to target behavior.

Figure 3 shows a final stage where the robot finished learning. The number of episodes (right) is envisaged (as was evident in the simulation in [14–16]) to show a pattern of convergence towards minimal number of step if the robot where left to run for a very long time. However, due to the time and physical constraints, this was difficult to do and the agent was let to run for a limited number of episodes. This comparison has been done by starting always form the same position the variation is due to the different learning stages of the agent parameters. A fifth order polynomial curve fitting is shown in Fig. 3 that illustrates possible convergence. It distinctively shows two main stages of learning that occurred in those episodes one that represents initial stage that used a costal behavior to reach the goal and the other trying to reach a better robust behavior that tries to avoid the walls. Our results show that out of 35(30 training + 5 testing) times. The agent confused the goal four times throughout the episodes. Hence, it can be said that the success rate is 31/35 ≈ 89 % for goal recognition. As opposed to many models the goal or the home is represented by just few snapshots taken for that location with the desired orientation of the robot. The model uses deep architecture for feature representation and an actor-critic update that mixes conjugate gradient with gradient update, which set its distinctive novelty. Also it is intended to show some other interesting properties of the model such as convergence and the relationship between deep feature learning and deep action learning in the future.

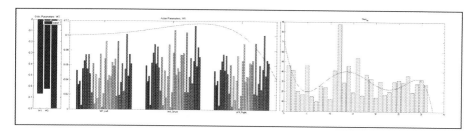

Fig. 3. To the left is the model learned parameters for the critic which uses conjugate gradient eligibility trace, while in the middle the parameters for the actor's different actions is shown where gradient eligibility trace is used; a tendency towards going forward then turning left is developed by the agent, which is what is expected when operating in the selected environment. To the right, the convergence to a stable number of steps is shown to be developing; the height shows number of steps needed for each episode to reach the goal.

References

1. Vardy, A., Moller, R.: Biologically plausible visual homing methods based on optical flow techniques. Connection Sci. **17**, 47–89 (2005)
2. Tomatis, N., et al.: Combining topological and metric: a natural integration for simultaneous localization and map building. In: Presented at Proceedings of the Fourth European Workshop on Advanced Mobile Robots (Eurobot) (2001)
3. Zeil, J.: Visual homing: an insect perspective, Current Opinion in Neurobiology. **22**(2), 285–293 (2012). ISSN 0959-4388
4. Sutton, R.S., Barto, A.: Reinforcement Learning, an introduction. MIT Press, Cambridge (1998)
5. Konda, V., Tsitsiklis, J.: Actor-Critic algorithms. In: Presented at NIPS 12 (2000)
6. Ziv, O., Shimkin, N.: Multigrid methods for policy evaluation and reinforcement learning. In: Presented at IEEE International Symposium on Intelligent Control, Limassol (2005)
7. Zhang, C., et al.: Efficient multi-agent reinforcement learning through automated supervision. In: Presented at International Conference on Autonomous Agents Estoril, Portugal (2008)
8. Bhatnagar, S., et al.: Incremental natural actor-critic algorithms. In: Presented at Neural Information Processing Systems (NIPS19) (2007)
9. Hinton, G., et al.: A fast learning algorithm for deep belief nets. Neural Comput. **18**(7), 1527–1554 (2006)
10. Coates, A., et al.: An analysis of single-layer networks in unsupervised feature learning. In: AISTATS 14 (2011)
11. Vincent, P., et al.: Extracting and composing robust features with denoising autoencoders. In: ICML (2008)
12. Andrew, Ng et al.: Tutorial in Deep Learning: Stanford University (2010). http://ufldl.stanford.edu/tutorial/
13. LeCun, Y., et al.: Learning methods for generic object recognition with invariance to pose and lighting. In: CVPR (2004)

14. Altahhan, A.: A robot visual homing model that traverses conjugate gradient TD to a variable λ TD and uses radial basis features. In: Mellouk, A. (ed.) Advances in Reinforcement Learning, pp. 225–254. InTech Education and Publishing, Vienna (2011)
15. Altahhan, A.: Robot visual homing using conjugate gradient temporal difference learning, radial basis features and a whole image measure. In: International Joint Conference on Neural Networks (IJCNN), Barcelona, Spain (2010). ISBN: 978-1-4244-6916-1
16. Altahhan, A., et al.: Visual robot homing using sarsa(λ), whole image measure, and radial basis function. In: International Joint Conference on Neural Networks (IJCNN), Hong Kong (2008)
17. Nocedal, J., Wright, S.: Numerical Optimization. Springer-Verlag, New York, 978-0-387-30303-1, 2nd Edition (2006)
18. Sutton, R.S., et al.: A new Q(lambda) with interim forward view and Monte Carlo equivalence. In: Proceedings of the 31 st International Conference on Machine Learning, Beijing, China, 2014. JMLR: W&CP vol. 32 (2014)

Heterogeneous Features Integration via Semi-supervised Multi-modal Deep Networks

Lei Zhao, Qinghua Hu[✉], and Yucan Zhou

School of Computer Science and Technology, Tianjin University, Tianjin, China
huqinghua@tju.edu.cn

Abstract. Multi-modal features are widely used to represent objects or events in pattern recognition and vision understanding. How to effectively integrate these heterogeneous features into a unified low-dimensional feature space has become a crucial issue in machine learning. In this work, we propose a novel approach which integrates heterogeneous features via an elaborate Semi-supervised Multi-Modal Deep Network (SMMDN). The proposed model first transforms the original data to high-level abstract homogeneous features. Then these homogeneous features are integrated into a new feature vector. By this means, our model can obtain abstract fused representations with lower-dimensionality and stronger discriminative ability. A Series of experiments are conducted on two object recognition datasets. Results show that our approach can integrate heterogeneous features effectively and achieve better performance compared to other methods.

Keywords: Deep neural network · Feature fusion · Pattern recognition

1 Introduction

With the development of sensor technology, multiple information from different channels can be obtained for one object. For example, information is described by images and text annotations in many social media websites, such as Flikcr, Google Picasa, etc. Meanwhile, more and more feature descriptors for one channel information (image or text) are proposed to extract high-level semantic information from the original low-level data. Different feature descriptors depict different aspects of the pattern's intrinsic structure. For an instance, in image processing, the Color Histogram depicts the color property of the image, HOG conveys the shape information and the feature of LBP extracts texture information from the original image.

Intuitively, heterogeneous features containing rich information can help to achieve better performance than single type of feature descriptors in many recognition tasks. However, these features usually have different structures, making it difficult or inefficient to use them in a classifier. What's more, they are always of high dimension which brings a drawback to computation efficiency. Therefore, how to integrate these heterogeneous features into new low-dimensional feature representations has become a crucial and attractive issue.

© Springer International Publishing Switzerland 2015
S. Arik et al. (Eds.): ICONIP 2015, Part IV, LNCS 9492, pp. 11–19, 2015.
DOI: 10.1007/978-3-319-26561-2_2

A simple way to integrate these high-dimensional heterogeneous features is concatenating all the features into a long feature vector and then utilizing conventional dimensionality reduction methods, e.g., PCA, ICA. However, it ignores the distinction among different types of features. Simply concatenation may even deteriorate the intrinsic structure of original features. To overcome this obstacle, some algorithms have been proposed.

Several methods are based on graph models. In paper [2,3], a shared common cluster indicator with non-negative constraint is constructed by non-negative matrix factorization (NMF), the different features are merged with the unsupervised spectral clustering. Cai [1] proposed another semi-supervised graph based approach for image classification. However, an intractable barrier for these graph based methods is the high computational complexity. Computation of these algorithms is infeasible when the dataset is very large.

Multiple Kernel Learning (MKL) based methods [6,7,12] are another commonly investigated methods for multi-modal learning. Considering each type of features as one modality, the MKL methods allocate one independent kernel for every feature modality. Then an ensemble kernel is learned to project all the features into an ensemble Reproducing Kernel Hilbert Space. A problem of these MKL based methods is that the base kernels for each modality should be specified manually. Since the base kernels can impact the final performance, how to select a proper kernel for each modality is a difficult and confusing problem.

Recently, deep learning has shown its power of learning latent feature representations. Some multi-modal deep learning models based on Deep Belief Net are also proposed [8,9]. However, these methods learn deep networks with two modalities, eg., image-text or video-audio pairs. In this paper, we propose a more flexible multi-modal learning model which integrates heterogeneous features. The proposed model is a Semi-supervised Multi-Modal Deep neural Networks (SMMDN) including multiple sub-networks and several top hidden layers. Treating each type of features as an independent modality, we allocate it a relatively independent sub-networks. The corresponding sub-networks of SMMDN will first transform the heterogeneous inputs into high-level abstract homogeneous representations. Then the top layers of SMMDN will integrate these homogeneous modality-free representations into a fused representation in a lower dimensional space, which is trained by another network.

In the following section, we will describe the architecture of the proposed model in detail.

2   Semi-Supervised Multi-Modal Deep Networks

2.1   Model Architecture

Figure 1 illustrates the complete structure of SMMDN. The whole model is decoupled into two subsections: Root Networks and Top Networks. As shown in Fig. 1, the root network comprises m sub-networks. We should note that all these sub-networks are different in terms of their inner layer structures. Different types of networks have distinct capability to extract features from raw data.

We should select appropriate deep neural networks dependent on the applications. After that, we introduce an auxiliary bridge layer to connect all sub-networks when jointly training them. These sub-networks are responsible for extracting high-level and modality-free representations from the input data. Once all the sub-networks are trained, batches of refined homogeneous feature representations can be extracted from the top hidden layers of these sub-networks. The Top Networks merge the refined homogeneous feature representations and project them into a shared low-dimensional feature space. In the following subsections, we introduce the two parts of SMMDN in detail and explain how to execute the training procedure.

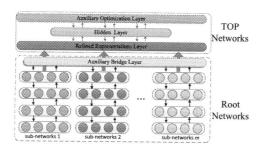

Fig. 1: Architecture of the SMMDN. The model consists of two parts: the Top Networks (*surrounded by the dotted box*) and the Root Networks (*boxed by dashed lines*).

2.2 Extracting Homogeneous Representations by Root Networks

A major problem of multiple features integration is the heterogeneity existing in the discriminative features. Fortunately, deep learning has demonstrated its potential in discovering latent hierarchical representations. In our model, we exploit this advantage of deep neural networks to eliminate the heterogeneity.

The intrinsic distribution structures of different modalities differ from each other. As to some simple structures, relatively shallow neural networks are able to extract their high-level latent representations. To the contrary, some modalities with complicated intrinsic structures may need deeper networks to extract their high-level representations. starting from this consideration, we design m heterogeneous sub-networks for the m modalities. They differ from each other in the number of hidden layers and hidden nodes. We denote the number of hidden layers of m-th modality as n_m. Figure 2 illustrates the detailed structure of the Root Networks.

There are two stages in training the Root Networks: unsupervised pre-training and supervised jointly fine-tuning, as shown in Fig. 2. Different from the methods of [8,9], in the pre-training stage, we train the sub-networks as Stacked Denoising Autoencoders (SDA) [13]. By this means, we can assign proper initialization to the connection weights between hidden layers. This unsupervised pre-training could avoid the networks from converging at local minimums. To find

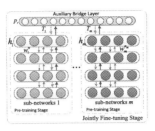

Fig. 2: Illustration of root networks. The training of root networks includes two stages: unsupervised pre-training and supervised fine-tuning. In stage of pre-training,we train every sub-networks (*bounded in grey dashed boxes*) independently. We introduce an auxiliary bridge layer in the fine-tuning stage to connect all the sub-networks and finetune all the sub-networks' parameters jointly.

the latent correlations across modalities, in the following fine-tuning stage we introduce an auxiliary bridge layer to connect all the sub-networks by utilizing the label information. As shown in Fig. 2, the auxiliary bridge layer is connected to the sub-networks top layer with the corresponding weights T_m.

In consideration of the correlations across modalities and to get homogeneous modality-free features, we let all the weights matrix T_m share the same weights matrix T. Let h_m denotes the top hidden layer of m-th modality and p_r denote the indicators of auxiliary bridge layer. We set the loss function as Eq. 1 where N denotes the number of training samples, $y^{(i)}$ denotes the label of sample $x^{(i)}$ and $h_j^{(i)}$ denotes the j-th modality's top hidden layer's output of sample $x^{(i)}$. For k-way classification problem, the probability that an input vector h_m belong to class i is denoted as Eq. 2, where b_{root} is a bias vector, T^{ℓ} denotes the ℓ-th row of matrix T. Actually, the loss is a logistical loss.

$$\mathcal{L} = -\sum_{j}^{m}\sum_{i}^{N}\log(P(Y = y^{(i)}|h_m^{(i)}, T, b_{root})) \tag{1}$$

$$P(Y = y^{(i)}|h_m, T, b_{root}) = \frac{exp(T^i h_m + b_{root_i})}{\sum_{\ell} exp(T^{\ell} h_m + b_{root_{\ell}})} \tag{2}$$

By minimizing the loss, we finely tune all the parameters of the Root Networks. Finally, we extract the abstract homogeneous features h_m from every sub-networks. Since we enforce the shared weighs T on all the connection weights T_m shown in Fig. 2. In fine-tuning stage, we fine-tune one sub-network with the remaining fixed and repeat this procedure until all the sub-networks are trained in every iteration. The auxiliary layer is just used for jointly finetuning all the Root Network, so it will be discarded when the fine tuning is done.

2.3   Feature Fusion with Top Networks

Since we have got a batch of homogeneous high-level abstract representations, the Top Networks will merge these representations with a non-linear transformation. Figure 3 demonstrates the detailed structure of the Top Networks.

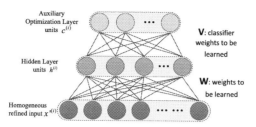

Fig. 3: Structure of the top networks. The homogeneous refined representations x' are projected into a low-dimensional feature space by the function $h = s(Wx' + b)$. The top auxiliary optimization layer is used for combining the supervised and unsupervised information to find the optimal weights W with back-propagation.

The homogeneous refined features will be concatenated as the input x' of the bottom layer. With the transformation matrix W, the input features will be projected into a new low-dimensional fused feature space. The final integrated feature h is extracted from the middle hidden layer. To take advantage of label information, we add an auxiliary optimization layer on the top of the Top Networks. This auxiliary could be a logistical classifier for our final classification task as well. This triple layer networks is used for the final feature integration. To train this networks, we set the final cost function $\mathcal{L}$ as following:

$$\mathcal{L} = \mathcal{L}_{dis} + \beta \mathcal{L}_{gen} + \lambda_1 \|W\|_{\ell_1} + \lambda_2 \|W\|_F^2 \tag{3}$$

$$\mathcal{L}_{dis} = -\sum_i^N \log(P(Y = y^{(i)} | h^{(i)}, V, b_{top})) \tag{4}$$

$$\mathcal{L}_{gen} = -\sum_i^N [x'^{(i)} \log \hat{x}'^{(i)} + (1 - x'^{(i)}) \log(1 - \hat{x}'^{(i)})] \tag{5}$$

There are three terms in the loss function: discriminative loss $\mathcal{L}_{dis}$, generative loss $\mathcal{L}_{gen}$ and the regularization. The discriminative loss let the integrated feature have strong discriminative ability. Similar to Eq. 2, we use the logistical loss here and formulated as Eq. 4. Beside the strong discriminative capability, we expect the fused features have better generative ability at the same time. The generative loss measures an average reconstruction error between x' and it's reconstruction $\hat{x}'$ with the transformation matrix of W. The small error mean that the fusion feature h has preserved most of the information from x'. We define this generative loss as Eq. 5. In Eq. 5, $\hat{x}' = s(W^T s(Wx + b) + b')'$ is the reconstruction of input x' where $s(\cdot)$ is the sigmoid function, b and b' are bias items. Now the weight matrix W is not only learned from the label information, but also from reconstructing the input x' unsupervisedly. Parameter β is introduced to balance loss $\mathcal{L}_{gen}$ and loss $\mathcal{L}_{dis}$. Moreover, we add two additional regularization items, which encourage sparsity and margin on weights W. This makes the model more robust. All these items form the final objective function. We use the gradient descent algorithm to minimize the objective function and obtain the optimal parameters of the Top Networks. The regularization item $\|W\|_{\ell_1}$ in Eq. 3 is not differentiable at 0. It brings a problem for the gradient descent method. So we set the following approximation to smooth the loss function:

$$\|W\|_{\ell_1} = \sum_{ij} \sqrt{W_{ij} + \sigma} \tag{6}$$

where σ is a pre-specified scalar with a very small positive value. We set it as 10^{-31} in our all experiments.

3   Experiments

3.1   Datasets and Experiment Setup

We evaluate our model on two public image classification tasks: NUS-Object[1] and AWA[2]. The NUS-Object dataset includes 30000 imgaes collected from Flickr. Text descriptions are attached to every image. These images are categorized into 31 classes. We utilize the published five types of image features: 64-dimension Color Histogram (CH) features, 144 dimension color auto-CORRelogram (CORR) features, 73 dimension Edge Direction Histogram (EDH) features, 225 dimension Wavelet Texture (WT) features, Block-wise Color Moments (CM) features. Besides, we get two types of text features from the attached text with the LDA method: Doc-Topic distribution of LDA Topic model with 31 topics (31 dimension LDA31) and 81 topics (81 dimension LDA81). Generally, a deep neural networks could converge well just with relatively large-scale training data for its numerous parameters. So we take 20000 samples randomly as train set, the rest 10000 images as test data.

The AWA dataset consists of 30475 images of 50 animal classes. We take all the published features: 2000 dimension Local Self-Similarity (LSS) features, 2000 dimension colorSIFT (RGSIFT) features, 2000 dimension SIFT features, 2000 dimension SURF features, 2688 dimension Color Histogram (CQ) features and 252 dimension Pyramid HOG (PHOG) features. We choose 9607 images of 10 classes and take 8000 of them as train set and the rest as test set.

We compare our model with some MKL based methods and simple SVM methods described in Sect. 1: (1) SVM with the concatenation of all the original features, (2) PCA+SVM, we first reduce the dimension of the concatenation of original features to 300 dimension, then SVM is applied to the low-dimensional projected features, (3) lMKL [5], (4) gMKL [11], (5) sMKL [10], (6) cabMKL [4]. Graph based methods couldn't process large-scale data with acceptable time cost, so we don't compare our model with these graph based approaches. The SVM in all our experiments is implemented with the LIBSVM[3] software package. For MKL methods, we test different kernel including gaussian, polynomial and

---

[1] http://lms.comp.nus.edu.sg/research/NUS-WIDE.htm.

[2] http://attributes.kyb.tuebingen.mpg.de/.

[3] http://www.csie.ntu.edu.tw/%7ecjlin/libsvm/.

linear kernel and select the best kernels according to the final classification accuracy for every modality. Our multi-modal deep neural networks is implemented with the deep learning library of Theano[4]. There are three hyper-parameters $(\beta, \lambda_1, \lambda_2)$ in our model as shown in Eq. 4. We tried a series of values in the range of $[10^{-5}, 1]$ for every hyper-parameter and select the better value according to the final classification accuracy on a small validation set.

3.2 Experimental Results

First we evaluate the discriminative ability of the integrated feature. In Table 1, we give the final classification accuracy with integrating multiple heterogeneous features on the two datasets by different methods. To judge the promotion of discriminative ability of the fused features, we also give the accuracy with single modality by SVM in Table 2.

Table 1: Classification accuracies produced with different methods

Methods	SVM	PCA+SVM	lMKL	gMKL	sMKL	cabMKL	Our Methods
NUS-Obejct	0.4563	0.4290	0.5733	0.6300	0.6145	0.5579	**0.6373**
AWA	0.6623	0.5893	0.3385	0.6459	0.6727	0.4306	**0.7119**

Table 2: Classification accuracies of SVM with different modalities

NUS-Object	Features	CH	CORR	EDH	WT	CM	LDA31	LDA81
	Acc	0.2426	0.3066	0.2894	0.3063	0.2857	0.5062	0.4564
AWA	Features	LSS	RGSIFT	SIFT	SURF	CQ	PHOG	
	Acc	0.4941	0.5022	0.4088	0.5271	0.3920	0.3765	

For dataset of NUS-Object, its seven types of features are extracted from two different channels: image and text. We noticed that the features from text (LDA31, LDA81) have much more discriminative ability than those features from image. However, if we simply concatenate all the features into a long feature vector and employ SVM, the classification accuracy is even lower than only using the LDA31 features. Because of the information loss of PCA, the method of PCA+SVM get even lower accuracy. These results verify the necessity of heterogeneous integrating especially when they come from quite different information channels. Though some MKL based methods performance well, our method outperform them and achieve better performance.

For dataset AWA, its six types of features are all from image. For SVM, PCA+SVM and most of the MKL methods, using the concatenation of all features can yields higher accuracy than using any single feature representations. However, our method get much higher classification accuracy compared with

---

[4] http://deeplearning.net/software/theano/index.html.

them. Results on two datasets confirm that our model can integrate heterogeneous features effectively and keep strong discriminative ability.

Though the discriminative ability is a very important criterion to evaluate the effectiveness of integration, outstanding fused features should preserve more information with less dimensionality from the original multiple features. So we verify our model's ability of dimensionality reduction. Figure 4 shows the variation of classification accuracy with changes in the fused feature's dimensionality. Many proposed deep architectures just exploit the discriminative loss, however we utilize the discriminative and generative information simultaneously. We also compared the experiment results with different loss function items in Eq. 3. The method with only discriminative loss is denoted by $\mathcal{L}_{dis}$ and shown with blue lines in Fig. 4. Red lines indicates the results of method with all the loss items in Eq. 3 and we denote it as $\mathcal{L}_{dis} + \mathcal{L}_{gen}$. It's clearly that our model can integrate all the original features into one fusion feature with very-low dimensionality while the fused features keep very strong discriminative ability. For the dataset of AWA, we can even integrate all its image features into a 10-dimension fused features without any classification accuracy loss nearly. Meanwhile, the experiment results show that with the introduced generative loss item $\mathcal{L}_{gen}$, our model can achieve better performance with same dimensionality. This fact confirms the integrating effectiveness of our model again.

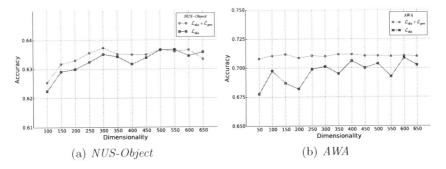

(a) *NUS-Object*   (b) *AWA*

Fig. 4: The classification accuracy changes with different dimensionality on the two Dataset (Color figure online).

4   Conclusion

We propose a novel deep neural networks based approach for heterogeneous features integration. Our model can integrate heterogeneous features from different sources into new fused features effectively. The fused feature have strong discriminative ability while low-dimensionality. All the experiment results confirm this effectiveness.

Acknowledgments. This work was supported in part by National Natural Foundation of China (No. 61222210) and 973 Program (2013CB329304). We gratefully acknowledge the support of NVIDIA Corporation with the donation of the Tesla K40 GPU used for this research.

References

1. Cai, X., Nie, F., Cai, W., Huang, H.: Heterogeneous image features integration via multi-modal semi-supervised learning model. In: ICCV 2013, pp. 1737–1744. IEEE (2013)
2. Cai, X., Nie, F., Huang, H., Kamangar, F.: Heterogeneous image feature integration via multi-modal spectral clustering. In: CVPR 2011, pp. 1977–1984. IEEE (2011)
3. Chen, H., Cai, X., Zhu, D., Nie, F., Liu, T., Huang, H.: Group-wise consistent parcellation of gyri via adaptive multi-view spectral clustering of fiber shapes. In: Ayache, N., Delingette, H., Golland, P., Mori, K. (eds.) MICCAI 2012, Part II. LNCS, vol. 7511, pp. 271–279. Springer, Heidelberg (2012)
4. Cortes, C., Mohri, M., Rostamizadeh, A.: Algorithms for learning kernels based on centered alignment. J. Mach. Learn. Res. **13**(1), 795–828 (2012)
5. Gönen, M., Alpaydin, E.: Localized multiple kernel learning. In: ICML 2008, pp. 352–359. ACM (2008)
6. Guillaumin, M., Verbeek, J., Schmid, C.: Multimodal semi-supervised learning for image classification. In: CVPR 2010, pp. 902–909. IEEE (2010)
7. Lin, Y.Y., Liu, T.L., Fuh, C.S.: Local ensemble kernel learning for object category recognition. In: CVPR 2007, pp. 1–8. IEEE (2007)
8. Ngiam, J., Khosla, A., Kim, M., Nam, J., Lee, H., Ng, A.Y.: Multimodal deep learning. In: ICML 2011, pp. 689–696 (2011)
9. Srivastava, N., Salakhutdinov, R.R.: Multimodal learning with deep boltzmann machines. In: NIPS 2012, pp. 2222–2230 (2012)
10. Subrahmanya, N., Shin, Y.C.: Sparse multiple kernel learning for signal processing applications. IEEE Trans. Pattern Anal. Mach. Intell. **32**(5), 788–798 (2010)
11. Varma, M., Babu, B.R.: More generality in efficient multiple kernel learning. In: ICML 2009, pp. 1065–1072. ACM (2009)
12. Vedaldi, A., Gulshan, V., Varma, M., Zisserman, A.: Multiple kernels for object detection. In: ICCV 2009, pp. 606–613. IEEE (2009)
13. Vincent, P., Larochelle, H., Bengio, Y., Manzagol, P.A.: Extracting and composing robust features with denoising autoencoders. In: ICML 2008, pp. 1096–1103. ACM (2008)

Multimodal Deep Belief Network Based Link Prediction and User Comment Generation

Feng Liu, Bingquan Liu$^{(\boxtimes)}$, Chengjie Sun, Ming Liu, and Xiaolong Wang

School of Computer Science and Technology, Harbin Institute of Technology,
Harbin, China
{fengliu,liubq,cjsun,mliu,wangxl}@insun.hit.edu.cn

Abstract. In social network services, the relationship among members can be represented as link network, and the link prediction problem is to infer the value of such links. By analysing the structure of link network, researchers have proposed several methods for solving link prediction. Nowdays, when some members label their link values, they also make comments, which have been seldom considered for link prediction. In this paper, by considering both the link network data and user comment data, we propose multimodal deep belief network based link prediction method, which outperforms other state-of-art methods. With the learned joint distribution of link network features and user comment features, our method could generate comment words properly.

Keywords: Link prediction · User comment generation · Social networks · Multimodal learning · Deep belief networks

1 Introduction

The classical link prediction problem is inferring new interactions among social members by a snapshot of the social network [6]. In this paper, the predicting task is to predict the link value of one user toward another. This problem can be solved by analysing the user's relations with other members as a traditional link prediction problem, which is shown in Fig. 1. In the condition that users can also make some comments when they label their link values, this problem can be also solved as a sentiment analysis problem, as shown in Table 1.

The link prediction problem can be solved by machine learning based methods. It treats this problem as a link value classification task [13]. Support vector machine is used to analysing how each feature effects the link value [9]. Deep belief network based approaches for link prediction are introduced in [7,8]. However, these methods only used the data from link network structure.

The sentiment analysis problem can be also treated as a sentiment polarity classification task [10]. Deep learning based sentiment analysis method achieved good performance in corpus from many different domains [2]. Deep belief network with fuzzy membership function based method is used to predict the sentiment polarity of user reviews [15].

© Springer International Publishing Switzerland 2015
S. Arik et al. (Eds.): ICONIP 2015, Part IV, LNCS 9492, pp. 20–28, 2015.
DOI: 10.1007/978-3-319-26561-2_3

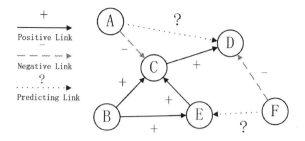

Fig. 1. Link network structure.

However, most researches solve the link prediction and sentiment analysis problem individually, or just try to combine their features for training linear classifier [14]. As both the link network (shown in Fig. 1) and user comments (shown in Table 1) could imply the concept of link value, it satisfies the properties of learning a good multimodal model [11]. Such as the multimodal models for image and text described the image are introduced in [5]. And multimodal model can represent the joint distribution of questions and answers [12].

In this paper, we assuming that both the link network and user comments imply the link value. And we proposed a multimodal model that can learn the joint distribution of link network features and user comment features. With two different learning processes, this model could be used to predict link values or generate user comments. The experiments are done on typical social dataset, and the results show that our method's link prediction performance outperforms other state-of-art methods. The user comment generation results show that our method could make up the missing comment properly.

Table 1. User comments

Link value	Comment words
(+)	Yea, I've seen him a lot. Good candidate
(−)	Sorry to hear you're getting bored of editing, but getting the admin tools is not the solution to that
(+)	No problems here. Great user who deserves the mop
(−)	Not enough admin area experience, as DGG says
(?)	Valuable contributions in the past and more anticipated
(?)	Sorry, no. Too little, not good enough yet

2 Models

2.1 Restricted Boltzmann Machine

A Restricted Boltzmann Machine(RBM) is a neural network that contains two layers. It has a single layer of hidden units h that are not connected with each other. And the hidden units have undirected, symmetrical connections w to a layer of visible units v. As shown in Fig. 2(a), $\{V_1, W_1, H_1\}$ constructs a RBM. It defines a probability distribution over v, h as

$$-\log P(v,h) \propto E(v,h) = -\sum_i v_i a_i - \sum_j h_j b_j - \sum_{i,j} v_i w_{ij} h_j \qquad (1)$$

where a_i is the bias of visible unit i, and b_j is the bias of hidden unit j. The parameters $\{w, a, b\}$ are usually trained by using the Contrastive Divergence(CD) learning procedure, which is introduced in [3].

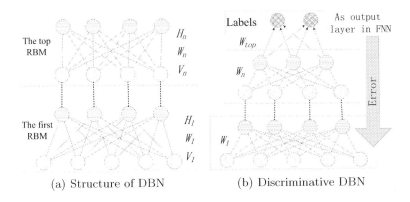

(a) Structure of DBN   (b) Discriminative DBN

Fig. 2. The structure of deep belief networks.

2.2   Deep Belief Network

The Deep Belief Network(DBN) is a multilayer, stochastic generative model that is created by learning a stack of RBMs, as shown in Fig. 2(a). When learning a DBN, only the first RBM is trained on the original samples by CD learning procedure. Then the second RBM is trained by the first RBM's hidden activation vectors which are generated from the original samples. Do that iteratively until the top RBM is learned. This greedy, layer-by-layer learning can be repeated as many times as desired [4]. If a sample vector is imputed to the first RBM of that DBN, the highly abstracted vector of that sample would be gotten from the top RBM's hidden layer.

2.3   Multimodal Deep Belief Networks

As introduced above, the DBN based on RBMs can represent the distribution of input samples. By using RBM's such ability, we designed a Multimodal Deep Belief Networks(MDBN) as shown in Fig. 3. Both the link network features and user comment features are represented by learned DBN. Then we join the represented features in the ith RBM as shown in figure, and learn another new DBN

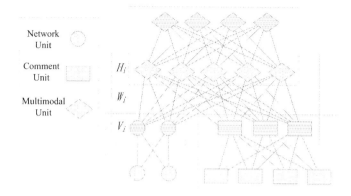

Fig. 3. Structures of multimodal deep belief networks.

by these features. The new DBN could represent the joint distribution of link network features and user comment features.

The reason of not joining the features in bottom RBM is because that the correlations between the link network features and user comment features may be highly non-linear. The $1st$ RBM may not model the joint distribution properly. As introduced in Sect. 2.2, well learned DBN could highly abstract the inputs, and the represented samples become linear classifiable. As a result, we design the MDBN that join the different data in higher RBM layers. It can be easier for the model to learn the higher-order correlations between link network data and user comments data.

3   Methodology

3.1   Link Network Structure Features

The link network is a directed graph. Denote $CNe(u, v)$ as common neighbours of user u and v. There are total 26 link network structure features, including 8 node features and 16 neighbour features.

The node features include the in and out degrees with sign values. Denote $d_{in}^{+}(u)$ for positive in-degree, $d_{in}^{-}(u)$ for negative in-degree, and collect 2 out-degree features. Then collect the degrees of v. The neighbour features includes statistical information from $CNe(u, v)$, such as the number of nodes w in $CNe(u, v)$ and the number of edges between w and u, v. Then select any node w from $CNe(u, v)$, whose edges could have any direction with any sign value connected with u and v. So there are 16 kinds of relationships of u, v and w.

3.2   User Comment Features

The text, such as comment, could be represented by the Bag of Words model(BOW). It treats each comment as the bag of its words, and represents text as a vector of words via the word dictionary. The word dictionary Dic contains all the appeared words, and the dictionary size is $lenDic$. The set of words appeared in comment from u to v is denoted as $W(u,v)$. Then build a word vector $w(u,v)$ with dimension $lenDic$, and set the ith position to '1' if the Dic's $ith\ word \in W(u,v)$, while all other positions are set to '0'.

The $lenDic$ is always vary large, but most of these words are only used few times. It makes word vector $w(u,v)$ very sparse. Because the number of first layer RBM's visible units is equal to $lenDic$. We use the 2000 top frequency words to build the dictionary and extract word vectors.

3.3 Discriminative Deep Belief Networks

In order to use the unsupervised learned DBN to solve link prediction problem, we added a layer of linear output units for class labels at the top as shown in Fig. 2(b). Each of the output unit stands for a class label, and the sample's label should be the output unit with the largest value. This output layer works as a linear Softmax classifier.

The Softmax classifier is learned with minimizing the cross-entropy loss error as $-\sum_i y_i \log p_i$ where y_i is the class label and p_i is the predicted link value. In order to get a better classification performance, the unsupervised learned DBN is fine-tuned when we update the output layer's weights by BP algorithm as a forward neural network shown in Fig. 2(b).

3.4 Reconstructive Deep Belief Networks

The discriminative DBN could be used to predict link values, but the DBN's reconstructive ability becomes vanished with the fine-tuning process by BP error from the label layer. In order to make the MDBN suitable for user comment generation by its reconstructive ability, we use another fine-tuning precess to adjust the unsupervised learned DBN as shown in Fig. 4.

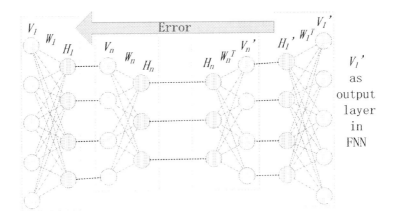

Fig. 4. Reconstructive deep belief networks.

This process stretches the whole DBN with generating visible unit activations by the top hidden layer H_n. Then use the reconstructed input sample V_1' as the output layer of a forward neural network and use BP algorithm slightly adjust the weights. This fine-tuning process makes the MDBN suitable for reconstructing the inputs. In order to generate user comment features, we use the link network features to get the activation of H_n. Then use it to get the activation of the visible comment units as shown in Fig. 3.

4 Experiments and Analysis

4.1 Experiment Setup

In our experiments, the link data of Wikipedia RfA prepared by [14] is used. To avoid the samples imbalance effects, we randomly selected 50000 balanced samples, using 40000 samples for training and the other 10000 for testing.

In order to make a more completely comparison with other state-of-art link prediction methods. We learned Support Vector based Classifier(SVC), which is widely used for link prediction [13], on the same dateset and features. We extracted features by programming in Python, train DBNs and MDBNs in Matlab, and the toolkit 'LIBLINEAR' by [1] is used for learning SVCs.

As the link network structure feature's dimension is 26, the DBN structure for link network is set as $1st$ RBM(26×30), $2nd$ RBM(30×50) and $3rd$ RBM(50×100). The DBN structure for user comment is set as $1st$ RBM(2000×2000), $2nd$ RBM(2000×1000) and $3rd$ RBM(1000×500). So the joint represented feature's dimension is $100(link) + 500(comment)$, and the top DBN's structure in MDBN is set as $1st$ RBM(600×600), $2nd$ RBM(600×800) and $3rd$ RBM(800×1000).

4.2 Link Prediction Results and Analysis

The link prediction results are shown in Table 2. The first column is the name of method. When introducing each method, the first row is the kind of classifier, and the second row is how the features processed. The Discriminative DBN could predict the link value directly. While the Reconstructive DBN without label output layer can only be used to get represent features from the top hidden layer, and need to learn another classifier by represent features.

Firstly, the models learned on multiple source of data outperform the ones learned on single source of data. Both the results of SVC and discriminative DBN have better accuracy than using only one kind of features. It shows that use the joint representation from different data spaces, which imply similarity concept, could improve models' performance.

Secondly, the classifiers with features represented by MDBNs have better performance than original features. It shows that MDBN could abstract the features properly, and transform them into a more linear classifiable space. The results show that SVC and discriminative DBN learned with the features represented by discriminative MDBN have the best performance.

Table 2. Experiment results (Accuracy %)

Methods	Link structure	Word vector	link + word
SVC with (original features)	79.96	82.65	85.96
SVC with represented features by (reconstructive MDBN)	80.43	85.14	86.26
Discriminative DBN by (discriminative MDBN)	82.53	85.22	88.50
SVC with represented features by (discriminative MDBN)	82.36	86.08	89.07

Thirdly, the features represented by discriminative MDBN has better performance than reconstructive MDBN. That is because discriminative DBN is tuned by the sample labels that can directly imply the meaning of samples. While the reconstructive DBN is tuned by a more unsupervised process. Anyway, the reconstructive DBN is more suitable for reconstructing user comments.

4.3 User Comment Generation Results and Analysis

The user comment generation results are shown in Table 3. In order to get a overview of user comments, we list out the top 20 frequency words from comments with positive(+) link and negative(−) links. The link network feature's dimension is 26 while user comment feature's is 2000. It is not sound to generate the user comments by only using the link network features. So we randomly lose half of the words in each comment, then generate the missing words by the link network features and the remaining words.

Table 3. Results of generated user comments(top 20)

Data	Link	Comment words
Original	(+)	the, to, and, a, i, of, good, in, with, is, that, be, this, no, not, for, have, an, admin, but
	(−)	the, to, and, of, a, i, in, is, not, that, but, this, for, with, be, have, you, per, an, as
Generated	(+)	good, this, candidate, an, see, and, my, user, tools, would, per, seems, are, for, reason, all, not, here, concerns, who
	(−)	this, per, not, an, and, for, are, experience, you, is, it, more, at, edits, rfa, be, would, see, also, sorry

Firstly, we could see that the positive comments and negative comments share many common words in the original data, The main different is the words 'good', 'not' and 'but'. The word 'good' appears more times in positive comments, while 'not' and 'but' are more likely to appear in negative comments. However, the article word is the most frequency in users' original comments.

Secondly, in the generated user comments, the positive comments and negative comments share less common words than the condition of original data. The word 'good' is the most frequency one and word 'but' is ranked top 280 in positive comments. In the negative comments, the word 'not' and 'but'(top 77) also have appeared more times than other words. We could also find that there are less article words in top rank.

Thirdly, by comparing the generate and original comments, we find that words with stronger sentiment polarity have higher rank in generate comments. It means that the visible unit standing for sentiment polarity words, such as 'sorry', are more likely to be active in the word generation process. And the top ranked generate comment words seems better to describe the link value.

5   Conclusion

In this paper, we proposed multimodal deep belief network based approaches for predicting link values and generating user comment words in social networks. By learning the joint distribution of link network data and user comment data for classification, the discriminative MDBN method's link prediction performance outperforms other state-of-art methods. By using the reconstructive MDBN method, we can get generated user comments that have a better description for link value than the original ones.

Acknowledgement. This work is supported by the National Natural Science Foundation of China (61272383, 61300114, and 61572151), Specialized Research Fund for the Doctoral Program of Higher Education (No. 20132302120047), the Special Financial Grant from the China Postdoctoral Science Foundation (No. 2014T70340), and China Postdoctoral Science Foundation (No. 2013M530156).

References

1. Fan, R.E., Chang, K.W., Hsieh, C.J., Wang, X.R., Lin, C.J.: Liblinear: a library for large linear classification. J. Mach. Learn. Res. **9**, 1871–1874 (2008)
2. Glorot, X., Bordes, A., Bengio, Y.: Domain adaptation for large-scale sentiment classification: a deep learning approach. In: Proceedings of the 28th International Conference on Machine Learning (ICML-2011), pp. 513–520 (2011)
3. Hinton, G.E.: Training products of experts by minimizing contrastive divergence. Neural Comput. **14**(8), 1771–1800 (2002)
4. Hinton, G.E.: To recognize shapes, first learn to generate images. Prog. Brain Res. **165**, 535–547 (2007)
5. Kiros, R., Salakhutdinov, R., Zemel, R.: Multimodal neural language models. In: Proceedings of the 31st International Conference on Machine Learning (ICML-14), pp. 595–603 (2014)
6. Liben-Nowell, D., Kleinberg, J.: The link-prediction problem for social networks. J. Am. Soc. Inf. Sci. Technol. **58**(7), 1019–1031 (2007)

7. Liu, F., Liu, B., Sun, C., Liu, M., Wang, X.: Deep learning approaches for link prediction in social network services. In: Lee, M., Hirose, A., Hou, Z.-G., Kil, R.M. (eds.) ICONIP 2013, Part II. LNCS, vol. 8227, pp. 425–432. Springer, Heidelberg (2013)

8. Liu, F., Liu, B., Sun, C., Liu, M., Wang, X.: Deep belief network-based approaches for link prediction in signed social networks. Entropy **17**(4), 2140–2169 (2015)

9. Liu, F., Liu, B., Wang, X., Liu, M., Wang, B.: Features for link prediction in social networks: A comprehensive study. In: 2012 IEEE International Conference on Systems, Man, and Cybernetics (SMC), pp. 1706–1711. IEEE (2012)

10. Pang, B., Lee, L.: Opinion mining and sentiment analysis. Found. Trends Inf. Retrieval **2**(1), 1–135 (2008)

11. Srivastava, N., Salakhutdinov, R.R.: Multimodal learning with deep boltzmann machines. In: Advances in neural information processing systems, pp. 2222–2230 (2012)

12. Wang, B., Liu, B., Wang, X., Sun, C., Zhang, D.: Deep learning approaches to semantic relevance modeling for chinese question-answer pairs. ACM Trans. Asian Lang. Inf. Process. (TALIP) **10**(4), 21–37 (2011)

13. Wang, P., Xu, B., Wu, Y., Zhou, X.: Link prediction in social networks: the state-of-the-art. Sci. China Inf. Sci. **58**(1), 1–38 (2014)

14. West, R., Paskov, S.H., Leskovec, J., Potts, C.: Exploiting social network structure for person-to-person sentiment analysis. Trans. Assoc. Comput. Linguist. **2**(1), 297–310 (2014)

15. Zhou, S., Chen, Q., Wang, X.: Fuzzy deep belief networks for semi-supervised sentiment classification. Neurocomputing **131**, 312–322 (2014)

Deep Dropout Artificial Neural Networks for Recognising Digits and Characters in Natural Images

Erik Barrow[✉], Chrisina Jayne, and Mark Eastwood

Coventry University, Coventry, West Midlands, UK
ab3065@coventry.ac.uk

Abstract. Recognising images using computers is a traditionally hard problem in computing, and one that becomes particularly difficult when these images are from the real world due to the large variations in them. This paper investigates the problem of recognising digits and characters in natural images using a deep neural network approach. The experiments explore the utilisation of a recently introduced dropout method which reduces overfitting. A number of different configuration networks are trained. It is found that the majority of networks give better accuracy when trained using the dropout method. This indicates that dropout is an effective method to improve training of deep neural networks on the application of recognising natural images of digits and characters.

Keywords: Character recognition · Natural images · Artificial neural network · Deep learning · Dropout network

1 Introduction

This paper investigates the application of Deep Neural Networks approach [1] and the Dropout Neural Networks method [2] to recognise and classify both digits and characters from a set of natural images [3].

Character recognition is a difficult problem to solve, as many characters can be similar to each other. With the addition of natural images the problem becomes harder, as many variances in the text can occur, such as font, colour, backgrounds, lighting, angle, and texture [3].

Neural networks have been previously applied to the MNIST dataset [4] for recognition of handwritten digits using a Deep Neural Network [4], producing an 0.35 % error rate. This is a low error rate, however all the images in the dataset have been pre-processed.

Deep Neural Networks [1] have been shown to be an effective machine learning technique allowing for more complex features to be learnt. Deep Neural Networks refer to Neural Network Architectures with a large number of hidden layers [1]. The breakthrough paper published in 2006 by Hinton et al. [1] introduced an unsupervised fast, greedy learning algorithm that finds a fairly good set of parameters quickly in deep networks with millions of parameters and

© Springer International Publishing Switzerland 2015
S. Arik et al. (Eds.): ICONIP 2015, Part IV, LNCS 9492, pp. 29–37, 2015.
DOI: 10.1007/978-3-319-26561-2_4

many hidden layers. This discovery enabled the pre-training of deep supervised multi-layer neural networks using the Restricted Boltzmann Machine (RBM) generative model for each layer [1,5]. The pre-training serves as an initialization of the neural network which is then fine-tuned with respect to a supervised criteria as usual. The unsupervised pre-training helps with initialization of the net into a more favorable region of the weight space [6] compared to the random initialization. Several studies have show that this leads to better generalization results [6–8].

More recently Neural Networks utilising the so called dropout method [2] have been shown to provide a better accuracy in classification problems, by reducing overfitting of the data and preventing a network becoming stuck in a local minima. Dropout is a method applied to neural networks that involves dropping neurons during training to cause other neurons to learn new features and prevent overfitting. Dropout based learning has previously been used successfully on the ImageNet dataset [9] of natural images. In this previous application [10] a Deep Convolutional Neural Network was used with dropout, and gained 37 % error on identifying the object in the image, and 17 % error in identifying the image within the top 5 predictions.

The purpose of this paper is to investigate the application of Deep Neural Networks with Dropout on the dataset of natural characters and to find out whether it can help solve the overfitting problem in this context. The results of the paper provide insight into the feasibility of using the dropout method for this problem, and provide recommendations for its practical application.

Motivation for this work comes from the various applications of deep neural networks and the dropout method to data sets of natural images [10] and handwritten digits [11], as well as the need for ways to improve accuracy of reading text in natural images.

This paper is organised into 5 sections. Section 1 introduces the paper, Sect. 2 describes deep neural networks and the dropout method, and Sect. 3 covers the data set and pre-processing used. Experiments and results are presented in Sect. 4, and Sect. 5 concludes.

2 Methodology

The methodology consists of Restricted Boltzmann Machines for pre-training of the network weights, and a Deep Neural Network trained with the dropout method for the classifier.

2.1 Restricted Boltzmann Machines

The first part of the neural network for this problem is a Restricted Boltzmann Machine (RBM). The RBM is used to pre-train the weights of a neural network to provide better results during later training of deep architectures [1].

A boltzmann machine is a system of random variables $\mathbf{v}, \mathbf{h}$ whose joint probability can be described by an energy function as follows:

$$P(\mathbf{v}, \mathbf{h}) = \frac{\exp(-E(\mathbf{v}, \mathbf{h}))}{Z} \tag{1}$$

where Z is a normalisation factor

$$Z = \sum_{\mathbf{h}} \sum_{\mathbf{v}} \exp(-E(\mathbf{v}, \mathbf{h}))$$

The RBM used here has an energy function:

$$E(\mathbf{v}, \mathbf{h}) = -\sum_i v_i b_i - \sum_j h_j c_j - \sum_{ij} v_i W_{ij} h_j \tag{2}$$

RBM is trained by forward propagating the input from one layer to the next. The output is then passed back to the layer that gave the inputs, essentially forward propagating in the reverse direction. The returned results represent a reconstruction of what the network thinks was the input into the system. This altered predicted input is then fed forward again and compared to the original inputs results to produce an error value [12].

The RBM minimises the energy function via approximate gradient descent. We use the enhanced gradient presented in [13]. This training is done one layer at a time. So the first layer will have all the input data run though it over a number of epochs and will then move onto training the next layer, using the outputs of the previous trained layer as the input.

For the energy function given in Eq. 2, the weight and bias gradients have the form:

$$\nabla_e w_{ij} = cov_d(v_i, h_j) - cov_m(v_i, h_j)$$

$$\nabla_e b_i = <v_i>_d - <v_i>_m - \sum_i <h_j>_{dm} \nabla_e w_{ij}$$

$$\nabla_e c_j = <h_j>_d - <h_j>_m - \sum_i <v_i>_{dm} \nabla_e w_{ij}$$

where $< \cdot >_d, < \cdot >_m$ denote averages over the data/model distributions respectively, and $< \cdot >_{dm} = \frac{1}{2} < \cdot >_d + \frac{1}{2} < \cdot >_m$. These values are multiplied by the learning rate λ to provide the parameter updates.

After training of the stacked RBMs, the weights are then used to initialise the Deep Neural Networks that will be used to further train the weights in a supervised manner.

2.2 Deep Neural Networks

A Deep Neural Network is a network that has many hidden layers (typically more than one hidden layer) which utilises initial weights pre-trained with the RBM method. To train the Deep neural network the traditional backpropagation algorithm is used [14]. The inputs are multiplied by their associated weight matrix and then summed before being put through a sigmoid function.

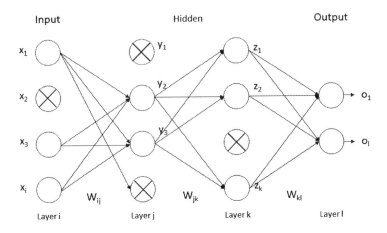

Fig. 1. A network using dropout during training

The result of the sigmoid activation function is then the output for our neuron that can be passed to the next layer as an input for the following layer [15].

When the input has forward propagated as far as the output layer the results of the prediction are compared against a list of targets for that input. The difference between the target and the actual input becomes an error value which is used to tweak the weights for that neuron. The error of that neuron is then passed back through back propagation as a factor for the neurons on the previous layer to take into account and update its own weights based on the error of all the neurons it was attached to.

This training is performed over a number of inputs and then repeated over a large number of epochs, stopping when the minimum has been found, or a maximum amount of epochs has been reached where more training would provide little improvement.

2.3 Dropout Method

Dropout is a relatively new method used in training neural networks and deep learning [2,10]. The idea of dropout is that we randomly select a number of neurons on each layer (excluding the output layer) to be switched off for one epoch. At the end of that epoch the neurons and their associated weights are switched back on, and another set of neurons are selected to be switched off for the next epoch (Fig. 1).

The idea of this is that turning off neurons that have been trained to recognise a certain feature will force a neuron that has learnt nothing or is slow in learning to speed up and try and replace the neurons that have been taken away. Doing this a number of times over a large amount of epochs helps to prevent over fitting of the data and getting stuck in a local minimum.

In a practical application, existing code can be edited to incorporate the dropout method. One of the simplest ways to do this is to select the neurons randomly at the beginning of each epoch, and then after taking a backup of all the weights set the weights associated with the off neurons to zero. This prevents the neurons that are turned off from having an impact on the training and also prevents them from being trained. The advantage of doing dropout this way is that you do not need to re-size and reshape the matrix to accommodate the change in neurons, although not reducing the size of weight matrix could effect speed on a large data set.

It is also necessary to modify the inputs of each node after training for the final prediction by multiplying the input by the proportion of nodes retained during training, so that the input of each node is similar to that of a network with a percentage of nodes switched off. If $\mathbf{x}^{(l)}$ are the node outputs at layer l, and p_l is the proportion of nodes retained during dropout for layer l, the node output during prediction is:

$$\mathbf{x}^{(l)} = sigmoid(p_l \mathbf{W}^{(l)} \mathbf{x}^{(l)} + \mathbf{b})$$

.

3   Data Set Description

The Chars74K image data set [3] is used to conduct experiments on the effectiveness of the dropout deep neural networks. The dataset consists of 74 thousand images in 64 classes (0–9, A-Z, a-z), with images obtained from three sources (natural images, hand drawn images, and synthesised characters). For the purpose of this application we use only the natural image part of this dataset, which provides an accurate simulation of the application of deep neural networks in a real world environment (Fig. 2).

The dataset in its original state is unsuitable for the application of a neural network due to the varying sizes of images and the wide variety of colours that introduce complexity into the recognition process as well as more features to be trained upon.

To prepare the dataset all the images were processed by first Gray scaling the image, and then resizing the image on its largest axis to 50 pixels. The smaller

Fig. 2. A sample of images from the data set

Fig. 3. A sample of images after processing

axis was then padded with pixels to match the average colour of the border pixels on one side to bring the image up to 50×50 pixels. This provides 2500 features to train our network upon.

The final step of preparation was to flip the colours of the Grey scale images so that the backgrounds were black and the number white. On a small number of instances the processed characters were too similar to the background and either became blanked out entirely in some cases, or in other cases where the background and text are the same colour the characters were not switched to white, as can be seen in Fig. 3.

These variations in the processed dataset were minimal to the correctly processed images and should provide minimal impact to the training of the network as it is.

The dataset was split into three subsets. A training set of 5705, a validation set of 1000 for use during training, and a test set of 1000 for testing after the training had finished.

4   Application and Results

The architecture of the Deep Neural Network used in our application is an input layer of 2500 neurons, at least 3 hidden layers, and an output layer of 62 neurons for all the classes. The network is initialised with weights produced by training Restricted Boltzmann machines on the hidden layers, leaving the output layer weights as randomly initialised.

Both Deep Neural Networks with and without the use of dropout were trained with the same architecture to provide a benchmark for the dropout method and prevent any bias.

The RBMs are trained over 10,000 iterations on each layer (stack), to initialise the weights for use with the Deep Neural Networks with and without dropout method.

A number of different configurations can be seen in results Table 1. These same configurations were used to create the 10 different networks. Each configuration has 1 network with and 1 network without dropout, both being initialised

Table 1. Test results with 10,000 iterations

Architecture	Error with dropout	Error without dropout
1000,500,500	45.7	46.9
1500,1000,500	46.4	47.3
1500,1000,500,250	46.6	49.1
500,500,500	45.4	45.3
500,500,1000	43.7	45.8
1500,1000,500,1000	46.6	46.0
500,500,200	45.7	48.3
800,500,500,200	45.4	47.6
500,500,500,500	42.9	44.6
800,500,500,500,250	44.1	46.8

with the same weights. The dropout networks all used the same drop limit of a maximum of 20 % on the input layer and 30 % on the hidden layers.

Table 1 shows that 8 out of 10 dropout based networks give better results over a cap of 10,000 iterations, with an average of 1.9875 % increase in accuracy over the best 8, and a 1.52 % increase over all 10. Where the results are worse for the networks with the dropout the difference is no more than 0.6 percent. The best trained dropout network gained an error of 42.9 % and the best network without only achieved 44.6 %, which is a increase of 1.7 % in accuracy.

Occasionally dropout networks can benefit from extra training as it can take up to double the iterations of a non dropout network to converge [10]. Because of this we gave the dropout networks an extra 5,000 iterations to run. Table 2 shows the results after these extra iterations, and we can see that most networks either improved or had already converged before, except one case where the error went up 0.1 %, which is negligible. For the two cases that had previously been worse for the dropout, one has now overtaken its competing neural network, where as the other has reduced the gap to 0.1 %, suggesting even more training time may eventually lead to the dropout network overtaking the non dropout. We have also reduced ther smallest dropout error to 42.7 %, and improved the average accuracy increase to 1.75 %.

Figure 4 shows the graph of validation error over time (iterations). From the left graph we can see that the network without dropout converged sooner at around 5000 epochs, whereas the dropout network continued to train past this point to eventually get a lower error. The right graph shows a zoomed image of a section of the graph, and we can see that the error on the normal network is very smooth and likely stuck in a local minima, whereas the dropout network error is a lot more jagged as it jumps in and out of minima. At the time of the non dropout network converging the dropout network had a worse validation error, however over further iterations it overtook the non dropout network.

Table 2. Test Results with 15,000 Iterations

Architecture	Error with dropout	Error without dropout
1000,500,500	45.7	46.7
1500,1000,500	46.4	47.5
1500,1000,500,250	46.4	49.1
500,500,500	44.4	45.3
500,500,1000	43.8	45.6
1500,1000,500,1000	46.0	45.9
500,500,200	45.7	48.3
800,500,500,200	44.8	48.1
500,500,500,500	42.7	44.7
800,500,500,500,250	44.1	46.8

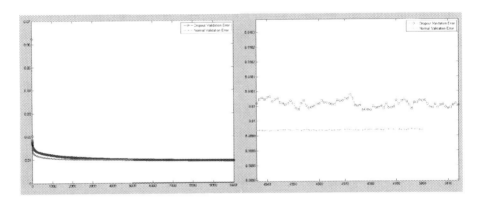

Fig. 4. Validation error during training for the [500,500,500,500] hidden layer architecture.

5   Conclusion

The Dropout method on deep neural networks has been shown to be effective at improving the results for image recognition of natural digits and characters. After extra iterations the average increase in accuracy was 1.75 %, with the best neural network providing 57.3 % accuracy. This indicates that our results are in line with the results reported in [3], which reported 55.26 % accuracy achieved with Multiple Kernel Learning experiments. The same subset of images used in [3] was selected from the dataset, but with 15 training examples per class and a balanced 15 test samples from each class. However our subsets were created using a random sample selection. The accuracy of our networks could be further improved in the future by increasing the maximum amount of iterations allowed and fine tuning the architecture of the neural network.

References

1. Hinton, G.E., Osindero, S., Teh, Y.W.: A fast learning algorithm for deep belief nets. Neural Comput. **18**(7), 1527–1554 (2006)
2. Srivastava, N., Hinton, G., Krizhevsky, A., Sutskever, I., Salakhutdinov, R.: Dropout: A simple way to prevent neural networks from overfitting. J. Mach. Learn. Res. **15**(1), 1929–1958 (2014)
3. de Campos, T.E., Babu, B.R., Varma, M.: Character recognition in natural images. In: Proceedings of the International Conference on Computer Vision Theory and Applications, Lisbon, Portugal, February 2009
4. LeCun, Y., Bottou, L., Bengio, Y., Haffner, P.: Gradient-based learning applied to document recognition. Proc. IEEE **86**(11), 2278–2324 (1998)
5. Bengio, Y.: Learning deep architectures for AI. Found. Trends Mach. Learn. **2**(1), 1–127 (2009)
6. Bengio, Y., Lamblin, P., Popovici, D., Larochelle, H., Montral, U.D., Qubec, M.: Greedy layer-wise training of deep networks. In: NIPS, MIT Press (2007)
7. Glorot, X., Bengio, Y.: Understanding the difficulty of training deep feedforward neural networks. In: Proceedings of the International Conference on Artificial Intelligence and Statistics (AISTATS 2010). Society for ArtificialIntelligence and Statistics. (2010)
8. Larochelle, H., Bengio, Y., Louradour, J., Lamblin, P.: Exploring strategies for training deep neural networks. J. Mach. Learn. Res. **10**, 1–40 (2009)
9. Russakovsky, O., Deng, J., Su, H., Krause, J., Satheesh, S., Ma, S., Huang, Z., Karpathy, A., Khosla, A., Bernstein, M., Berg, A.C., Fei-Fei, L.: ImageNet Large Scale Visual Recognition Challenge. Int. J. Comput. Vision (IJCV), 1–42 (2015). 10.1007/S11263-015-0816-y
10. Krizhevsky, A., Sutskever, I., Hinton, G.E.: Imagenet classification with deep convolutional neural networks. In: Advances in Neural Information Processing Systems, pp. 1097–1105 (2012)
11. Ciresan, D.C., Meier, U., Gambardella, L.M., Schmidhuber, J.: Deep, big, simple neural nets for handwritten digit recognition. Neural Comput. **22**(12), 3207–3220 (2010)
12. Salakhutdinov, R., Mnih, A., Hinton, G.: Restricted boltzmann machines for collaborative filtering. In: Proceedings of the 24th International Conference on Machine Learning, pp. 791–798. ACM (2007)
13. Cho, K.H., Raiko, T., Ilin, A.: Enhanced gradient for training restricted boltzmann machines. Neural Comput. **25**(3), 805–831 (2013)
14. Rumelhart, D.E., Hinton, G.E., Williams, R.J.: Learning representations by back-propagating errors. In: Cognitive Modeling 5 (1988)
15. Hinton, G., Deng, L., Yu, D., Dahl, G.E., Mohamed, A.R., Jaitly, N., Senior, A., Vanhoucke, V., Nguyen, P., Sainath, T.N., et al.: Deep neural networks for acoustic modeling in speech recognition: The shared views of four research groups. IEEE Signal Process. Mag. **29**(6), 82–97 (2012)

A Multichannel Deep Belief Network for the Classification of EEG Data

Alaa M. Al-kaysi[1][(⊠)], Ahmed Al-Ani[1][(⊠)], and Tjeerd W. Boonstra[2,3,4]

[1] Faculty of Engineering and Information Technology,
University of Technology Sydney, Ultimo, NSW 2007, Australia
Alaa.M.Dawood@student.uts.edu.au, Ahmed.Al-Ani@uts.edu.au
[2] School of Psychiatry, University of New South Wales, Sydney, Australia
t.boonstra@unsw.edu.au
[3] Black Dog Institute, Sydney, Australia
[4] Research Institute MOVE, VU University Amsterdam,
Amsterdam, The Netherlands

Abstract. Deep learning, and in particular Deep Belief Network (DBN), has recently witnessed increased attention from researchers as a new classification platform. It has been successfully applied to a number of classification problems, such as image classification, speech recognition and natural language processing. However, deep learning has not been fully explored in electroencephalogram (EEG) classification. We propose in this paper three implementations of DBNs to classify multichannel EEG data based on different channel fusion levels. In order to evaluate the proposed method, we used EEG data that has been recorded to study the modulatory effect of transcranial direct current stimulation. One of the proposed DBNs produced very promising results when compared to three well-established classifiers; which are Support Vector Machine (SVM), Linear Discriminant Analysis (LDA) and Extreme Learning Machine (ELM).

Keywords: Multichannel deep belief network · EEG classification · Transcranial direct current stimulation

1 Introduction

Deep Belief Networks (DBNs) have emerged recently as new learning algorithms composed of multilayer neural networks that have the ability of training high dimensional data. A deep belief network is a probabilistic, generative model which is constructed of multiple layers of hidden units. The undirected layers in the DBNs are called Restricted Boltzmann Machines (RBM) which are trained layer by layer. RBMs have been used in machine learning as a generative model for different types of data [1].

DBNs have mainly been applied to image classification due to the hierarchical structure of images, where edges can be grouped to form segments, which when grouped form objects [2]. They have also been successfully applied to a

© Springer International Publishing Switzerland 2015
S. Arik et al. (Eds.): ICONIP 2015, Part IV, LNCS 9492, pp. 38–45, 2015.
DOI: 10.1007/978-3-319-26561-2_5

number of other classification tasks, such as speech recognition and natural language processing [2]. In contrast, DBNs have not been fully explored for the classification of electroencephalogram (EEG) data, which records the electrical activity of the human brain from multiple electrodes on the scalp. EEG plays a vital role in understanding the functional state of the brain and diagnosing disorders. For example, EEG has been used to study the modulatory effect of transcranial direct current stimulation (tDCS) on changes in cortical activities [3] and the treatment of brain disorders [4].

Wulsin et al. evaluated the classification and anomaly measurement of EEG data using an autoencoder produced by unsupervised DBN learning [5]. Zheng et al. introduced a deep belief network to search the neural signature associated with positive and negative emotional categories from EEG data. They trained the DBN using different entropy features that were extracted from multichannel EEG data. They showed that the performance of DBN was better in higher individual frequency bands [6].

In addition to DBNs that have been proposed for the classification of single-stream data, multi-modality DBNs have also been investigated by a number of researchers. Ngiam et al. proposed a deep network to learn unsupervised features for text, image or audio. They showed that the performance of a bimodal DBN based autoencoder was worse in cross modality and shared representation tasks [2]. Srivastava and Salakhutdinov proposed a DBN of multimodal system for learning a joint representation of data. They used images and text bi-model data to demonstrate that Multimodal DBN (MDBN) can learn good generative attributes. [7]. Cheng et al. proposed MDBN as a new deep learning model to fit large images (e.g. 1024×768). First, they used Scale-invariant feature transform (SIFT) descriptor to reprocess the images and sent these images to MDBN to extract features. Also, they adapted the Markov sub-layer to reflect the neighboring relationship between the inputs [8].

In this paper, three implementations of DBN are presented; one is similar to the single-stream DBN, the second implements a single stream DBN for each channel and then combines the classification outcomes of the different channels using another DBN, and the third case utilises the concept of MDBN in proposing a new multi-stream (or multichannel) DBN. In the third case a "partial" DBN with no decision layer is constructed for each channel, and the last hidden layers of the partial DBNs are combined using higher level hidden and decision layers.

2 Deep Belief Networks

Deep learning is a set of algorithms in machine learning that is based on distributed representations. Hinton et al. proposed a Deep Belief Network as an efficient unsupervised learning algorithm to overcome the complexity of training deep generative model [9,10]. The principal operation of DBN depends on using hierarchical structure of multiple layers of Restricted Boltzmann Machines (RBM). The process of training DBNs involves the individual training of each

RBM one after another and then stack them on top of each other. Each two consecutive layers in DBNs are treated greedily as a Restricted Boltzmann Machine
[11]. RBM consists of a weight matrix w_{ij}, where i represents a visible node and
j a hidden node. Gaussian or Bernoulli stochastic visible units are usually used
in RBMs, while the hidden units are usually Bernoulli [10]. The energy function
of a joint configuration (v, h) of the visible and hidden units is described by [1]:

$$E(v, h) = \sum_{i \in visible} a_i v_i - \sum_{j \in hidden} b_j h_j - \sum_{ij} v_i h_j w_{ij} \qquad (1)$$

where $v_i h_j$ are the binary states of visible and hidden units i, j respectively, w_{ij}
is the weight between visible and hidden units and $a_i b_j$ are their biases. The
model parameters are composed of $[w, v, h]$, $a = [a_1, , a_V]^T$, $b = [b_1, , b_H]^T$. The
probability for every possible pair of a visible and a hidden victor was assigned
by the following function:

$$p(v, h) = \frac{1}{Z} e^{E(v,h)} \qquad (2)$$

where Z is the partition function that represents the summing of all possible
pairs of visible and hidden vectors:

$$Z = \sum_{v,h} e^{E(v,h)} \qquad (3)$$

In each RBM, there are direct connections between units of the two layers,
but there are no connections between units of the same layer, which leads to
an unbiased sample of the state of a hidden and a visible units. Updating the
visible and hidden units can be implemented using Gibbs sampling [1].

3   The Proposed Implementations of Multichannel DBN

Three DBN architectures have been implemented for the classification of multichannel EEG data, as shown in Fig. 1.

– The first architecture (Fig. 1(a)), is the traditional single-stream DBN that
   trains features extracted from all channels using a single network that consists
   of two or more RBMs. The process of training the RBM models is as follows:
   the hidden layer of the first RBM is the visible layer of the second RBM, and
   the hidden layer of the second RBM is the visible layer of the third RBM,
   and so on until arriving to the top RBM. This architecture is used to evaluate
   the ability of a single-stream DBN to extract high level information from the
   features of the different channels.
– The second architecture (Fig. 1(b)) consists of multiple DBNs, one for each
   EEG channel, each of which is trained in a way similar to that of the single-
   stream DBN. The labels of these DBNs are trained using another DBN to
   obtain the final decision. This architecture evaluates the ability of individual
   channels to correctly classify the data and the advantage of combining the
   classification results of individual channels.

– The third architecture, or the multi-stream DBN, (Fig. 1(c)) is implemented using partial DBN that is used to process individual channels. The top hidden layer of each of those partial DBN are combined using unified hidden layer(s) that is(are) followed by the decision layer. The rationale behind proposing this architecture is to extract "local" information from each channel using the partial DBNs, while the higher level information is extracted using the combined hidden layers that fuse the local information of the different channels.

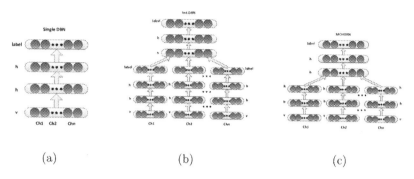

(a)   (b)   (c)

Fig. 1. (a) Single-stream DBN, (b) individual DBN for each channel and a top DBN to combine the channels' classification results, and (c) multi-stream (or multichannel) DBN

4   Data Set

In our experiments, we used data that was collected from patients suffering from depression. Depression is one of the significant mental disorders that affects many people around the world. According to the World Health Organization (WHO), depression is expected to be the major disability causing disease in the world by 2020 [12]. Transcranial direct current stimulation (tDCS) is a promising approach for treating depression [13,14]. tDCS treatment has only been recently established and it is important to study the effects of tDCS on brain functioning. In this study, we utilized data that was collected from twelve patients who were asked to go through three EEG recording sessions [4]. Each session lasted for 20 min. In the first session, baseline EEG was recorded for each patient. This was followed by one session of active and one session of sham tDCS that were randomly ordered among the twelve patients. Scalp EEG was recorded while patients were sitting still with their eyes closed using 62 electrodes placed according to the International 10/20 system, as shown in Fig. 3(a). In this paper, we attempt to differentiate between the three session types using the recorded EEG data. The data of each EEG session was sampled at $F = 2000\,\text{Hz}$. A Hanning window was applied and the power spectral density (PSD) was estimated using the Fourier transform. The average PSD in the five conventional frequency bands, which are Delta $(0.5 \leq \delta < 4)\,Hz$, Theta $(4 \leq \theta < 8)\,Hz$, Alpha $(8 \leq \alpha < 13)\,Hz$, Beta $(13 \leq \beta < 30)\,Hz$, Gamma $(30 \leq \gamma < 100)\,Hz$), were used as features.

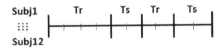

Fig. 2. Testing and training divisions

5 Experimental Results and Discussion

Two experiments were conducted, one aims at ranking channels based on their importance for this particular classification task, while the other aims at evaluating the three proposed multichannel DBN implementations and compare their performance with the benchmark classifiers of SVM, LDA and ELM. We divided the data of each session in to 10 segments, and used segments [5, 6, 9, 10] for testing, while the remaining segments were used to train the classifiers, as shown in Fig. 2. To increase the training data, we further divided each segment into 10 windows.

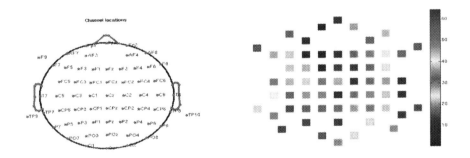

Fig. 3. (a) EEG electrode montage. (b) Classification accuracy for individual channels

We used a linear SVM, LDA and ELM classifiers to rank the performance of channels. The aim of using three classifiers is to identify channels that perform well for all three classifiers, and hence, reduce sensitivity to a particular classifier. We started by evaluating the performance of individual channels. In order to reduce fluctuations in the performance of channels, we considered evaluating subsets of four neighboring channels, and then for each channel we averaged the accuracy of the subsets that consist of that particular channel. For example, the accuracy of channel $C4$ is obtained by averaging the accuracy of subsets

Table 1. Classification error rates of the six classification methods for each of the selected channel subsets

Channels	Classification methods					
	SVM	LDA	ELM	DBN1	DBN2	DBN3
[F7,AF4]	0.507	0.534	0.243	0.396	0.361	0.264
[O1,AF3,FCz]	0.319	0.333	0.298	0.389	0.340	0.257
[TP9,CPz,POz]	0.423	0.416	0.312	0.359	0.403	0.299
[O1,TP9,P6,CPz]	0.340	0.354	0.305	0.375	0.389	0.257
[O1,O2,PO3,PO4,PO7,PO8]	0.465	0.423	0.312	0.382	0.306	0.278
[Fp2,TP9,AF3,AF4,CPz,POz]	0.306	0.305	0.229	0.340	0.340	0.243
[Fp1,Fp2,F7,F8,F9,F10,Fpz]	0.444	0.326	0.229	0.292	0.292	0.236
[Fp1,Fp2,AF3,AF4,AF7,AF8,Fpz]	0.333	0.263	0.277	0.271	0.243	0.215
[O1,O2,PO3,PO4,PO7,PO8,Oz,POz]	0.347	0.284	0.326	0.333	0.285	0.243
[P4,O2,P8,P2,PO4,P6,PO8,Oz]	0.368	0.368	0.263	0.368	0.278	0.257
[Cz,FC1,FC2,CP1,CP2,C1,C2,CPz]	0.410	0.314	0.256	0.361	0.312	0.271
[Fp1,F3,F7,F1,AF3,AF7,F5,F9,Fpz]	0.333	0.298	0.256	0.278	0.243	0.215
[Fp2,F4,F8,Fz,F2,AF4,AF8,F6,F10]	0.431	0.291	0.326	0.306	0.292	0.222
[P3,O1,P7,Pz,P1,PO3,P5,PO7,POz]	0.486	0.430	0.291	0.368	0.361	0.285
[C3,T7,FC5,CP5,TP9,FC3,CP3,C5,TP7]	0.424	0.416	0.333	0.319	0.299	0.229
[C4,T8,FC6,CP6,TP10,FC4,CP4,C6,TP8]	0.389	0.305	0.333	0.347	0.333	0.181
[Fp1,Fp2,F7,F8,AF7,AF8,F9,F10,Fpz]	0.396	0.222	0.259	0.299	0.222	0.167
[Fz,AF3,AF4,AF7,AF8,F5,F6,F9,F10,Fpz]	0.299	0.229	0.270	0.257	0.188	0.111
Average	0.390	0.340	0.284	0.336	0.305	0.235

($\{C4, FC2, C2, FC4\}$, $\{C4, CP6, CP4, C6\}$, $\{C4, CP2, C2, CP4\}$, $\{C4, FC6, FC4, C6\}$). The 62 channels were ranked based on their classification performance, as shown in Fig. 3(b). The color bar represents the rank order, where the best channels have dark blue color. The Figure indicates that channels located over the frontal midline perform relatively better than other regions of the brain, and hence, have the ability to differentiate between the three sessions of baseline, active tDCS and sham tDCS.

In the second experiment we constructed channel subsets of various sizes to have a better evaluation of the performance of the classifiers. These subsets are formed using frontal channels as well as channels of other regions of the brain. The channel combinations are listed in Table 1. Unlike the last experiment, each channel is used by itself in this experiment.

In order to evaluate the performance of the three DBN architectures, they were used to classify the subsets of selected channels. For the single-stream DBN shown in Fig. 1(a), the two hidden layers consist of 200 and 150 units. For the other two DBNs, the four hidden layers consist of 200, 150, 200 and 150 units. The models were trained for 1200 iterations. The obtained results of

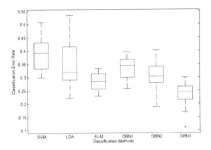

Fig. 4. Classification error of the six classification methods

the six classifiers shown in the table indicate that the ELM classifier performed noticeably better than both SVM and LDA. As for the DBN architectures, results obtained using the single-stream architecture was not found to produce very competitive results. The performance of the second DBN was slightly better than the first one, particularly for the large channel subsets. The multi-stream DBN on the other hand achieved smaller error rates in most channel subsets compared to the other five classifiers, especially for the larger ones. Figure 4 shows a box-plot of the obtained classification error.

The obtained results indicate that DBN needs to be carefully designed when applied to the classification of EEG data, as traditional or single-stream architecture may not produced the expected outcomes. The same is also applicable to combining the classification results of individual channel DBN classifiers. The multi-stream DBN proved to be the most appropriate architecture, as it attempts to extract local attributes from each channel and combine those at a higher level. It is worth mentioning that we have not attempted to optimize the computational cost of the DBN algorithms, and hence, their execution time was noticeably higher than the other three classifiers.

6   Conclusion

In this paper, we evaluated the effectiveness of DBN in the classification of EEG data. Three DBN architectures were presented that fuse the channel information at different levels. In order to evaluate the proposed DBNs, we utilized 62-channel EEG data that was collected from patients suffering from major depressive disorder and started the tDCS treatment. The classification task involved differentiating between baseline EEG, active tDCS and sham tDCS sessions, and the data of each session was represented using the power spectral density in five frequency bands. We first identified the most relevant channels for classification and found the frontal midline channels to be more influential. We have then evaluated the performance of the three DBNs and found the one the multi-stream DBN to outperform the other two DBN implementations as well as the SVM, LDA and ELM classifiers. The multi-stream DBN attempts to extract local attributes from the individual channels in the lower network layers, which are processed and then combined in the higher unified layer(s) to extract high level information that facilitates more accurate labeling of the data. These findings demonstrate the capacity of DBNs as a classification platform for multichannel EEG data. In future studies, we are planning to extend this work to differentiate patients that respond to the tDCS treatment from those who do not.

References

1. Hinton, G.: A practical guide to training restricted boltzmann machines. Momentum **9**(1), 926 (2010)
2. Ngiam, J., Khosla, A., Kim, M., Nam, J., Lee, H., Ng, A.Y.: Multimodal deep learning. In: Proceedings of the 28th International Conference on Machine Learning (ICML-11), pp. 689–696 (2011)
3. Jacobson, L., Ezra, A., Berger, U., Lavidor, M.: Modulating oscillatory brain activity correlates of behavioral inhibition using transcranial direct current stimulation. Clin. Neurophysiol. **123**(5), 979–984 (2012)
4. Powell, T.Y., Boonstra, T.W., Martin, D.M., Loo, C.K., Breakspear, M.: Modulation of cortical activity by transcranial direct current stimulation in patients with affective disorder. PloS one **9**(6), e98503 (2014)
5. Wulsin, D., Gupta, J., Mani, R., Blanco, J., Litt, B.: Modeling electroencephalography waveforms with semi-supervised deep belief nets: fast classification and anomaly measurement. J. Neural Eng. **8**(3), 036015 (2011)
6. Zheng, W.L., Zhu, J.Y., Peng, Y., Lu, B.L.: EEG-based emotion classification using deep belief networks. In: 2014 IEEE International Conference on Multimedia and Expo (ICME), pp. 1–6. IEEE (2014)
7. Srivastava, N., Salakhutdinov, R.: Learning representations for multimodal data with deep belief nets. In: International Conference on Machine Learning Workshop (2012)
8. Cheng, D., Sun, T., Jiang, X., Wang, S.: Unsupervised feature learning using markov deep belief network. In: 2013 20th IEEE International Conference on Image Processing (ICIP), pp. 260–264. IEEE (2013)
9. Do, V., Xiao, X., Chng, E.: Comparison and combination of multilayer perceptrons and deep belief networks in hybrid automatic speech recognition systems. In: Proceedings of Asia-Pacific Signal and Information Processing Association Annual Summit and Conference (APSIPA ASC) (2011)
10. Hinton, G.E., Salakhutdinov, R.R.: Reducing the dimensionality of data with neural networks. Science **313**(5786), 504–507 (2006)
11. Wu, Y., Cai, H.: A simulation study of deep belief network combined with the self-organizing mechanism of adaptive resonance theory. In: 2010 International Conference on Computational Intelligence and Software Engineering (CiSE), pp. 1–4. IEEE (2010)
12. Hosseinifard, B., Moradi, M.H., Rostami, R.: Classifying depression patients and normal subjects using machine learning techniques and nonlinear features from eeg signal. Comput. Methods Programs Biomed. **109**(3), 339–345 (2013)
13. Kalu, U., Sexton, C., Loo, C., Ebmeier, K.: Transcranial direct current stimulation in the treatment of major depression: a meta-analysis. Psychol. Med. **42**(09), 1791–1800 (2012)
14. Loo, C.K., Alonzo, A., Martin, D., Mitchell, P.B., Galvez, V., Sachdev, P.: Transcranial direct current stimulation for depression: 3-week, randomised, sham-controlled trial. Br. J. Psychiatry **200**(1), 52–59 (2012)

Deep Convolutional Neural Networks for Human Activity Recognition with Smartphone Sensors

Charissa Ann Ronao and Sung-Bae Cho[✉]

Department of Computer Science, Yonsei University,
Seoul, 120-749, South Korea
cvronao@sclab.yonsei.ac.kr, sbcho@cs.yonsei.ac.kr

Abstract. Human activity recognition (HAR) using smartphone sensors utilize time-series, multivariate data to detect activities. Time-series data have inherent local dependency characteristics. Moreover, activities tend to be hierarchical and translation invariant in nature. Consequently, convolutional neural networks (convnet) exploit these characteristics, which make it appropriate in dealing with time-series sensor data. In this paper, we propose an architecture of convnets with sensor data gathered from smartphone sensors to recognize activities. Experiments show that increasing the number of convolutional layers increases performance, but the complexity of the derived features decreases with every additional layer. Moreover, preserving the information passed from layer to layer is more important, as opposed to blindly increasing the hyperparameters to improve performance. The convnet structure can also benefit from a wider filter size and lower pooling size setting. Lastly, we show that convnet outperforms all the other state-of-the-art techniques in HAR, especially SVM, which achieved the previous best result for the data set.

Keywords: Human activity recognition · Deep learning · Convolutional neural network · Smartphone · Sensors

1 Introduction

Human activity recognition (HAR) is a classification task that makes use of time-series data from devices such as accelerometers and gyroscopes (as seen in Fig. 1), preprocess these signals, extract relevant and discriminative features from them, and finally, recognize activities by using a classifier. Especially those gathered from sensors, time-series data have a strong 1D structure in that they are very highly correlated to temporally nearby local readings [1]. Moreover, considering that people perform activities with different poses and styles, and a complex activity can be decomposed into more basic movements, time-series sensor data have inherent translation and hierarchical characteristics [2]. Therefore, it is vital to utilize a classifier that takes into consideration of these intrinsic properties of time-series sensor signals.

Deep learning, and in particular, convolutional neural networks (convnet), has been gaining a lot of attention in recent years due to its excellent performance in fields such as image and speech. In the same way, the ability of convnets to exploit the local

© Springer International Publishing Switzerland 2016
S. Arik et al. (Eds.): ICONIP 2015, Part IV, LNCS 9492, pp. 46–53, 2015.
DOI: 10.1007/978-3-319-26561-2_6

dependency and translation equivariance of data, together with its hierarchical feature extraction mechanism, is what makes it very suitable for use with time-series sensor signals [3]. In this paper, we exploit convnets to recognize activities using time-series data from smartphone sensors, investigate the effect of varying its architecture, and compare its performance with other state-of-the-art classifiers in the HAR domain.

This paper is organized as follows: we review the related work in Sect. 2, followed by a presentation of convnets with HAR and time-series sensor signals in Sect. 3. Section 4 presents and examines the experimental results, and we draw our conclusion in Sect. 5.

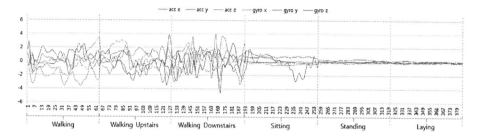

Fig. 1. One vector of accelerometer and gyroscope sensor data for every activity

2   Related Works

One of the most cited papers in HAR is by Bao and Intille, which proved that signals from accelerometers, especially those placed on the thigh, are very useful in recognizing different activities [4]. Years later, accelerometers in smartphones were utilized by Kwapisz et al. [5]. They placed the smartphone in the user's pocket and classified six different activities using classifiers such as J48 decision trees (DT), multilayer perceptrons (MLP), among others. In subsequent works, the accelerometer and gyroscope were found to be the lead sensors able to effectively recognize human activities [6]. Also, classifiers used in these latter works include naïve Bayes (NB) and k-nearest neighbors (KNN). Furthermore, Anguita et al. used a multi-class SVM to classify six different activities using 561 features extracted from both the accelerometer and gyroscope [7]. All of these classifiers, however, exhibited low performance in differentiating very similar activities such as walking upstairs and walking downstairs. We show that convnets are able to overcome this problem by exploiting the locally dependent and hierarchical characteristics of time-series sensor signals.

Among the deep learning techniques used with HAR and sensors, restricted Boltzmann machines (RBM) and sparse auto-encoders were the most common [8, 9]. However, both approaches are fully-connected neural networks which do not capture the local dependencies of time-series signals [1]. Convnets were finally applied to human activity recognition using sensor signals in [10]. However, this work only made use of a one-layered convnet architecture, which in turn did not exploit the hierarchical physiognomy of activities.

3 Human Activity Recognition with Convolutional Neural Networks

Human activities have inherent translation characteristics in that different people perform the same kind of activity in different ways, and that a fragment of an activity can manifest at different points in time [2]. Activities are also hierarchical in a sense that complex activities are composed of basic actions or movements prerequisite to the activity itself. Moreover, when using sensor signals to classify activities, it is important to take into account the temporal dependence of nearby readings [1]. Convolutional neural networks (convnet) exploit these data characteristics with its convolutional layer, which computes a mixture of nearby sensor readings, and pooling layer, which makes the representation invariant to small translations of the input [3]. A simple convnet architecture is illustrated in Fig. 2.

3.1 Convolutional Neural Networks

Convolutional neural networks perform convolution instead of matrix multiplication (as with fully-connected neural networks). Let $x_i^0 = [x_1, \ldots, x_N]$ as the accelerometer and gyroscope sensor data input vector, where N is the number of values per window. The output of the convolutional layer is:

$$c_i^{l,j} = \sigma \left(b_j + \sum_{m=1}^{M} w_m^j x_{i+m-1}^{0,j} \right),$$ (1)

where l is the layer index, σ is the activation function, b_j is the bias term for the j th feature map, M is the kernel/filter size, and w_m^j is the weight for the j th feature map and m th filter index. A summary statistic of nearby outputs are derived from $c_i^{l,j}$ by the pooling layer. The pooling operation used in this paper, max-pooling, is characterized by outputting the maximum value among a set of nearby inputs, given by

$$p_i^{l,j} = \max_{r \in R}(c_{i \times T+r}^{l,j}),$$ (2)

where R is the pooling size, and T is the pooling stride. Several convolutional and pooling layers can be stacked on top of one another to form a deep neural network architecture. These layers act as a hierarchical feature extractor; they extricate discriminative and informative representations with respect to the data, with basic to more complex features manifesting from bottom to top.

A simple softmax classifier is utilized to recognize activities, which is placed at the topmost layer. Features from the stacked convolutional and pooling layers are aligned/flattened to form feature vectors $p^l = [p_1, \ldots, p_I]$, where I is the number of units in the last pooling layer, as input to the softmax classifier:

$$P(c|p) = \text{argmax}_{c \in C} \frac{\exp(p^{L-1} w^L + b^L)}{\sum_{k=1}^{N_c} \exp(p^{L-1} w_k)},$$ (3)

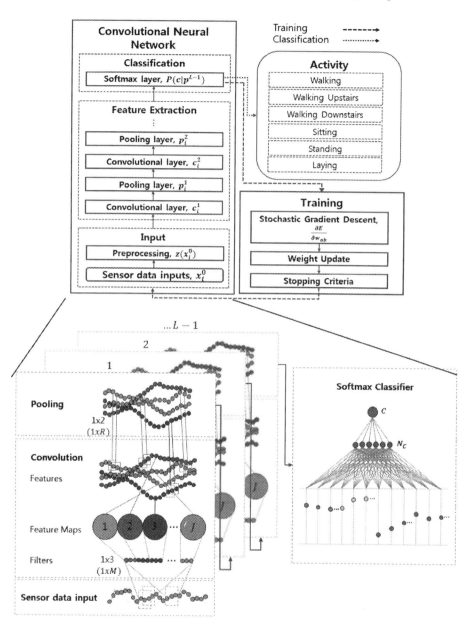

Fig. 2. Convolutional neural network architecture and training procedure

where c is the activity class, L is the last layer index, and N_C is the total number of activity classes.

Forward propagation is performed using Eqs. (1)-(3), which give us the error values of the network. Weight update and error cost minimization through training is done by stochastic gradient descent (SGD) on minibatches of sensor train data examples.

Backpropagation to adjust weights is done by computing the gradient of the convolutional weights:

$$\frac{\partial E}{\partial w_{ab}} = \sum_{i=0}^{N-M-1} \frac{\partial E}{\partial x_{ij}^{l}} y_{(i+a)}^{l-1},$$ (4)

where E is the error/cost function, $y_{(i+a)}^{l-1}$ is the nonlinear mapping function equal to $\sigma\left(x_{(i+a)}^{l-1}\right) + b^{l-1}$, and deltas $\frac{\partial E}{\partial x_{ij}^{l}}$ are equal to $\frac{\partial E}{\partial y_{ij}^{l}}\sigma'\left(x_{ij}^{l}\right)$. The forward and back propagation procedure is repeated until a stopping criterion is met, e.g., if a maximum number of epochs is reached, among others.

3.2   Convolutional Neural Network Architecture and Hyperparameters

Based on the equations above, we can clearly see that there is a large number of possible combination of settings for the convnet hyperparameters, resulting in different architecture configurations. To observe and assess the effects of varying the values of these hyperparameters to the performance of the network with respect to HAR sensor data, we incorporated greedy-wise tuning starting from the number of layers L (one-layer, L_1; two-layer, L_2; and three-layer, L_3), the number of feature maps J, the size of the convolutional filter M, and the pooling size R. We varied the number of layers from 1 to 3, the number of feature maps from 10 to 200 in intervals of 10 (the same number for all layers [13]), the filter size from 1x3 to 1x15, and pooling size from 1x2 to 1x15.

4   Experiments

The publicly-available HAR smartphone dataset from the UCI repository has been utilized for all our experiments. This dataset contains accelerometer and gyroscope data from 30 subjects performing 6 different activities, namely: walking, walking upstairs, walking downstairs, sitting, standing, and laying. Data from random 21 subjects were set aside for training and the remaining data from 9 subjects, for testing. The raw accelerometer and gyroscope xyz signals were standardized to have a mean of zero (subtracted by the mean and divided by the standard deviation), resulting in a vector of 128 z-score values for every activity example. This means that we perform 6-channel (acc and gyro xyz axes), 1D convolution on the input.

For every run, padding is used to perform 'full' convolution with the inputs in every layer, and ReLU activation function and max-pooling operation are used. We set the learning rate to 0.01, gradually increase momentum from 0.5 to 0.99, the weight decay to 0.00005, and maximum epochs to 5000, with an early stopping criterion. The model with the best score on the validation set is saved during the run [13].

Figure 3 shows the results of every run with increasing L and J. On validation data, it can be seen that there is a gradual increase in the performance gap after adding an additional layer. On the other hand, on test data, it can be observed that there is a noticeably bigger jump in performance with the addition of the second layer than with the addition of the third layer. This means that indeed, second layer features are much more complex than first layer ones, but difference in complexity between the second layer and the third layer features are not that great. Yet, we cannot deny that adding a third layer to the network still improves performance. We have also tried to add a fourth layer, but it didn't prove to be beneficial as it just lowered the performance.

The configuration that achieved the best accuracy on the test set is carried over for tuning the filter and pooling sizes, as seen in Fig. 4. All through the run, the best performance convnet would be of an L_3 architecture. With $J = 200$, $M = 11$, and $R = 2$, convnet achieved a performance of 92.60 % on test data. This shows that blindly increasing the hyperparameters as large as we can does not necessarily translate to better performance. Instead, it is more important to preserve the information passed from input to the convolutional layers; the product of the number of features and the number of values in the input should be roughly constant, or ensured to be maintained, with the addition of each layer. This is seen with the best J configurations, with $J(L_1) = 120$ and $J(L_2) = 130$ very close to the number of points in an activity example ($N = 128$), and the jump to $J(L_3) = 200$ covering for the addition of three convolutional layers from the input. Also, high performance were seen with filter sizes 1x9 to 1x14, a span of 0.18 to 0.28 s, indicating that there tends to be a wider correlation between nearby time readings that should be exploited (as compared to 1x3 (0.06 s), where only immediate neighboring readings are considered).

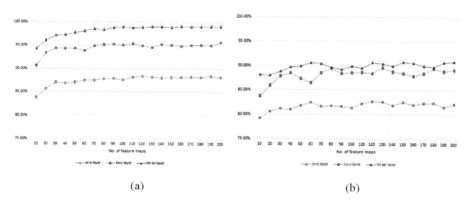

(a)   (b)

Fig. 3. Performance of different convnet architectures with increasing number of layers and feature maps on (a) validation data and (b) test data

Incorporating an MLP with 1000 units (with dropout) instead of only a softmax layer increases the performance by 1.2 %, and adopting an inverted pyramid architecture (96-192-192-1000-6, $M = 9$, and $R = 3$) with a tuned learning rate (0.02) gives us the best convnet performance of 94.79 %, as seen in Table 1. Comparing the confusion matrix of convnet with SVM's, it is apparent that convnet is better at discriminating moving activities than the latter, especially with very similar ones such as walking downstairs and walking upstairs, which achieved 100 % accuracy. However, stationary activities are a little hard for convnet to recognize, with Laying obtaining the worst classification rate.

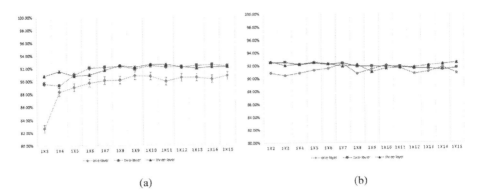

(a)   (b)

Fig. 4. Performance of convnet with $J(L_1) = 120, J(L_2) = 130, J(L_3) = 200$, and increasing (a) filter size, and with $M = 9, M = 14, M = 11$, and increasing (b) pooling size

Table 1. Confusion matrix of best convnet

		Predicted Class						
		W	WU	WD	Si	St	L	Recall
Actual Class	Walking	491	3	2				98.99%
	W. Upstairs		471					100.00%
	W. Downstairs			420				100.00%
	Sitting				436	34	21	88.80%
	Standing		1		24	496	11	93.23%
	Laying				43	23	471	87.71%
	Precision	100.00%	99.16%	99.53%	86.68%	89.69%	93.64%	**94.79%**

We further compare the best convnet with other algorithms, as seen in Table 2. Results show that convnet outperforms other state-of-the-art techniques, which are all using hand-crafted features. It is slightly better than SVM, which previously achieved the best performance in this data set. Using hand-crafted features with convnet also further improves performance, showing that convnet derives another layer of more discriminative features above the hand-crafted ones.

Table 2. Performance of best convnet compared to other algorithms

Algorithm	Accuracy
Naïve Bayes	76.63 %
J48 decision trees	83.63 %
Artificial neural network	91.08 %
Support vector machine	94.61 %
Convnet (L_3)	94.79 %
HCF + Convnet $(J\left(L_1\right) = 200)$	95.75 %

5   Conclusion

We have evaluated convolutional neural networks in recognizing activities using time-series, accelerometer and gyroscope sensor data. It is found that the complexity of derived features indeed increases with increasing convolutional layers, but the difference in complexity between adjacent layers decrease with each additional layer. Preserving the information passed from input to convolutional layers is also important, and a wider correlation between nearby readings should be exploited. Lastly, it was shown that convnet outperformed all other state-of-the-art algorithms in HAR, particularly SVM (which has previously achieved the best performance on the HAR dataset), exhibiting noticeably better performance in classifying moving activities than the latter.

Future studies will include examining and comparing the confusion matrices of other HAR algorithms with convnet, and the inclusion of frequency convolution.

Acknowledgements. This research was supported by the MSIP (Ministry of Science, ICT and Future Planning), Korea, under the ITRC (Information Technology Research Center) support program (IITP-2015-R0992-15-1011) supervised by the IITP (Institute for Information & communications Technology Promotion).

References

1. LeCun, Y., Bengio, Y.: Convolutional Networks for Images, Speech, and Time-Series, The Handbook of Brain Theory and Neural Networks, pp. 255–258 (1998)
2. Dobrucalı, O., Barshan, B.: Sensor-activity relevance in human activity recognition with wearable motion sensors and mutual information criterion. Inf. Sci. Syst. **264**, 285–294 (2013)
3. Bengio, Y., Goodfellow, I.J., Courville, A.: Deep Learning, Book in preparation for MIT Press (2015). http://www.iro.umontreal.ca/~bengioy/dlbook
4. Bao, L., Intille, S.S.: Activity recognition from user-annotated acceleration data. In: Ferscha, A., Mattern, F. (eds.) PERVASIVE 2004. LNCS, vol. 3001, pp. 1–17. Springer, Heidelberg (2004)
5. Kwapisz, J., Weiss, G., Moore, S.: Activity recognition using cell phone accelerometers. SIGKDD Explor. **12**(2), 74–82 (2010)
6. Wu, W., Dasgupta, S., Ramirez, E.E., Peterson, C., Norman, G.J.: Classification accuracies of physical activities using smartphone motion sensors. J. Med. Internet Res. **14**(5), 105–130 (2012)
7. Anguita, D., Ghio, A., Oneto, L., Parra, X., Reyes-Ortiz, J.L.: A public domain dataset for human activity recognition using smartphones. Eur. Symp. Artif. Neural Netw. (ESANN) **19**, 437–442 (2013)
8. Plotz, T., Hammerla, N.Y., Olivier, P.: Feature learning for activity recognition in ubiquitous computing. Int. Joint Conf. Artif. Intell. (IJCAI) **2**, 1729–1734 (2011)
9. Vollmer, C., Gross, H.-M., Eggert, J.P.: Learning features for activity recognition with shift-invariant sparse coding. In: Mladenov, V., Koprinkova-Hristova, P., Palm, G., Villa, A.E., Appollini, B., Kasabov, N. (eds.) ICANN 2013. LNCS, vol. 8131, pp. 367–374. Springer, Heidelberg (2013)
10. Zeng, M., Nguyen, L.T., Yu, B., Mengshoel, O.J., Zhu, J., Wu, P., Zhang, J.: Convolutional neural networks for human activity recognition using mobile sensors. In: International Conference on Mobile Computing, Applications and Services (MobiCASE) (2014)
11. Bengio, Y.: Practical Recommendations for Gradient-Based Training of Deep Architectures. arXiv:1206.5533v2 (2012)

Concentration Monitoring with High Accuracy but Low Cost EEG Device

Jun-Su Kang, Amitash Ojha, and Minho Lee[✉]

School of Electronics Engineering, Kyungpook National University,
1370 Sankyuk-Dong, Puk-Gu, Taegu 702-701, South Korea
{wkjuns,amitashojha,mholee}@gmail.com

Abstract. Concentration is an important part of our life especially during learning or thinking. Visually or auditory evoked concentration affects information processing in human brain. To understand the concentration process of humans, the underlying neural mechanism needs to be explored. EEG device is a promising device to understand underlying neural mechanism of various cognitive functions. In this paper, we propose an accurate concentration monitoring method using a low cost EEG device. Our low cost EEG device has two channel electrodes (FP1, FP2). Usually small channel EEG devices face filtering problem because commonly used filtering method, such as ICA, fails with less number of electrodes. In our work, we investigate effective filters for removing noises from raw data and suitable features for monitoring the concentration status with the low cost EEG device in real time. We collect EEG data from 10 participants for rest state with open eyes and concentration task state. For concentration task, Sudoku game is used. Using support vector machine, we successfully distinguish between rest state and concentration state over 88 % accuracy in real time.

Keywords: EEG · Concentration · Power spectral density · Support vector machine · Hurst exponent

1 Introduction

The basic idea of Brain-Computer Interface (BCI) is to interpret the user's thoughts to generate corresponding commands for controlling a computer or device. Several tools have been used to acquire brain signals such as electroencephalogram (EEG), functional Near Infra-Reds (fNIRs), Magnetoencephalography (MEG), Functional Magnetic Resonance Imaging (fMRI), electrocorticography (ECoG), etc. EEG is the most familiar among these devices because it is non-invasive, relatively cheap and safe to use. Moreover, there are several released commercial products to collect EEG signals such as Emotiv EPOC [1], Biopac B-alert [2], Quasar DSI system [3], etc.

Recently, several BCI studies related to cognitive science have been conducted [4]. One of the interesting studies is related to concentration detection and/or recognition. Concentration is a major research area in psychology and cognitive science but its application in technology and in real world faces several problems. For instance, in the case of EEG, one of the problems is the number of electrodes. Generally studies use over four or five electrodes [5, 6]. However, customers want to use cheap devices but

© Springer International Publishing Switzerland 2015
S. Arik et al. (Eds.): ICONIP 2015, Part IV, LNCS 9492, pp. 54–60, 2015.
DOI: 10.1007/978-3-319-26561-2_7

any cheap device usually has less number of electrodes with low sampling rate, which causes a filtering problem. Independent Component Analysis (ICA) based filtering is often used in EEG signal filtering. However, ICA based filtering requires a big number of electrodes. Using a low EEG device with small number of sensors, ICA cannot be used as a preprocessor.

In this study, we attempt to solve above mentioned problems for enhancing the possibility of concentration based applications using a low cost EEG device in real-world environment. In particular, we try to find concentration features using commercial two-channel EEG device from the frontal lobe (FP1, FP2). In order to improve the classification accuracy, we investigate several combination of different types of digital filter such as notch filter, band pass filter, median filter and average filter as a preprocessing, and try to compare the features consisting of frequency bands, combination of the frequency bands as well as hurst index to construct the concentration monitoring system with high accuracy but low cost.

The rest of the paper is organized as follows: in the next section, we explain our proposed method to signal processing and candidate features. In Sect. 3, we report the results of the experiment. In Sect. 4, we conclude our results and present our future plans.

2 Method

2.1 Overall Structure of the Proposed System

Figure 1 describes the overall structure of the proposed system. The system consists of EEG acquisition, filtering on time domain, filtering on feature domain and support vector machine (SVM) with radial basis function (RBF) kernel. The system uses the EEG signals obtained while participants play Sudoku game. To remove power noise, notch filter is applied at 60[Hz] followed by band pass filter based on 9th-order Butterworth type of 4-40[Hz]. Then, Short Time Fourier Transform (STFT) is used to acquire both the dynamic characteristic and time-frequency analysis of EEG signal while performing the concentration task. For removing eye movement effect on the power spectral density, the median filter and average filter are applied. For classifier, SVM with RBF kernel is used. The filtered power spectral density is used to classify user's status into rest state and concentration state.

2.2 Experiment Design

In this study, a low cost EEG device, developed by SOSO [7], is used (Fig. 2). The device has two channel EEG sensors located on the forehead (FP1, FP2). The electrode type is dry electrode type, which makes it simple to use. The sampling rate of this device is 256[Hz] with a high and low pass filter at 0.5[Hz] and 100[Hz].

The experiment includes discrimination of resting and concentrating states. In the beginning of the experiment, participants are asked to rest with closed eyes for 30 s and then with opened eyes for another 30 s. The concentration task (Sudoku game) starts after the rest session. We select the Sudoku task as it requires greater concentration

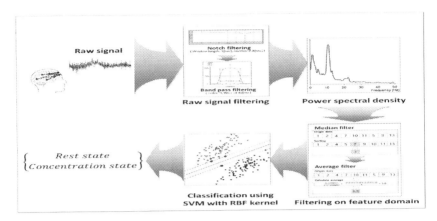

Fig. 1. Overall procedure of proposed system

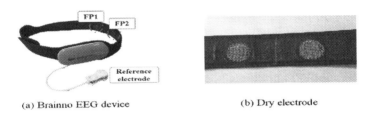

(a) Brainno EEG device   (b) Dry electrode

Fig. 2. Appearance of low cost EEG device

because the player needs to consider each row and column combination with each rectangle combination. After the concentration task, participants are asked to rest with opened and closed eyes for 30 s. For constructing the concentration monitoring system, we analyze the signals obtained during rest with opened eye and concentration task. Figure 3 shows the procedure of the study.

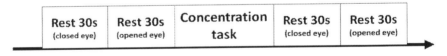

Fig. 3. Procedure of experiment in the study

2.3 EEG Signal Processing

As a preprocessing for the EEG signals, notch filter and band pass filter are used. The EEG device is wired type thus it requires line noise removal. We apply notch filter to remove the line noise. The stop band frequency is 60[Hz]. After notch filter, band pass filter is applied at frequency band of interest (4 ∼ 40[Hz]) (Fig. 4).

From the band pass filtered signals; we consider the Short-Time Fourier Transform (STFT) to time-frequency representation. STFT, the simplest time-frequency representation, is a two-dimensional representation created by computing the Fourier transform

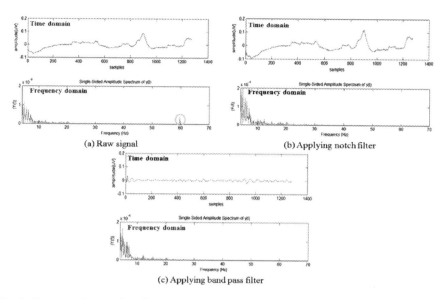

Fig. 4. Preprocessing results of raw EEG signals (a) raw signal (b) notch filtered signal (c) band pass filtered signal

using a sliding temporal window. By using the STFT, we can observe how the frequency of the EEG signals changes with time. As such the details of the resulting STFT are greatly influenced by the choice of window. In this study, the Hamming window with 1 s is used as window function in STFT. The overlap time is 0.8 s. However, even after we get the time-frequency representation from STFT, noise component such as blink still remains. To remove eye movement effect, we consider the median and average filters. The median filter is known as non-linear type filter and robust smoother [8]. From this median filter, we get the median value of the signals on each window. Average filter is also good to smooth the signal, which can remove the influences of noises after feature extraction. Figure 5 shows the effect of each filter.

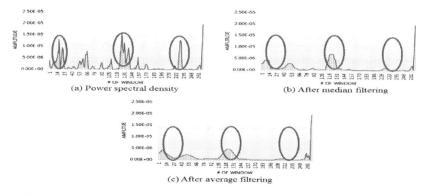

Fig. 5. Results of filtering on feature domain (a) power spectral density after band pass filtering (b) power spectral density after median filtering (c) power spectral density after average filtering

2.4   Candidate Features for Concentration Detection

For detecting user's concentration status from EEG signals, we test several concentration related features. In the field of EEG study, there are well-known six frequency bands. Table 1 shows the brief information of these frequency bands.

Table 1. EEG frequency bands and meaning of each bands

Band	Frequency [Hz]	Activity
Delta, δ	0.1 ~ 3.9	Deep sleep
Theta, θ	4 ~ 7.9	Stable
Alpha, α	8 ~ 12.9	Relaxed
SMR	12 ~ 14.9	Concentration
Beta, β	13 ~ 29.9	Nervous
Gamma, γ	30 ~ 50	More nervous

Alpha wave is a well-known feature of concentration [9]. Some researchers insist that β wave is also related to concentration [5]. Moreover, to improve ADHD patients, $\frac{SMR + mid\beta}{\theta}$ [10, 11] is popularly used. Therefore, to find the best features for our setup, we compare those features along with other possible candidate features such as high β, mid β and Hurst exponent (HE), etc. We include the HE because of its long-term dependency [12]. So, we test the HE for concentration detection, too. The range of HE is [0 1]. Equation (1) shows the mathematical expression for HE.

$$He = \frac{\log(R/S)}{\log(cT)} \tag{1}$$

where T is the duration of the sample data, R is distance between the maximum and minimum of cumulated sum and S is standard deviation.

3   Experimental Results

A total of 10 individuals participate in the experiment. Their average age is 25 years (22 ~ 29 years old). All participants are right-handed and without any obvious signs of medical or psychological diseases. All participants are native Korean speakers. The solving time of Sudoku game is distributed as 131[sec] ~ 500[sec]. Average solving time is 295.1[sec]($\pm$ 125.1[sec]).

Table 2 shows the results of classification in alpha band. This result is obtained from 10 cross validations. The results confirm that filtering on feature domain effectively improves the classification performance. In other words, the median filter and average filter effectively removes the eye movement noise making the signal dynamics simple.

Table 3 shows the classification result of features. High β power shows best performance among all features. Most subjects show the best classification accuracy at the

Table 2. Classification results for each filtering step

	BPF		BPF + median		BPF + median +avr.	
	Train (%)	Test (%)	Train (%)	Test (%)	Train (%)	Test (%)
Sub 1.	53.82	53.66	75.85	75.45	79.65	79.24
Sub 2.	61.03	60.93	84.09	83.94	85.56	85.47
Sub 3.	61.78	61.68	64.67	64.66	65.67	65.55
Sub 4.	57.34	56.89	77.09	77.05	84.52	84.50
Sub 5.	40.76	39.98	72.64	72.44	73.39	73.31
Sub 6.	45.77	45.84	75.03	74.98	76.88	76.83
Sub 7.	70.86	70.90	90.10	89.95	91.75	91.47
Sub 8.	69.25	68.91	77.15	76.60	76.50	76.25
Sub 9.	52.50	52.14	83.54	83.55	85.79	85.62
Sub 10.	39.46	39.44	76.88	76.87	86.70	86.70
Average	55.26	55.04	77.71	77.55	80.64	80.49

* avr.: average filter

Table 3. Classification results for features

	$\frac{SMR + mid\beta}{\theta}$		HE		High β	
	Train (%)	Test (%)	Train (%)	Test (%)	Train (%)	Test (%)
Sub 1.	69.81	69.66	59.28	59.07	87.93	87.77
Sub 2.	81.68	81.57	84.42	84.38	80.15	79.92
Sub 3.	75.10	74.95	81.72	81.58	90.50	90.46
Sub 4.	70.64	70.42	75.23	75.14	83.86	83.75
Sub 5.	69.79	69.43	69.05	69.00	66.37	66.15
Sub 6.	75.25	75.23	52.99	52.53	98.34	98.33
Sub 7.	80.94	80.84	80.08	79.67	93.69	93.66
Sub 8.	67.23	67.10	71.28	71.28	99.06	99.02
Sub 9.	83.00	82.91	76.53	76.43	94.67	94.55
Sub 10.	75.60	75.52	66.49	66.34	83.41	83.36
Average	74.90	74.76	72.79	72.66	88.77	88.67

high β feature. The high β band reflects the agitation or excitement [13]. The β band is well-known frequency band for focus or concentration. Our result also supports it.

4 Conclusion

The goal of this study is two-fold. First to test the performance of various candidate features ($\frac{SMR + mid\beta}{\theta}$, HE and high β, etc.) that are known to be related to concentration. Second to test various filters on EEG signals with less number of electrodes. For this we applied the STFT and SVM with RBF kernel. Results show that filtering on feature domain successfully improves the classification performance. Also, we found that the high β band shows the best result to identify the concentration state in current experimental condition. Based on our results we argue that the proposed method can be used to determine the concentration level for any commercial EEG device using small number of electrodes. In our future work, we would like to investigate the best features which are related to multiple intelligence [14].

Acknowledgements. This work was partly supported by the ICT R&D program of MSIP/IITP. [10041826, Development of emotional features sensing, diagnostics and distribution s/w platform for measurement of multiple intelligence from young children] (50 %) and Regional Specialized Industry R&D program funded by the Ministry of Trade, Industry and Energy (R0002982) (50 %).

References

1. Emotiv EPOC. https://emotiv.com/epoc.php
2. Biopac B-alert. http://www.biopac.com/
3. Quasar DSI system. http://www.quasarusa.com/products_dsi.htm
4. Gruzelier, J.H.: EEG-neurofeedback for optimising performance. I: a review of cognitive and affective outcome in healthy participants. Neurosci. Biobehav. Rev. **44**, 124–141 (2014)
5. Gola, M., Magnuski, M., Szumska, I., Wróbel, A.: EEG beta band activity is related to attention and attentional deficits in the visual performance of elderly subjects. Int. J. Psychophysiol. **89**(3), 334–341 (2013)
6. Tonin, L., Leeb, R., Sobolewski, A., del Millán, R.J.: An online EEG BCI based on covert visuospatial attention in absence of exogenous stimulation. J. Neural Eng. **10**(5), 056007 (2013)
7. SOSO. http://soso-g.co.kr/twfo/en/
8. Lan, T., Adami, A., Erdogmus, D., Pavel, M.: Estimating cognitive state using EEG signals. In: Signal Processing Conference, 2005 13th European. IEEE (2005)
9. Klimesch, W.: Alpha-band oscillations, attention, and controlled access to stored information. Trends Cogn. Sci. **16**(12), 606–617 (2012)
10. Lubar, J.O., Lubar, J.F.: Electroencephalographic biofeedback of SMR and beta for treatment of attention deficit disorders in a clinical setting. Biofeedback self-Regul. **9**(1), 1–23 (1984)
11. Lee, C., Kwon, J., Kim, G., Hong, K., Shin, D.-S., Lee, D.: A study on EEG based concentration transmission and brain computer interface application. In: The Institute of Electronics Engineers of Korea, pp. 41–46 (2009)
12. Kannathal, N., Acharya, U.R., Lim, C., Sadasivan, P.: Characterization of EEG—A comparative study. Comput. Methods Programs Biomed. **80**(1), 17–23 (2005)
13. Zhang, Y., Chen, Y., Bressler, S.L., Ding, M.: Response preparation and inhibition: the role of the cortical sensorimotor beta rhythm. Neuroscience **156**(1), 238–246 (2008)
14. Gardner, H.: Frames of mind: The theory of multiple intelligences. Basic books (2011)

Transfer Components Between Subjects for EEG-based Driving Fatigue Detection

Yong-Qi Zhang[1], Wei-Long Zheng[1], and Bao-Liang Lu[1,2(✉)]

[1] Center for Brain-like Computing and Machine Intelligence,
Department of Computer Science and Engineering,
Shanghai Jiao Tong University, Shanghai 200240, China
bllu@sjtu.edu.cn
[2] Key Laboratory of Shanghai Education Commission for Intelligent
Interaction and Cognitive Engineering, Shanghai Jiao Tong University,
Shanghai 200240, China

Abstract. In this paper, we first build up an electroencephalogram (EEG)-based driving fatigue detection system, and then propose a subject transfer framework for this system via component analysis. We apply a subspace projecting approach called transfer component analysis (TCA) for subject transfer. The main idea is to learn a set of transfer components underlying source domain (source subjects) and target domain (target subjects). When projected to this subspace, the difference of feature distributions of both domains can be reduced. Meanwhile, the discriminative information can be preserved. From the experiments, we show that the TCA-based algorithm can achieve a significant improvement on performance with the best mean accuracy of 77.56 %, in comparison of the baseline accuracy of 66.56 %. The improvement shows the feasibility and efficiency of our approach for subject transfer driving fatigue detection from EEG.

Keywords: EEG · Driving fatigue detection · Transfer learning · Domain adaptation

1 Introduction

Among all the driving accidents, driving fatigue is believed to be the most significant one [1]. When people feel tired, the ability to maintain essential vigilance and avoid accidents gets worse. In this case, to develop an efficient method for detecting drivers' fatigue during driving is an essential issue for many transportation safety researchers. Up till now, various methods have been studied to analyze fatigue status during driving. Among them, EEG is believed to be the most reliable one [2].

To present, most machine learning based Brain-Computer Interface systems [3] rely on a calibration session to train the models. This calibration is time-consuming for real world applications [4]. The intuitive approach is to train the classifiers on a set of collected data from the previous experiments and then

© Springer International Publishing Switzerland 2015
S. Arik et al. (Eds.): ICONIP 2015, Part IV, LNCS 9492, pp. 61–68, 2015.
DOI: 10.1007/978-3-319-26561-2_8

make prediction on the unseen data from a new subject. However, it becomes technically difficult as the nonstationary nature of EEG signals and the variance of environment enlarges the distribution difference between subjects. Traditional machine learning methods assume that the training data and test data follow the same distribution and they have the same feature space [5]. However, this assumption can not be always satisfied between different subjects for EEG-based driving fatigue detection.

Domain adaptation, one of the branches of transfer learning, is feasible to address this problem. Here, we introduce a domain adaptation method to driving fatigue detection across subjects. Let $X \in \mathcal{X}$ be the feature space and $y \in \mathcal{Y}$ be the corresponding drivers' fatigue labels. In this case, $\mathcal{X} = \mathbb{R}^{C \times d}$, where C is the number of samples and d is the number of feature dimensions. Let $P(X)$ be the marginal probability distribution of X. In our case, the source and target subjects share the same feature space, $\mathcal{X}_S = \mathcal{X}_T$, but the marginal distributions are different, $P(X_S) \neq P(X_T)$. The key assumption in most domain adaptation methods is that $P(Y_S|X_S) = P(Y_T|X_T)$.

The major problem for subject transfer is how to reduce the difference between the distributions of the source and target subject data. In this paper, we introduce a feature reduction based transfer learning method called transfer component analysis (TCA) [6] to subject transfer driving fatigue detection from EEG. Although the distributions of source domain and target domain in high dimensional space are different, there still exists a low dimensional manifold space where the distributions of both domains are similar [7]. TCA tries to learn a set of common transfer components underlying both domains. When projected to this subspace, the difference of feature distributions of both domains can be reduced, and at the same time, the discriminative information can be preserved. There exists a transformation function $\phi(.)$ such that $P(\phi(X_S)) \approx P(\phi(X_T))$ and $P(Y_S|\phi(X_S)) \approx P(Y_T|\phi(X_T))$. We demonstrate that TCA can be used to improve the performance of driving fatigue detection system for subject transfer from EEG.

This paper is organized as follows. In Sect. 2, a systematic description of feature extraction and TCA method is given. Section 3 describes the fatigue driving simulation experiment, the data collection and pre-processing, the fatigue measurement by PERCLOS rule, as well as the parameters we used for data processing. In Sect. 4, we compare the performance of the proposed method with that of the baseline. Finally, in Sect. 5, we make conclusion and future work.

2   Algorithm Description

2.1   Feature Extraction

The raw EEG data is firstly downsampled to 200 Hz in order to reduce computing complexity. After that, the EEG data is processed with a bandpass filter between 0.3 Hz and 60 Hz in order to filter the noise and artifacts. Each channel of EEG data is divided into 10 s segments without overlapping.

The existing studies indicate that differential entropy (DE) features can achieve better performance than conventional EEG features such as power spectral density [8]. Thus, we employ the DE features in this study. If a random variable obeys the Gaussian distribution $N(\mu, \sigma^2)$, the DE can simply be calculated by the following formulation,

$$
\begin{aligned}
h(X) &= -\int_{-\infty}^{\infty} \frac{1}{\sqrt{2\pi\sigma^2}} \exp \frac{(x-\mu)^2}{2\sigma^2} \log \frac{1}{\sqrt{2\pi\sigma^2}} \exp \frac{(x-\mu)^2}{2\sigma^2} dx \\
&= \frac{1}{2} \log 2\pi e \sigma^2.
\end{aligned}
\tag{1}
$$

For a fixed length EEG segment, DE is proven to be equivalent to the logarithm energy spectrum in a certain frequency band [9]. Then, we choose the differential entropy of each frequency band and each channel as features. The choice of frequency bands and channels is discussed in Sect. 3.4.

2.2 Transfer Component Analysis

The major computational problem in transfer learning is how to reduce the difference between the distributions of the source and target domain data. As the feature distributions of different subjects cannot be ignored, discovering a good feature representation across domains is crucial.

As mentioned in Sect. 1, we need to find a transformation $\phi(.)$ such that $P(\phi(X_S)) \approx P(\phi(X_T))$ and $P(Y_S|\phi(X_S)) \approx P(Y_T|\phi(X_T))$. Since we have no labeled data of target subject, $\phi(.)$ cannot be learnt through minimizing the distance between $P(Y_S|\phi(X_S))$ and $P(Y_T|\phi(X_T))$. Pan et al. [6] proposed an efficient approach called transfer component analysis to learn $\phi(.)$.

Let Gram matrices defined on the source domain, target domain and cross-domain data in the embedded space be $K_{S,S}$, $K_{T,T}$, $K_{S,T}$ and $K_{T,S}$, respectively. The kernel matrix K is defined as

$$
K = \begin{bmatrix} K_{S,S} & K_{S,T} \\ K_{T,S} & K_{T,T} \end{bmatrix} \in \mathbb{R}^{(n_1+n_2) \times (n_1+n_2)}.
\tag{2}
$$

By the virtue of kernel trick, the distribution distance can be written as $tr(KL)$, where $K = [\phi(x_i)^T \phi(x_j)]$, and L is defined as

$$
L_{ij} = \begin{cases} \frac{1}{n_1^2}, & x_i, x_j \in X_S \\ \frac{1}{n_2^2}, & x_i, x_j \in X_T \\ -\frac{1}{n_1 n_2}, & otherwise \end{cases}
\tag{3}
$$

A matrix $\widetilde{W} \in \mathbb{R}^{(n_1+n_2) \times m}$ transforms the empirical kernel map K to an m-dimension space (where $m \ll n_1 + n_2$). The resultant kernel matrix is

$$
\tilde{K} = (KK^{-1/2}\widetilde{W}(\widetilde{W}^T K^{-1/2}K)) = KWW^T K,
\tag{4}
$$

where $W = K^{-1/2}\widetilde{W}$. With the definition of $\widetilde{K}$ in (4), the distance between empirical means of the two domain X'_S and X'_T can be written as

$$Dist(X'_S, X'_T) = tr((KWW^T K)L) = tr(W^T KLKW). \tag{5}$$

A regularization term $tr(W^T W)$ is usually added to control the complexity of W, while minimizing (5).

Besides reducing the difference of the two distributions, $\phi(.)$ should also preserve the data variance which is related to the target learning task. From (4), the variance of the projected samples is $W^T KHKW$, where $H = I_{n_1+n_2} - \frac{1}{n_1+n_2}\mathbf{1}\mathbf{1}^T$ is the centering matrix, $\mathbf{1} \in \mathbb{R}^{n_1+n_2}$ is the column vector with all 1 s and $I_{n_1+n_2}$ is the identity matrix.

Therefore, the objective function of TCA is

$$\begin{aligned} \min_{W} \quad & tr(W^T KLKW) + \mu tr(W^T W) \\ s.t. \quad & W^T KHKW = I_m \end{aligned} \tag{6}$$

where $\mu > 0$ is a tradeoff parameter, and $I_m \in \mathbb{R}^{m \times n}$ is the identity matrix.

According to [6], the solutions W are the m leading eigenvectors of $(KLK + \mu I)^{-1}KHK$, where $m < n_1 + n_2$. The algorithm of TCA for subject transfer is summarized in Algorithm 1. The detailed descriptions of TCA can be refered to [6]. After obtaining the transformation matrix W, standard machine learning methods can be used in the subspace KW across domains.

Algorithm 1. TCA-based Subject Transfer

Input: Source domain data set $\mathcal{D}_S = \{(x_{S_i}, y_{src_i})\}_{i=1}^{n_1}$ and the target domain data set $\mathcal{D}_T = \{x_{T_j}\}_{j=1}^{n_2}$;
Output: New feature matrix Φ;
  1: Construct kernel matrix K from $\{x_{S_i}\}_{i=1}^{n_1}$ and $\{x_{T_j}\}_{j=1}^{n_2}$, matrix L and centering matrix H;
  2: Eigendecompose the matrix $(KLK + \mu I)^{-1}KHK$ and select the m ($m \ll n_1 + n_2$) leading eigenvectors to construct the transformation matrix W;
  3: **return** transformation matrix $\Phi = KW$.

---

3   Experimental Setup

3.1   Subjects and Procedure

In order to collect EEG data under different mental states, subjects are asked to drive in a car in a simulated driving environment. The experimental scene is shown in Fig. 1. A NeuroScan4.3 system is used to collect the original EEG data and the SMI eye tracking glasses are set for eye closure data. All the actions of each subject are recorded in each section in order to guarantee the appearance of both fatigue and wake period. Six healthy subjects of 20–23 years old in total have participated in this experiment. Before each session, the subjects have a short time to be acquainted with the driving simulation system. Each subject has performed three experiments on different days. Each of the experiments is carried out about 1 h and 30 min in a quiet and comfortable room with normal illumination.

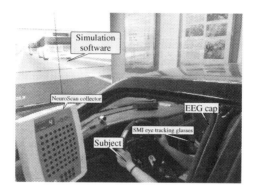

Fig. 1. The experimental scene

3.2 Data Collection and Pre-processing

For each experiment, a total of 62 EEG channels are recored and sampled at 1000 Hz. Then, the original data is down-sampled to 200 Hz to reduce computational complexity and filtered between 0 and 60 Hz to eliminate some artifacts. The 1h 30 min sequence of each experiment is divided with a 10 s windows, and thus get nearly 540 fragments for each experiment.

3.3 Feature Smooth

As driving fatigue is a relatively stable variable, the sudden changes of feature values are mostly caused by noisy and artifact, especially for the artifact caused by electromyography (EMG). Moving Average algorithm is used to choose the average value of a successive sequence as the value of a certain point. By using the Moving Average algorithm, the sudden changes caused by artifact can be removed.

3.4 Feature Extraction

Although we have 62 channels of EEG data collected, not all these channels are related to fatigue. Some unrelated signals may disturb the detection of fatigue. According to [10,11], occipital lobe region has strong relation with fatigue status. Figure 2(a) shows the chosen channels.

We extract the DE features in twenty-five frequency bands (1–2Hz, 2–4Hz,..., 48–50Hz), rather than the normal five frequency bands, with a 2048 point short-time fourier transform. The existing studies show that the increased number of

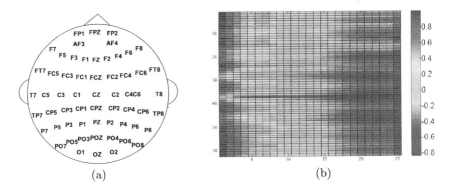

(a)   (b)

Fig. 2. The chosen channels in occipital lobe region (a), and the power spectrum in relation to 62 channels and 25 frequency bands (b).

frequency bands has a benefit of preserving more information of EEG signals [12]. From Fig. 2(b), we can see that, the power of adjacent frequency bands shows dynamic differences for different mental states. In contrast, the rough five frequency bands-separation will lose some important information.

3.5   Fatigue Measurement

Along with the EEG data, the eye closure data is recorded by the SMI eye tracking glasses. The glasses use infrared lighting sources to track the eye gaze and eye movements [13]. Then, the information of eye movement is used to calculate the proportion of time that a subject's eyes are closed over a specified period, namely the PERCLOS value. According to the existing work [14], we set the threshold to be 0.75. If the PERCLOS value is above 0.75, the subject is considered in fatigue state. Otherwise, it corresponds to normal state.

3.6   Detailed Parameters for Training

Here, we present the details about the parameters for training and the baseline for comparison. For baseline method of subject transfer for driving fatigue detection, a straightforward and intuitive method is to concentrate all the data of all the subjects and employ the leave-one-subject-out cross validation. We employ support vector machine with 'RBF' kernel as the classifiers.

For TCA, there are three parameters, kernel parameter σ, parameter μ, and the dimensionality of latent space D. We fix two parameters and adjust the remaining parameter one by one to find out the optimal three values. We first set $\mu = 1, m = 10$ and search for the best σ value in the range $[10^{-5}, 10^{5}]$. Afterwards, we set μ and m in the same manner as TCA. The range of searching for μ and m are $[10^{-3}, 10^{3}]$ and $[3, 220]$ respectively. After the searching process, the optimal parameter combination is $\sigma = 100, \mu = 2$ and $m = 7$.

4   Experiment Results

From our parameter searching results, we observe that TCA-based subject transfer algorithm is a parameter-stable algorithm. The valus of σ and μ influence little on the results. For the dimension m, this value has great effect on the result. If m is too big, the transferred feature matrix involves too much redundant features. Meanwhile, if m is too small, it contains little useful information. As a result, $m = 7$ is the best dimension for our driving fatigue detection problem.

From Fig. 3, we see that TCA-based SVM performs much better than conventional SVM for our 18 experiments. In experiment 1, 2, 8, TCA-based subject transfer does not show great improvements compared to the baseline as the noise in original EEG is strong. Even in experiment 8, there are two bad electrodes after the whole process. But for others, when the inherent structure of EEG data is good, the TCA gets more significant improvement. After eliminating the data of experiment 1, 2, 8, the improvement grows up to 10.46 % from the conventional SVM (66.56 %) to TCA-SVM (77.02 %).

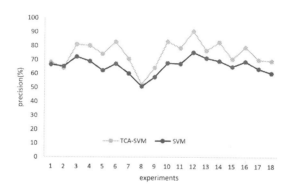

Fig. 3. The comparison of TCA-based framework and SVM

5   Conclusion and Future Work

In this paper, we have proposed a subject transfer framework for EEG-based driving fatigue detection via shared common components. We have introduced a domain adaptation method called TCA to address the structural and functional variability across subjects. This method can learn transfer components in a low-dimensional latent space from the source domain and target domain. Meanwhile, we present our experiment details for driving fatigue detection. The experimental results show that TCA can achieve a significant improvement of 10.46 % in accuracy compared to the conventional method.

Acknowledgments. This work was partially supported by the National Natural Science Foundation of China (Grant No. 61272248), the National Basic Research Program of China (Grant No. 2013CB329401), and the Science and Technology Commission of Shanghai Municipality (Grant No. 13511500200).

The authors would like to thank Prof. Sinno Jialin Pan, Nanyang Technological University, Singapore, for providing the source code of Tranfer Component Analysis.

References

1. Nobe, S., Wang, F.-Y., et al.: An overview of recent developments in automated lateral and longitudinal vehicle controls. In: 2001 IEEE International Conference on Systems, Man, and Cybernetics, vol. 5, pp. 3447–3452. IEEE (2001)
2. Cajochen, C., Khalsa, S.B.S., Wyatt, J.K., Czeisler, C.A., Dijk, D.-J.: EEG and ocular correlates of circadian melatonin phase and human performance decrements during sleep loss. Am. J. Physiol. Regul. Integr. Comp. Physiol. **277**(3), R640–R649 (1999)
3. Buch, E., Weber, C., Cohen, L.G., Braun, C., Dimyan, M.A., Ard, T., Mellinger, J., Caria, A., Soekadar, S., Fourkas, A., et al.: Think to move: a neuromagnetic brain-computer interface (bci) system for chronic stroke. Stroke **39**(3), 910–917 (2008)
4. Krauledat, M., Tangermann, M., Blankertz, B., Müller, K.-R.: Towards zero training for brain-computer interfacing. PLOS One **3**(8), 2967–2976 (2008)
5. Pan, S.J., Yang, Q.: A survey on transfer learning. IEEE Trans. Knowl. Data Eng. **22**(10), 1345–1359 (2010)
6. Pan, S.J., Tsang, I.W., Kwok, J.T., Yang, Q.: Domain adaptation via transfer component analysis. IEEE Trans. Neural Netw. **22**(2), 199–210 (2011)
7. Ben-David, S., Blitzer, J., Crammer, K., Pereira, F., et al.: Analysis of representations for domain adaptation. Adv. Neural Inf. Process. Syst. **19**, 137 (2007)
8. Duan, R.-N., Zhu, J.-Y., Lu, B.-L.: Differential entropy feature for EEG-based emotion classification. In: IEEE EMBS Conference on Neural Engineering, pp. 81–84. IEEE (2013)
9. Shi, L.-C., Bao-Liang, L.: Eeg-based vigilance estimation using extreme learning machines. Neurocomputing **102**, 135–143 (2013)
10. Kuzniecky, R.: Symptomatic occipital lobe epilepsy. Cortex **3**(12), 13 (1998)
11. Cheng, S.-Y., Hsu, H.-T.: Mental Fatigue Measurement Using EEG. INTECH Open Access Publisher, Rijeka (2011)
12. Ngoc, H.T., Nguyen, T.H., Ngo, C.: Average partial power spectrum density approach to feature extraction for EEG-based motor imagery classification. Am. J. Biomed. Eng. **3**(6), 208–219 (2013)
13. Ji, Q., Yang, X.: Real-time eye, gaze, and face pose tracking for monitoring driver vigilance. Real Time Imag. **8**(5), 357–377 (2002)
14. Gao, X.-Y., Zhang, Y.-F., Zheng, W.-L., Lu, B.-L.: Evaluating driving fatigue detection algorithms using eye tracking glasses. In: IEEE EMBS Conference on Neural Engineering, pp. 767–770. IEEE (2015)

A Proposed Blind DWT-SVD Watermarking Scheme for EEG Data

Trung Duy Pham, Dat Tran$^{(\boxtimes)}$, and Wanli Ma

Faculty of Education, Science, Technology and Mathematics,
University of Canberra, Canberra, Australia
{duy.pham,dat.tran,wanli.ma}@canberra.edu.au

Abstract. Copyright and integrity violations of digital medical data have become security challenges since the ever-increasing distribution of them between clinical centres and hospitals through the widespread usage of telemedicine, teleradiology, telediagnosis, and teleconsultation. Therefore, preserving authenticity and integrity of medical data including Electroencephalogram (EEG) has become a necessity. Watermark techniques have been thoroughly studied as a means to preserve the authenticity and integrity of the content of medical. Although there is a large volume of works on watermarking and stenography, not many researchers have addressed issues related to EEG data. This paper proposes a new approach that uses discrete wavelet transform (DWT) to decompose EEG signals and singular value decomposition (SVD) to embed watermark into the decomposed EEG signal. Based on the advantage of using the SVD technique, our proposed method achieved blind detection of watermark in which the receiver does not require the original EEG signal to retrieve the watermark. Experimental results show that the proposed EEG watermarking approach maintains the high quality of the EEG signal.

Keywords: EEG · Watermarking · DWT · SVD

1 Introduction

Electroencephalogram (EEG) is a recording of the electrical activity of human brain, usually acquired by a number of electrodes placed on the scalp. In the past decade, there has been tremendous growth in EEG based research activities, e.g. automated EEG analysis for diagnosis of neurological diseases, and brain computer interface (BCI) [1]. EEG provides an insight on the human brain: it can detect abnormalities, diagnose mental disorders like dementia, epileptic seizures and psychiatric disorders. EEG is also used in telemedicine and brain-computer interface (BCI) applications. The widespread emergence of computer networks and the popularity of electronic managing of medical records have made it possible for digital medical data to be shared across the world for services such as telemedicine, teleradiology, telediagnosis, and teleconsultation [2]. However, there are multiple danger zones like copyright and integrity violations of digital

© Springer International Publishing Switzerland 2015
S. Arik et al. (Eds.): ICONIP 2015, Part IV, LNCS 9492, pp. 69–76, 2015.
DOI: 10.1007/978-3-319-26561-2_9

objects [3]. It is well known that the integrity and confidentiality of medical data including EEG data are critical issues for ethical as well for legal reasons.

Watermarking techniques have been thoroughly studied as a means to achieve proof of ownership and transaction tracking [4]. Data watermarking is relatively old branch in the domain of stenography (data hiding). Its main purpose is to trace the watermarked data as that data is used. Data watermarking can be used to protect data ownership priority, copyright assertion and securely sending data through unsecured communication channels such as the Internet or the wireless channels.

EEG signals over long periods of time can be enormous in size, and can be used as a host to carry other biomedical information watermarked inside them. In this scenario, the watermarking algorithm has to preserve the main features of the EEG signal. Moreover, it must guarantee that diagnosis of the EEG signal can be done directly without removing the watermark. The watermark must thus ideally be invisible on a trace. Although there is a large volume of works on watermarking and stenography, not many researchers have addressed issues related to medical data such as EEG or electrocardiogram (ECG).

Kong and Feng [5] describe three popular watermarking techniques applied to EEG signals: Benders Patchwork, Van Schyndels LSB, and Chens QIM watermarking with regard to their ability to verify EEG signal integrity after noise contamination resulting from communication. The authors of [6] describe a spread spectrum watermarking scheme that embeds robust and imperceptible watermarks into ECG signals. However, such a scheme addresses security considerations only during communication of the data rather than over the course of sharing it. The authors of [7] propose an LSB watermarking scheme in support of proof-of-ownership for ECG signals. However, LSB watermarks provide poor robustness to malicious alterations.

The wavelet transform is a powerful mathematical tool that has been used in different research areas. Multi-scale capabilities of the wavelet transform that highlight the local and global characteristics of the signal make it being an efficient tool in particular for watermarking applications [8]. Kasmani and Naghsh-Nilchi [9] proposed a combination of DWT and DCT to embed the binary watermark. They performed 3-level DWT decomposition and then applied DCT to embed the watermark. Results showed a good watermark recovery against many attacks but this scheme suffers from high time complexity. Moreover, it had a non-blind detection. Recently the SVD transform has been widely used in robust image watermarking due to its attractive mathematical features [10]. The main idea in SVD-based watermarking methodology is to embed the watermark into the singular values by applying the SVD onto whole or small blocks of the cover image [11].The advantage of using SVD technique is to achieve blind detection of watermark. Another advantage of using SVD technique is that there is no constraint on size of matrices, they can be either square or rectangle [12]. In spite of the robust performance of SVD based watermarking techniques, they cannot outperform the robustness of frequency-based methods against different attacks [13]. Therefore, the best approach to enhance the robustness of SVD-based methods is to employ this transform along with the frequency based transforms like DWT.

In addition, watermarking schemes can be classified in to two schemes: non-blind scheme and blind scheme. Non-blind scheme requires presence of original content during watermarking detection, while blind scheme does not. Non-blind one suffers from two distinct disadvantages [3]:

1. Security compromise: Attacker may claim the ownership by inserting another watermark in the cover object since non-blind detection does not guarantee unequivocal claims of ownership by the content creator.
2. Practical application constraints: During detection for every watermarking application, non-blind scheme can not ensure the presence of original content.

With the development in watermarking research, blind schemes are matching the performance criteria of non-blind schemes. In this paper, we propose the blind watermarking scheme based on DWT and SVD. The rest of paper is organized as follows. We first present the proposed EEG watermarking approach in Sect. 2. Section 3 presents experiments and results. We conclude the paper and present our future work in Sect. 4.

2   The Proposed EEG Watermarking Approach

The SVD mathematical technique provides an elegant way for extracting algebraic features from a 2-D matrix. The main properties of the matrix of the singular values can be exploited in EEG watermarking. The proposed approach is presented in Fig. 1 including two phases: watermarking embedding and watermarking extraction.

2.1   EEG Watermarking Embedding

The steps of the EEG watermarking embedding are summarized as follows:

1. The 1-D EEG signals is transformed into a 2-D matrix
2. Watermark W is decomposed using SVD

$$W = U_W S_W V_W^T \tag{1}$$

3. Haar wavelet is applied to decompose the 2-D matrix into sub-bands: LL, HL, LH and HH.
4. Applying SVD to HH band

$$H = U_H S_H V_H^T \tag{2}$$

5. Replacing the singular values of the HH band with singular values of the watermark
6. Applying inverse SVD to obtain the modified HH band

$$H' = U_H S_W V_H^T \tag{3}$$

7. The watermarked matrix in 2-D format is obtained using inverse DWT
8. The 2-D watermarked matrix is transformed again into a 1-D EEG signal

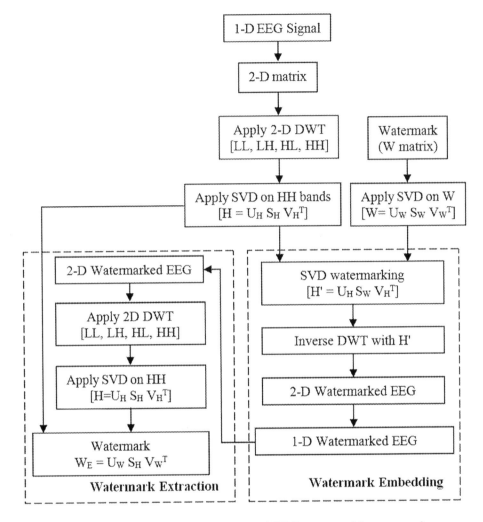

Fig. 1. The flow chart of the proposed EEG watermarking approach

2.2 EEG Watermarking Extraction

The EEG watermarking extraction includes:

1. The 1-D watermarked EEG signals is transformed into a 2-D matrix
2. Using Haar wavelet, decompose the 2-D Watermarked EEG in to four sub-bands: LL, HL, LH and HH.
3. Applying SVD to HH band.

$$H = U_H S_H V_H^T \qquad (4)$$

4. Extracting the singular values from HH band.

5. Constructing the watermark using singular values and orthogonal matrices U_W and V_W obtained using SVD of original watermark

$$W_E = U_W S_H V_W^T \tag{5}$$

The EEG watermarking extraction phrase does not require original cover signal at the receiver, thus the proposed approach constitutes a blind watermarking scheme.

3 Experiments and Results

3.1 Dataset

The dataset was used in this research is the DEAP dataset (Dataset for Emotion Analysis using Electroencephalogram, Physiological and Video Signals) which is an open database proposed by Koelstra et al. [14]. It includes the EEG and multiple peripheral physiological signals of 32 subjects who were subject to music video stimuli. During the experiments, EEG signals were recorded with 512 Hz sampling frequency, which were down-sampled to 128 Hz and filtered between 4.0 Hz and 45.0 Hz. In this section, several experiments are carried out to test the performance of the proposed DWT-SVD watermarking approach for some EEG signal channels. All the experimentation and testing is performed in Windows 8 platform. MATLAB version 2014 is used for the implementation of the proposed algorithm.

3.2 Results

The proposed EEG watermarking embedding and extraction process are performed and the results are shown in Fig. 2.

These figures reveal that the proposed EEG watermarking does not degrade the quality of the watermarked EEG signals. The watermarked signals are very similar to the cover signal and the differences are not evident to naked eye. We use Peak-Signal-to-Noise ratio (PSNR) as a metric to check perceptual similarity between original EEG and watermarked EEG. PSNR is the ratio of maximum amplitude of the cover EEG signal to the mean squared deviation between two signals. Higher the value of PSNR, better is the quality. PSNR represents a measure of peak error and expressed in terms of the logarithmic decibel units as follows:

$$PSNR = 20 log_{10} \frac{max[x_c]}{\sqrt{\frac{1}{N} \sum_{n=1}^{N} [x_c - x_w]^2}} \tag{6}$$

where N is total number of samples. x_c is the amplitude of the cover signal and x_w is the amplitude of the watermarked signal. According to Chen et al. [15], PSNR above 40 dB indicates a good perceptual fidelity. The PSNR (in dB) of the watermarked EEG are shown in Table 1, all of them are higher than 40 dB, thus this indicates that diagnosability is not lost and degradation to the overall signal is acceptable

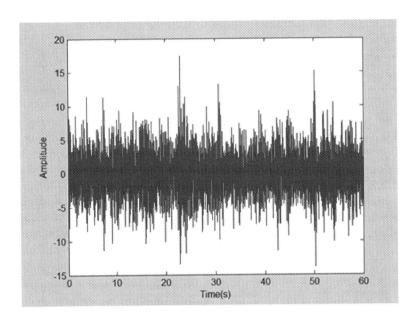

Original EEG signal

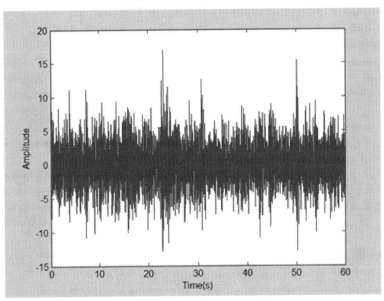

Watermarked EEG signal

Fig. 2. Original EEG signal and watermarked EEG signal in Fp1 channel, subject 1

Table 1. PSNR of watermarked EEG channels

EEG channel	PNSR (in dB)
Fp1	57.533981
F3	54.062241
C3	55.179020
P7	56.187211
Oz	56.757695
T8	47.890337

4 Conclusion and Future Work

We presented the watermarking scheme for the purpose of ensuring the integrity and privacy of the EEG signals. We propose a basic watermarking which is based on cascading DWT with SVD mathematical technique. The EEG signal is transformed into a 2-D format and the singular values of the resulting 2-D matrix are used for watermark embedding. The proposed scheme is based on the idea of replacing singular values of the HH band with the singular values of the watermark. The results show that hiding the watermark in the cover signal in such a way that the diagnosability is not lost and degradation to the overall signal is acceptable. In the future, we intend to develop the reliably and secure scheme for secret patient information transmission in EEG cover signal without much loss in the diagnosability.

References

1. Nidermeyer, E., Silva, F.L.D.: Electroencephalography: Basic Principles, Clinical Applications and Related Fields, 5th edn. Lippincott Williams and Wilkins, Philadelphia (2005)
2. Bhatnagar, G., Jonathan, W.U.Q.M.: Biometrics inspired watermarking based on a fractional dual tree complex wavelet transform. Future Gener. Comput. Syst. **29**(1), 182–195 (2013)
3. Gupta, A.K., Raval, M.S.: A robust and secure watermarking scheme based on singular values replacement. Sadhana **37**(4), 425–440 (2012)
4. Cox, I., et al.: Digital Watermarking and Steganography, 2nd edn. Morgan Kaufmann Publishers Inc., San Francisco (2008)
5. Kong, X., Feng, R.: Watermarking medical signals for telemedicine. IEEE Trans. Inf. Technol. Biomed. **5**(3), 195–201 (2001). ISSN: 1089–7771
6. Kaur, S. et al.: Digital watermarking of ECG data for secure wireless commuication. In: 2010 International Conference on Recent Trends in Information, Telecommunication and Computing, pp. 140–144 (2010)
7. Ibaida, A., Khalil, I., van Schyndel, R.: A low complexity high capacity ECG signal watermark for wearable sensor-net health monitoring system, in Computing in. Cardiology **2011**, 393–396 (2011)
8. Mousavi, S.M., et al.: Watermarking techniques used in medical images: a survey. J. Digit. Imag. **27**, 714–729 (2014). doi:10.1007/s10278-014-9700-5

9. Kasmani, S.A., Naghsh-Nilchi, A.: A new robust digital image watermarking technique based on joint DWT-DCT transformation. In: 2008 Third International Conference on Convergence and Hybrid Information Technology ICCIT, vol. 2(1), pp. 539–544 (2008)
10. Lei, B.Y., Soon, I.Y., Li, Z.: Blind and robust audio watermarking scheme based on SVDDCT. Signal Process. **91**(8), 1973–1984 (2011)
11. Lai, C.-C.: An improved SVD-based watermarking scheme using human visual characteristics. Opt. Commun. **284**(4), 938–944 (2011)
12. Liu, R., Tan, T.: A SVD-based watermarking scheme for protecting rightful ownership. IEEE Trans. Multimed. **4**(1), 121–128 (2002). ISSN: 1520–9210
13. Tsai, H.-H., Jhuang, Y.-J., Lai, Y.-S.: An SVD-based image watermarking in wavelet domain using SVR and PSO. Appl. Softw. Comput. **12**(8), 2442–2453 (2012)
14. Koelstra, S., Muhl, C., Soleymani, M., Lee, J.S., Yazdani, A., Ebrahimi, T., Pun, T., Nijholt, A., Patras, I.: DEAP : a database for emotion analysis using physiological signals. IEEE Trans. Affect. Comput. **3**, 18–31 (2012)
15. Chen, T.S., Chang, C.C., Hwang, M.S.: A virtual image cryptosystem based upon vector quantization. IEEE Trans. Image Process. **7**(10), 1485–1488 (1998)

A Study to Investigate Different EEG Reference Choices in Diagnosing Major Depressive Disorder

Wajid Mumtaz[1], Aamir Saeed Malik[1(✉)],
Syed Saad Azhar Ali[1], and Mohd Azhar Mohd Yasin[2]

[1] Center for Intelligent Signal and Imaging Research (CISIR),
Universiti Teknologi PETRONAS (UTP), 32610 Bandar Seri Iskandar, Perak, Malaysia
aamir_saeed@petronas.com.my
[2] Department of Psychiatry, Hospital Universiti Sains Malaysia (HUSM),
65000 Kota Bharu, Kelantan, Malaysia
mdazhar@kb.usm.my

Abstract. Choice of an electroencephalogram (EEG) reference is a critical issue during measurement of brain activity. An appropriate reference may improve efficiency during diagnosis of psychiatric conditions, e.g., major depressive disorder (MDD). In literature, various EEG references have been proposed, however, none of them is considered as gold-standard [1]. Therefore, this study aims to evaluate 3 EEG references including infinity reference (IR), average reference (AR) and link-ear (LE) reference based on EEG data acquired from 2 groups: the MDD patients and healthy subjects as controls. The experimental EEG data acquisition involved 2 physiological conditions: eyes closed (EC) and eyes open (EO). Originally, the data were recorded with LE reference and re-referenced to AR and IR. EEG features such as the inter-hemispheric coherences, inter-hemispheric asymmetries, and different frequency bands powers were computed. These EEG features were used as input data to train and test the logistic regression (LR) classifier and the linear kernel support vector machine (SVM). Finally, the results were presented as classification accuracies, sensitivities, and specificities while discriminating the MDD patients from a potential population of healthy controls. According to the results, AR has provided the maximum classification efficiencies for coherence and power based features. The case of asymmetry, IR and LE performed better than AR. The study concluded that the reference selection should include factors such as underlying EEG data, computed features and type of assessment performed.

Keywords: EEG measurements · Infinity reference · Average reference · Link-ear reference · Major depressive disorder

1 Introduction

Choice of electroencephalogram (EEG) reference is a critical issue while measuring brain activity. It has been established that there is no constant, zero reference in the human body [2]. Hence, due to factors such as the effects of volume conduction and a non-zero activity at reference electrodes, the EEG measurements may be confounded

© Springer International Publishing Switzerland 2015
S. Arik et al. (Eds.): ICONIP 2015, Part IV, LNCS 9492, pp. 77–86, 2015.
DOI: 10.1007/978-3-319-26561-2_10

with errors. This significantly hampers the physiological interpretations associated with the EEG derived measures. For example, the EEG coherence describes the functional relationship between 2 EEG signals acquired from different brain locations. However, reference related issues caused variations of measurements and have posed problems during coherence estimation. As a result, the functional connectivity between different brain regions may be misleading [3].

In this paper, three different EEG references have been considered: the infinity reference (IR), the average reference (AR) and the link-ear (LE) reference. The IR is considered as having a static ground and theoretically provides zero-reference [4, 5]. In addition, it is considered to be located at a far location when compared with other recording electrodes and hence provides the least interference to these recording electrodes. The AR computes an average of all the electrodes and subtracted the value from individual electrodes over the scalp [6]. The AR works well for dense net array of electrodes than small number of electrodes. Moreover, the LE montage considers the reference electrodes to be placed on ear lobes, both on the left and right sides. A physical connection between the 2 ear lobe references forces their potential difference to be the same. However, for LE reference, the change in the resistance of different electrodes caused a change in noise levels.

Studies based on EEG references have reported differences in the computed measures due to different EEG references, e.g., bias in the computed head surface integral [7]. In addition, differences in the computation of the frequency spectrum based on EEG data sets are reported [8]. The interpretations associated with these measures are critical for clinical applications such as diagnosing major depressive disorder (MDD). Therefore, in this study, we hypothesized that the abnormalities, associated with the condition MDD, have associated aberrant EEG recordings that may not be found commonly in analyzing healthy controls. Based on this hypothesis, we can discriminate the MDD patients from healthy controls. For this purpose, we have selected coherence, asymmetry, and power as features. They are also reported to have effects due to difference reference choices. According to our literature review, no one has performed such comparisons involving EEG references for the MDD patients. However, most recently, there are few related studies [9–11]. In contrast, this paper focuses on the assessment of the 3 reference selections while evaluating EEG-based feature performances as discriminants of the MDD patients from controls.

The paper is organized as follows. The Sect. 2 illustrated the EEG-based measures. The recruitment of the study participants and the experiment design is provided in Sect. 3. Sections 4 and 5 provides as brief explanation about proposed method and the results. Finally, the paper provides the conclusion in Sect. 6.

2   Proposed EEG Measures

2.1   Inter-hemispheric Asymmetry

Asymmetry is an EEG-based measure used to compute the changes in signal power between the left and right hemispheres [12]. The relative EEG signal power is computed in the left and right hemisphere as described in Eqs. 1 and 2,

$$W'_{Lmn} = \sum_{f=f_1}^{f_2} s_{Lmn} \Big/ \sum_{f=0.5Hz}^{30Hz} s_{Lmn} \tag{1}$$

$$W'_{Rmn} = \sum_{f=f_1}^{f_2} s_{Rmn} \Big/ \sum_{f=0.5Hz}^{30Hz} s_{Rmn} \tag{2}$$

where, the f_1 and f_2 are the lower and higher frequencies of the selected EEG frequency bands, respectively. The s_{Lmn} and s_{Rmn} are the left and right hemispheric power spectral densities. The interhemispheric EEG asymmetry is calculated for each study subject based on Eq. 3 described:

$$A_{mn}(f_1, f_2) = \frac{W'_{Lmn} - W'_{Rmn}}{W'_{Lmn} + W'_{Rmn}} \times 100 \tag{3}$$

where A_{mn} is the inter-hemispheric asymmetry at frequencies f_1 and f_2.

2.2 Inter-hemispheric Coherence

Coherence is a measure that evaluates functional connections among different brain locations. The coherence values are high when the signals at 2 electrodes have similar behavior. It means that those electrodes might have the functional connections among each other. Power spectral density is used to compute the coherence according to the formula give above. The inter-hemispheric coherences were computed based on a mathematical formula (Eq. 4) [12]:

$$C_{xy}(f_1, f_2) = \frac{\left(\sum_{f=f_1}^{f_2} s_{xy} \right)^2}{\sum_{f=f_1}^{f_2} s_{xx}(f) \cdot \sum_{f=f_1}^{f_2} s_{yy}(f)} \tag{4}$$

where s_{xy} is the power cross-spectral density of 2 signals; s_{xx} and s_{yy} are the power spectral densities of each signal. The calculations for coherence performed separately for individual frequency bands: delta (0.5-4 Hz), theta (4-8 Hz), alpha (8-12 Hz), beta (12-20 Hz) and gamma (> 20 Hz).

2.3 Power Computation Based on Welch Periodogram Method

EEG signal powers are computed based on welch periodogram method with Hanning window. According to the method, the EEG signal is segmented into 8 segments with 50 % overlap. The spectrum is computed for each segment, and for power computation, the computed values are averaged over all segments to finally get the power of that EEG signal.

3 Participant Recruitment and Experiment Design

3.1 Study Participants

For data acquisition, the experimental procedure was approved by the ethics committee, hospital Universiti Sains Malaysia (HUSM) and well informed to all the study participants. For this study, 2 groups of study participants were recruited: (1) MDD patients (n = 33, mean age: 40.33, std. ± 12.86) and (2) well-matched healthy subjects (n = 19 mean age: 39.8, std. ± 15.6) as controls. Both groups were asked to sign the consent forms before the actual experiment commencement. The details of experiment performance and basic operation of EEG equipment was explained to each participant. The MDD patients have met the diagnosis criteria for MDD, termed as DSM-IV [13]. MDD patients were recruited from the outpatient clinic, HUSM. The healthy subjects are well-matched group and properly screened for the absence of any psychiatry problem in the history or in family.

3.2 Experiment Design

Figure 1 describes the experimental data acquisition scheme. The EEG data were recorded in a quiet environment while the participants were asked to sit in a comfortable chair in a semi-recumbent position. A wearable EEG cap with 24 electrodes was used. Out of those 24 electrodes, 19 electrodes were used to record EEG data, other 3 were used for ECG and remaining 2 for reference purpose. Only EEG data were used for analysis. Placement of the EEG electrodes followed standard 10-20 international electrode placement system. The Brain master discovery software was used for preprocessing and storing the datasets

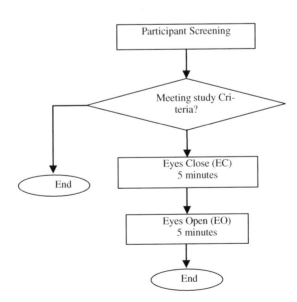

Fig. 1. EEG data acquisition scheme

on a computer disk. For this purpose, the sampling rate was set at 256 Hz, and a band-pass filter within the range 0.5-70 Hz was performed with an additional 50 Hz notch filter for power line noise filtration. The EEG data were amplified by the Brain master 24 E amplifier. The duration of EO and EC sessions was five (5) minutes each. At first hand, the EEG data were recorded with LE reference. For the computation of EEG features, the data were re-referenced from LE to AR and finally to the IR.

4 Data Analysis

4.1 EEG Data Preprocess

EEG data preprocessing involved either correction of the EEG data from artifacts or removing/deleting the EEG segment culminated with artifacts. The correction of the data is preferable as it safeguards useful information which might be lost by deleting the data with artifacts. In this study, the EEG data were cleaned with a standard tool named, brain electrical source analysis (BESA) [14]. Moreover, this correction of EEG data from artifacts involves modeling the artifact topographies. The successful modeling of the artifacts in the data further helped during data correction. The artifact topographies were learned from the data recordings itself. The process named surrogate filtering was employed for this whole process [15].

4.2 EEG Analysis

The EEG analysis process was based on machine learning (ML) concepts: features extraction, selection and classification with 10 fold cross validation. Feature extraction stage involved 3 proposed EEG-based measures: inter-hemispheric asymmetries, coherences, and power of different frequency bands of EEG signals. The feature space matrix was formed with EEG-based measures extracted over the whole scalp. Feature extraction may have involved redundant or irrelevant features, therefore, a selection of most significant and relevant features was performed. The selection of most useful features can effectively reduce the computation time and improves the efficiencies of classifier models. The features selection were performed based on ranking them according to receiver operating characteristics (ROC) criterion and then selection the top ranked features only among a list of all the features ranked accordingly. The feature selection was followed by classification stage. The classification was performed with the reduced set of selected features.

In this study, the 2 possible classes were MDD patients and Controls. Therefore, classifiers such as logistic regression (LR) and support vector machine (SVM) with linear kernel were selected. The LR classifier is considered as a simple classifier model that can easily model the independent EEG derived measures with the dependent outcome classes: MDD patients Vs controls. It is used to model epidemiologic data with good efficiency while using a logistic model between the dependent and independent variables [16]. Specifically, during conditions such as the number of data points for training and testing are less. The secondly selected classification model was SVM with linear kernel. The SVM is considered as a high-efficiency classifier that can classify the feature space according to a hyper-plan learned from the test data set. It is efficient because the boundary which is the called hyper-plan is most optimally selected providing maximal separation between the 2 classes. For this purpose, the data points that make the hyper-plan are termed as support vectors.

4.3   Validation

A 10-fold cross-validation was performed in combination with a Monte-Carlo iteration to achieve a certain confidence in the results. In the Monte-Carlo simulations, 100 times iterations of 10-fold cross validation were performed with random permutations. Finally, an average of 100 values was performed to achieve the classification accuracies, sensitivities, and specificities.

5   Results

5.1   Calculations of Accuracy, Sensitivity, and Specificity

In this paper, both the classification models utilized supervised learning procedures. The performance metrics used for the success of training and testing were the accuracies, sensitivities, and the specificities. These metrics were computed based on the formulas provided in Eqs. 5 to 6 respectively. The sensitivity of a classification model corresponds to the percentage of true cases (TP) which are correctly classified as cases defined by Eq. (6). The specificity of a classification model refers to the percentage of true non-cases (TN) which are correctly classified as non-cases as described by Eq. (7). The accuracy of a classification model illustrates the percentage of correctly classified cases and non-case among all the example points as depicted in Eq. (5).

$$Accuracy = \frac{TP + TN}{TP + TN + FP + FN} \tag{5}$$

$$Sensitivity = \frac{TP}{TP + FN} \tag{6}$$

$$Specificity = \frac{TN}{TN + FP} \tag{7}$$

5.2   Low Dimensional Representation

To visualize the distribution of the extracted features on a 2D plain, principal component analysis (PCA) was performed on the data matrixes of coherence (Fig. 2), inter-hemispheric asymmetry (Fig. 3) and power (Fig. 4). Performing PCA to visualize data distribution is a standard process and provides information about a spread of the data. In case, an artifact is detected, e.g., an outlier can be removed from the data. In our study, we have also performed the standardization of feature space matrix with the z-score standardization.

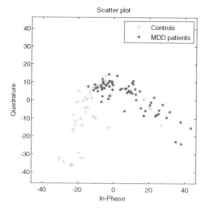

Fig. 2. Clustering for average reference (Coherence)

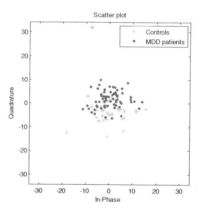

Fig. 3. Clustering for average reference (Asymmetry)

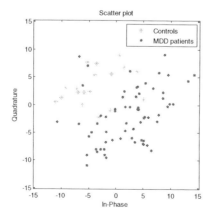

Fig. 4. Clustering for average reference (Power)

5.3 Classification Results

Table 1 shows results during classification with LR model. A comparison of the different references is provided for coherence, asymmetry, and power. In case of coherence, the AR (accuracy = 91.4 %) has performed better than IR (accuracy = 86.4 %) and LE (accuracy = 83.72 %) references. Also, the IR shows better performance than LE reference. In case of asymmetry, LE (accuracy = 82.46 %) and IR (accuracy = 80.8 %) performed better than AR (accuracy = 77.8 %). Both IR and LE have approximately equal performances. In case of power, AR (accuracy = 89.36 %) performed better than both IR (accuracy = 82 %) and LE (accuracy = 86.7 %).

Table 1. Perfomance of logistic regression classification

		Coherence	Asymmetry	Power
IR	Acc.	86.4 %	80.8 %	82 %
	Sen.	70.8 %	72.8 %	72.5 %
	Spec.	95.4 %	85.9 %	87.8 %
AR	Acc.	91.4 %	77.8 %	89.36 %
	Sen.	82.5 %	65.8 %	100 %
	Spec.	96.9 %	85 %	83.57 %
LE	Acc.	83.72 %	82.46 %	86.7 %
	Sen.	71.66 %	74.16 %	84.16 %
	Spec.	90.7 %	86.90 %	88.88 %

Table 2. Performance of support vector machine classification

		Coherence	Asymmetry	Power
IR	Acc.	86 %	80 %	85.75 %
	Sen.	70 %	60 %	78.57 %
	Spec.	95 %	90 %	90 %
AR	Acc.	91 %	83 %	90.5 %
	Sen.	85.7 %	76 %	82.85 %
	Spec.	93.8 %	86.5 %	94.6 %
LE	Acc.	84.5 %	84 %	88.5 %
	Sen.	71.4 %	75 %	85.7 %
	Spec.	91.5 %	89.2 %	89.2 %

Table 2 shows results during classification with SVM model. A comparison of the different references is provided for coherence, asymmetry, and power. In case of coherence, the AR (accuracy = 91 %) has performed better than IR (accuracy = 86 %) and LE (accuracy = 84.5 %) references. Also, the IR shows better performance than LE reference. In case of asymmetry, LE (accuracy = 84 %) and AR (accuracy = 83 %) performed better than IR (accuracy = 80 %). Both AR and LE have approximately equal performances. In case of power, AR (accuracy = 90.5 %) performed better than both IR (accuracy = 85.75 %) and LE (accuracy = 88.5 %).

6 Conclusions

This paper emphasizes following main points, (a) the choice of reference is a critical issue during any measurement, (b) EEG merely collects brain electric activity, which may or may not relate to the organ's physiological activity, and thus (c) the reference choice has a particular role in reducing confusions in presentation, representation, and interpretation of the acquired data.

In context of a suitable EEG reference, this study has provided the comparisons among 3 references based on their discriminatory performances for MDD patients and controls. The coherence, inter-hemispheric asymmetry, and power were selected as features. Based on the results, it is concluded that the choice of EEG reference for diagnosing the MDD patients depends on the type of features being used. The case of coherence and power, the AR would a preferable choice. For asymmetry, both the IR and LE can be preferred over AR reference. A further validation of this result is cautioned because of few limitations such as small sample sizes. The effects of medication on brain cannot be ignored, although the study MDD patients were medication-free for 2 weeks. These results further motivate to explore effects of other references to the same features and will be considered for future research into the same area.

Acknowledgement. This research is supported by the HiCoE grant for CISIR (0153CA-005), Ministry of Education (MOE), Malaysia, and NSTIP strategic technologies programs, grant number (12-INF2582-02), in the Kingdom of Saudi Arabia.

References

1. Nunez, P.L.: REST: A good idea but not the gold standard. Clin. Neurophysiol. Official J. Int. Fed. Clin. Neurophysiol. **121**(12), 2177 (2010)
2. Geselowitz, D.B.: The zero of potential. IEEE Eng. Me. Biol. Mag. Q. Mag. Eng. Med. Biol. Soc. **17**(1), 128–132 (1997)
3. Nunez, P.L., et al.: EEG coherency: I: statistics, reference electrode, volume conduction, Laplacians, cortical imaging, and interpretation at multiple scales. Electroencephalogr. Clin. Neurophysiol. **103**(5), 499–515 (1997)
4. Yao, D.: A method to standardize a reference of scalp EEG recordings to a point at infinity. Physiol. Meas. **22**(4), 693 (2001)
5. Zhai, Y., Yao, D.: A study on the reference electrode standardization technique for a realistic head model. Comput. Methods Programs Biomed. **76**(3), 229–238 (2004)

6. Bertrand, O., Perrin, F., Pernier, J.: A theoretical justification of the average reference in topographic evoked potential studies. Electroencephalogr. Clin. Neurophysiol./Evoked Potentials Section **62**(6), 462–464 (1985)
7. Junghöfer, M., et al.: The polar average reference effect: a bias in estimating the head surface integral in EEG recording. Clin. Neurophysiol. **110**(6), 1149–1155 (1999)
8. Yao, D., et al.: A comparative study of different references for EEG spectral mapping: the issue of the neutral reference and the use of the infinity reference. Physiol. Meas. **26**(3), 173 (2005)
9. Xu, P., et al.: Recognizing mild cognitive impairment based on network connectivity analysis of resting EEG with zero reference. Physiol. Meas. **35**(7), 1279 (2014)
10. Fried, S.J., Smith, D.M., Legatt, A.D.: Median nerve somatosensory evoked potential monitoring during carotid endarterectomy: does reference choice matter? J. Clin. Neurophysiol. **31**(1), 55–57 (2014)
11. Mumtaz, W., et al.: Review on EEG and ERP predictive biomarkers for major depressive disorder. Biomed. Signal Process. Control **22**, 85–98 (2015)
12. Hinrikus, H., et al.: Electroencephalographic spectral asymmetry index for detection of depression. Med. Biol. Eng. Compu. **47**(12), 1291–1299 (2009)
13. Association, A.P.: Diagnostic and statistical manual of mental disorders: *DSM-IV-TR®*. 2000: American Psychiatric Pub
14. Berg, P., Scherg, M.: Dipole modelling of eye activity and its application to the removal of eye artefacts from the EEG and MEG. Clin. Phys. Physiol. Meas. **12**(A), 49 (1991)
15. Hoechstetter, K., Berg, P., Scherg, M.: BESA research tutorial 4: Distributed source imaging. BESA Research Tutorial, 1–29 (2010)
16. Hosmer Jr., D.W., Lemeshow, S.: Applied logistic regression. John Wiley & Sons, New York (2004)

Prosthetic Motor Imaginary Task Classification Based on EEG Quality Assessment Features

Sherif Haggag$^{(\boxtimes)}$, Shady Mohamed, Omar Haggag, and Saeid Nahavandi

Centre for Intelligent Systems Research, Deakin University, Victoria, Australia
{shaggag,shady.mohamed,saeid.nahavandi}@deakin.edu.au

Abstract. Brain Computer Interface (BCI) plays an important role in the communication between human and machines. This communication is based on the human brain signals. In these systems, users use their brain instead of the limbs or body movements to do tasks. The brain signals are analyzed and translated into commands to control any communication devices, robots or computers. In this paper, the aim was to enhance the performance of a brain computer interface (BCI) systems through better prosthetic motor imaginary tasks classification. The challenging part is to use only a single channel of electroencephalography (EEG). Arm movement imagination is the task of the user, where (s)he was asked to imagine moving his arm up or down. Our system detected the imagination based on the input brain signal. Some EEG quality features were extracted from the brain signal, and the Decision Tree was used to classify the participant's imagination based on the extracted features. Our system is online which means that it can give the decision as soon as the signal is given to the system (takes only 20 ms). Also, only one EEG channel is used for classification which reduces the complexity of the system which leads to fast performance. Hundred signals were used for testing, on average 97.4 % of the up-down prosthetic motor imaginary tasks were detected correctly. This method can be used in many different applications such as: moving artificial limbs and wheelchairs due to it's high speed and accuracy.

1 Introduction

Some people suffer from an injury in the spinal cord or paralysis. These injuries lead those people not to be capable of communicating or connecting with the outside world as normal people do during their daily life. They are literally 'locked in' their bodies, as they can't do any motor activity. Technology plays a very important role to find another way of communication and control for them. The role is to push the current research towards the brain signals as it can help those people in finding a way out [1]. Non-invasive electroencephalogram (EEG) is one of the best ways to understand the behavior of the human brain, and researchers are now focusing on translating the EEG signal to a set of commands that can control a robot or a machine via BCI applications in real time and with high accuracy [2].

© Springer International Publishing Switzerland 2015
S. Arik et al. (Eds.): ICONIP 2015, Part IV, LNCS 9492, pp. 87–94, 2015.
DOI: 10.1007/978-3-319-26561-2_11

Nowadays, the Brain computer interface (BCI) research is very competitive due to the huge demand [3,4]. This demand is due to the huge desire for improving the human lifestyle. Disabled people can do most of the daily activities using their brain via BCI applications [5,6]. Brain machine interface mainly translates the brain signal to an action based on the brain behavior and the extracted features [7]. Robotics is one of the famous fields using BCI, but it is used in games [8], military [9], hospitals [10] and many other fields and gradually it will be used in every single field in our life.

Brain activities are the main source of information for the BCI. There are many different ways to record these activities [11,12]. One way to record the brain activities is EEG (electroencephalography) where a number of electrodes are placed on the human scalp [13]. Another way for recording the brain activities is Subdural Electrode Recording but it is so complex process and a lot of experience is needed as it is an invasive procedure, which means that the electrodes are physically placed on the brain itself [14]. EEG is the most famous method for recording human's brain behavior [15], the reason behind that is that EEG is a very safe, simple and non-invasive procedure. These reasons made most of the people use EEG signal as an input to the BCI applications [16].

Translating the EEG signal to a command is based mainly on two steps. First step is to extract the important features from the signal. Secondly, the signal is classified to a specific action group based on the extracted features [17]. These features are extracted from the motor cortex area which is responsible for any body movements. C3, C4 and Cz are the main three motor cortex area channels [18].

BCI research is very competitive and is growing these days. Researchers are now focusing on feature extraction methods and classification techniques, their main target is to improve the classification accuracy which will leads to improve the accuracy of the BCI applications.

Recording the cortical brain signals is the initial step. Feature extraction is the second step where a meaningful and a unique features are extracted from the input EEG signal [19], this will help to classify the signal to the correct action group [20]. The most famous feature extraction techniques are the standard deviation (SD), variance, Fast Fourier transforms (FFT) [21], wavelet transform (WT) [22] and power spectral density (PSD) [23]. It is very important to select a suitable feature extraction technique which will match with the classification method, otherwise the accuracy and the speed will be very low.

The most famous feature extraction methods are Fourier Transform (FT) and short Time Fourier Transform (STFT). But in case using EEG signal in real time applications, STFT and FT time-frequency trade off which will lead to inefficient signal localization with various spectral-temporal characteristics. Wavelet transform takes time to be performed which will make it hard to be implemented in real time applications. PCA and ICA are usually used after extracting the features to reduce their number, but our method doesn't need any feature reduction method as it already gives a small number of features (Fig. 1).

Raw EEG

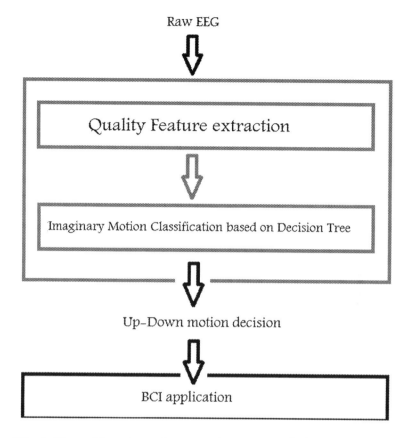

Fig. 1. The red box shows the steps of our method (Color figure online)

Finally, the classification method decides which action fired the brain signal. Many different classifiers were used, such as Quadratic Discriminant Analysis (QDA) [24], Naive Bayesian Classifier (NBC) [25], Support Vector Machine (SVM) [26,27], K-Nearest Neighbor (KNN) [28], Linear Discriminant Analysis (LDA) [29] and Hidden Markov Model(HMM) [36]. Most of these classifiers take some time to classify the signal (offline algorithms) which will not be the perfect case for any BCI application. LDA doesn't modify the original dataset's location but supports more class separability and draws a decision area between the each class. Moreover, LDA assumes that the covariance of classes is the same which does not happen in all the cases. On the other hand, QDA discard that assumption but it only gives good results when points are normally distributed. KNN has an assumption that the probabilities of a class are locally nearly constant. Also, the performance of KNN extremely relies on choosing an appropriate k value. When k is big number, small classes will be destroyed by bigger classes. On the other hand, if k is very small, there won't be any advantage to use KNN algorithm. Moreover, KNN is sensitive to the data structures and the noise level in the data. Also it needs huge amount of memory and it has high computational cost. SVM is affected by noise and it assumes that there is only two classes in the data. Also, it assumes that all attributes in a given class are independent. NBC also ignores interaction between attributes within individuals of the same class. For all these reasons, a classifier is needed with these properties: fast, online and doesn't depend on any data assumptions.

2   Methodology

The main idea is to extract the most significant features from the input EEG signal. Then the signal is classified into different groups based on the extracted features. The features are extracted based on some properties in the input signal.

2.1   Signal Recording

The EEG data was recorded from a healthy people. The person was asked to do one of two opposite prosthetic motor imaginary task (up or down), and his cortical potentials were recorded at the same time. During the recording, a visual feedback is shown when cortical potentials are slow (Cz-Mastoids). When the cursor moves downward, this indicates the occurrence of cortical positivity and it moves upward when cortical negativity occurs. The time frame of one trial is 6 s. The target was visually shown by a clear goal shown at the top or bottom of the screen during each trial, this shows the negativity or positivity during each trial. The visual feedback appears 2 s late after the beginning of each trial. The data is recorded by 256 Hz (sample rate) and the action is recorded within 3.5 s which means 896 samples per trial (for each channel). PsyLab EEG8 is used as an amplifier and computer boards PCIM-DAS1602/16 bit is used as A/D-converter. The amplitude of the input signal range from -1000 μ V to 1000 μ V, and it's sampling rate is 256 S/s. Six channels were recorded which are A2-Cz, A1-Cz, 2 cm parietal of C3, 2 cm frontal of C3, 2 cm parietal of C4 and 2 cm frontal of C4 [30].

2.2   Feature Extraction

Our main idea is to extract the most significant features from the input EEG signal. Then the signal is classified into a specific group based on the extracted features. The features are extracted based on some quality feature which is used before in our previous work. These features are the number of spikes, maximum amplitude, minimum amplitude and the duration between each spike. These features were used beside other features before in our previous work for EEG quality assessment and it gave a very good results in [31–35].

After recording the EEG signal, the features were calculated from the input signal to be classified in the next step.

2.3 Classification Model

A fast and online classification model is needed to be used by the BCI applications. Classification and Regression Tree is an accurate and fast classification algorithm [37]. Decision tree was build for predicting up and down imaginary response as a function of the input features predictors. The tree is a binary tree where each non-terminal node is split based on the values of the input training signals.

The next step is to use 20 signals from each motor imaginary task to build a Decision tree.

3 Results

The EEG data was recorded while a person is doing a prosthetic motor imaginary task, this task is to imagine moving an arm up or down. The features were extracted from the signal using the Quality features which was explained in Sect. 2 II. Using the EEG quality features enables us to get a really good classification accuracy by using only one channel (from the recorded 6 channels). These features were used by the Decision tree classifier.

Our work was also compared to the most popular feature extraction method which is the wavelet coefficients. On the other hand, it was compared with the most common classifiers, which are used in classifying the up-down brain motor imaginary task. It was compared with LDA (Linear Discriminant Analysis), QDA (Quadratic Discriminant Analysis), KNN(K-Nearest Neighbour), SVM (Support Vector Machine), NBC (Naive Bayesian Classifier) and HMM (Hidden Markov Model).

The classification accuracy is significantly improved by using EEG quality features followed by DT classifier. Twenty signals were used to build the DT classification model and 100 signals were used for testing. Our model accuracy was 97.4±0.5 % which means that the accuracy was improved by 9 %. Table 1 shows a comparison between different methods accuracy.

Table 1. Comparison between the most common classifiers used using wavelet coefficients as a feature extraction and our method. Accuracy is calculated with and without using the feature reduction step using PCA.

Classifiers	Accuracy (%)	Accuracy (reduced features using PCA) (%)
LDA	79.29	81.14
QDA	79.29	80.71
KNN	76.43	81.43
SVM (Linear)	80.71	80.71
SVM (RBF kernel)	82.14	82.14
Naïve Bayesian	80	81.43
HMM	89.71	89.72
DT	97.1	97.4

4   Conclusion and Future Work

This paper proposed a new classification method for prosthetic motor imaginary task based on EEG signals. The method showed a huge accuracy improvement. Quality features were used to represent the signal in a meaning way. Then Decision Tress was used to build a classification model that classifies the input EEG signal to up or down motor imaginary task group in real time. The classification accuracy was improved by 9 % compared to all the all famous methods. Our method can be used to control robots by brain signal in real time, with high accuracy and using only one EEG channel.

References

1. Machado, S., Araújo, F., Paes, F., Velasques, B., Cunha, M., Budde, H., Basile, L.F., Anghinah, R., Arias-Carrión, O., Cagy, M., et al.: EEG-based brain-computer interfaces: an overview of basic concepts and clinical applications in neurorehabilitation. Rev. Neurosci. **21**(6), 451–468 (2010)
2. Wolpaw, J.R., Birbaumer, N., Heetderks, W.J., McFarland, D.J., Peckham, P.H., Schalk, G., Donchin, E., Quatrano, L.A., Robinson, C.J., Vaughan, T.M., et al.: Brain-computer interface technology: a review of the first international meeting. IEEE Trans. Rehabil. Eng. **8**(2), 164–173 (2000)
3. Soe, A.K., Nahavandi, S., Khoshmanesh, K.: Neuroscience goes on a chip. Biosens. Bioelectron. **35**(1), 1–13 (2012)
4. Zheng, H., Zhang, J., Nahavandi, S.: Learning to detect texture objects by artificial immune approaches. Future Gener. Comput. Syst. **20**(7), 1197–1208 (2004)
5. Zickler, C., Di Donna, V., Kaiser, V., Al-Khodairy, A., Kleih, S., Kübler, A., Malavasi, M., Mattia, D., Mongardi, S., Neuper, C., et al.: BCI applications for people with disabilities: defining user needs and user requirements. In: Assistive Technology from Adapted Equipment to Inclusive Environments, AAATE. Assistive Technology Research Series, vol. 25, pp. 185–189 (2009)
6. Daly, J.J., Wolpaw, J.R.: Brain-computer interfaces in neurological rehabilitation. Lancet Neurol. **7**(11), 1032–1043 (2008)
7. Leuthardt, E.C., Schalk, G., Wolpaw, J.R., Ojemann, J.G., Moran, D.W.: A brain-computer interface using electrocorticographic signals in humans. J. Neural Eng. **1**(2), 63 (2004)
8. Gürkök, H., Nijholt, A., Poel, M.: Brain-computer interface games: towards a framework. In: Herrlich, M., Malaka, R., Masuch, M. (eds.) ICEC 2012. LNCS, vol. 7522, pp. 373–380. Springer, Heidelberg (2012)
9. Kotchetkov, I.S., Hwang, B.Y., Appelboom, G., Kellner, C.P., Connolly Jr., E.S.: Brain-computer interfaces: military, neurosurgical, and ethical perspective. Neurosurg. Focus 28(5), E25 (2010)
10. Shyu, K.K., Chiu, Y.J., Lee, P.L., Lee, M.H., Sie, J.J., Wu, C.H., Wu, Y.T., Tung, P.C.: Total design of an FPGA-based brain-computer interface control hospital bed nursing system. IEEE Trans. Industr. Electronics **60**(7), 2731–2739 (2013)
11. Wolpaw, J.R., Birbaumer, N., McFarland, D.J., Pfurtscheller, G., Vaughan, T.M.: Brain-computer interfaces for communication and control. Clin. Neurophysiol. **113**(6), 767–791 (2002)

12. Schalk, G., McFarland, D.J., Hinterberger, T., Birbaumer, N., Wolpaw, J.R.: BCI 2000: a general-purpose brain-computer interface (BCI) system. IEEE Trans. Biomed. Eng. **51**(6), 1034–1043 (2004)
13. Guger, C., Edlinger, G., Harkam, W., Niedermayer, I., Pfurtscheller, G.: How many people are able to operate an EEG-based brain-computer interface (BCI)? IEEE Trans. Neural Syst. Rehabil. Eng. **11**(2), 145–147 (2003)
14. Felton, E.A., Wilson, J.A., Williams, J.C., Garell, P.C.: Electrocorticographically controlled brain-computer interfaces using motor and sensory imagery in patients with temporary subdural electrode implants: report of four cases. J. Neurosurg. **106**(3), 495–500 (2007)
15. Linkenkaer-Hansen, K., Nikouline, V.V., Palva, J.M., Ilmoniemi, R.J.: Long-range temporal correlations and scaling behavior in human brain oscillations. J. Neurosci. **21**(4), 1370–1377 (2001)
16. Teplan, M.: Fundamentals of EEG measurement. Meas. Sci. Rev. **2**(2), 1–11 (2002)
17. Burke, D.P., Kelly, S.P., de Chazal, P., Reilly, R.B., Finucane, C.: A parametric feature extraction and classification strategy for brain-computer interfacing. IEEE Trans. Neural Syst. Rehabil. Eng. **13**(1), 12–17 (2005)
18. Ball, T., Schreiber, A., Feige, B., Wagner, M., Lücking, C.H., Kristeva-Feige, R.: The role of higher-order motor areas in voluntary movement as revealed by high-resolution EEG and FMRI. Neuroimage **10**(6), 682–694 (1999)
19. Al-Fahoum, A.S., Al-Fraihat, A.A.: Methods of EEG signal features extraction using linear analysis in frequency and time-frequency domains. International Scholarly Research Notices 2014 (2014)
20. Lotte, F., Congedo, M., Lécuyer, A., Lamarche, F.: A review of classification algorithms for EEG-based brain-computer interfaces. J. Neural Eng. **4**, R1–R13 (2007)
21. Schiff, S.J., Aldroubi, A., Unser, M., Sato, S.: Fast wavelet transformation of EEG. Electroencephalogr. Clin. Neurophysiol. **91**(6), 442–455 (1994)
22. Chen, G.: Automatic EEG seizure detection using dual-tree complex wavelet-fourier features. Expert Syst. Appl. **41**(5), 2391–2394 (2014)
23. Zhou, S.M., Gan, J.Q., Sepulveda, F.: Classifying mental tasks based on features of higher-order statistics from EEG signals in brain-computer interface. Inf. Sci. **178**(6), 1629–1640 (2008)
24. Farquhar, J., Hill, N.J.: Interactions between pre-processing and classification methods for event-related-potential classification. Neuroinformatics **11**(2), 175–192 (2013)
25. Berta, R., Bellotti, F., De Gloria, A., Pranantha, D., Schatten, C.: Electroencephalogram and physiological signal analysis for assessing flow in games. IEEE Trans. Comput. Intell. AI Games **5**(2), 164–175 (2013)
26. Yoo, J., Yan, L., El-Damak, D., Altaf, M.A.B., Shoeb, A.H., Chandrakasan, A.P.: An 8-channel scalable EEG acquisition SoC with patient-specific seizure classification and recording processor. IEEE J. Solid-State Circuits **48**(1), 214–228 (2013)
27. Bhattacharyya, S., Khasnobish, A., Konar, A., Tibarewala, D., Nagar, A.K.: Performance analysis of left/right hand movement classification from EEG signal by intelligent algorithms. In: 2011 IEEE Symposium on Computational Intelligence, Cognitive Algorithms, Mind, and Brain (CCMB), pp. 1–8. IEEE (2011)
28. Kim, M.K., Kim, M., Oh, E., Kim, S.P.: A review on the computational methods for emotional state estimation from the human EEG. Comput. Math. Methods Med. **2013**, 1–13 (2013)
29. Rodrguez-Bermdez, G., García-Laencina, P.J., Roca-González, J., Roca-Dorda, J.: Efficient feature selection and linear discrimination of EEG signals. Neurocomputing **115**, 161–165 (2013)

30. Birbaumer, N., Perelmouter, J., Taub, E., Flor, H.: A spelling device for the paralysed. Nature **398**(6725), 297–298 (1999)
31. Haggag, S., Mohamed, S., Bhatti, A., Haggag, H., Nahavandi, S.: Neuron's spikes noise level classification using hidden markov models. In: Loo, C.K., Yap, K.S., Wong, K.W., Beng Jin, A.T., Huang, K. (eds.) ICONIP 2014, Part III. LNCS, vol. 8836, pp. 501–508. Springer, Heidelberg (2014)
32. Zhou, X., Garcia-Romero, D., Duraiswami, R., Espy-Wilson, C., Shamma, S.: Linear versus mel frequency cepstral coefficients for speaker recognition. In: 2011 IEEE Workshop on Automatic Speech Recognition and Understanding (ASRU), pp. 559–564 (2011)
33. Haggag, S., Mohamed, S., Bhatti, A., Gu, N., Zhou, H., Nahavandi, S.: Cepstrum based unsupervised spike classification. In: 2013 IEEE International Conference on Systems, Man, and Cybernetics (SMC), pp. 3716–3720. IEEE (2013)
34. Haggag, S., Mohamed, S., Haggag, H., Nahavandi, S.: Hidden markov model neurons classification based on mel-frequency cepstral coefficients. In: 2014 9th International Conference on System of Systems Engineering (SOSE), pp. 166–170. IEEE (2014)
35. Haggag, S., Mohamed, S., Bhatti, A., Haggag, H., Nahavandi, S.: Neural spike representation using cepstrum. In: 2014 9th International Conference on System of Systems Engineering (SOSE), pp. 97–100. IEEE (2014)
36. Haggag, S., Mohamed, S., Haggag, H., Nahavandi, S.: Prosthetic motor imaginary task classification using single channel of electroencephalography. In: SMC IEEE International Conference, Hong Kong (2015)
37. Breiman, L., Friedman, J., Stone, C.J., Olshen, R.A.: Classification and Regression Trees. CRC Press, Boca Raton (1984)

Enhancing Performance of EEG-based Emotion Recognition Systems Using Feature Smoothing

Trung Duy Pham[1], Dat Tran[1(✉)], Wanli Ma[1], and Nga Thuy Tran[2]

[1] Faculty of Education, Science, Technology and Mathematics,
University of Canberra, Canberra, Australia
{duy.pham,dat.tran,wanli.ma}@canberra.edu.au
[2] The Department of Information Technology and Mathematics,
Hanoi Medical University, Hanoi, Vietnam
tranthuynga@hmu.edu.au

Abstract. Electroencephalography (EEG) has been used recently in emotion recognition. However, the drawback of current EEG-based emotion recognition systems is that the correlation between EEG and emotion characteristics is not taken into account. There are the differences among EEG features, even with the same emotion state in adjacent time because EEG extracted features usually change dramatically, while emotion states vary gradually or smoothly. In addition, EEG signals are very weak and subject to contamination from many artefact signals, thus leading to an accuracy reduction of emotion recognition systems. In this paper, we study on feature smoothing on EEG-based Emotion Recognition Model to overcome those disadvantages. The proposed methodology was examined on two useful kinds of features: power spectral density (PSD) and autoregressive (AR) for two-level class and three-level class using DEAP database. Our experimental results showed that feature smoothing affects to both the feature sets, and increases the emotion recognition accuracy. The highest accuracies are 77.38 % for two-level classes and 71.75 % for three-level classes, respectively in valence space.

Keywords: EEG · Emotion recognition · Feature smoothing · Saviztky-Golay

1 Introduction

Emotions play an important role on most of dimensions of cognition such as attention, memory and decision making. Emotion recognition is a step towards aiding people such as in care taking and designing brain-computer interfaces. Nowadays, automatic emotion recognition has been one of the most popular research topics in the fields of computer vision, speech recognition, brain-machine interface, and computational neuroscience [1].

Through many studies, EEG signals have been proven to provide informative characteristics in response to the emotional states [2–4]. In addition, human EEG signals emerge as a potential recognition for human emotion with advantages [5]

© Springer International Publishing Switzerland 2015
S. Arik et al. (Eds.): ICONIP 2015, Part IV, LNCS 9492, pp. 95–102, 2015.
DOI: 10.1007/978-3-319-26561-2_12

such as: (1) Brain activities have direct information about emotion, (2) EEG signals can be measured at any moment and are not dependent on other activities of user such as speaking or generating a facial expression, and (3) Different recognition techniques can be used. However, there still exist some limitations on EEG-based emotion recognition framework. Firstly, those works did not consider the correlation between EEG and emotion characteristics [6]. While emotional states usually change gradually, features extracted directly from EEG data always have strong fluctuations and contain some information unrelated to the emotion task. This leads to significant differences among EEG features, even with the same emotion state in adjacent time. In addition, the existence of noise is an unpleasant phenomenon in any real signal analysis technique and EEG is not an exception. EEG signals are very weak and subject to the contamination from many artefact signals. As a result, extracting relevant brain activity from EEG data remains a challenging task due to its low signal-to-noise ratio (SNR). The activity of neuronal processes is typically much smaller than total variance of the EEG data [7]. Consequently, many applications based on EEG signals may not be analysed correctly, including EEG-based emotion classification.

To overcome these shortcomings, we should smooth the EEG features to reduce the large correlative differences between EEG and emotion characteristics, and to minimize the presence of artefacts in the EEG features. As emotional states are time-dependent, the features extracted from EEG components are also time-dependent. Additionally, smoothers not only reduce destructive noise and short-term components of a signal, but also enhance the classifier performance so that the EEG signal will have a unique characteristic given a particular emotion.

It has been found in the literature the feature smoothing method has been used in order to increase the accuracy and efficiencies of EEG-based emotion classifiers [8,9]. Ruo-Nan Duan [8] and Jia-Yi ZhuThe [9] used Linear Dynamical Systems (LDS) as feature smoothing method for different entropy (DE) features, and it has a good effect. However, the effectiveness of the feature smoothing methods has not examined on different kinds of features. The feature smoothing methods also need to be validated on other feature extraction methods.

Currently, varieties of feature extraction methods have been applied to extract features from EEG signals for emotion recognition [10]. The most widely used method is spectra analysis using Fourier Transform [11]. Koelstra et al. [12] proposed a system based on power spectral features from EEG signals, Fisher criterion for feature selection and Nave Bayes classifier; they achieved the average recognition accuracy of 57.6 % for two classes of valence using DEAP database. With this database, using Bayesian weighted-log-posterior function and perception convergence algorithm based on Fast Fourier transform feature extraction for emotion classification, Yoon and Chung obtained the accuracy of the valence is 70.9 % for the two-level class and 55.4 % for three-level class [13].

Besides spectral features, autoregressive (AR) modelling is one of the prominent parametric methods. Hatamikia et al. found that AR features are efficient to recognize emotional states from EEG signals. The classification accuracies of the valence based on AR features, sequential forward feature selection (SFS) method and K-nearest neighbor (KNN) classifier are 72.33 % and 61.10 % for two classes and for three classes, respectively. This has been the highest achieved accuracies using DEAP database [14].

In this paper, we propose to take advantage of feature smoothing technique in EEG-based emotion recognition to improve the recognition performance. The proposed model is evaluated on the DEAP database. Our experimental results showed that the proposed model increases the emotion recognition accuracy. Compared with previous studies using the DEAP database, classification accuracy from our proposed method is highest. We obtained 77.38 % and 71.75 % for two-level and three-level classes, respectively in valence space.

The rest of the paper is organized as follows. In Sect. 2 we propose the EEG-based emotion recognition system with feature smoothing technique. Section 3 describes experiments and results. We conclude the paper with a conclusion and future work in Sect. 4.

2 The EEG-based Emotion Recoginition Model with Feature Smoothing

Figure 1 shows the proposed EEG-based emotion recognition system with feature smoothing including two phases: training phase and recognition phase. In the training phase, the labelled EEG data, i.e. the data from a known emotion, are recorded. After collecting the data, the EEG signals are preprocessed and extracted features. Before building a model for that emotion, these features are smoothed to minimize the presence of artefacts and noise in the EEG features. In addition, feature smoothing helps remove the rapid fluctuations in the feature sequence as emotional state varies smoothly. The Moving Average and Saviztky-Golay are examples for feature smoothing technique. In the recognition phase, an unknown EEG data will be compared with the built emotion models and the recognised emotion is the label of the best matching model.

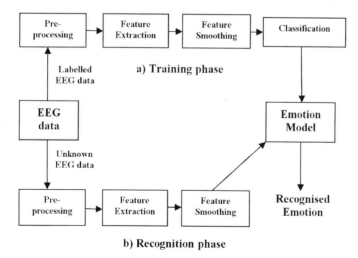

Fig. 1. The proposed EEG-based emotion recognition system with feature smoothing

3   Experiments and Results

3.1   Dataset

The dataset was used in this research is the DEAP dataset (Dataset for Emotion Analysis using Electroencephalogram, Physiological and Video Signals) which is an open database proposed by Koelstra et al. [18]. It includes the EEG and multiple peripheral physiological signals of 32 subjects (16 women and 16 men with the average age of 26.9) who were subject to music video stimuli. After the presentation of each stimulus, participants rated its content in terms of arousal, valence, likability, dominance (on scale from 1 to 9) and familiarity (on scale 1 to 5) by using self-assessment manikins (SAM) questionaire. During the experiments, EEG signals were recorded with 512 Hz sampling frequency, which were down-sampled to 128 Hz and filtered between 4.0 Hz and 45.0 Hz.

3.2   Feature Extraction

In our experiments, we use Power spectral density (PSD) and Autoregressive (AR) to extract the features. We categorized emotional valance in two-level class and three-level class, according to the unnormalized SAM-rating values on a scale of 1–9. Table 1 shows the conditions for categorizing the emotional class. Then the performance of the proposed emotion recognition methodology was examined in both the two-level and three-level classes. Each epochs of 60 s were split for training and testing from 32 EEG channels, the remaining other channels were unused. The channel signal in each epochs was used to extract features, and these features were merged together to make a single feature vector.

Table 1. Sam rating for each emotional class

	Two-level class		Three-level class		
	postive	negative	postive	neutral	negative
SAM rating (S_R)	$S_R \geq 5$	$S_R \leq 5$	$S_R \geq 6$	$4 \leq S_R \leq 6$	$S_R \leq 4$

We test the proposed methodology on two sets of features: PSD features and AR features to evaluate the effectiveness of feature smoothing technique on different kinds of features. Firstly, all of the 32 EEG channel data were spectrally analysed using fast Fourier transform with a non-overlap window of 2 seconds for each subject and each stimulus. Then the spectral power density in theta band (4–7 Hz), alpha band (8–15 Hz), beta band (15–30 Hz), and gamma band (31–45 Hz) were computed for each window. Hence, the feature vector, $x = (x_1, x_2, \ldots, x_D)^T$, is defined as a vector of D band powers, in our experiment, the number of original features is D $= 32 \times 4 = 128$. For AR features extraction, we also divided the EEG signal of each channel in to 2 s non-overlapping window and to ensure the assumption of stationary and AR coefficients were extracted for each window. 11th AR coefficients of the 11th order AR model with Burg's lattice-based method were extracted for each electrode and there are $11 \times 32 = 352$ features in total.

3.3   Feature Smoothing

After feature extraction, we used the Savitzky-Golay smoothing method to smooth out EEG features. As Saviztky-Golay smoother proposed a power method of data smoothing based on local least-squares polynomial approximation, Savitzky and Golay were interested in smoothing of noisy data and they demonstrated that least-squares smoothing reduces noise while maintaining the shape and height of waveform peaks. Therefore, this property of the Saviztky-Golay smoother has been found to be attractive in ECG processing [15]. To design Savitzky-Golay smoother, we should define the order of the polynomial N and the frame size M, which represents the half width of the approximation interval [16]. Furthermore, the order of the polynomial must be strictly less than the frame size [17]. In this research, we used $N = 17$ and $M = 33$, because these values gave a good balance between smoothing and denoising techniques [18].

3.4   Experimental Results

The Support Vector Data Description (SVDD) method was used for training models. In order to obtain reliable classification result, we randomly divided the trials into training set and testing set with ratio 6:4. The parameters were selected using 5-fold cross validation on training set and the best parameters found were used to train models on the whole training set and test on a separate test set. In all experiments, to validate the effect of features smoothing, we compared the classification results with or without using the feature smoothing technique. Table 2 shows the classification performance obtained by SVDD classifiers. It can be seen that almost all the classification accuracies are improved, which are significant improvement compared to the case when smoothing method is not employed. These results indicate that using feature smoothing techniques can effectively improve classification performance.

In addition, the feature smoothing method is more effective on AR features than on PSD features. It has improved the classification accuracies by 6.03 % for two-level class and 6.53 % for three-level class with AR features. While the proposed method has only improved by 0.94 % for two-level class and 4.01 % for three-level class with PSD features.

Compared with previous studies using DEAP database, classification accuracy from our proposed method is higher. We obtained 77.38 % and 71.75 % compared with the highest achieved accuracies until now are 74.20 % and 61.10 % for two and three classes of valence levels, respectively [14].

Figures 2–5 show the classification performance of two-level and three-level classes with AR features and PSD features. In general, the feature smoothing technique is efficient when smoothing the feature sequence. But for some subjects, the smoothed feature sequence decreases the accuracies of classifiers for the reason that the smoothing causes information lost in some degree. This problem could be solved by adjusting the parameters of smoothers to find the best parameters. Finding the best parameters will be investigated in further studies.

Table 2. Sam rating for each emotional class

	Two-level class		Three-level class	
	PSD Features	AR Features	PSD Features	AR Features
Non Feature Smoother	74.68 %	71.77 %	67.74 %	63.00 %
Feature Smoother	75.62 %	**77.38 %**	**71.75 %**	69.53 %

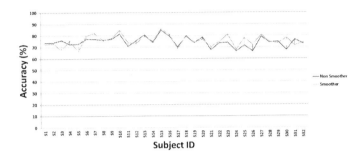

Fig. 2. The classification performance of two-level class with PSD features

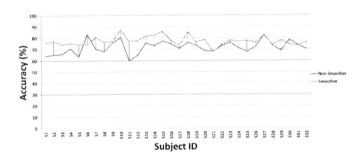

Fig. 3. The classification performance of two-level class with AR features

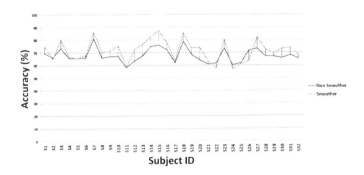

Fig. 4. The classification performance of three-level class with PSD features

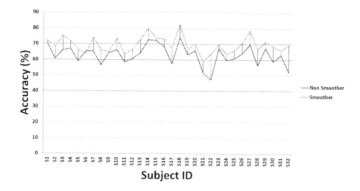

Fig. 5. The classification performance of three-level class with AR features

4 Conclusion and Future Work

We have proposed an emotion recognition system using EEG data with feature smoothing technique. The results showed that the feature smoothing techniques are good effective on AR features and PSD features. The feature smoothing technique is more effective on AR features than on PSD features. The results also revealed that the classification accuracy using the proposed method is higher than the best one in the literature until now which were examined on the DEAP database for emotion recognition using EEG signals. It can be concluded that the proposed feature extraction method showed very promising performance and is a very efficient method. The experimental results confirmed that our proposed method can be implemented in emotion recognition systems for EEG signals to improve its performance and reliability. In near future, we will experiment our proposed EEG-based emotion recognition system with feature smoothing technique on other large datasets. In addition, finding the best parameters of smoother or the different feature smoothing technique will be investigated to enhance the performance of EEG-based emotion recognition systems.

References

1. Cowie, R., Douglas-Cowie, E., Tsapatsoulis, N., Votsis, G., Kollias, S., Fellenz, W., Taylor, J.G.: Emotion recognition in human-computer interaction. Signal Process. Mag. **18**(1), 32–80 (2001)
2. Coan, J., Allen, J.: Frontal EEG asymmetry as a moderator and mediator of emotion. Biol. Psychol. **67**, 7–50 (2004)
3. Petrantonakis, P., Hadjileontiadis, L.: A novel emotion elicitation index using frontal brain asymmetry for enhanced EEG-based emotion recognition. IEEE Trans. Inf. Technol. Biomed. **15**, 737–746 (2011)
4. Li, X., Hu, B., Zhu, T., Yan, J., Zheng, F.: Towards affective learning with an EEG feedback approach. In: Proceedings of the1st ACM Interna-tional Workshop on Multimedia Technologies for Distance Learning, pp. 33–38 (2009)

5. Pham, T.D., Tran, D.: Emotion recognition using the emotiv EPOC device. In: Huang, T., Zeng, Z., Li, C., Leung, C.S. (eds.) ICONIP 2012, Part V. LNCS, vol. 7667, pp. 394–399. Springer, Heidelberg (2012)

6. Wang, X.-W., Nie, D., Lu, B.-L.: Emotional state classification from EEG data using machine learning approach. Neurocomput. **129**, 94–106 (2014)

7. Dyrholm, M., Dyrholm, M., Parra, L.C., Parra, L.C.: Smooth bilinear classification of EEG. IEEE Eng. Med. Biol. Soc. **1**, 42–49 (2006)

8. Duan, R.-N., Zhu, J.-Y., Lu, B.-L.: Differential entropy feature for EEG-based emotion classification. In: IEEE/EMBS Conference on Neural Engineering (NER), pp. 81–84 (2013)

9. Zhu, J.-Y., Zheng, W.-L., Peng, Y., Duan, R.-N., Lu, B.-L.: EEG-based emotion recognition using discriminative graph regularized extreme learning machine. In: IJCNN 2014, pp. 525–532, July 2014

10. Murugappan, M., Ramachandran. N., Sazali. Y., Rizon, M.: Inferring of human emotion states using multichannel EEG. Int. J. Soft Comput. Appl. (IJSCA), EURO Journals (2009)

11. Schaaff, K., Schultz, T.: Towards emotion recognition from electroencephalographic signals. In: 3rd International Conference on Affective Computing and Intelligent Interaction and Workshops, ACII 2009, pp. 1–6, September 2009

12. Koelstra, S., Muhl, C., Soleymani, M., Lee, J.S., Yazdani, A., Ebrahimi, T., Pun, T., Nijholt, A., Patras, I.: DEAP : a database for emotion analysis using physiological signals. IEEE Trans. Affect. Comput. **3**, 18–31 (2012)

13. Yoon, H.J., Chung, S.Y.: EEG-based emotion estimation using bayesian weighted-log-posterior function and perceptron convergence algorithm. Comput. Biol. Med. **43**(12), 2230–2237 (2013)

14. Hatamikia, S., Maghooli, K., Nasrabadi, A.M.: The emotion recognition system based on autoregressive model and sequential forward feature selection of electroencephalogram signals. J. Med. Signals Sens. **4**(3), 194–201 (2014)

15. Pandia, K., Revindran, S., Cole, R., Kovacs, G., Giaovangrandi, L.: Motion artifact cancellation to obtain heart sounds from a single chest-wworn accelerometer. In: Proceedings of ICASSP-2010, pp. 590–593 (2010)

16. Sanei, S., Chambers, J.: EEG Signal Processing. Wiley-Interscience, Chichester (2007)

17. Shete, V.V., Sonar, S., Charantimath, A., Elgendelwar, S.: Detection of K-complex in sleep EEG signal with matched filter and neural network. Int. J. Eng. Res. Technol. **1**, 1–4 (2012)

18. Al-Kadi, M.I., Reaz, M.B.I., Ali, M.A.M., Liu, C.Y.: Reduction of the dimensionality of the EEG channels during scoliosis correction surgeries using a wavelet decomposition technique. Sens. **14**(7), 13046–13069 (2014)

Intelligent Opinion Mining and Sentiment Analysis Using Artificial Neural Networks

Keith Douglas Stuart[1]([✉]) and Maciej Majewski[2]

[1] Department of Applied Linguistics, Polytechnic University of Valencia,
Camino de Vera, s/n, 46022 Valencia, Spain
`kstuart@idm.upv.es`
[2] Faculty of Mechanical Engineering, Koszalin University of Technology,
Raclawicka 15-17, 75-620 Koszalin, Poland
`maciej.majewski@tu.koszalin.pl`

Abstract. The article formulates a rigorously developed concept of opinion mining and sentiment analysis using hybrid neural networks. This conceptual method for processing natural-language text enables a variety of analyses of the subjective content of texts. It is a methodology based on hybrid neural networks for detecting subjective content and potential opinions, as well as a method which allows us to classify different opinion type and sentiment score classes. Moreover, a general processing scheme, using neural networks, for sentiment and opinion analysis has been presented. Furthermore, a methodology which allows us to determine sentiment regression has been devised. The paper proposes a method for classification of the text being examined based on the amount of positive, neutral or negative opinion it contains. The research presented here offers the possibility of motivating and inspiring further development of the methods that have been elaborated in this paper.

Keywords: Computational linguistics · Natural language processing · Neural networks · Opinion mining · Sentiment analysis

1 Introduction

Intelligent opinion mining and sentiment analysis tasks concern the use of natural language processing, text analysis and computational linguistics to identify and extract subjective information in a selected set of texts. It involves the use of methods for predicting the orientation of subjective content in opinionated text documents, with various applications in many areas including opinion-oriented information-seeking systems for data mining, Web mining, and text mining.

Sentiment analysis analyzes people's opinions, sentiments, evaluations, appraisals, attitudes, and emotions towards entities such as products, services, organizations, individuals, issues, events, topics, and their attributes. Opinion mining represents a large problem space and includes many different tasks: opinion extraction, sentiment mining, subjectivity analysis, affect analysis, emotion

S. Arik et al. (Eds.): ICONIP 2015, Part IV, LNCS 9492, pp. 103–110, 2015.
DOI: 10.1007/978-3-319-26561-2_13

analysis, review mining. Opinions may be direct or indirect (explicit/implicit) and comparative.

Sentiment analysis tries to measure subjectivity and opinion in text, usually by capturing speaker/writer evaluations (positive, negative or neutral) and the strength of these evaluations (the degree to which the word, phrase, sentence, or document in question is positive or negative). The task of automatically classifying the polarity (whether the expressed opinion is positive or negative) of texts (technically, large amounts of unstructured data) at the document, sentence, or feature/aspect level can be a challenging task. It is these kinds of linguistic subtleties that make automatic classification difficult and result in low accuracy rates (60 %-70 %) of automated systems when sentences express both negative and positive opinions and express implicit negativity. There is a need for better modeling of compositional sentiment [1]. And at the sentence level, this means more accurate calculation of the overall sentence sentiment of the sentiment-bearing words, the sentiment shifters, and the sentence structure.

Texts contain ideational and interpersonal meaning: facts and opinions. The construing of experience through interpersonal meaning is a very human act. Interpersonal meanings are often expressed in texts through opinions and points of view, which are given about a wide range of topics from commercial products to presidential speeches. What makes sentiment analysis difficult is how natural languages are used for communicative purposes. Words in natural languages are dynamic contributors to a process of meaning creation which is strongly affected by the context of use. Hence, a word is a dynamic variable whose value may change depending on the context in which it is used. Words enter into relationships with other words. These in turn form phrases/clauses that form sentences that form texts.

Sentiment analysis is the computational study of interpersonal meanings: the computational extraction of emotional expression in text. There are many advantages to be derived from information systems that are able to extract information about people's appraisal, opinion, sentiment about a product, individual, event, organization, or topic (for example, security reasons). Sentiment analysis is the task of finding people's opinions about specific entities in text, a technique to classify people's opinions. Research in sentiment analysis has focused mainly on two problems: detecting whether the text is subjective or objective, and determining whether the subjective text is positive or negative. An important concept in sentiment analysis is semantic orientation which refers to the sentiment polarity (positive or negative) and sentiment strength of words, phrases, or texts [2]. It is often the goal of sentiment analysis to find the semantic orientation of texts [3,4].

This research presents a rigorously developed concept of opinion mining and sentiment analysis shown in Fig. 1. The concept consists of a general processing scheme, using artificial intelligence methods and techniques, for opinion analysis and classification. It also contains a conceptual method for processing natural-language text [5]. The aim of the concept is to enable a variety of analyses of the subjective content of texts. The system contains the following subsystems:

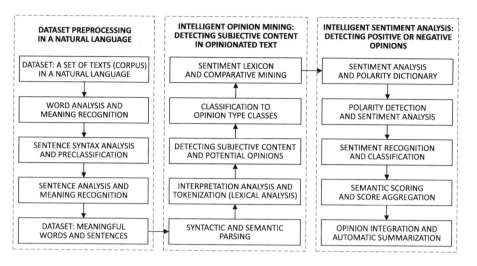

Fig. 1. Proposed concept of opinion mining and sentiment analysis

natural language dataset preprocessing subsystem, intelligent opinion mining subsystem for detecting subjective content in opinionated texts, intelligent sentiment analysis for detecting positive or negative opinions.

2 The State of the Art

Sentiment analysis has been investigated mainly at the following levels: document level, sentence level, and feature/aspect level [1].

The task at the document level is to classify whether a whole opinion document expresses a positive or negative sentiment. This task is commonly known as document-level sentiment classification. This level of analysis assumes that each document expresses opinions on a single entity. Therefore, it is not applicable to documents which evaluate or compare multiple entities.

The task at sentence level determines whether each sentence expresses a positive, negative, or neutral opinion. Neutral usually means no opinion. This level of analysis is closely related to subjectivity classification [2,3], which distinguishes sentences (called objective sentences) that express factual information from sentences (called subjective sentences) that express subjective views and opinions.

Both document level and sentence level analyses do not discover what exactly people like and do not like. Aspect level performs finer-grained analysis. Aspect level was earlier called feature level (feature-based opinion mining and summarization). Instead of looking at language constructs (documents, paragraphs, sentences, clauses or phrases), aspect level directly looks at the opinion itself. It is based on the idea that an opinion consists of a sentiment (positive or negative) and a target (of opinion). An opinion without its target being identified is of limited use. Realizing the importance of opinion targets also helps to understand the sentiment analysis problem better. In many applications, opinion targets are described by entities and their different aspects. Therefore, the goal of this level of analysis is to discover sentiments on entities and their aspects. Based on this level of analysis, a structured summary of opinions about entities and their aspects can be produced, which turns unstructured text to structured data and can be used for all kinds of qualitative and quantitative analyses. Both document level and sentence level classifications are already highly challenging. The aspect-level is even more difficult. It consists of several sub-problems [4].

3   Description of the System

This research proposes an intelligent system for opinion mining and sentiment analysis presented in Fig. 2. The system consists of a processing scheme, using artificial hybrid neural networks, for natural language processing, opinion mining, sentiment analysis and classification. It uses a method for processing natural-language texts [5–7]. The aim of the system is to enable a variety of analyses of the text and its subjective content.

The proposed system contains many specialized modules and it is divided into the following subsystems: a subsystem analyzing natural-language texts [8,9] - for the purpose of sentiment analysis - in search of dictionary words and meaningful sentences, and a subsystem for intelligent sentiment analysis for detecting positive or negative opinions. In this system, artificial intelligence methods [10,11] allow interpretation of a natural language, resulting in the detection of potential subjective contents, analyses of opinions and sentiments in examined texts, as well as classification of opinion type classes. The system is equipped with several adaptive intelligent layers for word analysis and recognition, sentence syntax analysis, sentence segment analysis, sentence recognition, syntactic parsing, semantic parsing, interpretation analysis [12,13], sentence tokenization, polarity detection and classification, sentiment analysis and classification.

This research also proposes a methodology based on hybrid neural networks for detecting subjective content and potential opinions, which is presented in Fig. 3. It involves intelligent opinion mining (detecting subjective content in examined text). It also contains a method allowing to classify to opinion type and sentiment score classes. The hybrid neural network consists of a modified probabilistic neural network [14] combined with a single layer classifier. The inputs of the network comprise binary images of potential opinion samplers. The consecutive bits represent evaluative words (the words which have the greatest impact on sentiment scores) which were scored on a scale of intensity in an appropriate manner. The output provides the detection of potential opinions and the classification of opinion type classes. Figure 3 shows the hybrid neural network architecture (A) and its details: (Fig. 3B) neural classifier, (Fig. 3C) neuron of the pattern layer, (Fig. 3D) neuron of the output layer.

The method uses modified hybrid multilayer probabilistic neural networks to recognize the opinion type and find its score. The network is a pattern classifier. It uses learning files containing patterns of possible opinion type classes. The network allows for detection of any potential combination of a subjective content with similar meanings but different lexico-grammatical patterns. It becomes an effective tool for solving classification problems of subjective content text structures, where the objective is to assign cases of opinions to one of a number of discrete opinion type classes. It offers a way to interpret the network's structure in the form of a probabilistic density function. The overall results also include text characteristics, statistical analysis, checking opinion occurrences, and validating sentiments.

The presented system also consists of a methodology based on a regression neural network [15] allowing to determine sentiment regression (Fig. 4A). Inputs of the network comprise sentiment scores (vectors) for each utterance. The network's output produces sentiment regressions. The sentiment analysis is modeled with a linear combination of sentiment scores to produce an opinion approximation. Another method based on a Hamming neural network proposes a

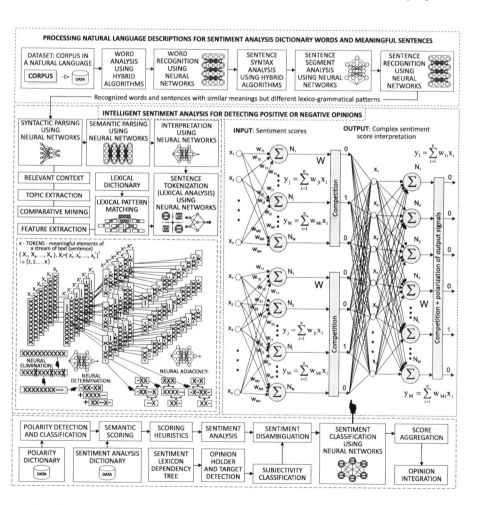

Fig. 2. Proposed intelligent system for opinion mining and sentiment analysis

classification process of the examined text depending on the amount of positive, neutral or negative opinion it contains (Fig. 4B). The inputs of the network comprise binary sentiment score images representing sentiment tone intensity. The outputs provide binary images of the recognized sentiment/opinion classes. Because of the binary input signals, the Hamming neural network [5] is chosen for the recognition of normalized sentiment scores and sentiment tone intensities, which directly realizes the one-nearest-neighbour classification rule.

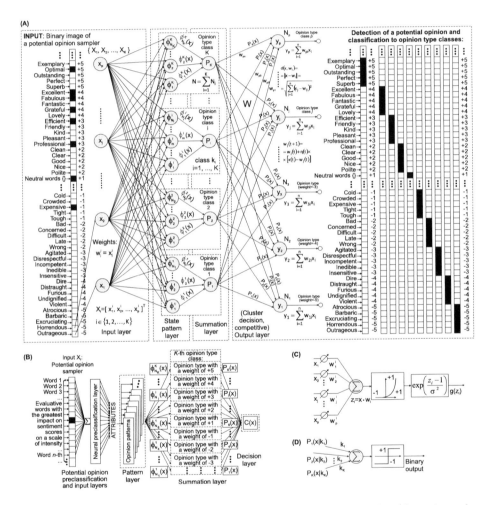

Fig. 3. Proposed hybrid neural networks for intelligent opinion mining (detecting subjective content in examined text): (A) neural network architecture for detection of a potential opinion in a text and its classification of opinion type classes, (B) neural classifier, (C) neuron of the pattern layer, (D) neuron of the output layer

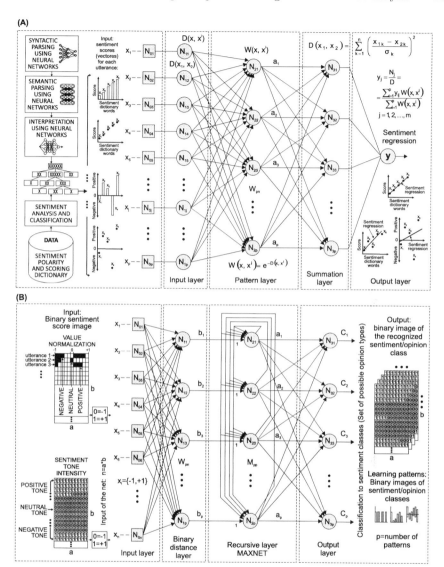

Fig. 4. Proposed neural networks for sentiment and opinion analysis: (A) methodology based on a regression neural network for determining sentiment regressions, (B) method based on a Hamming neural network for recognition of sentiment/opinion classes.

4   Conclusions and Perspectives

In the past few years, opinion mining and sentiment analysis has attracted a great deal of attention from both academia and industry due to many challenging research problems and a wide range of applications. The research presented in this article has proposed a concept of intelligent opinion mining and sentiment analysis which enables a variety of analyses of the subjective content of texts. Methods based on the proposed concepts can be extended in order to provide an even greater number of possible experimental applications.

References

1. Feldman, R.: Techniques and applications for sentiment analysis. Commun. ACM **56**(4), 82–89 (2013)
2. Taboada, M., Brooke, J., Tofiloski, M., Voll, K., Stede, M.: Lexicon-based methods for sentiment analysis. Comput. Linguist. **37**(2), 267–307 (2011)
3. Mohammad, S.M., Turney, P.D.: Crowdsourcing a word-emotion association lexicon. Comput. Intell. **29**(3), 436–465 (2013)
4. Chen, H., Zimbra, D.: AI and opinion mining. IEEE Intell. Syst. **25**(3), 74–80 (2010)
5. Majewski, M., Zurada, J.M.: Sentence recognition using artificial neural networks. Knowl. Based Syst. **21**(7), 629–635 (2008)
6. Kacalak, W., Stuart, K.D., Majewski, M.: Intelligent natural language processing. In: Jiao, L., Wang, L., Gao, X., Liu, J., Wu, F. (eds.) ICNC 2006. LNCS, vol. 4221, pp. 584–587. Springer, Heidelberg (2006)
7. Kacalak, W., Stuart, K., Majewski, M.: Selected problems of intelligent handwriting recognition. In: Melin, P., Castillo, O., Ramírez, E.G., Kacprzyk, J., Pedrycz, W. (eds.) IFSA 2007. Advances in Soft Computing, vol. 41, pp. 298–305. Springer, Cancun (2007)
8. Stuart, K.D., Majewski, M.: Selected problems of knowledge discovery using artificial neural networks. In: Liu, D., Fei, S., Hou, Z., Zhang, H., Sun, C. (eds.) ISNN 2007, Part III. LNCS, vol. 4493, pp. 1049–1057. Springer, Heidelberg (2007)
9. Stuart, K., Majewski, M.: A new method for intelligent knowledge discovery. In: Castillo, O., Melin, P., Ross, O.M., Cruz, R.S., Pedrycz, W., Kacprzyk, J. (eds.) IFSA 2007. Advances in Soft Computing, vol. 42, pp. 721–729. Springer, Heidelberg (2007)
10. Stuart, K.D., Majewski, M.: Artificial creativity in linguistics using evolvable fuzzy neural networks. In: Hornby, G.S., Sekanina, L., Haddow, P.C. (eds.) ICES 2008. LNCS, vol. 5216, pp. 437–442. Springer, Heidelberg (2008)
11. Stuart, K.D., Majewski, M.: Evolvable neuro-fuzzy system for artificial creativity in linguistics. In: Huang, D.-S., Wunsch II, D.C., Levine, D.S., Jo, K.-H. (eds.) ICIC 2008. LNCS (LNAI), vol. 5227, pp. 46–53. Springer, Heidelberg (2008)
12. Stuart, K.D., Majewski, M., Trelis, A.B.: Selected problems of intelligent corpus analysis through probabilistic neural networks. In: Zhang, L., Lu, B.-L., Kwok, J. (eds.) ISNN 2010, Part II. LNCS, vol. 6064, pp. 268–275. Springer, Heidelberg (2010)
13. Stuart, K.D., Majewski, M., Trelis, A.B.: Intelligent semantic-based system for corpus analysis through hybrid probabilistic neural networks. In: Liu, D., Zhang, H., Polycarpou, M., Alippi, C., He, H. (eds.) ISNN 2011, Part I. LNCS, vol. 6675, pp. 83–92. Springer, Heidelberg (2011)
14. Specht, D.F.: Probabilistic neural networks. Neural Netw. **3**(1), 109–118 (1990)
15. Specht, D.F.: A general regression neural network. IEEE Trans. Neural Netw. **2**(6), 568–576 (1991)

Mining Top-k Minimal Redundancy Frequent Patterns over Uncertain Databases

Haishuai Wang[1], Peng Zhang[1], Jia Wu[1,2(✉)], and Shirui Pan[1]

[1] Centre for Quantum Computation & Intelligent Systems,
University of Technology Sydney, Ultimo, NSW 2007, Australia
Haishuai.Wang@student.uts.edu.au,
{Peng.Zhang,Jia.Wu,Shirui.Pan}@uts.edu.au
[2] Department of Computer Science, China University of Geosciences,
Wuhan 430074, China

Abstract. Frequent pattern mining from uncertain data has been paid closed attention due to most of the real life databases contain data with uncertainty. Several approaches have been proposed for mining high significance frequent itemsets over uncertain data, however, previous algorithms yield many redundant frequent itemsets and require to set an appropriate user specified threshold which is difficult for users. In this paper, we formally define the problem of top-k minimal redundancy probabilistic frequent pattern mining, which targets to identify top-k patterns with high-significance and low-redundancy simultaneously from uncertain data. We first design uncertain pattern correlation based on Pearson correlation coefficient, which considers pattern uncertainty. Moreover, we present a new algorithm, UTFP, to mine top-k minimal redundancy frequent patterns of length no less than minimum length min_l without setting threshold. We further propose a set of strategies to prune and reduce search space. Experimental results demonstrate that the proposed algorithm achieves good performance in terms of finding top-k frequent patterns with low redundancy on probabilistic data. Our method represents the first research endeavor for probabilistic data based top-k correlated pattern mining.

Keywords: Top-k · Frequent patterns · Uncertain · Redundancy

1 Introduction

Uncertain frequent pattern mining is a crucial data mining task due to the wide applications, such as sensor network monitoring, moving object tracking, and protein interaction data. One fruitful research direction on frequent pattern mining is *support* measurement of patterns, that is pattern significance. In this case, the definitions of uncertain frequent pattern mining can be categorized into two classes: *expected support-based frequent patterns* [1,3] that uses the expectation of the support, and *probabilistic frequent patterns* [2,11] that considers the probability that the support of a pattern is no less than specified

© Springer International Publishing Switzerland 2015
S. Arik et al. (Eds.): ICONIP 2015, Part IV, LNCS 9492, pp. 111–119, 2015.
DOI: 10.1007/978-3-319-26561-2_14

threshold. The common framework among this two definitions is to consider the *support* of a pattern as a discrete random variable. Tong et al. [4] clarified the relationship of the two uncertain frequent pattern definitions and proved the two definitions can be integrated together. Despite the different frequentness metrics are employed, both of them hold the downward closure property, which leads to generate an exponential number of short result patterns. As a remedy, the other line of research work aims to eliminate *redundancy* among discovered patterns, such as pattern summarization [6] and pattern representative [7], which only consider the correlation between pattern and its sub-patterns with distance measure.

These pervious works over uncertain data only focus on significance or redundancy separately, whereas many real applications aim to discover high-significance and low-redundancy patterns. Furthermore, users may only interested in the k most frequent and low-redundancy result patterns, and in practice, setting an appropriate support threshold is often difficult, if not impossible.

Therefore, it is imperative for mining a small probabilistic pattern set with high-significance and low-redundancy without user specified threshold settings. A straightforward approach to solve the above problems is to combine the currently research works of frequent pattern mining methods and pattern correlation calculation methods. While this jointly approach needs two steps: generating the whole pattern set and finding top-k target patterns, which is computational and spatial inefficient.

Motivated by the fact that correlations between two patterns measures the similarity of the pairs occurrence distributions, in this paper, we formally define *Combined Pattern Significance and Redundancy* (*CPRS*), an criterion to assess the utility of each pattern by jointly considering the pattern significance and redundancy. Afterwards, we integrates the calculation of *CPRS* into the frequent pattern mining process. To obtain a set of top-k frequent patterns with weak correlations, we derive two effective pruning rules to reduce the unpromising itemsets. Our experimental results show that our approach identifies the top-k long frequent patterns with weak correlations.

The remainder of the paper is structured as follows. Section 2 introduces the preliminaries and problem definition. Section 3 presents the proposed approaches. The experiment results and conclusion are demonstrated in Sects. 4 and 5, respectively. Section 6 surveys related work.

2   Preliminaries and Problem Definition

Definition 1 *(Pattern Significance)*. A significance measure S is the degree of usefulness of a pattern [9]. For uncertain data, S can be expected support, confidence, coherence, and other attribute values. In this paper, we use support random variables of patterns as significance measure.

Assuming the existence probabilities of patterns X_1, X_2 in the ith transaction t_i are p_1^i, p_2^i, and the corresponding support random variables are $\hat{X}_j^i$ $(j = 1, 2)$, which follows Bernoulli distribution, that is $\hat{X}_j^i \sim Bern(p_j^i)$, $j = 1, 2$.

As $\hat{X}_j$ follows Poisson binomial distribution, the mean value and variance of $\hat{X}_j$ are

$$u_j = \sum_{i=1}^{n} p_j^i, \; \sigma_j^2 = \sum_{i=1}^{n} (p_j^i(1 - p_j^i)), \; j = 1, 2$$

The covariance of $\hat{X}_1$ and $\hat{X}_2$ is $Cov(\hat{X}_1, \hat{X}_2) = \sum_{i=1}^{n} \sum_{j=1}^{n} Cov(\hat{X}_1^i, \hat{X}_2^j)$.

As $\hat{X}_1^i$, $\hat{X}_2^j$ are independent with respect to transaction i and j such that $i \neq j$, then the covariance $Cov(\hat{X}_1^i, \hat{X}_2^j) = 0$. The larger of $Cov(\hat{X}_1^i, \hat{X}_2^j)$, the more stronger correlation of $\hat{X}_1^i$ and $\hat{X}_2^j$.

We resort to *Pearson's Correlation Coefficient* [5,8] (PCC) to define the uncertain pattern correlation. Thus, we have the PCC of $\hat{X}_1$ and $\hat{X}_2$, $\Phi(\hat{X}_1, \hat{X}_2)$, as shown in the following definition.

Definition 2 *(PCC for Uncertain).* Pearson's Correlation Coefficient between two support random variables $\hat{X}_1$ and $\hat{X}_2$ with respect to pattern X_1 and X_2 is defined in Eq. (1).

$$\Phi(\hat{X}_1, \hat{X}_2) = \frac{E(\hat{X}_1, \hat{X}_2) - E(\hat{X}_1)E(\hat{X}_2)}{\sqrt{E(\hat{X}_1^2) - (E(\hat{X}_1))^2}\sqrt{E(\hat{X}_2^2) - (E(\hat{X}_2))^2}} = \frac{\sum_{i=1}^{n} \sum_{j=1}^{n} Cov(\hat{X}_1^i, \hat{X}_2^j)}{\sigma_{\hat{X}_1} \sigma_{\hat{X}_2}}$$

$$= \frac{\sum_{i=1}^{n} Cov(\hat{X}_1^i, \hat{X}_2^i)}{\sigma_{\hat{X}_1} \sigma_{\hat{X}_2}} = \frac{\sum_{i=1}^{n} (p_{(1,2)}^i - p_1^i p_2^i)}{\sqrt{\sum_{i=1}^{n} (p_1^i(1 - p_1^i)) \sum_{j=1}^{n} (p_2^j(1 - p_2^j))}} \tag{1}$$

where $p_{(1,2)}^i$ is the existence probability of union set of X_1 and X_2 in the ith transaction t_i. $\sum_{i=1}^{n} \sum_{j=1}^{n} Cov(\hat{X}_1^i, \hat{X}_2^j) = \sum_{i=1}^{n} Cov(\hat{X}_1^i, \hat{X}_2^i)$ due to $Cov(\hat{X}_1^i, \hat{X}_2^j) = 0$.

Definition 3 *(Pattern Pair Redundancy).* Assume the significance of pattern X, Y are S_X and S_Y respectively, then the redundancy of patterns X and Y, $PPR(X, Y)$, is denoted in Eq. (2).

$$PPR(X, Y) = |\Phi(S_X, S_Y)| \tag{2}$$

Note that this redundancy definition focuses on the strong or weak correlation of two patterns, so this definition just considers positive correlation, and pattern redundancy should satisfy $0 \leqslant PPR(X, Y) \leqslant 1$.

Definition 4 *(Pattern Set Redundancy).* Given a set of k patterns $\mathcal{P}^k = \{p_1, p_2, \cdots, p_k\}$, let $PSR(\mathcal{P}^k)$ denotes the redundancy of $\mathcal{P}^k$, we have

$$PSR(\mathcal{P}^k) = \sum_{i=1}^{k} \sum_{j=i+1}^{k} PPR(p_i, p_j), \; p_i, p_j \in \mathcal{P}^k \tag{3}$$

Definition 5 *(Combined Pattern Significance and Redundancy).* Let $CPSR(X)$, *pattern absolutely significance of* X, be an evaluation function which measures the significance of a pattern $X \in \mathcal{P}^k$ and the redundancy of $\mathcal{P}^k$. Assume patterns in $\mathcal{P}^k$ are all independent, we have

$$CPSR(X) = S_X(C - PSR(\mathcal{P}^k)). \tag{4}$$

where constant $C = \binom{k}{2}$. *Note that* $0 \leqslant PPR(X, Y) \leqslant 1$, *therefore, we have* $0 \leqslant PSR(X, Y) \leqslant \binom{k}{2}$ *from Definition 4, but* $0 \leqslant S_X \leqslant 1$, *so we use constant* C *to tradeoff the influence of* S_X *and* $PSR(\mathcal{P}^k)$. *From the formula, we have* $CPSR$ *is larger for high* S_X *and low* PSR.

Definition 6 *(Top-k Frequent Patterns with Minimum Redundancy and Length Constraint).* A pattern X is a top-k frequent pattern with minimum redundancy and length is no less than min_l if there are no more than $(k-1)$ patterns which length at least min_l whose $CRSR$ is higher than that of X.

3   Proposed Method

In the previous section, we define the top-k minimum redundancy frequent patterns whose length is at least min_l. In this section, we propose an efficient algorithm, UTFP, for mining top-k frequent patterns with minimal redundancy and length is no less than minimal length over uncertain data. We resort to UH-Mine [1] as the basic mining algorithm. This section firstly sketches out the UH-Mine algorithm, and two strategies, which are used to pruning and reducing search space. Then we present the overall framework for the proposed algorithm.

UH-Mine: UH-Mine [1] is expected support-based algorithm and extended from the H-Mine [10]. UH-Mine is also based on the divide-and-conquer framework and the depth-first search strategy, and can be outlined as follows. Initially, it scans the database once to find all expected support-based frequent items and remove the infrequent items from database. Then, it builds a head table which contains all the frequent items. After that, the transformed database is stored in a structure, UH-Struct, which contains label, appearing probability and a pointer for each item. After building the global UH-Strut, UH-Mine uses the depth-first strategy to build the head table where each items in global head table as the prefix. Then, recursively building the head tables where different itemsets are prefix and generating all expected support-based frequent itemsets.

An example of UH-Mine is shown in Fig. 1. The algorithm firstly builds the global UH-Struct of Table 1 which contains item A, B, C and their expected support by descending order, and uses hyper-linked array to store each transactions that length is no less than min_l. Figure 1(B) shows the head table where item A is the prefix.

Strategy 1. During the UH-Struct construction, if there are less than min_l distinct items in a transaction T_i, T_i can be skipped, because the items in T_i cannot be used as a pattern with minimum length min_l.

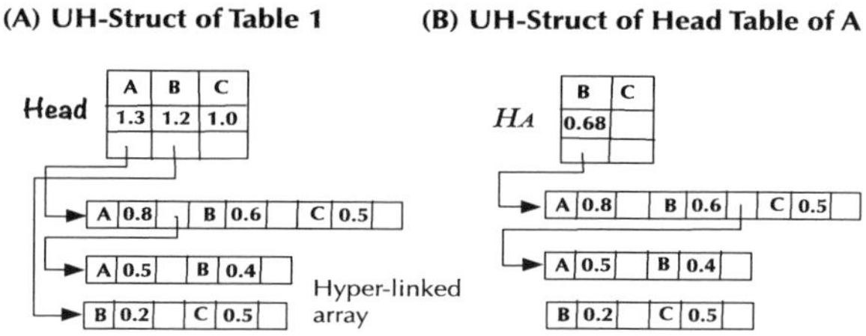

(A) UH-Struct of Table 1   **(B) UH-Struct of Head Table of A**

Fig. 1. (A) is the global UH-Struct generated from Table 1, and the UH-Mine algorithm uses the depth-first strategy to build the head table in (B) where item A is the prefix.

Algorithm 1. UTFP (DB, min_l, k)

---

1 threshold $min\_CPSR = 0$;
2 $\mathcal{P}^k, \mathcal{P}^{k-1} = \{\phi\}$, $new\_\mathcal{P}^k = \{\phi\}$;
3 Get patterns p with min_l from UH-Mine;
4 **forall the** p **do**
5 $\quad$ **if** $\mathcal{P}^k.size \leq k$ **then**
6 $\quad\quad$ $\mathcal{P}^k \leftarrow p$;
7 $\quad\quad$ **if** $\mathcal{P}^k.size = k$ **then**
8 $\quad\quad\quad$ Get pattern p, p^{min}, whose $CPSR$ is minimum in $\mathcal{P}^k$ and $CSPR(p^{min})$;
9 $\quad\quad\quad$ $min\_CPSR \leftarrow CPSR(p^{min})$;
10 $\quad$ **else**
11 $\quad\quad$ $new\_\mathcal{P}^k \leftarrow \mathcal{P}^k$;
12 $\quad\quad$ $\mathcal{P}^{k-1} \leftarrow \mathcal{P}^k.remove(p^{min})$;
13 $\quad\quad$ **forall the** $p^i \in \mathcal{P}^{k-1}$ **do**
14 $\quad\quad\quad$ Get the redundancy of p^i and p, $R(p^i, p)$;
15 $\quad\quad\quad$ $PSR^* = PSR^* + R(p^i, p)$;
16 $\quad\quad$ $CPSR(p) = S_p(C_k^2 - PSR^*)$;
17 $\quad\quad$ **if** $CPSR(p) > min\_CPSR$ **then**
18 $\quad\quad\quad$ $new\_\mathcal{P}^k \leftarrow \mathcal{P}^{k-1}.add(p)$;
19 $\quad\quad\quad$ Get $CPSR(p^{min})$ in $\boldsymbol{new\_\mathcal{P}^k}$;
20 $\quad\quad\quad$ $min\_CPSR \leftarrow CPSR(p^{min})$;

21 **return** $new\_\mathcal{P}^k$;

Strategy 2. During the mining stage, if the length of current header table is l_h, and the current hyper-link in transaction t is t_i, the length from t_i to the end of t is l_r. If $l_h + l_r < min\_l$, the transaction t can be skipped.

For example, assume the current header table is H_{abd}, and the minimal length is $min\_l = 6$. When scanning a transaction $t : \{a, b, c, d, e, f\}$, due to $l_h = 3, l_r = 2$, and $l_h + l_r = 5 < min\_l$, so this transaction should be skipped.

Framework of the UTFP Algorithm: We use UH-Mine algorithm and follow the two strategies to generate frequent patterns whose length is no less than

Table 1. An example of uncertain database.

TID	Transactions
T1	A:0.8 B:0.6 C:0.5
T2	A:0.5 B:0.4
T3	B:0.2 C:0.5

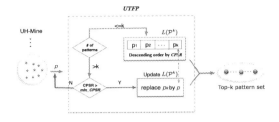

Fig. 2. The overall framework of UTFP.

min_l, and initialize the threshold min_CPSR to 0 (steps 1–4). During pattern generation, at the beginning, the discovered patterns are added to a list of patterns $L(\mathcal{P}^k)$ which are ordered by descending $CPSR$ of patterns (Definition 5) until the kth pattern. The list $L(\mathcal{P}^k)$ is used to maintain the top-k patterns. Once k patterns are found, computing the minimal $CPSR$ in $L(\mathcal{P}^k)$, and the current threshold min_CPSR is raised to the value (steps 5–10). The raised min_CPSR is used to prune the search space when searching for the following patterns. Thereafter, once a pattern is generated, if the $CPSR$ of the new patter meets the min_CPSR threshold, then the tail pattern in $L(\mathcal{P}^k)$ is removed and the new pattern is inserted into the list, and simultaneous updating the current min_CPSR. Repeat the process until there is no pattern generation (steps 11–21). Finally, the algorithm retruns the result patterns based on Definition 6. The overall framework of $UTFP$ is shown in Fig. 2.

Example: Assuming there are five patterns p_1, p_2, p_3, p_4, p_5, and the generation sequence is $p_1 \prec p_2 \prec p_3 \prec p_4 \prec p_5$. We also assume $k = 3, min\_l = 2$. After generating patterns p_1, p_2, p_3, we add them to a list $L(\mathcal{P}^k)$ and compute the redundancy of each two patterns in $L(\mathcal{P}^k)$ by Definition 3. Then, by Eqs. (3) and (4), we can get the pattern set redundancy $PSR(\mathcal{P}^k)$ and the $CPSR$ of each pattern in $L(\mathcal{P}^k)$ respectively, after that, sorting $L(\mathcal{P}^k)$ by $CPSR$ with descending order and setting the threshold min_CPSR to the minimal $CPSR$ value (steps 1–10 in Algorithm 1). When pattern p_4 is generated, calculating the pattern set redundancy of p_4 with first $(k-1)$ patterns in $L(\mathcal{P}^k)$ and $CPSR(p_4)$. If $CPSR(p_4)$ is no less than the current min_CPSR, the pattern of tail of $L(\mathcal{P}^k)$ is removed and p_4 is added to $L(\mathcal{P}^k)$, otherwise, continue these steps with pattern p_5 (steps 11–21 in Algorithm 1). Finally, the algorithm returns the result pattern set with 3 patterns following Definition 6.

4   Experimental Results

To validate the performance of our method, we use the deterministic database, Retail dataset, which can be downloaded from the Frequent Itemset Mining(FIMI) Dataset Repository[1]. The number of transactions and items in the

---

[1] http://fimi.cs.helsinki.fi/data

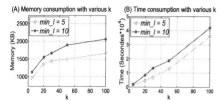
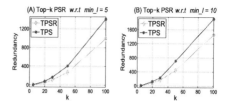

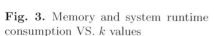

Fig. 3. Memory and system runtime consumption VS. k values

Fig. 4. Top-k pattern set redundancy of TPSR and TPS.

retail dataset is 88162 and 16470 respectively, and the average length is 10.3. To bring uncertainty into the dataset, we synthesize an existential probability for each item based on a Gaussian distribution with the mean of 0.9 and the variance of 0.125. The algorithm is implemented in Java. The system is a Linux Ubuntu server with 8*2.9 GHz CPU and 32 G memory.

In Fig. 3, we report the memory and system runtime consumptions $w.r.t$ different top-k values and minimal length constraint. We set two different minimal length constraints, $min\_l = 5$ and $min\_l = 10$, respectively. Figure 3(A) shows the memory consumption with various k, the algorithms consume more memory with the increasing of k or min_l, mainly because UTFP holds the k patterns in memory until finishing the mining process, so more patterns or longer patterns need more memory space to store. On the other hand, if there are more patterns or longer patterns, the calculation of redundancy and minimal threshold consume more computation time, as the results demonstrated in Fig. 3(B).

We compare the redundancy of mined top-k patterns which with both high-significance and low-redundancy (TPSR), and high-significance only (TPS), as shown in Fig. 4. Figure 4(A) and (B) demonstrate Top-k Pattern Set Redundancy (PSR) with length is no less than 5 and 10 respectively. The redundancy of TPS is much higher than that of TPSR. The redundancy increases with increasing minimal length constraint under same k value both in TPSR and TPS.

5 Conclusion

In this paper, we studied top-k frequent pattern mining and pattern correlation over uncertain data, and proposed the UTFP algorithm, which aims to find top-k probabilistic frequent patterns with minimum redundancy and length constraint. To address the data uncertainty issue, we define the concept of probabilistic pattern redundancy with pearson's correlation coefficient, as well as a measure combines pattern significance and redundancy. To dynamically raise the threshold, prune unpromising itemsets, and reduce search space, we further present two strategies and an UH-Mine based algorithm to find top-k frequent patterns, which length is no less than minimal length threshold, with minimal redundancy from uncertain data. Our experimental results demonstrate that the proposed method effectively discovers the set of top-k probabilistic frequent patterns with high-significance and low-redundancy without setting thresholds, and the redundancy of result set is less than that only considers significance.

6   Related Work

Existing probabilistic frequent pattern mining can be roughly categorized into two classes: expected support-based patterns mining and probabilistic-based pattern mining. The first class focus on using the expectation of support as the frequentness measure, a pattern is frequent only if its expected support is no less than a specified minimum expected support, such as UApriori [3], UFP-growth [1] and UH-Mine [1], while the second class uses frequency probability as the metric, which means the probability that a pattern appears no less than a specified minimum support times, such as DP [2] and DC [11]. However, the anti-monotonicity still hold in uncertain data mining, which leads to generate an exponential number of patterns.

A variety of definitions have been proposed to reduce the number of patterns, such as maximal patterns [12], frequent closed patterns [14] and nonderivable patterns [13]. While all frequent patterns can be recovered from maximal patterns, the loss of support information is unacceptable in some circumstances. Therefore, to minimize the information losing, maximize user requirement and speed up response time, there are some works to mine top-k frequent patterns in mining determinate data [15, 16] which is unnecessary to set up minimal support threshold by users. But to the best of our knowledge, no existing works/studies exist for mining top-k frequent patterns in uncertain databases.

Acknowledgement. This work was supported by the Australian Research Council (ARC) Discovery Project under Grant No. DP140102206.

References

1. Aggarwal, C.C., Li, Y., Wang, J.: Frequent pattern mining with uncertain data. In: SIGKDD, pp. 29–38 (2009)
2. Bernecker, T., Kriegel, H.P., Renz, M., Verhein, F., Zuefle, A.: Probabilistic frequent itemset mining in uncertain databases. In: SIGKDD (2009)
3. Chui, C.-K., Kao, B., Hung, E.: Mining frequent itemsets from uncertain data. In: Zhou, Z.-H., Li, H., Yang, Q. (eds.) PAKDD 2007. LNCS (LNAI), vol. 4426, pp. 47–58. Springer, Heidelberg (2007)
4. Tong, Y., Chen, L., Yurong, Y., Yu, P.S.: Mining frequent itemsets over uncertain databases. In: VLDB, pp. 1650–1661 (2012)
5. Zhou, W., Xiong, H.: Volatile correlation computation: a checkpoint view. In: KDD, pp. 848–856 (2008)
6. Liu, C., Chen, L., Zhang, C.: Summarizing probabilistic frequent patterns. In: KDD, pp. 527–535 (2013)
7. Liu, C., Chen, L., Zhang, C.: Mining probabilistic representative frequent patterns from uncertain data. In: SDM, pp. 73–81 (2013)

8. Pan, S., Zhu, X.: Continuous top-k query for graph streams. In: CIKM, pp. 2659–2662 (2012)
9. Xin, D., Cheng, H., Yan, X., Han, J.: Extracting redundancy-aware top-k patterns. In: KDD, pp. 444–453 (2006)
10. Pei, J., Han, J., Lu, H., Nishio, S., Tang, S., Yang, D.: H-mine: hyper-structure mining of frequent patterns in large databases. In: ICDM, pp. 441–448 (2001)
11. Sun, L., Cheng, R., Cheung, D.W., Cheng, J.: Mining uncertain data with probabilistic guarantees. In: SIGKDD, pp. 273–282 (2010)
12. Bayardo Jr., R.: Efficiently mining long patterns from databases. In: SIGMOD, pp. 85–93 (1998)
13. Calders, T., Goethals, B.: Mining all non-derivable frequent itemsets. In: Elomaa, T., Mannila, H., Toivonen, H. (eds.) PKDD 2002. LNCS (LNAI), vol. 2431, pp. 74–85. Springer, Heidelberg (2002)
14. Pasquier, N., Bastide, Y., Taouil, R., Lakhal, L.: Discovering frequent closed itemsets for association rules. In: Beeri, C., Bruneman, P. (eds.) ICDT 1999. LNCS, vol. 1540, pp. 398–416. Springer, Heidelberg (1998)
15. Tzvetkov, P., Yan, X., Han, J.: TSP: mining top-k closed sequential patterns. Knowl. Inf. Syst. 7(4), 438–457 (2005)
16. Han, J., Wang, J., Lu, Y., Tzvetkov, P.: Mining top-k frequent closed patterns without minimum support. In: ICDM, pp. 211–218 (2002)

Exploring Social Contagion in Open-Source Communities by Mining Software Repositories

Zakariyah Shoroye, Waheeb Yaqub, Azhar Ahmed Mohammed,
Zeyar Aung, and Davor Svetinovic$^{(\boxtimes)}$

Institute Center for Smart and Sustainable Systems,
Department of Electrical Engineering and Computer Science,
Masdar Institute of Science and Technology, P.O. Box 54224, Abu Dhabi, UAE
{zshoroye,wyaqub,amohammed,zaung,dsvetinovic}@masdar.ac.ae

Abstract. The emergence of data mining has helped improve our understanding of social contagion in networks. The magnitude of contagion in
networks such as Facebook and Twitter has been studied in detail. Study
of social contagion in software development networks can provide interesting findings in order to increase return on investment and improve
quality of software. For example, developers could be incentivised and
the time to start an open-source projects optimized by analyzing social
contagion in online repositories. In this study, open-source repositories'
data was analyzed and it was observed that highly followed developers
tend to attract more contributors to a project. Also, the number of commits was aggregated on a yearly basis to provide insight into the question
of the best time to start a project. GitHub online repository data was
collected since its inception until 2014. The number of commits in the
online repository was found to follow the "power law". By considering
only large projects, a correlation between the number of followers a user
has and the contagion rate of their commits was observed. Understanding these questions and social contagion can help software companies to
leverage on the open-source community and improve their own internal
social networks.

Keywords: Software repositories · Social network analysis · Social
contagion

1 Introduction

Collaborative software development is gaining momentum in the competitive
market as companies have to innovate faster and cut costs. Recently a new
trend in software development is emerging in which companies across industries
are collaborating to develop common code bases, which can then be extended.
In March 2014, Linux Foundation carried out an invitation-only survey to the
companies that included Cisco, Fujitsu, HP, IBM, Intel, Google, and Samsung
among others [1]. Exactly 686 software developers took part in this survey from

© Springer International Publishing Switzerland 2015
S. Arik et al. (Eds.): ICONIP 2015, Part IV, LNCS 9492, pp. 120–127, 2015.
DOI: 10.1007/978-3-319-26561-2_15

organizations that make 500 million dollars or more in annual revenue. The findings of the study showed that open-source development is no longer driven by the love for programming but by money. 35 % of the developers still contribute to open-source projects in their free time but 44 % of the developers admitted that the job requirements were the reason they started contributing [1,2]. Thus, the recent trend is that companies encourage their staff to contribute to open-source projects. As such, commercial development interest and influence in open-source projects is on the rise.

This approach of collaborative development is a win-win situation for both the organizations and the developers. Half of the managers surveyed were of the opinion that collaborative development gives them the flexibility to innovate and help transform the industry. Cloud computing, mobile devices, the Internet of Things, software-defined-networking were identified as the top five that would see increasing use of collaborative development practices [1]. The Linux Foundation report emphasizes the use of open-source collaboration and hence it is imperative to answer the following questions:

– When is the best time to start a new open-source project?
– Who should be given the incentive to start new open-source project?

The study of the social contagion in the open-source development networks can help us understand and answer such questions. The answers can in turn help software companies to leverage the open-source community and improve their own internal social networks.

In this study, we try to answer these questions by analyzing GitHub [3] data. GitHub is a web-based Git repository hosting service that offers source code management (SCM) and reversion control of projects in distributed fashion. As of 2014, GitHub is the largest open-source projects host [4]. Our study showed that the best starting time for a project is between June and August. We also found that there is a correlation between the followers a user has and the number of such followers that contributed to a particular project after their contribution. This makes such developers potential targets for incentivisation.

2   Related Work and Research Method

Social networks have been widely studied by researchers for its various characteristics such as contagion, geography, centrality, etc. For example, Ugander et al. [5] studied the structural diversity in social contagion on the Facebook network. The studies on the veracity of GitHub [3] repository data for research purposes has shown that although it is rich in data, it has some underlying issues which developers have to be aware of [6]. These include the presence of discrepancies in the pull commit data and that a lot of the projects are small and personal. Social networks are being more and more leveraged in software development [7].

GitHub data is either accessible through its application program interface (API) which limits the number of queries requested per hour to 60 records if using an unauthenticated account and 5,000 records for an authenticated account.

Given the scale of data required to be accessed by us, 5,000 records every hour was merely not enough. Therefore we obtained all of GitHub using GHTorrent [8]. The total size of the data was 40 GB.

GHTorrent provides archives of close to 600 million rows of MySQL data which can be downloaded as a *.sql* file that can be used to build a native database. GHTorrent provides data in a very organized manner dividing it into several tables. The two tables used in this study are *commits* and *followers*. The size of *commits* table is 39 GB with almost 1.5 billion records and that of *followers* table is 397 MB with 5 million records.

First, we tired to categorize the repositories based on their respective sizes. As of 2014, there are 16.7 million repositories in GitHub with 3.4 million users. These have varying number of commits with some having zero commits and some having more than 400,000 commits. Contrary to the method used by Kolassa et al. [9], in which the number of contributors was chosen as a representative of the size of GitHub repository, we chose the number of commits to represent the size of the project. This is because there can be some projects with few authors but a large number of commits. Furthermore, while Kolassa et al.'s method cannot observe the disparity between the two projects with the same number of developers but a notably different number of commits, our proposed commit-based method can do so. As we mentioned before, large software companies are adapting open-source and collaborative approach to develop projects in order to limit the cost of projects, etc. Later we will show that large projects are the ones that fall in the long tail of the "power law distribution" where the number of commits is the representative of project size.

Before trying to fit the data to any distribution, it can be assumed that the empirical data follows normal (a.k.a. Gaussian) distribution. The normal distribution is a natural guess since it is heavily used in natural sciences [10]. From the "central limit theorem" it is deduced that if we take small sequence of small independent random variables then their average will be distributed across normal distribution. On the other hand a study on the distribution of web links showed something different from central limit theorem [10]. The number of web pages that have m links is proportional to $\frac{1}{m^2}$. Such extreme observations are better captured using the power law distribution.

Since population in large cities, biological extinction, genetic networks or the World Wide Web all follow the power law [11,12] and so does the sizes of commits in GitHub [13], we hypothesize that the number of commits would also follow the power law. We are interested in the projects that fall in the heavy tail of the power distribution. We believe that these projects represent the scale and size of repositories similar to projects that companies are willing to collaborate on. The total number of commits for more than 6.5 million repositories was used as an input to fit the power law distribution. We followed the method introduced by Clauset et al. in [14] to fit the empirical data to power law distribution. In [14,15], it was shown that when testing empirical data, creating the cumulative distribution function and fitting the resulting function to the linear form by least-squares linear regression is subject to systematic and potentially large errors. To avoid such errors incurred in [9,13], the steps listed below were followed for analysing the data.

1. Estimate the x_{min} and α using bootstrap method to get handle on parameter uncertainty.
2. Compute the goodness of fit between the empirical data and the power law using goodness-of-fit test, which generates a p-value that represents the soundness of the hypothesis. These tests depend on the distance between distribution of empirical data and the hypothesized model.
3. Eliminate the possible competing distributions such as exponential or log normal.

3 Results and Discussions

3.1 Power Law Distribution

The histograms in Figs. 1 and 2 depict the uncertainty in estimated parameters of x_{min} and α. Based on the bootstrapping method proposed by Clauset et al. [14], we ran and estimated the parameters for 100 simulations. The frequency of parameters x_{min} and α was 384 and 2.372, respectively. Based on the hypothesis that data is generated from the power law distribution and $p = 0.82 > 0.1$, we cannot rule out the power law. The p-values examined for other distributions were less than that of the power law. Hence, we found that the power law is the best fit for this data.

The data was fitted to the power law distribution. Based on this fitting, we were able to categorize the size of projects based on its commits size. This helps to avoid some of the shortcomings of using the GitHub data as discussed in Kalliamvakou et al. [6].

3.2 Best Time to Start New Projects

Close to three million projects were launched on GitHub during its five years of existence: 2008 until 2013. We limited the retrieved data until the end of 2013 because complete data for year 2014 was not available at the time of conducting our experiments. Figure 3 shows the monthly commits activity of these projects. The values were normalized to scale uniformly across all years. The plot shows a gradual increase in the commit activity with a slight decrease in the month of December. The decrease in the commit activities in December (except for 2011) is possibly because of the festive season. Figure 3 shows that the best time to start the project on GitHub would be from June to August as across most of the years the commit activity continues to increase during this time. This can be attributed to the fact that the summer vacation in most parts of the world starts during that period and consequently developers find more free time. It would thus be imperative to start the project at the peak times as, based on basic probability studies, it is most likely going to get the much needed attention.

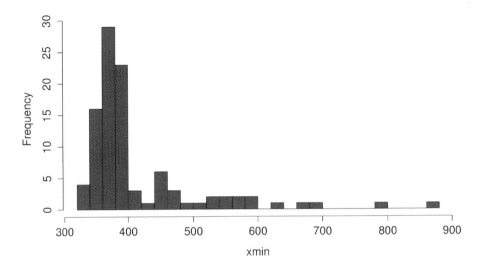

Fig. 1. Frequency of different values of x_{min}. Y-axis represents the frequency of each possible X-min values from the bootstrapped method. X-axis represents the minimum possible values after which the power law holds.

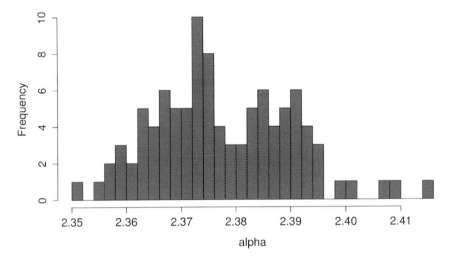

Fig. 2. Frequency of different values of alpha. Y-axis represents the frequency of each possible α values from the bootstrapped method.

3.3   Social Contagion in Open-Source Software Development

We were also able to identify those who should be given an incentive to start a project because of their high contagion rate. During the analysis, we considered only the large projects using the criteria established above using the power law. We analyzed 443 projects that had more than 10,000 commits. This is because

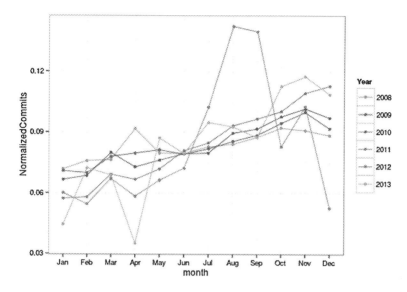

Fig. 3. Normalized number of commits for all months from 2008 to 2013.

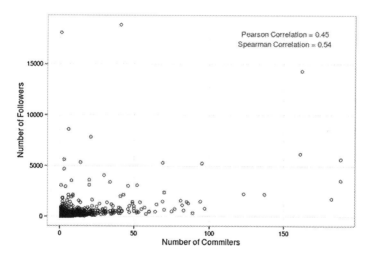

Fig. 4. Correlation of number of followers vs. number of followers who are committed.

smaller projects might skew the data as the probability of a developer follower committing in the same project as him decreases if only a few developers are partaking in the project. The threshold of 10,000 commits was thus chosen. We were able to compare over 70,000 unique developers. Figure 4 shows the correlation between the number of followers a user has and the followers who committed in a project after she has committed. We found a Spearman correlation of 0.54 and a Pearson correlation of 0.45. Cohen provides the following guidelines for

Pearson correlation in social sciences: *small* if the value is 0.1, *moderate* if it is 0.3 and *large* if it is 0.5 [16,17]. Hence, we found that the contagion factor is highly prevalent in the GitHub open-source community. This indicates that the developers are likely to participate in projects where their followees have earlier participated. Thus, coding can be considered as contagious in the follower network of GitHub.

4   Limitations and Future Work

In this study only two variables were considered: the number of commits and the number of followers. Our assumptions can be made stronger by considering other variables such as different criteria on the size of the project, bi-directional following, time-relative following trends, etc. Another interesting direction would be to analyze the behavior of the GitHub users in correlation with their behavior on the other social networking sites, e.g., Twitter. A good question to answer would be how many of the followers on Twitter have joined GitHub. We used the number of commits made as a measure of contribution to the project. The use of other actions like pull requests might yield additional insights. Our preliminary results show that there exists a correlation between the number of commits made by the followers and the total number of followers. The correlation however does not show causation and hence a causal analysis is necessary.

5   Conclusion

In this study, we tried to answer two interesting questions derived from the Linux Foundation report [1]. The answers to these questions might help companies decide when is the best time to start a project and whom to incentivise to start and lead the project. Our preliminary analysis showed that starting a project between June and August is probably optimal to leverage free time and social contagion within open-source community networks. We also found that there is a correlation between the followers a user has and the number of such followers that commit to a particular project after she committed. This is an evidence of the existence of the social contagion within the development networks. This preliminary study provides us with sufficient support for extending the study of the social contagion in the open-source development communities, especially in the early stages of development, e.g. [18,19], and its further incorporation within our main research initiative [20].

References

1. Linux Foundation. Collaborative development trends report. Technical report (2014)
2. Vaughan-Nichols, S.J.: The new open source motivation, Show me the money (2014)

3. GitHub: Build software better, together (2014)
4. Gousios, G., Vasilescu, B., Serebrenik, A., Zaidman, A.: Lean GHTorrent: GitHub data on demand. In: Proceedings of the 11th Working Conference on Mining Software Repositories (MSR), pp. 384–387. ACM (2014)
5. Ugander, J., Backstrom, L., Marlow, C., Kleinberg, J.: Structural diversity in social contagion. Proc. Natl. Acad. Sci. **109**(16), 5962–5966 (2012)
6. Kalliamvakou, E., Gousios, G., Blincoe, K., Singer, L., German, D.M., Damian, D.: The promises and perils of mining GitHub. In: Proceedings of the 11th Working Conference on Mining Software Repositories (MSR), pp. 92–101. ACM (2014)
7. Adepetu, A., Ahmed, K.A., Al Abd, Y., Al Zaabi, A., Svetinovic, D.: Crowdrequire: a requirements engineering crowdsourcing platform. In: AAAI Spring Symposium: Wisdom of the Crowd (2012)
8. Gousios, G.: The GHTorrent dataset and tool suite. In: Proceedings of the 10th Working Conference on Mining Software Repositories (MSR), pp. 233–236. ACM (2013)
9. Kolassa, C., Riehle, D., Salim, M.A.: A model of the commit size distribution of open source. In: van Emde Boas, P., Groen, F.C.A., Italiano, G.F., Nawrocki, J., Sack, H. (eds.) SOFSEM 2013. LNCS, vol. 7741, pp. 52–66. Springer, Heidelberg (2013)
10. Easley, D., Kleinberg, J.: Networks, Crowds, and Markets: Reasoning About a Highly Connected World. Cambridge University Press, Cambridge (2010)
11. Bak, P.: How Nature Works. Oxford University Press, Oxford (1997)
12. Barabási, A.-L., Albert, R.: Emergence of scaling in random networks. Sci. **286**(5439), 509–512 (1999)
13. Arafat, O., Riehle, D.: The commit size distribution of open source software. In: Proceedings of the 42nd Hawaii International Conference on System Sciences (HICSS), pp. 1–8. IEEE Press (2009)
14. Clauset, A., Shalizi, C.R., Newman, M.E.J.: Power-law distributions in empirical data. SIAM Rev. **51**(4), 661–703 (2009)
15. Goldstein, M.L., Morris, S.A., Yen, G.G.: Problems with fitting to the power-law distribution. Eur. Phys. J. B: Condens. Matter Complex Syst. **41**(2), 255–258 (2004)
16. Cohen, J.: Statistical Power Analysis for the Behavioral Sciences. Routledge Academic, New York (2013)
17. Cohen, J.: A power primer. Psychol. Bull. **112**(1), 155–159 (1992)
18. Svetinovic, D., Berry, D.M., Day, N.A., Godfrey, M.W.: Unified use case state-charts: case studies. Requir. Eng. **12**(4), 245–264 (2007)
19. Svetinovic, D.: Architecture-level requirements specification. In: STRAW, pp. 14–19 (2003)
20. Svetinovic, D.: Strategic requirements engineering for complex sustainable systems. Syst. Eng. **16**(2), 165–174 (2013)

Data Mining Analysis of an Urban Tunnel Pressure Drop Based on CFD Data

Esmaeel Eftekharian[1], Amin Khatami[2]([✉]),
Abbas Khosravi[2], and Saeid Nahavandi[2]

[1] School of Mechanical Engineering, Shiraz University,
Mollasadra Street, Shiraz, Iran
e_eftekharian@shirazu.ac.ir
[2] Centre for Intelligent Systems Research,
Deakin University, Geelong, VIC 3216, Australia
{skhatami,abbas.khosravi,saeid.nahavandi}@deakin.edu.au

Abstract. An accurate estimation of pressure drop due to vehicles inside an urban tunnel plays a pivotal role in tunnel ventilation issue. The main aim of the present study is to utilize computational intelligence technique for predicting pressure drop due to cars in traffic congestion in urban tunnels. A supervised feed forward back propagation neural network is utilized to estimate this pressure drop. The performance of the proposed network structure is examined on the dataset achieved from Computational Fluid Dynamic (CFD) simulation. The input data includes 2 variables, tunnel velocity and tunnel length, which are to be imported to the corresponding algorithm in order to predict presure drop. 10-fold Cross validation technique is utilized for three data mining methods, namely: multi-layer perceptron algorithm, support vector machine regression, and linear regression. A comparison is to be made to show the most accurate results. Simulation results illustrate that the Multi-layer perceptron algorithm is able to accurately estimate the pressure drop.

Keywords: CFD · Tunnel · Data mining · Pressure drop

1 Introduction

Tunnel ventilation is one of the most important issues in tunnel projects, as poorly ventilated tunnels can put passengers and drivers into danger. A precise estimation of tunnel pressure drop due to vehicles inside the tunnel leads to a better prediction of required ventilation equipment resulting in providing people inside the tunnel with fresh air. CFD simulation is one of the most reliable approaches of simulating flow field in industrial as well as academic projects. To assess the accuracy of CFD results, comparisons were made between the results obtained from experiment and those achieved from CFD simulation and these comparisons proved the accuracy and reliability of CFD simulation [1, 15]. Some researchers compared drag coefficient of a moving vehicle achieved from CFD simulation with that presented by 1D model [17]. It was shown that

© Springer International Publishing Switzerland 2015
S. Arik et al. (Eds.): ICONIP 2015, Part IV, LNCS 9492, pp. 128–135, 2015.
DOI: 10.1007/978-3-319-26561-2_16

1D model is successful to predict drag coefficient of an isolated moving vehicle with a good precision, nevertheless, it fails to represent a correct estimation when two consecutive cars are moving in a tunnel. In the same vein, it was shown that 1D model is unable to accurately predict the impacts of wind on tunnel ventilation [6]. The impacts of tunnel curvature on the drag coefficient of a moving vehicle in a curved tunnel were also investigated [16]. Moreover, estimation of drag coefficient of vehicles in unsteady condition for an urban tunnel by using optimization technique was presented [9]. Furthermore, using experimental data obtained from a wind tunnel, Watkins and Vino (2008) [18] showed that the presence of one car near by another one affects drag coefficient of both of them. In addition, the airflow and pollution level investigation of an urban tunnel in traffic jam were carried out [3,4]. The present literature review represents that few studies have been carried out regarding estimation of pressure drop due to combination of cars in traffic jam in a tunnel. On the other side, it is generally accepted that data mining approach is a strong and accurate tool for predicting patterns in engineering problems. This method is to be used to represent the correlation between pressure drop due to vehicles and tunnel velocity and length based on CFD simulation data.

This paper concentrates on proposing a new intelligent method for prediction of pressure drop of vehicles in one-way tunnels as a function of tunnel length and tunnel average air velocity. Generally, there are two methods for computing tunnels pressure drop, namely 1D (one dimensional) approach and CFD simulation; the former incorporates in oversimplifying assumptions in flow field inside the tunnel resulting in considerable errors, however, it involves a few computational costs; as for the latter, although it involves in huge computational costs, it enjoys very accurate results. More information about the procedure with which the tunnel pressure drop is calculated based on 1D model can be found in [2]. Difficulties such as high computational costs of CFD simulation motivate us to develop a technique so that based on the results achieved from CFD simulation, tunnel pressure drop can be predicted as fast as 1D model with an acceptable degree of accuracy. Hence, different methods in data mining approach are investigated in order to improve prediction of tunnel pressure drop based on CFD data (35 data in 3 dimensions). In fact, tunnel pressure drop in a tunnel in traffic congestion is a function of tunnel average air velocity and the tunnel length.

The rest of this article is organised as follow: in the second section we briefly explain the techniques used in this study, The experimental results in section three are followed by the conclusion section at the end.

2   Methodology

This section is allocated to describe the methods utilized in this study to conduct the research in order to propose a new prediction system for pressure drop due to cars in traffic jam inside the tunnels. The following items explain three different data mining techniques utilized in this study to predict this system:

- Artificial Neural Network: Artificial Neural Networks (ANNs) is recognized as a set of instrumental techniques which can learn functions based on real and discrete input data. Back-propagation algorithm (BP) is one of them to be taken into account. Using gradient recent, this method, categorized in supervised learning tool, sets interior parameters in a way resulting in the best prediction. It also is a general statement of the Widrow-Hoff learning rule 1 to multiple-layer networks and non-linear differentiable conversion functions [5]. ANN is also powerful in detecting errors in the training dataset. Multilayer perceptron (MLP) has been frequently used by researchers working in this area. It was proved that the MLP, BP algorithm for training the network according to [11,12], is a general function approximator [7]. The MPL tries to fit input data with anticipated outputs. Figure 1 shows a schematic view of MPL structure. As can be seen in the Fig. 1, it consists of several layers of nodes and each of them is linked to the following one. From the second layer onwards, the layers follow a non-linear activation function [7]. The MPL is used in this study as it is very helpful method for those data which are not linearly separable. Feed-forward neural network commonly classifies neurons into some separate layers. Neurons in the first layer are responsible to send signals to other layers through connections. Feed-forward network shown schematically in Fig. 1 is vastly used in practical applications [10]. A multilayer feed-forward neural network consists of an input layer, at least a hidden layer and one output layer.

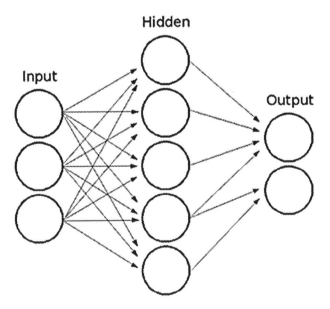

Fig. 1. A schematic view of typical MLP.

– Linear Regression: Linear regression is one of the simplest approximation for prediction of patterns of the data. For instance, to estimate x, it calculates the following linear expression.

$$x = w_0 + w_1 a_1 + ... + w_k a_k \qquad (1)$$

where w_i and a_i are the ith weight and the ith point respectively. The procedure of this prediction technique is consists of three main steps as following:
 • step 1: Calculating weights for training data.
 • step 2: Predicting value for first training instance, $a^{(1)}$:

$$Predicted\ value = \sum_{j=0}^{k} w_j a_j^{(1)} \qquad (2)$$

 • step 3: Minimizing the following squared error function on training data in order to obtain the optimal weights:

$$\sum_{i=1}^{n} (x^{(i)} - \sum_{j=0}^{k} w_j a_j^{(i)})^2 \qquad (3)$$

– Support Vector Regression: SVM regression is somehow an SVM applied for regression problems. The major purpose of SVM regression is transformation of input space of training data to higher order space by implementing a non-linear mapping. As SVM regression performs in high-dimensional space, it can predict new functions in target space; this way, it uses non-linear functions to provide an estimation of an unknown function [13].

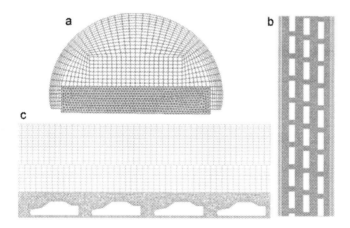

Fig. 2. Different views of the tunnel. (a) The tunnel cross section. (b) Top view of the tunnel. (c) Side view of the tunnel.

3   Experimental Result

The model and data are obtained from CFD simulation. Five tunnels which are 200, 400, 600, 890 and 1200 m long with the cross section area of 94 $m2$ in traffic congestion are modelled, and by allocating different velocities, ranging from 1 to 7, as inlet boundary condition, in each case the tunnel pressure drop is computed. For all tunnel lengths and velocities, pressure drop is also calculated where there is no car in the tunnels. The difference between the two obtained pressure drop results is the pressure drop due to vehicles only. It should be mentioned that in all cases, zero amount of gage pressure is used as the pressure outlet boundary condition. Figure 2 shows three views of a tunnel used in this study.

Table 1. The dataset used in this study

Length(m)	velocity(m/s)	PCFD	1D Error	Length(m)	velocity(m/s)	PCFD	1D Error
200	1	0.9	32.22	600	5	58.57	19.94
200	2	3.47	29.3	600	6	84.16	19.77
200	3	7.96	30.59	600	7	114.61	19.81
200	4	13.81	28.91	890	1	3.42	17.89
200	5	21.57	28.87	890	2	13.73	18.29
200	6	31.05	28.85	890	3	30.93	18.39
200	7	42.3	28.93	890	4	55	18.41
400	1	1.61	22.91	890	5	85.5	17.99
400	2	6.42	22.39	890	6	127.04	20.52
400	3	14.39	22.15	890	7	176.93	26.57
400	4	25.68	22.47	1200	1	4.61	17.87
400	5	39.98	22.17	1200	2	18.4	17.69
400	6	57.58	22.19	1200	3	41.35	17.59
400	7	78.48	22.29	1200	4	73.47	17.54
600	1	2.35	20.22	1200	5	114.8	17.54
600	2	9.39	20.07	1200	6	165.34	17.56
600	3	21.04	19.77	1200	7	225.2	17.61
600	4	37.39	19.75				

Table 1 shows the pressure drop achieved from CFD simulation and that obtained from 1D model for different tunnel lengths and tunnel average air velocity. As mentioned previously, CFD simulation provides accurate results, however, the model proposed [2] incorporates in errors. Table 4 demonstrates that the mean absolute error of [2] is 21.85 %.

The dataset used here is divided into two training and test parts; 82 % for train, and 18 % for test data. We use Weka toolbox to analyse the data. It is a collection of machine learning techniques for data mining tasks [8]. The aforementioned data mining techniques are utilized and compared together to present the best model for predicting the pressure drop. The methods are compared in terms of mean absolute error defined by equation.

$$Mean\ absolute\ error = \frac{1}{n}\sum_{i=1}^{n}|f_i - y_i| \qquad (4)$$

where n in the number of dataset, f_i and y_i is the ith prediction and the true values.

Table 2. A comparison among the techniques.

	Multi perceptron	Linear regression	SVM(Regression)
Mean absolute error	8.83	22.42	21.82
Correlation coefficient	0.996	0.852	0.838

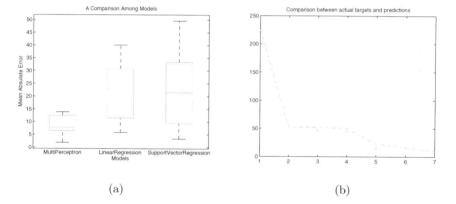

(a)   (b)

Fig. 3. (a) A comparison between three data mining techniques. (b) The comparison between actual and prediction tested data for the last fold of cross validation.

A 10-fold cross validation technique is utilized to evaluate our algorithm. As can be seen in Table 2, multi-layer perceptron algorithm performs better compared to other methods. It improves about 13 % accuracy in terms of mean absolute error to SVR and linear regression. This table also compares the results in terms of correlation coefficient which is the relation between real and predicted value on the dataset. Figure 3a illustrates the Box plot graph of the above experiments. The functional boxplot is an informative data exploratory visualization tool for descriptive statistics describing the major attributes of a collection of information [14] (Table 3).

Therefore, according to aforementioned results, the multi-perceptron technique is the best model in this regard. Figure 3 shows this optimal network with the lowest error on the basis of all performances of the network models in the validation set. Table 4 also demonstrates the superiority of our model compared to 1D model proposed [2].

Table 3. The optimal multi perceptron network with the lowest error

Parameters	Description
Network type	feed-forward back-propagation
Number of input nodes	13
Hidden layer	one layer
Number of hidden nodes	15 nodes
Momentum term	0.6
Learning rate	0.5
Number of training epochs	500

Table 4. A comparison between the proposed technique and the model introduced in [2].

	Multi perceptron	1D Model proposed [2]
Mean absolute error	8.83	21.85

4   Conclusion

Prediction of tunnel pressure drop in critical conditions is a challenging issue which in turn is of prime importance in designing ventilation systems for tunnels. The available methods are involved in either high computational costs (CFD simulation) or considerable errors (1D method). In this study, it is tried to present a new model based on data mining being capable of predicting tunnel pressure drop with a fairly good degree of precision and speed. Different data mining methods are used to predict pattern between the data achieved from CFD simulation for tunnel pressure drop due to vehicles. It is shown that of three data mining approaches used in this study, multi-layer perceptron algorithm presents more accurate results with mean absolute error of about 8 %, while this value for 1D model, a conventional method frequently used by engineers for predicting pressure drop in a tunnel in traffic jam, is almost 21 %.

References

1. Angelis, W., Drikakis, D., Durst, F., Khier, W.: Numerical and experimental study of the flow over a two-dimensional car model. J. Wind Eng. Ind. Aerodyn. **62**(1), 57–79 (1996)
2. Association, W.R., et al.: Road tunnels: vehicle emissions and air demand for ventilation. In: PIARC Committee on Road Tunnels Operation (C4) (2012)
3. Bari, S., Naser, J.: Simulation of smoke from a burning vehicle and pollution levels caused by traffic jam in a road tunnel. Tunn. Undergr. Space Technol. **20**(3), 281–290 (2005)
4. Bari, S., Naser, J.: Simulation of airflow and pollution levels caused by severe traffic jam in a road tunnel. Tunn. Undergr. Space Technol. **25**(1), 70–77 (2010)

5. Demuth, H.B., Beale, M.H.: Neural network toolbox for use with MATLAB: computation, visualization. In: Programming-User's Guide. MathWorks, Incorporated (2000)
6. Eftekharian, E., Dastan, A., Abouali, O., Meigolinedjad, J., Ahmadi, G.: A numerical investigation into the performance of two types of jet fans in ventilation of an urban tunnel under traffic jam condition. Tunn. Undergr. Space Technol. **44**, 56–67 (2014)
7. Gybenko, G.: Approximation by superposition of sigmoidal functions. Math. Control Signals Syst. **2**(4), 303–314 (1989)
8. Hall, M., Frank, E., Holmes, G., Pfahringer, B., Reutemann, P., Witten, I.H.: The weka data mining software: an update. ACM SIGKDD Explor. Newsl. **11**(1), 10–18 (2009)
9. Jang, H.M., Chen, F.: On the determination of the aerodynamic coefficients of highway tunnels. J. Wind Eng. Ind. Aerodyn. **90**(8), 869–896 (2002)
10. Philippe, D.: Neural network models (1997)
11. Rosenblatt, F.: The perceptron: a probabilistic model for information storage and organization in the brain. Psychol. Rev. **65**(6), 386 (1958)
12. Rumelhart, D.E., Hinton, G.E., Williams, R.J.: Learning internal representations by error propagation. Technical report, DTIC Document (1985)
13. Smola, A.J., Schölkopf, B.: A tutorial on support vector regression. Stat. Comput. **14**(3), 199–222 (2004)
14. Sun, Y., Genton, M.G.: Functional boxplots. J. Comput. Graph. Stat. **20**(2), 316–334 (2011)
15. Vega, M.G., Díaz, K.M.A., Oro, J.M.F., Tajadura, R.B., Morros, C.S.: Numerical 3D simulation of a longitudinal ventilation system: memorial tunnel case. Tunn. Undergr. Space Technol. **23**(5), 539–551 (2008)
16. Wang, F., Wang, M., He, S., Deng, Y.: Computational study of effects of traffic force on the ventilation in highway curved tunnels. Tunn. Undergr. Space Technol. **26**(3), 481–489 (2011)
17. Wang, F., Wang, M., Wang, Q., Zhao, D.: An improved model of traffic force based on CFD in a curved tunnel. Tunn. Undergr. Space Technol. **41**, 120–126 (2014)
18. Watkins, S., Vino, G.: The effect of vehicle spacing on the aerodynamics of a representative car shape. J. Wind Eng. Ind. Aerodyn. **96**(6), 1232–1239 (2008)

MapReduce-based Parallelized Approximation of Frequent Itemsets Mining in Uncertain Data

Jing Xu, Xiao-Jiao Mao, Wen-Yang Lu, Qi-Hai Zhu,
Ning Li, and Yu-Bin Yang[(✉)]

State Key Laboratory for Novel Software Technology,
Nanjing University, Nanjing, China
yangyubin@nju.edu.cn

Abstract. In recent years, frequent itemsets mining in uncertain data has drawn increasingly attractions from data mining communities. Currently, frequent itemsets mining algorithms in uncertain data mainly use frequent itemsets defined based on the expected support rather than the probabilistic support since the computational complexity is prohibitively high. To address this issue, various approximation algorithms for mining the probabilistic frequent itemsets have been proposed. However, the existing approximation algorithms are not adequately effective when the uncertain data is very large or extremely dense or sparse. In this paper, we propose a parallelized approximation algorithm, which is capable of mining probabilistic frequent itemsets on large-scale, dense or sparse uncertain data, based on the MapReduce platform. Experimental results are illustrated and analyzed to demonstrate the computational effectiveness of our algorithm.

Keywords: Uncertain data mining · Probabilistic frequent itemset · Mapreduce

1 Introduction

Data mining in uncertain data has become an attractive research topic in recent years. As the popularity of new applications, the dataset, such as the user locations obtained by GPS and indicated as places with probability rather than a precise one, is usually inherently uncertain. Another example is the satellite image data, in which an object may appear with a probability instead of an absolute one. In addition, some artificial noise may be added into the data to prevent from the disclosure of personal information, which also makes the data uncertain [1]. Consequently, mining frequent itemsets in uncertain data has been attracting increasingly research attentions [2–7]. For uncertain data, an itemset's support value indicating its occurrence is a random variable rather than a fixed value. Therefore, the definition of "frequent itemset" is different from that in certain data. Uncertain frequent itemset has two different definitions: (1) the *expected support-based* frequent itemset [4], and (2) the *probabilistic* frequent itemset [1, 2]. Most priori researches believe that the two definitions should be studied respectively [1, 2, 4]. In this paper, we focus on the definition of "probabilistic frequent itemsets". We design and implement a parallelized approximation algorithm, named as Parallel Normal Distribution based UApriori (PNDUA), on MapReduce platform.

© Springer International Publishing Switzerland 2015
S. Arik et al. (Eds.): ICONIP 2015, Part IV, LNCS 9492, pp. 136–144, 2015.
DOI: 10.1007/978-3-319-26561-2_17

Moreover, to further analyze the efficiency of PNDUA, experiments are carried out and comparisons with state-of-the-art methods are made.

2 Related Work

Given an uncertain dataset T containing N transactions, let V be a set of distinct items in T. Each transaction t has a set of items existing in V, in which each item v is assigned with a random variable $P(v,t) \in (0,1]$, representing the probability that v appears in t. Suppose that $X = \{x_1, x_2, \ldots, x_n\}$ is a non-empty subset of V, and X is a n-itemset if it has n items. The number of transactions containing X is a random variable $sup(X)$, we may have $P_i(X) = P(x_1, t_i) \times P(x_2, t_i) \times \ldots \times P(x_n, t_i)$. The expected support of X is then indicated as: $esup(X) = \sum_{i=1}^{N} P_i(X)$.

Definition 2.1 *Probabilistic Frequent Itemset.* By setting a minimum expected support ration *minsup* and a user-specified parameter τ, X is defined as a probabilistic frequent itemset if the probabilistic frequentness of X is no less than τ, which is denoted as follow: $P\{sup(X) \geq N \times minsup\} \geq \tau$.

Bernecker et al. [2] proposed a probabilistic frequent itemsets mining algorithm in uncertain data by integrating dynamic programming techniques into the basic UApriori algorithm. The time complexity of Dynamic Programming-based UApriori algorithm for mining each itemset is $O(N^2 \times minsup)$, where N is the number of transactions in uncertain data. Because computing the exact probabilistic frequentness is very complicated, the quick approximation of the support of each itemset can improve the efficiency. The basic idea behind the existing two representative approximation algorithms is that the support of an itemset in uncertain data is a random variable following the Poisson Binomial distribution. When uncertain datasets are large enough [10], Poisson distribution and Normal distribution can be used to approximately compute the support of an itemset. Wang et al. [5] used the cumulative distribution function (cdf) of the Poisson distribution to approximate the confidence that an item is frequent, based on which they proposed a Poisson distribution approximation algorithm for probabilistic frequent itemsets mining. In addition, based on the Normal distribution, approximation algorithm called ND [6] has also been proposed to efficiently obtain the probabilistic frequent itemsets by calculating their expectations and variances. Afterwards, the confidence probability of itemset X is efficiently computed as:

$$P(X) \approx \Phi(\frac{N \times minsup - 0.5 - esup(x)}{\sqrt{Var(X)}}) \tag{1}$$

where $\Phi(.)$ is the cdf of standard Normal distribution, and $Var(X)$ is the variance of the support of itemset X. Because the computational complexity of calculating an itemset's expectation and variance is only $O(N)$. With the development of cloud computing, Carson et al. [9] is the first work proposing a parallel algorithm for frequent itemset mining in uncertain data based on MapReduce. However, their work is about the expected support-based frequent itemsets mining rather than the probabilistic frequent itemset mining.

3   MapReduce-based Parallel Approximation Algorithm

Given the minimum support ratio *minsup* and the parameter τ, PNDUA computes the expectation and variance of each itemset in parallel. Then, the probabilistic frequentness is calculates according to Eq. (1). The process of PNDUA is composed of five steps, among which **Step 1** illustrated as below is the main step responsible for counting the probability of frequent itemset. All algorithms in this paper are based on the MapReduce framework, which contains both Map procedure and Reduce procedure.

Algorithm PNDUA, Step 1: Counting the probability of frequent itemset

---

1: **Procedure:** Map(*key*, *value*=transaction *t*)

2:   float *val* = 0f;

3:   **for** each item $x \in$ transaction *t* **do**

4:      *val* = (*1-prob(x,t)*)**prob(x,t)*;

5:      Output(*x*, *prob(x,t)*+" "+*val*);

6:   **end for**

7: **End Procedure**

8: **Procedure:** Reduce(*key, Iterable values*)

9:   int *totalSum* = 0, *expSup* = 0, *varSum* = 0;

10:   **for** each value in values **do**

11:     *expSup* = *expSup* + *value.prob(x,t)*;

12:     *varSum* = *varSum* + *value.val*;

13:     *totalSum* ++;

14:   **end for**

15:   float *fre*=0f;

16:   $fre = 1 - Normal(totalSum - 0.5 - expSup) / \sqrt{valSum}$;

17:   **if** $totalSum \geq min\_sup \times N$ **and** $fre \geq \tau$

18:     Output(*key, expSup*);

19: **End Procedure**

If the results of **Step 1** are not empty, PNDUA algorithm continues to perform **Step 2**. Otherwise, it will exit. **Step 2** generates the candidates of frequent 2-itemset. The results obtained from **Step 1** are sorted and combined to form the final candidates of frequent 2-itemset. The results are also stored in HDFS (Hadoop Distributed File System). Afterwards, **Step 3** is launched to generate frequent *k*-itemset, which is described as follows:

Algorithm PNDUA, Step 3: generating frequent k-itemset

1: **if** the candidates of frequent k-itemset is empty **then** stop the iteration

    else

2:     **Procedure:** Map(key, $value$=transaction);

3:     load the candidates of frequent k-itemset presented as list of candidates

4:     **for** each candidate in the list of candidates **do**

5:       int $count$=0;

6:       float $probSum$=1f;

7:       **for** each item x in candidate **do**

8:         **if** $x \in$ transaction t **then**

9:           $count$++;

10:           $probSum$=$probSum$*$prb(x,t)$;

11:         **else** break;

12:       **end for**

13:       **if** $count <$ candidate.length **then**

14:         continue;

15:       **else**

16:         val=(1-$probSum$)*$probSum$;

17:         Output(candidate, $probSum$+" "+val);

18:     **end for**

19:   **End Procedure**

20:   **Procedure**: Reduce(key, $Iterable\ values$)

21:     The process is as same as the Reduce Procedure in **Step 1**

22:   **End Procedure**

If the frequent k-itemset generated by **Step 3** is not empty, PNDUA algorithm continues to perform **Step 4**. Otherwise, it will exit. **Step 4** generates the candidate frequent $(k + 1)$-itemset by adopting the optimization method [8] based on the probabilistic frequent k-itemset achieved by **Step 3**, which greatly reduces the time of the "union" operation [3] and the memory cost.

4 Experimental Results

Our experiments are carried out on a 13-node computing cluster, one of which worked as master and the others worked as slaves. All slave-nodes are of the same hardware settings: Intel Core i5 4-core CPU (3.1 GHz) and 4 GB RAM. The master-node uses a Pentium Dual-Core E5800 CPU (3.2 GHz) and 1 GB RAM. All nodes run on a 32-bit

Ubuntu 11.04 operating system and the stable version of Apache Hadoop 1.0.4. The PNDUA algorithm is implemented with Java.

As for the datasets used in the experiments, we choose the benchmarks from FIMI repository containing two dense datasets "accidents" and "connect" and one sparse dataset "kosarak". Table 1 shows the characteristics of the original datasets. The datasets mentioned above are all certain data. We applied a Gaussian distribution on the datasets to calculate a probability for each item to finally generate the uncertain datasets used in our experiments, same as the literature [2–7] usually did. We then replicated the above three datasets to generate 1 GB, 2 GB, 3 GB and 4 GB large-scale datasets respectively in order to better evaluate the performance of our algorithm on large-scale data. We obtain he executed jar of PD and ND from Reference [7], which was implemented using Microsoft's Visual C ++ 2010. PD and ND were executed on a PC with Intel Core i5 4-core CPU (3.1 GHz) and 4 GB RAM. Since Dynamic Programming-based UApriori cannot handle large-scale uncertain data, we also implemented its Mapreduce-based version named as PDPA.

Table 1. The characteristics of original datasets

Dataset	Num of Trans.	Num of Items	Ave. Len.	Density
accidents	340,183	468	33.8	0.072
connect	67,557	129	43	0.333
kosarak	990,002	41,270	8.1	0.00019

4.1   Performance Analysis

We evaluate the performance of PNDUA algorithm using widely used measures including *speedup*, *sizeup* and *scaleup*. For the details of the measures please refer to [11]. As for *speedup* evaluation, we fixed the size of dataset and varied the cluster size as 1, 3, 9 and 12 nodes respectively. Figure 1 illustrates the *speedup* value achieved by PNDUA on datasets "accidents", "connect" and "kosarak" respectively. It shows that the *speedup* values achieved by our parallel algorithm are the best on 4 GB-scale for all three datasets. On the other hand, the *speedup* performances of "accidents" and "connect" are better than that of "kosarak". Table 3 shows that both "accidents" and "connect" are very dense datasets, but "kosarak" is very sparse which includes a large amount of distinct items and transactions with very short average length. The above characteristics of "kosarak" make **Step 3** of PNDUA algorithm fail to generate as many frequent *k*-itemsets as possible. As a result, its *speedup* performance tends to be worse than the other two datasets.

As for *sizeup* evaluation, we set the size of cluster as 3, 6, 9 and 12 nodes respectively. Figure 2 shows the *sizeup* performance achieved by PNDUA algorithm on "accidents", "connect" and "kosarak" respectively. It illustrates that PNDUA is able to handle data growth elegantly at all configurations on all three datasets. With the increase of the number of nodes in the cluster, the increase of *sizeup* value b slow. As for *scaleup* evaluation, we increase the size of data and the number of nodes in the cluster by a same factor *p*. For all three datasets, we replicated them to achieve the scale

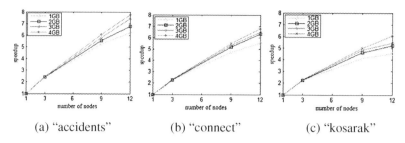

Fig. 1. speedup of PNDUA

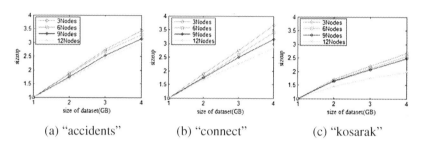

Fig. 2. sizeup of PNDUA

of 1 GB, 2 GB, 3 GB and 4 GB. We set the cluster's size as 3, 6, 9 and 12 nodes. The ideal *scaleup* performance can be achieved if the *scaleup* value keeps equal to 1, shown as the blue dot line in Fig. 3(a), (b) and (c) respectively. We can see from the figures that PNDUA scales up quite well on all datasets by maintaining a *scaleup* value higher than 73 %.

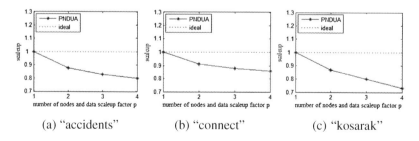

Fig. 3. scaleup of PNDUA

4.2 Performance Comparisons

Since PNDUA is an approximation algorithm, we need to evaluate its accuracy of frequent itemsets mining as well, which is generally measured by Precision and Recall [12]. Assume that the results obtained from PNDUA (approximation) are denoted as *appr* and the results obtained from PDPA (baseline) are expressed as *exact*. Afterwards, the *precision* and *recall* can be defined as:

$$precision = \frac{|appr \cap exact|}{|appr|}, \quad recall = \frac{|appr \cap exact|}{|exact|} \tag{2}$$

Tables 2, 3 and 4 demonstrate the comparisons of the running time between PNDUA, PD, ND, and PDPA on different large-scale uncertain datasets. When the scale of three datasets is 1 GB, PD and ND always need more time than PNDUA to get the results and both of them failed due to the out-of-memory crash. But PNDUA is able to return the results in acceptable time when the scale of datasets is larger than 1 GB. Although PDPA can always return the results, it is quite time-consuming. Table 5 shows the precision (*P*) and recall (*R*) of PNDUA compared with PDPA on all three datasets with different *minsup* settings. Overall, PNDUA is able to achieve a comparable mining accuracy while keeping overwhelmed advantages over other methods on the mining efficiency. Moreover, from the above analysis, we may also conclude that PNDUA is able to achieve better accuracy if a dataset is denser.

Table 2. Running times of accidents on PD, ND, PDPA and PNDUA, minsup = 0.15, $\tau = 0.9$

Size of dataset	Running time (s)			
	PD	ND	PDPA	PNDUA
1 GB	1,202	1,297	91,800	1,009
2 GB	Out of memory	Out of memory	146,909	1,656
3 GB	Out of memory	Out of memory	195,039	2,363
4 GB	Out of memory	Out of memory	268,304	3,039

Table 3. Running times of connect on PD, ND, PDPA and PNDUA, minsup = 0.7, $\tau = 0.9$

Size of dataset	Running time (s)			
	PD	ND	PDPA	PNDUA
1 GB	1,248	1,380	56,700	963
2 GB	Out of memory	Out of memory	96,390	1,618
3 GB	Out of memory	Out of memory	129,843	2,110
4 GB	Out of memory	Out of memory	168,940	2,900

Table 4. Running times of kosarak on PD, ND, PDPA and PNDUA, minsup = 0.003, $\tau = 0.9$

Size of dataset	Running time (s)			
	PD	ND	PDPA	PNDUA
1 GB	1,412	1,873	3,592	721
2 GB	Out of memory	Out of memory	5,388	1,045
3 GB	Out of memory	Out of memory	7,039	1,272
4 GB	Out of memory	Out of memory	9,469	1,419

Table 5. Accuracies of PNDUA on "accidents", "connect" and "kosarak"

Accidents			Connect			Kosarak		
Minsup	P	R	Minsup	P	R	Minsup	P	R
0.15	1	1	0.65	1	1	0.002	0.98	0.99
0.2	0.99	1	0.7	1	1	0.003	0.97	1
0.25	1	1	0.75	1	1	0.004	0.99	1
0.3	1	1	0.8	1	1	0.005	1	1
0.35	1	1	0.85	1	1	0.006	1	1

5 Conclusions

In this paper, we propose a parallelized approximation algorithm called PNDUA, which is capable of mining probabilistic frequent itemsets on large-scale, dense or sparse uncertain data, based on the MapReduce platform. PNDUA successively generates frequent itemsets in four steps. Experimental results are illustrated and analyzed to demonstrate the computational effectiveness of our algorithm. In the future, we will continue our work to provide theoretical analysis for our algorithm.

Acknowledgement. This work is supported by the Program for New Century Excellent Talents of MOE China (Grant No.NCET-11-0213), the Natural Science Foundation of China (Grant Nos. 61273257, 61321491), and the Program for Distinguished Talents of Jiangsu (Grant No. 2013-XXRJ-018).

References

1. Xia, Y., Yang, Y.R., Chi, Y.: Mining association rules with non-uniform privacy concerns. In: ACM DMKD, pp. 27–34 (2004)
2. Bernecker, T., Kriegel, H.-P., Renz, M., Verhein, F., Züfle, A.: Probabilistic frequent itemset mining in uncertain databases. In: KDD, pp. 119–128 (2009)
3. Aggarwal, C.C., Li, Y., Wang, J., Wang, J.: Frequent pattern mining with uncertain data. In: KDD, pp. 29–38 (2009)
4. Chui, C.-K., Kao, B., Hung, E.: Mining frequent itemsets from uncertain data. In: Zhou, Z.-H., Li, H., Yang, Q. (eds.) PAKDD 2007. LNCS (LNAI), vol. 4426, pp. 47–58. Springer, Heidelberg (2007)
5. Wang, L., Cheng, R., Lee, S.D., Cheung, D.W.-L.: Accelerating probabilistic frequent itemset mining: a model-based approach. In: CIKM, pp. 429–438 (2010)
6. Calders, T., Garboni, C., Goethals, B.: Approximation of frequentness probability of itemsets in uncertain data. In: ICDM, pp. 749–754 (2010)
7. Tong, Y., Chen, L., Cheng, Y., Yu, P.S.: Mining frequent itemsets over uncertain databases. PVLDB 5(11), 1650–1661 (2012)
8. Lin, M.-Y., Lee, P.-Y., Hsueh, S.-C.: Apriori-based frequent itemset mining algorithms on MapReduce. In: ICUIMC 2012, art. 76 (2012)
9. Leung, C.K.-S., Hayduk, Y.: Mining Frequent patterns from uncertain data with MapReduce for big data analytics. In: Meng, W., Feng, L., Bressan, S., Winiwarter, W., Song, W. (eds.) DASFAA 2013, Part I. LNCS, vol. 7825, pp. 440–455. Springer, Heidelberg (2013)

10. Cam, L.L.: An approximation theorem for the Poisson binomial distribution. Pac. J. Math. **10**(4), 1181–1196 (1960)
11. Li, N., Zeng, L., He, Q., et al.: Parallel implementation of apriori algorithm based on MapReduce. In: IEEE SNPD, pp. 236–241(2012)
12. Michael, K.B., Fredric, C.G.: The relationship between recall and precision. JASIS **45**(1), 12–19 (1994)

A MapReduce Based Technique for Mining Behavioral Patterns from Sensor Data

Md. Mamunur Rashid[1]([✉]), Iqbal Gondal[1,2], and Joarder Kamruzzaman[1,2]

[1] Faculty of Information Technology, Monash University, Melbourne, Australia
{md.rashid,iqbal.gondal,joarder.kamruzzaman}@monash.edu
[2] ICSL, Federation University, Ballarat, Australia

Abstract. WSNs generate a large amount of data in the form of streams, and temporal regularity in occurrence behavior is considered as an important measure for assessing the importance of patterns in WSN data. A frequent sensor pattern that occurs after regular intervals in WSNs is called regularly frequent sensor patterns (RFSPs). Existing RFSPs techniques assume that the data structure of the mining task is small enough to fit in the main memory of a processor. However, given the emergence of the Internet of Things (IoT), WSNs in future will generate huge volume of data, which means such an assumption does not hold any longer. To overcome this, a distributed solution using MapReduce model has not yet been explored extensively. Since MapReduce is becoming the de-facto model for computation on large data, an efficient RFSPs mining algorithm on this model is likely to provide a highly effective solution. In this work, we propose a regularly frequent sensor patterns mining algorithm called RFSP-H which uses MapReduce based framework. Extensive performance analyses show that our technique is significantly time efficient in finding regularly frequent sensor patterns.

Keywords: Wireless sensor networks · Data mining · MapReduce · Knowledge discovery · Regularly frequent sensor pattern

1 Introduction

Due to the recent advancements in technology, the deployment of large-scale WSNs is increasing day-by-day in many application domains, such as habitat monitoring, object tracking, environment monitoring, military, disaster management, as well as smart environments [1]. These applications generate huge volume of dynamic, geographically distributed and heterogeneous data. Also WSNs will be the large integral part of the emerging IoT. Data mining techniques play a vital role to efficiently analyse and facilitate automated or human induced strategic decisions.

Recently, extracting knowledge from sensor data has received a great deal of attention by the data mining community. Knowledge discovery in WSN has been used to extract information about the surrounding environment, that are deduced from the data reported by sensor nodes [2], and behavioral patterns

© Springer International Publishing Switzerland 2015
S. Arik et al. (Eds.): ICONIP 2015, Part IV, LNCS 9492, pp. 145–153, 2015.
DOI: 10.1007/978-3-319-26561-2_18

about sensor nodes, which are evolved from the meta-data describing sensors' behaviors. In literature different behavioral pattern mining techniques such as sensor association rules [3] and associated sensor patterns [4] have been successfully used on sensor data where patterns are extracted regarding the sensor nodes rather than the area monitored. Both techniques work based on the occurrence frequency of a pattern and reflect only the number of epochs in the sensor database which contain that pattern.

Another important criterion for identifying the interestingness of frequent patterns is the characteristic of occurrence, i.e., whether they occur regularly, irregularly, or mostly at specific time interval in the sensor database. A frequent pattern that occurs after regular intervals in WSNs is called regularly frequent sensor patterns (RFSPs). In a WSN, among all frequently occurred patterns, the user may be interested only on the regularly occurred patterns compared to other patterns that frequently occurred in specific time duration. In analysis of sensor network data, finding the occurrence regularity can be very necessary to provide significant information in network monitoring. Regularly frequent sensor patterns also can identify a set of temporally correlated sensors. This knowledge can be helpful to overcome the undesirable effects (e.g., missed reading) of the unreliable wireless communications.

Traditional frequent pattern mining methods fail to discover such RFSPs because they only focus on the high frequency pattern. Recently, Rashid et al. [5] have introduced a method of discovering regularly frequent sensor patterns that follow a temporal regularity in their occurrence characteristics. For regularly frequent sensor patterns mining, they used a tree structure, called a RSP-tree (Regularly Frequent Sensor Pattern tree), which captures the database contents in a highly compact manner with one database scan.

Existing RFSPs mining methods have been proposed for single processor and main memory based machine. However, enormous amount of data will be generated from scenarios like IoT which is basically composed of sensor deployed everywhere [6]. These limited hardware resources are not capable of handling large sensor data for mining and analyzing, and thereby suffer from scalability problems. To handle such bottleneck more efficient approaches (besides the serial processing approach) are needed to process terabyte or petabyte of information.

To handle Big Data in general, some researchers proposed the use of MapReduce [7], which mines the search space with parallel or distributed computing. With the emergence of cloud computing environment, the MapReduce framework gained popularity because of its ease of parallel processing, capability of distributing data and load balancing. It assumes a data centric method of distributed computing with the principle of moving computation to data. It uses a distributed file system that is particularly optimized to improve I/O performance while handling Big Data. Hadoop is an open source implementation of MapReduce framework. MapReduce on Hadoop needs only to share and pass the support of individual candidate pattern rather using the whole dataset itself. Therefore, inter-processor communication cost is low compared to the traditional distributed environments. Although recently a few works based on

MapReduce have been proposed to mine frequent patterns from transactional database [7–9] and DNA sequence dataset [10], but the impact of MapReduce to discover behavioral sensor patterns from Big Sensor Data is not investigated yet. In this paper, we propose RFSP-H (regularly frequent sensor pattern mining on Hadoop), a distributed mining technique over MapReduce. Extensive performance study with diverse datasets shows that our proposed technique is very efficient in discovering RFSPs over large sensor data streams.

2   RFSPs Mining Problem in Wireless Sensor Networks

Let $S = \{s_1, s_2, ..., s_n\}$ be the set of sensors in a specific WSN. We assume that the time is divided into equal-sized slots $t = \{t_1, t_2, ..., t_q\}$ such that $t_{j+1} - t_j = \lambda, j \in [1, q-1]$ where λ is the size of each time slot. A set $P = \{s_1, s_2, ..., s_p\} \subseteq S$ is called a pattern of a sensors.

An epoch is a tuple $e(e_{ts}, Y)$ such that Y is a pattern of the event detecting sensors that report events within the same time slot and e_{ts} is the epoch's time slot. A sensor database SD shown in Table 1, is a set of epochs $E = \{e_1, e_2, ..., e_m\}$ with $m = |SD|$, i.e., total number of epochs in SD. If $X \subseteq Y$, it is said that X occurs in e and denoted as $e_j^X, j \in [1, m]$. Let $E^X = \{e_j^X, ..., e_k^X\}$, where $j \leq k$ and $j, k \in [1, m]$ be the ordered set of epochs in which pattern X has occurred in SD. Let e_s^X and e_t^X, where $j \leq s < t \leq k$ be the two consecutive epochs in E^X. The number of epochs or time difference between e_t^X and e_s^X, can be defined as a period of X, say p^X, i.e., $p^X = \{e_t^X - e_s^X\}$. Let $P^X = \{p_1^X, ..., p_s^X\}$ be the set of periods for patterns X. For simplicity in period computation, assume the first and last epochs in SD as *null* with $e_f = 0$ and $(e_l = e_m)$ respectively.

Definition 1 (*Regularity of Patten X*): For a given E^X, let P^X be the set of all periods of X, i.e., $P^X = \{p_1^X, p_2^X ..., p_n^X\}$, where n is the total number of periods in P^X. Then the average period value of pattern X is represented as, $\bar{p}^X = \sum_{k=1}^N \frac{p_k^X}{n}$ and its variance as $\sigma^X = \sum_{k=1}^N \frac{(p_k^X - \bar{p}^X)^2}{n}$. The regularity of X can be denoted as $Reg(X) = \sigma^X$ (variance of periods for pattern X).

Definition 2 (*Support of a pattern X*): The number of epochs in a SD that contain X is called the support of X in SD and is denoted as $Sup(X) = |E^X|$, where $|E^X|$ is the size of E^X.

Definition 3 (*Regularly frequent sensor Pattern*): A pattern is called a regularly frequent pattern if it satisfies both of the following two conditions: (i) its support is no less than a user-given minimum support threshold, say, *min_sup* and (ii) its regularity is no greater than a user-given maximum regularity threshold say, *max_var*.

Given a SD, *min_sup* and *max_var* constraints, the objective is to discover the complete set of interesting patterns in SD having than support no less *min_sup* and regularity no more than *max_var*.

3   RFSPs Mining Using MapReduce Model

The two key functional programming primitives in MapReduce are 'Map' and 'Reduce'. The input data are read and divided into several segments, and assigned to different nodes. Each node executes the map function on each segment. The map function takes a pair of keys, value data, and returns a list of intermediate $< key, value >$ pairs.

map: $(key_1, value_1) \rightarrow$ list of $(key_2, value_2)$,

where (i) key_1 &key_2 are keys in the same or different domains, and (ii) $value_1$ & $value_2$ are the corresponding values in some domains. Then, these pairs are shuffled and sorted. Each node then executes the Reduce function to the set of intermediate pairs with the same key. Typically, the Reduce function produces output pairs by performing a merging operation. The reduce function 'reduces' - joining, totaling, compressing, sifting, or transforming - the list of values associated with a given key ($\forall$ k keys) and returns a list of k values.

reduce: $(key_2, listofvalue_2) \rightarrow$ list of $(value_3)$,

where (i) key_2 is a key in some domains, and (ii) $value_2$ & $value_3$ are the corresponding values in some domains. All the output pairs are finally sorted based on their key values. Examples of MapReduce applications include the construction of an inverted index and the word counting of a document.

Table 1. A sensor database (SD)

TS	Epoch	TS	Epoch
1	$s_1 s_2 s_5 s_6$	5	$s_2 s_3 s_4$
2	$s_1 s_2 s_3 s_5$	6	$s_3 s_4 s_5$
3	$s_1 s_2 s_3 s_5$	7	$s_2 s_3 s_4$
4	$s_1 s_2 s_5 s_6$	8	$s_4 s_5 s_6$

Table 2. I/O scheme for proposed model

I/O	Map	R-1	R-2
Input:	k: TS	k: can-s.set	f-s.set
key/value (k/v) pairs	v: s.set	v: sup	reg
Output:	k: can-s.set	k: f-s.set	RFSP
key/value (k/v) pairs	v: sup	v: sup	sup

Hadoop is an open source implementation of MapReduce [7]. The Hadoop system comprises of two primary segments: (i) Hadoop Distributed File System (HDFS) and (ii) the MapReduce. The Hadoop runtime framework coupled with HDFS deals with the subtle elements of parallelism and concurrency to give simplicity of parallel programming with strengthened unwavering quality. In a Hadoop cluster two types of nodes are accessible: master node and slave node, where a master node controls a collection of slave nodes on which the Map and Reduce functions run in parallel. The master node assigns a job to a slave node that has any empty job slot. Hadoop permits the MapReduce structure to viably assign computing jobs to an array of storage nodes where data files reside and prompts a high aggregate bandwidth over the whole Hadoop cluster. An input data file is fed to Map functions that reside on the Hadoop distributed file system

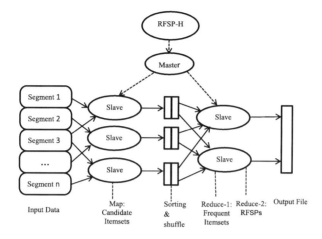

Fig. 1. Proposed MapReduce framework

on a cluster. Hadoop's HDFS parts the data file into even-sized sections, which are dispersed to a pool of slave nodes for further MapReduce handling.

Table 3. Candidate sensorset for segment 1 from slave 1

can-s.set	sup	can-s.set	sup
s_2	2	$s_2 s_4$	2
s_3	3	$s_3 s_4$	3
s_4	3	$s_3 s_5$	2
s_5	3	$s_2 s_3 s_4$	2
s_6	1	$s_3 s_4 s_5$	1
$s_2 s_3$	2	–	–

Table 4. Candidate sensorset for segment 2 from slave 2

can-s.set	sup	can-s.set	sup
s_1	4	$s_2 s_3$	2
s_2	4	$s_2 s_5$	3
s_3	3	$s_3 s_5$	1
s_5	3	$s_2 s_6$	2
s_6	1	$s_1 s_2 s_3$	2
$s_1 s_2$	2	$s_1 s_2 s_6$	2
$s_1 s_5$	4	$s_2 s_3 s_5$	2
$s_1 s_6$	2	$s_1 s_2 s_3 s_5$	2

The sensor database (SD) is split into smaller segments automatically after being stored on the HDFS. The Hadoop components perform job execution, file staging, and workflow information storage and used the files to replace the database automatically. We used Hadoop-based parallel FP-Growth proposed in [9] and we follow the data placement strategy indicated in [11]. Figure 1 indicates the workflow of the proposed MapReduce framework where the RFSPs are generated at reduce phase-2. Table 2 shows the input/output schemes of the proposed framework.

After splitting the sensor database into smaller segments, the master node assigns task to each slave node. Each slave node scans the epochs in the smaller data segments as $<TS, s.set>$ and the balanced FP-growth [9] was applied to produce output as $<can\text{-}s.set, sup>$ pairs, where $s.set$ and $can\text{-}s.set$ represents sensorset and candidate sensorset respectivly. These values are to be stored in the local disk as intermediate results and the HDFS will perform the sorting or merging operations to produce results as $<can\text{-}s.set, sup>$ pairs. For this, we used two levels of pruning techniques, local pruning, and the global pruning, using two minimum support thresholds - $local\_min\_sup$ and $global\_min\_sup$. The local pruning is applied on the Map phase in each segment and the global pruning is applied in the Reduce phase. For this purpose we modified the balanced FP-growth [9] using MapReduce library function using in Java.

Then the master node will assigns the slave nodes for reduce operation. In reduce phase-1 a slave node takes input as $<can\text{-}s.set, sup>$ pairs and generates output as $<f\text{-}s.set, sup>$ pair where $f\text{-}s.set$ represents frequent sensorset. In reduce phase-2 a slave node takes input as $<f\text{-}s.set, sup>$ pairs, then checks the regularity criteria among the frequent patterns and writes regularly frequent sensor patterns on the output files as $<RFSP, sup>$ pairs and stores in the local disks. Here phase incorporation of two sub-phases is possible because, the size of produced dataset is less compared to the map phase.

Consider the min_sup is 3 and max_var is 1.1. Assume the SD in Table 1 is divided into two equal segments: TS 1 to 4 are in the first and TS 5 to 8 are in the second segment. Let master node has assigned segment 01 to slave node 1 and segment 02 to slave node 2. Mapper maps the $<TS, s.set>$ pairs and generate output as $<can\text{-}s.set, sup>$ pairs and store them as intermediate values in the local disk of slave node 1 and 2 and inform the master node. Tables 3 and 4 show the results of the Map phase. These values are fed into the Reduce phase after sorting and merging operation.

The master node assigns reduce task to idle slave nodes. In reduce phase 1, slave nodes take input as $<can\text{-}s.set, sup>$ pairs, share the support value of the candidate sensorset with other slave nodes, find out the complete set of frequent sensorset as $<f\text{-}s.set, sup>$ pairs, and write the result on the local disk of slave node 3 is shown in Table 5. In reduce phase 2 the idle slave node take the $<f\text{-}s.set, sup>$ pairs as input, apply the constraint indicated in definition 2, and generates $<RFSP, sup>$ pairs as output. The final output is shown in Table 6.

Table 5. Frequent s.set from slave 3

f-s.set	sup	f-s.set	sup
s_1	4	$s_1 s_2$	4
s_2	6	$s_1 s_5$	4
s_3	5	$s_2 s_3$	4
s_4	5	$s_2 s_5$	3
s_6	3	$s_3 s_5$	3

Table 6. Final output from slave 4

f-s.set	Reg	RSFP	f-s.set	Reg	RSFP
s_1	1.44	×	$s_1 s_2$	1.44	×
s_2	0.693	✓	$s_1 s_5$	1.44	×
s_3	0.556	✓	$s_2 s_3$	1.04	✓
s_4	3.04	×	$s_2 s_5$	1.44	×
s_6	1.5	×	$s_3 s_5$	0	✓

4   Experimental Results

We used Hadoop version 1.2.1, running on a cluster with five nodes: one master (3.8 GHz processor, 8 GB RAM) and 4 slaves (2.63 GHz processor 4 GB RAM). We configured HDFS on Ubuntu-14.04.1. The MapReduce based RFSPs mining technique as illustrated in Sect. 3 are implemented using Java on MapReduce library functions. We apply the RFSP-H algorithm on *T10I4D100*, *kosarak* and *Connect-4* datasets downloaded from [12], whose characteristics are summarized in [4].

The first experiment shows the runtime of RFSP-H for varying *max_var* values. We kept record of the running time of RFSP-H for *max_var* values that varied between 0.15 % to 0.45 % for *T10I4D100K* dataset, with *min_sup* fixed at 1 %. The result is shown in Fig. 2(a). For *kosarak* and *connect-4* datasets the result are shown in Fig. 2(b-c) where the *min_sup* was fixed at 3 % and 0.7 %, respectively. As expected, the runtime increases as the *max_var* value increases. This is because an increasing *max_var* value also increases the number of candidate patterns for regularly frequent patterns. Then, the mining operation was performed by varying *min_sup* threshold, where the *max_var* was fixed at 0.15 %, 20 % and 40 %, respectively for the above datasets. The results are shown in Fig. 3, where the runtime decreases as the *min_sup* value increases. The reason is that, with the increase of the *min_sup* value, the number of candidate patterns numbers deceases.

In the second experiment, we demonstrate how RFSP-H's runtime varies with the number of slave nodes. Here we fixed the *min_sup* and *max_var* values at 1 % and 0.15 %, 3 % and 20 %, 7 % and 40 % respectively for *T10I4D100K*, *kosarak* and *Connect-4* datasets. We varied the number of slaves nodes from 2

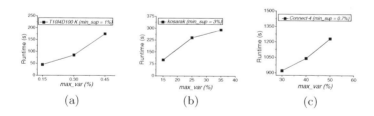

Fig. 2. Runtime of RFSP-H by varying *max_var*.

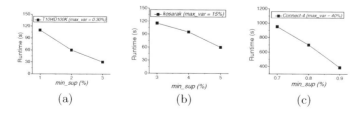

Fig. 3. Runtime of RFSP-H by varying *min_sup*.

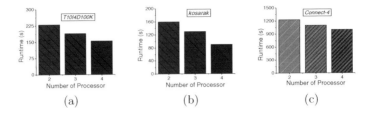

Fig. 4. Relationship between the runtime and the number of slave nodes.

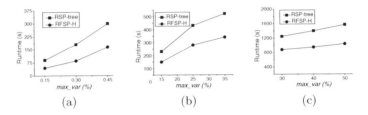

Fig. 5. Runtime comparison RFSP-H v/s RSP-tree. (a) *T10I4D100K*, (b) *Kosarak* and (c) *Connect-4*

to 4, and recorded the runtime for each dataset. Figure 4 shows that the runtime reduces significantly with increasing number of slave nodes.

Finally, we compared the runtime of RFSP-H with RSP-tree [5] by varying the *max_var* threshold. Although, RSP-tree shows good performance, this technique uses single processor to process large data. In this experiment, for RFSP-H the number of processor was fixed at 4 for each dataset. Figure 5 reveals that RFSP-H significantly outperforms RSP-tree in runtime. The reason is that RFSP-H uses MapReduce based parallel process with less inter-processor communications which enables the RFSP-H to mine the patterns more efficiently.

5   Conclusion

The key contribution of this paper is to provide a MapReduce based regularly frequent sensor pattern mining algorithm, called RFSP-H over sensor data streams. We evaluated the performance of RFSP-H over diverse datasets. Extensive performance analyses show that RFSP-H is very efficient for regularly frequent sensor pattern mining and significantly better than the existing method.

References

1. Mahmood, A., Shi, K., Khatoon, S., Xiao, M.: Data mining techniques for wireless sensor networks: a survey. Int. J. Distrib. Sensor Netw. **2013**, 24 (2013). doi:10.1155/2013/406316
2. Duarte, M.F., Hu, Y.H.: Vehicle classification in distributed sensor networks. J. Parallel Distrib. Comput. **64**(7), 826–838 (2004)

3. Boukerche, A., Samarah, S.A.: Novel algorithm for mining association rules in wireless ad-hoc sensor networks. IEEE Trans. Para. Dist. Sys. **19**(7), 865–877 (2008)
4. Rashid, M.M., Gondal, I., Kamruzzaman, J.: Mining associated patterns from wireless sensor networks. IEEE Trans. Comput. PP(99) (2014)
5. Rashid, M.M., Gondal, I., Kamruzzaman, J.: Regularly frequent patterns mining from sensor data stream. In: Lee, M., Hirose, A., Hou, Z.-G., Kil, R.M. (eds.) ICONIP 2013, Part II. LNCS, vol. 8227, pp. 417–424. Springer, Heidelberg (2013)
6. Chen, F., Deng, P., Wan, J., Zhang, D., Vasilakos, A.V., Rong, X.: Data mining for the internet of things: literature review and challenges. Int. J. Distrib. Sensor Netw. **2015**, 14 (2015)
7. Dean, J., Ghemawa, S.: MapReduce: simplified data processing on large clusters. Commun. ACM **55**(1), 107–113 (2008)
8. Leung, C.K.-S., Hayduk, Y.: Mining frequent patterns from uncertain data with MapReduce for big data analytics. In: Meng, W., Feng, L., Bressan, S., Winiwarter, W., Song, W. (eds.) DASFAA 2013, Part I. LNCS, vol. 7825, pp. 440–455. Springer, Heidelberg (2013)
9. Zhou, Z. J., and Feng, S.: Balanced parallel FP-Growth with MapReduce. In: IEEE Youth Conference on Information Computing and Telecom (YC-ICT), pp. 243–246 (2010)
10. Karim, M.R., Jeong, B.S., Choi, H.J.: A MapReduce framework for mining maximal contiguous frequent patterns in large DNA sequence datasets. IETE Tech. Rev. **29**(2), 162–168 (2012)
11. Xie, J., Majors, J., Qin, X.: Improving MapReduce performance through data placement in heterogeneous hadoop clusters. In: IEEE International Symposium, IPDPSW (2010)
12. Frequent itemset mining repository. http://fimi.cs.helsinki.fi/data/

A Methodology for Synthesizing Interdependent Multichannel EEG Data with a Comparison Among Three Blind Source Separation Techniques

Ahmed Al-Ani[1]([⊠]), Ganesh R. Naik[1], and Hussein A. Abbass[2]

[1] Faculty of Engineering and Information Technology,
University of Technology Sydney, Ultimo, NSW 2007, Australia
{ahmed.al-ani,ganesh.naik}@uts.edu.au
[2] School of Engineering and Information Technology, University of New South Wales,
Canberra, ACT 2600, Australia
h.abbass@adfa.edu.au

Abstract. In this paper, we introduce a novel method for constructing synthetic, but realistic, data of four Electroencephalography (EEG) channels. The data generation technique relies on imitating the relationships between real EEG data spatially distributed over a closed-circle. The constructed synthetic dataset establishes ground truth that can be used to test different source separation techniques. The work then evaluates three projection techniques – Principal Component Analysis (PCA), Independent Component Analysis (ICA) and Canonical Component Analysis (CCA) – for source identification and noise removal on the constructed dataset. These techniques are commonly used within the EEG community. EEG data is known to be highly sensitive signals that get affected by many relevant and irrelevant sources including noise and artefacts.

Since we know ground truth in a synthetic dataset, we used differential evolution as a global optimisation method to approximate the "ideal" transform that need to be discovered by a source separation technique. We then compared this transformation with the findings of PCA, ICA and CCA. Results show that all three techniques do not provide optimal separation between the noisy and relevant components, and hence can lead to loss of useful information when the noisy components are removed.

Keywords: Synthetic multichannel EEG · Artefact removal · PCA · ICA · CCA · Optimal projection · Differential evolution

1 Introduction

The Electroencephalography (EEG) is a well established modality for recording brain's electrical activities. In addition to the various signal sources arising

© Springer International Publishing Switzerland 2015
S. Arik et al. (Eds.): ICONIP 2015, Part IV, LNCS 9492, pp. 154–161, 2015.
DOI: 10.1007/978-3-319-26561-2_19

from the cerebral, scalp EEG is also influenced by a number of noise and arte-fact sources. These include scalp muscles, eye movements and blinks, breathing, heart beat, and electrical line noise. Proper interpretation of EEG data is very important as artefact and noise may bias the neurological interpretation [1–4].

There has been attempts to extract relevant information from the EEG sig-nals using classical signal processing techniques, including adaptive supervised filtering, parametric and nonparametric spectral estimation, time frequency analysis, and higher-order statistics. Unfortunately, all techniques face difficulties arising from the spectrum overlap of brain signals with artifacts. For instance, most of these techniques have been reported to fail in completely eliminating ocular artifacts [5–7].

Multivariate techniques, such as Principal Component Analysis (PCA), Inde-pendent Component Analysis (ICA) and Canonical Component Analysis (CCA) have been widely used in identifying the relevant sources and denoising the EEG data. The aforementioned techniques are commonly used in processing EEG data [6–8]. However, some important information might be lost while applying the above procedures to EEG data. Identification and removal of exact location of noisy components are of great interest to both engineers and clinicians. However, the number of sources, whether of cerebral origin or artefacts, that contribute to the recorded scalp signal as well as their combination methodology are not known. Thus, one possible approach to overcome this limitation is the use of synthetic EEG data to model and project the sources.

This research reports on the ability of the three techniques – PCA, ICA and CCA – in re-constructing the relevant sources using synthetic data. A novel method for constructing synthetic data of four EEG channels is presented with the aim of imitating relationships between real EEG channels, spatially distrib-uted with varying distances. Five measures are presented to evaluate the per-formance of these methods, where the first three evaluate the reconstruction of sources without eliminating noisy components, while the remaining two evaluate the information loss and noise elimination ability of these methods when remov-ing noisy component(s). A differential evolution (DE) algorithm is utilized to search for the optimal transformation, as the synthetic sources are known a pri-ori, to estimate the difference in performance of the three multivariate methods to that of the "best" projection for each of the five evaluation measures.

The paper is organized as follows. Section 2 describes the construction of synthetic data. Evaluation of the three projection techniques is presented in Sect. 3, and a conclusion is given in Sect. 4.

2   Construction of Synthetic Data

In the construction of synthetic data, we consider the case of limited number of EEG channels (four in particular), and presume that there are five brain sources, four of them are local, one for each channel, while the fifth one is global. We also presumed that there are two noise sources, similar to the synthetic data described in [4]. The seven synthesised sources are described as shown in Table 1. We also

Table 1. Synthesized sources

Source	Equation	Description
1	$14 \sin(2\pi \times 4t) + 52 \sin(2\pi \times 22t)$	Delta and Beta
2	$23 \sin(2\pi \times 7t) + 70 \sin(2\pi \times 19t)$	Theta and Beta
3	$16 \sin(2\pi \times 5t) + 43 \sin(2\pi \times 11t)$	Theta and Alpha
4	$44 \sin(2\pi \times 9t) + 56 \sin(2\pi \times 47t)$	Alpha and Gamma
5	$34 \sin(2\pi \times 6t) + 24 \sin(2\pi \times 45t)$	Theta and Gamma
6	$144 \sin(2\pi \times 31t) + 337 \sin(2\pi \times 51t)$	EMG artefact
7	$282 \sin(2\pi \times 28t) + 246 \sin(2\pi \times 49t)$	EMG artefact

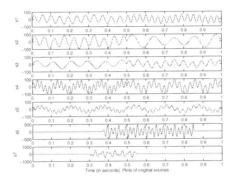

Fig. 1. The seven original signal sources of the synthetic data

decided to change the frequency of the second component for each of the five EEG sources in the second half of the signal and make the EMG artefacts active during limited portions of the signal. Figure 1 shows the seven sources, which are sampled at 256 Hz.

In order to define relationships between the four channels, we studied a real EEG dataset that consists of 64 channels according to the montage shown in Fig. 2. The four synthesized channels are considered to form a rectangle shape with different lengths of its horizontal and vertical sides, as described in Table 2. The real EEG dataset was used to calculate the correlation values between the six pairs of the four channels, i.e., {ch1, ch2}, {ch1, ch3}, {ch2, ch3}, {ch1, ch4}, {ch2, ch4} and {ch3, ch4}.

The nine cases listed in the table have different number of channel combinations, which are: 38, 33, 26, 22, 18, 13, 10, 4 and 2 respectively, as the smaller the rectangle the larger the number of possible combinations that can be formed using the 64 channels. In general, the obtained results indicate that the longest the rectangle side the smaller the correlation between channels. We also noticed that vertical channels tend to have smaller correlation compared to horizontal ones with similar distance (i.e., larger inter-lobe differences compared to intra-lobe), and that the lower horizontal channels (parietal/occipital) usually have higher correlation than their corresponding higher ones (frontal/central).

Table 2. Considered scenarios based on distances between the four channels

Case	D-h	D-v	Example {Ch1, Ch2, Ch3, Ch4}	Median correlation between channels {1,2}, {1,3}, {2,3}, {1,4}, {2,4}, {3,4}
1	1	1	{F5, F3, FC5, FC3}	{0.84, 0.78, 0.73, 0.71, 0.76, 0.83}
2	2	1	{F5, F1, FC5, FC1}	{0.65, 0.79, 0.57, 0.52, 0.75, 0.65}
3	2	2	{F5, F1, C5, C1}	{0.64, 0.52, 0.36, 0.30, 0.42, 0.66}
4	3	2	{F5, DZ, C5, CZ}	{0.30, 0.55, 0.27, 0.23, 0.42, 0.38}
5	4	2	{F5, F2, C5, C2}	{0.20, 0.57, 0.19, 0.20, 0.47, 0.20}
6	4	3	{F5, F2, CP5, CP2}	{0.24, 0.24, 0.13, 0.22, 0.18, 0.21}
7	5	3	{F5, F4, CP5, CP4}	{0.22, 0.28, 0.18, 0.27, 0.19, 0.52}
8	6	4	{F5, F6, P5, P4}	{0.29, 0.10, 0.27, 0.37, 0.13, 0.59}
9	7	4	{F5, F8, P5, P8}	{0.15, 0.07, 0.25, 0.37, 0.07, 0.54}

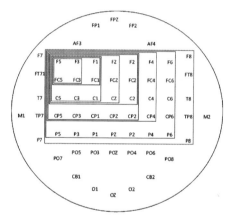

Fig. 2. EEG montage that shows distribution of 64 channels with examples of different distances between the four selected channels (corners of the rectangles)

We considered one of the two synthetic noise sources to be close to the upper left corner of the rectangle (e.g., close to the left frontal channels), while the other has a stronger effect on the lower right corner of the rectangle (e.g., close to the right parietal/occipital channels). The observed synthetic signals of the four channels are formed using a weighted sum of the seven sources. We fixed the weight of the local source of each channel to 1.0 and varied the weights of EMG artefacts based on their distances from each of the four channels. The objective was to search for an appropriate weight value for the other local and global sources to achieve correlation values that are close to the ones listed in Table 2. The weight search that was implemented using Differential Evolution (DE) [9,10] was restricted by an upper limit. The obtained weight matrices are listed in Table 3. The nine obtained sets shown in Fig. 3 demonstrate that the varying level of correlation between the four signals with the first set (Fig. 3(a)) having

Table 3. Weigh matrices for each of the nine cases. Weights that are in bold font were fixed, while the remaining ones were optimized given that they do not exceed an upper limit

	W1				W2				W3			
S1	**1.000**	0.500	−0.500	−0.500	**1.00**	−0.450	0.450	0.450	**1.000**	0.450	−0.450	−0.353
S2	0.224	**1.000**	0.494	0.112	0.450	**1.000**	−0.183	0.370	−0.056	**1.000**	−0.450	−0.450
S3	0.500	−0.500	**1.000**	−0.127	0.348	−0.070	**1.000**	−0.304	0.234	0.019	**1.000**	−0.410
S4	0.500	0.379	0.440	**1.000**	−0.450	0.224	0.450	**1.000**	−0.45	0.240	0.450	**1.000**
S5	−0.122	0.110	−0.092	0.074	0.143	−0.281	−0.066	0.047	−0.248	−0.250	0.087	−0.134
S6	**0.500**	**0.400**	**0.350**	**0.300**	**0.500**	**0.350**	**0.350**	**0.250**	**0.500**	**0.350**	**0.250**	**0.200**
S7	**0.020**	**0.030**	**0.040**	**0.050**	**0.020**	**0.040**	**0.04**	**0.070**	**0.020**	**0.040**	**0.070**	**0.150**

	W4				W5				W6			
S1	**1.000**	−0.400	−0.400	−0.400	**1.000**	−0.300	−0.300	−0.300	**1.000**	−0.250	−0.250	−0.250
S2	−0.400	**1.000**	−0.230	−0.400	−0.300	**1.000**	−0.300	−0.300	−0.250	**1.000**	−0.250	−0.250
S3	−0.400	−0.400	**1.000**	0.400	−0.280	−0.300	**1.000**	−0.172	−0.250	−0.250	**1.000**	0.250
S4	−0.400	0.198	−0.400	**1.000**	−0.300	0.300	−0.300	**1.000**	−0.250	0.016	−0.216	**1.000**
S5	−0.500	0.500	0.500	0.219	−0.750	0.750	0.418	0.070	−0.750	0.750	0.750	−0.266
S6	**0.500**	**0.300**	**0.250**	**0.170**	**0.500**	**0.250**	**0.250**	**0.130**	**0.500**	**0.250**	**0.150**	**0.100**
S7	**0.020**	**0.050**	**0.070**	**0.200**	**0.020**	**0.055**	**0.070**	**0.250**	**0.020**	**0.055**	**0.100**	**0.300**

	W7				W8				W9			
S1	**1.000**	−0.250	−0.250	0.250	**1.000**	−0.200	−0.200	0.200	**1.000**	−0.026	−0.150	0.150
S2	−0.250	**1.000**	−0.250	−0.250	−0.200	**1.000**	0.066	−0.200	−0.150	**1.000**	0.019	−0.150
S3	−0.250	−0.250	**1.000**	0.250	−0.200	−0.200	**1.000**	0.200	−0.150	−0.150	**1.000**	0.150
S4	−0.069	−0.059	0.250	**1.000**	0.200	−0.077	0.200	**1.000**	0.150	−0.150	0.150	**1.000**
S5	−0.750	0.672	0.551	0.094	−0.750	0.294	0.230	−0.603	−0.750	0.750	0.380	−0.732
S6	**0.500**	**0.200**	**0.150**	**0.070**	**0.500**	**0.150**	**0.050**	**0.020**	**0.500**	**0.100**	**0.050**	**0.020**
S7	**0.020**	**0.060**	**0.100**	**0.350**	**0.020**	**0.065**	**0.150**	**0.450**	**0.020**	**0.070**	**0.150**	**0.500**

the maximum correlation between the four signals that are noticeably affected by the first noise source. The effect of the first noise source was kept constant for the fist signal and gradually decreased for the remaining three signals, especially the fourth one, as the distance between the channels increased. In contrast, the effect of the second noise source gradually increased for the fourth signal.

3   Evaluation of the Projection Techniques

It is important to find the optimal reconstruction matrices as their projections will serve as a baseline to evaluate the performance of PCA, ICA and CCA. Finding these matrices will also be helpful for the development of future projection techniques. Two objectives will be defined and a search mechanism using DE will be utilized for this purpose. The first objective is to search for the weight reconstruction matrix that maximises correlation with the original sources. We will first attempt to maximise the correlation with the four local sources only, and then with all five EEG sources. Those two implementations will be named DE1 and DE2. The second objective is to dedicate one component to the two noise sources, i.e., maximise correlation with them, while the remaining three components are to be dedicated to the EEG sources (one version (DE3) for the four local sources, and another one (DE4) for all five EEG sources). This will enable the removal of noisy component similar to how ICA, PCA and CCA are

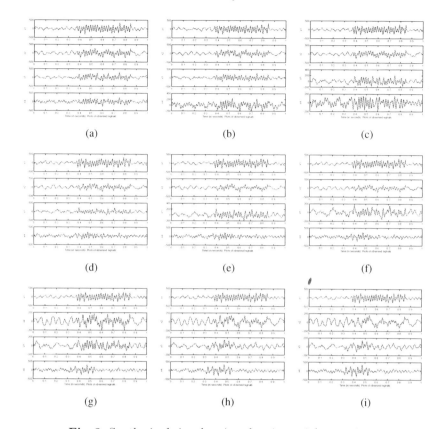

Fig. 3. Synthesised signals using the nine weight matrices.

usually utilized in EEG processing. We used five measures for evaluating the performance of ICA, PCA and CCA. The first three for evaluating the transformation without removing any component, and the other two evaluate the estimated noise free signals after removing noisy component(s). The five measures are:

1. Distance of reconstructed sources from the original local sources, i.e., distance between $Corr(S_r, S_l)$ and $Corr(S_l, S_l)$, where $Corr(S_r, S_l)$ is the cross correlation matrix between the reconstructed sources, S_r, and the original local sources, S_l, while $Corr(S_l, S_l)$ is the autocorrelation matrix of the original local sources. Note that we manually arranged the components of ICA, PCA and CCA to achieve maximum cross correlation between S_r and S_l.
2. Correlation between the reconstructed sources and the two noise sources, $Corr(S_r, S_n)$, where S_n represents the two noisy sources.
3. Correlation between the reconstructed sources and the original global source, $Corr(S_r, S_g)$, where S_g is the global source.

Table 4. Performance of ICA, PCA, CCA and the four DE-optimized benchmark methods

	Method	Case 1	Case 2	Case 3	Case 4	Case 5	Case 6	Case 7	Case 8	Case 9	Mean
Measure 1	ICA	0.1624	0.2305	0.1836	0.2448	-	0.2922	0.2564	0.2229	0.2171	0.2262
	PCA	0.2378	0.3341	0.2350	0.3201	0.2495	0.2562	0.2539	0.2179	0.2136	0.2576
	CCA	0.3087	0.2119	0.1929	0.2131	0.2251	0.2388	0.2569	0.2743	0.2589	0.2423
	DE1	0.0996	0.1710	0.1453	0.1854	0.1869	0.1886	0.1804	0.1208	0.1129	**0.1545**
	DE2	0.1131	0.1708	0.1446	0.1920	0.1994	0.1963	0.1852	0.1318	0.1347	0.1631
Measure 2	ICA	0.2804	0.3203	0.3488	0.3536	0.4433	0.3479	0.3337	0.3181	0.3252	0.3412
	PCA	0.2738	0.3144	0.3291	0.3533	0.3667	0.3829	0.4264	0.4273	0.4341	0.3676
	CCA	0.2729	0.2907	0.3301	0.3526	0.3695	0.3889	0.3673	0.3622	0.3701	0.3449
	DE1	0.2391	0.0712	0.1588	0.0964	0.1412	0.1476	0.1658	0.3419	0.3612	0.1915
	DE2	0.2075	0.0682	0.1605	0.0904	0.1170	0.1377	0.1702	0.3287	0.3365	**0.1796**
Measure 3	ICA	0.0946	0.1700	0.0965	0.2925	0.1710	0.3872	0.3288	0.2612	0.3692	0.2412
	ICA	0.0990	0.1713	0.1079	0.2613	0.3413	0.4189	0.3116	0.2450	0.3752	0.2591
	ICA	0.0954	0.1704	0.1130	0.2905	0.3628	0.4467	0.3743	0.2876	0.4024	0.2826
	ICA	0.0988	0.1714	0.1140	0.3378	0.4308	0.5435	0.4716	0.2646	0.3490	0.3091
	ICA	0.0981	0.1745	0.1137	0.3598	0.4467	0.5722	0.4981	0.2939	0.4296	**0.3319**
Measure 4	ICA	0.1268	0.1511	0.1591	0.1630	0.1397	0.1229	0.1202	0.0671	0.4641	0.1682
	PCA	0.0697	0.1145	0.1019	0.0922	0.1342	0.1661	0.2033	0.1192	0.3433	0.1494
	CCA	0.0702	0.0588	0.1119	0.1216	0.1505	0.1690	0.0335	0.0838	0.2148	0.1127
	DE3	0.0081	0.0303	0.0180	0.0071	0.0068	0.0113	0.0013	0.0047	0.0013	**0.0099**
	DE4	0.0089	0.0211	0.0091	0.0364	0.0139	0.0067	0.0089	0.0121	0.0070	0.0138
Measure 5	ICA	0.2837	0.2945	0.3051	0.3388	0.3352	0.3781	0.3478	0.3546	0.3537	0.3324
	PCA	0.3398	0.3742	0.3400	0.3835	0.3715	0.3855	0.3733	0.3332	0.3332	0.3594
	CCA	0.3531	0.3066	0.2908	0.3198	0.3271	0.3236	0.3741	0.3682	0.3627	0.3362
	DE3	0.1730	0.1258	0.0906	0.0294	0.0506	0.0754	0.1102	0.1887	0.1878	**0.1146**
	DE4	0.1836	0.1287	0.0917	0.0342	0.0489	0.0804	0.1104	0.1966	0.1938	0.1187

4. Error remaining in the reconstructed signals after removing noisy component(s), which is calculated using $Corr(Sig_R, S_n)$, where Sig_R are the signals after removing noisy components, while S_n is the noisy sources.
5. Information loss due to removal of noisy component(s), which is estimated using $Corr(S_{rn}, [S_l, S_g])$, where S_{rn} is(are) the reconstructed noisy component(s) that is(are) removed to obtain an estimate of noisy-free signals.

Results of the ICA, PCA, CCA and the four DE optimized methods, which represent the upper limits to compare with, are shown in Table 4. The first three measures shown in the table indicate that the three methods of ICA, PCA and CCA could not reach the optimal performance in terms of correlation with the local and global EEG sources as well as reducing the influence of noise. The table shows that the performance of ICA was slightly better than CCA and noticeably better than PCA for the first two measures, but slightly worse than them for the third measure. On the other hand, the performance of CCA was slightly better than ICA and PCA in reconstructing the global source. Please note that measure 1 could not be calculated for ICA in the fifth case, as it only managed to find two components and could not converge when calculating the third one. For the fourth and fifth measures that deal with removing noisy component(s),

CCA tends to perform slightly better than ICA and PCA in terms of removing the noise sources and not losing relevant EEG information, however, all three methods are not optimal as removal of noisy sources led to noticeable loss of information.

4 Conclusion

In this paper, we proposed a new approach for the construction of synthetic EEG signals using limited number of channels. The construction aimed at mimicking relationships between real EEG channels. The distances were varied between channels, and hence, varying their cross-correlation. The projections obtained using PCA, ICA and CCA were evaluated using five measures and compared to a best approximate projection that was obtained using a differential evolution algorithm. Results reveal that all three projection techniques are not optimal in terms of reconstructing the original sources. Also, the removal of noisy component(s) led to a loss of relevant information for all three methods. These findings motivate the need for more research on the reconstruction of relevant EEG sources. The proposed synthetic dataset construction method can serve as one of the testbeds for evaluating EEG source separation techniques.

References

1. Platt, B., Riedel, G.: The cholinergic system, eeg and sleep. Behav. Brain Res. **221**(2), 499–504 (2011)
2. Mirowski, P., Madhavan, D., LeCun, Y., Kuzniecky, R.: Classification of patterns of eeg synchronization for seizure prediction. Clin. Neurophysiol. **120**(11), 1927–1940 (2009)
3. Simon, M., Schmidt, E.A., Kincses, W.E., Fritzsche, M., Bruns, A., Aufmuth, C., Bogdan, M., Rosenstiel, W., Schrauf, M.: Eeg alpha spindle measures as indicators of driver fatigue under real traffic conditions. Clin. Neurophysiol. **122**(6), 1168–1178 (2011)
4. Goh, S.K., Abbass, H.A., Tan, K.C., Al-Mamun, A.: Decompositional independent component analysis using multi-objective optimization. Soft Comput. **19**, 1–16 (2015)
5. Cichocki, A., Amari, S.I.: Adaptive Blind Signal and Image Processing: Learning Algorithms and Applications, vol. 1. Wiley, New York (2002)
6. Sweeney, K.T., McLoone, S.F., Ward, T.E.: The use of ensemble empirical mode decomposition with canonical correlation analysis as a novel artifact removal technique. IEEE Trans. Biomed. Eng. **60**(1), 97–105 (2013)
7. Urigüen, J.A., Garcia-Zapirain, B.: Eeg artifact removalstate-of-the-art and guidelines. J. Neural Eng. **12**(3), 031001 (2015)
8. Mammone, N., Foresta, F.L., Morabito, F.C.: Automatic artifact rejection from multichannel scalp eeg by wavelet ica. IEEE Sensors J. **12**(3), 533–542 (2012)
9. Sarker, R., Elsayed, S., Ray, T.: Differential evolution with dynamic parameters selection for optimization problems. IEEE Trans. Evol. Comput. **18**(5), 689–707 (2014)
10. Guo, S.M., Yang, C.C.: Enhancing differential evolution utilizing eigenvector-based crossover operator. IEEE Trans. Evol. Comput. **19**(1), 31–49 (2015)

Analysing the Robust EEG Channel Set
for Person Authentication

Salahiddin Altahat$^{(\boxtimes)}$, Girija Chetty, Dat Tran, and Wanli Ma

Faculty of ESTeM, University of Canberra, Canberra, Australia
{salah.altahat,girija.chetty,dat.tran,wanli.ma}@canberra.edu.au,
http://www.canberra.edu.au

Abstract. In this paper, we present the findings on the EEG channel selection and its impact on the robustness for EEG based person authentication. We test the effect of the enhancement threshold value (T_e), EEG frequency rhythms, mental task and the person identity on the selected EEG channels. Experimental validation of the work with publicly available EEG dataset, showed that the idle mental task provides the highest accuracy rates compared to other considered mental tasks. Moreover, we noticed that imaginary movement tasks provide better accuracy than actual movement tasks. Also for the frequency rhythm effect, the combined frequency rhythms increase the authentication accuracy better than using a single rhythm, so no single rhythm contains all the related identity information. Also for the T_e value, we found that the less T_e we consider, the more EEG channels to be included. Further, for the final part of this work, we tested if the selected channel are person specific. As a result, we found that EEG channel set, if selected for each person differently does enhance the authentication accuracy.

Keywords: EEG · Authentication · Reduced channel set · Mental task · EEG band

1 Introduction

One of the main challenges of any successful biometric modality is its immunity against spoofing. This explains the increasing interest in using EEG in biometrics [14]. Recent reports show the ability to spoof face recognition security systems [18], and also to spoof fingerprint security systems [7]. Further, the recent emergence of long distance iris scanner [1] open the door for iris security system spoofing. The spoofing problem may be extended to emerging biometrics including hand print and voice biometrics. Enhancing the fundamental approaches of acquisition of these biometric modalities may reduce the spoofing threat, such as adding liveness detection procedures. However, the effectiveness of these procedures is directly related to the cost and to the degree of user inconvenience [35]. Bioelectric signals like EEG may provide a better solution for the spoofing problem by using them solely or in conjunction with other biometric modalities, either as liveness detectors or fused to make multimodal biometric system.

© Springer International Publishing Switzerland 2015
S. Arik et al. (Eds.): ICONIP 2015, Part IV, LNCS 9492, pp. 162–173, 2015.
DOI: 10.1007/978-3-319-26561-2_20

EEG has shown promising cues related to human identity [6,38] and many recent studies propose different approaches in using EEG as a biometric modality for identity verification [20,23–26,28–31,33]. Also, recent years have witnessed a revolution in the technology that measures the EEG signal, and it has become more usable, cheap, and does not require lab environment. This may be due to widely used applications of EEG in Brain Computer Interface (BCI) for both entertainment or medical application such as helping paralyzed people [40].

EEG has many advantages *e.g.* confidentiality, high immunity to forgery and promising recognition accuracy in identifying people. However, it still faces serious problems, and among these problems are: the high noise content in the EEG signal, large dependency on mental task, high signal variation between EEG recording sessions and the cumbersome procedure of EEG electrodes placement on the client's scalp [37]. Reducing the number of required EEG channels implies less number of EEG electrodes are needed, and this will reduce the degree of user inconvenience. Moreover, reducing the number of electrodes will lessen and the complexity of the EEG recording system, this will lead to a smaller size and lower cost of the EEG recording system.

EEG is known to be highly sensitive to the user's mental task [19]. This may still be acceptable for an authentication problem, where the cooperation of the users is assumed for them to be authenticated, but this will highly affect its usability as a biometric modality in identification problem where the users' cooperation should not be assumed. This work is an extension to our previous work in [5] where a novel approach was proposed for channel selection for person authentication based on the channel stability in different mental tasks. In the previous work [5] we studied stable channel set for a sample size of 106 people. This was based on a novel measure, called stability index S_i, which ranks the EEG channels based on their similarity in different mental tasks. Having the EEG channels ordered in a table based on their S_i values, then by applying a feature selection approach based on Sequential Forward Selection algorithm, our approach assures that each selected channel fits well with the previously selected and more stable channels. In the Sequential Forward Selection approach we arbitrarily set the enhancement threshold T_e to a value of 1 %, which means that any added channel that did not increase the authentication accuracy by at least 1 % will be excluded from the feature vector. The final result was eight-channel set that produce Half Total Error Rate (HTER) of 14.69 %. In this work we extend this work further, and investigate the effect of the T_e value on the number of selected channels, where, rather than using 1 %, we use different values from 15 % to 0 % and examine its impact on the number of selected channels. Also, we analyze the effect of EEG frequency bands on these channels (considering the 1 % value of T_e), and verify if a particular EEG frequency band provide more identity related information than the others. Further, we analyze the effect of the mental task on the selected channels, Lastly, we examined if using person specific channel set may provide better accuracy.

In this work the signal Power Spectral Density (PSD) was considered as a feature, as it was noted in [22] that the EEG signal periodogram (which is a method of estimating the PSD) can lead to better or similar performances than more elaborated features such as parameters of autoregressive (AR) models and wavelets.

2   Related Work

In examining the effect of mental task on the EEG biometric accuracy, [20] performed EEG recording from nine participants using 32 electrodes. Three mental tasks were considered: imagining a repetitive left hand movement, imagining a repetitive right hand movement and generating words beginning with the same random letter. In their experiment, the results were highly affected by the experiment settings like the number of Gaussian mixtures used in the classification stage. However, they found that there are mental tasks more appropriate for authentication than others. In their work the left hand movement task gives the best average accuracy.

In [12] the researchers perform EEG recording from six participants, two EEG channels and three mental tasks: relax (baseline), mathematical multiplication and reading. In their case, Multiplication task gives the best average accuracy 97.5 %. In [39] the researchers used data for nine participants with 59 electrodes. Two mental tasks: imagine moving left index and imagine moving right index. In their study they found the task of imagining moving the left index was more distinctive, which is closely related to the result found in [20]. Furthermore, they trained the Neural Network (NN) classifier to produce two main classes of outputs: the mental task, and the participants' identity. It was noted that training the classifier for both outputs gives better result than training the classifier for one output only i.e. the person identity.

In [8] the researchers used EEG data for three participants, six channels and four motor imagery mental tasks to imagine moving: left hand, right hand, foot and tongue. They studied the effect of mental task and the frequency band on the identification accuracy. Also in their results, the left hand imaginary movement results in less accuracy than right hand imaginary movement in the μ band and in the β band, but it produce significantly better accuracy in the γ band. The best achieved accuracy was reached with the tongue imaginary mental task in the γ band (90.6 %).

In [4] the researchers perform EEG recording from 10 male participants while resting with eyes open and eyes closed in 5 separate sessions conducted over a course of 2 weeks. They found that the idle mental task with eyes closed gives slightly better accuracy than idle mental task with eyes open, however the difference is not significant, and they refer it to the fact that idle mental task with eyes closed may contains less artifact than idle mental task with eyes open. Also they noticed that including four channels in the feature vector gives better accuracy that using one or two EEG channel.

In [16] the researchers perform EEG recording using 54 electrodes for 9 participants during the resting state with open eyes (OE) and closed eyes (CE) mental task. They found that the mental task of CE provides better accuracy than the OE mental task which complies with what was found in [4]. They also infer this result due to two facts: EEG OE mental task contains more artefact like eye movements than EEG CE task, and there are certain distinctive traits existing in α rhythm which appears more clearly when resting CE mental state that contain identity related information. Also, the extensive review in [9] highlighted this result.

Many studies linked the frequency band effect with the mental task as shown above, since it is known that related mental tasks activate certain frequency bands and therefore these bands becomes dominant in the EEG signal. Also for Brain Computer Interface (BCI) applications, the link between mental task and its related spectrum is emphasized by the fact that EEG spectrum is used as a feature to distinguish mental tasks, and in [17] it was found that there are certain frequency bands that are more appropriate to detect different mental tasks, such as: baseline, multiplication, letter composing, geometric figure rotation and counting. They used sum of weighted power spectrum as a feature and found that γ band (30–100) Hz provide the best accuracy for mental task detection, and it was shown that their result is superior as compared to the use of multivariate AR model as a feature. Further, the change of EEG power spectrum of the mental task is used to diagnose mental illness e.g. Alzheimer Disease (AD) [32] and the neuropsychiatric disorders [13].

In [15] the researchers perform EEG recording using 56 electrodes to record EEG data for 45 participants during the resting CE mental task for the person identification problem. In their study, the effect of used frequency band was analysed for identification accuracy. Their results show that δ and α as a separate bands contain more person specific information than the other bands, despite the fact that δ has not been reported before or used in biometric application. However, combining δ, θ, α and β bands provides the highest accuracy. Also, it is noted that γ band was not considered in their work although a higher accuracy rates for γ band only was reported in [3] but under different mental task (visually evoked potential only). Moreover, these results still related to one specific mental task (idle CE), and the effect of different mental tasks was not considered.

To the best of our knowledge, no study had so far analyzed the frequency band that provides best biometric accuracy over different mental tasks, and we focus on this aspects in this work. Also in this study, we extend our work in [5] to verify the effect of the enhancement threshold T_e value which was set initially to 1 % in our previous study. By changing the T_e threshold value, we are trying to verify the effect of EEG neighbouring channel correlation [21] and if there are any dominant or key EEG channels that often enhance the authentication accuracy regardless of the previously chosen electrodes. Also, we tried to verify if there is a maximum limit of the number of channels to be considered, and the enhancement of recognition accuracy becomes insignificant when adding any further channels.

Finally, since it was reported in [42] that the EEG channels that contain identity related information might be person specific, we also examined if using person specific channel subset may improve the authentication accuracy.

3   Proposed Method

3.1   Dataset

We used the publicly available data described in [36], and was downloaded from [2]. This data was selected since it contains large number of participants and contains 6 mental tasks, which makes it more appropriate to measure EEG channel stability in different mental tasks. The data contains EEG recordings for 109 persons with the following mental tasks: Task a: Idle OE, Task b: Idle CE, Task c: Open and close left or right fist, Task d: Imagine opening and closing left or right fist, Task e: Open and close both fists or both feet and Task f: Imagine opening and closing both fists or both feet. The dataset contains recording of 64 EEG channels. The left earlobe and mastoid electrodes were used as reference and ground electrodes respectively. The data sampling rate is 160 samples/s. Three subjects out of 109 were disregarded because they contain data that was sampled at 128 sample/s. So in total we used EEG data for 106 subjects. Task a, b, c, d, e and f samples represent roughly 50 %, 4 %, 12 %, 11 %, 12 % and 11 % of the total samples number respectively.

3.2   Preprocessing and Feature Extraction

The channel recording during each mental task was separated from other tasks based on the recorded annotation. So for each participant of the 106, the data for each channel was separated to 6 parts. Then, the data was segmented to one second segments. This segment length was chosen because there was not much interest in very low frequencies, as the frequency range 4–52 Hz was found to be more useful, and we considered frequency bins of size 8 as a feature to reduce the size of the feature vector and increase the learning speed of the model due to limited number of data samples. Hence, there was no need to have high resolution spectrum. For the situations requiring testing of frequencies rhythms effect on the authentication accuracy, different bin size was used to accommodate the frequency rhythm range. Also, the frequency range 4–52 Hz was suggested, as it was found in [14], that combining EEG rhythms (θ (4–8) Hz, α (8–15) Hz and β (15–31) Hz) gives optimal result for recognition. Further, other similar studies use only the γ rhythm (30–50) Hz as in [3] and claim maximum identification accuracy. So we considered merging θ, α, β and γ rhythms for better results. No filtration was performed to reject EEG artifacts and all samples were considered. Moreover, for cases where we tested the δ band authentication accuracy, the filtration was considered from 1–52 Hz to include the δ rhythm (1–4) Hz.

For a single channel, every segment have 6 features represent the frequency bins $(52 - 4)/8$. The features were normalized to the sum of all PSD bins in the same segment. When multiple channels are considered, their features were combined together. The frequency range and resolution was changed when considering different EEG rhythms.

3.3   Channel Selection Criteria

In this step we consider the data for 50 participants to avoid over fitting. For the same mental task, channel and person, the EEG feature vectors were assumed to have Gaussian distribution. In order to select stable channel subset among the available 64 channels, two values were measured. The first is the average Mahalanobis distance between the means of the feature vectors distributions of the mental tasks for the same channel and person. So the Mahalanobis distance between the distribution mean of *Task a* was measured against distribution mean of *Task b*, *Task c*, *Task d*, *Task e* and *Task f* for the same channel and person. This was repeated for all the other five tasks. This resulted in 30 distances per channel per person. Averaging these 30 distances over the 50 persons resulted in the "within person distance", which will be referred to as DW_i, where i refers to channel number. The distance, DW_i, measures how far from each other the feature vectors distributions cluster during different mental tasks within the same person for some channel i. The second measure was the Mahalanobis distance between the means of the distributions of the same mental task and channel but for different persons. So the Mahalanobis distance was measured between the mean of *Task a* distribution against the other means for the same task and channel for different persons. This was repeated for all the other five tasks. This resulted in 49 distances per task per channel per person. Then we averaged over all 6 tasks for all 50 persons to find the "between persons distance". This we refer to as DB_i, where i refers to the channel number. DB_i measures how far from each other the EEG feature vectors cluster using the channel i during the same mental task but in different persons.

The channels' stability S_i was measured as the difference between DB_i and DW_i as shown in Eq. 1

$$S_i = DB_i - DW_i \tag{1}$$

Higher values of S_i indicates high separability for the related channel i. After finding the channel stability for all channels, they were ranked based on their stability value. In order to find the best channel subset, we run Sequential Forward Selection algorithm on the channels based on their stability value. Sequential Forward Selection is a simple greedy search algorithm, and hence it does not guarantee finding the global minimum. The change of person authentication half total error rate (HTER) measure for the training set (50 persons) was set as an objective function. So, the channels that have higher stability were prioritized to be used in person authentication problem, as they give better between person separation. But this does not exclude other channels from being used as they may have different information that are related to person identity. Any added channel that did not improve the recognition accuracy by at least T_e value was considered non-informative, and thus was not considered in the feature vector. The non-informative channels contain redundant information that exists in more stable channels that were included previously in the feature vector. The threshold T_e was set to 1% empirically in our previous work [5], in this work we will tune the T_e value to verify its effect on the selected channel set and the related authentication accuracy. The details of the authentication experiment are described in Sect. 4.

4   Experimental Details

For the person authentication experiments, we built a person authentication framework based on Gaussian Mixture Model (GMM) classifier. Among the EEG data of 50 persons, the data during all mental tasks for the first 30 persons were used to build a Universal Background Model (UBM). The number of used mixtures in the GMM was set to 8 based on the best results of multiple trials. The remaining data of 20 persons were used in client/imposter testing. All samples from one mental task only (*Task a*) were used in training to build the client model, and all the remaining samples during the other five mental tasks were used in client/imposter testing. *Task a* represents around 50 % of the total samples for each participant. In order to verify this channel set, we run the authentication experiment using the optimized channel set considering all the data for all 106 persons. In this final testing, the data of the first 60 participants were used to build the UBM model. The remaining 46 participants were used in client /imposter testing. The Gaussian mixtures were trained considering diagonal covariance, and the clients' mixtures' means were adapted using Maximum a Posteriori with the prior UBM model means as described in [34]. The authentication was made such that the ratio between the probability density function (pdf) of a test feature vector measured by the client model and the pdf of the same feature vector measured by the background model should be greater than predefine threshold. Assuming independence for all feature vectors and a uniform class distribution, the likelihood that a certain feature vector $\bar{x} = x_1, x_2, x_3, ..., x_m$ belongs to a specific class λ which has n mixtures is measured by probability density function shown in Eq. 2.

$$p(\bar{x}|\lambda) = \sum_{i=1}^{n} k_i N(\tilde{\mu}_i, \tilde{\Sigma}_i) \tag{2}$$

where, N: normal distribution with $\tilde{\mu}_i$ multivariate mixture mean and $\tilde{\Sigma}_i$ is multivariate mixture covariance matrix. k_i: the probability of the ith mixture in the GMM model.

In order to test the EEG rhythms effect, the above experiment was repeated considering only δ, θ, α, β and γ rhythms. In testing the mental task suitability for person authentication, only the samples that belong to the task of interest was divided into training and testing (80 % - 20 %) ratio.

For the person specific channels, the stability ranking was different for each participant. Therefore there was no Sequential Forward Selection applied for these channels, rather we perform the testing considering first one to five channels and compare the results with the first one to five averaged channel. Also the person specific testing setup was based on using the Maximum Likelihood (ML) frame work rather than Maximum a Posteriori (MAP) and no UBM was created and mean adaptation.

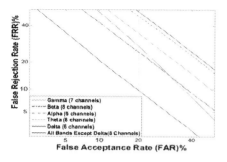

Fig. 1. DET curve for different frequencies.

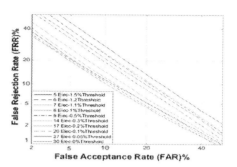

Fig. 2. DET curve for different T_e values.

Fig. 3. DET curve for single task only.

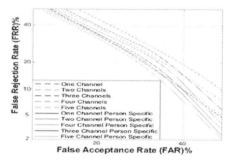

Fig. 4. Person specific channels DET curve.

5   Results and Discussion

Figure 3 shows the Detection Error Tradeoff (DET) curve of the accuracy of the selected channels using one mental task only. The idle task provides the best authentication accuracy, then the imaginary movement tasks share almost the same accuracy and finally the actual movement tasks comes with the least accuracy. This may be due to the idea that idle task has less amount of noise compared to imaginary movement mental tasks. Also, the imaginary movement mental task has less noise component than the actual movement, thus the better accuracy. This also is different from the results shown in [8] where it was found to imagine moving the tongue more appropriate than idle mental task.

Figure 1 shows the DET curve of channel accuracy considering EEG rhythms δ, θ, α, β and γ separately, and compared to the result with the combined EEG rhythms. The figure shows that no single EEG rhythm contains sufficient identity related information, rather it is spread among all these rhythms. Also, it can be noticed that δ and α rhythms contains very close amount of identity related information which complies with the results found in [15], but the results here shows that these two bands contains least amount of identity related information which negates the findings in [15]. This may be due to the fact that δ and α rhythms are known to be dominant in idle mental task which was the only task used in [15], and suppressed in other mental tasks which we have in this study. Interestingly in our work, θ rhythm provides higher accuracy rates than δ. Also initial EEG biometric studies refer to α rhythm only as the one that contains the best identification accuracy e.g. [27,31]. Also θ rhythm provide better accuracy results than δ and α which also can be related to the nature of the used mental tasks, as it is known that θ rhythm may be present in repetitive tasks [10], which is the case of the dataset in hand. γ and β rhythms share the highest accuracy rates, again this may be related to the mental tasks in the used dataset. It is known that γ is active in visual processing mental task [41], and the switching between tasks in the dataset collection was based on a visual cue. Also the β is known to be active during mental tasks that require visual attention [11], which is the case in this EEG dataset. So since the mental tasks we have in this dataset is related to processing and visual attention, this may explain the high accuracy for β and γ bands.

Figure 2 shows the DET curve using different T_e values. The best DET curve achieved when using 27 channels, however the accuracy improvement from 20 channels was very small. This indicate either 20 channels may be the optimal option for using EEG in biometrics, or this result may be due to lack of training data for EEG channels number more than 20 since training the GMM model for a larger feature vector will require extra data which is not available. Also it was noted that channels O2, Iz and TP8 were able to enhance the authentication accuracy in all test cases. The last comparison in Fig. 4 shows clearly that using person specific channels produced better accuracy. This agrees with the result shown in [42] even for different mental tasks.

6   Conclusion and Future Work

This work is an extension for our previous work [5]. In this work we analyzed our previous results of the stable channel set, and how it is affected by the mental task, frequency rhythm, the enhancement threshold value and whether using person specific channel set can improve the accuracy. The results shows that these factors affects the selected channels number as well as their locations. Also some results appeared to be different from previous work, i.e. the EEG rhythm that provide best recognition accuracy. This may be linked to the dataset in hand and the different mental tasks used during the EEG recording. Using different mental tasks may lead to different results. Also, in the dataset used, the EEG recording considered the left earlobe and mastoid electrodes as reference and ground respectively. Selecting a different reference electrode may lead to a different result.

References

1. Long-range iris scanning is here. http://www.theatlantic.com/technology/archive/2015/05/long-range-iris-scanning-is-here/393065/. Accessed: 04 June 2015

2. Physionet. http://www.physionet.org/pn4/eegmmidb/. Accessed: 30 September 2014

3. Roodaki, A.A., Rezatofighi, S.H., Misaghian, K., Setarehdan, S.K.: Fisher linear discriminant based person identification using visual evoked potentials. In: 9th International Conference on Signal Processing, ICSP 2008, pp. 1677–1680, October 2008

4. Abdullah, M.K., Subari, K.S., Loong, J.L.C., Ahmad, N.N.: Analysis of effective channel placement for an eeg-based biometric system. In: IEEE EMBS Conference on Biomedical Engineering and Sciences (IECBES), pp. 303–306. IEEE (2010)

5. Altahat, S., Wagner, M., Marroquin, E.M.: Robust electroencephalogram channel set for person authentication. In: IEEE International Conference on Acoustics, Speech and Signal Processing (ICASSP), pp. 997–1001. IEEE (2015)

6. Anokhin, A., Steinlein, O., Fischer, C., Mao, Y., Vogt, P., Schalt, E., Vogel, F.: A genetic study of the human low-voltage electroencephalogram. Hum. Genet. **90**(1–2), 99–112 (1992)

7. Arthur, C.: iphone 5s fingerprint sensor hacked by Germany's chaos computer club (2013). Accessed: 30 September 2014

8. Bao, X., Wang, J., Hu, J.: Method of individual identification based on electroencephalogram analysis. In: International Conference on New Trends in Information and Service Science, NISS 2009, pp. 390–393. IEEE (2009)

9. Campisi, P., Rocca, D.L.: Brain waves for automatic biometric based user recognition (2014)

10. Collins, G.F.: Cosmopsychology: The Psychology of Humans as Spiritual Beings. Xlibris US (2009)

11. Gola, M., Magnuski, M., Szumska, I., Wróbel, A.: Eeg beta band activity is related to attention and attentional deficits in the visual performance of elderly subjects. Int. J. Psychophysiol. **89**(3), 334–341 (2013)

12. Hema, C.R., Paulraj, M.P., Kaur, H.: Brain signatures: a modality for biometric authentication. In: International Conference on Electronic Design, ICED 2008, pp. 1–4. IEEE (2008)

13. Hidasi, Z., Czigler, B., Salacz, P., Csibri, É., Molnár, M.: Changes of eeg spectra and coherence following performance in a cognitive task in Alzheimer's disease. Int. J. Psychophysiol. **65**(3), 252–260 (2007)

14. Rocca, D.L., Campisi, P., Sole-Casals, J.: Eeg based user recognition using bump modelling. In: 2013 International Conference of the Biometrics Special Interest Group (BIOSIG), pp. 1–12, September 2013

15. Rocca, D.L., Campisi, P., Scarano, G.: Eeg biometrics for individual recognition in resting state with closed eyes. In: 2012 BIOSIG-Proceedings of the International Conference of the Biometrics Special Interest Group (BIOSIG), pp. 1–12. IEEE (2012)

16. La Rocca, D., Campisi, P., Scarano, G.: Stable EEG features for biometric recognition in resting state conditions. In: FernÁndez Chimeno, M., Fernandes, P.L., Alvarez, S., Stacey, D., Solé-Casals, J., Fred, A., Gamboa, H. (eds.) BIOSTEC 2013. CCIS, vol. 452, pp. 313–330. Springer, Heidelberg (2014)

17. Liu, H., Wang, J., Zheng, C., He, P.: Study on the effect of different frequency bands of eeg signals on mental tasks classification. In: 27th Annual International Conference of the Engineering in Medicine and Biology Society. IEEE-EMBS 2005, pp. 5369–5372. IEEE (2006)
18. Maatta, J., Hadid, A., Pietikainen, M.: Face spoofing detection from single images using micro-texture analysis. In: 2011 International Joint Conference on Biometrics (IJCB), pp. 1–7, October 2011
19. Malmivuo, J., Plonsey, R.: Bioelectromagnetism: Principles and Applications of Bioelectric and Biomagnetic Fields. Oxford University Press, Oxford (1995)
20. Marcel, S., Del, J., Millán, R.: Person authentication using brainwaves (eeg) and maximum a posteriori model adaptation. IEEE Trans. Pattern Anal. Mach. Intell. **29**(4), 743–752 (2007)
21. Menon, V., Freeman, W.J., Cutillo, B.A., Desmond, J.E., Ward, M.F., Bressler, S.L., Laxer, K.D., Barbaro, N., Gevins, A.S.: Spatio-temporal correlations in human gamma band electrocorticograms. Electroencephalogr. Clin. Neurophysiol. **98**(2), 89–102 (1996)
22. del José, R., Millán, J., Franzé, M., Mouriño, J., Cincotti, F., Babiloni, F.: Relevant eeg features for the classification of spontaneous motor-related tasks. Biol. Cybern. **86**(2), 89–95 (2002)
23. Palaniappan, R.: Method of identifying individuals using vep signals and neural network. IEEE Proc. Sci. Measure. Technol. **151**(1), 16–20 (2004)
24. Palaniappan, R.: Electroencephalogram signals from imagined activities: a novel biometric identifier for a small population. In: Corchado, E., Yin, H., Botti, V., Fyfe, C. (eds.) IDEAL 2006. LNCS, vol. 4224, pp. 604–611. Springer, Heidelberg (2006)
25. Palaniappan, R., Mandic, D.P.: Eeg based biometric framework for automatic identity verification. J. VLSI Sig. Process. Syst. Sig. Image Video Technol. **49**(2), 243–250 (2007)
26. Poulos, M., Rangoussi, M., Alexandris, N.: Neural network based person identification using eeg features. In: IEEE International Conference on Acoustics, Speech, and Signal Processing, vol. 2, 1117–1120 (1999)
27. Poulos, M., Rangoussi, M., Alexandris, N.: Neural network based person identification using eeg features. In: Proceedings of IEEE International Conference on Acoustics, Speech, and Signal Processing, vol. 2, pp. 1117–1120. IEEE (1999)
28. Poulos, M., Rangoussi, M., Alexandris, N., Evangelou, A.: On the use of eeg features towards person identification via neural networks. Med. Inform. Internet Med. **26**(1), 35–48 (2001)
29. Poulos, M., Rangoussi, M., Alexandris, N., Evangelou, A.: Person identification from the eeg using nonlinear signal classification. Meth. Inf. Med. **41**(1), 64–75 (2002)
30. Poulos, M., Rangoussi, M., Chrissikopoulos, V., Evangelou, A.: Parametric person identification from the eeg using computational geometry. In: The 6th IEEE International Conference on Electronics, Circuits and Systems, Proceedings of ICECS 1999, vol. 2, pp. 1005–1008, September 1999
31. Poulos, M., Rangoussi, M., Chrissikopoulos, V., Evangelou, A.: Person identification based on parametric processing of the eeg. In: The 6th IEEE International Conference on Electronics, Circuits and Systems, Proceedings of ICECS 1999, vol. 1, pp. 283–286 (1999)
32. Rajna, P., Csibri, É., Pal, I., Szelenberger, W.: Task related difference eeg spectrum-a new diagnostic method for neuropsychiatric disorders. Med. Hypotheses **61**(3), 390–397 (2003)

33. Ravi, K.V.R., Palaniappan, R.: A minimal channel set for individual identification with eeg biometric using genetic algorithm. In: International Conference on Computational Intelligence and Multimedia Applications, vol. 2, pp. 328–332, December 2007
34. Reynolds, D.A., Quatieri, T.F., Dunn, R.B.: Speaker verification using adapted gaussian mixture models. Digital Sig. Process. **10**(13), 19–41 (2000)
35. Sabarigiri, B., Suganyadevi, D.: The possibilities of establishing an innovative approach with biometrics using the brain signals and iris features. Res. J. Recent Sci. **3**(3), 60–64 (2013). 2277:2502
36. Schalk, G., McFarland, D.J., Hinterberger, T., Birbaumer, N., Wolpaw, J.R.: Bci 2000: a general-purpose brain-computer interface (bci) system. IEEE Trans. Biomed. Eng. **51**(6), 1034–1043 (2004)
37. Singh, Y., Singh, S., Ray, A.: Bioelectrical signals as emerging biometrics: Issues and challenges. ISRN Sig. Process. **2012**, 13 (2012)
38. Stassen, H.H.: The similarity approach to EEG analysis. Meth. Inf. Med. **24**(4), 200–212 (1985)
39. Sun, S.: Multitask learning for eeg-based biometrics. In: 19th International Conference on Pattern Recognition, ICPR 2008, pp. 1–4. IEEE (2008)
40. Tangermann, M., Müller, K.-R., Aertsen, N.B., Braun, C., Brunner, C., Leeb, R., Mehring, C., Miller, K.J., Müller-Putz, G.R., Nolte, G., Pfurtscheller, G., Preissl, H., Schalk, G., Schlögl, A., Vidaurre, C., Waldert, S., Blankertz, B.: Review of the bci competition iv. Frontiers Neurosci. **6** (2012)
41. Tzelepi, A., Bezerianos, T., Bodis-Wollner, I.: Functional properties of sub-bands of oscillatory brain waves to pattern visual stimulation in man. Clin. Neurophysiol. **111**(2), 259–269 (2000)
42. Yeom, S.-K., Suk, H.-I., Lee, S.-W.: Person authentication from neural activity of face-specific visual self-representation. Pattern Recogn. **46**(4), 1159–1169 (2013)

Automatic Brain Tumor Segmentation in Multispectral MRI Volumetric Records

László Szilágyi[1,2,3]($\boxtimes$), László Lefkovits[2], Barna Iantovics[4], David Iclănzan[2], and Balázs Benyó[1]

[1] Department of Control Engineering and Information Technology,
Budapest University of Technology and Economics, Budapest, Hungary
lszilagyi@iit.bme.hu
[2] Sapientia University of Transylvania, Tîrgu-Mureş, Romania
[3] Canterbury University of Christchurch, Christchurch, New Zealand
[4] Department of Informatics, Petru Maior University of Tîrgu-Mureş,
Tîrgu-Mureş, Romania

Abstract. The aim of this study was to establish a multi-stage fuzzy *c*-means (FCM) framework for the automatic and accurate detection of brain tumors from multimodal 3D magnetic resonance image data. The proposed algorithm uses prior information at two points of the execution: (1) the clusters of voxels produced by FCM are classified as possibly tumorous and non-tumorous based on data extracted from train volumes; (2) the choice of FCM parameters (e.g. number of clusters, fuzzy exponent) is supported by train data as well. FCM is applied in two stages: the first stage eliminates the most part of non-tumorous tissues from further processing, while the second stage is intended to accurately extract the tumor tissue clusters. The algorithm was tested on 13 selected volumes from the BRATS 2012 database. The achieved accuracy is generally characterized by a Dice score in the range of 0.7 to 0.9. Tests have revealed that increasing the size of the train data set slightly improves the overall accuracy.

Keywords: Image segmentation · Fuzzy clustering · Tumor detection · Magnetic resonance imaging

1 Introduction

The early detection of brain tumors is utmost important as it can save human lives. The accurate segmentation of brain tumors is also essential, as it can assist the medical staff in the planning of treatment and intervention. The manual segmentation of tumors requires plenty of time even for a well-trained expert.

Research supported by the Hungarian National Research Funds (OTKA), Project no. PD103921. The work of L. Lefkovits was supported by The Sectorial Operational Program Human Resources Development POSDRU/159/1.5/S/137516 financed by the European Social Found and by the Romanian Government.

S. Arik et al. (Eds.): ICONIP 2015, Part IV, LNCS 9492, pp. 174–181, 2015.
DOI: 10.1007/978-3-319-26561-2_21

A fully automated segmentation and quantitative analysis of tumors is thus a highly beneficial service. However, it is also a very challenging one, because of the high variety of anatomical structures and low contrast of current imaging techniques which make the difference between normal regions and the tumor hardly recognizable for the human eye [1]. Recent solutions, usually assisted by the use of prior information, employ various image processing and pattern recognition methodologies like: combining multi-atlas based segmentation with non-parametric intensity analysis [2], AdaBoost classifier [3], level sets [4], active contour model [5], graph cut distribution matching [6], diffusion and perfusion metrics [7], confidence guided discriminative classifier [8], 3D blob detection [9], and support vector machine [10].

The main goal of our research work is to build a reliable procedure for brain tumor detection from multimodal MRI records, based on semi-supervised clustering algorithms, using the MICCAI BRATS data set that contains several dozens of image volumes together with ground truth provided by human experts. As a first step, in the current paper we present preliminary results achieved using a two-stage fuzzy c-means clustering cascade algorithm and 13 selected volumes from the above mentioned data set.

2   Materials and Methods

2.1   BRATS Data Sets

Brain tumor image data used in this work were obtained from the MICCAI 2012 Challenge on Multimodal Brain Tumor Segmentation [11]. The challenge database contains fully anonymized images originating from the following institutions: ETH Zürich, University of Bern, University of Debrecen, and University of Utah. The image database consists of multi-contrast MR scans of 30 glioma patient, out of which 20 have been acquired from high-grade (anaplastic astrocytomas and glioblastoma multiforme tumors) and 10 from low-grade (histological diagnosis: astrocytomas or oligoastrocytomas) glioma patients. For each patient, multimodal (T1, T2, FLAIR, and post-Gadolinium T1) MR images are available. All volumes were linearly co-registered to the T1 contrast image, skull stripped, and interpolated to 1 mm isotropic resolution. All images are stored as signed 16-bit integers, but only positives values are used. Each image set has a truth image which contains the expert annotations for "active tumor" and "edema".

In our application, each voxel in a volume is represented by a four-dimensional feature vector:

$$\mathbf{x} = [\log(x^{(\mathrm{T1})}), \log(x^{(\mathrm{T2})}), \log(x^{(\mathrm{T1C})}), \log(x^{(\mathrm{FLAIR})})]^T.$$

Those voxels which have zero intensity in any of the channels are neglected. Most of these are voxels outside the volume of interest, but there are a few others as well. The intensity information of voxels of a whole volume are collected in a large matrix whose number of rows is 4, while the number of columns is the number of actual voxels, somewhere between 1 and 2 million. This matrix will represent the input data for the FCM cascade. We also store the position of each voxel, so that we are able to localize them after clustering. The log values of intensities are likely to be in the range between 3 and 8.

2.2   The FCM Cascade

The FCM cascade algorithm has the main goal of effectively separating homogeneous areas in the volume. Further on, based on decision support built upon prior information, it also distinguishes tumor tissues from normal brain tissues.

The conventional FCM algorithm provides the input data set a fuzzy partitioning into a previously set number (c) of clusters, based on the minimization of a quadratic objective function that contains a parameter referred to as fuzzy exponent. This parameter $m > 1$ regulates the fuzzyness of the partition. The smaller the value of the fuzzy exponent is, the closer the partition will be to the crisp one [12]. FCM usually places the cluster prototypes in areas where lots of input vectors are in the neighborhood. In order to provide an accurate clustering, we should know in advance the exact number of such accumulation points in the 4D color space, and be able to initialize a cluster prototype in the proximity of each. Obviously this is not the case. Even if we could do this somehow, clustering would represent a huge computational burden because of the hundreds of clusters and millions of input vectors. To avoid this case, we propose to perform FCM in two stages. The first stage will help us get rid of the most part of the input data, those vectors which are far from all tumor tissue intensities that we have stored in a predefined atlas. In the second stage we only have voxels of intensities that are close to those of tumor patterns. Cluster prototypes obtained in the second stage are individually analyzed, to separate tumor clusters from normal ones.

So in the first stage, fuzzy c-means is applied to the whole set of n voxels in the volume. The number of clusters c is typically varied between 6 and 20. When this first FCM stage is ready, the voxels are categorized into c clusters, each represented by its cluster prototype $\mathbf{v}_i$, $i = 1 \ldots c$. These cluster prototypes are individually investigated by the decision support system which decides whether they are suspected of containing tumor tissues or not. Those clusters whose centroid vector is distant from tumor intensities, are excluded from the further stages. The decision support at this stage usually keeps up to 3 of the clusters obtained in the first FCM. The remaining n' voxels, or more precisely, the collection of feature vectors that represent them, will serve as input data for the second stage of the FCM cascade. Fuzzy c-means clustering is applied again, this time with c' clusters, also varied between 6 and 20. Final cluster prototypes are checked again by the decision support system. This time all clusters having their prototypes in the proximity of tumor intensities will be declared tumor tissues (positive), while all others negative. The fuzzy exponent may vary from stage to stage within the interval $[1.5, 2.0]$, that is why we denoted by m and m' the exponent applied in the first and second stage, respectively.

2.3   FCM Initialization

Especially in multi-dimensional problem, the FCM algorithm is highly sensitive to prototype initialization. Basically it is advised to attempt placing the initial cluster prototypes far from each other, possibly in accumulation areas of input vectors. Randomly chosen input vectors can produce high variability among different runs. So a deterministic rule is needed for a stable solution.

Our algorithm applies FCM clustering in each dimension d on the scalar data $\log(x_1^{(d)}), \log(x_2^{(d)}) \ldots \log(x_n^{(d)})$ to produce three clusters, whose cluster prototypes $\nu_1^{(d)}$, $\nu_2^{(d)}$, and $\nu_3^{(d)}$ will serve as grid points in the given dimension d. The 4D grid formed this way defines $3^4 = 81$ grid points, which are all treated as potential cluster seeds for the 4D clustering problem. Let us denote these potential seeds by $\mathbf{w}_i$, $i = 1 \ldots 81$. Out of these, we choose initial cluster prototypes those ones, which have lowest average square distance from input vectors, computed with the formula $\Delta(\mathbf{w}_i) = \sum_{k=1}^{n} ||\mathbf{x}_k - \mathbf{w}_i||^2$.

2.4   Decision Support

After the first stage of the FCM cascade, it is necessary to separate those clusters that are likely to contain tumor tissues from the other clusters. After the second stage of the cascade, a decision has to be taken, which are the tumor tissue clusters and which are declared negative. These decisions rely on prior information extracted from train volumes. A series of previously performed FCM runs have established clusters that were declared tumor tissues or negative ones based on the available ground truth. In the testing phase, a k-nearest neighbors algorithm is employed to decide, whether the extracted clusters contain tumor tissues or not. Ground truth information extracted from the test volume is never used by the decision support algorithm.

2.5   Evaluation of Accuracy

The Jaccard index (JI) is a normalized score of accuracy, computed as $JI = \frac{TP}{TP+FP+FN}$, where TP stands for the number of true positives, FP for the false

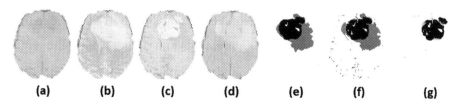

(a)   **(b)**   **(c)**   **(d)**   **(e)**   **(f)**   **(g)**

Fig. 1. Volumetric segmentation results presented in a single slice: (a)–(d) the four input channels; (e) ground truth with tumor shown in black and edema in grey; (f) and (g) results after first and second FCM stage, respectively (TP: black, FN: light grey, FP: dark grey).

Table 1. Dice scores achieved in unsupervised and various supervised settings. US stands for unsupervised processing mode, while SS1...SS12 indicate the semi-supervised processing with the number of train volumes varying between 1 and 12.

Scenario	HG01	HG02	HG03	HG04	HG07	HG11	HG14	HG15
US best	0.8925	0.7576	0.8984	0.7878	0.7476	0.7557	0.7855	0.8468
US average	0.8575	0.5318	0.8564	0.7746	0.7188	0.4480	0.7215	0.8256
SS1 average	0.8701	0.6801	0.8770	0.7786	0.7306	0.6910	0.7262	0.8359
SS2 average	0.8696	0.6872	0.8758	0.7769	0.7299	0.6956	0.7247	0.8367
SS3 average	0.8710	0.6889	0.8773	0.7773	0.7307	0.6946	0.7278	0.8371
SS4 average	0.8744	0.6912	0.8794	0.7774	0.7323	0.6974	0.7276	0.8380
SS5 average	0.8757	0.6929	0.8807	0.7775	0.7338	0.6988	0.7279	0.8382
SS6 average	0.8763	0.6940	0.8817	0.7773	0.7347	0.7002	0.7288	0.8381
SS7 average	0.8770	0.6957	0.8827	0.7769	0.7352	0.7007	0.7294	0.8382
SS8 average	0.8771	0.7001	0.8833	0.7763	0.7357	0.7006	0.7309	0.8384
SS9 average	0.8781	0.7093	0.8848	0.7758	0.7364	0.7005	0.7320	0.8385
SS10 average	0.8791	0.7207	0.8867	0.7756	0.7374	0.7012	0.7334	0.8387
SS11 average	0.8792	0.7349	0.8885	0.7754	0.7381	0.7022	0.7358	0.8391
SS12 average	0.8789	0.7491	0.8873	0.7749	0.7395	0.7016	0.7361	0.8368

positives, and FN for false negatives. Further on, the Dice score (DS) can be computed as $DS = \frac{2TP}{2TP+FP+FN} = \frac{2JI}{1+JI}$. Both indices score 1 in case of an ideal clustering, while a fully random result is indicated by a score close to zero.

3   Results and Discussion

Thirteen volumes from the BRATS 2012 data set were selected for evaluation. These underwent the proposed FCM cascade algorithm using various settings regarding fuzzy exponent and number of classes in each stage, namely $c, c' \in \{6, 7, \ldots 20\}$, and $m, m' \in \{1.5, 1.6, \ldots, 2.0\}$. All these variants sum up to 8100 tests performed for each volume. The average and maximum Dice score for each volume was extracted, to be used as reference during the evaluation of the semi-supervised algorithm. Figure 1 presents a slice of a successfully segmented volume.

In order to decide the optimal number of clusters for each processing stage, we propose to employ a semi-supervised learning scenario, based on the following terms. The 13 volumes are separated into train and test data. In a given evaluation, each volume can be either train or test volume, never both. The number of train volumes, denoted by p, can vary from $p_{\min} = 1$ to $p_{\max} = 12$. At the evaluation of a certain test volume's segmentation, there are $p_{\max}!/(p!(p_{\max} - p)!)$ different possibilities to choose exactly p train volumes. Each of these cases yield a different Dice score. Average Dice scores were extracted for each test volume and each possible number of train volumes.

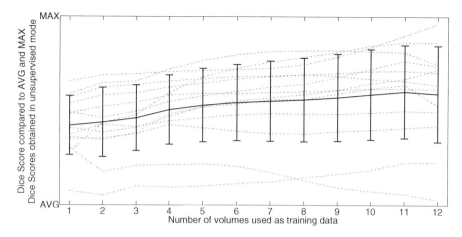

Fig. 2. Evolution of average Dice score obtained for individual test volumes, plotted against the number of train volumes (dashed lines). Average and standard deviation of the thirteen curves obtained for individual test volumes. As the number of train volumes grows, the achieved Dice score slightly tends towards the maximum.

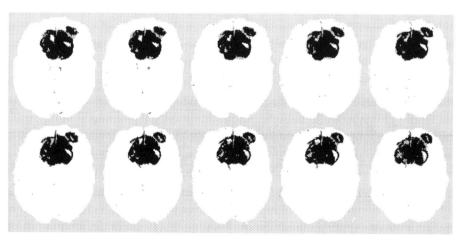

Fig. 3. Consecutive slices of a detected tumor. Black pixels are true positives, dark grey pixels represent false positives, light grey ones indicate false negatives.

Table 1 summarizes the obtained Dice scores in various circumstances. The first two rows exhibit the reference DS values (average and maximum) obtained in case of unsupervised processing. Semi-supervision induced by a given single volume, say HG01, means that we choose that parameter setting (m, m', c, and c') which led to the best DS for volume HG01 in unsupervised processing, and applied it at the processing of the other 12 volumes. DS values shown in the SS1 row were obtained for each volume by choosing the average DS out of the 12 outcomes. Whichever is the test volume, each of the 12 others can be the inducing train volume in SS1 mode.

In semi-supervision induced by two volumes (SS2) processing mode, the parameter setting is chosen such a way that the average DS of the train volumes is greatest. This setting is applied to the other 11 volumes. Each test volume receives $12!/(10!2!) = 66$ different optimal settings from the different combinations of train volumes. Out of the 66 Dice scores were computed the values given in row SS2 of Table 1. Further rows SS3 to SS12 were obtained analogously.

The dashed curves in Fig. 2 present the evolution of relative DS achieved for each individual test volume, compared to the average (AVG) and maximum (MAX) Dice score obtained in unsupervised mode, plotted against the number of train volumes that induced the semi-supervised processing. Figure 2 also shows the global average and standard deviation of the relative Dice scores indicated by the dashed curves.

Figure 3 exhibits a tumor detected by the proposed algorithm, in consecutive slices of volume HG15. True positive pixels represented by black seem to have captured the boundary of the tumor accurately in most places. False negatives (light grey) and false positives (dark grey) are also present in a considerable proportion. So there are still challenges to improve the implemented algorithm.

As it was initially suspected, the parameters that assure best segmentation accuracy for train volumes can indeed give useful support in choosing the right parameters for the test volume. Parameters provided by the train data lead to better accuracy than the expected accuracy via randomly chosen parameters. However, not every train volume adds to the success of every test volume. Figure 2 reveals the slight improvement achieved by adding more and more volumes to the train set. Besides the initial goal of including further dozens of volumes in our study, there are several possible directions for further development of the proposed system:

1. There is a strong need to establish whether all four channels are necessary for an accurate tumor detection. If one of the channels does not really contain useful data, its presence not only hardens the computational burden, but also hinders c-means clustering models in providing accurate partitions.
2. Obviously not all image volumes share the same properties, not all of them are clustered best with the same parameter settings. A brief evaluation of the intensity histograms in different channels of the test data, and matching with certain previously established templates could be effective in determining, which train volumes would give best parameter setting for the test volume.
3. The current system neglects the handling of intensity non-uniformity effects. However, the effective compensation of such phenomena is usually performed simultaneously with image segmentation [13,14].
4. The tumor detection process should be extended to the identification and separation of neighboring edema regions as well.

4   Conclusion

In this paper we proposed an automatic tumor detection and segmentation algorithm employing a fuzzy c-means clustering cascade algorithm and fusing it with decision support based on prior information. The proposed algorithm was validated using 13 selected MRI volumes originating from the BRATS 2012 data set. Preliminary evaluation results revealed the ability of the algorithm to extract accurate information concerning the presence and position of the tumor. Further development directions enumerated in Sect. 3 are likely to improve the benchmarks of the algorithm.

References

1. Gordillo, N., Montseny, E., Sobrevilla, P.: State of the art survey on MRI brain tumor segmentation. Magn. Res. Imag. **31**, 1426–1438 (2013)
2. Asman, A.J., Landman, B.A.: Out-of-atlas labeling: a multi-atlas approach to cancer segmentation. In: 9th IEEE International Symposium on Biomedical Imaging, pp. 1236–1239. IEEE Press, New York (2012)
3. Ghanavati, S., Li, J., Liu, T., Babyn, P.S., Doda, W., Lampropoulos, G.: Automatic brain tumor detection in magnetic resonance images. In: 9th IEEE International Symposium on Biomedical Imaging, pp. 574–577. IEEE Press, New York (2012)
4. Hamamci, A., Kucuk, N., Karamam, K., Engin, K., Unal, G.: Tumor-Cut: segmentation of brain tumors on contranst enhanced MR images for radiosurgery applicarions. IEEE Trans. Med. Imag. **31**, 790–804 (2012)
5. Sahdeva, J., Kumar, V., Gupta, I., Khandelwal, N., Ahuja, C.K.: A novel content-based active countour model for brain tumor segmentation. Magn. Res. Imag. **30**, 694–715 (2012)
6. Njeh, I., Sallemi, L., Ayed, I.B., Chtourou, K., Lehericy, S., Galanaud, D., Hamida, A.B.: 3D multimodal MRI brain glioma tumor and edema segmentation: a graph cut distribution matching approach. Comput. Med. Imag. Graph. **40**, 108–119 (2015)
7. Svolos, P., Tsolaki, E., Kapsalaki, E., Theodorou, K., Fountas, K., Fezoulidis, I., Tsougos, I.: Investigating brain tumor differentiation with diffusion and perfusion metrics at 3T MRI using pattern recognition techniques. Magn. Res. Imag. **31**, 1567–1577 (2013)
8. Reddy, K.K., Solmaz, B., Yan, P., Avgeropoulos, N.G., Rippe, D.J., Shah, M.: Confidence guided enhancing brain tumor segmentation in multi-parametric MRI. In: 9th IEEE International Symposium on Biomedical Imaging, pp. 366–369. IEEE Press, New York (2012)
9. Yu, C.P., Ruppert, G., Collins, R., Nguyen, D., Falcao, A., Liu, Y.: 3D blob based brain tumor detection and segmentation in MR images. In: 11th IEEE International Symposium on Biomedical Imaging, pp. 1192–1197. IEEE Press, New York (2012)
10. Zhang, N., Ruan, S., Lebonvallet, S., Liao, Q., Zhou, Y.: Kernel feature selection to fuse multi-spectral MRI images for brain tumor segmentation. Comput. Vis. Image Underst. **115**, 256–269 (2011)
11. Menze, B.H., Jakab, A., Bauer, S., Kalpathy-Cramer, J., Farahani, K., Kirby, J., et al.: The multimodal brain tumor image segmentation benchmark (BRATS). IEEE Trans. Med. Imag. **34**(10), 1993–2024 (2015)
12. Bezdek, J.C.: Pattern Recognition with Fuzzy Objective Function Algorithms. Plenum, New York (1981)
13. Vovk, U., Pernuš, F., Likar, B.: A review of methods for correction of intensity inhomogeneity in MRI. IEEE Trans. Med. Imag. **26**, 405–421 (2007)
14. Szilágyi, L., Szilágyi, S.M., Benyó, B.: Efficient inhomogeneity compensation using fuzzy c-means clustering models. Comput. Meth. Prog. Biomed. **108**, 80–89 (2012)

Real-Time EEG-based Human Emotion Recognition

Mair Muteeb Javaid, Muhammad Abdullah Yousaf,
Quratulain Zahid Sheikh, Mian M. Awais[⊠], Sameera Saleem,
and Maryam Khalid

LUMS, D.H.A, Lahore Cantt., Lahore 54792, Pakistan
{15100159,15100067,14100146,awais,sameera.saleem,
maryam.khalid}@lums.edu.pk

Abstract. Recognition of user felt emotion is an exciting field because visual, verbal and facial communications can be falsified more easily than 'inner' emotions. Non-invasive EEG-based human emotion recognition entails the classification of discrete emotions using EEG data. These emotions can be defined by the arousal-valence dimensions. We performed real-time emotion classification for four categories of emotional states, namely: pleasant, sad, happy and frustrated. Higuchi's Fractal Dimension was applied on EEG data and used as a feature extraction method and Support Vector Machine was used for classification. This paper documents a comparative study of classification accuracy achieved by collecting raw EEG data from 3 electrode locations vs. collection from 8 electrode locations.

Keywords: EEG · Emotion recognition · Arousal-valence · SVM

1 Introduction

Human emotions are an important part of everyday life, as they assist in understanding human behavior. As electroencephalogram devices have become more accessible commercially, it can be used in bio-feedback applications' research with the motivation to create an application which will directly respond to the mood of the user instead of using buttons or touch screens to provide input.

These applications can be related to entertainment, E-learning, virtual worlds, and cyber world [3–9]. Several measures have been used for emotion recognition which are explained below:

EMG has been used to detect a startle response magnitude which is caused by reflexive myoneural (junction between nerve fibre and muscle it supplies) response. It is however limited in detecting only one portion of the emotional states for example the arousal state.

Behavioral measurement monitors the face muscles or body movement using EMG or video recording in order to predict the emotional state of the subject but the emotions are restricted to pre-defined emotional states.

Autonomous measurement can detect emotion-induced physiological responses caused by the nervous system when exposed to physiologically arousing stimulus.

© Springer International Publishing Switzerland 2015
S. Arik et al. (Eds.): ICONIP 2015, Part IV, LNCS 9492, pp. 182–190, 2015.
DOI: 10.1007/978-3-319-26561-2_22

These include Skin Conductance Responses (SCRs) and Heart Rate Variability (HRV). But these will recognize only a small subspace of our emotional states.

Non-invasive sensor modalities have specific characteristics regarding spatiotemporal resolution and mobility [10]. These include Magnetoencephalography(MEG), functional Magnetic Resoance Imaging (fMRI) and Electroencephalography (EEG). fMRI has been previously used in finding cortical structures in the brain related to emotional states. MEG has been used to identify emotion states using precise timing and sensitive spatiotemporal resolution. EEG captures immediate responses of brain activity with very sensitive temporal resolutions and is cost-effective as well as mobile. Hence EEG has more potential to be used in online interactive applications.

Electroencephalography is a non-invasive method to record electrical activity of the brain by placing electrodes on the scalp. Brain rhythms are rhythmic fluctuations of brain's electrical activity and reflect the state of the brain. The five major brain rhythms associated with emotions are Delta (1.5–4 Hz), Theta (4–7.5 Hz), Alpha (8–13 Hz), Beta (13–30 Hz) and Gamma (30 Hz and above).

The format of this paper is as follows: first we will introduce the techniques applied by researches in emotion classification. Then we will discuss our own experimental design which includes subject selection and experimental protocol. This leads us to the experimental procedure of our classification technique that covers how EEG data was acquired, feature extraction method and finally our supervised classification of emotional states. Finally the results section compares the two approaches of using different number of electrodes.

2 Background

Liu et al. [1] proposed a fractal Dimension threshold-based classifier and quantified emotions on the arousal-valence dimensions. Auditory clips from IADS database were used as stimulus. They conducted two experiments using Emotiv headset, with sets of stimuli belonging to IADS and a self-collected set. Recorded EEG data was labelled with it's corresponding arousal and valence level using Self-Assessment Manikin (SAM). Three electrode channel locations were utilized; AF3, F4 and FC6. Higuchi's Fractal Dimension was employed for feature extraction. Valence and arousal values associated to a dataset were converted to discrete values whose combination provided an emotional state. One of the disadvantages of this technique was that the emotion recognition scheme had to be first trained for each individual subject. In another paper by Sourina et al.

[2] similar feature extraction method was implemented using Higuchi's Fractal Dimension and Box-counting dimension and the resultant feature matrix was fed into Support Vector Machine for classification using arousal and valence binary models. They achieved a minimum accuracy of 70 % and a maximum accuracy of 100 % for both arousal and valence level recognition.

Lin et al. uses 12 symmetric electrode pairs to calculate power spectral difference of symmetric electrode pairs. Three different schemes of multi-class SVM were applied which included all-together (a single model which classified four emotional states), one-against-one (k[k-1]/2 models were created for k classes of emotions) and model-based (arousal-valence model based nested binary classifiers were created). The best performance was achieved using the one-against-one scheme whereby 92.57 % accuracy was achieved to distinguish all four emotional states: Joy, Pleasure, Angry and Sadness. Bos et al. utilized 3 electrodes Fpz, F3 and F4 in their research whereby feature extraction was done using the combination of power spectra of the five rhythms: delta, theta, alpha, beta and gamma which amounted to a total of 1000 features. PCA was applied to reduce the feature space from 1000 to about 25 dimensions. Binary linear Fisher's Discriminant Analysis was performed for emotion level classification of valence and arousal level. A performance rate of 92.3 % was achieved.

Rahman et al. compared the classification performance as a result of using four different feature extraction methods which included Higuchi's Fractal Dimension, Minkowski Bouligand, Fractional Brownian Motion and Gaussian Mixture Models (GMM). Support Vector Machine and K-Nearest Neighbor were used and a hierarchical model was constructed to separate the four emotional states: positive, negative, clam and excited. It was concluded that GMM performed with the highest accuracy of about 82 %.

3 Real-Time Emotion Recognition

3.1 Experimental-Design

Two sessions were performed. A training session was conducted where stimuli was shown to each subject corresponding to each emotional state and using that, a hierarchical model was created. The second session consisted of live collection of EEG data from a subject and real-time recognition of the emotion.

Stimulus Selection. After surfing through a wide range of online resources, 32 audio-visual stimuli were shortlisted for the experimental procedure. Stimuli were gathered using tags associated with the specific emotion available online and survey conducted among university students. A large portion of the time was spent in collecting a generic database which induces the same set of emotions in different individuals. It should be noted that stimulus for the emotional state of frustration was highly subjective among the subjects. It should be noted that unlike works discussed in previous section, we did not have stimuli from IADS or IAPS database at our disposal.

Subject Demographics. We experimented on eight subjects, one female and seven males, aged between 17– 28 years, with education ranging from high school to post-graduate, and fields of specialization varying from engineering sciences to accounting and finance.

Experimental Protocol. Following guidelines were used for the experiment protocol:

1. Subject should be exposed to the stimulus for sufficient time for the stimulus to induce the required emotion in the subject's brain. Therefore our subjects were exposed to the stimulus for 3–4 min.
2. Subject is NOT to be informed of the nature of stimuli we show him/her, in order to preserve the element of surprise and avoid leading the subject into answering in a biased manner.

3. Stimulus should not cause multiple emotions to be felt by the subjects. It should have a single emotional theme.
4. Stimulus should not be biased and should elicit the same emotions irrelevant of the age group or sex of the subject.
5. Cooling off period i.e. time interval between each stimuli exposure should be introduced so that the subject can move his/her eyes or other parts of the body.
6. During the time interval in which the stimulus is being played, the subject should be in a relaxed body position and try to prevent any muscle tension or movement. Specific actions such as eye movement, jaw clenching, tongue rolling, teeth grinding and contraction of muscles should be avoided.
7. The OpenBCI Kit has wires connected to the RFDuino board. Care should be taken when handling it. In order to avoid distractions in the environment in the form of people talking or passersby, a custom made cardboard box was made which covered the user and the laptop/PC on which the stimulus was being shown.
8. Every subject was required to fill a Self-Assessment Manikin (SAM) after being exposed to a stimulus. This was used to validate whether the subject consciously reports the same emotion that the experiment stimulus aimed to induce. In cases where we got disparaging results, the data file was not used any further, and instead another stimulus was shown to the participant. This was an important change which helped us improve the models that were generated after training.

Data Acquisition. OpenBCI Kit was used to extract EEG signals from the subject with a sampling rate of 250 Hz. Two different electrode locations' were used to obtain EEG data. We used the electrode locations of AF3, F4 and FC6 in case of 3 electrodes and Fp1, Fp3, F3, F4, T3, T4, P3 and P4 for 8 electrode locations attached bilaterally. A conductive paste was applied to each of the electrodes before placing them on the scalp. It was ensured that there was no contact of hair with the electrodes to avoid noise. An EEG data buffer with a size of 1024 and an overlapping window of 99 % was kept. The buffer fed the data into the emotion recognition algorithm at fixed intervals. Each time new data arrived, previous buffer data was removed and filled with new incoming data. We gathered a total 192 data sets, 96 each for 3 and 8 electrodes placements.

3.2   Emotion Recognition Algorithm

Preprocessing. The brain waves which are involved in determining human emotions are delta (1–3 Hz), theta (4–7 Hz), alpha (8–13 Hz), beta (14–30 Hz) and gamma (31–50 Hz). The raw data was filtered by a notch filter to remove any background noise. Preprocessing also involved band-pass filtering in the range of 1.5–50 Hz to obtain data relevant to the five rhythms. The pre-processed data is divided into segments using a sliding window of 1024 with 99 % overlapping between segments.

Feature Extraction. For feature selection, we used the fractal dimension values from each electrode and combined them in a feature vector. For 3 electrode paradigm, we used [FDAF3 FDF4] as a feature vector for valence recognition and [FDFC6] feature vector for arousal level recognition. In the case of 8 electrodes (4 electrode bilateral pairs), we used all 8 electrode FD values in the feature vector for arousal and valence recognition. We used Higuchi's Fractal Dimension algorithm to calculate fractal dimension as explained below.

Higuchi's Fractal Dimension. Let X (1), X (2),..., X (N) be a finite set of time series samples, a new time series is constructed as follows:

$$X_k^m : X(m), X(m+k), X(m+2k), \ldots + X\left(m + \left[\frac{N-m}{k}\right].k\right) \tag{1}$$

where m = 1, 2,..., k, m is the initial time and k is the interval time. Then, k sets of $L_m(k)$ are calculated as follows:

$$L_m(k) = \frac{\left\{\left[\sum_{i=1}^{\left[\frac{N-m}{k}\right]}|X(+ik) - X(m+(i-1).k)|\right]\frac{N-1}{\left[\frac{N-m}{k}\right].k}\right\}}{k} \tag{2}$$

<L(k)> denotes the average value over k sets of Lm(k) and the relationship exists as follows:

$$L(k) \propto k^{-D} \tag{3}$$

Finally, the fractal dimension can be obtained by logarithmic plotting between different k and it's associated <L (k)>.

Classification. James A. Russell introduced the bipolar model of affect (i.e. emotion as represented in language). It was concluded by Russell that pleasure is the bipolar opposite of displeasure and arousal of sleepiness. This bipolar affective space can represent variance in 42 commonly used scales e.g. scales of happiness, elation, anger, fear,, anxiety and depression. We will use this bipolar model otherwise known as arousal-valence model in our classification algorithm.

Using the feature vector for each segment that we created in the preprocessing stage, feature matrices for valence and arousal models are created where each row corresponds to one of the labels: high arousal or low arousal for arousal model, negative valence or positive valence for valence model. In SVM, a hyper plane is constructed in n-dimensional space in order to separate distinct classes of data. Hierarchical classification was performed where SVM is chosen to create two binary models; arousal and valence. This dimensional model defines an emotional state in spatial region using the basic dimensions of arousal and valence. In this way, discrete emotions can be defined on the arousal-valence axes. As our feature space has very few dimensions, radial basis function kernel is used to project the feature space to a higher dimension in order to separate non-linear EEG data. Four emotional states were recognized using the two models of arousal and valence:- positive/low arousal (pleasant), negative/low arousal (sad), positive/high arousal(happy) and negative/high arousal (frustrated). Performance

Table 1. Performance Comparison of 8 electrodes usage vs. 3 electrodes usage

# of EEG channels	Channel Locations	Feature Matrix Format (FD denotes fractal Dimension Value)	Classification accuracy for arousal model	Classification accuracy for valence model	Avg. Emotional State Classification Accura
3	AF3, F4, FC6	[FDAF3, FDF4] For valence model, [FDFC6] for arousal model	59.1 %	68.39 %	51.94 %
8	Fp1, Fp3, F3, F4, T3, T4, P3 and P4	[FDFp1, FDFp3, FDF3, FDF4, FDT3, FDT4, FDP3, FDP4] for both arousal and valence model	87.62 %	83.28 %	77.38 %

of the emotional states was determined by the correct classification of emotion label by the two models of the valence and arousal label associated with each EEG data segment (Table 1).

4 Comparison and Results

A comparative study was performed between using 3 and 8 electrode locations. The table on the previous page summarizes the performance results for the two experimental setups. It can be deduced from the results that using 8 electrodes, the performance has improved significantly. The minimum performance accuracy in the case of the four emotional states has increased to 75 %. In the case of arousal level recognition, accuracy improved from 59.17 % to 87.62 % and for valence level, it increased from 68.39 % to 83.28 % when using 8 electrodes. The average emotional states' classification accuracy also rises from 51.94 % to 77.38 % (Figs. 1, 2 and 3).

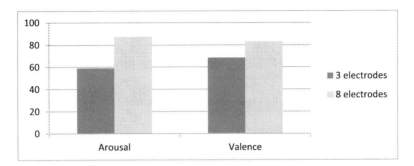

Fig. 1. Performance of valence and arousal levels for both 3 and 8 electrodes

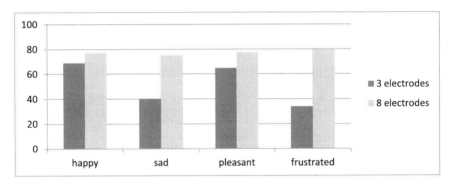

Fig. 2. Performance of valence and arousal model combined in predicting four emotional states for both 3 and 8 electrode

5 Software and Tools

We used Emotiv Headset and OpenBCI for data acquisition. We used MATLAB for pre-processing of data and LIBSVM library of MATLAB in the classification step.

6 Conclusion

What we have demonstrated is that using greater number of electrodes, it is possible to significantly increase the performance of human emotion detection using a hierarchical classification model based on the valence-arousal model. Future work can include the following undertakings; use the optimal number of electrodes for this type of study, explore alternative classification techniques like Fuzzy C-means clustering (where extracted EEG data can be labeled as belonging to more than one emotion category) and involve equal number of male and female subjects to remove any gender bias. We faced some major challenges in the course of our project. The first issue was regarding the stimulus where it was very challenging to gather audio or audio-visual database which could be generalized for a group of people even within the same age-group.

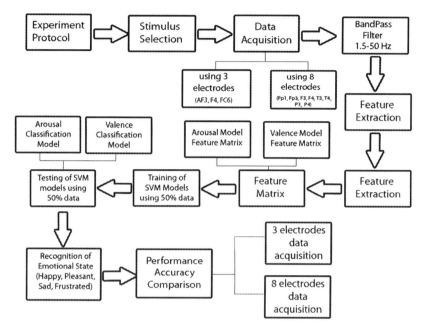

Fig. 3. Pipeline for EEG-based emotion recognition

Number of electrodes available were fewer and it has been proven in past research that using for example 12 pairs of electrodes can improve the results significantly [2].

References

1. Liu, Y., Sourina, O., Nguyen, M.K.: Real-time EEG-based Human Emotion Recognition and Visualization
2. Lin, Y.P., Wang, C.H., Wu, T.L., Jeng, S.K., Chen, J.H.: EEG-based emotion recognition in music listening: A comparison of schemes for multiclass support vector machine. In: ICASSP, IEEE International Conference on Acoustics, Speech and Signal Processing - Proceedings, Taipei, pp. 489–492 (2009)
3. Bos, D.O.: EEG-based emotion recognition (2006). http://hmi.ewi.utwente.nl/verslagen/capitaselecta/CS-Oude_Bos-Danny.pdf
4. Russell, J.A.: Affective space is bipolar. J. Pers. Soc. Psychol. **37**, 345–356 (1979)
5. Higuchi, T.: Approach to an irregular time series on the basis of the fractal theory
6. Liu, Y., Sourina, O., Nguyen, M.K.: Real-time EEG-based Human Emotion Recognition and Visualization
7. Lin, Y.P., Wang, C.H., Wu, T.L., Jeng, S.K., Chen, J.H.: EEG-based emotion recognition in music listening: A comparison of schemes for multiclass support vector machine. In: ICASSP, IEEE International Conference on Acoustics, Speech and Signal Processing - Proceedings, Taipei, pp. 489–492 (2009)

8. Bos, D.O.: EEG-based emotion recognition (2006). http://hmi.ewi.utwente.nl/verslagen/capitaselecta/CS-Oude_Bos-Danny.pdf
9. Russell, J.A.: Affective space is bipolar. J. Pers. Soc. Psychol. **37**, 345–356 (1979)
10. Higuchi, T.: Approach to an irregular time series on the basis of the fractal theory

Dynamic 3D Clustering of Spatio-Temporal Brain Data in the NeuCube Spiking Neural Network Architecture on a Case Study of fMRI Data

Maryam Gholami Doborjeh$^{(\boxtimes)}$ and Nikola Kasabov

Knowledge Engineering and Discovery Research Institute,
Auckland University of Technology, Auckland, New Zealand
{mgholami,nkasabov}@aut.ac.nz

Abstract. The paper presents a novel clustering method for dynamic Spatio-Temporal Brain Data (STBD) on the case study of functional Magnetic Resonance Image (fMRI). The method is based on NeuCube spiking neural network (SNN) architecture, where the spatio-temporal relationships between STBD streams are learned and simultaneously the clusters are created. The clusters are represented as groups of spiking neurons inside the NeuCube's spiking neural network cube (SNNc). The centroids of the clusters are predefined by spatial location of the brain data sources used as input variables. We illustrate the proposed clustering method on an fMRI case study STBD recorded during a cognitive task. A comparative analysis of the clusters across different mental activities can reveal new findings about the brain processes under study.

1 Introduction

The human brain is acting as a complex information processing machine [1,2] that processes data through communications between billions of neurons. In order to analyze STBD, suitable models are needed to trace such complex patterns and to understand processes that generate data. In this paper we use the NeuCube framework [3] as a rich model for modelling and understanding of STBD. We propose a new method for dynamic clustering of STBD in a NeuCube model and illustrate it on fMRI data.

2 The Proposed NeuCube-Based Spiking Neural Network Methodology for Learning, Visualization and Clustering of STBD

2.1 Spiking Neural Networks for Modelling STBD

Spiking neural networks model operation is based on neuron synaptic states that incorporate spiking time. SNNs have become popular computational methods for complex spatio/spectro temporal data analysis [3]. Their neuromorphic highly parallel hardware implementations are advancing very fast [4].

© Springer International Publishing Switzerland 2015
S. Arik et al. (Eds.): ICONIP 2015, Part IV, LNCS 9492, pp. 191–198, 2015.
DOI: 10.1007/978-3-319-26561-2_23

2.2   The NeuCube Architecture [3]

NeuCube is an evolving spiking neural network (eSNN) architecture for STBD learning, modelling, knowledge extraction, and for the analysis of the brain processes that generated the data [3]. The NeuCube architecture consists of several modules:

An Input encoding module for transforming spatio-temporal data into spike trains; A 3D SNN cube (SNNc) module for unsupervised learning of STBD; An output classification/regression module for supervised learning of data; Optimization module; Visualization and knowledge extraction module. Based on the NeuCube architecture we propose here a new clustering method schematically represented in Fig. 1.

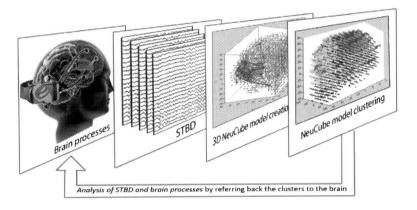

Fig. 1. The steps performed as part of the STBD clustering procedure

2.3   3D Dynamic Neuronal Clustering in a NeuCube SNN Model

The proposed STBD clustering method includes the following steps:

Step 1: STBD recording, i.e. for a given problem, relevant STBD is recorded.
Step 2: STBD encoding and mapping into a 3D SNNc as part of a NeuCube model.
Step 3: Unsupervised learning in the 3D SNNc and cluster evolution based on the spiking activity and connectivity in SNNc.
Step 4: Analysis of the connectivity of the trained 3D SNNc as dynamic spatio-temporal clusters of the STBD.
Step 5: Mapping SNNc clusters into brain regions.
Step 6: Functional analysis of the dynamic interaction between SNNc clusters to understand brain functional dynamics.

At Step 2, STBD is encoded into spike trains using an appropriate encoding algorithm such as Threshold-Based Representation method (TBR). The spike trains are then entered into a 3D SNNc which is created with a suitable size to map the brain template relevant to the data, such as Talairach [5], MNI [6], or voxel coordinates of individual brain data. Input neurons are allocated in the SNNc to enter input variable spike trains. The SNNc is initialized using a small world connectivity rule [7].

At Step 3, spike trains are learned in a SNNc using Spike-Timing Dependent Plasticity (STDP) learning rule [8]. During the learning procedure, spatio-temporal relationships within data are captured in the form of evolving connection weights and simultaneously neurons are clustered into homogeneous groups with respect to the neurons' activation similarity. Formation of the clusters is based on the intensity of spike communication within the SNNc as a similarity measure. The similarity measure is not only based on the spatial information of the data, but also on the temporal similarity as spike-time relationships. The results are dynamic 3D clusters containing the most evoked neurons by the corresponding cluster's centroid. The clusters represent spiking neurons with similar spiking activity at each time, reflecting dynamic spatio-temporal brain processes. The proposed clustering method differs significantly from the existing methods for clustering, such as evolving clustering (e.g. DENFIS [9] which method deals with static data and no temporal relationships between input vectors are learned) or self-organizing maps [10].

3 Application of the Proposed Method on a Benchmark fMRI STBD

To demonstrate our clustering method, the known STAR/PLUS fMRI data [11] is used to study how neurons in the SNNc are clustered based on the spatio-temporal similarity measure of the neurons' activation against different sentence polarities.

3.1 FMRI Data Acquisition Description

In STAR/PLUS fMRI data, the whole brain volume is recorded every 500 ms, while a cognitive task (pictures/sentences stimuli matching) performed by a subject. When the subject reads a sentence, the brain activity patterns are performed differently depending on the sentence polarity (affirmative vs. negative). In this experiment, fMRI STBD is divided into two time series corresponding to the two classes: Class 1, reading affirmative sentences; Class 2, reading negative sentences. In this study, the most activated voxels against sentence stimuli are selected using Signal-to-Noise Ratio (SNR) feature selection as reported in Table 1.

Table 1. The 20 more activated ROIs are presented in decreasing order of SNR values. In bracket is the number of voxels located in every brain region

Activated brain region (number of selected voxels)					
1	'LT' (3)	4	'LDLPFC' (6)	7	'RDLPFC'(1)
2	'LOPER' (3)	5	'RT'(2)	8	'RSGA' (1)
3	'LIPL'(1)	6	'LSGA'(1)	9	'RIT'(1)

3.2 FMRI Data Mapping, Learning and Visualization in a SNNc

The whole fMRI set of voxels is spatially mapped into the SNNc. The spatio-temporal patterns of the 20 pre-selected voxels are encoded to sequences of spikes using TBR and later transferred to the SNNc via 20 spatially allocated input neurons. During the NeuCube training, the post synaptic potential of each neuron at time t, PSPi(t), increases by every input spike received from all pre-synaptic neurons [8,12]. Once the PSPi (t) exceeds a firing threshold, neuron emits a spike. Based on the STDP learning rule, if neuron spikes first and then spikes, the connection weight between these neurons increases, otherwise it decreases. After the NeuCube unsupervised learning is completed, the spatio-temporal relationships between fMRI input streams are reflected on the created neuronal connections (Fig. 2). These connections are generated differently for each of the two stimuli revealing the fact that the subject is performing differently when processing affirmative versus negative sentences. Figure 2 shows that more and stronger neural connections are created in the left hemisphere (LDLPFC and LT) than in the right hemisphere (RDLPFC and RT) while reading negative sentences.

3.3 Dynamic Cluster Evolution in a NeuCube Model on the fMRI Case Study STBD

During the learning procedure, 3D neuronal clusters are evolved when new input fMRI vectors are entered and learned in the SNNc. Step-wise visualizations of the clusters evolution reveals differences of spike-time relationships between fMRI voxel activity patterns against sentence polarities. Figure 3 shows the process of cluster creation over time for 16 selected time points during unsupervised learning of fMRI data.

During the NeuCube learning, there are step-wise changes in the evolution of the clusters. When a SNNc was training with fMRI data of affirmative sentences, the first produced clusters corresponded to the RDLPFC brain region after the 3rd fMRI time frame was learned. In the case of negative sentences, the first created neuronal clusters related to RDLPFC and LOPER regions and were created after the 3rd fMRI data frame was learned. Figure 4 illustrates how the size of the cluster is changing in terms of the number of neurons belong to each cluster.

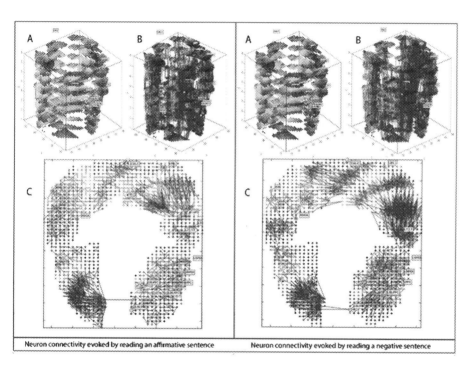

Fig. 2. Mapping of 5062 fMRI voxels to the SNNc and the spatio-temporal connections which evolved around 20 input voxels against affirmative /negative sentence presentation. Blue lines represent positive connection weights while red lines represent negative weights. The brighter the color of a neuron, the stronger its activity is with neighboring neurons. Thickness of the lines also identifies the neuron's enhanced connectivity. (a) 3D visualization of the initial connections before the NeuCube training process; (b) Connections after the NeuCube training process; (c) 2D visualization of the spatio-temporal connections after the NeuCube training (Color figure online)

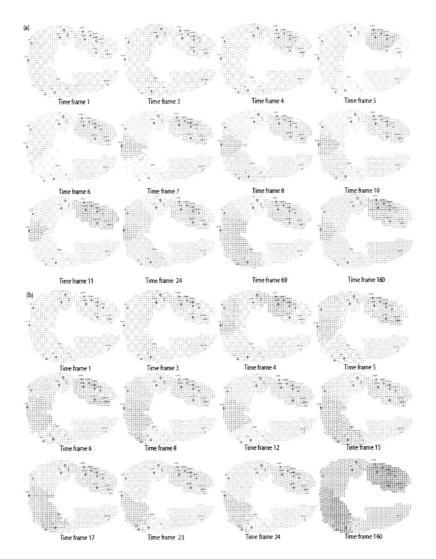

Fig. 3. Clusters' centroids are predefined and labeled by different colors. Before training the SNNc by fMRI data streams, the clusters had no members. They evolved with respect to the neurons spiking activities generated during the training process (a) Step-wise visualization of the dynamic neuronal cluster evolution corresponding to 20 voxels (clusters' centroids), while the subject was reading affirmative sentences. The first neuronal cluster is created at the 3rd time frame of the fMRI data during the unsupervised learning in a SNNc and is associated with the RDLPFC brain region; (b) Dynamic neuronal cluster evolution while the subject was reading negative sentences. The first neuronal clusters are created at the 3rd time frame of the learning procedure associated with the RDLPFC and LOPER brain regions (Color figure online)

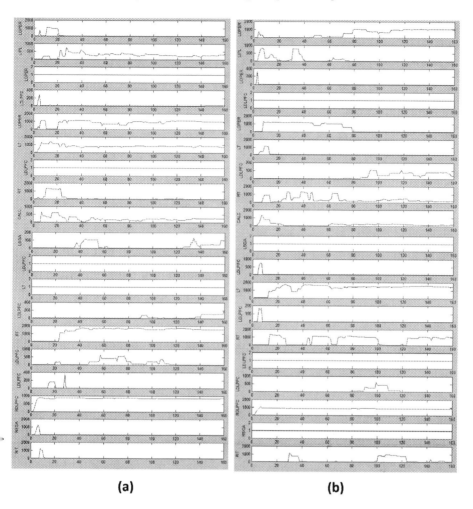

(a)   (b)

Fig. 4. Changings of the neuronal cluster size during their creation in the NeuCube SNNc while training on fMRI STBD against (a) affirmative versus (b) negative sentences

4   Conclusion

In this study, a generic SNN methodology was proposed as a novel method for 3D dynamic neuronal clustering of STBD. The method is based on the following scheme: Brain processes STBD 3D NeuCube model creation 3D NeuCube model clustering Analysis of STBD Analysis of spatio-temporal brain processes.

The method is illustrated on a benchmark fMRI STBD. The clustering procedure uses STDP learning rule as part of unsupervised learning in a 3D SNN NeuCube -based model. The proposed method for dynamic clustering of STBD is illustrated on fMRI data, but can be applied on EEG and other STBD to study dynamic functional processes in the brain during cognitive tasks.

Acknowledgment. The research is supported by the Knowledge Engineering and Discovery Research Institute of the Auckland University of Technology (www.kedri.aut.ac.nz). The Authors would like to acknowledge the following participants and researchers that have contributed to the realization of this study: Dr Enmei Tu, Lei Zhou, Joyce D'Mello.

References

1. Doborjeh, M.G., Capecci, E., Kasabov, N., Engineering, K.: Temporal brain data with a NeuCube evolving spiking neural network model, pp. 73–80 (2014)
2. Kasabov, N., Capecci, E.: Spiking neural network methodology for modelling, classification and understanding of EEG spatio-temporal data measuring cognitive processes. Inf. Sci. (Ny) **294**, 565–575 (2014)
3. Kasabov, N.K.: NeuCube: a spiking neural network architecture for mapping, learning and understanding of spatio-temporal brain data. Neural Netw. **52**, 62–76 (2014)
4. Indiveri, G., Linares-Barranco, B., Hamilton, T.J., van Schaik, A., Etienne-Cummings, R., Delbruck, T., Liu, S.C., Dudek, P., Hfliger, P., Renaud, S., Schemmel, J., Cauwenberghs, G., Arthur, J., Hynna, K., Folowosele, F., Saighi, S., Serrano-Gotarredona, T., Wijekoon, J., Wang, Y., Boahen, K.: Neuromorphic silicon neuron circuits. Front. Neurosci. **5**, 1–23 (2011)
5. Talairach, J., Tournoux, P.: Co-planar Stereotaxic Atlas of The Human Brain. 3-Dimensional Proportional System: An Approach to Cerebral Imaging. Thieme Medical, New York (1988)
6. Brett, M., Christoff, K., Cusack, R., Lancaster, J.: Using the talairach atlas with the MNI template. Neuroimage **13**, 85 (2001)
7. Tu, E., Kasabov, N., Othman, M., Li, Y., Worner, S., Yang, J., Jia, Z.: NeuCube (ST) for spatio-temporal data predictive modelling with a case study on ecological data (2014)
8. Song, S., Miller, K.D., Abbott, L.F.: Competitive Hebbian learning through spike-timing-dependent synaptic plasticity. Nat. Neurosci. **3**, 919–926 (2000)
9. Kasabov, N.K., Song, Q.: DENFIS: dynamic evolving neural-fuzzy inference system and its application for time-series prediction. IEEE Trans. Fuzzy Syst. **10**, 144–154 (2002)
10. Deboeck, G., Kohonen, T.: Visual Explorations in Finance: with Self-Organizing Maps. Springer Finance, xlv, 258. Springer, New York (1998)
11. StarPlus fMRI data. http://www.cs.cmu.edu/afs/cs.cmu.edu/project/theo-81/www/
12. Kasabov, N.: To spike or not to spike: a probabilistic spiking neuron model. Neural Netw. **23**, 16–19 (2010)

Vigilance Differentiation from EEG Complexity Attributes

Junhua Li[1], Indu Prasad[1], Justin Dauwels[2], Nitish V. Thakor[1,3],
and Hasan AI-Nashash[1,4(✉)]

[1] Singapore Institute for Neurotechnology (SINAPSE), Centre for Life Sciences,
National University of Singapore, Singapore, Singapore
juhalee@nus.edu.sg, {indu.iitbbs,sinapsedirector}@gmail.com
[2] School of Electrical and Electronic Engineering,
Nanyang Technological University, Singapore, Singapore
jdauwels@ntu.edu.sg
[3] Department of Biomedical Engineering, Johns Hopkins University, Baltimore, USA
[4] Department of Electrical Engineering, College of Engineering,
American University of Sharjah, Sharjah, UAE
hnashash@aus.edu

Abstract. Vigilance is an ability to maintain concentrated attention on a particular event or target stimulus. Monitoring tasks require certainly high vigilance to properly detect rare occurrence or accurately respond to stimulation. Changes in vigilance can be reflected by EEG signal, so vigilance levels can be classified based on features extracted from EEG. Up to now, power spectral density was commonly employed as features to differentiate between vigilance levels in majority of previous studies. To the best of our knowledge, multifractal attributes for vigilance differentiation have not been exploited, and their feasibility still need to be investigated. In this study, we first extracted multifractal attributes based on wavelet leaders, and then selected statistically significant distinct attributes for the following classification (two vigilance levels). According to the results, classification accuracy was improved with increase of time window used for feature extraction. When time window was increased to 50 s, an averaged accuracy of 91.67 % was achieved, and accuracies for all subjects were higher than 85 %. Our results suggest that multifractal attributes are promising for vigilance differentiation.

Keywords: Vigilance classification · Sustained attention · Complexity attribute · Self-similarity · Feature selection

1 Introduction

In our daily life, vigilance plays an important role on interactive activities. High vigilance facilitates to perform stressful and hard tasks, and to reduce the rate of occurrence of mistakes. In contrast, vigilance decrement could lead to serious consequences, such as traffic accidents. A few psychophysiological signals

S. Arik et al. (Eds.): ICONIP 2015, Part IV, LNCS 9492, pp. 199–206, 2015.
DOI: 10.1007/978-3-319-26561-2_24

have been utilized to estimate vigilance levels. EEG is the most common among them because it is directly recorded from the brain. In EEG studies, power spectral density is usually extracted to represent ongoing brain states [1–6]. For instance, J.N. Gu et al. extracted features of power spectral density and classified vigilance states by Gaussian mixture model [1]. Power spectral density in different bands was also investigated in [2–4,6]. Before vigilance classification, principle component analysis was usually used to reduce dimension of features [2,4,7]. In general, components corresponding to first several largest variances are selected and features are projected onto the space spanned by these components. In [7], the authors proposed to calculate correlation between features and vigilance index for feature selection. By this way, the retained components are more relevant to vigilance. All papers mentioned above utilized power spectral density as features to estimate vigilance level, but vigilance information may also be encoded in EEG by other representations. As indicated in [8], EEG signal recorded under motor imagery has multifractal characteristics. The multifractal attributes can be used to successfully decode different motor imageries. In this case, both power spectral density and multifractal attributes encode relevant information of motor imagery. In this paper, we explore multifractal attributes in EEG recordings and investigate feasibility of vigilance differentiation. To this end, a monotonous task was designed to induce vigilance decrement with time. Multifractal attributes were compared between high vigilance and low vigilance states, and were used to classify vigilance levels.

2   Subjects and Experimental Environment

Twelve healthy volunteers (age: $20 \sim 31$) participated in the study. All of them self-reported no history of mental illness, neurological illness, and they were not on medication. Their sights were normal or corrected-to-normal by wearing glasses. All subjects gave their written consent before conducting the experiment. The experiment was approved by the Institutional Review Board (IRB) of the National University of Singapore.

Subject was seated in a comfortable armchair with an appropriate height. A simulated factory scene (see Fig. 1) was presented in front of the subject by a 21-inch computer screen. The distance from the screen to the plane of the subject's eyes was 50 cm. The subject was instructed to detect a rarely occurring intruder (marked by red circle in the Fig. 1) from civilians who were randomly walking in the scene (marked by blue circles). The intruder sneaked from an arbitrarily appeared location to another location and disappeared on reaching that location. Subject needed to detect the intruder by pressing the key 'Q'. During the experiment, 62 channels EEG, 2 channels EOG, and 1 channel ECG were recorded as referenced to the linked mastoids. The impedance of all electrodes was kept below 10 $k\Omega$. In addition, eye tracker was also used to track eyes' behaviour, but the data analysis is beyond the scope of this paper.

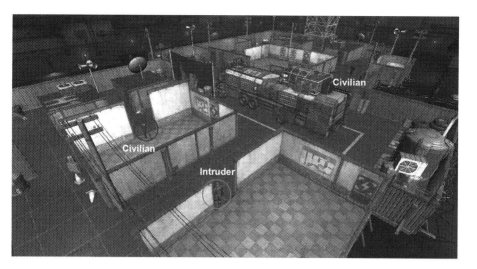

Fig. 1. Simulated factory scene. A few civilians are randomly walking in the scene (as indicated by blue circles). Intruder wearing military uniform infrequently appears at random location (marked by red circle) and then sneaks to another location (Color figure online).

3   Methodology

In this experiment, subjects performed saccades from one point to another point to detect the intruder. This results in obvious artefacts corrupting the EEG signals. At the first step, we should remove the effects of artefacts, such as EOG and EMG. Figure 2 shows the schematic of our methodology for vigilance classification. Least mean squares (LMS) was employed to remove EOG by subtracting weighted EOG signal from EEG signal [9]. Then, EEG signal was decomposed by canonical correlation analysis (CCA) to obtain components. Subsequently, EEG was projected onto a few maximally auto-correlated components to suppress the effect of EMG [10]. EEG signal plots at the upper right part of Fig. 2 show examples before and after artefacts removal.

After removing artefacts, we extracted multifractal attributes for each channel by the following procedure. Let $x(t)$ stand for time series of one channel in a sliding time window. $x(t)$ is first transformed into a time-scale representation (wavelet coefficients) by discrete wavelet transform (DWT). Because multifractal spectrum can be more accurately estimated based on wavelet leaders than based on the wavelet coefficients themselves according to the suggestion in [11,12], we adopted wavelet leaders in this study. Wavelet leader $L_x(j, k)$ is defined as the greatest value of adjacent wavelet coefficients, which includes wavelet coefficients in the time interval $[k - 1, k + 1]$ at scale 2^j and all smaller scales. Hölder exponent h is defined as the largest power that $x(t)$ is expanded into a polynomial at t_0 with error no more than the product of a constant and the power of $|t - t_0|$.

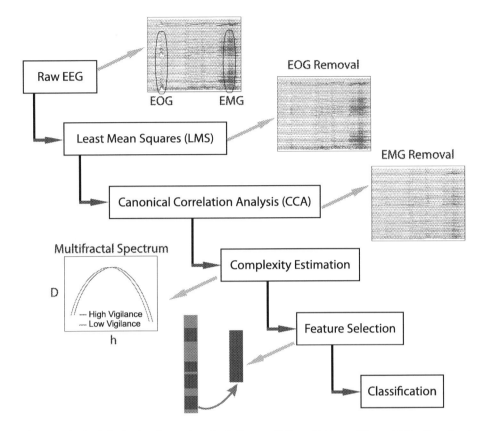

Fig. 2. Methodological schematic of vigilance differentiation. The middle flowchart shows each step to perform classification. The upper right subplots illustrate examples of artefacts removal. The bottom left subplot and diagram show multifractal spectrum and feature selection, respectively.

Under mild regularity condition, wavelet leaders exactly reproduce the Hölder exponent h of $x(t)$ at t_0 in the limit of fine scales ($2^j \to 0$) [12].

$$L_x(j, k) \leq C2^{jh}, \tag{1}$$

It means $h(t_0)$ is the supremum of all values h. The formula (1) shows that the wavelet leader structure functions $S^L(j, q)$ possess power law behaviour with respect to scales in the limit $2^j \to 0$ [12].

$$S^L(j, q) = \frac{1}{n_j} \sum_{k=1}^{n_j} L_x(j, k)^q = F_q 2^{j\zeta(q)}, \tag{2}$$

where n_j is the number of wavelet leaders $L_x(j, k)$ available at the scale 2^j, F_q is coefficient. $\zeta(q)$ is scaling exponents that indicate signal complexity. The higher order in $\zeta(q)$, the more complex the EEG signal is. It has been proven that the

Legendre transform of the scaling exponents $\zeta(q)$ provide an upper bound for the multifractal spectrum $D(h)$ [12],

$$D(h) \leq \min_{q \neq 0}(1 + qh - \zeta(q)). \tag{3}$$

For the most commonly used multifractal models [13], the inequality (3) becomes equality,

$$D(h) = \min_{q \neq 0}(1 + qh - \zeta(q)). \tag{4}$$

Hence, the multifractal spectrum $D(h)$ can be drawn with respect to the scaling exponents $\zeta(q)$ and Hölder exponents h. An example of multifractal spectrum is shown in the bottom left subplot in the Fig. 2. Two curves respectively represent multifractal spectra of high and low vigilance states. For further information about multifractal spectrum, readers could refer to [11–13]. In order to reduce the number of features, we did not use original multifractal spectrum as features, only three attributes are extracted from multifractal spectrum to be used as features. They are the location of maximum c_1, width c_2, and asymmetry c_3 of multifractal spectrum. The physical meaning of c_1 is self-similarity of signal over time. All attributes of all channels are assembled to form a feature vector

$$f = [c_{1,1}\ c_{1,2}\ c_{1,3}\ \cdots\ c_{n,1}\ c_{n,2}\ c_{n,3}], \tag{5}$$

where n is the number of channels.

The dimension of features is still high if attributes extracted from all channels are used, so we select a subset of features according to statistical evaluation (see the following pseudocode). Feature selection is performed for each individual. Then, selected features are normalized and fed into an SVM classifier with radial basis function (RBF) kernel [14]. Performance is evaluated by 4-fold cross-validation.

Feature Selection Procedure

```
Input: All Extracted Features
Output: Selected Features
   begin
    repeat
      channel = channel+1;
        repeat
          attribute = attribute+1;
          if difference is significant (p<0.05)
             keep current feature;
          else
             remove current feature;
          end
        until attribute = 3
    until channel = 62
   end
```

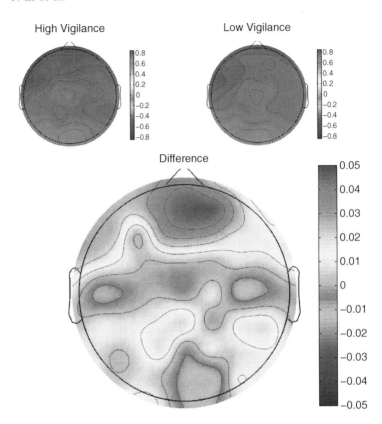

Fig. 3. Topographies at the upper row are averaged multifractal attributes (self-similarity) for the high and low vigilance states, respectively. Difference topography at the bottom is obtained by subtracting self-similarity values of low vigilance state from that of high vigilance state (Color figure online).

4   Results

Figure 3 shows the difference topography of multifractal attribute between high vigilance and low vigilance conditions. Areas in red color mean self-similarity value for the high vigilance condition is bigger than that of the low vigilance condition. The inverse situation is marked by blue color in the topography. The most dominant regions between high and low vigilances are located on the frontal and occipital cortices. A higher self-similarity is observed on the frontal cortex during high vigilance state. It becomes lower on the occipital cortex. Interestingly, the dominant spatial regions found in our study are matched with the findings with spectral power study, such as alpha band on occipital cortex is related to vigilance [15].

High vigilance and low vigilance portions are selected from the whole recording according to the protocol setting. Each portion is 12 min long, and is further partitioned into segments using specific time window length. Time window

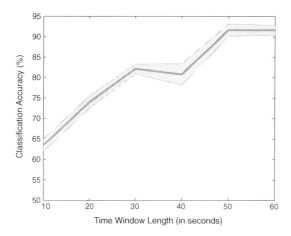

Fig. 4. The grand accuracies in different time window lengths. Horizontal axis represents window length while vertical axis represents classification accuracy. Bold green line stands for averaged accuracy across subjects. Standard error is shown as shaded region (Color figure online).

length is a crucial parameter in multifractal analysis. Different window lengths used for extracting multifractal attributes result in different classification accuracies. Therefore, we used different window lengths from 10 s to 60 s with an incremental step of 10 s. The results are shown in the Fig. 4. Grand accuracies averaged across all subjects are shown as bold green line. From the figure, we can see that classification accuracy is generally improved with increase in window length. The highest accuracy of 91.67 % is reached when window length is 50 s. Accuracies for every subject are higher than 85 %. Accuracies in 7 out of 12 subjects exceed 90 %. Variance over subjects are shown as shaded region. The largest inter-subject variance occurs at the window length of 40 s. The variances for the rest of window lengths are small.

4.1   Conclusion

In this study, we exploited a new representation of multifractal attributes for vigilance degree, rather than conventional representation of power spectral density. Three attributes derived from the location of maximum, width, asymmetry of multifractal spectrum are extracted as features. Then, the number of features was reduced by feature selection according to statistical evaluation. Classification accuracies demonstrate that multifractal attributes are feasible to differentiate vigilance levels. We further explored the spatial distribution of self-similarity attribute, and found that the most distinct difference between high and low vigilance levels is appeared on the frontal and occipital cortices.

Vigilance differentiation could be utilized to construct a passive brain computer interface (BCI) for monitoring the alertness of people. It could be integrated into an active BCI, such as motor-imagery wheelchair system [16], to assess mental state while people are manipulating by an active BCI.

Acknowledgments. This paper is supported by the Singapore Ministry of Defence, Singapore (Grant No. 9011103788).

References

1. Gu, J.N., Liu, H.J., Lu, H.T., Lu, B.L.: An integrated hierarchical gaussian mixture model to estimate vigilance level based on EEG recordings. In: Lu, B.L., Zhang, L., Kwok, J. (eds.) ICONIP 2011, Part I. LNCS, vol. 7062, pp. 380–387. Springer, Heidelberg (2011)
2. Li, W., He, Q.C., Fan, X.M., Fei, Z.M.: Evaluation of driver fatigue on two channels of EEG data. Neurosci. Lett. **506**, 235–239 (2012)
3. Trejo, L.J., Kubitz, K., Rosepal, R., Kochavi, R.L., Matthews, B.L., Montgomery, L.D.: EEG-based Estimation and Classification of Mental Fatigue Leonard, pp. 1–44 (2009)
4. Yu, Z.E., Kuo, C.C., Chou, C.H., Yen, C.T., Chang, F.: A machine learning approach to classify vigilance states in rats. Expert Syst. Appl. **38**, 10153–10160 (2011)
5. Li, J., Struzik, Z., Zhang, L., Cichocki, A.: Feature learning from incomplete EEG with denoising autoencoder. Neurocomputing **165**, 23–31 (2015)
6. Lin, C.T., Chuang, C.H., Huang, C.S., Tsai, S.F., Lu, S.W., Chen, Y.H., Ko, L.W.: Wireless and wearable EEG system for evaluating driver vigilance. IEEE Trans. Biomed. Circ. Syst. **8**, 165–176 (2014)
7. Shi, L.-C., Lu, B.-L.: EEG-based vigilance estimation using extreme learning machines. Neurocomputing **102**, 135–143 (2012)
8. Li, J., Cichocki, A.: Deep learning of multifractal attributes from motor imagery induced EEG. In: Loo, C.K., Yap, K.S., Wong, K.W., Teoh, A., Huang, K. (eds.) ICONIP 2014, Part I. LNCS, vol. 8834, pp. 503–510. Springer, Heidelberg (2014)
9. He, P., Wilson, G., Russell, C.: Removal of ocular artifacts from electroencephalogram by adaptive filtering. Med. Biol. Eng. Comput. **42**, 407–412 (2004)
10. De Clercq, W., Vergult, A., Vanrumste, B., Van Paesschen, W., Van Huffel, S.: Canonical correlation analysis applied to remove muscle artifacts from the electroencephalogram. IEEE Trans. Biomed. Eng. **53**, 2583–2587 (2006)
11. Jaffard, S., Lashermes, B., Abry, P.: Wavelet leaders in multifractal analysis. In: Wavelet Analysis and Applications, pp. 201–246 (2007)
12. Wendt, H., Abry, P.: Multifractality tests using bootstrapped wavelet leaders. IEEE Trans. Sig. Process. **55**, 4811–4820 (2007)
13. Wendt, H., Abry, P.: Bootstrap for multifractal analysis. In: Proceedings of 2006 IEEE International Conference Acoustic Speech Signal Process, vol. 3, pp. 38–48 (2006)
14. Vapnik, V.: The Nature of Statistical Learning Theory. Springer, New York (1995)
15. Dockree, P.M., Kelly, S.P., Foxe, J.J., Reilly, R.B., Robertson, I.H.: Optimal sustained attention is linked to the spectral content of background EEG activity: greater ongoing tonic alpha (10 Hz) power supports successful phasic goal activation. Euro. J. Neurosci. **25**, 900–907 (2007)
16. Li, J., Liang, J., Zhao, Q., Li, J., Hong, K., Zhang, L.: Design of assistive wheelchair system directly steered by human thoughts. Int. J. Neural Syst. **23**, 1350013 (2013)

Robust Discriminative Nonnegative Patch Alignment for Occluded Face Recognition

Weihua Ou[1]([✉]), Gai Li[2], Shujian Yu[3], Gang Xie[1], Fujia Ren[1],
and Yuanyan Tang[4]

[1] School of Mathematics and Computer Science,
Guizhou Normal University, Guiyang, China
ouweihua_biac@hust.edu.cn
[2] Department of Electronics and Information Engineering,
Shunde Polytechnic, Foshan, China
[3] Department of Electrical and Computer Engineering,
University of Florida, Gainesville, USA
yusjlcy9011@ufl.edu
[4] Faculty of Science and Technology, University of Macau, Macau, China
yytang@umac.mo

Abstract. Face occlusion is one of the most challenging problems for robust face recognition. Nonnegative matrix factorization (NMF) has been widely used in local feature extraction for computer vision. However, standard NMF is not robust to occlusion. In this paper, we propose a robust discriminative representation learning method under nonnegative patch alignment, which can take account of the geometric structure and discriminative information simultaneously. Specifically, we utilize linear reconstruction coefficients to characterize local geometric structure and maximize the pairwise fisher distance to improve the separability of different classes. The reconstruction errors are measured with weighted distance, and the weights for each pixel are learned adaptively with our proposed update rule. Experimental results on two benchmark datasets demonstrate the learned representation is more discriminative and robust than most of the existing methods in occluded face recognition.

Keywords: Face recognition · Nonnegative patch alignment · Geometric structure · Discriminative information · Iteratively reweighted scheme

1 Introduction

Face recognition under partial occlusions is commonly encountered in real applications, and is also viewed as the most challenging topic in face recognition [1]. Ordinary occlusions include different accessories (e.g., sunglasses, scarf. hats, etc.), or even a face occluded with hands or other faces. Occlusions can severely degrade the performance of most sophisticated face recognition systems. Robustness to these occlusions is therefore essential to practical face recognition.

© Springer International Publishing Switzerland 2015
S. Arik et al. (Eds.): ICONIP 2015, Part IV, LNCS 9492, pp. 207–215, 2015.
DOI: 10.1007/978-3-319-26561-2_25

An effective solution is to find the robust low-dimensional representation which is insensitive to occlusions.

Holistic approaches, such as linear discriminative analysis (LDA) [2] and principal component analysis (PCA) [3], can achieve satisfactory performance for face recognition in a controlled environment. However, the extracted global features make them sensitive to occlusions. Obviously, local features are less likely to be corrupted with occlusions. A lot of local feature-based methods have been proposed to solve the problem of occlusions [4–6]. Among them, nonnegative matrix factorization (NMF) [7] is one of the most emerging and promising areas of research in this direction, which is consistent with the psychological intuition of combing parts to form a whole. Learning parts-based representation based on NMF provides a new way for robust face recognition under occlusions, and many related methods have been proposed in this direction. They can be categorized into two main classes. The first class aims at improving the robustness of NMF. For example, $\ell_{2,1}$-norm [8], ℓ_1-norm [9], earth mover's distance metric [10] and correntropy induced metric (CIM-NMF) [11] are employed as the error function, respectively. With respect to these methods, although CIM-NMF proposed in [11] can achieve the best performance, the local geometric structure is not considered in their model. The second class pays attention to the geometric structure and supervised information. For example, Cai *et al.* [12] proposed the graph-regularized NMF (GNMF), in which the geometric structure is encoded via k-NN graph. But it is difficult to find the suitable graph and associated parameters. Yang *et al.* [13] proposed non-negative embedding (NGE) within the graph embedding framework. NGE preserves the favorite similarities and unfavored similarities via intrinsic graph and penalty graph. However, the graph in NGE becomes unreliable for the noisy data and unreliable graph will result in indistinctive local structure. In the framework of patch alignment, Guan *et al.* [14] proposed a non-negative patch alignment (NPA). NPA shows that intrinsical differences for various NMFs lie on the patches that they build. It provides systemical framework to understand different nonnegative matrix factorization algorithms and presents a new way to develop a new one. However, NPA still assumes that the approximation errors follow Gaussian distribution.

In this paper, we propose a robust discriminative nonnegative patch alignment (RD-NPA) for occluded face recognition, which takes into account geometric structure and supervised information simultaneously. Specifically, for each sample, we obtain the locally linear representation within its k near neighbors, which is preserved in the low-dimensional space. To improve the separability of different classes, we maximize the margin of different classes based on weighted pairwise fisher criteria. We utilize weighted distance to measure the approximation error, which adaptively learns weight for each pixel. We then propose an iteratively reweighted update scheme to solve the objective function and obtain a simple multiplicative rule. Experimental results on two benchmark datasets demonstrate the learned low-dimensional representation is more robust to occlusion, large magnitude noises, compared with most of the existing methods.

The rest of paper is organized as follows. We present the robust discriminative nonnegative patch alignment (RD-NPA) in Sect. 2, followed by the algorithm implementation details in Sect. 3. The experiments are conducted in Sect. 4 prior to conclusions in Sect. 5.

2 Robust Discriminative Nonnegative Patch Alignment (RD-NPA)

Nonnegative patch alignment framework [14] is a general dimensionality reduction method, which contains part optimization and whole alignment. In this framework, different part optimization can be designed according to specific objective. In this section, we build part optimization in Sect. 2.1 and whole alignment in Sect. 2.2, respectively. After that, we present the objective function for RD-NPA in Sect. 2.3.

2.1 Part Optimization

Given a sample $x_i \in X = [x_1, x_2, \cdots, x_N] \in \mathbb{R}^{m \times N}$, we group the other samples into two classes based on the label of x_i: 1) samples in the same class as x_i, denoted by X^s; and 2) samples in classes which are different from x_i, denoted by X^d. We use x_i and its k_1 nearest neighbors in X^s to form the with-in class patch $X_i^w = [x_i, x_{i^1}, x_{i^2}, \cdots, x_{i^{k_1}}]$, and then use x_i and its k_2 nearest neighbors in X^d to form the between-class patch $X_i^b = [x_i, x_{i_1}, x_{i_2}, \cdots, x_{i_{k_2}}]$. The corresponding low-dimensional representation of X_i^w and X_i^b are denoted by $H_i^w = [h_i, h_{i^1}, h_{i^2}, \cdots, h_{i^{k_1}}]$ and $H_i^b = [h_i, h_{i_1}, h_{i_2}, \cdots, h_{i_{k_2}}]$, respectively.

Based on the locally linear assumption [15], x_i can be linearly represented by its neighbors X_i^w, i.e., $x_i \approx s_{i_0} x_i + s_{i_1} x_{i^1} + s_{i_2} x_{i^2} + \cdots + s_{i_{k_1}} x_{i^{k_1}}$, where $s_{i_0} = 0$. The reconstruction coefficients $s_i = [s_{i_1}, s_{i_2}, \cdots, s_{i_{k_1}}]$ can be obtained by minimizing the following objective function in real application.

$$s_i^w = arg \min_{s_i} \left\| x_i - \sum_{n=1}^{k_1} s_{i_n} x_{i^n} \right\|_2^2 \tag{1}$$

We expect that such linear reconstruction relationship s_i^w can be preserved in low-dimensional space, i.e., $\min \|h_i - H_i^w s_i^w\|_2^2$. Thus, the part optimization on with-in class patch X_i^w can be formulated as, $\min_{H_i^w} \text{Tr} \left(H_i^w L_i^w H_i^{wT} \right)$, where $L_i^w = cc^T$, $c^T = \left[1, -s_i^{k_1} \right] \in \mathbb{R}^{k_1+1}$, $s_i^{k_1} = [s_{i^1}^w, \cdots, s_{i^{k_1}}^w] \in \mathbb{R}^{k_1}$.

For the between-class patch X_i^b, we expect that if x_i and the neighbors x_{i_j} are "close", then the corresponding distance between h_i and h_{i_j} are as large as possible. According to the weighted pairwise fisher criteria [16], we formulate the part optimization on patch X_i^b as follows,

$$\max \sum_{j=1}^{k_2} \omega_{i,i_j} \|h_i - h_{i_j}\|_2^2 \tag{2}$$

where $\omega_{i,i_j} = 1/2(d_{i,i_j})^2 erf(d_{i,i_j}/2\sqrt{2})$, $d_{i,i_j} = \|\boldsymbol{x}_i - \boldsymbol{x}_{i_j}\|_2$ is the distance between $\boldsymbol{x}_i$ and $\boldsymbol{x}_{i_j}$, and $erf(x) = \frac{2}{\sqrt{\pi}} \int_0^x \exp(-t^2)dt$. With simple algebraic reformulation, the part optimization can be represented as

$$\max_{H_i^b} \mathrm{Tr}\left(H_i^b L_i^b H_i^{b^T}\right) \tag{3}$$

where $L_i^b = \begin{bmatrix} \boldsymbol{e}_{k_2} \\ -I_{k_2} \end{bmatrix} \mathrm{diag}(\boldsymbol{\omega}_i) \begin{bmatrix} \boldsymbol{e}_{k_2} \\ -I_{k_2} \end{bmatrix}^T$, $\boldsymbol{e}_{k_2} = [1, 1, \cdots, 1] \in \mathbb{R}^{k_2}$ is a row vector, and I_{k_2} is a $k_2 \times k_2$ identity matrix, $\boldsymbol{\omega}_i = \left[\omega_{i,i_1}, \cdots, \omega_{i,i_{k_2}}\right]^T$.

2.2   Whole Alignment

According to the whole alignment strategy [14], we obtain the following two objective functions

$$\min_H \mathrm{Tr}\left(HL^w H^T\right), \max_H \mathrm{Tr}\left(HL^b H^T\right) \tag{4}$$

where $L^w = S_i^w L_i^w S_i^{w^T}$ and $L^b = S_i^b L_i^b S_i^{b^T}$ are the alignment matrices of within-class and between-class, respectively. $S_i^w \in \mathbb{R}^{N \times (k_1+1)}$ and $S_i^b \in \mathbb{R}^{N \times (k_2+1)}$ are the selection matrices for the with-in class patch and the between-class patch, which are defined below:

$$(S_i^w)_{pq} = \begin{cases} 1, & \text{if } p = F_i^w(q) \\ 0, & \text{otherwise} \end{cases}, (S_i^b)_{jk} = \begin{cases} 1, & \text{if } j = F_i^b(k) \\ 0, & \text{otherwise} \end{cases}$$

where $F_i^w = [i, i^1, \cdots, i^{k_1}]$ and $F_i^b = [i, i_1, \cdots, i_{k_2}]$ are the set of indices on with-class patch X_i^w and between-class patch X_i^b, respectively.

As suggested in [14], we obtain the whole alignment, $\min_H \mathrm{Tr}\left(HLH^T\right)$, where $L = (L^b)^{-\frac{1}{2}} L^w \left((L^b)^{-\frac{1}{2}}\right)^T$.

2.3   Objective Function of RD-NPA

According to the discussion in [17], the representation fidelity measured by l_2-norm or l_1-norm of the reconstruction error is not suitable for the occlusions. Therefore, we utilize weighted distance to measure the representation fidelity and obtain the objective function of RD-NPA as follows:

$$\min_{W \geq 0, H \geq 0} \left\{\|X - WH\|_P^2 + \lambda \mathrm{Tr}\left(HLH^T\right)\right\}, \tag{5}$$

where $\|X - WH\|_P^2 = \sum_{i,j} P_{ij}\left(X_{ij} - (WH)_{ij}\right)^2$, and $P_{ij} \in [0, 1]$ is the weight for pixel.

3   Algorithm for Robust Discriminative Nonnegative Patch Alignment

For problem (5), we can solve it by recursively optimizing one with the others fixed.

Optimize P for given W and H: Given W and H, problem (5) can be solved separately with respect to P_{ij}. Motivated by [17], for $E_{ij} = X_{ij} - (WH)_{ij}$, the estimation of P_{ij} is given by: $P_{ij} = g_{\mu,\delta}(X_{ij} - (WH)_{ij})$, where $g_{\mu,\delta}(e) = 1/(1 + \exp(\mu(e^2) - \mu\delta))$.

Optimize W for given H and P: Given H and P, the problem (5) can be solved as follows by optimizing each row of W separately,

$$\mathcal{L}(W,\Theta) = \sum_{i=1}^{m} (X_{i,*} - W_{i,*}H) A_i (X_{i,*} - W_{i,*}H)^T + \mathrm{Tr}(\Theta^T W),$$

where $A_i = diag(P_{i,*}) \in \mathbb{R}^{N \times N}$, $\Theta = [\theta_{ik}] \in \mathbb{R}^{m \times r}$ is the Lagrange multipliers for the non-negative constraints $W \geq 0$. Setting the partial derivatives of $\frac{\partial \mathcal{L}(W,\Theta)}{\partial W_{ik}}$ to zero and utilizing the KKT conditions $\theta_{ik} W_{ik} = 0$, we can get following equation for W_{ik},

$$\left(-2 \left(X_{i,*} A_i H^T\right)_k + 2 \left(W_{i,*} H A_i H^T\right)_k + \theta_{ik}\right) W_{ik} = 0.$$

This equation leads to the update rule below for W_{ik}:

$$W_{ik} = W_{ik} \frac{\left((X \odot P) H^T\right)_{ik}}{\left(((WH) \odot P) H^T\right)_{ik}}. \tag{6}$$

Optimize H for given W and P: In this case, the problem (5) is equivalent to minimize following objective function:

$$\mathcal{L}(H,\Psi) = \sum_{j=1}^{n} \left\{(X_{*,j} - WH_{*,j})^T B_j (X_{*,j} - WH_{*,j})\right\} + \mathrm{Tr}(\Psi^T H + \lambda H L H^T)$$

where $B_j = diag(P_{*,j}) \in \mathbb{R}^{m \times m}$, $\Psi = [\psi_{kj}] \in \mathbb{R}^{r \times N}$ is the Lagrange multipliers for the non-negative constraints $H \geq 0$. Setting the partial derivatives of $\frac{\partial \mathcal{L}(H,\Psi)}{\partial H_{kj}}$ to zero and utilizing the KKT conditions $\psi_{kj} H_{kj} = 0$, we can get following equation for H_{kj},

$$\left(-2 \left(W^T B_j X_{*,j}\right)_k + 2 \left(W^T B_j W H_{*,j}\right)_k H_{kj}\right) + (\psi_{kj} + 2\lambda(HL)_{kj}) H_{kj} = 0 \tag{7}$$

By separating L into two parts, i.e., $L = L^+ - L^-$, $L_{ij}^+ = (|L_{ij}| + L_{ij})/2$, $L_{ij}^- = (|L_{ij}| - L_{ij})/2$, and with some simple calculus, Eq. (7) leads to the update rule for H_{kj}:

$$H_{kj} = H_{kj} \frac{\left(W^T(X \odot P) + \lambda H L^-\right)_{kj}}{\left(W^T(WH \odot P) + \lambda H L^+\right)_{kj}}. \tag{8}$$

4   Experiments

In this section, we conduct experiments to demonstrate the effectiveness and robustness of the proposed method on CMU-PIE and ORL datasets. The following representative algorithms are re-implemented for comparison purpose: PCA [3], NMF [7], LDA [2], GNMF [12], $L21$-NMF [18], CIM-NMF [11], and NDLA [14].

4.1   Data and Parameter Setting

For CMU-PIE, 42 images of resolution 32×32 at pose 27 with different illumination conditions are selected for each person. For ORL dataset, all the 400 images are selected. 20 % of the images from each dataset are randomly selected and are corrupted with salt & pepper noise or occluded by white block. For salt & pepper noise, the noise level is 10 %. For the occlusion, the block size is 10×10 pixels and the position is random.

The subspace dimension r ranges from $\{30, 60, 90, 120, 150\}$. For PCA and LDA, the nearest neighbor rule was used for classification. For the other methods, we adopt the algorithm proposed in [10] to test algorithms performance. We set $k_1 = 5, k_2 = 20, \lambda = 0.0001$ for RD-NPA and NDLA. The μ and δ are set according to [17]. For GNMF, the neighbor size k is set to 5 and λ is set to 0.001 according to [12]. For NDLA, we utilize multiplicative updating rule by separating the whole alignment matrix L into two parts according to [14].

4.2   Experimental Results

As shown in Table 1, all the NMF-based methods can achieve better performance with the increment of subspace dimension r. On the CMU-PIE dataset, our proposed RD-NPA can always get the best results, and the best recognition rate is 99.12 % for the salt & pepper noises. On the contrary, the best recognition rate achieved with the well-known PCA is lower than 70 %. According to Table 2, our proposed method can effectively improve the performance of robust face recognition in the presence of occlusion on the ORL dataset. Both RD-NPA and CIM-NMF can achieve satisfactory performance, while PCA failed in this dataset.

Figure 1 shows reconstruction images selected from the test set of CMU-PIE with $r = 150$. For NMF and NDLA, we can clearly observe the occlusion block in the reconstruction images, as marked by red box in Fig. 1(a) and (b). Thus, for NMF and NDLA, both the subspace and the associated occlusions are learnt, which will seriously affect the representation ability of the associated basis. However, it can be seen that the faces reconstructed by RD-NPA are more reliable compared with the other methods.

Learning parts-based representation is the basic motivation for NMF. Figure 2 shows the associated basis vectors learned by different methods. It is clear to see that the basis vectors learned by RD-NPA are much sparser and clearer than those learned by other methods. Many basis vectors learned by

Table 1. Recognition rates with increase of subspace dimension r on CMU-PIE dataset.

Method	Salt & pepper (%)						Occlusion (%)					
	30	60	90	120	150	Avg	30	60	90	120	150	Avg
PCA	40.19	51.90	58.24	59.50	64.87	54.94	47.29	58.58	65.43	67.68	69.14	61.62
LDA	76.57	80.42	–	–	–	–	75.99	85.31	–	–	–	–
NMF	82.61	87.19	89.63	92.25	92.30	88.80	73.24	80.29	82.09	83.61	83.93	80.63
L21-NMF	87.03	92.11	93.77	94.08	94.19	92.24	73.90	80.69	82.21	82.95	83.66	80.68
GNMF	82.78	87.53	90.16	92.31	92.42	89.04	72.71	79.89	81.85	83.61	83.98	80.41
CIM-NMF	79.00	88.32	91.19	92.83	93.87	89.04	77.94	87.58	90.49	91.98	93.27	88.25
NDLA	84.96	90.35	91.71	93.88	94.72	91.12	76.18	85.22	86.40	87.80	89.70	85.06
RD-NPA	**91.25**	**95.44**	**98.35**	**98.80**	**99.12**	**96.60**	**89.21**	**95.16**	**96.65**	**97.81**	**98.46**	**95.46**

Table 2. Recognition rates with increase of subspace dimension r on ORL dataset.

Method	Salt & pepper (%)						Occlusion (%)					
	30	60	90	120	150	Avg.	30	60	90	120	150	Avg.
PCA	19.58	18.33	25.83	25.83	20.00	21.91	25.42	26.67	25.42	21.67	22.08	24.25
LDA	64.17	–	–	–	–	–	55.42	–	–	–	–	–
NMF	83.75	82.50	83.75	86.25	83.75	84.00	73.33	75.00	71.57	77.18	71.67	73.75
L21-NMF	81.67	85.83	83.75	85.00	84.17	84.08	71.87	74.38	77.08	72.50	78.33	74.83
GNMF	82.78	87.53	90.16	92.31	92.42	89.05	70.42	77.08	76.67	77.50	74.58	75.25
CIM-NMF	85.42	84.17	84.58	83.75	82.08	84.00	77.94	87.58	90.49	91.98	93.27	88.25
NDLA	72.92	79.58	77.50	76.67	80.42	77.42	74.58	73.33	74.17	77.92	77.50	75.50
RD-NPA	**90.83**	**92.75**	**93.75**	**93.92**	**93.50**	**93.15**	**91.25**	**93.75**	**93.92**	**93.75**	**95.00**	**93.53**

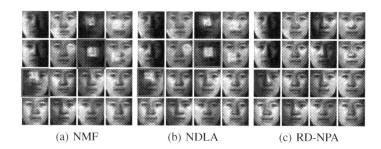

(a) NMF   (b) NDLA   (c) RD-NPA

Fig. 1. Reconstruction images. (a)–(c) are reconstructed by NMF, NDLA, and RD-NPA, respectively.

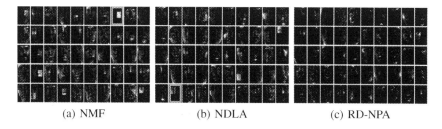

(a) NMF   (b) NDLA   (c) RD-NPA

Fig. 2. The learned basis vectors on the CMU-PIE dataset with occlusion size 10×10.

NMF and NDLA are partially occluded, as shown in Fig. 2(a) and (b) marked by red box. This suggests that RD-NPA can learn a better parts-based and compact representation even in the presence of occlusions.

5   Conclusion

In this paper, we propose a robust discriminative nonnegative patch alignment (RD-NPA) for face recognition under occlusions. It incorporates the local geometric structure and employs discriminative information simultaneously. We propose locally linear embedding preservation to characterize local geometric structure and maximize the weighted pairwise fisher distance of different classes to improve the separability of different classes. The weights are learned adaptively via iteratively reweighted scheme. Experimental results demonstrate that the learned representation is more robust to large magnitude noises and occlusion than most of the existing methods.

Acknowledgements. This work is supported in part by the National Natural Science Foundation of China (No.61402122), the 2014 Ph.D. Recruitment Program of Guizhou Normal University and the Outstanding Innovation Talents of Science and Technology Award Scheme of education department in Guizhou Province (Qianjiao KY word [2015]487).

References

1. Ekenel, H.K., Stiefelhagen, R.: Why Is facial occlusion a challenging problem? In: Tistarelli, M., Nixon, M.S. (eds.) ICB 2009. LNCS, vol. 5558, pp. 299–308. Springer, Heidelberg (2009)
2. Belhumeur, P.N., Hespanha, J.P., Kriegman, D.: Eigenfaces vs. fisherfaces: recognition using class specific linear projection. TPAMI **19**(7), 711–720 (1997)
3. Turk, M., Pentland, A.: Face recognition using eigenfaces. In: CVPR, pp. 586–591. IEEE (1991)
4. Tan, X., Chen, S., Zhou, Z.H., Zhang, F.: Recognizing partially occluded, expression variant faces from single training image per person with som and soft k-nn ensemble. TNN **16**(4), 875–886 (2005)
5. Li, S.Z., Hou, X.W., Zhang, H.J., Cheng, Q.S.: Learning spatially localized, parts-based representation. In: CVPR, pp. 1–6. IEEE (2001)
6. Jia, H., Martinez, A.M.: Support vector machines in face recognition with occlusions. In: CVPR, pp. 136–141. IEEE (2009)
7. Lee, D.D., Seung, H.S.: Learning the parts of objects by non-negative matrix factorization. Nat. **401**(6755), 788–791 (1999)
8. Huang, J., Nie, F., Huang, H., Ding, C.: Robust manifold nonnegative matrix factorization. ACM TKDD **8**(3), 11 (2014)
9. Guan, N., Tao, D., Luo, Z., Shawe-Taylor, J.: Mahnmf: manhattan non-negative matrix factorization. Mach. Learn. **1**(5), 11–43 (2012)
10. Sandler, R., Lindenbaum, M.: Nonnegative matrix factorization with earth mover's distance metric for image analysis. TPAMI **33**(8), 1590–1602 (2011)

11. Du, L., Li, X., Shen, Y.: Robust nonnegative matrix factorization via half-quadratic minimization. In: ICDM, pp. 201–210. IEEE (2012)
12. Cai, D., He, X., Han, J., Huang, T.: Graph regularized nonnegative matrix factorization for data representation. TPAMI **33**(8), 1548–1560 (2011)
13. Yang, J., Yang, S., Fu, Y., Li, X., Huang, T.: Non-negative graph embedding. In: CVPR, pp. 1–8. IEEE (2008)
14. Guan, N., Tao, D., Luo, Z., Yuan, B.: Non-negative patch alignment framework. TNN **22**(8), 1218–1230 (2011)
15. Roweis, S.T., Saul, L.K.: Nonlinear dimensionality reduction by locally linear embedding. Sci. **290**(5500), 2323–2326 (2000)
16. Loog, M., Duin, R.P.W., Haeb-Umbach, R.: Multiclass linear dimension reduction by weighted pairwise fisher criteria. TPAMI **23**(7), 762–766 (2001)
17. Yang, M., Zhang, L., Yang, J., Zhang, D.: Regularized robust coding for face recognition. TIP **22**(5), 1753–1766 (2013)
18. Kong, D., Ding, C., Huang, H.: Robust nonnegative matrix factorization using l21-norm. In: Proceedings of the 20th International Conference on Information and Knowledge Management, pp. 673–682. ACM (2011)

Single-Image Expression Invariant Face Recognition Based on Sparse Representation

Ya Su[✉] and Mengyao Wang

University of Science and Technology Beijing, Beijing, China
suyacv@163.com

Abstract. Face recognition under expression variation has been paid few attentions and remains a difficult problem. This is because the non-linear shape variation makes it infeasible to match two images linearly. This paper believes that expression variations affects the face recognition in three aspects, i.e., alignment error, shape change, and occlusion. Based on this observation, this paper solves this problem using a shape-constrained sparse representation (SSR) framework. The proposed method presumes that expression variations have dramatic effect on shape, along with minor effect on texture such as occlusion. It has two contributions. First, SSR introduces a shape-constrained texture matching (STM) algorithm to solve the alignment error and the shape changes. This strategy is able to eliminate shape changes as long as the two objects have similar textures. This is different from state-of-the-art algorithms which directly match two faces using learnt measures beforehand. Second, SSR matches the two aligned images based on the sparse representation theory. As a result, an image can be sparsely represented by the image with the same identity, even if the texture has been partly occluded. Extensive experiments show that SSR obtains the state-of-the-art performance robustly.

Keywords: Sparse representation · Expression variation · Face recognition

1 Introduction

Although been presented more than ten years, face recognition across different expressions has been paid relatively few attentions. Recent researches on the problem can be classified into two categories [9,12]. The first category is the discriminant analysis method. These methods perform face recognition by a discriminant model. Wang et al. [12] modeled this problem by linear discriminant analysis. Mohammadzade and Hatzinakos [9] improved the performances by using new expression images synthesized from gallery subjects for training.

The other category of methods regard the texture of an individual invariant to expression variation. Thus, it is feasible to perform recognition by transforming the probe image and the gallery image into the same expression. For example, Li et al. [7] separated texture and shape information and projecting them into

© Springer International Publishing Switzerland 2015
S. Arik et al. (Eds.): ICONIP 2015, Part IV, LNCS 9492, pp. 216–223, 2015.
DOI: 10.1007/978-3-319-26561-2_26

separate PCA spaces. Lee and Kim [6] proposed to transform the probe image with an arbitrary expression to its corresponding neutral expression face.

Although much research has been conducted in this field, facial recognition under expression variation remains to be a difficulty. To solve the problem, following obstacles should be addressed.

- How to deal with **alignment error**? Since different expressions produce difference appearances which is the most important evidence of landmark definition, it is not easy to define the same landmarks.
- **Shape Change.** Along with expression varies, shapes of the same person changes a lot. As a result, difference between two pictures of the same person is larger than that of distinct person. An following question is how to align an image that it can be recognized using its expressionless version.
- Given an image with aligned shape, Euclidean distance does not work when matching to another image. It is caused by changes of textures along with shape variation, such as eyes. This difficulty can be regarded as texture disappearance or **occlusion**.

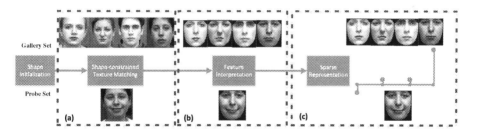

Fig. 1. Flow chart of the framework

This paper addresses the expression invariant face recognition problem based on a shape-constrained sparse representation (SSR) framework. It has following two contributions. First, to overcome the shape change caused by expression, a shape-constrained texture matching algorithm (STM) is proposed. It deals with both alignment error and shape change simultaneously. Second, the sparse representation technique is introduced to solve the occlusion problem.

2 Shape-Constrained Sparse Representation

It has been investigated that sparse representation is powerful in face recognition problem (SRC) when faces are well aligned [13]. Then, Wagner et al. [11] proposed an extension to SRC to eliminate the image misalignment by introducing an affine transformation.

Motivated by the success of SRC, we proposed an expression invariant face recognition framework based on sparse representation. The proposed framework is illustrated in Fig. 1.

2.1   Shape-Constrained Texture Matching

Suppose that T is a probe image subject to an unknown expression and I denotes a gallery image with the same identity with T. STM assumes that I and T has the similar texture but different shape.

Formulation of STM. We formulate STM as an image matching problem. Suppose we have a shape model and two images, a template image T with given shape and an input image I. The goal of STM is to find the transformation of I such that the shape-transformation difference between T and $I(W(x,p))$ is minimized, i.e.,

$$\min_p \sum_x \|I(W(x,p)) - T(x)\|_2, \tag{1}$$

where W denotes the transformation of I with parameters p, and $\|.\|_2$ means $L2$ norm. The summation is over the mesh on T.

It has been discussed that either $L1$ norm or $L2$ norm is appropriate for image matching [11]. They have the conclusion that if there is only small noise between the two images, $L1$ norm or $L2$ norm have similar performance. On the other hand, $L1$ norm is better in case of serious occlusion. Since this paper only deals with expression change, $L2$ norm is preferred.

STR is different from AAM [4] in that there is no texture variation. As a result, only shape parameters are needed to be optimized.

Solution to STM. To obtain the solution of the optimization problem (1), it is necessary to parameterize the shape transformation. For this purpose, we make use of a linear shape model based on statistical of shape variation [2]. Typically, given a shape being defined by the coordinates of the v vertices that make up the face:

$$s = (x_1, y_1, x_2, y_2, \ldots, x_v, y_v)^T. \tag{2}$$

Then, PCA is conducted and a shape s can be expressed as a base shape s_0 plus a linear combination of n shape vectors s_i [8]:

$$s = W(s_0, p) = s_0 + \sum_{i=1}^{n} p_i s_i, \tag{3}$$

where $S = [s_0, s_1, \ldots, n]$ is the shape model, the coefficients p_i can be considered as the shape parameters. As a result, we have a linear transformation model parameterized by p.

Based on the shape parameterization, we can further solve (1) based on the Lucas Kanade algorithm [1]. Particularly, we linearize the problem (1) about p and obtain

$$\min_p \|I(W(x,p)) + J\triangle p - T(x)\|_2, \tag{4}$$

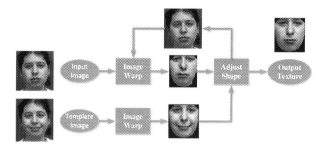

Fig. 2. Face alignment using STM algorithm.

where J is the Jacobian of $I(W(x, p))$ that

$$J = \frac{\partial}{\partial p} I = \nabla I \frac{\partial W}{\partial p}, \tag{5}$$

where ∇I denotes the gradient of I, and $\frac{\partial W}{\partial p}$ can be computed based on shape parameterization (3) [8].

The closed form solution of Eq. (4) for $\triangle p$ is:

$$\triangle p = H^{-1} \sum_x J^T (T(x) - I(W(x, p))), \tag{6}$$

where H is the Gauss-Newton approximation to the Hessian matrix that

$$H = J^T J. \tag{7}$$

Solving (4) is an iterative procedure, c.f. Fig. 2.

2.2   Feature Interpretation

Finally, given a probe/template image and the initialized shape, all gallery images as well as the probe image are warped to the mean shape. The resulted shape-free patches are shown in Fig. 1(b). This can be done based on Delauney triangulation which partitions the convex hull of the control points into a set of triangles [2].

Subsequently, each shape-free patch is concatenated to a vector, v. Finally, we arrange the gallery images of the same class as a matrix $A_i = [v_{i,1}, v_{i,2}, \ldots, v_{i,n_i}] \in \mathfrak{R}^{m \times n_i}$, where n_i denotes the number of samples in the i-th class. Thus, samples of all the k classes can be concatenated as

$$A = [A_1, A_2, \ldots, A_k] = [v_{1,1}, v_{1,2}, \ldots, v_{k,n_k}]. \tag{8}$$

Since the shapes are obtained using the texture matching criterion, shape differences due to expression are suppressed to some extent. Therefore, these textures are convenient to use for recognition.

2.3   Sparse Texture Representation (STR)

Sparse representation [13] has been proved to be robust to the choice of feature, occlusion and corruption. Therefore, it is utilized for our recognition procedure, c.f. Fig. 1(c).

According to the theory of Sparse representation [13], any sample of i-th class can be represented as the linear combination of gallery samples. In our case, the texture of the probe image can be represented sparsely that

$$y = A\alpha \in \Re^m, \tag{9}$$

where $\alpha = [0, \ldots, 0, \alpha_{i,1}, \alpha_{i,2}, \ldots, \alpha_{i,n_i}, 0, \ldots, 0]^T \in \Re^n$ is a coefficient vector whose entries are zero except those associated with the ith class, c.f., Fig. 1(c). The optimal α can be formulated as an optimization problem that

$$\alpha^* = \arg\min_x \|\alpha\|_1 + \|e\|_1$$
$$s.t. \quad y = A\alpha + e. \tag{10}$$

This is a L1-minimization problem, which can be solved using Augmented Lagrange Multiplier (ALM) algorithm [14]. Finally, we recognize the identity of y by

$$\mathrm{id}(y) = \mathrm{id}(\arg\max_i \|\alpha_i\|) \tag{11}$$

Algorithm 1. Shape-constrained Sparse Representation

Input: Gallery images $\{(I_i, c_i)\}_{i=1}^N$, a probe image T, a shape model S, an initialized
    shape s
1: **for** each gallery image I_i
2:    obtain the transformation for each gallery images by
    solving (4)
3: **end**
4: find α by solving the L1-minimization problem (10).
Output: $\mathrm{id}(y) = \mathrm{id}(\arg\max_i \|\alpha_i\|)$

---

3   Experiments

In this section, evaluations are conducted to demonstrate the performance of the proposed expression invariant face recognition method. To this purpose, Multi-PIE [5] is involved. CMU Multi-PIE contains images of 337 subjects across simultaneous variation in pose, expression, and illumination.

In the proposed algorithm, facial landmarks are automatically initialized by adaboost based face detection [10] and ASM model [3]. Examples of the initialized landmarks by ASM and human are given in Fig. 3. It can be found that the landmarks given by ASM are far from accurate labeling. As a result, it challenges algorithms for not only expression variation

Table 1. Face recognition on CK.

Methods	Accuracy reported (%)
Wang et al. [12]	0.67
ESP [9]	0.84
ESP-SEE [9]	0.87
The proposed SSR	0.88

Fig. 3. Illustration of two initialization methods. The left picture is labeled by human. The middle one is initialized by ASM. The right one is shifted by 10 pixels.

We first compare the proposed algorithm on the CK database with three state-of-the-art algorithms, i.e., Wang et al. [12], ESP [9], ESP+SEE [9]. It is notably that all algorithms for comparison require the accurate image alignment, e.g., eye localization [9] or facial landmarks [6,12]. The neural faces of all subjects are taken as gallery and expressions of utmost intensity are regarded as probe. The recognition results are shown in Table 1. We can find that SSR obtains the best performance. It is remarkable that SSR outperforms other algorithms without accurate initialization.

3.1 Shape Change

It is interesting to verify how shape changes affect the recognition result, i.e., the influence of shape initialization and STR. On the one hand, we investigate the influence of the initialization of landmark points. Given the localization of an probe image, s, the shape of a probe image is initialized by shifting s by a number of pixels, which is illustrated in Fig. 3. Thus, the initialization can be considered to be disturbed by noise. The recognition results are shown in Fig. 4. It shows that the accuracy remains a high accuracy of 88 % when the shift is less than 5 pixels. While the shift is up to 10 pixels, the accuracy reduces to 61 %.

On the other hand, we compare the proposed algorithm with SRC [11] to study the impact of STR. SRC assumes that faces of the same person are under different affine transformations. As a result, when shape change caused by expression occurs, SRC is not applicable. For this comparison, we used the similar criterion with SRC to perform this evaluation, i.e., neutral faces are taken as gallery and smile faces as probe. To highlight the expression variation, we only involved frontal illumination. The results shown in Table 2 indicate that the shape transformation has greatly improved the accuracy of SRC. This is because SSR is more powerful than SRC to eliminate the influence of shape variation.

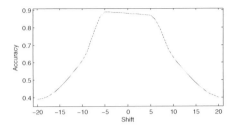

Fig. 4. Recognition accuracy for disturbed initialization.

Table 2. Face recognition on Multi-PIE.

Methods	Accuracy reported
Wagner et al. [11]	0.65
The proposed SSR	0.81

3.2 Expression Impact

Finally, we investigated the influence of different expression. For this purpose, neutral expression is considered as gallery and all the other five expressions in Multi-PIE database are involved as probe, i.e., smile, surprise, squint, disgust, and scream. The results are shown in Table 3. It can be found that the smile and the squint obtain the highest accuracy. This may be because these two expressions have the least change in shape with neutral face. Scream face is the most difficult to recognize. This is reasonable because the shape changes the most under scream.

Table 3. Face recognition of All Expressions on Multi-PIE.

Smile	Surprise	Squint	Disgust	Scream
0.81	0.78	0.81	0.77	0.75

4 Conclusion

This paper proposed an expression invariant face recognition algorithm, shape-constrained sparse representation (SSR). Different from traditional algorithms which need accurate facial landmarks, SSR only requires a coarse initialization for facial shape. For this purpose, SSR finds the shape correspondence between probe and gallery images by the shape-constrained texture matching (STM). Furthermore, SSR utilizes the sparse texture representation technique to robustly classify the probe image. Extensive experiments indicate that SSR outperforms the state-of-the-art algorithms, without accurate initialization.

References

1. Baker, S., Matthews, I.: Lucas-kanade 20 years on: a unifying framework. IJCV **56**(3), 221–255 (2004)
2. Cootes, T.F., Edwards, G.J., Taylor, C.J.: Active appearance models. In: Burkhardt, H., Neumann, B. (eds.) ECCV 1998. LNCS, vol. 1407, pp. 484–498. Springer, Heidelberg (1998)
3. Cootes, T.F., Taylor, C.J.: Active shape models - smart snakes. In: Hogg, D., Boyle, R. (eds.) BMVC, pp. 266–275. Springe, Heidelberg (1992)
4. Gross, R., Matthews, I., Baker, S.: Generic vs. person specific active appearance models. IVC **23**(12), 1080–1093 (2005)
5. Gross, R., Matthews, I., Cohn, J., Kanade, T., Baker, S.: Multi-pie. In: AFGR, pp. 1–8 (2008)
6. Lee, H.-S., Kim, D.: Expression-invariant face recognition by facial expression transformations. PRL **29**(13), 1797–1805 (2008)
7. Li, X., Mori, G., Zhang, H.: Expression-invariant face recognition with expression classification. In: CRV, p. 77 (2006)
8. Matthews, I., Baker, S.: Active appearance models revisited. IJCV **60**(2), 135–164 (2004)
9. Mohammadzade, H., Hatzinakos, D.: Projection into expression subspaces for face recognition from single sample per person. IEEE Trans. AC **4**(1), 69–82 (2013)
10. Viola, P., Jones, M.: Robust real-time face detection. IJCV **57**(2), 137–154 (2004)
11. Wagner, A., Wright, J., Ganesh, A., Zhou, Z., Mobahi, H., Ma, Y.: Toward a practical face recognition system: Robust alignment and illumination by sparse representation. PAMI **34**(2), 372–386 (2012)
12. Wang, J., Plataniotis, K., Lu, J., Venetsanopoulos, A.: On solving the face recognition problem with one training sample per subject. Pattern Recogn. **39**(9), 1746–1762 (2006)
13. Wright, J., Yang, A.Y., Ganesh, A., Sastry, S.S., Yi, M.: Robust face recognition via sparse representation. PAMI **31**(2), 210–227 (2009)
14. Yang, A., Sastry, S., Ganesh, A., Ma, Y.: Fast l1-minimization algorithms and an application in robust face recognition: a review. In: ICIP, pp. 1849–1852 (2010)

Intensity-Depth Face Alignment Using Cascade Shape Regression

Yang Cao[1] and Bao-Liang Lu[1,2(✉)]

[1] Department of Computer Science and Engineering,
Center for Brain-like Computing and Machine Intelligence,
Shanghai Jiao Tong University,
800 Dongchuan Road, Shanghai 200240, China
bllu@sjtu.edu.cn
[2] Key Laboratory of Shanghai Education Commission for Intelligent Interaction
and Cognitive Engineering, Shanghai Jiao Tong University,
800 Dongchuan Road, Shanghai 200240, China

Abstract. With quick development of Kinect, depth image has become an important channel for assisting the color/infrared image in diverse computer vision tasks. Kinect can provide depth image as well as color and infrared images, which are suitable for multi-model vision tasks. This paper presents a framework for intensity-depth face alignment based on cascade shape regression. Information from intensity and depth images is combined during feature selection in cascade shape regression. Experimental results show that this combination improves face alignment accuracy notably.

Keywords: Face alignment · Depth image · Cascade shape regression

1 Introduction

Face alignment is to detect key points such as eye corner, mouth corner and nose tip on human face, and is an important step for many face related vision tasks like face tracking and face recognition. A variety of models are proposed to tackle face alignment problem. Notable ones are Active Shape Model (ASM) [8], Active Appearance Model (AAM) [7], Constrained Local Model (CLM) [9], Explicit Shape Regression [4] and Deep Convolutional Network [14]. Besides, numerous improvements for these models have been proposed.

Among those different models, the family of cascade regression models is the leading one. Generally, cascade regression models solve face alignment problem with many stages of regressions (in a cascade manner). Based on the basic structure, a lot of models are proposed. Cao et al. [4] use two-level boosted regression, shape indexed features and fast correlation-based feature selection. They achieve remarkable results. Asthana et al. [1] make the learning parallel and incremental, so that the model can automatically adapt to data. Burgos et al. [3] model occlusion explicitly to locate occluded regions and improve performance for occluded faces. Chen et al. [5] deal with face detection and face alignment jointly, which improves performance on both problems.

© Springer International Publishing Switzerland 2015
S. Arik et al. (Eds.): ICONIP 2015, Part IV, LNCS 9492, pp. 224–231, 2015.
DOI: 10.1007/978-3-319-26561-2_27

Most face alignment studies focus on intensity image. Unlike intensity image, each pixel in a depth image represents the distance between the point in the scene and the camera. Intuitively, information from intensity image and depth image is complementary. For example, eyelid and eyebrow are very distinct on intensity image because of their darkness; tip of nose is very distinct on depth image because of its shape. However, only a few researchers try to combine them in face alignment tasks [2,6].

Both [2] and [6] are based on CLM. Nevertheless, this approach has some limitations. The first one is that it is not optimal because intensity information and depth information do not always contribute equally to face alignment in the whole face region. Moreover, this approach is based on CLM, but recent studies show that cascade shape regression achieves higher performance. In this paper, we address those two problems in a novel way.

2 Framework

We adapt cascade shape regression model to depth image, and use a novel approach to combine intensity image and depth image. Figure 1 is an overview of our framework.

2.1 Problem Description

Suppose that I is an image (intensity-depth image in our case), which contains human face, R is a rectangle, which gives face region in I, and $Y = [x_1, y_1, \ldots, x_N, y_N]^T$ is the ground truth of facial landmarks. The task of face alignment is computing an estimation $\hat{Y}$ from only I and R, which minimizes

$$\|Y - \hat{Y}\|_2. \tag{1}$$

2.2 Framework Structure

Since computing $\hat{Y}$ in one-shot is difficult, almost all face alignment models work in a cascade manner. That is, let $\hat{Y}_1$ (which is usually the average landmarks

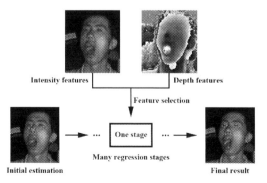

Fig. 1. The workflow of our framework.

positions translated into R) be the initial estimation, and compute a better estimation $\hat{Y}_2$ from I and $\hat{Y}_1$. Repeat this for several times and we can get the final result $\hat{Y}$.

Our framework also works in this manner. We use the two-level structure proposed in [4]. In our framework, there are T stages. Therefore we begin with $\hat{Y}_1$, and get $\hat{Y}_2, \hat{Y}_3, \ldots, \hat{Y}_t, \ldots$, iteratively, until $\hat{Y}_T$. $\hat{Y}_T$ will be our final result $\hat{Y}$.

2.3   Feature

Doing regression on I and $\hat{Y}_t$ directly is hard. We must extract features from I and $\hat{Y}_t$.

Shape Indexed Features. Shape indexed features mean that features are extracted relative to landmarks. For face alignment, shape indexed features are more robust against pose variation. In [4], a feature is the difference between intensity values of two pixels. The pixel position is generated by taking an offset to a certain facial landmark. Therefore, pixels positions are relative to facial landmarks, which makes them invariant in different poses. In [3], linear interpolation between two landmarks is used as the position of a pixel, which makes it more robust.

In our framework, we also use shape indexed features. For intensity image, we directly use the difference between intensity values of two pixels, whose positions are randomly generated, as a feature. We generate many such features and then do feature selection. For depth image, we use normalized depth features.

Normalized Depth Features. Shape indexed features can effectively deal with pose variation problem for intensity image. But for depth image, shape indexed features are not enough, since the depth value of each pixel will change dramatically during pose variation. To make information of depth image more robust against pose variation, we propose normalized depth features.

As in the intensity image case, we use the difference between depth values of two pixels as a feature. To make it invariant under pose variation, we compensate each value of pixel according to the pose.

If we use a plane to approximate the human face, the plane can be expressed by

$$X\beta = z, \tag{2}$$

where $X = [1, x, y]^T$ is the location of the pixel, β is the parameter of the plane and z is the depth of the pixel.

Suppose $\boldsymbol{X} = [\boldsymbol{x_1}, \boldsymbol{x_2}, \ldots, \boldsymbol{x_m}]^T$, $\boldsymbol{x_k} = [1, x_k, y_k]^T$ is a landmark on the depth image, and $\boldsymbol{z}$ is the corresponding depth values. Then their relationship can be expressed by

$$\boldsymbol{X}\beta = \boldsymbol{z}, \tag{3}$$

which can be solved by linear regression. Using normal equation, the result is

$$\beta = (\boldsymbol{X}^T \boldsymbol{X})^{-1} \boldsymbol{X}^T \boldsymbol{z}. \tag{4}$$

Having got the parameter β of the face plane, we can compensate the depth value of a pixel by

$$z' = z - \alpha X \beta, \tag{5}$$

where z is the original depth value, z' is the compensated depth value, and X is the location of the pixel. We also use attenuation parameter α for the compensation, because the estimated face pose may not be very accurate.

The estimation of β is an iterative process. As the estimation of landmarks becomes more accurate during face alignment, the estimation of β will also become more accurate, which in turn contributes to the estimation of landmarks.

Feature Selection and Fusion. Using the feature extraction method described above, we can randomly (randomly select a landmark and randomly choose an offset) generate a huge feature pool for each image of training data, which contains not only intensity features but also depth features. We use correlation-based (Pearson Correlation) feature selection method proposed in [4],

$$j_{\text{opt}} = \arg\min_j \text{corr}(\boldsymbol{Y}\boldsymbol{v}, \boldsymbol{X}_j), \tag{6}$$

where $\boldsymbol{X}$ is a matrix to represent all randomly generated features of all the training data in current external-stage, $\boldsymbol{X}_j$ is a column vector which represents the jth feature of all the training data, $\boldsymbol{Y}$ is a matrix in which each row represents the disparity between current estimation of landmarks positions and true landmarks positions, and $\boldsymbol{v}$ is a vector drawn from unit Gaussian to project $\boldsymbol{Y}$ into a vector. Note that $\boldsymbol{v}$ is necessary here, so that we can compute the correlation.

In our model, each stage contains K internal-stages [4]. In the training of each internal-stage (a fern [10]), we use the feature selection method described above to select F features for that fern. Both intensity features and depth features are considered. Therefore, one fern can use intensity features and depth features at the same time, which effectively combine the best part of two sources of features. Figure 2 gives an example of such a fern. Further more, we use a parameter ρ to control the ratio between intensity features and depth features, and use cross validation to select the best value for ρ.

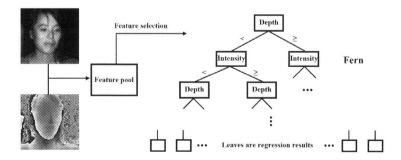

Fig. 2. Feature fusion. A fern can contain both intensity features and depth features at the same time.

2.4   Initial Estimation

An accurate initial estimation of facial landmarks positions can greatly contribute to face alignment. We use k-means on training data to get representative initial estimations of facial landmarks positions. Multiple initializations [4] are used in testing.

In a tracking scenario, a reasonable initial estimation is the face alignment result on previous frame. However, using that estimation directly will lead to model drift, since the model is trained with certain initial estimations obtained by k-means as described above. To solve drift problem, we use Procrustes method [8] to align initial estimations to face alignment result on previous frame, and then use those transformed initial estimations.

3   Experiments

We conduct experiments on two datasets to evaluate our model. Shape Root-Mean Square (RMS) error normalized w.r.t the inter-ocular distance of the face is used in the evaluation. In each experiment, we create three models: intensity, depth, intensity-depth. The only difference among those three models is feature. Depth model only uses depth features, intensity model only uses intensity features, and intensity-depth model is allowed to use both intensity features and depth features. All of the parameters are determined by cross validation on training data. To make the comparison fair, even though intensity-depth model is allowed to use both kinds of features, the numbers of total features which these three models are allowed to use are equal.

3.1   FRGC

FRGC (Face Recognition Grand Challenge) Version 2.0 database [11] is originally designed for face recognition, but some researchers [12,13] annotated those face images with 68 landmarks. It consists 4950 face images with color and depth information. We conduct this experiment in a similar manner (with minor difference) as [6], so that the results can be compared.

From Fig. 3, we can see that the performance of intensity model and intensity-depth model is on the same level, and both are much higher than that of the depth model. We also observe that the performance of our model is better than that of [6]. The average landmark errors of our intensity model, depth model and intensity-depth model are 0.0211, 0.0408 and 0.0209 respectively.

3.2   LIDF

The experiment on FRGC validates our model. However, the improvement by combining intensity feature and depth feature is very small. We believe that it is because of performance saturation (faces in FRGC are almost frontal and not

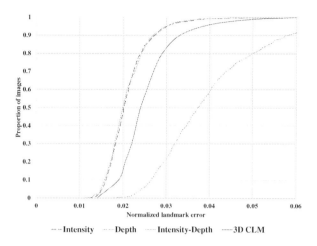

Fig. 3. Landmark error curves of our three models and 3D CLM [6] on FRGC.

challenging enough). To better evaluate the effect of combining intensity feature and depth feature, we create a more difficult dataset. Our labeled infrared-depth face (LIDF) dataset is acquired in a lab environment. It consists of 17 subjects (10 males and 7 females), each subject has 9 poses, and each pose has 6 expressions. Infrared image and depth image are taken simultaneously and are perfectly aligned (using Microsoft Kinect One). So we have 918 infrared-depth images in total. 15-points manually labeled landmarks are provided. This dataset is publicly available[1]. We use images from first 7 male subjects and first 4 female subjects as training set, and the rest images as testing set (Fig. 4).

From Fig. 5 we can see that the performance of intensity-depth model is notably higher than both intensity model and depth model. The average landmark errors of our intensity model, depth model and intensity-depth model are 0.0329, 0.0380 and 0.0305 respectively. Some alignment results obtained by our method are shown in Fig. 6.

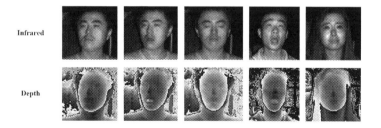

Fig. 4. Some images in LIDF.

---

[1] Available soon on http://bcmi.sjtu.edu.cn/resource.html.

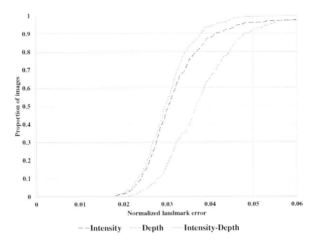

Fig. 5. Landmark error curves of our three models on LIDF.

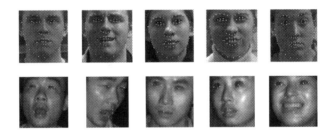

Fig. 6. Some alignment results of our intensity-depth model (first row: FRGC, second row: LIDF).

4   Conclusion

We proposed a face alignment model for intensity-depth image based on cascade regression model. Depth features were made to be robust against pose variation. Intensity and depth features were effectively combined during feature selection. Our intensity-depth model got 0.9 % error reduction on FRGC and 7.3 % error reduction on LIDF over intensity model, which indicated that higher performance could be achieved by combining intensity image and depth image.

Acknowledgments. This work was partially supported by the National Natural Science Foundation of China (Grant No. 61272248), the National Basic Research Program of China (Grant No. 2013CB329401), and the Science and Technology Commission of Shanghai Municipality (Grant No. 13511500200).

References

1. Asthana, A., Zafeiriou, S., Cheng, S., Pantic, M.: Incremental face alignment in the wild. In: IEEE Conference on Computer Vision and Pattern Recognition (CVPR), pp. 1859–1866. IEEE (2014)
2. Baltrusaitis, T., Robinson, P., Morency, L.: 3D constrained local model for rigid and non-rigid facial tracking. In: IEEE Conference on Computer Vision and Pattern Recognition (CVPR), pp. 2610–2617. IEEE (2012)
3. Burgos-Artizzu, X.P., Perona, P., Dollár, P.: Robust face landmark estimation under occlusion. In: IEEE International Conference on Computer Vision (ICCV), pp. 1513–1520. IEEE (2013)
4. Cao, X., Wei, Y., Wen, F., Sun, J.: Face alignment by explicit shape regression. Int. J. Comput. Vis. (IJCV) **107**(2), 177–190 (2014)
5. Chen, D., Ren, S., Wei, Y., Cao, X., Sun, J.: Joint cascade face detection and alignment. In: Fleet, D., Pajdla, T., Schiele, B., Tuytelaars, T. (eds.) ECCV 2014, Part VI. LNCS, vol. 8694, pp. 109–122. Springer, Heidelberg (2014)
6. Cheng, S., Zafeiriou, S., Asthana, A., Pantic, M.: 3D facial geometric features for constrained local model. In: IEEE Conference on Image Processing (ICIP). IEEE (2014)
7. Cootes, T.F., Edwards, G.J., Taylor, C.J.: Active appearance models. IEEE Trans. Pattern Anal. Mach. Intell. (TPAMI) **23**(6), 681–685 (2001)
8. Cootes, T.F., Taylor, C.J., Cooper, D.H., Graham, J.: Active shape models-their training and application. Comput. Vis. Image Underst. (CVIU) **61**(1), 38–59 (1995)
9. Cristinacce, D., Cootes, T.F.: Feature detection and tracking with constrained local models. In: British Machine Vision Conference (BMVC), vol. 2, p. 6 (2006)
10. Dollár, P., Welinder, P., Perona, P.: Cascaded pose regression. In: IEEE Conference on Computer Vision and Pattern Recognition (CVPR), pp. 1078–1085. IEEE (2010)
11. Phillips, P.J., Flynn, P.J., Scruggs, T., Bowyer, K.W., Chang, J., Hoffman, K., Marques, J., Min, J., Worek, W.: Overview of the face recognition grand challenge. In: IEEE Conference on Computer Vision and Pattern Recognition (CVPR), vol. 1, pp. 947–954. IEEE (2005)
12. Sagonas, C., Tzimiropoulos, G., Zafeiriou, S., Pantic, M.: 300 faces in-the-wild challenge: the first facial landmark localization challenge. In: IEEE International Conference on Computer Vision Workshops (ICCVW), pp. 397–403. IEEE (2013)
13. Sagonas, C., Tzimiropoulos, G., Zafeiriou, S., Pantic, M.: A semi-automatic methodology for facial landmark annotation. In: IEEE Conference on Computer Vision and Pattern Recognition Workshops (CVPRW), pp. 896–903. IEEE (2013)
14. Sun, Y., Wang, X., Tang, X.: Deep convolutional network cascade for facial point detection. In: IEEE Conference on Computer Vision and Pattern Recognition (CVPR), pp. 3476–3483. IEEE (2013)

Bio-Inspired Hybrid Framework
for Multi-view Face Detection

Niall McCarroll[1(✉)], Ammar Belatreche[1], Jim Harkin[1], and Yuhua Li[2]

[1] Intelligent Systems Research Centre, University of Ulster, Derry, Northern Ireland
mccarroll-n1@email.ulster.ac.uk,
{a.belatreche,jg.harkin}@ulster.ac.uk
[2] School of Computing, Science and Engineering, University of Salford, Manchester, UK
y.li@salford.ac.uk

Abstract. Reliable face detection in completely uncontrolled settings still remains a challenging task. This paper introduces a novel hybrid learning strategy that achieves robust in-plane and out-of-plane multi-view face detection through the enhanced implementation of the hierarchical bio-inspired HMAX framework using spiking neurons. Through multiple training trials, separate pools of neurons are trained on different face poses to extract features through feed-forward unsupervised STDP. The trained neurons are then processed by an additional STDP mechanism to generate a streamlined repository of broadly tuned multi-view neurons. After unsupervised feature extraction, supervised feature selection is implemented within the hybrid framework to reduce false positives. The hybrid system achieves robust invariant detection of in-plane and out-of-plane rotated faces that compares favourably with state-of-the-art face detection systems.

Keywords: Multi-view face detection · Spiking neural networks · STDP · Hybrid learning · Hierarchical object detection · HMAX

1 Introduction

Face detection is an important computer vision task as it forms the first essential stage in all face processing algorithms such as face recognition and facial expression tracking. Despite significant research into reliable multi-view face detection, it remains a challenging problem as the appearance of a face will vary greatly depending on size, head pose, orientation, illumination, facial expression and occlusions [1].

Serre et al. [2] argue that there is merit in exploring biologically representative solutions that behave adaptively through learning, so as to adequately account for the well documented superior performance of human and primate object detection. This rapid detection of objects, achieved in the brain through largely feed-forward computations, has motivated groundbreaking work on sparse spiking neural networks (SNNs) that can process substantial amounts of data through a relatively small sequence of spikes [3].

This paper presents an extended bio-inspired hierarchical SNN framework that adopts a hybrid approach to learning, combining a bottom-up unsupervised Spike-Timing Dependent Plasticity (STDP) feature extraction phase that builds up low-level

© Springer International Publishing Switzerland 2015
S. Arik et al. (Eds.): ICONIP 2015, Part IV, LNCS 9492, pp. 232–239, 2015.
DOI: 10.1007/978-3-319-26561-2_28

edge features to higher level diagnostic face features, followed by a supervised feature selection phase that provides feedback to the framework in an effort to reduce false positives. After hybrid training a new cohesive, robust and efficient multi-view face detection system is formed by merging the small but expressive set of multi-view neurons. This new approach is characterised by the following novel features:

Multiple training trials using different pools of neurons on separate multi-view face datasets, each neuron pool learning a specific face pose to address invariance.
Streamlining the system by harnessing the broadly-tuned behaviour of neurons.
Extending a hierarchical SNN framework with an advanced STDP-based feature extraction module that selects the most diagnostic neurons for a particular face pose.
Introducing hybrid learning by implementing an additional supervised learning mechanism that handles error in the system by reducing false positives.

The remainder of this paper is organised as follows: Sect. 2 provides a review of bio-inspired face detection. Section 3 discusses the multi-trial training scheme, the extended STDP streamlining strategy, concluding with the implementation of a supervised mechanism to improve detection accuracy. Section 4 provides detailed results and analysis with final conclusions outlined in Sect. 5.

2 Bio-inspired Approaches to Face Detection

In SNNs the timing of spiking activity is the key information processing parameter which has the potential to be more computationally powerful than traditional rate-based coding [4]. Having demonstrated that primates can analyse and categorise complex visual stimuli within 100–150 ms, Thorpe et al. [5, 6] developed an encoding scheme in which rapid visual categorisation could be achieved through a temporal pattern generated by the asynchronous firing of single spiking neurons across a population–known as Rank Order Coding (ROC). Employing their four-layer feed-forward 'SpikeNet' neural network simulator [6], they examined the possibility that robust face detection could be achieved through ROC. The detection system performed comparatively well in relation to the more classical learning techniques, with a best performance of 96.3 % accuracy on what was deemed an 'easy' database of 400 exclusively frontal faces.

2.1 Unsupervised and Supervised Learning in Face Detection

STDP provides a useful learning mechanism for feature extraction in SNNs by adjusting synaptic strength as a function of the relative timing of the spikes of pre- and post-synaptic neurons. Introducing lateral inhibitory competition leads to self-organisation within a network with each neuron learning to be optimally responsive to different input patterns or features. This self-organising behaviour was used to train a multi-neural network to distinguish between nine separate head poses in [7]. Post training, specific neurons began to show a clear preference for specific head poses with 94.7 % of previously unseen examples being correctly identified.

There is a significant body of evidence that instruction-based learning is exploited by the brain and that these supervised learning mechanisms control information representations in sensory networks such as pattern recognition. In [8] the authors introduce a supervised system that adopts an image encoding scheme inspired by the HMAX model [9]. The multi-layer model was tested on a pattern recognition task of handwritten digits. The supervised learning layer tunes synaptic weights in such a way that four pools of 20 neurons respond to specific patterns collectively with either a 1 (firing) or a 0 (no firing) based on majority vote. The supervised network achieved 93.7 % accuracy on a 500 image training set which drops to 79 % accuracy on a testing set.

2.2   Hierarchical Frameworks for Robust Face Detection

To achieve robust face detection and recognition that is invariant to view point, it becomes necessary to move away from the holistic (global template) approach and to view a face as a combination of local features. A hierarchical approach with a build-up of simple features (e.g. oriented edges processed by primary visual cortex V1 neurons), to more complex features (e.g. facial features processed by Infero Temporal Cortex (ITC) neurons), is more biologically representative of what happens in the ventral stream of the visual cortex [9, 10].

To achieve this combinatorial approach [10] introduces a four layer, feed-forward spiking adaptation of the original HMAX framework [9]. The hierarchical system tends to learn salient features as they correspond to the earliest spikes from the most strongly activated neurons. The higher order visual cells gradually become selective to frequently occurring feature combinations (eyes, nose, mouth, head) that are consistently present in the face training set. The system detected 96.5 % of faces in a test set of 217 images but is should be noted that all subjects were in an upright frontal pose.

2.3   Challenge

While it is accepted that rapid object detection is a predominantly feed-forward process, there is a wealth of experimental evidence that feedback loops occur extensively throughout the visual cortex [10]. Modelling these feedback loops provides an opportunity to introduce supervisory signals to reduce the level of error and improve detection accuracy. As the visual input will be initially too complex for the system to deal with directly, an unsupervised learning phase will reduce the dimensions into clusters of salient features. After this feature extraction phase, supervised learning will perform a feature selection process resulting in a streamlined multi-view face detection system that is optimised in terms of efficiency and accuracy.

3   Methodology

The multi-view face dataset that was sourced and generated in [11] was used for training and testing in this study. As HMAX neurons exhibit robust invariance to faces rotated within a range from -15° to +15° of their trained rotation, this allowed training and testing datasets with rotation steps of 30° to be used for both the 12 different in-plane rotated face sets (0°, ± 30°, ± 60°, ± 90°, ± 120°, ± 150° and 180°) and the 6 different out-of-plane rotated face sets (0°, ± 30°, ± 60° and ± 90°). The training and testing groups contained 500 face images each for the 12 different in-plane and 6 different out-of-plane face poses. Consequently, the entire in-plane training and testing sets are composed of 6,000 (12 × 500) face images, totalling 12,000 images for in-plane faces. The out-of-plane training and testing populations consist of 3,000 images (6 × 500) each, totalling 6,000 images for out-of-plane faces. The negative (non-face) test set that was sourced and generated in [11] was used as the 9,000 negative image set for the merged system testing phase.

3.1   Unsupervised STDP Feature Extraction

Using the pseudo SNN architecture introduced in [10], spiking activity is simplified through a manipulation of intensity-to-latency values at each layer of the network. The feed-forward network uses a four layer hierarchy (S1-C1-S2-C2) and operates a time-to-first-spike scheme (Fig. 1).

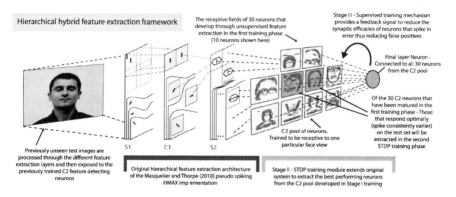

Fig. 1. Hybrid adaptation of the spiking HMAX model [10]. The different layers implement the build-up of selectivity to simple features (oriented edges) which are then pooled together to form more complex diagnostic face features. The model includes an additional layer of STDP facilitated feature extraction which acts to filter out the less optimally responsive neurons for a particular face pose. Stage III introduces a supervisory signal that responds to false positives by reducing the synaptic efficacy of neurons that consistently spike in error with non-face images.

Edge detection is performed at the first layer (S1 cells) by convolving five scaled versions of the input face image with Gabor filters at four preferred orientations (22.5°, 67.5°, 112.5° & 157.5°). Implementing a winner-takes-all approach, only the earliest spike which corresponds to the best matching orientation at a given location is propagated.

Unsupervised learning is achieved through a simplified STDP mechanism which cultivates the C1-S2 connections leading to the development of S2 receptive fields formed from combined salient edges that are responsive to intermediate-complexity features (e.g. full faces, heads or sizeable sections of a face with salient features). A lateral inhibition process is also in place to promote competition amongst the C2 cells, preventing them from learning the same feature. The maximum response of the next layer of C2 cells simply takes the first spike emitted by the S2 cells over all positions and scales leading to scale and position invariant feature detection (Fig. 1). Full implementation details of this SNN architecture can be found in [10].

18 different pools of neurons develop selectivity to different multi-view face features through individual targeted training sessions. This allows each of the 18 individual neuron pools to 'focus' on one particular set of facial features that are consistently present in faces of that particular pose. In each session a different pool of 30 neurons is trained on one of the 18 different face poses. The decision to increase the number of neurons per face pose to 30 from the smaller number of 10 in the original experimental setup [10] is due to the fact that three face datasets are being used instead of one. It was decided to triple the number of neurons to account for any increase in within-class variability that is introduced with new faces being learned.

3.2   Hybrid Feature Extraction and Selection

As a consequence of competitive learning only a subset of each pool of 30 neurons develop receptive fields that are responsive to diagnostic and recognisable facial features with the rest developing less diagnostic features that contribute little or indeed negatively (higher false positive rates) to the overall detection accuracy. This presents an opportunity to optimize these pools in terms of efficiency (reduced number of neurons) and detection accuracy performance (reduced false positives) for each particular face orientation. This was achieved in an unsupervised way in [11] by extending the HMAX framework to include a second simplified STDP feature extraction module. This extended, feed-forward streamlining process achieved a significant 26.8 % improvement in detection accuracy across 9000 multi-view faces and 9000 non-face images. However, there remained a significant amount of error due to false positives (non-faces being detected as faces). The false positive count was 2,250 (25 %) for the 9,000 non-face test set which is at an unacceptably high level.

The new hybrid system extends the framework by first applying a new STDP module that accounts for the relative timing of spikes, and not just their order. In previous implementations of the spiking HMAX framework [10, 11] weight change in the STDP module did not depend on the exact timing of a pre or post-synaptic spike, only on the order in which they occurred. All pre-synaptic neurons that spike before the post-synaptic neuron are rewarded equally regardless of their timing as well as being punished equally regardless of the time they spike after the post-synaptic neuron.

This simplified STDP mechanism was utilized because the pixel intensity to latency conversion used to model spiking activity for the large number of S1 cells compressed the entire spike wave making the influence of relative timing negligible. However, at the C2 layer there are a significantly smaller number of neurons (30 C2 neurons per face pose pool) with a greater range of spike timings. As a result, the relative timing of each individual neuron's output spike becomes important for the adjusting of connection strengths based on the responsiveness to input stimuli. The new STDP module is sensitive to both the positive and negative inter-spike interval (ISI) between action potentials of pre- and post-synaptic neurons, with short intervals producing maximal plasticity change and longer intervals generating little or no change in synaptic strength.

The rationale for introducing this extended STDP module is that Masquelier's [10] simplified STDP module strips out a lot of neurons (from 30 down to as few 7 on average). Many of these neurons make a valuable contribution to positive face detection but because they are not as responsive with lower spike counts they are removed in the original simplified STDP process. Implementing the new relative timing STDP module retains a lot more useful neurons (an average of 21 across each pool).

With the second stage of unsupervised feature extraction complete, the next phase was to further extend the HMAX framework to include a new supervised feature selection module to filter out the worst performing neurons from each face pose pool. This is implemented using a final layer of the network that contains a single neuron, to which the subset of 30 C2 neurons filtered out in the second phase of STDP extraction form synaptic connections with. These remaining synaptic connections were assigned random weights with a mean 0.8 and standard deviation 0.05 as defined in [10] (Fig. 1).

The 18 neurons pools are exposed in separate trials to the test sets of face images and non-face images. Throughout the training process neurons that respond in error (spiking when a non-face pattern is presented) are punished by a supervisory signal by having their synaptic efficacies reduced. Weights are maintained within the range 0–1 and soft bound is implemented to ensure that weight changes tend towards zero when the weight approaches either of the bounds. Supervised training continues with the penalty for error increasing incrementally until a threshold is met where false positive rates and false negative rates result in the best overall detection accuracy. After training, the neurons with synaptic weights stabilized at 0 are inactive and filtered out of the neuron pools leaving a subset of the best performing C2 neurons.

The outcome of the combined unsupervised (relative timing STDP) followed by supervised training for the novel hybrid system is a set of 18 streamlined neuron pools (each different pool has been trained and optimised for one of the 18 different face poses). Cohesive multi-view face detection systems were formed by merging the 18 individual streamlined pools of neurons together to form a system composed of 148 neurons. A 540 neuron system is also formed from the 18 pools of 30 C2 neurons generated from the first training phase, as well as a 124 neuron multi-view system composed of neurons extracted by the unsupervised STDP framework discussed in [11]. Face detection performance comparisons are made between the three multi-view systems by exposing each to 9000 multi-pose face images (formed by combining the sets of 500 images of each of the 18 different face poses) and the extended set of 9,000 non-face images. The spike responses for faces and non-faces were recorded and the overall detection accuracy for each multi-view system was calculated as:

$$Accuracy = \frac{\#\ True\ Positives\ +\ \#\ True\ Negatives}{Total\ Population\ Size}$$

The detection performance for the three merged multi-view face detection systems are discussed in Sect. 4.

4   Results & Discussion

Table 1 shows that the new multi-view hybrid system exhibits the best performance achieving both the highest true positive rate and lowest false positive rate resulting in an overall detection accuracy of 93.10 % across all rotations compared to the substantially lower detection performance of 59.50 % for the 540 neuron multi-view detection system. This 93.10 % detection rate also represents a significant 6.8 % improvement on the multi-view detection accuracy of the unsupervised system presented in [11]. This improvement is due to the significant reduction in false positives achieved through supervised training. The false positive rate after hybrid supervised training is 11.6 % compared to a much higher false positive rate of 25 % for the unsupervised system. The combined process of streamlining through relative timing STDP, followed by error reduction through supervised learning, filters out the erratic and random spiking neurons, whilst retaining the useful neurons that contribute the most to positive face detection.

Table 1. Results from the merged multi-view system tests on the 9,000 multi-view face and 9,000 non-face datasets. Column II lists the number of multi-view neurons that each system is composed of, Column III lists the TP (true positive) rate. Column IV lists the FP (false positive) rate and Column V lists the overall detection accuracy of each merged system

System Description	# neurons	TP Rate	FP Rate	Detection Accuracy
Original 30 Neuron System	540	99.01 %	98.66 %	59.50 %
Unsupervised System	124	97.60 %	25.00 %	86.30 %
Hybrid System	148	97.73 %	11.60 %	93.10 %

The proposed system has achieved significant incremental improvement compared to the original [10] and extended [11] HMAX spiking frameworks that it builds upon. However, it is also useful to benchmark the system against state-of-the-art commercial and research systems. The current system was trained on 9,000 face images and compares favourably to Google's commercial face detection system – Picasa 3.9 [12] which has been trained on billions of faces and an extended version of the Viola Jones boosted cascade scheme that incorporates rotated Haar features [13].

The extended Viola Jones scheme struggles with in-plane rotated faces achieving an overall detection rate of only 61.49 %. Picasa also struggles with in-plane rotated faces once they rotate away from the frontal upright pose. Overall, across all 18 face poses, Picasa only achieves an overall detection accuracy of 68.14 %. The new hybrid system, using only 148 neurons or features for in-plane and out-of-plane face detection, achieves a significantly improved detection rate of 93.10 % when tested on all 9,000 face images and 9,000 non-face images.

5   Conclusions & Future Work

The novel hybrid system proposed in this study successfully combines bottom-up unsupervised learning for information reduction through feature extraction, with a supervised learning phase that filters these new formed features in a way that rejects those that introduce error into the system. A true positive detection rate of 97.7 % is an excellent result given that the 9,000 faces were presented in 18 different multi-view poses but there still remains an unacceptable level of 11.6 % false positive error. The complex behaviour of spiking neurons provides opportunity to investigate further biological mechanisms of feedback control, including spike-threshold adaptation and threshold dynamics relating to spike latency that would further enhance feature selectivity leading to greater accuracy within the multi-layer face detection framework.

References

1. Zhang, C., Zhang, Z.: A survey of recent advances in face detection. MSR-TR-2010-66. Microsoft Research (2010)
2. Serre, T., Oliva, A., Poggio, T.: A feedforward architecture accounts for rapid categorization. Proc. National Acad. Sci. US Am. **104**(15), 6424–6429 (2007)
3. Van Rullen, R., Gautrais, J., Delorme, A., Thorpe, S.: Face processing using one spike per neuron. BioSystems **48**(1–3), 229–239 (1998)
4. Maass, W.: Networks of spiking neurons: The third generation of neural network models. Neural Netw. **10**(9), 1659–1671 (1997)
5. Thorpe, S., Fize, D., Marlot, C.: Speed of processing in the human visual system. Nature **281**(6582), 520–522 (1996)
6. Thorpe, S.: Ultra-Rapid scene categorization with a wave of spikes. In: Bülthoff, H.H., Lee, S.-W., Poggio, T.A., Wallraven, C. (eds.) BMCV 2002. LNCS, vol. 2525, pp. 335–351. Springer, Heidelberg (2002)
7. Weidenbacher, U., Neumann, H.: Unsupervised learning of head pose through spike-timing dependent plasticity. In: André, E., Dybkjær, L., Minker, W., Neumann, H., Pieraccini, R., Weber, M. (eds.) PIT 2008. LNCS (LNAI), vol. 5078, pp. 123–131. Springer, Heidelberg (2008)
8. Yu, Q., Tang, H., Tan, K., Li, H.: Rapid feedforward computation by temporal encoding and learning with spiking neurons. IEEE Trans. Neural Netw. Learn. Syst. **24**(10), 1539–1553 (2013)
9. Serre, T., Wolf, L., Bileschi, S., Riesenhuber, M., Poggio, T.: Robust object recognition with cortex-like mechanisms. IEEE Trans. Pattern Anal. Mach. Intell. **29**(3), 411–426 (2007)
10. Masquelier, T., Thorpe, S.: Learning to recognize objects using waves of spikes and spike timing-dependent plasticity. In: International joint conference on neural networks (IJCNN), Barcelona (2010)
11. McCarroll, N., Belatreche, A., Harkin, J., Li, Y.: Bio-inspired hierarchical framework for multi-view face detection and pose estimation. accepted for publication In: International joint conference on neural networks (IJCNN), Killarney (2015)
12. Google Picasa 3.9. http://picasa.google.com/
13. Lienhart, R., Kuranov, A., Pisarevsky, V.: Empirical analysis of detection cascades of boosted classifiers for rapid object detection. In: Michaelis, B., Krell, G. (eds.) DAGM 2003. LNCS, vol. 2781, pp. 297–304. Springer, Heidelberg (2003)

Convolutional Neural Networks Considering Robustness Improvement and Its Application to Face Recognition

Amin Jalali, Giljin Jang, Jun-Su Kang, and Minho Lee[✉]

School of Electronics Engineering, Kyungpook National University,
1370 Sankyuk-Dong, Puk-Gu, Daegu, 702-701, South Korea
{max.jalali,gjang7,wkjuns,mholee}@gmail.com

Abstract. This paper proposes a novel activation function to promote robustness to the outliers of the training samples. Data samples in the decision boundaries are weighted more by adding the derivatives of the sigmoid function outputs to avoid drastic update of the network weights. Therefore, the network becomes more robust to outliers and noisy patterns. We also present appropriate backpropagation learning algorithm for the convolutional neural networks. We evaluate the performance improvement by the proposed method on a face recognition task, and proved that it outperformed the state of art face recognition methods.

Keywords: Convolutional neural network · Back propagation · Robustness in cost function · Deep learning · Gradient descent

1 Introduction

Face recognition from visual images has been a challenging problem by many researchers due to posture varieties, occlusions, variations in shape and color, illumination variations in training and testing conditions, and shading. The conceptual structure of convolutional neural networks (CNN) was presented by Yann LeCun and Yoshua Bengio [1, 2] for image processing and recognition tasks. CNN was biologically inspired by the nerve cells in the visual striate cortex of a cat called receptive filed [3]. The CNN conducts feature extraction as well as classification with least preprocessing of the input images.

LeCun [4] presented the fundamental structure of CNNs for digit recognition and face detection, which exhibited better recognition results than probability density function approaches. Simrad [5] proposed the implementation of more productive methodology of subsampling layers in the operation of the convolutional layers instead of separated subsampling layer. In recent face recognition research, Khalajzadeh et al. [6] proposed a progressive CNN approach and the standard backpropagation as the learning algorithm. However, these approaches have lower accuracy for recognition rate.

In this study, we used the robustness in the cost function of the CNN to extract the features and classify the images of FERET dataset. The samples of the dataset include illumination, occlusion, and distinctive postures. Therefore to achieve better image labeling and classification, we utilize a penalty term to be considered in the cost function

© Springer International Publishing Switzerland 2015
S. Arik et al. (Eds.): ICONIP 2015, Part IV, LNCS 9492, pp. 240–245, 2015.
DOI: 10.1007/978-3-319-26561-2_29

of the training algorithm to diminish the affectability of input patterns over the output error to ensure the robustness. To make robust error cost function, the derivative of the activation function of the neurons are added in cost function of the error back propagation-learning algorithm. Through in the training process, as the error flows back to the first layer, the weights are adjusted considering the Hebb's guideline. This helps data samples in the decision boundaries to be weighted more by adding the derivatives of the sigmoid function outputs and create better features which prevents drastic update of the network weights and causes general exploitation. On the other hand, to gain less training time, the fused convolution-subsampling approach is applied instead of convention method [5]. This method has achieved higher accuracy recognition rate compared with other state of art approaches.

The remainder of this paper is organized as follows. Section 2 presents the proposed method. Experimental results and benchmarking of results with existing works are discussed in Sect. 3. Ultimately, the final section concludes the work.

2 Proposed Method

Convolutional neural network [1] is made out of several feature map arrays followed by a classifier toward the end. Each layer performs feature extraction and the last layer does classification. Its invariance to scaling, deformation, and shifting make it as an applicable structure. In proposed configuration the combination of the convolution and subsampling operations [5, 7] is implemented. This makes the architecture simpler, which leads to a design that has decreased configuration complexity, higher generalization capacity with lower number of trainable parameters, and quicker execution.

The expanded problem complexities like pose, occlusion, and illumination variations are typically spoken to as disturbance on training data. To obtain generalized robust classification, the effect of outliers with nonsensically high sensitivities has to be lessen. Particularly for the classifier networks, the low sensitivity of output values ensures valid classification even though the input values are changed by noise or other unknown reasons. In proposed method the derivatives of the activation function are applied to the standard output error as an extra term with relative significance factor, and the total error is minimized by the back propagation. In proposed approach, the new cost error $(\widetilde{E})$ [9] is presented as:

$$(\widetilde{E}) = E_o + \sum_{l=1}^{L-1} \Upsilon_l \, E_h^l = \frac{1}{M} \sum_{s=1}^{M} E_o^s + \frac{1}{M} \sum_{s=1}^{M} \sum_{l=1}^{L-1} \Upsilon_l E_h^{sl} \qquad (1)$$

$$E_o^s = \frac{1}{2N_o} \sum_i \left(t_i^s - y_i^s \right)^2 \qquad (2)$$

$$E_h^{sl} = \frac{1}{N_l} \sum_{j_l}^{N_l} \overline{f' \left(h_{j_l}^{sl} \right)} \qquad (3)$$

where E_o^s and E_h^{sl} are the output error and the additional penalty term defined from the lth hidden layer for the sth training pattern, respectively. t_i^s and y_i^s are target and actual output values of the ith output neuron for the sth stored pattern, respectively, and $\left(\widehat{h^{sl}_{J_l}}\right)$ is the corresponding post-synaptic value for the j_l^{th} element at the lth hidden-layer. $f'()$ is the derivative of sigmoid activation function to measure the sensitivity of the output. Here M, N_o, and N_l are the number of stored patterns, number of output neurons, and number of neurons at the lth hidden layer, respectively, and the errors are normalized with these numbers. A constant Y_l represents relative significance of the hidden-layer error E_h^l over output error E_o.

We derived appropriate back propagation algorithm to train the structure by tuning the weights of kernels. The weights are updated as follows:

$$(w^{(l)}_{j_l j_{l-1}})_{new} = (w^{(l)}_{j_l j_{l-1}})_{old} - \eta \quad * \quad \triangle w^{(l)}_{j_l j_{l-1}} \tag{4}$$

where $w^{(l)}_{j_l j_{l-1}}$, is the weights of the kernels, η is the learning rate and $\triangle w^{(l)}_{j_l j_{l-1}}$, is the derivative of error with regard to weights which is an interconnection between the jth element in l^{th} hidden-layer and the element in $l + 1^{th}$ hidden-layer. $\triangle w_{j_l j_{l-1}}$, is calculated for each layer (l) as:

$$\triangle w^{(l)}_{j_l j_{l-1}} = -\eta_l \frac{\partial \widetilde{E}}{\partial w^l_{j_l j_{l-1}}} = \eta_l h^{l-1}_{j_{l-1}} \delta^l_{j_l} \tag{5}$$

where

$$\delta^l_{j_l} = \left(\sum_{j_{l+1}=1}^{N_{l+1}} \delta^{l+1}_{j_{l+1}} w^{(l+1)}_{j_{l+1} j_l} + \frac{Y_l}{MN_l} h^l_{j_l}\right) f'\left(\widehat{h^l_{J_l}}\right) \tag{6}$$

Where, $\delta^l_{j_l}$ is the total error (E) over the neural activations h_{j_l} at the lth layer. As indicated by Eqs. (5) and (6), there are two parts of weight updates, i.e., the back propagated error and gradient of the hidden neuron penalty term. Data samples in the decision boundaries are weighted more by adding the additional term, i.e. $\frac{Y_l}{MN_l} h^l_{j_l} h^{l-1}_{j_{l-1}} f'\left(\widehat{h^l_{J_l}}\right)$. This follows the Hebbian learning rule [8]. The Hebbian term is also multiplied by the derivative of activation function, i.e. $f'\left(\widehat{h^l_{J_l}}\right)$, which avoids drastic update of the network weights. The theoretical justification is represented in [9].

Hebb's principle [8] can be described as a method for deciding how to change the weights between neurons. The weight between two neurons increments if the two neurons activate simultaneously, and decreases if they initiate independently. Nodes that have a tendency to be either both positive or both negative at the same time have solid positive weights, while those that have a tendency to be inverse have solid negative weights.

We applied the proposed robust cost function to the first layer of the CNN and for the other layers just used the simple error cost function. This leads to faster learning and better performance. By using the chain rule for an L-layer CNN, the values of weights are tuned through the all layers of the structure. The comparison studies is performed on FERET dataset.

To calculate $\Delta w^{(l)}_{j_l j_{l-1}}$ for each layer, the effects of error, which is propagated back from previous layers need to be considered. So in the first term of δ^l_{ji}, the multiplication of weights of previous layers and the derivative of the output of each hidden node are mandatory. This is actually the chain rule for L-layer structure and comes from following.

$$\frac{\partial y_i}{\partial x_k} = \sum_{j_1,\ldots,j_{L-1}} W^{(L)}_{ij_{L-1}} W^{(L-1)}_{j_{L-1}j_{L-2}} \cdots W^{(1)}_{j_1 j_k} \widehat{f'^{(y_i)}_L} \; \widehat{f'^{h^{L-1}_{j_{L-1}}}_{L-1}} \; \widehat{f'^{h^1_{j_1}}_1} \tag{7}$$

where x_k and y_i denote the kth element of the input vector and i^{th} element of the output vector, respectively, and $W^{(L-1)}_{j_{L-1}j_{L-2}}$ is the interconnection weights between the $(L-1)^{th}$ and $(L-2)^{th}$ layers. $f'^{(y_i)}_L$, indicates the derivate of activation function at L^{th} hidden layer, and $h^{L-1}_{j_{L-1}}$ is the feature map's values at $(L-1)^{th}$ layer.

3   Experimental Results

The experiments were carried out using conventional cost function and that with the proposed cost function; then, compared with other approaches [10, 11] for FERET dataset. The total number of used samples for FERET dataset is 1100 including 50 subjects each 22 samples. We used 850 of the dataset samples for training and 250 for the test phase. All the samples were chosen randomly and equally distributed for test and train phase. Table 1 presents the detailed structure of conventional CNN and that CNN with the proposed robust cost function.

Table 1. The structure specifications of conventional and proposed CNN

The structure of conventional and proposed CNN	
No. FM: L0/L1/L2/L3/L4	1/4/16/120/50
Type of layer: L1/L2/L3/L4	CS/CS/FC/FC
FM size: L0/L1/L2/L3/L4	29/13/5/1/1
K size: K1/K2/K3/K4	5/5/5/1
SF size: S1/S2/S3/S4	2/2/1/1

In the following tables the abbreviated terms are as follows: No. FM: number of feature maps, L: Layer, CS: Convolution-Subsampling, FC: Fully connected, FM size: feature map size, K size: Kernel size, SF size: Subsampling size.

For preprocessing steps, we used image intensity adjustment which maps the intensity values in grayscale image to new values such that 1% of data is saturated at low and high intensities. This increase the contrast of the output image. Moreover, we applied contrast-limited adaptive histogram equalization (CLAHE) [12] method to enhance the contrast of the image. MATLAB is utilized to prepare the samples in preprocessing stage. Table 2 shows the accuracy rate regarding to change of the Lambda value.

Table 2. The accuracy rate regarding to value of Lambda

CNN	Robust CNN		
	Lambda: 0.6	Lambda: 0.75	Lambda: 0.85
88.8 %	92 %	90.4 %	89.2 %

Table 3 shows the accuracy of the proposed method compared with the other state of art methods. The results show the superior performance of the proposed method in dealing with the illumination, occlusion, and pose variations.

Table 3. The performance of different methods on FERET dataset

Reference/year	Method	No. of subjects	Accuracy
Rinky et al., 2012 [12]	DWT	35	88 %
Shih et al., 2005 [11]	Fisherface	20	90.8 %
Proposed	CNN	50	88.8 %
Proposed	Robust CNN	50	92 %

The proposed cost error was utilized for each layer of the CNN structure and also to the entire layers of the structure. The best performance is obtained from the analysis in which the proposed method was applied only to the first layer. This is because the training process does a hierarchical procedure of forward and backward propagation from the first layer to the last and vice versa. Otherwise by applying the proposed approach just to one layer other than the first or the entire layers, would cause disorder, irregularity, and divergence in the chain of tuning the network weights. The reason is that there are interconnections between different parts of the network including feature maps and the kernels. It is recommended that the weights tuning to be performed layer wise beginning from the last layer and propagates to the first.

4   Conclusion

In this study, new robust cost function for convolutional neural networks is proposed. This robust error contains back propagated error and gradient of the hidden neuron penalty term which promote robustness to the outliers of the training samples. The error value is used to update the values of the weights during the backpropagation process.

The proposed cost function leads to salient feature extraction. This occurs when the new learning algorithm incorporates two popular learning algorithms, i.e., the back-propagation and Hebbian learning principle. The Hebbian term is also multiplied by the derivative of activation function. This additional term gives more weights to the data samples in the decision boundaries which avoids drastic update of the network weights when it feedbacks the small variations of the output error to the input layer. The proposed method presents better performance in compare to other state of art methods.

Acknowledgements. The research was partly supported by 'Basic Science Research Program', through the National Research Foundation of Korea (NRF) funded by the Ministry of Science, ICT and future Planning (2013R1A2A2A01068687) (50%) and by the Industrial Strategic Technology Development Program (10044009) funded by the Ministry of Trade, Industry and Energy (MOTIE, Korea) (50 %).

References

1. LeCun, Y., Kavukcuoglu, K., Farabet, C.: Convolutional networks and applications in vision. In: Proceedings of 2010 IEEE International Symposium on Circuits and Systems (ISCAS). IEEE (2010)
2. LeCun, Y., Bengio, Y.: Convolutional networks for images, speech, and time series. In: The Handbook of Brain Theory and Neural Networks, vol. 3361, pp. 310 (1995)
3. Bouchain, D.: Character recognition using convolutional neural networks. In: Institute for Neural Information Processing, vol. 2007 (2006)
4. LeCun, Y., Huang, F.J., Bottou, L.: Learning methods for generic object recognition with invariance to pose and lighting. In: Proceedings of the 2004 IEEE Computer Society Conference on Computer Vision and Pattern Recognition, CVPR 2004. IEEE (2004)
5. Simard, P.Y., Steinkraus, D., Platt, J.C.: Best practices for convolutional neural networks applied to visual document analysis. In: 2013 12th International Conference on Document Analysis and Recognition. IEEE Computer Society (2003)
6. Khalajzadeh, H., Mansouri, M., Teshnehlab, M.: Hierarchical structure based convolutional neural network for face recognition. Int. J. Comput. Intell. Appl. 12(03), (2013)
7. Mamalet, F., Garcia, C.: Simplifying convnets for fast learning. In: Villa, A.E., Duch, W., Érdi, P., Masulli, F., Palm, G. (eds.) ICANN 2012, Part II. LNCS, vol. 7553, pp. 58–65. Springer, Heidelberg (2012)
8. Hebb, D.O: The Organization of Behavior. Wiley, New York (1949, 1968)
9. Soo-Young, L., Dong-Gyu, J.: Merging back-propagation and hebbian learning rules for robust classifications. Neural Netw. Official J. Int. Neural Netw. Soc. 9(7), 1213–1222 (1996)
10. Shih, F., Fu, C., Zhang, K.: Multi-view face identification and pose estimation using B-spline interpolation. Inf. Sci. 169(3), 189–204 (2005)
11. Rinky, B., Mondal, P., Manikantan, K., Ramachandran, S.: DWT based feature extraction using edge tracked scale normalization for enhanced face recognition. Procedia Technol. 6, 344–353 (2012)
12. Zuiderveld, K.: Contrast limited adaptive histogram equalization. In: Graphics Gems IV. Academic Press Professional, Inc. (1994)

Weighted-PCANet for Face Recognition

Jiawen Huang[(⊠)] and Chun Yuan

Graduate School at Shenzhen, Tsinghua University,
Shenzhen 518055, Guangdong, China
huangjw13@mails.tsinghua.edu.cn, yuanc@sz.tsinghua.edu.cn

Abstract. Weighted-PCANet, a novel feature learning method is proposed to face recognition by combining Linear Regression Classification model (LRC) and PCANet construction. The sample specific hat matrix is used to handle different images in feature extraction stage. After appropriate adaption, the performance of this new model outperform than various mainstream methods including PCANet for face recognition on Extended YaleB dataset. Particularly, various experiments testify the robustness of weighted-PCANet while dealing with less training samples or corrupted data.

Keywords: Convolutional neural network · Principal component analysis · Linear regression classification · Feature extractor · Face recognition

1 Introduction

Manifold learning methods are known to play an important role in face recognition systems [14]. The original image space is such a high-dimensional space that the feature extraction stage is necessary. Lower-dimensional vectors in the face space are the goal of many mainstream recognition algorithms. Over the last several decades, considerable efforts have been devoted to designing appropriate feature extractor, such as the Local Binary Patterns (LBP) [1] features, which use one of the best performing texture depicters in face recognition.

Furthermore, another kind of popular feature depicter is sparse coding. Sparse Representation-based Classification (SRC) [15], a powerful tool in distinguishing signal categories which lie on different subspaces, is based on a reconstructive perspective. An improved method, Superposed Sparse Representation based Classification (SSRC) [4], in which the intra-class differences are based on the P+V model, casts the recognition problem as finding a sparse representation of the test image in terms of a super-position of the class centroids. The Relaxed Collaborative Representation (RCR) [16] model, effectively exploits the similarity and distinctiveness of features, with each feature vector coded on its associated dictionary of coding vectors to address the similarity among features.

What's more, deep neural networks (DNNs), utilizing two or more successive layers of low-level feature extractors and a following supervised classifier to extract high-level features, is a representation of high-level feature descriptors. Particularly, Convolution Neural Networks (CNNs) [9], the key concepts

© Springer International Publishing Switzerland 2015
S. Arik et al. (Eds.): ICONIP 2015, Part IV, LNCS 9492, pp. 246–254, 2015.
DOI: 10.1007/978-3-319-26561-2_30

of which are local receptive fields and tied weights, has been the most popular model for face recognition because of its excellent ability in numeral complex datasets [6–8]. The utility of local receptive fields makes CNNs more efficient in training and more stable in transformation data and weight-tying observably reduces the number of learnable parameters.

However, despite the superior success of these deep network architectures, experience of weight-tuning and extra training tricks are necessary for these deep networks. Therefore, some researchers have exerted themselves to work on simplifying the training process.

Tiled Convolution Neural Networks (TCNN) [11] base on a statistical method Topographic Independent Component Analysis (TICA), obtains better robustness and competitive performance than CNNs. Its application in LFW dataset [18] even obtains the state-of-the-art record. Random weight-TCNN [12] analysis and demonstrate the importance of convolution pooling architectures, which can be inherently frequency selective and translation invariant, even with random weights. Even though the random-weight TCNN has already immensely simplified CNNs, the performance and efficiency cannot be ensured at the same time. Then another simple deep network appeared. PCANet [2] combines principal component analysis (PCA) with deep neural networks, earning new records for many classification tasks including several face recognition tasks.

Many variants of PCA have been presented since PCA is widely used for dimensionality reduction in computer vision fields, especially for face recognition technology, such as LRC [10], bases on a class-specific hat matrix.

Based on this concept, an efficient neural network model weighted-PCANet, attempting to more detail analyses in sample feature extraction is proposed for face recognition in this paper. When extracting features, the importance of different training examples is also considered because the information for face recognition in every training image is not the same and this may influence the performance of feature extracting. Taking the advantage of the workedout resolution matrix of LRC and the efficiency of cascading construction in PCANet, the weighted-PCANet improved the ability of handling small-amount of training set and corrupted inputs including many kinds of transformation, such as noises and shelters. After combining the LRC idea and the PCANet neural network model suitably, this new model outperforms many current methods including PCANet, in Extended YaleB dataset [5], a benchmark database in face recognition, reducing 20 % of the error rate.

2   Algorithm

2.1   Sample Specific Hat Matrix

Since the information for feature extracting in every training sample is different, a sample specific hat matrix H is set to depict more details about the whole training set, which is denoted as follow,

$$H_i = S_i(S_i^T S_i)^{-1} S_i^T \tag{1}$$

where S_i is defined as the feature extracted from the i_{th} sample. The weight matrix is developed from LRC [10] algorithm, which is a modified feature extractor of PCA. Assume that there are N training samples. The matrix F contains all feature vectors from N samples $F = [S_1, S_2, ..., S_N]$. In order to adapt the LRC algorithm to unsupervised extractor, the hat matrix in LRC is changed in the light of feature extraction from all image samples no matter which class it came from. Assume that y_i is the feature matrix of S_i after projection.

$$y_i = \beta_i S_i + e \tag{2}$$

where $_i$ is the projection parameter and e is an error vector whose components are independent random variables with mean zero. The goal of feature extraction is to find a set of suitable $\widetilde{\beta}_i$ to minimize the residual errors e. In hence, the projection coefficients can be solved through the least-square estimation in LRC and can be written as a matrix form in (4)

$$\widetilde{\beta}_i = (S_i^T S_i)^{-1} S_i^T y \tag{3}$$

Then the feature matrix y_i and the sample specific matrix H_i can be denoted as

$$y_i = S_i(S_i^T S_i)^{-1} S_i^T x, H_i = S_i(S_i^T S_i)^{-1} S_i^T \tag{4}$$

where x is the input of previous layer. And y_i can be expressed as $y_i = H_i x$

There are two advantages of this adaptation:

1. The abandon of the idea of extraction features between different classes through class specific hat matrix in original LRC keeps the feature extraction stage an unsupervised process. The importance of unlabeled extractors is well-known to the explosive-growth data nowadays. At the same time, this convert keeps more information from the input data, and after training the filters becomes more suitable for the whole training set and more powerful for the testing set.
2. Since the neural network model of PCANet has already blocked the raw images, the grouping in original LRC is unnecessary in weighted-PCANet. Every training sample in weighted-PCANet is treated as a set of block-size input data, making this operation a kind of grouping of the raw samples.

2.2   Weighted-PCANet

Propose there are N image samples in the dataset for training. For randomness, these samples are out-of-order firstly. Every sample X, is taken into $b = k \times k$ overlapping patches, and then the mean of every patch is subtracted, denoted as $\bar{X}_i$. The constructed matrix of a training sample X is defined as

$$\bar{X} = [\bar{X}_1, \bar{X}_2, ..., \bar{X}_N] \tag{5}$$

The target of feature extractors is to find a suitable set of feature F, which is a concise representation of original data. So a family of orthonormal filters A is needed to minimize the reconstruction error as follows,

$$min\|\bar{X} - AF\|^2, s.t.D(F) = I_L \tag{6}$$

where $D(F)$ is the variance of F. As mentioned in Sect. 2.1, a modified method of PCA can be used to resolve this optimization problem. Since the raw image is blocked, the original LRC is already inappropriate for deep neural network models. Therefore, a sample specific hat matrix is designed in (7),

$$\bar{Y} = mean(\bar{X}) \tag{7}$$

where $\bar{Y}$ is the mean block of $\bar{X}$. So the optimization problem is changed into

$$min\|\bar{Y} - AF\|^2, s.t.D(F) = I_L \tag{8}$$

The solution is known as described in Subsect. 2.1. The filters are therefore expressed as

$$W = mat(q(\bar{Y}_i(\bar{Y}_i^T\bar{Y}_i)^{-1}\bar{Y}_i^T)) = mat(q(H)) \tag{9}$$

where $mat_{k_1,k_2}(x)$ is a function that maps $x \in R^{k_1,k_2}$ to a matrix $W \in R^{k_1 \times k_2}$ and q denotes the principal eigenvectors of the sample specific matrix. The leading principal eigenvectors capture the main variation of all the mean-removed training patches. Then, the output for this layer is a convolution result like this,

$$O_i^l = I_i^{l-1} * W_j^l \tag{10}$$

Of course, similar to CNNs, multiple layers of filters can be stacked to extract higher level features. Two layers are set to extract features after various experiments. A pooling layer is set to follow these convolution layers like the traditional CNNs. In pooling layer, the outputs from the last convolution layer are binarized by a Heaviside like function.

$$H_i^l = \begin{cases} 1 & if O_i^l > 0 \\ 0 & else \end{cases} \tag{11}$$

The order of eigenvalues is reflected by output value multiple a set of exponents of constant 2 because the value of H_i^l is 1 or 0.

$$T_i^l = \sum_{j=1}^{L} 2^{j-1} H_i^l \tag{12}$$

Each of the L_1 images T, $l = 1, ..., L_1$, is partitioned into B blocks. For the feature of face image is not the same in every position of the picture, the blocks is set to be non-overlapping. The histogram of the decimal values in each block is computed, and then concatenated into one vector, denoted as $BHist(T_i^l)$. After this process, the feature of the input sample X is then defined to be the set of block-wise histograms,

$$TF = [BHist(T_i^1), BHist(T_i^2), ..., BHist(T_i^L)] \tag{13}$$

The whole feature extraction procedure of weighted-PCANet is shown in Sect. 2.2.

Algorithm 1. Weighted-PCANet

Input: The training sample x

OutPut:Feature vector as (13)

Step 1: Data normalization: block the images and subtract the block mean. $\bar{X} = [\bar{X}_1, \bar{X}_2, ..., \bar{X}_N]$ by (5)

Step 2: Convolution layer. Change $\bar{X}$ to the image mean block $\bar{Y}$ by (10). Utilize sample specific matrix (1) and (9) to compute the l_{th} matrix W, and the output of this layer is $O_i^l = \{I_R^{l-1} * W_j^l\}$ by (10)

Step 3: Convolution layer. Repeat Step 2.

Step 4: Pooling layer. Binarized the output by (11) and hist them as (12) and (13).

2.3   Comparison with CNNs and PCANet

Weighted-PCANet shares the main construction characteristics With classical CNNs [9] as a cascading neural network, including convolution layers and pooling layers. The advantage of CNNs is taken in the design of weighted-PCANet, such as the utility of local vision fields. However, the traditional back propagation (BP) algorithms in CNNs are replaced with a solved optimization problem. There are two parts in BP algorithm: information transmission and forward error back-propagation, which is extensively utilized in traditional Artificial Neural Networks (ANNs). The common point in all ANN models is to find the optimal solution for the loss function iteratively. Especially in CNNs, the kernel filters are randomly set at the very first time and fine-tuned over and over again by stochastic gradient descent method in the process of forward error back-propagation. Not only does the iteration refinement last so long, but also there are many problems about the gradient descent method such as gradient diffusion and convergence to the local optimum which need abundant training experience. Instead, there isnt any forward error back-propagation process in weighted-PCANet. Moreover, the information transmission in it is only conducted for one time. Since the weight matrix of filters is set at a time based on sample specific matrix for all training data. Last but not the least, a worked-out optimization problem in convolution layer simplifies the theoretical analysis of deep neural networks which is an urgent need when the performance of it refreshed new record again and again and the application of it is more and more common.

While comparing with PCANet, in feature extracting stage, the significance of every training sample is different in weighted-PCANet. As shown in Fig. 1, even from the same individual, the global information for face recognition in two samples is far different. Furthermore, LRC is an improved feature depicter of PCA especially in face recognition. So after being elaborately designed to adapt block-calculation in cascading neural network, the performance of this model will outperform the original PCANet without the increase of time.

Fig. 1. Different training samples from Extended YaleB database

3 Experiments

3.1 Dataset and Experiment settings

The Extended Yale B dataset consists of 2414 frontal-face images of 38 human subjects, respectively 64 images for every individual. The cropped and normalized 192×168 face images were captured under various laboratory-controlled lighting conditions. The images are resized to 32×32 pixels. 8 samples are randomly picked for training and the rest for testing. The parameters of weightedPCANet and PCANet-2 [2] are set as follow, 2 convolution layers and 8 filters for each layer. The overlapping block size in convolution layers is 5×5, and the non-overlapping block size in pooling layer is 4×3. The whole model is followed by a linear-SVM classifier [3]. One layer weight-PCANet is also set to prove the excellent ability of multi-layer learning. The setting of CNNs-2 is the same as the famous LeNet-2 [9], 2 layers with filter size 5×5; 20 channels for the first convolution layer and 50 channels for the second convolution layer. Each convolution layer is followed by a 2×2 max-pooling layer. The output layer is a softmax classifier and the whole model is iterated for 500 times. In random-weight TCNNs [12], 20 maps is set for the first layer and 50 maps for second, window size is 5×5 and untied pooling size is 1×1. The model also uses a linear-SVM for classification [3].

3.2 Experimental Results and Analysis

A summary of results is reported in Table 1.Weighted-PCANet obtains the best performance in all testing tasks. Although the training samples are only seventh of the whole data set, weighted-PCANet can get excellent performance because of the painstaking design of feature extraction with sample specific matrix. As analyzed before, when the amount of samples of per individual is small and the difference between every sample is global such as illumination, weighted-PCANet can extract more precise features than other feature depicters. Compare with CNNs-2, the efficiency of the weighted-PCANet is remarkable. Since the use of time without iteration in weighted-PCANet is hundredth of traditional

CNNs-2, the accuracy of weighted-PCANet super outperforms this 500-iteration CNNs-2. Random-weight TCNNs [12] is used to compare with weighted-PCANet because it is also without iteration. However, the performance is barely satisfactory. The performance of weighted-PCANet also outperforms PCANet, LRC and their variants. Especially, the error rate of weighted-PCANet cuts down more than 40 % of LRC and at the same time weighted-PCANet reduces 20 % of the error-rate of PCANet. These results proved that the design of weighted-PCANet applies the LRC in PCANet is suitable for face recognition. The outstanding performance of LBP fea-ture and weighted-PCANet also merits our attentions that these simple methods can also get excellent performance in benchmark dataset.

Table 1. The Error Rates of the Recognition on Extended YaleB Database(%)

Algorithms	Error rate
CNNs-2	96.60
random-weighted TCNNs-2	77.54
LBP [14]	6.91
SVM [13]	23.79
CRC [17]	28.51
RCR [16]	28.11
SRC [15]	41.53
SSRC [4]	40.09
LRC [10]	46.14
PCANets-2 [2]	7.09
weighted-PCANet-1	10.89
weighted-PCANet-2	**5.88**

To prove the analysis mentioned before, the corrupted test images are used for further test tasks, including two levels of contiguous occlusion 10 percent and 20 percent, by replacing a randomly located square block of each image with an unrelated image and multiplicative-Gaussian-noised test image. See Fig. 2 for example. Because the accuracy of CNNs-2 and random-weight TCNNs-2 are too low in the raw test task, they are ignored in these corrupted image tasks. The results are shown in Table 2.

Fig. 2. Corrupted testing images (Left to Right) : raw, 10 % occlusion, 20 % occlusion and noisy image.

Table 2. The Error Rates on Corrupted Test Set(%)

Algorithms	Noised	10 %	20 %
LBP [14]	12.44	8.58	12.73
SVM [13]	24.08	23.91	25.17
CRC [17]	29.72	32.60	53.92
RCR [16]	30.41	49.77	34.45
SRC [15]	41.70	46.03	51.96
SSRC [4]	35.25	40.04	46.49
LRC [10]	46.66	49.77	53.69
PCANets-2 [2]	12.21	7.50	9.62
weighted-PCANet-1	16.24	11.00	13.77
weighted-PCANet-2	**10.77**	**5.88**	**7.14**

It is apparent that weighted-PCANet outperforms other methods even though the images are corrupted seriously. Furthermore, the performance of weighted-PCANet even better than PCANet while in the original testing task. It is a powerful certificate that the new weighted-PCANet model is more robust than PCANet while the input data is contaminate cause the design of combination with LRC for detail conducting in training stage.

4 Conclusion

In this paper, a novel feature learning method by exploiting PCA-Net is proposed to face recognition. The sample specific hat matrix is used to handle different image in training set, which is capable of learning complementary, hierarchical representations. The idea of LRC is applied in PCA-Net after appropriate adaption. Experimental results show the effectiveness of weighted-PCANet for face recognition on Extended YaleB dataset. Particularly, weighted-PCANet outperforms LRC when images are different in illumination and more meticulous than PCANet in information conduct to training samples.

References

1. Ahonen, T., Hadid, A., Pietikainen, M.: Face description with local binary patterns: application to face recognition. IEEE Trans. Pattern Anal. Mach. Intell. **28**(12), 2037–2041 (2006)
2. Chan, T.H., Jia, K., Gao, S., Lu, J., Zeng, Z., Ma, Y.: PCANet: a simple deep learning baseline for image classification? arXiv preprint (2014). arXiv:1404.3606
3. Chang, C.-C., Lin, C.-J.: LIBSVM: a library for support vector machines. ACM Trans. Intell. Syst. Technol. (TIST) **2**(3), 27 (2011)
4. Deng, W., Hu, J., Guo, J.: In defense of sparsity based face recognition. In: 2013 IEEE Conference on Computer Vision and Pattern Recognition (CVPR), pages 399–406. IEEE (2013)

5. Georghiades, A.S., Belhumeur, P.N., Kriegman, D.J.: From few to many: illumination cone models for face recognition under variable lighting and pose. IEEE Trans. Pattern Anal. Mach. Intell. **23**(6), 643–660 (2001)
6. Hinton, G.E., Srivastava, N., Krizhevsky, A., Sutskever, I., Salakhutdinov, R.R.: Improving neural networks by preventing co-adaptation of feature detectors. arXiv preprint (2012). arXiv:1207.0580
7. Krizhevsky, A., Sutskever, I., Hinton, G.E.: Imagenet classification with deep convolutional neural networks. In: Advances in Neural Information Processing Systems, pp. 1097–1105 (2012)
8. Lawrence, S., Giles, C.L., Tsoi, A.C., Back, A.D.: Face recognition: a convolutional neural-network approach. IEEE Trans. Neural Netw. **8**(1), 98–113 (1997)
9. LeCun, Y., Bottou, L., Bengio, Y., Haffner, P.: Gradient-based learning applied to document recognition. Proc. IEEE **86**(11), 2278–2324 (1998)
10. Naseem, I., Togneri, R., Bennamoun, M.: Linear regression for face recognition. IEEE Trans. Pattern Anal. Mach. Intell. **32**(11), 2106–2112 (2010)
11. Ngiam, J., Chen, Z., Chia, D., Koh, P.W., Le, Q.V., Ng, A.Y.: Tiled convolutional neural networks. In: Advances in Neural Information Processing Systems (2010)
12. Saxe, A., Koh, P.W., Chen, Z., Bhand, M., Suresh, B., Ng, A.Y.: On random weights and unsupervised feature learning. In: Proceedings of the 28th International Conference on Machine Learning (ICML-11), pp. 1089–1096 (2011)
13. Suykens, J.A.K., Vandewalle, J.: Least squares support vector machine classifiers. Neural Process. Lett. **9**(3), 293–300 (1999)
14. Turk, M., Pentland, A.: Eigenfaces for recognition. J. Cogn. Neurosci. **3**(1), 71–86 (1991)
15. Shankar Sastry, S., Wright, J., Ganesh, A., Ma, Y., Yang, A.: Robust face recognition via sparse representation. IEEE Trans. Pattern Anal. Mach. Intell. **31**(2), 210–227 (2009)
16. Yang, M., Zhang, L., Zhang, D., Wang, S.: Relaxed collaborative representation for pattern classification. In: 2012 IEEE Conference on Computer Vision and Pattern Recognition (CVPR), pp. 2224–2231. IEEE (2012)
17. Zhang, L., Yang, M., Feng, X.: Sparse representation or collaborative representation: which helps face recognition? In: 2011 IEEE International Conference on Computer Vision (ICCV), pp. 471–478. IEEE (2011)
18. Zhu, Z., Luo, P., Wang, X., Tang, X.: Deep learning identity-preserving face space. In: 2013 IEEE International Conference on Computer Vision (ICCV), pp. 113–120. IEEE (2013)

Feature Extraction Based on Generating Bayesian Network

Kaneharu Nishino[✉] and Mary Inaba

Graduate School of Information Science and Technology,
The University of Tokyo, Tokyo, Japan
{nishino.kaneharu,mary}@ci.i.u-tokyo.ac.jp

Abstract. Networks used in Deep Learning generally have feedforward architectures, and they can not use top-down information for recognition. In this paper, we propose Bayesian AutoEncoder (BAE) in order to use top-down information for recognition. BAE constructs a generative model represented as a Bayesian Network, and the networks constructed by BAE behave as Bayesian Networks. The network can execute inference for each stochastic variable through belief propagation, using both bottom-up information and top-down information. We confirmed that BAE can construct small networks with one latent layer and extract features in 3×3 pixel input data as latent variables.

1 Introduction

Recently, Deep Learning has demonstrated high performance in object recognition tasks [3,4]. There are two types: methods that produce generative models, and methods using auto-encoders. Both of them use multi-layer networks and extract features of input data. Nodes in the higher layers represent more abstract features, and they are effective for object recognition.

One of the major methods of making generative models is Deep Belief Network (DBN) [2]. DBN is trained so that the network tends to generate training data, and acquires the factors which generate the data as its hidden variables. In recognition, DBN uses these factors as features of the input data. DBN calculates feature values from bottom to top as does feedforward. On the other hand, methods using auto-encoders [8,11] use multi-layer neural networks. The networks are trained in order to reconstruct input data from the output of their latent layers. In recognition, these networks behave as feedforward networks. In these networks, the data flows only from the input layer to the output layer.

In brains, there are not only feedforward connections from the lower level to the higher level but also feedback connections from higher to lower. As the reason for the feedback connections, some studies propose that the information processing model of brains is based on a Bayesian Network (Bayes Net) [5,9]. When using belief propagation of Bayes Net, input information flows top-down and bottom-up. For example, the Bayes Net can predict and attend to local information such as words and characters from their global context.

© Springer International Publishing Switzerland 2015
S. Arik et al. (Eds.): ICONIP 2015, Part IV, LNCS 9492, pp. 255–262, 2015.
DOI: 10.1007/978-3-319-26561-2_31

In this paper, we propose the Bayesian AutoEncoder (BAE). This is a method that constructs a generative model represented as a Bayes Net, and extracts features as latent variables. Networks constructed through BAE can handle both feedforward and feedback flows of information.

2   Related Work

2.1   Deep Belief Networks

Deep Belief Network (DBN) is one of the major technologies of Deep Learning using a generative model. DBN extracts hidden variables through training Restricted Boltzmann Machines (RBMs) [1].

RBM is an undirected probabilistic graphical model represented by a complete bipartite graph with one hidden layer h and one visible layer v. Given values for visible layer h, RBM can calculate posterior distributions over hidden nodes. RBM can also calculate the distributions over visible variables v, and RBM adjust its parameters in order to maximize the probability that it will generate training data for v. As a result, RBM extracts hidden nodes as factors to better generate training data.

DBN is produced by stacking RBMs. The higher RBM is trained by using the values which are sampled from the hidden layer of a lower trained RBM as values of visible nodes. The higher RBM extracts hidden variables as factors for posterior distributions for the lower hidden variables. By stacking multiple layers, DBN extracts variables representing complex concepts, which are invariant to small changes in the input data. Using these invariant variables as features for classification, DBN yields high performance. However, each RBM only calculates each hidden layer distribution from each vector given to a visible layer, so the activities of variables depend on only the lower layer information.

2.2   Bayesian Networks

To handle top-down information, we adopted what is called *belief propagation* on Bayesian Networks (Bayes Nets) [7].

A Bayes Net is a directed probabilistic graphical model, and it is suitable for representing causality between random variables. If variables have a conditional relationship, their nodes have links. Exchanging information called *messages* along links, Bayes Nets give the posterior distributions of any variables when any set of variables is observed. This inference using messages is *belief propagation*. A Bayes Net gives exact posterior distributions if the network has a tree structure, and often gives approximate posteriors even if the network has loops [10].

A belief propagation on a Bayes Net uses not only bottom-up information, but also top-down information. We considered whether generative models of training data represented by a Bayes Net might exhibit high performance in object recognition tasks.

3   Proposal for a Bayesian AutoEncoder

In order to handle the feedback information in recognition, we propose a Bayesian AutoEncoder (BAE) to construct a generative model as a Bayes Net.

BAE first constructs a two layer Bayes net as a directed complete bipartite graph containing a visible layer of child variable nodes and a latent layer of parent variable nodes. BAE adjusts the parameters of the network, and cuts some links so that the network can predict input data. The parent nodes come to represent features of input data. The variables are binary variables, which are either true (T) or false (F) (existent or non existent) respectively of corresponding features. When the network has been trained sufficiently, BAE stacks a new latent layer over the parent variables, completely connects between them, and adjusts parameters between them in the same way. Thus BAE constructs a multi-layer Bayes Net as a generative model of the input data. The constructed network recognizes each pattern corresponding to each of its variable nodes from the given data through belief propagation.

BAE trains each layer of the network so that it can predict each part of the input for the layer from the rest of the input (where the input is bottom-up information for the layer) inspired by predictive coding [6]. BAE adjusts each link parameter in order to minimize the difference between the top-down message inference for each variable and the bottom-up message inference.

During training, links that are not necessary for prediction weaken. When links are weaker than the threshold, BAE cuts these links and reduces its calculation of messages sent on them. Thus BAE changes the network structure.

4   Implementation of Bayesian AutoEncoder

In this section, we describe the implementation of BAE. We implemented BAE in order to construct the Bayes Net with one parent layer and one child layer. Both of the parent variables and child variables are binary variables, which have either true (T) state or false (F) state. In this implement, each child variable node has a visible variable node (called the *input node*) as its child, which has a gradient value. The links between the child nodes and the input nodes are fixed, and BAE adjusts links between the parent layer and the child layer.

Below we present a definition of the parameters of the network, and the belief propagation in the network. In addition, we define the loss function that is used to update the parameters.

4.1   Definitions and Assumptions for the Parameters

A Bayes Net needs parameters which denote conditional probabilities between parents and children. However, BAE trains networks from a complete bipartite graph, and therefore each child node needs $O(2^n)$ parameters where n is the number of parents. To reduce the parameters, we make the assumption that only one parent node can have a T state among parent variables sharing child variables with the parent variable. Given this assumption, each child node in networks constructed through BAE has $2n+2$ parameters where n is the number of parents. We define the parameters of a child node v which has links to parent nodes $u_1, u_2, \dots$ as:

$$p_{v,u_i} = b_{v,u_i} p(v = T | u_i = T), p'_{v,u_i} = b_{v,u_i} p(v = F | u_i = T) \tag{1}$$

$$p_v = p(v = T | u_1 = F, \dots), p'_v = p(v = F | u_1 = F, \dots) \tag{2}$$

where $b_{v,u} = p_{v,u} + p'_{v,u}$, and we assign $b_{v,u}$ as the link strength.

4.2   Belief Propagation in BAE

Networks constructed through BAE execute belief propagation by exchanging information called messages along links between nodes. The messages are calculated at each node from messages the node receives based on the formula below. Messages are updated repeatedly.

The message $m_{d,v}$ passed to a child node v from its input node d is given by:

$$m_{d_v} = \frac{x}{x_{max} - x} \cdot \frac{p(v = T)}{p(v = F)} = \frac{p(d = x | v = T)}{p(d = x | v = F)} \tag{3}$$

where x denotes the input value for d, and x_{max} denotes the max value of x. The prior probability for v is $p(v = T)$ and is acquired through the expectation of $p(v = T | d = x)$, where $p(v = T | d = x) = x/x_{max}$. The message $m_{u,v}$ passed to a child node v from its parent node u is given by:

$$m_{u,v} = \frac{p(u = T)}{p(u = F)} \prod_{v' \in V \setminus v} m_{v',u} \tag{4}$$

where V denotes a set of child nodes of u, and $p(u)$ denotes the prior probability of u. However, in this implementation, $p(u = T)$ is fixed to 0.5. The message $m_{v,u}$ passed to a parent node v from its child node u is given by:

$$m_{u,v} = n_{u,v}^{p_{v,u} + p'_{v,u}} \tag{5}$$

$$n_{u,v} = \frac{1}{p_{v,u} + p'_{v,u}} \frac{p_{v,u} m_{d,v} + p'_{v,u}}{p_v m_{d,v} + p'_v + \sum_{u' \in U \setminus u} p_{v,u'} m_{d,v} m_{u',v} + p'_{v,u'}} \tag{6}$$

where U denotes a set of parent nodes of v. We use Eq. (5) so that, if a child node v has a stronger link to a parent node u, messages from u is more affected by the message $m_{v,u}$.

4.3   Definition of Loss Function

Here we define a loss function to update parameters. BAE updates the parameters of a child node v in order to minimize difference between the inference for v

from messages from its parent nodes U, and the inference from a message from its input node d. Therefore, the error function L used in BAE is defined as a cross entropy between the two inference such as:

$$L = -\frac{1}{r+1} \log \frac{1}{q+1} - \left(1 - \frac{1}{r+1}\right) \log \left(1 - \frac{1}{q+1}\right) \qquad (7)$$

where :

$$q = \frac{p_v + \sum_{u' \in U} p_{u',v} m_{u',v}}{p'_v + \sum_{u' \in U} p'_{u',v} m_{u',v}}, r = \frac{p(v = T)}{p(v = F)} m_{d,v} \qquad (8)$$

The value $q/(q+1)$ is the probability of $v = T$ calculated from messages from U, and $r/(r+1)$ is the probability of $v = T$ calculated from messages from d. BAE minimizes L through a stochastic gradient method to update the parameters.

Through the training, a link strength $b_{v,u} = p_{v,u} + p'_{v,u}$ changes. If $b_{v,u}$ is less than its threshold, the link between u and v is cut. In this implementation the threshold is half of the mean of the link strength of all links of the child node v.

5 Experiments for Feature Extraction Through BAE

To confirm that the network trained through BAE extracts features as the parent variables from its input data, we carried out several experiments. In these experiments, we constructed small networks through BAE using 3×3 pixel data as its input, and confirmed extracted features visually. As a result, the network structure changed, and latent variables were found to correspond to features in the input data.

5.1 Configurations

This network is composed of two layers, the child variable node layer and the parent variable node layer. Each layer has 9 nodes. Each child node has a input node which can have input values from 0 to 20. The parameters $p_{v,u}$ are initialized randomly from 0 to 1, and parameters $p'_{v,u}$ are given through $p'_{v,u} = 1 - p_{v,u}$. The values p_v, p'_v are initialized as 0.5. Messages are updated 10 times for each input datum, and the parameters are updated for every two message updates.

5.2 Input Data

As input, we used nine dimensional data corresponding to a 3×3 pixel image. Each pixel has a value 0.1 or 19.9. The value of 0.1 is black, and 19.9 is white. Pixels are indexed as shown in Table 1. We call triad pixels (1,2,3), (4,5,6) and (7,8,9) rows, and triad (1,4,7), (2,5,8) and (3,6,9) columns.

Table 1. Index of pixels

1	2	3
4	5	6
7	8	9

We prepared three groups of data for input: horizontal, vertical, and mixed. We independently chose two states, one 25 % ON and the other 75 % OFF for each row in the horizontal, for each column in the vertical, and for each row and column in the mixed. The background of each input is 0.1 (black).

Horizontal: Each row is in two independently states. ON rows had the pixel value set to 19.9. OFF rows do not change their pixel values, and each pixel in OFF rows remains at 0.1. This input group has 8 patterns as shown in Fig. 1(a). Vertical: Each column is in two independently states. This group has 8 patterns as shown in Fig. 1(b). Mixed: We chose for each row and column two independently states. Each ON row or column is set to 19.9 for each pixel value. The pixels which are in the OFF rows and in the OFF columns remain at 0.1. This group has 48 patterns as shown in Fig. 1(c).

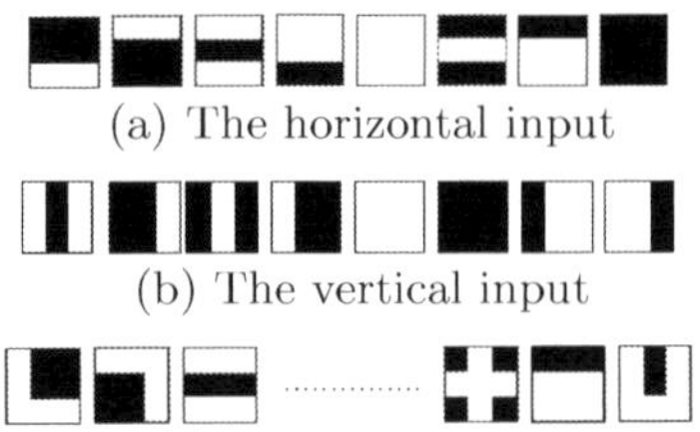

(a) The horizontal input

(b) The vertical input

(c) The mixed input

Fig. 1. The three groups of input data: the horizontal input (a), the vertical input (b), and the mixed input (c).

5.3 Learned Network Structure

The parameters learned in the experiment are shown in Table 2. These parameters are updated using the horizontal input.

In this table, the link parameters of each parent node between each child node are in each cell. The higher parameter is $p_{v,u}$, and the other is $p'_{v,u}$. The 'c' in the darker cells indicates their link is cut.

Table 2 shows that many links have been cut. This demonstrates that BAE can change the network structure. We visualize the links of each parent variable in the next section.

Table 2. Parameters adjusted through BAE using the horizontal input.

\ child parent	1	2	3	4	5	6	7	8	9
1	c	c	c	0.1 3.0	c	c	c	c	c
2	3.6 0.0	3.6 0.0	0.3 1.5	c	c	c	c	c	c
3	2.4 0.0	2.4 0.0	1.7 1.4	c	c	c	c	c	c
4	c	c	c	c	c	0.6 0.6	2.0 0.0	1.7 0.8	
5	c	c	c	c	c	c	1.9 0.8	2.0 0.1	c
6	c	c	c	c	c	c	0.6 0.5	2.1 0.0	1.4 0.9
7	c	c	c	2.2 0.6	c	2.9 0.0	c	c	c
8	c	c	c	c	2.4 1.5	2.8 0.0	c	c	c
9 deleted	c	c	c	c	c	c	c	c	c

Visualized Links of Parent Nodes. Fig. 2 visualizes how parent nodes are connected to child nodes; each box corresponds to a parent node, 3×3 pixels in the box corresponds to child nodes, and their color corresponds to the parameters of the links. If the link is stronger, the pixel is more opaque: when $b_{v,u}$ is a higher value, the pixel corresponding to the child v is more opaque in the box corresponding to the parent u. When $b_{v,u}$ is lower, the pixel becomes more transparent, and becomes red because the background is red. When a conditional probability $p_{v,u}/b_{v,u}$ is near 1, the pixel is near white, and when the probability is near 0, the pixel becomes black. Figure 2(a), (b) and (c) show links updated using the horizontal input, the vertical input and the mixed input.

In Fig. 2(a), the boxes have horizontal patterns. This denotes that the parents are activated (the parent variables have a high posterior probability) specifically

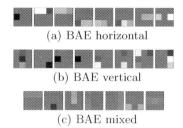

(a) BAE horizontal

(b) BAE vertical

(c) BAE mixed

Fig. 2. Extracted links of parent nodes, using the horizontal (a), the vertical (b), and the mixed input (c).

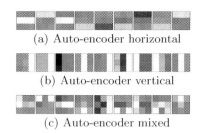

(a) Auto-encoder horizontal

(b) Auto-encoder vertical

(c) Auto-encoder mixed

Fig. 3. Weights learned by auto-encoders, from the horizontal (a), the vertical (b) and the mixed input (c).

through the horizontal input. Boxes in Fig. 2(b) have vertical patterns. These parents became specific to the vertical input. In Fig. 2(c), each parent node has links to only one row or column from the mixed input.

From these results, we can conclude that the network links reflect patterns in its input data. From the horizontal input, the parent nodes have links to the pixels of only one horizontal row, and they are activated specifically when horizontal rows are the input. A case in which the input is vertical is the same. Patterns in the input data are extracted as parent variables. Also when the input is the mixed, patterns in the input data are extracted as rows and columns.

Results of the Conventional Auto-Encoder. For comparison, we have also tried the same feature extraction through the auto-encoder. The neural networks have an input layer possessing 9 nodes, and one latent layer possessing 9 nodes. The neural networks were trained without any regularizations. We visualized the weight parameters in Fig. 3. If a weight parameter is 0, the pixel is gray, when a weight parameter falls, the pixel becomes darker, and when a weight parameter becomes larger, the pixel becomes brighter. Figure 3(a) is from the horizontal input, 3(b) is from the vertical input, and 3(c) is from the mixed input.

Comparison Between BAE and Auto-Encoder. The auto-encoder is superior to BAE in feature extraction from the horizontal input and the vertical input, but not from the mixed input.

Observing Fig. 2(a) and (b), some parent nodes have only two children, and the parameters of parents which have three children vary. This is not a positive result: each parent should have three children, and its parameter should be nearly the same. The results of auto-encoder are better. On the other hand, for the feature extraction from the mixed input, BAE is superior to auto-encoder. The auto-encoder did not extract its features as shown in Fig. 3(c): the patterns of the mixed input should horizontal rows and vertical columns. From Fig. 3(c) shows that the auto-encoder did not separate each column and row. BAE separated them as shown in Fig. 2(c).

We have considered the reason for this. The auto-encoder used in this experiment did not have any regularization, but in the network constructed by BAE, the parent nodes sharing the same child nodes suppressed each other so that only one parent node had a high prior probability (because of the parameter assumption), and this worked as regularization.

6   Concluding Remarks

In this paper we proposed the Bayesian AutoEncoder as a method to construct a generative model represented as a Bayesian Network for training data, and confirmed that BAE can extract patterns of data as the network structure.

In the experiment using mixed input, the results of BAE were better than that of auto-encoder. We concluded that this is because of the regularization of BAE, and we need to carry out more experiments to confirm that this regularization is valid for other input data.

Our eventual goal is to construct multi-layer networks from training data through BAE. If BAE can extract more abstract features with a multilayer network, the network can execute inference for each variable using, not only bottom-up, but also top-down information. By using top-down information, the recognition of an environment through the networks can affect the recognition of local objects, and this will be beneficial for object recognition in our opinion.

References

1. Hinton, G.: Boltzmann machines. Scholarpedia
2. Hinton, G., Salakhutdinov, R.: Reducing the dimensionality of data with neural networks. Science **313**(5786), 504–507 (2006)
3. Krizhevsky, A., Sutskever, I., Hinton, G.E.: Imagenet classification with deep convolutional neural networks. In: Pereira, F., Burges, C., Bottou, L., Weinberger, K. (eds.) Advances in Neural Information Processing Systems, vol. 25, pp. 1097–1105. Curran Associates, Inc. (2012). http://papers.nips.cc/paper/4824-imagenet-classification-with-deep-convolutional-neural-networks.pdf
4. Le, Q.: Building high-level features using large scale unsupervised learning. In: 2013 IEEE International Conference on Acoustics, Speech and Signal Processing (ICASSP), pp. 8595–8598, May 2013
5. Lee, T.S., Mumford, D.: Hierarchical bayesian inference in the visual cortex (2002)
6. Olshausen, B.A., Field, D.J.: Emergence of simple-cell receptive field properties by learning a sparse code for natural images. Nature **381**, 607–609 (1996)
7. Pearl, J.: Bayesian networks: a model of self-activated memory for evidential reasoning. In: Cognitive Science Society, pp. 329–334 (1985)
8. Ranzato, M., Poultney, C.S., Chopra, S., LeCun, Y.: Efficient learning of sparse representations with an energy-based model. In: NIPS, pp. 1137–1144. MIT Press (2006)
9. Shon, A.P., Rao, R.P.: Implementing belief propagation in neural circuits. Neurocomputing **65–66**, 393–399 (2005). Computational Neuroscience: Trends in Research 2005
10. Yedidia, J.S., Freeman, W., Weiss, Y.: Constructing free-energy approximations and generalized belief propagation algorithms. IEEE Trans. Inf. Theory **51**(7), 2282–2312 (2005)
11. Bengio, Y., Lamblin, P., Popovici, D., Larochelle, H.: Greedy layer-wise training of deep networks, pp. 153–160 (2007)

Neural Population Coding of Stimulus Features

David Iclănzan[(⊠)] and László Szilágyi

Sapientia Hungarian University of Transylvania, Tîrgu-Mureş, Romania
david.iclanzan@gmail.com

Abstract. While empirical evidence suggest that the brain can represent and operate on probability distributions, it is not clear how multivariate dependencies can be detected and represented by neural circuits. Based on previous work and the principle of entropy distillation, the paper introduces a massively parallel connectionist machine whose spiking behavior adapts to the statistical distribution of binary inputs. Experimental results confirm that the network is able to accurately capture the joint probability distribution of the inputs and it is able to represent even higher-order features lacking pairwise correlations.

Keywords: Neural coding · Entropy distillation · Higher-order features

1 Introduction

Behavioral and physiological evidence imply that the brain is able to represent and work with probability distributions [1]. Despite advancements [2,3], it is not yet clear how complex distributions might be encoded and operated upon.

In this work we address aspects from the first part of this issue: encoding and decoding statistical dependencies arising from binomial probability distributions.

Neural encoding describes the mapping function from stimulus to response, while the decoding reconstructs certain aspects of the input, from the neural activity it evokes.

In our proposed approach, even complex multivariate features can be encoded by a massively parallel network, where a firing neuron unequivocally identifies the dependent variables, while the spike-count rate encodes the strength of the dependency.

The novel foundation of the proposed network, is based on the principle of entropy distillation and exploitation of imbalances, principles and techniques that were developed and used in linear cryptanalysis.

Namely, the piling-up lemma is an analytical tool used in linear cryptanalysis, introduced by Matsui [4] in order to construct linear approximation to the effect of block chippers. A statistical independence test can be constructed on the converse of this lemma [5]. This converse was successfully used in the area of machine learning, linkage detection [6,7].

In these previous works, an exhaustive search was performed to find groups of variables that violates the independence converse, therefore uncovering the

© Springer International Publishing Switzerland 2015
S. Arik et al. (Eds.): ICONIP 2015, Part IV, LNCS 9492, pp. 263–270, 2015.
DOI: 10.1007/978-3-319-26561-2_32

statistical dependencies. The computation of order k entropy distillations (statistical independence test between k variables) are independent, therefore the process can be readily mapped to a massively parallel network.

2   Materials and Methods

2.1   Entropy Distillation and Statistical Independence

The statistical independence testing method is derived starting from the Piling-up Lemma [4], which is stated in the following:

Lemma 21. *Given n independent random binary variables $X_1, X_2, \ldots, X_n$, the bias $\epsilon = P(X = 0) - 1/2$ of the sum $X = X_1 \oplus X_2 \oplus \ldots \oplus X_n$ is given by:*

$$\epsilon = 2^{n-1} \prod_{i=1}^{n} \epsilon_i \tag{1}$$

where $\oplus$ denotes the exclusive OR (XOR) operation and ϵ_i are the biases of the terms X_i i.e. $\epsilon_i = P(X_i = 0) - 1/2$.

By denoting the imbalance (or correlation [5]) of an expression with $c = 2\epsilon = 2P(X = 0) - 1$, the Eq. 1 can be reduced to $c = \prod_{i=1}^{n} c_i$.

While the Piling-up lemma is usually used for calculating the imbalance or correlation of linear approximations, its converse offers a natural criterion that one can use to verify statistical independence of linear approximations: *all* (not just pairwise) linear combinations of a linear approximation must be of a small magnitude [5]. Formally:

Theorem 22. *Let $n \geq 2$ be an integer. The binary variables $X_1, X_2, \ldots, X_n$, with imbalances $c_i = c(X_i)$, $i = 1 \ldots n$ are statistically independent, if and only if for all index sets $I \subset \{1, 2, \ldots n\}$,*

$$c(\oplus_{i \in I} X_i) = \prod_{i \in I} c_i \tag{2}$$

For the proof refer to pp. 332 in [5].

If we encode the binary states $\{FALSE, TRUE\}$ as $\{-1, 1\}$ instead of $\{0, 1\}$, XOR is equivalent with the negated multiplication. The product $-(X_1 * X_2)$ outputs $TRUE$ i.e. 1, only when both inputs differ (one is -1, the other is 1). With such an encoding only the computation of the biases change to $\epsilon = (p(X = -1) - p(X = 1))/2$.

For ease of computation, in these paper we use the $\{-1, 1\}$ encoding and the negated multiplication.

2.2   Entropy Distillation Networks

The proposed entropy distillation network operation principle is the following:

(i) The combinations of binary inputs are "shuffled" n-parallel and propagated through a multilayer network of neurons in a feed-forward manner, an operation that normally increases the entropy at each step i.e. an entropy distillation is performed, the imbalances quickly approach 0.

(ii) If present, smaller or larger deviations from the expected entropy, are accumulated over time by each neuron.

(iii) If the accumulated bias exceeds a certain threshold, the neurons fire, signaling a possible underlying statistical dependence.

(iv) Higher spike-count rates correspond to stronger statistical dependencies.

Let $2 \leq k \leq n$ be the maximal order of interactions (size of features) we wish to cover. Then the entropy distillation network will contain k layers, where the first layer is just a copy of the n inputs. We denote such a network as $EDN(n, k)$.

If the biases ϵ_i of the input variables $X_1, X_2, \ldots, X_n$ are known or can be measured, reliably approximated, the expected imbalances can be computed as per Eq. 2 as in previous work [7], making the comparison with the empirically observed imbalances precise.

As we wish to propose a plausible, very simple connectionist model for detecting and signaling/encoding multivariate features, here we neglect the computation of the expected higher order imbalances and assume that these values are 0.

The simplifying assumption that higher order imbalances are exactly 0, introduces some noise and stochastic spiking behavior.

However, these effects, especially in higher index layers, are usually very small, as imbalances are multiplied 2. Note that when multiplying, if any imbalance is 0 (at least one of the binary variables is unbiased), the entire product will be 0.

Even when all imbalances are relatively high, because these numbers are subunitary, their product is subject to exponential decay. For example, if every variable has an imbalance $c(X_i) = 0.25$, the expected imbalances in layer 4 are already close to 0: $0.25^4 = 0.00390625$, in layer 5 this value being $0.25^5 = 0.0009765625$.

2.3   Synthetic Datasets

We test the behavior of the network using synthetically generated data-sets from 3 probability distributions P_1, P_2, P_3:

1. P_1 is simply white noise, used to test the baseline activity of the networks.

2. P_2 defines dependencies over six variables $X_1, X_2, \ldots, X_6$. X_1 and X_2 are moderately direct correlated, $cov(X_4, X_5) = 0.5$. X_6 is direct correlated with both X_4 and X_5, $cov(X_4, X_6) = 1/3$ and $cov(X_5, X_6) = 2/3$. This distribution is used to test the activity of the network in the presence of both strong and moderate pairwise and higher order features.

3. P_3 defines dependencies over eight variables $X_1, X_2, \ldots, X_8$. The number of ones in the variable subset $\{X_2, X_4, X_6, X_8\}$ is always even, while in the subset $\{X_1, X_3, X_5, X_7\}$ is always odd. This distributions lacks pairwise correlations, it is used to test the activity of the network in the presence of only higher order features.

In each distribution, the probability mass function of the variables is unbiased $(P(X = -1) = P(X = 1) = 0.5)$.

3   Results and Discussion

In order to test the activity of EDN-s under different types of inputs, we performed 4 experiments, each averaged over 100 trials. In each trial, the networks were presented with 10000 newly generated samples, according to the tested distributions.

3.1   Noise

In the first experiment we tested the baseline activity of an $EDN(6,6)$ with different spiking threshold values $\{32, 64, 128\}$.

The results are presented in Table 1, where the 2nd column contains the theoretical number of spikes according to a random walk, the 3rd and 4th column contain the experimentally determined average value and the standard deviation. Columns $6 - 8$ repeat the same prediction and the measurements on a single neuron level.

Table 1. $EDN(6,6)$ predicted and average activity on white noise input.

spike_threshold	E_{TOT}	A_{TOT}	STD_{TOT}	E_N	A_N	STD_N
32	556.6406	547.5200	17.8385	9.7656	9.6056	0.2434
64	139.1602	131.1900	9.1439	2.4414	2.3016	0.1174
128	34.7900	25.0800	4.0940	0.6104	0.4400	0.0639

As observed, the empirical data follows closely the theoretical baseline activity model, there are no highly spurious activities, the standard deviation of the activity level at the neuronal level is very tight.

3.2   Pairwise and Higher Order Features

In the 2nd experiment series we tested the activity of an $EDN(6,6)$ on samples coming from P_2 distribution. To reduce the baseline activity, we set the spiking threshold value to 128.

The measured average neural spiking activity A_{TOT} was 312.2600, almost 9th-fold of the expected one, with a standard deviation of 6.5314, signaling strong statistical dependencies in the inputs. The mean average neural activity was $A_N = 5.4782$ with a comparatively very high $STD_N = 14.5562$. This suggest that the neural activity is not uniform, some neurons spike with an order of magnitude higher frequency. Closer look revealed that around 93 % of neural activity is related to 7 neurons, with spike counts between 20 and 70, while the other neurons spiked maximally 2 times.

An example of spiking rate activity from one run, is depicted in Fig. 1, where the 7 always active neurons can be identified by their lighter colors. The highly active neurons and the variable subset they account for are (in descending order of average activity): L2_1:[1 2], L2_15:[5 6], L4_6:[1 2 5 6], L2_13:[4 5], L4_4:[1 2 4 5], L2_14:[4 6].

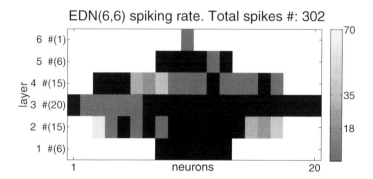

Fig. 1. $EDN(6,6)$ activity on 10000 samples from P_2. Lighter colors correspond to higher spiking rates, black colored neurons did not spike (Color figure online).

The data confirms, that the EDN accurately detects the underlying statistical dependencies, and the spike rates are proportional with the strength of these dependencies.

3.3 Just Higher Order Features

In experiment 3, a bigger network, $EDN(8,8)$ was tested on samples coming from P_3 distribution, lacking pairwise dependencies. Again, a spiking threshold value 128 was used.

As the size of this network is around 4 times bigger than $EDN(6,6)$ (255 vs. 63 neurons), the expected average total activity is $E_{TOT} = 150.7568$.

The measured average neural spiking activity A_{TOT} was 342.1500, with $STD_{TOT} = 18.1149$, signaling less activity. The mean average neural activity was $A_N = 1.3741$, again with a comparatively very high $STD_N = 8.4792$. This time, there were 3 neurons, all with spike counts between 77 and 79, while the other neurons spiked maximally 2 times on average.

As expected from the dependencies in P_3, the highly active 3 neurons and the variable subset they account for are: L4_21:[1 3 5 7], L4_50:[2 4 6 8], L8_1:[1 2 3 4 5 6 7 8].

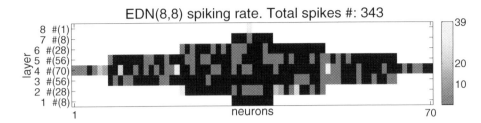

Fig. 2. $EDN(8,8)$ activity on 10000 samples from alternating P_2 and P_3 distributions. Lighter colors correspond to higher spiking rates, black colored neurons did not spike (Color figure online).

Again, the EDN accurately captures multivariate dependencies, even in the case of lacking pairwise dependencies.

3.4   Changing Features

In the last experiment we tested, how an $EDN(8,8)$ reacts to dynamically changing distributions. Here, in each run, the first 2500 samples from the total of 10000, came from P_2, then the next 2500 from P_3, followed again by 2500 samples from P_2, while the last 2500 came from P_3 again.

The averages and standard deviations over the 100 runs were: $A_{TOT=366.7100}$, $STD_{TOT} = 14.4273$, $A_N = 1.4727$, $STD_N = 5.4588$. In Fig. 2, from the one run spiking rate activity, we can observe that there are 10 neurons with high spiking rates.

Figure 3 depicts a partial population code, resulted from the joint activities of these 10 most active neurons. The time-dependent firing rate is computed over 100 ticks, i.e. if a spike appeared between times t and $t + 100$, the corresponding pixel is black.

Once again, the EDN correctly identifies the correct statistical dependencies and the firing pattern quickly adapts even in the case of dynamically changing features.

4   Conclusions

The paper demonstrated that a very simple connectionist machine, called entropy distillation network, based on simple operations like multiplication of inputs, accumulation of imbalances and a threshold, can accurately detect and encode complex probability distributions, containing pairwise and higher order dependencies.

The neural code carries both spatial and temporal information. The neural decoding of the multivariate features is straightforward: by looking at which neurons had a high spike rate, the exact dependent variables are exposed, as every neuron accounts for one unique variable subset. The spike-count rate of

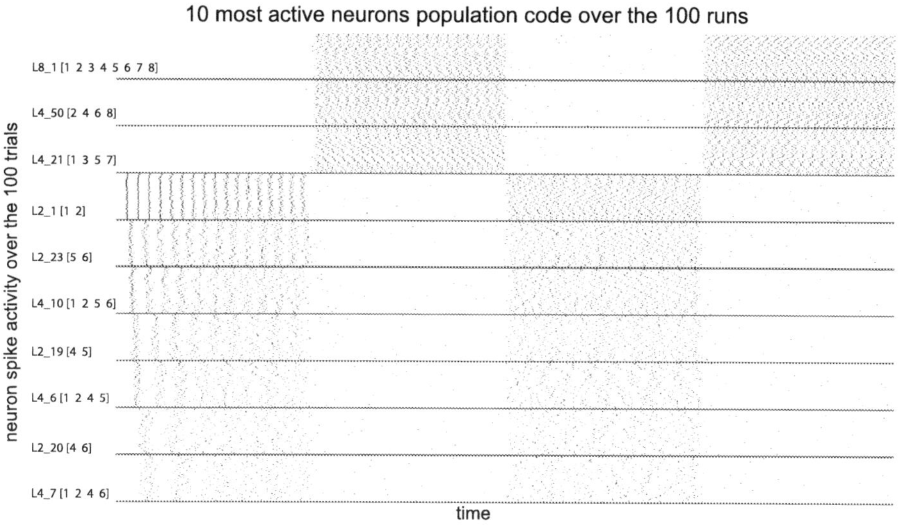

10 most active neurons population code over the 100 runs

Fig. 3. Spiking activity of 10 neurons over 10000 samples from alternating P_2 and P_3 distributions. For each neuron, all 100 runs are depicted vertically.

the neuron encodes the probability of that feature being manifested. Also, by looking at when the spiking activity was more prevalent, the part of the input set containing the feature is also revealed.

Future work will investigate stochastic firing, where the threshold on the imbalance is replaced with a spiking stochastic likelihood function. Also, in more general models, by connecting the axon recurrently or hierarchically into another network, the spike signal could have a dynamic affect, influencing synaptic plasticity.

Acknowledgments. We acknowledge the support of the Sapientia Institute for Research Programs (KPI). The second author was also supported by the Hungarian National Research Funds (OTKA), Project no. PD103921.

References

1. Pouget, A., Beck, J.M., Ma, W.J., Latham, P.E.: Probabilistic brains: knowns and unknowns. Nature Neurosci. **16**(9), 1170–1178 (2013)
2. Shadlen, M.N., Newsome, W.T.: Noise, neural codes and cortical organization. Curr. Opin. Neurobiol. **4**(4), 569–579 (1994)
3. Chechik, G.: Spike-timing-dependent plasticity and relevant mutual information maximization. Neural Comput. **15**(7), 1481–1510 (2003)
4. Matsui, M.: Linear cryptanalysis method for DES cipher. In: Helleseth, T. (ed.) EUROCRYPT 1993. LNCS, vol. 765, pp. 386–397. Springer, Heidelberg (1994)
5. Hermelin, M., Nyberg, K.: Dependent linear approximations: the algorithm of biryukov and others revisited. In: Pieprzyk, J. (ed.) CT-RSA 2010. LNCS, vol. 5985, pp. 318–333. Springer, Heidelberg (2010)

6. Iclănzan, D.: A multi-parent search operator for bayesian network building. In: Coello, C.A.C., Cutello, V., Deb, K., Forrest, S., Nicosia, G., Pavone, M. (eds.) PPSN 2012, Part I. LNCS, vol. 7491, pp. 246–255. Springer, Heidelberg (2012)
7. Iclanzan, D.: Higher-order linkage learning in the *ecga*. In: Proceedings of the Fourteenth International Conference on Genetic and Evolutionary Computation Conference, pp. 265–272. ACM (2012)

Extending Local Features with Contextual Information in Graph Kernels

Nicolò Navarin[✉], Alessandro Sperduti, and Riccardo Tesselli

Department of Mathematics, University of Padova, Padua, Italy
{nnavarin,sperduti}@math.unipd.it, rtessell@studenti.math.unipd.it

Abstract. Graph kernels are usually defined in terms of simpler kernels over local substructures of the original graphs. Different kernels consider different types of substructures. However, in some cases they have similar predictive performances, probably because the substructures can be interpreted as approximations of the subgraphs they induce. In this paper, we propose to associate to each feature a piece of information about the context in which the feature appears in the graph. A substructure appearing in two different graphs will match only if it appears with the same context in both graphs. We propose a kernel based on this idea that considers trees as substructures, and where the contexts are features too. The kernel is inspired from the framework in [7], even if it is not part of it. We give an efficient algorithm for computing the kernel and show promising results on real-world graph classification datasets.

Keywords: Graph kernels · Kernel-based methods · Structured data · Classification

1 Introduction

In many application domains data can be naturally represented in a structured form, e.g. in Chemoinformatics [1] or in natural language processing [3]. For this reason, in the last few years an interest in machine learning techniques applicable to data represented in structured (non-vectorial) form arose [9,14]. When dealing with machine learning for graph-structured data, kernel methods are one of the most popular approaches to follow. It just suffices to use a kernel for graphs together with any kernelized learning algorithm (e.g. SVM, SVR, KPCA, ...) and the user has a powerful, ready-to-use learning algorithm with strong theoretical bounds on its generalization performance. The predictive performance of the resulting learning procedure strongly depends on the particular kernel choice. The design of efficient graph kernels is not a trivial task, because several graph operations (e.g. the graph isomorphism) are not efficiently computable. The idea is to design kernels that are the most expressive as possible, in order to have a small information loss. Several alternatives have been proposed in literature. However it is difficult to state a priori which kernel will perform better in a specific task, because most of the existing kernels consider different approximations of the same local structures. In this paper, we propose a method to

© Springer International Publishing Switzerland 2015
S. Arik et al. (Eds.): ICONIP 2015, Part IV, LNCS 9492, pp. 271–279, 2015.
DOI: 10.1007/978-3-319-26561-2_33

enrich the feature space of a kernel with contextual information, i.e. we attach to a feature a piece of information about the topology of the graph in which that feature appeared. We apply this idea to the ODD kernel [7], and we define as the context of a feature another feature from the same kernel. We give an efficient algorithm for the kernel computation, and experimentally evaluate our proposal on five real-world datasets.

2   Definitions and Notation

Let $G = (V_G, E_G, L_G)$ be a graph, where V_G is the set of vertices (or nodes), $E_G \subseteq \{(v_i, v_j) | v_i, v_j \in V_G\}$ is the set of edges and $L_G : V_G \to \Sigma$ is a labeling function mapping each vertex to an element in a fixed alphabet Σ.

A graph is undirected if $(i, j) \in E_G \implies (j, i) \in E_G$, otherwise it is directed. A walk $w(u, v)$ in a graph is a sequence of nodes $v_1, \ldots, v_n$ s.t. $(v_i, v_{i+1}) \in E_G$ and $v_1 = u, v_n = v$. The length of a walk $|w(u, v)|$ is defined as the number of edges in such walk. A cycle is a walk where $v_1 = v_n$. A graph is acyclic if it does not contain cycles. A DAG is a directed acyclic graph. A path is a walk with no repeated nodes, i.e. where $\forall_{i=1}^n \forall_{j=1}^n, i \neq j \implies v_i \neq v_j$. A shortest path between two vertices $sp(u, v) \in V_G$ is a path with the minimum length that starts from u and ends in v. Note that the shortest paths are not unique, but their length $|sp(u, v)|$ is. $n\_sp(u, v)$ is a function returning the number of such shortest paths.

A rooted DAG D is a DAG in which one vertex r has been designated as the root. The root have no incoming edges, i.e. $\nexists u \in E_D, (u, r) \in E_D$. The function $r(D)$ returns the root of a rooted DAG.

A (rooted) tree is a rooted DAG where for each node there exists exactly one path connecting the root node to it. The children $children(v)$ of a node $v \in V_T$ in a tree are all the nodes $u \in V_T$ s.t. $(v, u) \in E_T$. The number of children, or out-degree, of a vertex v is $\rho(v)$. Similarly we can say that v is a parent of u. $ch_i(v, G)$ is the function retuning the i-th child of $v \in V_G$ (according to a particular order).

A proper subtree rooted at $u \in V_T$ of a tree T is the subtree that comprehends u and all its descendants. We will refer to it as $\overset{u}{\triangle} \in T$. We define $T_j(v, G)$ as a function returning the tree-visit of a graph G, rooted at v and limited at height j. Note that this tree-visit is the shortest-path tree between v and any $u \in V_G$ s.t. $|sp(u, v)| \leq j$. Moreover, we denote with $T(v, G)$ the tree-visit at the maximum possible height, i.e. $T_\infty(v, G) = T_{diam(G)}(v, G)$ where $diam(G)$ is the diameter of a graph, i.e. the length of the longest shortest path between two vertices.

A DAG-visit of a graph G, $DAG_j(v, G)$, is defined as the DAG of the shortest paths of length up to j. The main difference between $DAG(v, G)$ and $T(v, G)$ is that the number of nodes in the former is bounded by $|V_G|$ while in the latter it is not. We assume the nodes in $T_j(v, G)$ or $DAG_j(v, G)$ to be ordered according to the lexicographic order between the node labels (in case two nodes have the same label, the ordering is recursively induced from the children). Such an ordering has been proven to be well-defined in [7] for DAGs. Since trees are a special case of DAGs, the ordering relation is well-defined on trees as well. For ease of notation, when clear from the context, the link to the graph G will be omitted from the above-mentioned functions.

3   Graph Kernels

Most of the existing graph kernels are members of the R-convolution kernels framework [11]. The idea of this framework is to decompose the original structure into a set of simpler structures, where a (efficient) kernel is already defined. For example, the all-subgraphs graph kernel [10] has a feature associated to each possible graph. However, this kernel also happens to be NP-complete. An approach to reduce the computational complexity of the resulting kernel is to restrict the set of considered substructures of the graph. Different substructures raise different kernels. For example, in literature kernels based on random walks [13], shortest paths [2], subtree-patterns [12], subtrees [7] or pairs of small rooted subgraphs [5] have been proposed. The main drawback of these kernels is that they consider only local substructures of the original graphs, whose size is bounded to some limit due to computational complexity. For this reason, in some cases they have similar predictive performances [8], probably because the different substructures can be interpreted as different, but still similar, approximations of small subgraphs of the original graph. Enlarging the substructures to let the kernel consider a larger amount of information will increase the computational burden. We recall that the main challenge while designing graph kernels is the trade-off between the efficiency and the expressive power of the kernel. Among the available graph kernels, the NSPDK [5] is the most related to the proposed kernel. Specifically, in the RKHS of NSPDK, every feature represents a couple of small rooted subgraphs S_1 and S_2 of a certain diameter (radius) r, at a certain distance, i.e. where $|sp(r(S_1), r(S_2))| = d$. In a sense, S_1 can be seen as a context for S_2 and vice versa.

Let us define a set of Ordered Decomposition DAGs of a graph G limited to the maximum (user-specified) depth h as $ODD_G = \{DAG_h(v, G)|v \in V_G\}$, where we recall that the nodes in each DAG are ordered according to a recursive relation looking at the labels of a node and all its descendants. The $ODDK$ kernel [7] is defined as:

$$ODDK(G_1, G_2) = \sum_{\substack{OD_1 \in ODD_{G_1} \\ OD_2 \in ODD_{G_2}}} \sum_{j=1}^{h} \sum_{l=1}^{h} \sum_{\substack{v_1 \in V_{OD_1} \\ v_2 \in V_{OD_2}}} C_{ST}(r(T_j(v_1)), r(T_l(v_2)))$$

where $C_{ST}()$ is a function that defines the subtree kernel, i.e. a kernel that counts the number of shared proper subtrees between two trees. This kernel allows to obtain an explicit feature space representation ϕ [6]. Let us define a total ordering between all the possible labeled trees that appear from the kernel application on the dataset. Then each feature $\phi_i(G)$ represents the frequency of the i-th tree in the RKHS of the ODD kernel.

4   Adding Contexts to Graph Kernels

The graph kernels described in the previous section extracts local patterns of the graph as features, i.e. the feature itself does not bring any information regarding where it has appeared within the graph. The idea we propose in order to increase the expressiveness of a kernel, while preserving efficiency, is to enrich the local features (e.g. the features extracted by the ODD kernel) with their contextual information. The contextual information we are interested in is a description of the topology of the graph around the extracted feature. Thus, a substructure that appears in two different graphs will match if and only if it appears within the same context in both graphs. Considering contextual information, we obtain kernels that are more sparse. In some cases, the resulting kernel may be more discriminative with respect to the original one. However, in other cases it may be too much sparse to obtain good performance. In the latter case, it can be beneficial to add the contribution of the new kernel to the original one. In our experiments, we will implement both these variants. Note that, with our proposed approach, the computation of the contributions of the contextualized kernel and of the original kernel can be performed efficiently at the same time.

Fixed a feature of the original graph kernel, we want the following property to hold:

$$\sum_{c \in Contexts(f)} \phi_{f \circ c}(G) = \phi_f(G),$$

where $\phi_f(G)$ is the frequency of a feature f in the RKHS of the original kernel, and $\phi_{f \circ c}$ is the frequency of f appearing within the context c. From the formula it is clear that for each feature we need to consider also the empty context($\varnothing$-context), i.e. the situation in which a feature does not appear in any particular context e.g. because it has reached the maximum allowed dimension and we have no information about its context in the original graph.

In the remaining of this section, we will introduce our proposed kernel instantiating the context idea to the ODD kernel. As a feature represents a substructure, in the same way we can represent a context for a feature as a substructure of the graph, that incorporates the feature. Therefore, contexts and features can share the same representation and so it is possible that a context for a given feature can be a feature itself. To compute the contextualized features we only need to combine a feature with other features representing the context in which the first feature appears in the graph.

The first important difference between the proposed Tree Context Kernel (TCK) and ODDK is that, for technical reasons, the former is defined over tree-visits while the latter over DAG-visits. Note that the nodes of a tree-visit $T(v, G)$ of a graph G can grow exponentially in its size, while if we consider a DAG-visit $DAG(v, G)$, each node in the original graph can appear at most once, thus limiting the size of the resulting structure to at most $|V_G|$ nodes. However, in the next section we will provide an efficient implementation that does not need to store in memory the tree-visits, but only the DAG-visits. The Tree Context Kernel can be defined as:

$$TCK(G_1, G_2) = \sum_{\substack{v_1 \in V_{G_1} \\ v_2 \in V_{G_2}}} \sum_{i=1}^{h} \sum_{j=1}^{h}$$

$$[\delta(T_i(v_1), T_j(v_2)) + \sum_{\substack{\overset{u_1}{\triangle} \in T_i(v_1) \\ \overset{u_2}{\triangle} \in T_j(v_2)}} \delta(\overset{u_1}{\triangle}, \overset{u_2}{\triangle}) \sum_{l=1}^{\rho(u_1)} C_{ST}(ch_l(u_1), ch_l(u_2))]$$

where we recall that:

$$C_{ST}(v_1, v_2) = \begin{cases} \lambda \cdot K_L(v_1, v_2) & \text{if } v_1 \text{ and } v_2 \text{ are leaves} \\ \lambda \cdot K_L(v_1, v_2) \prod_{j=1}^{\rho(v_1)} C_{ST}(ch_j(v_1), ch_j(v_2)) & \text{if } \rho(v_1) = \rho(v_2) \\ 0 & \text{otherwise} \end{cases}$$

and δ is the Kronecker's delta function. We recall that $C_{ST}(v_1, v_2), v_1 \in T_1, v_2 \in T_2$ is a function that counts the common proper subtrees of two trees. The function depends on T_1 and T_2. We decided to follow the original definition of [4] omitting that dependency for ease of notation.

The kernel is positive semidefinite because it is a composition of positive semidefinite kernels, defined over the ordered tree visits $T_i(v, G) = T(DAG_i(v, G))$ that are well defined as shown in [7].

Intuitively, this kernel matches two subtree features $\overset{u_1}{\triangle} \in T_i(v_1, G_1), 0 \le i \le h$ and $\overset{u_2}{\triangle} \in T_j(v_2, G_2), 0 \le j \le h$ in one of the following cases:

- both v_1 and v_2 are the root nodes of the tree visit, i.e. $u_1 = v_1$ and $u_2 = v_2$;
- u_1 and u_2 occur within the same context in both trees, i.e. their parents generate the same proper subtree.

5 Efficient Implementation

Algorithm 1 shows the pseudocode to decompose a graph G into its explicit (sparse) feature vector ϕ. We will denote with f the map that stores the keys of the local subtree features, i.e. $f_{u,d}, u \in V_G, d \in \{0, \ldots, h\}$ is the key of the subtree rooted in u of height d. Similarly, $size$ is a map that stores the size of each feature, i.e. $size_{u,d}$ is the number of nodes that compose the feature $f_{u,d}$. Let κ be a perfect hash function from strings to integers. Such a function can be implemented with an incrementally-built hashmap that associates an unique id to each string. Alternatively, a normal hashing function can be used if we tolerate some clashes. We define reserved special symbols "[", "]", "#" and "○" that do not have to appear in the labels of the graphs and they are needed to encode subtree features into strings. In the following, we will discuss the most sensitive steps of the algorithm. In line 6 the nodes of the DAG-visit are traversed in a reverse topological order, ensuring that every node will be

Algorithm 1. An algorithm for computing the explicit feature space representation of a graph G according to the kernel TCK_{ST} with maximum (user-specified) height h and weight factor λ

1: $\phi = [0, \ldots, 0]$ ▷ Explicit feature space represented as sparse vector
2: **for all** $v \in V_G$ **do**
3:     $D \leftarrow DAG_h(v, G)$
4:     $f = \{\}$ ▷ dictionary that stores the features related to a node u and height d
5:     $size = \{\}$ ▷ dictionary that stores the size of each feature
6:     **for all** $u \in$ REVERSETOPOLOGICALORDER(D) **do**
7:         **for** $d \leftarrow 0, \ldots, diam(D) - |sp(v, u)|$ **do**
8:             **if** $d = 0$ **then**
9:                 $f_{u,0} \leftarrow \kappa(L(u))$
10:                $size_{u,0} \leftarrow 1$
11:            **else**
12:                $(S_1, \ldots, S_{\rho(u)}) \leftarrow$ SORT$(f_{ch_1(u),d-1}, f_{ch_2(u),d-1}, \cdots, f_{ch_{\rho(u)}(u),d-1})$
13:                $f_{u,d} \leftarrow \kappa(L(u)\lceil S_1 \# S_2 \# \cdots \# S_{\rho(u)} \rfloor)$
14:                $size_{u,d} \leftarrow 1 + \sum_{i=1}^{\rho(u)} size_{ch_i(u),d-1}$
15:                **for all** $ch \in children(u)$ **do**
16:                    $\phi_{f_{ch,d-1} \circ f_{u,d}} \leftarrow \phi_{f_{ch,d-1} \circ f_{u,d}} + n\_sp(v,u) \cdot \lambda^{\frac{size_{ch,d-1}}{2}}$
17:            **if** $u = v$ **then**
18:                $\phi_{f_{u,d} \circ \varnothing} \leftarrow \phi_{f_{u,d} \circ \varnothing} + \lambda^{\frac{size_{u,d}}{2}}$
19: **return** ϕ

---

processed before its parent. In line 7, for each node u of the current DAG-visit D, we consider all the heights for the feature generation. Note that when $d = 0$, $f_{u,0}$ is a feature (proper subtree) of the tree $T_{|sp(v,u)|}(v)$ and when $d = diam(D) - |sp(v, u)|$, $f_{u,d}$ is a feature of $T_{diam(D)}(v)$, where $diam(D) \leq h$. Notice that if D is unbalanced and we are considering a node u whose $|sp(v, u)|$ is not maximum, then we are considering many times the feature associated to u at its maximum height. In lines 12–14, the local feature related to the current node and height is generated. The hashed feature values of the children of the current node at height $d - 1$ are sorted, generating a feature of height d and inducing an order on the children of every node that is the lexicographic order over the hash values of the corresponding features. This step allows us not to define any particular ordering on the nodes of D. Then the extracted feature is encoded and finally it is hashed. Lines 15–16 generate the contextualized features and increment their frequency in ϕ according to a weight term multiplied by $n\_sp(v, u)$. This multiplication allows us to compute the statistics related to the tree-visit while working on the smaller (in terms of number of nodes) corresponding DAG-visit. Notice that $n\_sp(v, u)$ is efficiently computed during the creation of $DAG_h(v, G)$ in a top-down fashion without any additional cost. Finally, lines 17–18 increment the value corresponding to the feature with empty context $\phi_{f_{u,d} \circ \varnothing}$. This implementation returns the explicit sparse feature vector ϕ, therefore in order to compute the kernel function between two graphs is sufficient to compute the dot product between the two feature vectors.

Table 1. Accuracy results of the proposed kernels and the considered baselines, in nested 10-fold cross validation.

Kernel/dataset	CAS	GDD	NCI1	AIDS	CPDB
$NSPDK$	$83.6_{\pm0.34}$	$74.09_{\pm0.91}$	$83.46_{\pm0.46}$	$82.71_{\pm0.66}$	$76.99_{\pm1.15}$
WL	$83.33_{\pm0.37}$	$75.29_{\pm1.33}$	$84.41_{\pm0.49}$	$82.02_{\pm0.4}$	$76.36_{\pm1.4}$
$ODDK$	$83.53_{\pm0.21}$	$76.99_{\pm0.36}$	$85.31_{\pm0.26}$	$\mathbf{82.99_{\pm0.50}}$	$78.44_{\pm0.76}$
TCK	$83.53_{\pm0.32}$	$\mathbf{79.35_{\pm0.45}}$	$\mathbf{85.78_{\pm0.22}}$	$82.88_{\pm0.39}$	$76.96_{\pm0.96}$
$TCK + ODDK$	$\mathbf{83.94_{\pm0.26}}$	$78.03_{\pm0.56}$	$85.48_{\pm0.182}$	$82.97_{\pm0.5}$	$\mathbf{78.89_{\pm0.98}}$

6   Experimental Results

We measured the predictive performance of TCK and other state-of-the-art kernels on the following real-world datasets: AIDS, CAS, CPDB, GDD and NCI1. Each dataset represents a binary classification problem and is composed by labeled graphs with no self-loops. The AIDS, CAS, CPDB and NCI1 datasets are collections of chemical compounds represented as graphs, with nodes labeled according to the atom type and edges that represent the bonds. The GDD dataset is composed by proteins represented as graphs, where the nodes represent amino acids and two nodes in a graph are connected by an edge if they are less than $6\,\text{Å}$ apart. The largest datasets are CAS and NCI1 with more than 4000 graphs, and the smallest is CPDB with 684 instances.

Since we cannot know in advance whether the sparsity is beneficial for a particular task, we choose to test two versions of the proposed kernel. The first version (TCK) considers only contextualized features, while the second version $(TCK + ODDK)$ combines TCK with the base (non-contextualized) kernel, $ODDK$ in our case. Note that $TCK + ODDK$ can be computed with a slight modification of Algorithm 1, thus the computational complexities of the two versions of the proposed kernels are the same. We compare the proposed kernels with the NSPDK kernel [5], the Fast Subtree Kernel (FS) [12], and the original version of the ODDK based on the subtree kernel [7]. To assess the predictive performances of the different kernels, we used a *nested* 10-fold cross validation: within each of the 10 folds, another 10-fold cross validation is performed over the corresponding training set in order to select the best parameters for the current fold. Thus, the parameters are optimized on the training dataset only. The whole process has been repeated 10 times using different random data splits. The parameter space for both versions of TCK and $ODDK$ was restricted to the following values: $h = \{1, 2, \ldots, 10\}$ and $\lambda = \{0.1, 0.5, 0.8, 0.9, \ldots, 1.5, 1.8\}$. The parameter h of the FS kernel were restricted to $h = \{1, 2, \ldots, 10\}$ and for the NSPDK the values $h = \{1, 2, \ldots, 8\}$ and $d = \{1, 2, \ldots, 7\}$ were considered. The SVM solver had the C parameter ranging in $C = \{10^{-4}, 10^{-3}, \ldots, 10^3\}$.

Table 1 reports the averaged accuracy results of our experiments with the corresponding standard deviations. At a first glance, it is clear that in almost all the considered datasets, one of the two proposed kernels is the better performing among all the considered kernels, with the only exception of the AIDS dataset.

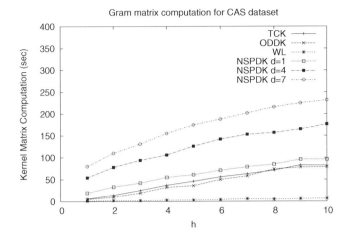

Fig. 1. Copmuptational time (in seconds) required for the Gram matrix computation of the considered kernel, with different parameters.

Looking at the results in more detail, in two datasets (GDD, NCI1) both versions of TCK perform better than the others. If we consider the CAS dataset, the performance of the worst of the proposed kernels is comparable with the better kernel among the baselines (NSPDK). In the CPDB dataset the worst of the proposed kernels is worse than the best kernel among the baselines (ODDK), but it is still competitive, such as in AIDS dataset, where the proposed kernels are very close to the best one. Let us finally anlyze the computational requirements of our proposed kernel. Figure 1 reports the computational times required for the Gram matrix computation of the kernels considered in this section on the CAS dataset. The execution times of the proposed kernel are very close to the ones of the original kernel. The situation is similar for other datasets, and thus the corresponding plots are omitted. The results presented in this section suggest that the introduction of contextualized features is a promising approach, and that in principle also other kernels can benefit from such an extension.

7   Conclusions and Future Work

In this paper, we proposed a technique to incorporate context information in the kernels that allow for an explicit feature space representation. In particular, we defined a relationship between the explicit features where one feature can be considered as the context of another one. We applied our idea to the $ODDK$ kernel, and slightly modified the kernel definition in order to provide an efficient algorithm for the computation of the contextualized kernel. We evaluated the predictive performance of the resulting kernel (in two variants) over five real-world datasets, and the proposed approach shows promising results. As future works, we plan to apply the contextualization idea to other state-of-the art graph kernels, as well as to kernels for other discrete structures.

Acknowledgments. This work was supported by the University of Padova under the strategic project BIOINFOGEN.

References

1. Aggarwal, C.C.: Managing and Mining Graph Data. Advances in Database Systems, vol. 40. Springer, Boston (2010)
2. Borgwardt, K., Kriegel, H.-P.: Shortest-path kernels on graphs. In: Fifth IEEE International Conference on Data Mining (ICDM 2005), pp. 74–81. IEEE, Los Alamitos (2005)
3. Collins, M., Duffy, N.: Convolution kernels for natural language. In: Dietterich, T.G., Becker, S., Ghahramani, Z. (eds.) NIPS, pp. 625–632. MIT Press, Cambridge (2001)
4. Collins, M., Duffy, N.: Convolution kernels for natural language. Adv. Neural Inf. Process. Syst. **14**, 625–632 (2001)
5. Costa, F., De Grave, K.: Fast neighborhood subgraph pairwise distance kernel. In: Joachims, J.F., Thorsten (eds.) Proceedings of the 27th International Conference on Machine Learning (ICML 2010), pp. 255–262. Omnipress (2010)
6. Da San Martino, G., Navarin, N., Sperduti, A.: A memory efficient graph kernel. In: The 2012 International Joint Conference on Neural Networks (IJCNN). IEEE, June 2012
7. Da San Martino, G., Navarin, N., Sperduti, A.: A tree-based kernel for graphs. In: Proceedings of the Twelfth SIAM International Conference on Data Mining, pp. 975–986 (2012)
8. Da San Martino, G., Navarin, N., Sperduti, A.: Exploiting the ODD framework to define a novel effective graph kernel. In: Proceedings of the European Symposium on Artificial Neural Networks, Computational Intelligence and Machine Learning (2015)
9. Da San Martino, G., Sperduti, A.: Mining structured data. IEEE Comput. Int. Mag. **5**(1), 42–49 (2010)
10. Gärtner, T., Flach, P.A., Wrobel, S.: On graph kernels: hardness results and efficient alternatives. In: Schölkopf, B., Warmuth, M.K. (eds.) COLT/Kernel 2003. LNCS (LNAI), vol. 2777, pp. 129–143. Springer, Heidelberg (2003)
11. Haussler, D.: Convolution kernels on discrete structures. Technical report, Department of Computer Science, University of California at Santa Cruz (1999)
12. Shervashidze, N., Schweitzer, P., van Leeuwen, E.J., Mehlhorn, K., Borgwardt, K.M.: Weisfeiler-Lehman graph kernels. J. Mach. Learn. Res. **12**, 2539–2561 (2011)
13. Vishwanathan, S., Borgwardt, K.M., Schraudolph, N.N.: Fast computation of graph kernels. In: Schölkopf, B., Platt, J., Hoffman, T. (eds.) Advances in Neural Information Processing Systems, vol. 19, pp. 1449–1456. MIT Press, Cambridge (2007)
14. Vishwanathan, S.V.N., Schraudolph, N.N., Kondor, R., Borgwardt, K.M.: Graph kernels. J. Mach. Learn. Res. **11**, 1201–1242 (2010)

Classification of Alzheimer's Disease Based on Multiple Anatomical Structures' Asymmetric Magnetic Resonance Imaging Feature Selection

Yongming Li[1(✉)], Jin Yan[1], Pin Wang[1], Yang Lv[1],
Mingguo Qiu[2], and Xuan he[1]

[1] Communication Engineering of Chongqing University,
Chongqing 400030, China
{Yongminli,lvyang}@cqu.edu.cn,
{994321134,61266584,704720981}@qq.com
[2] Department of Medical Imaging of School of Biomedical Engineering,
The Third Military Medical University, Chongqing 400038, China
qiumg_2002@sina.com

Abstract. The quality of magnetic resonance imaging (MRI) features is key to the classification of Alzheimer's disease. However, relevant research has as yet paid little attention to asymmetric MR features. In this paper, the asymmetric MR features of multiple anatomical structures are extracted. The MR feature types include volume feature and several kinds of texture features. Subsequently, the extracted features are selected based on the wrapper feature selection method with chain-like agent genetic algorithm (CAGA) and support vector machine (SVM). Finally, the selected asymmetric MR features are used for classification of Alzheimer's disease. Experimental results show that the extracted features have apparent asymmetrical characteristics. The asymmetric volume feature of single anatomical structure can have better discrimination capability than the whole volume feature of same anatomical structure. Single selected asymmetric MR feature has displayed a superior discrimination capability in regards to three conditions of Alzheimer's disease. The improvement is very apparent compared to before feature selection and the p-value-based feature selection method. In conclusion, this proposed method offer a new kind of feature type and can improve the classification rate for diagnosis of Alzheimer's disease.

Keywords: Magnetic resonance imaging(MRI) · Alzheimer's disease · Asymmetric MR feature · Image feature selections · Chain-like agent genetic algorithm(CAGA) · Support vector machine(SVM)

1 Introduction

Alzheimer Disease (AD) is a common neurodegenerative disease, which can cause serious damage to the central nervous system. A key step in the prevention and cure for AD is early, non-invasive diagnosis [1]. Currently, Magnetic Resonance Imaging

© Springer International Publishing Switzerland 2015
S. Arik et al. (Eds.): ICONIP 2015, Part IV, LNCS 9492, pp. 280–289, 2015.
DOI: 10.1007/978-3-319-26561-2_34

(MRI) is a widely used non-invasive detection technique. Additionally, MRI can precisely describe quantitative changes in structural and functional lesions of different brain tissues and can also suggest morphological changes and metabolite concentrations that have occurred in these tissues [2].

Relevant research reports show that during early stages of AD, these anatomical structures undergo some variations in shape and texture [3–5]. Neural research reports on the brain show that brain asymmetry exists between the left and right parts of the brains of normal individuals. When starting from early lesion of AD, the asymmetry existed between the left and right parts of the brains will change accordingly [6, 7].

Tianzi Jiang et al. have found that the asymmetry of the volume of the hippocampus changes a lot during AD [8]. Tsai. K.-J et al. found the similar case on a transgenic AD mouse model. With the progression on neuro-degeneration, the level of transthyretin (TTR) transcripts on the left side of the brain became significantly lower than that of the right part and this lowering of TTR transcript is accompanied with a higher Abeta level in the left part of the brain [9]. Jong Hun Kim et al. found the same case with multiple anatomical structures [10]. Markus Donix et al. found that, during early stages of AD, the cortex thickness on the left part of brain decreased [11].

The study reports above show that asymmetry is an important characteristic of the neurodegeneration process caused by AD and classification of brain asymmetry can provide help in the early diagnosis of AD. Regrettably, there are some limitations to the studies mentioned above. (1) Little attention is paid to multiple anatomical structures for asymmetric MR feature extraction. (2) Number and types of extracted features are little. Only several shape features are considered and asymmetric texture features are not considered. (3) Since there is such a small number of extracted feature types, feature selection is not considered. (4) Asymmetry features just are used for studing their relationship with the different conditions of AD. Connections between early diagnosis and asymmetry features for AD are not found. (5) The asymmetry features are rarely applied in the human body.

Based on the analysis above, in this paper, one novel asymmetric MR feature selection algorithm for AD diagnosis is proposed. For convenience, the algorithm is called multi-anatomical structures' multiple asymmetric MR features selection algorithm for AD diagnosis (MASMR_FF). Firstly, four major anatomical structures are selected for feature extraction. Secondly, each anatomical structure is divided into two parts-left part and right part. Thirdly, according to each part of each anatomical structure, several kinds of MR features are extracted. Subsequently, CAGA- based wrapper feature selection is performed on the extracted features and feature selection is completed. Finally, classification on the AD samples with the selected features is performed.

2 Feature Selection Algorithm Based on Multiple Anatomical Structures' asymmetric MR Features (MASMR_FS)

The figure below shows the flowchart of classification of Alzheimer's disease based on MASMR_FS algorithm (Fig. 1).

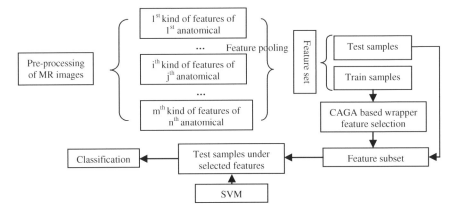

Fig. 1. Flow chart of classification of AD based on MASMR_FS algorithm

2.1   Pre-processing of Brain MRI Samples

Pre-processing of brain MRI samples includes: (1) Image registration: select a relatively standard brain MR image to be the reference image; then, operate a non-linear affine conversion to other brain MR images; finally, operate registration step using reference image as basis. This procedure is usually carried out by program SPM8 in MATLAB 2012b integrated development environment and include brain atlas space. (2) Image Segmentation: first, remove irrelevent tissues such as skull, nueroglia, subcutaneous fat, muscles, connective tissue and so on; then, extract 4 tissues (anatomical structures): white matter (WM), gray matter(GM), cerebrospinal fluid (CSF) and hippocampus (HPC); finally, divide each anatomical structure into two parts-left part and right part. MIPAV software are applied (http://www.softpedia.com/get/Science-CAD/MIPAV.shtml).

Figure 2 shows the results of major steps of the pre-processing as described above. Figure 3 shows the four anatomical structures after pre-processing, as described above. From the Figs. 2 and 3, it can be seen that every major step can obtain satisfactory results. This reflects that each step is controllable and can provide the confident images for following feature extraction.

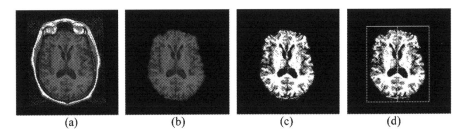

Fig. 2. Results of major steps in pre-processing: (a) After image registration; (b) After peeling skull; (c) After segmentation; (d) Left and right parts after division

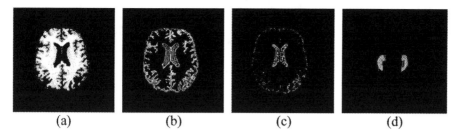

(a)   (b)   (c)   (d)

Fig. 3. Results of an MR image after tissue segmentation. (a) Grey matter, (b) White matter, (c) Cerebrospinal fluid, (d) Hippocampus

2.2 Multiple Types of MR Feature Extraction from Segmented Anatomical Structures

As lesions occur during AD, some anatomical structures will have some variations in shape and texture. This paper focuses on four anatomical structures for extracting shape and texture features; every anatomical structure is divided into two parts-left part and right part, comprising eight parts in total. For each part, the volume feature and two categories of texture features including the gray level co-occurrence matrix features and run length features are selected.

The gray level co-occurrence matrix features contain the following 5 features: Energy, Contrast, Inverse Difference Moment(IDM), Entropy and Correlation; The run length features contain the following 10 features: Gray-Level Nonuniformity (GLN), Run Length Nonuniformity(RLN), Long Run Emphasis(LRE),Short Run Emphasis, (SRE), Low Gray-Level Run Emphasis(LGRE), High Gray-Level Run Emphasis (HGRE), Short Run Low Gray-Level Emphasis(SRLGE), Short Run High Gray-Level Emphasis(SRHGE), Long Run Low Gray-Level Emphasis(LRLGE), Long Run High Gray-Level Emphasis(LRHGE). The total 15 texture features are extracted.

Thus in total there are 128 features extracted. Since the features are from the left part and right part of the anatomical structures separately, they reflect the asymmetric characteristics of the same anatomical structures within the same brain. For clarification, the extracted features are called asymmetric features.

2.3 Wrapper Feature Selection Algorithm Based on Chain-like Agent Genetic Algorithm (CAGA) and SVM (MASMR_FS Algorithm)

Since there are many extracted features, if all the features are used for classification, there will be dimensional disaster in terms of computational cost and correction rate. Therefore, it is necessary to perform feature selection, thereby selecting an optimal feature subset for image classification of AD.

In consideration of the capability to remove redundancy and the large number of extracted features, this paper adopts chain-like agent genetic algorithm (CAGA) based feature selection algorithm, which can avoid local optima with global optimization for a large feature space. The following table is flowchart of wrapper feature selection algorithm with CAGA.

2.3.1   Caga

Feature selection is performed by a former ensemble model combining chain-like agent genetic algorithm (CAGA), which was carried out previously by the authors [12]. This algorithm has higher levels of stability and a higher accuracy compared with other GA algorithms. CAGA is conducted as the following steps: (1) Dynamic neighborhood competition selection operator; (2) Neighborhood adaptive crossover operator; (3) Adaptive mutation operator; (4) Adaptive termination.

2.3.2   Svm

For wrapper feature selection mode, the fitness value of agents in CAGA is identified from SVM's classification accuracy.

SVM is a kind of machine learning method. It maps the samples in input space into high dimensional feature space with kernel function. The mapping is changed to convex quadratic programming problem and the objective is to find optimal hyperplane to correctly classify all the training samples. Gaussian radial basis function (RBF) is used as kernel function in this paper.

2.4   Classification of AD Based on SVM

Based on the above chapter, the optimized feature subsets are obtained via feature selection algorithm, then SVM is used to classify all brain MRI samples to realize diagnosis. Since the samples are limited, in order to verify effectiveness of the MASMR_FS algorithm, leave-k-out cross validation method is used. The main parameters of SVM are similar as those in Sect. 2.3.2 (Table 2).

3   Experimental Results and Analysis

3.1   Experimental Conditions

The image samples are from the public dataset for AD-ADNI (http://adni.loni.usc.edu/). In order to remove the bias in terms of different number of different classes, the same number of CTL (NC), MCI and ADs are chosen, and the name is 60. In the samples, the male and female samples are chosen randomly, but the proportion of two genders is close to 1:1 to remove the bias in terms of different number of different genders. The acquisition mode of samples is T2 weight gradient echo (T2*GRE). The relevant information can be found in Table 1.

For CAGA, the population size is 150, iteration number is 50, initial crossover probability is 0.8, initial mutation probability is 0.05. RBF kernel function is used for SVM. Leave-k-out cross validation method is adopted. Repetition time for same experiment is 30.

Table 1. The major procedure of MASMR_FS algorithm

Start

Ensemble n types of feature sets as to be selected feature sets $fea()$;

Proceed sample processing using $fea()$ so as to gain sample feature matrix $data\_fea()_{m \times n}$, where m represents sample amount, n represents feature amount; As for $data\_fea()_{m \times n}$,it is randomly being divided into sample training feature matrix $data\_fea\_train()$ and sample testing matrix $data\_fea\_test()$;

A chain like agent structure sets up on account of CAGA, the length of each agent is n, represents feature number as n, among these agent, one of them its length is zero indicate this feature has not been chosen, 1 indicated chosen one. The amount of all agent is P.

According to no. i agent , $data\_fea\_test()$ and $data\_fea\_train()$ are being cut off separately, and then generate the corresponding sample training feature matrix $data\_fea\_train()_{ithAgent}$ and sample testing feature matrix $data\_fea\_test()_{ithAgent}$, then put $data\_fea\_train()_{ithAgent}$ into SVM to operate training procedure, after this put $data\_fea\_test()_{ithAgent}$ into to output the testing accuracy results as agent adaptations;

Finish all agent adaptation calculations;

Genetic operation(Neighborhood competition selection, self-adaption crossing, self-adaption mutation);

Verdict whether it meet determine condition; if so , go to the next step; if not, return;

Stop Output the optimal agent, which is the corresponding optimal feature subset with prior classification accuracy.

Table 2. Basic information of samples

Class	Gender proportion	Age	MMSE score	CDR value
NC(CTL)	27/33	76.1 ± 6.7	24-30	0
MCI	34/26	74.7 ± 6.1	20-30	0.5 or 1
AD	28/32	74.8 ± 6.9	20-26	1 or 2

3.2 Discrimination Power of Single Asymmetric Feature

In order to clearly show the discriminative abilities of the asymmetric feature, the volume features of left and right parts of hippocampus are chosen for comparison with the volume feature of whole hippocampus. For each class, 30 samples are selected randomly and are plotted on the graph below (Fig. 4). The '*'refers to CTL samples, the '□' refers to MCI samples, the '+' refers to AD sample.

From the figure above, it can be seen that the three kinds of samples (CTL, MCI, AD) are not separable according to the volume feature of the whole hippocampus.

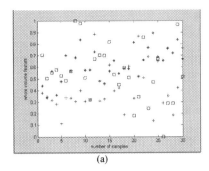 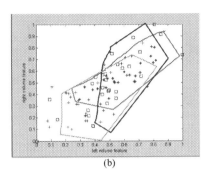

(a)   (b)

Fig. 4. Discrimination power of volume feature of hippocampus: (a) whole volume feature; (b) left and right volume features

But, according to the input of left and right volume features, the three kinds of samples can be distinguished to a extent To a certain extent, the three kinds of samples cluster into a specific area. Please see the Fig. 4-(b).

3.3   Comparison Before and After Feature Selection

This group of experiments aims to compare the MASMR_FS algorithm with the asymmetric features without selection. The volume features of the four anatomical structures are extracted, merged together and referred to as 'fea_vw'. The texture features of the four anatomical structures are extracted, merged together and referred to as 'fea_tw'. The full group of extracted features is referred to as 'fea_vw' + 'fea_tw'. In this process, 'leave one out' cross validation method is used. The process is repeated 10 times (Table 3).

Table 3. Comparison of classification performance before and after feature selection

	Features with MASMR_FS		Fea_vw + fea_tw		Fea_vw		Fea_tw	
	ADvsCTL	CTLvsMCI	ADvsCTL	CTLvsMCI	ADvsCTL	CTLvsMCI	ADvsCTL	CTLvsMCI
amount	61	62	128	128	4	4	124	124
accuracy	**0.9500**	**0.8344**	0.8167	0.5833	0.7500	0.4333	0.7333	0.6000
sensitivity	**0.9267**	**0.8044**	0.8000	0.6333	0.7333	0.4333	0.6667	0.6333
specificity	**0.9733**	**0.8644**	0.8333	0.5333	0.7667	0.4333	0.8000	0.5667

Firstly, from this table we can see that the number of features with MASMR_FS has reduced greatly, which reduces the complexity of classification dimension and reduces time cost. Secondly, through the MASMR_FS method, classification accuracy has been significantly improved. Moreover, it can be seen that volume features used alone or texture features used alone lead to lower classification accuracy than a combination of volume and texture features. This shows that the combination of shape and texture features is a superior method for AD classification. Additionally, the accuracy of the fea_vw and fea_tw methods is similar. Since the volume feature is easier to extract, shape features are more widely used in AD–related research in comparison to the texture feature.

3.4   Comparison with P-Value Based Method

Hypothesis-testing based on P-value is often used for feature selection. This method is widely used in medical diagnosis. In terms of feature selection, P-value based method ignores redundancy among features, so improvements to classification accuracy are limited. Table 4 shows comparison between P-value based method and MASMR_FS algorithm.

Table 4. Comparison between P-value Based method and MASMR_FS algorithm

	MASMR_FS		P-value based $P < 0.05$		P-value based $P < 0.01$	
	ADvsCTL	CTLvsMCI	ADvsCTL	CTLvsMCI	ADvsCTL	CTLvsMCI
amount	61	62	64	12	55	9
accuracy	**0.950**	**0.834**	0.850	0.600	0.783	0.550
sensitivity	**0.927**	**0.804**	0.833	0.600	0.767	0.533
specificity	**0.973**	**0.864**	0.867	0.600	0.800	0.567

From this table, it can be seen that classification accuracy is improved significantly with MASMR_FS compared to the P-value based feature selection method. Moreover, when the threshold of P-value is relatively lower, the number of selected features is fewer, but the classification accuracy falls a lot. The possible reason for this is that some features without strong relevance to labels, but with low correlation to other features, were removed incorrectly.

3.5   Classification Ability of Selected Features

This part of the experiment was conducted in order to test the classification ability of those features chosen by feature selection. Most of the features selected have strong monotonicity for CTL, MCI and AD. This represents that the features can be used to monitor the pathogenic progress from CTL to AD via MCI. The figure below shows some of these feature values. Please note the values shown have undergone normalization (Fig. 5).

In the 4 histograms above, the vertical axes represent the feature values after normalization, the horizontal axis represents the group and the confident coefficient can be seen on the top of each bar. In regards to the value of the features, monotonicity is apparent. With the progress from CTL to AD via MCI, the values of the features gradually become higher or lower. This demonstrates that the features have strong discriminatory ability, through which they are able to classify the three conditions, thereby achieving diagnosis. There is a large difference between the three different conditions, which means that the classification accuracy is stable.

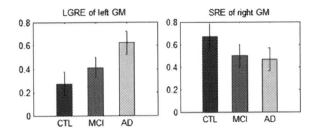

Fig. 5. Monotonicity of several features after feature selection

4   Conclusion

Compared to the former research on brain asymmetry in relation to AD, this paper considers the shape and texture features for more than one anatomical structure simultaneously. One feature selection algorithm is introduced and used for classification of AD involving human body. Experimental results show that the asymmetric feature analysis has a clear advantage over holistic feature analysis in terms of anatomical structure and feature type. Moreover, the proposed MASMR_FS algorithm can identify asymmetric features with high quality and the selected features demonstrate a superior discrimination ability for CTL, MCI and AD. The classification accuracy is higher than the traditional p-value based method. Future works should attempt to apply the MASMR_FS algorithm to other modalities, including fMRI, PET, and so on.

Acknowledgments.   This research is funded by NSFC (No: 61108086), CSTC (2012jjA0612), Innovation Ability Training Foundation of Chongqing University (CDJZR13160008, CDJZR 155507), Postdoctoral fund (2013M532153) and the Youth Training Project of Army Medical Technology (13QNP120).

References

1. Selkoe, D.J.: Preventing Alzheimer's disease. J. Sci. **337**, 1488–1492 (2012)
2. Khedher, L., Ramírez, J., Górriz, J.M., et al.: Early diagnosis of Alzheimer's disease based on partial least squares, principal component analysis and support vector machine using segmented MRI images. J. Neurocomputing **151**(1), 139–150 (2015)
3. Schmitter, D., Roche, A., Maréchal, B., et al.: An evaluation of volume-based morphometry for prediction of mild cognitive impairment and Alzheimer's disease. J. NeuroImage: Clinical. **7**, 7–17 (2015)
4. Orta-Salazar, E., Cuellar-Lemu, C.A.S., Díaz-Cintra, S., et al.: Cholinergic markers in the cortex and hippocampus of some animal species and their correlation to Alzheimer's disease. J. Neurol. **29**(8), 497–503 (2014)
5. Chincarini, A., Bosco, P., Calvini, P., et al.: Local MRI analysis approach in the diagnosis of early and prodromal Alzheimer's disease. J. NeuroImage. **58**(2), 469–480 (2011)

6. Toga, A.W., Thompson, P.M.: Thompson. Mapping brain asymmetry. J Nature Rev. Neurosci. **4**(1), 37–48 (2003)
7. Derflingera, S., Sorg, C., Gaser, C.: Grey-matter atrophy in Alzheimer's disease is asymmetric but not lateralized. J. J. Alzheimer's Dis. **25**(2), 347–357 (2011)
8. Shi, F., Liu, B., Zhou, Y., Yu, C., Jiang, T.: Hippocampal volume and asymmetry in mild cognitive impairment and Alzheimer's disease: Meta-analyses of MRI studies. J. Hippocampus. **19**(11), 1055–1064 (2009)
9. Tsai, K.-J., Yang, C.-H., Lee, P.-C., et al.: Asymmetric expression patterns of brain transthyretin in normal mice and a transgenic mouse model of Alzheimer's disease. J. Neurosci. **159**(2), 638–646 (2009)
10. Kim, J.H., Lee, J.W., Kim, G.H., et al.: Cortical asymmetries in normal, mild cognitive impairment, and Alzheimer's disease. J. Neurobiol. Aging. **33**(9), 1959–1966 (2012)
11. Donix, M., Burggren, A.C., Scharf, M., et al.: APOE associated hemispheric asymmetry of entorhinal cortical thickness in aging and Alzheimer's disease. J. Psychiatry Res. Neuroimaging. **214**(3), 212–220 (2013)
12. Li, Y., Zeng, X., Han, L., Wang, P.: Two coding based adaptive parallel co-genetic algorithm with double agents structure. J. Eng. Appl. Artif. Intell. **23**(4), 526–542 (2010)

Novel Feature Extraction and Classification Technique for Sensor-Based Continuous Arabic Sign Language Recognition

Mohammed Tuffaha[1], Tamer Shanableh[1(✉)], and Khaled Assaleh[2]

[1] Department of Computer Science and Engineering,
American University of Sharjah, Sharjah, UAE
{b00054842,tshanableh}@aus.edu
[2] Department of Electrical Engineering,
American University of Sharjah, Sharjah, UAE
kassaleh@aus.edu

Abstract. This paper proposes a novel approach to continuous Arabic Sign Language recognition. We use a dataset which contains 40 sentences composed from 80 sign language words. The dataset is collected using sensor-based gloves. We propose a novel set of features suitable for sensor readings based on covariance, smoothness, entropy and uniformity. We also propose a novel classification approach based on a modified polynomial classifier suitable for sequential data. The proposed classification scheme is modified to take into account the context of the feature vectors prior to classification. This is achieved through the filtering of predicted class labels using median and mode filtering. The proposed work is compared against a vision-based solution. The proposed solution is found to outperform the vision-based solution as it yields an improved sentence recognition rate of 85 %.

Keywords: Sign language recognition · Feature extraction · Sensor-based gloves · Pattern classification

1 Introduction

Sign language is the term used to describe the language that the deaf community uses to communicate together or with the hearing society. Sign language recognition systems are used to translate sign language into text or speech. Sign Language recognition systems have been developed for many languages including but not limited to English, Chinese, Korean and Arabic. Sign language recognition systems are divided into 2 categories based on data collection:

- Vision-based systems: Data is collected using one or more cameras. Typically such systems require high computational complexity and are not too accurate. An example of which is reported in [1].
- Glove-based systems: Data is collected using sensor-based gloves, an example of which is reported in [2]. It is more accurate than vision-based systems and is not affected by background motion, colors and light intensity.

© Springer International Publishing Switzerland 2015
S. Arik et al. (Eds.): ICONIP 2015, Part IV, LNCS 9492, pp. 290–299, 2015.
DOI: 10.1007/978-3-319-26561-2_35

Moreover, sign language recognition systems can be used for recognizing sign language alphabet, words or sentences. The latter is used in this paper and it is also referred to as contentious sign language recognition.

Three sensor-based gloves were used for Arabic sign language recognition [3]; the PowerGlove, DT Data Gloves and CyberGlove. With the use of PowerGlove, the work in [4] developed an Arabic sign language recognition system using Support Vector Machine (SVM) classifier. While in [2] it was reported that DG5-VHand 2.0 data gloves is suitable for Sign Language Recognition as it contains flex sensors and accelerometers without the need for motion detectors. The classification system is based on a method of accumulated differences to eliminate the temporal dependencies in the data with low-complexity. This word-based system worked on two modes; user dependent mode produced accuracy of 95.3 % and user independent mode achieved 92.5 % accuracy. Using the same gloves, a user-dependent recognition system was proposed by [5], the system works on continuous Arabic Sign Language (ArSL).

In [6], the CyberGlove and Flock of birds 3D motion tracker were used for American Sign Language (ASL) recognition using neural networks. The algorithm is designed to recognize one-handed words. The system is trained and tested on a set of 50 words for single and multiple users. In [7], CyberGlove are also used for continuous single handed American Sign Language (ASL) recognition. Classification was done using a two layer Conditional Random Field (CRF), Support Vector Machines (SVM) and Bayesian network (BN). CyberGlove is also used in [8] to develop a user independent two handed Chinese Sign Language recognition for isolated and continuous signs.

In this work we propose a sentence-based Arabic sign language recognition system using sensor-based gloves. Both hands are used for collecting data from sign language sentences ranging from 3 to 7 words each. We propose a novel set of features suitable for sensor-based sign language data. We also propose a novel alternative for existing classification techniques based on polynomial classifier.

2 The Dataset

In this section, the dataset used for training and testing will be described. The current work makes use of data collected earlier by [5]. The collected data made use of the DG5-VHand 2.0 data gloves that give 8 sensor readings per hand; 5 flex sensors for each finger and 3 3D-accelerometers.

The dataset used was created in collaboration with Sharjah City for Humanitarian Services [9]. The dataset consists of 40 sentences built from 80 words. Each sentence is composed of 3 to 7 words. The sentences are the same as the ones reported in [10]. Each of the 40 sentences was repeated 10 times performed by a single user. Again, a sequence of sensor readings make up a sign language word and a sequence of words make up a sign language sentence.

3   Data Collection and Labeling

A sensor based approach is to be used and compared against previous work to demonstrate results. The process start by collecting two sets of data one from sensory gloves and one from a camera. The camera input is used to label the sensory input manually.

Below is a descriptive diagram showing the process of data collection and labeling as used in [5]. The process starts by signing a new sentence, from which 2 attributes are acquired; sensor readings and record video. The sensor readings are manually labeled from the video and store the labels in a database. This process is illustrated in Fig. 1.

4   Proposed Feature Extraction

This section describes the proposed process by which meaningful features are extracted from a set of raw data. The feature extraction is performed on a window of sensor readings. This is important to put each sensor reading in its right context. In this case the raw data is 16 sensors from 2 DG5-VHand 2.0 data gloves divided as follows: 2×5 flex sensors for each finger and 2×3 3D accelerometer sensors to determine the rotation and position of the hand in 3D space.

Consider vector f_i containing the 16 sensor readings as one vector at a single point in time. The matrix of readings F_i will be expressed as $F_i = [f_1 f_2 .. f_N]^T$ where N is the number of vectors in a single window depending on time variant of the window and the number of readings taking per second. Typically 30 sensor readings per seconds are captured.

The raw data proved to contain some noise and are not very representative of the actual input so other statistical data are obtained from the feature vectors and added to make a new enriched FVs. In the beginning, mean and standard deviation were computed. Mean (μ) provides the average which eliminates to some extent the noise from the original feature vectors shown in the equation below.

$$\mu = \frac{1}{T} \sum_{k=1}^{T} f_k \qquad (1)$$

While the standard deviation measures the dispersion from the average sample. In this case, an estimate of the standard deviation called sample standard deviation (s) is used to reduce complexity, it is calculated using the equation below.

$$s = \left(\frac{1}{T-1} \sum_{k=1}^{T} (f_k - \mu)^2 \right)^{1/2} \qquad (2)$$

For the sign language data, this is done on a window of feature vectors, where the size of the window is varied via trial and error to find achieve highest performance.

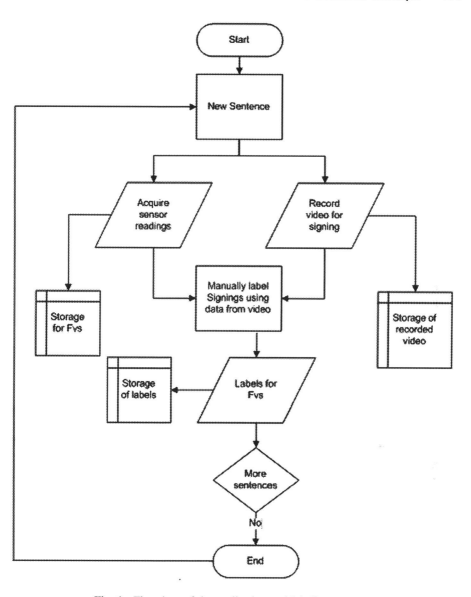

Fig. 1. Flowchart of data collection and labeling process.

For a predefined window size, the mean and standard deviation calculations are show below.

$$\mu_i = \frac{1}{\omega} \sum_{k=i-\frac{\omega-1}{2}}^{i+\frac{\omega-1}{2}} f_k \tag{3}$$

$$s_i = \left(\frac{1}{\omega - 1} \sum_{k=i-\frac{\omega-1}{2}}^{i+\frac{\omega-1}{2}} (f_k - \mu)^2 \right)^{1/2} \tag{4}$$

Where ω is an odd number denoting the window size and i is the current feature from a set of features.

Then other statistical features are added to further enrich the feature vectors such as the covariance, the entropy and the uniformity.

The covariance shows how much the features change with respect to each other. Mathematically, covariance is calculated using the equation below.

$$cov(x, y) = \frac{\sum_{i=1}^{n} (x_i - \mu_x)(y_i - \mu_y)}{n - 1} \tag{5}$$

In this work we compute the covariance of sensor readings for a window of size ω. We use the upper triangular values of the covariance matrix as feature variables.

Where x is the independent variable, y is the dependent variable, n is the number of points in the sample, μ_x is the mean of variable x and μ_y is the mean of variable y. The final feature vector consists of 200 values divided as follows: 16 raw sensor readings, 16 values for each of the window-based mean, standard deviation, entropy and uniformity and 120 values for the window-based covariance.

Since the data features have different scales, normalization is necessary to set a common range before classification for it to be successful. In this case, z-score normalization is used, the end result will have a mean of zero and unit variance.

5   Proposed Classification Solution

The proposed classification solution is based on the Polynomial classifier which was successfully used for classifying isolated sign language words [1]. Hence we start with a brief review of the polynomial classifier.

A Polynomial classifier is a supervised classifier technique which nonlinearly expands a sequence of input vectors to a higher dimension and maps them to a desired class labels.

Training a Pth order polynomial network is done in two stages. Stage one is expanding the training feature vectors through polynomial expansion. Stage two is linearly mapping the polynomial-expanded vectors to class labels by minimizing an objective criterion. Polynomial expansion of an M-dimensional feature vector $\mathbf{x} = [x_1\ x_2 \ldots x_M]$ is achieved by combining the vector elements with multipliers to form a set of basis functions, $\mathbf{p(x)}$. The elements of $\mathbf{p(x)}$ are the monomials of the form $\prod_{j=1}^{M} x_j^{k_j}$, where k_j is a positive integer, and $0 \leq \sum_{j=1}^{M} k_j \leq P$. For class i the sequence of feature vectors $\mathbf{X}_i = [\mathbf{x}_{i,1}\ \mathbf{x}_{i,2}\ \cdots,\ \mathbf{x}_{i,N_i}]^T$ is expanded into:

$$\mathbf{V}_i = [\,\mathbf{p}(\mathbf{x}_{i,1}) \quad \mathbf{p}(\mathbf{x}_{i,2}) \quad \cdots \quad \mathbf{p}(\mathbf{x}_{i,N_i})\,]^T \tag{6}$$

Where $\mathbf{X}_i$ is a $N_i \times M$ matrix and $\mathbf{V}_i$ is a $N_i \times O_{M,\mathrm{p}}$ matrix.

Expanding all the training feature vectors results in a global matrix for all K classes obtained by concatenating all the individual $\mathbf{V}_i$ matrices such that $\mathbf{V} = [\,\mathbf{V}_1 \quad \mathbf{V}_2 \quad \cdots \quad \mathbf{V}_K\,]^T$.

For each class i, the training objective is to find an optimum weight vector obtained by minimizing the distance between the class labels $\mathbf{y}_i$ and a linear combination of the polynomial expansion of the training feature vectors $\mathbf{V}\,\mathbf{w}_i$ such that

$$\mathbf{w}_i^{opt} = \arg\min_{\mathbf{w}_i} \|\mathbf{V}\,\mathbf{w}_i - \mathbf{y}_i\|_p \tag{7}$$

The class labels for the ith class, $\mathbf{y}_i$, is a column vector comprised of ones and zeros such as $\mathbf{y}_i = [\mathbf{0}_{N_1}, \mathbf{0}_{N_2}, \ldots, \mathbf{0}_{N_{i-1}}, \mathbf{1}_{N_i}, \mathbf{0}_{N_{i+1}}, \ldots, \mathbf{0}_{N_k}\,]^T$.

In the identification stage we are given a sequence of N_c feature vectors $\mathbf{X}_c$ and we are required to determine its class c as one of the enrolled classes in the set $\{1, 2, \cdots, K\}$. This is done by two steps: first, expand $\mathbf{X}_c$ into its polynomial basis terms $\mathbf{V}_c = [\,\mathbf{p}(\mathbf{x}_{c,1}) \quad \mathbf{p}(\mathbf{x}_{c,2}) \quad \cdots \quad \mathbf{p}(\mathbf{x}_{c,N_c})\,]^T$, and second, evaluate the output sequences against all K models $\{\mathbf{w}_i^{opt}\}$ to obtain a set of score sequences $\{\mathbf{s}_i\}$ such as

$$\mathbf{s}_i = \mathbf{V}_c \mathbf{w}_i^{opt} \tag{8}$$

The elements of the score sequence $\mathbf{s}_i$ represent the individual scores of each feature vector in the vector sequence $\mathbf{X}_c$. The class of the sequence $\mathbf{X}_c$ is determined by maximizing $\mathbf{s}_i$ such as

$$c = \arg\max_i(\mathbf{s}_i) \tag{9}$$

The proposed classifier is based on the above polynomial classifier of Pth order. Each feature vector is labeled according to the sign language word it belongs to. First, the classification weights vector is generated and then multiplied by the individual feature vectors to find the corresponding class label. We propose to use the context of the predicted label before arriving to the final classification result. We examine the predicted labels of the surrounding feature vectors and use a majority vote to decide on the class label of the current feature vector. Moreover, in Eq. (9) above, instead of computing the max score, we compute the first 3 maximum scores for each feature vector in the context window and then apply the majority vote. Figure 2 shows a block diagram for the classification process.

6 Experimental Results

In our dataset, the individual words are labeled by giving all feature vectors making up a word the same label. That is, each feature vector is labeled separately. There is a total of 82 classes where 80 are for the words and 2 are for the start and end of the sentences.

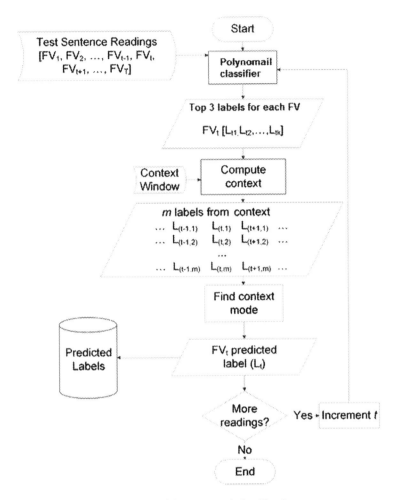

Fig. 2. Flowchart of the proposed classification system

Following the experimental setup of [5], the data is divided into 70 % training and 30 % testing in a round robin fashion.

In the results to follow, the sentences are considered correctly classified if all the words of the sentence are recognized successfully. Hence, the recognition rate is calculated by dividing the number of correctly recognized sentences by the total number of sentences. The word recognition rate on the other hand is found using the equation below [11].

$$Rate_{word} = 1 - \frac{D + S + I}{N} \tag{10}$$

Where D is the number of deleted words, I is the number of inserted words, S is the number of substituted words, and N is the total number of the words.

We use three classification measures. The first is the accuracy of classifying each feature vector individually, we refer to it as "class (%)". The second is the accuracy of classifying words based on the equation above, which is a sequences of feature vectors, we refer to it as "word (%)". The third and last is the accuracy of classifying each sentence correctly, which is a sequence of words, we refer to this as "sentence (%)". The tests are run on a Windows 7 machine with intel-i7 4790 k 3.5 GHz processor (quad core) and RAM of 16 GB.

In Table 1, we present the classification results using raw sensor features with window-based mean and standard deviation (total of 48 features). Polynomial classifier was implemented from 2nd to 7th order. In this experiment the FVs window ω is 43 and context width is 32.

Table 1. Classification results using raw sensor features with window-based mean and standard deviation

Order	Class (%)	Word (%)	Sentence (%)	Time
2nd	66.94	44.38	33.89	13 s
6th (peak)	81.6	70.85	63.33	39 s

The same results are repeated with the addition of the upper diagonal window-based covariance of the features (total of 168 features). The results are presented in Table 2. In this experiment the FVs window ω is 63 and context width is 24.

Table 2. Classification results using the features in Table 1 and the upper diagonal window-based covariance.

Order	Class (%)	Word (%)	Sentence (%)	Time
2nd	82.72	77.06	71.11	62 s
5th (peak)	87.48	86.67	82.22	142 s

The entropy and uniformity can also be added to the above feature set. This brings up the total features to 200 variables. The classification results of which are reported in Table 3. In this experiment the FVs window ω is 75 and context width is 22.

Table 3. Classification results using the features in Tables 1 and 2 and the entropy and uniformity of features

Order	Class (%)	Word (%)	Sentence (%)	Time
2nd	84.37	79.80	76.39	75 s
5th (peak)	87.77	88.95	85.00	135 s

The results in the above tables show that the proposed feature variables are suitable for sensor-based sign language recognition. The results also show that the proposed feature extraction and classification approach are not computationally expensive. For instance in Table 3, at a second order polynomial, it takes an average of 0.6 s to classify a sign language sentence.

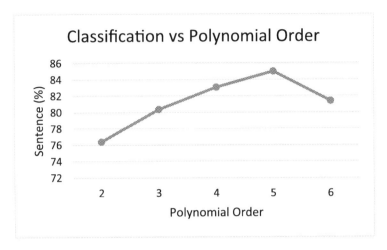

Fig. 3. Classification vs polynomial order

Figure 3 plots the sentence classification rate as a function of the polynomial classification order from 2 to 6. It is shown that the classification rates peaks at the 5th order.

Lastly, the proposed sensor-based solution is compared to [10] in terms of classification accuracy. As mentioned, the work in [10] used the same set of sentences with similar experimental setup. Features are based on Discrete Cosine Transform of window-based accumulated image differences. Hidden Markov Models are used for classification. The results are show in Table 4.

Table 4. Classification rates of proposed and reviewed solutions.

	Proposed	Reviewed [10]
Sentence recognition rate	85 %	75.6 %

This result is expected because sensor-based data is more accurate than vision-based data. Sensor readings are specific to hand movements whereas in sign language videos hand movements need to be segmented out. Segmentation techniques are typically not accurate and therefore the features are not an exact representation of the sign language sentences.

7   Conclusion

We proposed to modify the polynomial classifier to work with sequential data. This is implemented using a window-based feature extraction approach and through the use of statistical filtering of the predicted labels. We also proposed a new set of window-based features based on covariance, entropy, uniformity, smoothness and skewness. These features enhanced the classification accuracy. Lastly, we showed that the proposed system is computationally attractive and more accurate than existing vision-based solutions.

Acknowledgement. The authors gratefully acknowledge the American University of Sharjah for supporting this research through grant FRG14-2-26.

References

1. Shanableh, T., Assaleh, K.: User-independent recognition of Arabic sign language for facilitating communication with the deaf community. Digit. Signal Process. **21**, 535–542 (2011)
2. Assaleh, K., Shanableh, T., Zourob, M.: Low complexity classification system for glove-based arabic sign language recognition. In: Huang, T., Zeng, Z., Li, C., Leung, C.S. (eds.) ICONIP 2012, Part III. LNCS, vol. 7665, pp. 262–268. Springer, Heidelberg (2012)
3. Mohandes, M., Deriche, M., Liu, J.: Image-based and sensor-based approaches to arabic sign language recognition. IEEE Trans. Hum. Mach. Syst. **44**, 551–557 (2014)
4. Mohandes, M., A-Buraiky, S., Halawani, T., Al-Baiyat, S.: Automation of the Arabic sign language recognition. In: Proceedings of the 2004 International Conference on Information and Communication Technologies: From Theory to Applications, pp. 479–480 (2004)
5. Tubaiz, N., Shanableh, T., Assaleh, K.: Glove-based continuous Arabic sign language recognition in user-dependent mode. IEEE Trans. Hum. Mach. Syst. **PP**(99), 1–8 (2015). doi:10.1109/THMS.2015.2406692
6. Oz, C., Leu, M.C.: American sign language word recognition with a sensory glove using artificial neural networks. Eng. Appl. Artif. Intell. **24**, 1204–1213 (2011)
7. Kong, W.W., Ranganath, S.: Towards subject independent continuous sign language recognition: a segment and merge approach. Pattern Recogn. **47**, 1294–1308 (2014)
8. Gao, W., Fang, G., Zhao, D., Chen, Y.: A Chinese sign language recognition system based on SOFM/SRN/HMM. Pattern Recogn. **37**, 2389–2402 (2004)
9. Sharjah City for Humanitarian Services (SCHS). http://www.schs.ae/indexs.aspx
10. Assaleh, K., Shanableh, T., Fanaswala, M., Amin, F., Bajaj, H.: Continuous Arabic sign language recognition in user dependent mode. J. Intell. Learn. Syst. Appl. **2**, 19–27 (2010)
11. Starner, T., Weaver, J., Pentland, A.: Real-time American sign language recognition using desk and wearable computer based video. IEEE Trans. Pattern Anal. Mach. Intell. **20**(12), 1371–1375 (1998)

Weighting Based Approach for Semi-supervised Feature Selection

Khalid Benabdeslem[1(✉)], Mohammed Hindawi[2], and Raywat Makkhongkaew[1]

[1] University of Lyon1, 43 Bd du 11 Novembre 1918, 69622 Villeurbanne, France
khalid.benabdeslem@univ-lyon1.fr
[2] Computer Science Department, Zirve University, Kizilhisar Campus,
27260 Gaziantep, Turkey

Abstract. Semi-supervised feature selection has become more important as the number of features has increased in partially labeled data sets. In this paper we present a feature weighting-based model to address this problem. Our proposal is based on a semi-supervised clustering paradigm that can rank features according to their relevance from high-dimensional data. We propose an adaptation of the constrained K-Means algorithm to semi-supervised feature selection by an embedded approach. Experiments are provided on several known data sets for validating our proposal. The results are promising and competitive with several representative methods.

Keywords: Semi-supervised feature selection · Embedded approach · K-Means · Weighting approach

1 Introduction

The general problem of feature selection is well addressed in the literature by the data mining and machine learning communities. The aim of this task is to remove both irrelevant and redundant features in order to improve performance, decrease the complexity and improve the interpretability of learning algorithms [1]. Feature selection is well studied in both supervised and unsupervised contexts in several works [2,3].

Recently, learning from both labeled and unlabeled data has given rise to considerable interest. Thus, the semi-supervised feature selection became more important and more adaptive to real-world applications where labeled data are difficult and costly to obtain. In addition, the task becomes more challenging with the so called "small-labeled-sample problem" in which the amount of data that is unlabeled may be much larger than the amount of labeled data [4]. On the one hand, supervised feature selection algorithms require a large amount of labeled training data. Such algorithms provide insufficient information about the structure of the target concept, and thus can fail to identify the relevant features that are discriminative to different classes. On the other hand, unsupervised feature selection algorithms ignore label information, which may lead to performance deterioration.

© Springer International Publishing Switzerland 2015
S. Arik et al. (Eds.): ICONIP 2015, Part IV, LNCS 9492, pp. 300–307, 2015.
DOI: 10.1007/978-3-319-26561-2_36

Currently, the supervision information offered by the labeled part of data is transformed into background knowledge to be integrated into the feature selection process, along with the geometric structure exploited from the unlabeled part of data. Such background information is generally expressed by pairwise constraints, that specify that two instances are to be in the same class if both have the same label (must-link constraint) or in different classes if not (cannont-link constraint). Several recent works have attempted to exploit pairwise constraints or other prior information in feature selection. The authors in [5] proposed an efficient algorithm, called SSDR, which can simultaneously preserve the structure of original high-dimensional data and the pairwise constraints specified by users. The main problem with this method is that although the use of variance is very important to preserve the feature locality, the proposed objective function is independent of this parameter. In addition, the similarity matrix used in the objective function uses the same value for all pairs of data which are not related by any constraint. The same authors proposed a constraint score based method [6] which evaluates the relevance of features according to constraints only. The method carries out with little supervision information in labeled data ignoring the unlabeled data regardless of how large it is. The authors in [7] proposed to solve the problem of semi-supervised feature selection by a simple combination of scores computed on both labeled and unlabeled data. This method (called C4) tries to find a consensus between an unsupervised score and a supervised one (by multiplying both scores). The combination is simple, but can dramatically bias the selection of the features having best scores for the labeled part of data and bad scores for the unlabeled part and vice-versa. More recently, we proposed in [8] a constrained laplacian score (CLS) for semi-supervised feature selection by a more developed combination between a Laplacian score and a Constraint score. In the same way, we improved CLS in [9] by another method (CSFS), which exploits a constraint selection procedure during the feature selection process. Our recent work can be found in [10] with a theoretical analysis of CSFS and a new graph-based procedure to reduce redundancy.

The approach we introduce in this paper is an embedded method called wCKM (weighted constrained K-Means). It is based on semi-supervised clustering by a K-Means algorithm [11,12]. We propose the incorporation of a feature weighting procedure in the objective function of the constrained K-Means. The feature weights produced by wCKM measure the relevance of the associated features.

The rest of the paper is organized as follows: In Sect. 3, we describe our proposed weighting approach for semi-supervised feature selection. It is based on two important concepts, which are weighting and constrained clustering. An empirical study is given in Sect. 3. Finally, we conclude in Sect. 4 by a summary and an insight into future research based on our proposed approach.

2   Weighting Based Feature Selection: wCKM

In the context of semi-supervised learning, the data set $X = \{X_1, ..., X_n\}$ consists of two subsets depending on the label availability: $\{X_1, ..., X_L\}$ for which the labels $\{Y_1, ..., Y_L\}$ are provided, and $\{X_{L+1}, ..., X_{L+U}\}$ whose labels are not given. Each X_i is characterized by m features $x_{i1}, x_{i2}, ..., x_{im}$. Here, each label $Y_i \in \{1, 2, ..., C\}$ where C is the number of different labels, and $L + U = n$.

Before describing our semi-supervised feature selection method, we construct the different constraint sets (Ω_{ML} and Ω_{CL}) from the labeled part of data. Ω_{ML} contains pairs of instances that have the same label and Ω_{CL} contains those having different labels. Consequently, $|\Omega_{ML} \cup \Omega_{CL}| = \frac{L(L-1)}{2}$.

The idea behind our proposal is to associate to each feature j a weight w_j in a new objective function $\gamma(u, v, w)$ to be minimized, where the variable u is used for cluster-assignment, v for centroid-updating and w for feature-weighting. The goal is to assign a higher weight to a dimension along which the distance between instances and centroids is smaller.

$$\min_{u,v,w} \gamma(u, v, w) = \sum_{l=1}^{K} \sum_{i=1}^{n} \sum_{j=1}^{m} u_{il}^2 w_j^\beta (x_{ij} - v_{lj})^2 + \vartheta_{ML} + \vartheta_{CL} \qquad (1)$$

s.t. Eqs. (2), (3) and (4):

$$\sum_{l=1}^{K} u_{il} = 1; \; i = 1..n \qquad (2)$$

$$\sum_{j=1}^{m} w_j = 1; \; w_j \in [0, 1] \qquad (3)$$

$$u_{il} \in [0, 1]; i = 1..n \; and \; l = 1..K \qquad (4)$$

where β is a parameter for feature weights and:

$$\vartheta_{ML} = \sum_{(X_i, X_r) \in \Omega_{ML}} \sum_{k=1}^{K} \sum_{(l=1, l \neq k)}^{K} u_{ik} u_{rl} \qquad (5)$$

and

$$\vartheta_{CL} = \sum_{(X_i, X_r) \in \Omega_{CL}} \sum_{k=1}^{K} u_{ik} u_{rk} \qquad (6)$$

The optimization of the problem in Eq. (1) aims at partitioning X into K clusters whose centroids are $\{v_1, v_2, ..., v_K\}$. The obtained partition is constrained by the labeled part of the data, via ϑ_{ML} and ϑ_{CL} defined in Eqs. (5) and (6) respectively. These terms represent the penalty costs of constraint violation in Ω_{ML} and Ω_{CL} respectively. They control the influence given to the extracted constraints during the assignment phase of the algorithm, which is assured by u and constrained by Eqs. (2) and (4).

Note that β cannot be equal either to zero or to one. Indeed, if $\beta = 0$, the weighting is removed and so the feature selection cannot be performed. If $\beta = 1$, w would disappear because of the following derivations for solving the problem. Thus, to solve the optimization problem under these assumptions for β, we minimize Eq. (1) by solving the following three minimization problems:

- **Optimization: O1**: Minimizing $\gamma(u, v, w)$ with respect to v for calculating the centroids of clusters.
- **Optimization: O2**: Minimizing $\gamma(u, v, w)$ with respect to u for calculating the cluster-membership values of instances.
- **Optimization: O3**: Minimizing $\gamma(u, v, w)$ with respect to w for measuring the weights of features.

O1 represents the centroid updating procedure in the process and can be easily solved, providing the solution:

$$v_{lj} = \frac{\sum_{i=1}^{n} u_{il}^2 x_{ij}}{\sum_{i=1}^{n} u_{il}^2}; l = 1..K \ and \ j = 1..m \tag{7}$$

O2 represents the assignment step in the process. We use Lagrange multipliers to solve this problem and obtain:

$$u_{il} = \frac{\frac{1}{\sum_{j=1}^{m} w_j^{\beta}(x_{ij} - v_{lj})^2}}{\sum_{t=1}^{K} \frac{1}{\sum_{j=1}^{m} w_j^{\beta}(x_{ij} - v_{tj})^2}} + \frac{1}{2\sum_{j=1}^{m} w_j^{\beta}(x_{ij} - v_{lj})^2}$$

$$\times \left[\frac{\sum_{t=1}^{K} \frac{\sum_{(X_i,X_r)\in\Omega_{ML}} \sum_{(k=1,k\neq t)}^{K} u_{rk} + \sum_{(X_i,X_r)\in\Omega_{CL}} u_{rt}}{\sum_{j=1}^{m} w_j^{\beta}(x_{ij} - v_{tj})^2}}{\sum_{t=1}^{K} \frac{1}{\sum_{j=1}^{m} w_j^{\beta}(x_{ij} - v_{tj})^2}} \right.$$

$$\left. - \sum_{(X_i,X_r)\in\Omega_{ML}} \sum_{(k=1,k\neq l)}^{K} u_{rk} - \sum_{(X_i,X_r)\in\Omega_{CL}} u_{rl} \right] \tag{8}$$

Note that the instance assignments represented by Eq. (8) are made in a "fuzzy" manner where each instance has K membership values $w.r.t$ Eqs. (2) and (4).

O3 represents the feature weighting procedure in the process. The solution of this problem allows us to update the relevance of features in an embedded manner. To solve this problem, we can also considering the minimization via a Lagrange multiplier and obtaining:

$$w_j = \frac{1}{\sum_{k=1}^{m} \left(\frac{\sum_{l=1}^{K} \sum_{i=1}^{n} u_{il}^2 (x_{ij} - v_{lj})^2}{\sum_{l=1}^{K} \sum_{i=1}^{n} u_{il}^2 (x_{ik} - v_{lk})^2} \right)^{\frac{1}{\beta-1}}} \tag{9}$$

Note that the weights defined by Eq. (9) depend on the labeling constraints only indirectly through the cluster-membership values. Thus, the semi-supervised feature selection represented by this equation implicitly combines the geometrical structure from the unlabeled data with the supervision information of the labeled data. Subsequently, we can summarize all the above mathematical developments in Algorithm 1.

3 Results

In this section, we present an empirical study for evaluating the performance of our proposal over several high-dimensional data sets and compare it with several

---

Algorithm 1. wCKM

Input: Data set X of dimension $(n \times m)$, β, K
Output: Weighted and ranked features

1: Construct the constraint sets $(\Omega_{ML}$ and $\Omega_{CL})$ from the labeled part of X
2: Initialize the cluster-membership values to zero
3: Randomly choose initial centroids $v_1, v_2, ..., v_K$ from X
4: Randomly generate initial weights $w_1, w_2, ..., w_m$
5: **repeat**
6:    Calculate the cluster-membership values using Eq. (8)
7:    Update the cluster-centroids using Eq. (7)
8:    Update the feature weights using Eq. (9)
9: **until** Convergence
10: Rank the features $\{j\}$ according to their weights $\{w_j\}$ in a descending order.

---

representative methods. All data sets are available in http://featureselection.asu. edu/datasets.php. They are of different types, text data ("PCMAC", "RELATHE"); microarray data ("TOX-171") and face-image data ("PIX10P"). The whole of the data sets information is detailed in Table 1 in which the last column (S) represents the percentage of supervision.

Table 1. Data sets

Data sets	n	m	C	S
PCMAC	1943	3289	2	0.3 %
RELATHE	1427	4322	2	0.4 %
TOX-171	171	5748	4	7 %
PIX10P	100	10000	10	30 %

The data sets are high-dimensional, with different numbers of classes and little supervision; for evaluating the performance of wCKM and comparing it with other semi-supervised feature selection methods: sSelect [4], C4 [7], CLS [8] and CSFS [9].

3.1   Experimental Setting

To simulate the small labeled sample context, we set L, the number of labeled data, by randomly selecting 3 instances per class; the remaining instances are used as unlabeled data. The portion of the supervised information is very small for each data set (see the last column $(S = \frac{3C}{n})$ in Table 1).

The parameter β should be carefully tuned according to the problem at hand. If underestimated, the optimization of the objective function could be non-convex and having a local minimizer would be against our optimization procedure to have a local minimizer. On the other hand, if overestimated, the effect of feature weighting would disappear. For this reason, we set in the experiment:

$$\beta = percentile(\{\frac{log(\frac{1}{||x_i - \mu||^2})}{10}, i = 1..n\}, 10) \tag{10}$$

where μ is the centroid of the used data set. This usually results in good learning performance.

After feature selection, a SVM classifier with an RBF kernel (using LIBSVM package [13]) is employed for obtaining classification accuracy. The classifier is tuned via 10-fold cross-validation on training data set by repeating the process for 20 times on 20 different partitions of the data.

3.2 Comparison of Feature Quality on High-Dimensional Data

In this section, we assess the performance of wCKM and compare it with the above cited methods. This comparison concerns the classification accuracy that we present in both Fig. 1 and Table 2.

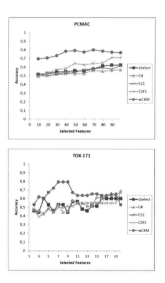
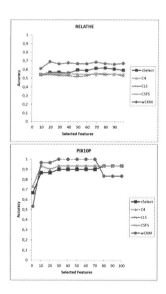

Fig. 1. Performance on classification accuracy vs. different number of selected features

Figure 1 plots the curves of the whole algorithms for classification accuracy vs. the different number of selected features. This figure indicates that in most cases wCKM outperforms the other methods, especially for text data sets in which the noise is important. It can be shown that in particular, the performances of C4 is the worst. The performance of C4 is weak for small-labeled data and relatively good for the high-labeled ones. We estimate that this is because C4 naively combines (by multiplying) two scores from both labeled and unlabeled

Table 2. Classification accuracy (SVM in %).

Datasets	sSelect	C4	CLS	CSFS	wCKM
PCMAC	56.3±3.75	53.68±2.41	55.26±3.71	60.55±7.45	**72.9±3.96**
RELATHE	58.26±2.52	54.77±0.51	53.79±1.19	55.02±0.47	**63.46±3.05**
TOX-171	53.16±5.87	53.99±7.73	51.78±5.91	51.78±5.91	**65.90±2.25**
PIX10P	89.8±3.63	92.56±2.95	92.76±2.5	92.77±2.6	**93.6±1.11**

data. wCKM seems to more efficiently combine the labeled and unlabeled parts of data than the other constraint based semi-supervised methods. This shows that the combination is more efficient using an embedded approach than that using a filter one, in which the relevance of features is independently measured. However, wCKM does not perform very well for Face datasets, in which the number of clusters and the number of generated constraints are both high. This is because the algorithm tends to simultaneously optimize both the proximity between instances and their closest centroids on one hand, and the violation of constraints on the other hand. Moreover, It is worth mentioning that the classification accuracy of wCKM generally increases at the beginning (with a small number of features), but such an increase lessens at the end when the effect of irrelevant features begins to appear.

Table 2 compares the averaged accuracy under different numbers of selected features. From this table and Fig. 1, we can find that, the performance of wCKM is almost always better than that of sSelect, C4 and CLS, and is comparable to CSFS. More specifically, wCKM is superior to the other methods on all datasets except those with high values of both K and l (as in Image datasets).

From Table 2, we can calculate differences between averaged accuracies among algorithms. We can see that in terms of accuracy gains, wCKM is 9.42 % better than sSelect, 7.75 % better than C4, 7.82 % better than CLS and 6.73 % better than CSFS. This observation suggests that the label information is more adopted for semi-supervised feature selection with our method than the others. This is also consistent with our understanding that the *embedded* character of the method has an important role in feature selection compared to the filter based approaches.

Hence, with wCKM, the label information is explicitly learned by the minimization of constraint violation in the associated objective function. This minimization is simultaneously performed with the minimization of the weighted distance between the data instances and their closest prototypes in the different clusters. Both minimizations are required to provide relevant features in a semi-supervised context.

4   Conclusion

In this paper, we proposed an embedded approach for semi-supervised feature selection. We presented wCKM, a new method based on constrained K-Means algorithm which calculates feature weights automatically. We showed that the optimization of the associated objective function is based on both, minimization of proximity between instances and their closest centroids, and minimization of constraint violation. Experimental results on several high dimensional data sets are performed by measuring both, classification and clustering accuracies. We showed that with only a small number of constraints, the proposed algorithm outperforms other semi-supervised features selection methods. For future works, it would be interesting to extend the proposed method to deal with regression problems in which the supervised part of data contains continuous values.

References

1. Guan, Y., Dy, J., Jordan, M.: A unified probabilistic model for global and local unsupervised feature selection. In: Proceedings of the Twenty Eight International Conference on Machine Learning (2011)
2. Guyon, I., Elisseeff, A.: An introduction to variable and feature selection. J. Mach. Learn. Res. **3**, 1157–1182 (2003)
3. Dy, J., Brodley, C.E.: Feature selection for unsupervised learning. J. Mach. Learn. Res. **5**, 845–889 (2004)
4. Zhao, Z., Liu, H.: Semi-supervised feature selection via spectral analysis. In: Proceedings of SIAM International Conference on Data Mining, pp. 641–646 (2007)
5. Zhang, D., Zhou, Z., Chen, S.: Semi-supervised dimensionality reduction. In: Proceedings of SIAM International Conference on Data Mining (2007)
6. Zhang, D., Chen, S., Zhou, Z.: Constraint score: a new filter method for feature selection with pairwise constraints. Pattern Recogn. **41**(5), 1440–1451 (2008)
7. Kalakech, M., Biela, P., Macaire, L., Hamad, D.: Constraint scores for semi-supervised feature selection: a comparative study. Pattern Recogn. Lett. **32**(5), 656–665 (2011)
8. Benabdeslem, K., Hindawi, M.: Constrained laplacian score for semi-supervised feature selection. In: Gunopulos, D., Hofmann, T., Malerba, D., Vazirgiannis, M. (eds.) ECML PKDD 2011, Part I. LNCS, vol. 6911, pp. 204–218. Springer, Heidelberg (2011)
9. Hindawi, M., Allab, K., Benabdeslem, K.: Constraint selection based semi-supervised feature selection. In: Proceedings of International Conference on Data Mining, pp. 1080–1085 (2011)
10. Benabdeslem, K., Hindawi, M.: Efficient semi-supervised feature selection: constraint, relevance and redundancy. IEEE Trans. Knowl. Data Eng. **26**(5), 1131–1143 (2014)
11. Wagstaff, K., Cardie, C., Rogers, S., Schroedl, S.: Clustering with instance level constraints. In: Proceedings of the Seventeenth International Conference on Machine Learning, pp. 1103–1110 (2001)
12. Bilenko, M., Basu, S., Mooney, R.: Integrating constraints and metric learning in semi-supervised clustering. In: Proceedings of the Twenty First International Conference on Machine Learning, pp. 11–18 (2004)
13. Chang, C.C., Lin, C.J.: LIBSVM: a library for support vector machines. ACM Trans. Intell. Syst. Technol. **2**, 27:1–27:27 (2011). http://www.csie.ntu.edu.tw/~cjlin/libsvm

Path Planning with Slime Molds: A Biology-Inspired Approach

Masafumi Uemura[1]([⊠]), Haruna Matsushita[1],
and Gerhard K. Kraetzschmar[2]

[1] Kagawa University,
2217-20 Hayashi-cho, Takamatsu, Kagawa 761-0396, Japan
s14g456@stmail.eng.kagawa-u.ac.jp, haruna@eng.kagawa-u.ac.jp
[2] Bonn-Rhein-Sieg University,
Grantham-Allee 20, 53757 Sankt Augustin, Germany
gerhard.kraetzschmar@fh-bonn-rhein-sieg.de

Abstract. This paper proposes an Artificial Plasmodium Algorithm (APA) mimicked a contraction wave of a plasmodium of physarum polucephalum. Plasmodia can live using the contracion wave in their body to communicate to others and transport a nutriments. In the APA, each plasmodium has two information as the wave information: the direction and food index. We apply the APA to 4 types of mazes and confirm that the APA can solve the mazes.

1 Introduction

Of late years, many algorithms which mimicked various creatures have proposed. Major examples are follows: Particle Swarm Optimization (PSO) [1]: based on the swarm behavior such as fish and bird schooling in nature, and Bees Algorithm (BA) [2,3]: based on the food foraging behavior of swarms of honey bees. They can perform smart in the nature making full use of their memory and own ability although they are small.

By the way, the creature, which we focus on in this study is the plasmodium of Physarum polucephalum. This plasmodium is multinuclear and unicellular organism, and it does not have any differentiated organ. Thereby, the plasmodium senses environment, decides and moves using the whole body. When the plasmodium is cultivated, their body is getting bigger like $5m^2$. However, no matter how big their body become, they can behave like in a body in spite of unicellular organism. In addition, the results that this highly homogenized plasmodium can solve the maze and find the shortest path were reported [4–6]. This is indeed true and was experimentally proved.

In this study, we consider about how the plasmodium solves the maze and find a shortest path. Furthermore, based on our consideration, we propose a new algorithm modeled on the plasmodium, called "Artificial Plasmodium Algorithm (APA)", and confirm that it can solve the maze. This is the first approach to simulate a behavior of the plasmodium mathematically.

© Springer International Publishing Switzerland 2015
S. Arik et al. (Eds.): ICONIP 2015, Part IV, LNCS 9492, pp. 308–315, 2015.
DOI: 10.1007/978-3-319-26561-2_37

2 Plasmodium in Nature

2.1 Behavior

The plasmodium is a large unicellular organism with a lot of nucleuses and network of flowing protoplasm. The plasmodium can move to flow the protoplasm as its direction switches back and forth periodically known as shuttle streaming. The shuttle streaming is caused by a contraction and relation oscillation of the plasmodium which is gathered in a food source. When one of the plasmodium's cell is stimulated, the cell generates a vibration phase and changes dynamic pattern to propagate the vibration phase to all cells by a *contraction wave*. Thus, the plasmodium is gathered at comfortable place (i.e. nutriment, warmth and humidity) or escape from harsh conditions (i.e. coldness and drying).

2.2 Application of Real Physarum Polycephalum to Solve a Maze

It had been reported that the true plasmodium of physarum polycephalum can solve and find a shortest path in the maze by T. Nakagaki et al. [4–7]. The process of true experimentation shown in Fig. 1 [4].

The plasmodium propagate from original sites along pathways, avoiding walls, and they merge into a single cell (a). After two pieces of nutrients are placed as a start and exit point in the maze, the network of pronounced tube is approximated (b). Protoplasmic tubes are connected all the path between two points and only the shortest path are remained gradually with time progress (c).

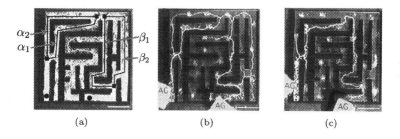

      (a)   (b)   (c)

Fig. 1. Process of the plasmodial maze-solving [4]. (a) Initial state. (b) Intermediate state. (c) Final state.

3 Artificial Plasmodium Algorithm (APA)

We propose an artificial plasmodium algorithm (APA) mimicked the contraction wave of the true plasmodium. The contraction wave is related with the shuttle streaming and it has the information of propagation direction. The wave sources are frontal part of the plasmodium and any food sources. When there are two food sources, the contraction wave propagates between two sources, and it is switched from one food source to another every several waves.

In our algorithm, the search space is defined as a maze which divided into discrete space of arbitrary size of $x \times y$ cells. The maze is composed of pathways and walls. Plasmodia can propagate and exist on such pathways. Each cell has a propagation direction to neighbors which are located on the four points of the compass: north, south, east and west. During process of the contraction wave, the contraction wave transmits transport direction to each cell from frontal part or food sources. After the wave, each cell has and knows the direction to the destination. In order to find the path, they trace their own direction. There are four processes until finding a path shown in Fig. 2.

1. Initialization.
2. Placement of two foods in the maze.
3. Spread of contraction waves.
4. Search for the path.

Fig. 2. Pseudo code of APA

3.1   Initialization

Initial state of APA, all the pathways of the maze are covered by the plasmodia, namely, M plasmodia exist in the maze.

3.2   Placement of Two Foods in the Maze

We put two food sources as the start and the exit points on the pathways of the maze. The food sources are distinguished that respective food sources have a food index like food "1" and "2".

3.3   Spread of Contraction Waves

A contraction wave starts to spread between two food sources to find a path. Figure 3 shows the flowchart of spread of contraction waves and the example is shown in Fig. 4(a)–(d). Each plasmodium i $(i = 1, 2, \cdots, M)$ has two information: a direction and the food index. We call these two information "wave information", and they are decided by the contraction wave. The direction comes in 4 types (north, south, east and west) and it denotes which neighbor the contraction wave came from. Note that decision of the direction by the contraction wave is once. Moreover, the food index comes in 2 types, and it denotes which food source the contraction wave came from. For example, the wave from food "1" has the food index "1".

First, the contraction wave spreads into circumference of the two food sources. This is to say, the four plasmodia in the neighborhood of each food source have the food index and the direction of each food source (as shown in Fig. 4(a)).

Then, the plasmodium i, which have no wave information, is chosen randomly. When one or more plasmodia, which have the wave information exist in the neighborhood of i, the plasmodium i obtains the food index and the direction. The food index of i is copy of that of the neighbor, and the direction information of i is a direction of the neighbor viewed from i. We repeat these process. In this way, the contraction wave gradually spreads into all the plasmodia.

When all the plasmodia have the wave information, the contraction wave stops propagating.

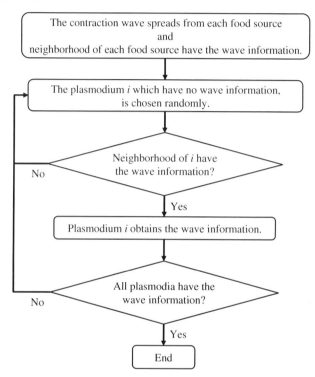

Fig. 3. Flowchart of spread of the contraction waves

3.4 Search for the Path

After the contraction wave finishes propagating, two plasmodia are chosen randomly with fulfilled following rules:

1. Two plasmodia have different food indexes, respectively.
2. Two plasmodia are located in the neighborhood of each other.

In Fig. 4(c), two plasmodia with ellipse shape are possible candidacy. After two plasmodia are chosen, the two plasmodia become a way point between two food sources and pursue the each direction to food sources as a path, with using the direction information obtained by Sect. 3.3 shown in Fig. 4(d).

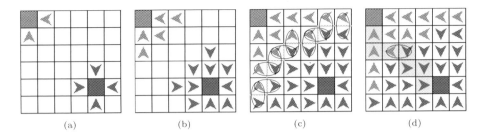

Fig. 4. The extraction of process of the contraction wave in APA. (a) The contraction wave spreads from two food sources, then the neighborhood of each food source have the wave information. (b) The contraction wave gradually spreads into all the plasmodia. (c) Two plasmodia are chosen randomly. (d) The path is appeared pursuing the each direction from the two chosen plasmodia.

4   Computer Simulation

4.1   Netlogo

In this study, we simulate the APA by using a Netlogo. The Netlogo which was made by Uri Wilensky in 1991 is the programming environment to simulate a nature and a social phenomenon. The Netlogo is suitable for modeling the complicated system which is changing every time.

4.2   Setting of Mazes

We use 4 types of the mazes shown in Fig. 5. All the mazes are set as 100 cells length × 100 cells width and composed pathways (*wall height* = 0) and walls (*wall height* = 1). Red circles in each maze indicate the start and exit point. There is only one operating route between two points in (a) Maze A (b) Maze B (c) Maze C. Especially in (d) Maze D, it is same as the maze, which was used in the real experimentation, and it has several routes including the shortest path between the start and exit.

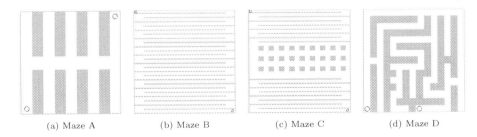

(a) Maze A   (b) Maze B   (c) Maze C   (d) Maze D

Fig. 5. Mazes used in the simulations. Two red circles indicate the start and the exit points, respectively. (a) Maze A (b) Maze B and (c) Maze C just have one route between two points. In (d) Maze D, there are 4 routes between two points (Color figure online).

Fig. 6. Result of the simulation. The proposed APA found the path between two food sources in the mazes.

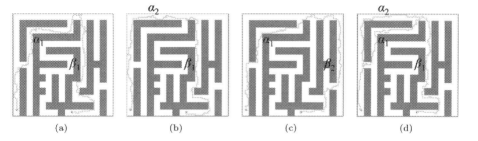

Fig. 7. Found routes of the simulation results. (a) The shortest path: $\alpha_1 + \beta_1$. (b) $\alpha_2 + \beta_1$. (c) $\alpha_1 + \beta_2$. (d) $\alpha_1 + \alpha_2 + \beta_1$.

5   Results

Simulation results of the maze (a) Maze A (b) Maze B and (c) Maze C are shown in Fig. 6 respectively. We can see from this figure, the proposed APA found the optimal route between two food sources in each maze (Fig. 7).

Table 1 shows a simulation result of 100 trials of (d) Maze D. The discovery rate means the percentage of the number of times that found any routes including the shortest path between two food sources. We can see that APA found the routes with a probability of 91 %. In addition, found routes are also shown in

Table 1. Simulation results of (d) Maze D

Number of trials	Discovery rate	
100	91 %	
	Routes	Discovery rates
	$\alpha_1 + \beta_1$	72 %
	$\alpha_2 + \beta_1$	12 %
	$\alpha_1 + \beta_2$	6 %
	$\alpha_1 + \alpha_2 + \beta_1$	1 %

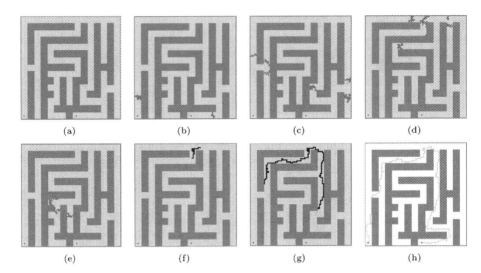

Fig. 8. Simulation using APA to Maze D. (a) Initial state ($t = 0$). (b) $t = 10$. (c) $t = 30$.
(d) $t = 60$. (e) $t = 90$. (f) $t = 120$. (g) $t = 190$. (h) Simulation results ($t = 300$).

Table 1. Route $\alpha_1 + \beta_1$ is the shortest path and was found with 72 %. Other
routes are not shortest path, but the APA can find with the highest probability.
Figure 8(a)–(h) shows the simulation process of APA to the Maze D. The two
food sources are placed in Fig. 8(a). Then the contraction wave spreads into all
the plasmodia in Fig. 8(b)–(e), and the simulation result is Fig. 8(h).

6   Conclusions

We have proposed artificial plasmodium algorithm (APA) mimicked the contraction wave of the plasmodium of physarum polucephalum. In the APA, each
plasmodium has two information as the wave information: the direction and
food index. The contraction wave spreads from the two food sources and when
the plasmodium received the contraction wave, the plasmodium can know the
direction of either food source. After the contraction wave finishes propagating,
two plasmodiums are chosen randomly and pursued each direction to the food
sources as a path.

We have applied APA to 4 types of the mazes. From these results, we can
say that the proposed APA found the optimal route in any mazes. In addition,
the found routes are not only the shortest path but also non-optimal routes. To
elucidate why and how non-optimal routes are made and to apply the APA to
more complicated mazes are remaining as a future works.

Acknowledgments. This work was supported by KAKENHI 24700226.

References

1. Kennedy, J., Eberhart, R.C.: Particle swarm optimization. In: Proceedings of IEEE International Conference on Neural Network, pp. 1942–1948 (1995)
2. Pham, D.T., Ghanbarzadeh, A., Koc, E., Rahim, S., Zaidi, M.: The Bees Algorithm - A Novel Tool for Complex Optimisation Problems. Cardiff University, UK, Manufacturing Engineering Centre (2005)
3. Karaboga, D.: An idea based on honey bee swarm for numerical optimization in Technical report TR06 (2005)
4. Nakagaki, T., Yamada, T., Toth, A.: Maze-solving by an amoeboid organism. Nature **407**, 470 (2000)
5. Nakagaki, T.: Smart behavior of true slime mold in labyrinth. Res. Microbiol. **152**, 767–770 (2001)
6. Nakagaki, T., Yamada, T., Toth, A.: Path findingby tube morphogenesisin an amoeboid organism. Biophys. Chem. **92**, 47–52 (2001)
7. Tero, A., Kobayashi, R., Nakagaki, T.: A mathematical model for adaptive transport network in path finding by true slime mold. J. Theor. Biol. **244**, 553–564 (2007)

Convolutional Neural Network
with Biologically Inspired ON/OFF ReLU

Jonghong Kim, Seonggyu Kim, and Minho Lee[✉]

School of Electronics Engineering, Kyungpook National University,
1370 Sankyuk-Dong, Puk-Gu, Taegu 702-701 South Korea
jonghong89@ee.knu.ac.kr, {kimsg99550,mholee}@gmail.com

Abstract. This paper proposes a modification of convolutional neural network (CNN) with biologically inspired structure, retinal structure and ON/OFF rectified linear unit (ON/OFF ReLU). Retinal structure enhances input images by center surround difference of green and red, blue and yellow components and creates positive results and negative results like ON/OFF visual pathway of retina to make totally 12 feature channels. This ON/OFF concept also adopted to each convolutional layer of CNN and we call this ON/OFF ReLU. Conventional ReLU only passes positive results of each convolutional layer so it loses negative information such as how much it was negative and also loses learning chance if results are saturated to zero but proposed model uses both positive and negative information so that additional learning chance also exist through negative results. Experimental results show how much the negative information and retinal structure improves the performance of CNN with public data.

Keywords: Object recognition · Neural network · Deep learning · Biologically inspired structure

1 Introduction

The concept of receptive field penetrates through recent computational theories of visual processing such as saliency map [1, 2] and convolutional neural network (CNN) [3]. The CNN is one of the popular deep network which is more suitable for image recognition than other deep models. The CNN structure consists of simple and complex cells, inspired by primary visual cortex receptive field structure which revealed by Hubel and Wiesel [4]. The saliency map consists of biologically inspired center surround difference structure [5]. The center surround difference is well known retinal structure but ON/OFF paths of visual information have not much considered as important characteristic before from engineering implementation point of view. Julijana Gjorgjieva et al. [6] evaluated the benefits of visual pathway splitting. They have shown that ON/OFF pathway increases information and reduces required spikes for the same information compared with ON/ON path way [7]. The shape of ReLU is similar with frequency response of neural spikes likewise sigmoid is similar with graded potential. Namely, it is consequentially derived that ON/OFF ReLU carries more information than conventional ReLU. Kaiming He et al. [8] proposed a generalized rectified linear unit which is called

© Springer International Publishing Switzerland 2015
S. Arik et al. (Eds.): ICONIP 2015, Part IV, LNCS 9492, pp. 316–323, 2015.
DOI: 10.1007/978-3-319-26561-2_38

parametric rectified linear unit (PReLU). PReLU has negative information with adaptively learned ratio and specific learning rule. Comparison result of ReLU and PReLU shows that PReLU is better for every image scale [8], so this result also supports the importance of negative information. Therefore, in this paper, we modified CNN with biologically inspired aspects to make CNN much similar with brain visual processing. First, we applied retinal structure for the pre-processing. This structure creates red, green, blue and yellow component from original color image and then, with center surround difference, it creates four positive features and four negative features so it is totally 12 features, i.e. we can expand amount of information and enhance features. Second, we propose a new modification of ReLU, which we call ON/OFF rectified linear unit (ON/OFF ReLU). ON/OFF ReLU consist of ON ReLU and OFF ReLU. ON ReLU is same as conventional ReLU but OFF ReLU is reversed ReLU which is activated by negative input signal. ON and OFF ReLU are not separated but work as pair. It expands each convolutional result as double and this negative information gives additional learning chance to make model learn faster and get good result.

Proposed model uses Caltech 101 dataset which consists of 101 different object image classes. Each object category contains about 40 to 800 images of size 300×200.

2 Proposed Method

2.1 Modeling Retinal Structure

To implement brain like visual processing model, we need to consider biological retinal structure at first. Retina consists of cone cells, rod cells, horizontal cells, bipolar cells, amacrine cells, ganglion cells, etc. Cone cells perceive color whereas rod cells perceive only intensity. In detail, cone cells also perceive intensity in bright illumination and rod cell do not work at that time. This phenomenon is related to power of illumination. Therefore, from the engineering implementation point of view, conventional RGB input data is enough information to build the retinal structure because we only use bright color images. Information from cone cells goes next to horizontal cells. Horizontal cells operate center surround difference. There are two significant operations which we call color opponency, one is red/green opponency, and other is blue/yellow opponency. The red and green information is important for the visual perception because most cone cells in retina are L cells (for red) and M cells (for green) so segregating that information should be considered at first. In contrast, S cone cells which perceive blue color are rarely found from fovea compare with other cone cells but effect of each S cell is very significant and yellow can be regarded as representation of the power of illumination so blue/yellow opponency indicates how much the blue component is included in color. For the engineering point of view, operation of horizontal cell is similar with Gaussian filtering which is low pass filter so this makes it possible to consider the surrounding information of particular cone cell. The size of horizontal Gaussian filter is about 10 times larger than center size on the basis of biological facts [9]. Center surround difference operation is similar with difference of Gaussian but unlike difference of Gaussian for the single color, center surround difference uses different color for each center and surround. For example, for the red center, green surround case, red center signal excites response whereas green center signal exhibits response and green surround signal excites response whereas red surround signal exhibits response.

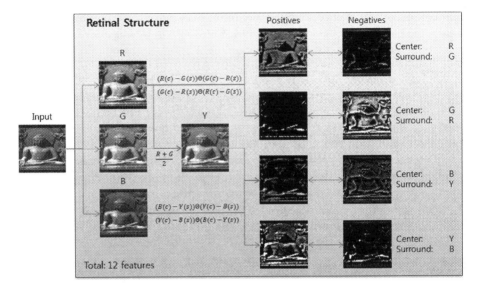

Fig. 1. Retinal structure for pre-processing (Color figure online)

These signals pass through bipolar cells. From this step, signals start to separate to ON path and OFF path. When the photo receptor is illuminated, ON bipolar cell is activated and when the light is turned off, OFF bipolar cell is activated [10]. Until this step, all the responses are graded potential but ON and OFF ganglion cells generate neural spikes whose frequency is proportional to graded potential of bipolar cells. This proportional frequency response is similar to ReLU. The action potential is transmitted to visual cortex, which is the motivation of CNN.

Figure 1 shows retinal structure process with an example. Input color image is separated to RGB channels and Y component is computed by averaging R and G. Each center surround difference is computed by Eq. (1).

$$CS(A, B) = \{(A(c) - B(s)) \, \Theta \, (B(c) - A(s))\} \tag{1}$$

$CS(A, B)$ is center surround difference of argument feature A and B. $A(c)$ means the center feature of A and $A(s)$ means the surround feature of A, it is also same for B. Θ means center surround difference operator [11]. After this operation, results are separated to positives and negatives which only include positive value and negative value.

2.2   ON/oFF Rectified Linear Unit

Rectified linear unit (ReLU) which is proposed by Xavier Glorot et al. [12] is inspired by biological facts that early part of our frequency response of neural spikes is similar with the rectifier. For the visual perception process, these neural spikes are found firstly

ON ganglion cells of the retina as mentioned above section. Because of the relative easiness of investigating eye than other nerve system, structure of retina is well known and one of the important things is discovery of ON/OFF visual pathway [13]. This separation can generate more visual information [6] but from the engineering point of view, this effect not that much considered before. In this paper, we tried to consider both of these two, one is positive information and the other is negative information.

For the implementation, existance of negative part of the ReLU is not recommended because of regularization. Separation of information can avoid this problem and this structure still has negative information. Figure 2 shows the structure of ON/OFF ReLU. Conventional neural network considers only one activation function but to separate information, we used same weighted sum result for different activation function. This different information creates doubled number of results. Because of this unusual change, we need to redefine forward propagation and back propagation formulas more carefully.

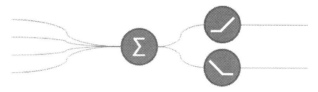

Fig. 2. ON/OFF ReLU structure. Weighted sum results pass through ON/OFF ReLU. Result of ON ReLU is positive information and OFF ReLU is negative information.

ON ReLU is calculated as same as conventional ReLU by Eq. (2) but because OFF ReLU is the reversed ReLU, it is calculated by Eq. (3).

$$y_{jOn} = y_j = max\left(0, \textstyle\sum_{i=1}^{n} w_{ij}x_i + w_{0j}\right) \tag{2}$$

$$y_{jOff} = y_{m+j} = max\left(0, -\left(\textstyle\sum_{i=1}^{n} w_{ij}x_i + w_{0j}\right)\right) \tag{3}$$

y_{jOn} is jth output value of ON ReLU which is same as conventional ReLU and the argument is weighted sum whereas y_{jOff} is output value of OFF ReLU, where n is number of previous nodes and w_{0j} means bias. If the number conventional ReLU results was m, then number of ON/OFF ReLU results is doubled 2 m so order of result is j and m + j. These formulas itself are not difficult to apply but after this forward propagation, doubled number of results come out. At this point, number of weights seems the problem, but the number of weights doesn't need to be changed. The purpose of the OFF ReLU is generation of extra information and the information comes from ignored negative part of ReLU. The only change is the number of weights of next level. Because of doubled number of input, the number of weights also needs to be doubled.

Another difficulty is computation of backpropagation. The problem comes when we calculate the so-called delta which is back-propagated factor to update weights in gradient descent method. Equation (4) shows how the delta can be calculated with ON/OFF ReLU.

$$\delta_j = \phi'_{On}\left(\sum_{i=1}^{n} w_{ij}x_i + w_{0j}\right) \sum_{k=1}^{l} w_{jk}\delta_k$$
$$+ \phi'_{Off}\left(\sum_{i=1}^{n} w_{ij}x_i + w_{0j}\right) \sum_{k=1}^{l} w_{(m+j)k}\delta_k \tag{4}$$

$$\phi'_{On}(u) = \begin{cases} 1 & u > 0 \\ 0 & u \le 0 \end{cases}$$
$$\phi'_{Off}(u) = \begin{cases} 0 & u \ge 0 \\ -1 & u < 0 \end{cases} \tag{5}$$

δ_j is jth delta, ϕ_{On} is ON ReLU, ϕ_{Off} is OFF ReLU, l is number of kth output and w is weight between two nodes according to its subscript. The first term of delta formula indicates well known delta rule and the second term is additional negative information. Because OFF ReLU result is another output, weights are also learned differently. Even though they are learned differently, those two outputs come from same weighted sum so we need to combine ON and OFF term.

Delta calculation requires derivative of activation function. Equation (5) shows the differential results.

2.3   Modification of Convolutional Neural Network

CNN structure which is inspired by primary visual cortex of brain is the most suitable deep neural network model for image recognition task. The operation itself is similar with conventional image processing but significant difference is that it learns kernels by itself. Usually it is expected that well trained kernels reflect the common characteristics of training samples. In this paper, we expected that if we can use negative information, the negative kernels can reflect the characteristics which should not be included in the images in same category.

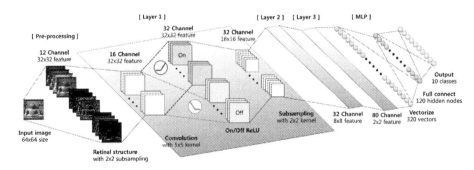

Fig. 3.   Overall structure of proposed model

Figure 3 shows how the convolutional network is constructed. Primary part is the preprocessing with retinal structure. Input image size is fixed with 64 × 64 and 12 features from retinal structure are subsampled with 2 × 2 max pooling to make 32 × 32 size features. These features pass through as an input of first layer of CNN. First convolution result is 16 channel features with the 32×32 as same size of input. Convolution operation in CNN is same with the weighted sum of neural network so after this it passes through ON/OFF ReLU. ON/OFF ReLU doubles the number of input so result is totally 32 features. Then do the subsampling with 2×2 size max pooling to make 16×16 size features. This operation performs consequently through layer 2 and layer 3. Result of layer 2 is 8×8 size 32 channel features and result of layer 3 is 2×2 size 80 channel features. This feature map is vectorized to 320 input nodes and it passes through MLP with 120 hidden nodes. Final number of class is 10 in this paper.

3 Experimental Results

The experiment has taken with Caltech-101 image dataset which includes 101 image categories. Each image category includes about 30–800 image samples but only few of specific categories such as faces has hundreds of samples and most of the categories includes about 40 image samples. For the experiment, 10 image categories are selected, chair, camera, anchor, lobster, dolphin, dragonfly, beaver, binocular, sunflower and garfield. For the training, augmented 2000 samples from 20 samples of the each category are used and 10 samples of each category are used for the testing. The training has performed with stochastic gradient descent without batch and 50 % dropout is used. The learning rate is 5×10^{-7} for the convolutional layers and initial weight is ± 0.05 for all the weights. The dropout only applied to hidden layer of MLP, first the model trained without dropout with 10^{-5} learning rate and after training until it over fits, retrain the model with dropout and 4×10^{-5} learning rate.

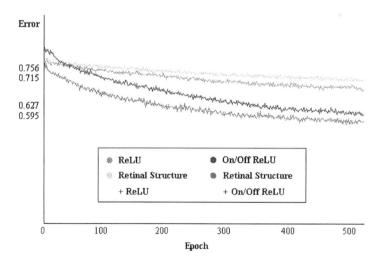

Fig. 4. The convergence of models with different settings

Figure 4 shows the convergence of different model settings. For the comparison, each model has trained with same epoch. Proposed model shows best performance compare with other settings which do not include or partly include the proposed model.

Table 1 shows the performance comparison result among different model settings and PReLU [8] which is primarily developed modified ReLU. Proposed model gives significantly better accuracy gain as 10 %. Even if we do not consider the preprocessing, the accuracy gain is still significant as 7 %. The absolute accuracy comparison is not possible because of the differences of target data and model settings.

Table 1. Performance comparison result

Model setting	Accuracy		Accuracy gain
	ON/OFF ReLU	PReLU [8]	
ReLU	44 %	66.18 %	–
Retinal structure + ReLU	46 %	–	2 %
ON/OFF ReLU	51 %	–	7 %
Retinal structure + ON/OFF ReLU	54 %	–	10 %
PReLU	–	67.36 %	1.2 %

4   Conclusions

In this paper, we proposed biologically inspired structures, retinal structure and ON/OFF ReLU for CNN. Retinal structure consists of center surround difference structure which generates additional feature for more accurate learning. ON/OFF ReLU consists of ON ReLU which is same as conventional ReLU and OFF ReLU which is reversed ReLU for negative information. Because OFF ReLU generates additional negative information as same size of positive information, we redefined forward and back propagation formulas. Experimental result presents the importance of negative information and it is significant compare with former method [8]. Our further work will be more detailed and variety of measurement to investigate how the OFF ReLU affects performance and efficiency of neural networks.

Acknowledgement. This work was supported by the Industrial Strategic Technology Development Program (10044009) funded by the Ministry of Trade, Industry and Energy (MOTIE, Korea) and was supported by Regional Specialized Industry R&D program funded by the Ministry of Trade, Industry and Energy (R0002982).

References

1. Itti, L., Koch, C.: Computational modelling of visual attention. Nat. Rev. Neurosci. **2**(3), 194–203 (2001)
2. Ban, S.-W., Lee, I., Lee, M.: Dynamic visual selective attention model. Neurocomputing **71**(4), 853–856 (2008)
3. LeCun, Y., Bottou, L., Bengio, Y., Haffner, P.: Gradient-based learning applied to document recognition. Proc. IEEE **86**(11), 2278–2324 (1998)
4. Hubel, D.H., Wiesel, T.N.: Receptive fields, binocular interaction and functional architecture in the cat's visual cortex. J. Physiol. **160**(1), 106 (1962)
5. Werblin, F.S., Dowling, J.E.: Organization of the retina of the mudpuppy, Necturus maculosus. II. Intracellular recording. J. Neurophysiol. **32**(3), 339–355 (1969)
6. Gjorgjieva, J., Sompolinsky, H., Meister, M.: Benefits of pathway splitting in sensory coding. J. Neurosci. **34**(36), 12127–12144 (2014)
7. Pitkow, X., Meister, M.: Decorrelation and efficient coding by retinal ganglion cells. Nat. Neurosci. **15**(4), 628–635 (2012)
8. He K., Zhang X., Ren S. and Sun J.: Delving deep into rectifiers: surpassing human-level performance on imagenet classification (2015). arXiv preprint arXiv:1502.01852
9. Dacey, D., Packer, O.S., Diller, L., Brainard, D., Peterson, B., Lee, B.: Center surround receptive field structure of cone bipolar cells in primate retina. Vision Res. **40**(14), 1801–1811 (2000)
10. Schiller, P.H.: Parallel information processing channels created in the retina. Proc. Natl. Acad. Sci. **107**(40), 17087–17094 (2010)
11. Kim, B., Okuno, H., Yagi, T., Lee, M.: Implementation of visual attention system using artificial retina chip and bottom-up saliency map model. In: Lu, B.-L., Zhang, L., Kwok, J. (eds.) ICONIP 2011, Part III. LNCS, vol. 7064, pp. 416–423. Springer, Heidelberg (2011)
12. Glorot, X., Bordes, A., Bengio, Y.: Deep sparse rectifier neural networks. In: International Conference on Artificial Intelligence and Statistics (2011)
13. Hartline, H.K.: The response of single optic nerve fibers of the vertebrate eye to illumination of the retina. Am. J. Physiol. **121**, 400–415 (1938)

Multivariate Autoregressive-based Neuronal Network Flow Analysis for In-vitro Recorded Bursts

Imali T. Hettiarachchi[1]($\boxtimes$), Asim Bhatti[1], Paul A. Adlard[2],
and Saeid Nahavandi[1]

[1] Centre for Intelligent Systems Research, Deakin University, Geelong, Australia
{imali.hettiarachchi,asim.bhatti,saeid.nahavandi}@deakin.edu.au
[2] Synaptic Neurobiology Laboratory,
The Florey Institute of Neuroscience and Mental Health, Melbourne, Australia
paul.adlard@florey.edu.au

Abstract. Neuroscientific studies of *in vitro* neuron cell cultures has attracted paramount attention to investigate the behaviour of neuronal networks in response to different environmental conditions and external stimuli such as drugs, optical and electrical stimulations. Microelectrode array (MEA) technology has been widely adopted as a tool for this investigation. In this work, we present a new approach to estimate interconnectivity of neural spikes using multivariate autoregressive (MVAR) analysis and Partial Directed Coherence (PDC). The proposed approach has the potential to discover hidden intra-burst causal connectivity patterns and to help understand the spatiotemporal communication patterns within bursts, pre and post stimulations.

Keywords: Multi electrode array · Bursts · Partial directed coherence · Multivariate autoregressive modelling

1 Introduction

Studies of *in vitro* cultured neuronal networks have gained more and more attention to study individual neuron and neuronal network behaviour under different conditions such as, drug exposure and electrical stimulation. Toxicity determination of drugs is one of the major applications based on *in vitro* neuronal cultures [1], which has opened new doors to study pre and post drug behaviour of neuronal networks. The electrophysiology-based communication of an *in vitro* cultured neuronal network can be measured using microelectrode arrays (MEAs). In *in vitro* cultures the MEAs act as an interface through which neural signals are recorded or electrical signals are delivered to *in vitro* cultures of neurons.

P.A. Adlard—PAA is supported by an ARC Future Fellowship. In addition, the Florey Institute of Neuroscience and Mental Health acknowledge the strong support from the Victorian Government and in particular the funding from the Operational Infrastructure Support Grant.

© Springer International Publishing Switzerland 2015
S. Arik et al. (Eds.): ICONIP 2015, Part IV, LNCS 9492, pp. 324–331, 2015.
DOI: 10.1007/978-3-319-26561-2_39

The fundamental unit of communication of neurons is the action potential. This phenomenon is caused by exchange of positive ions across the cell membrane, that generates a sharp change in voltage in the extracellular environment referred to as a spike. Neurons emit action potentials in two different modes, such as single spiking and high frequency bursting. During single spiking, action potentials are emitted irregularly, where on the other hand bursting neurons emit a cluster of action potentials (bursts). Bursting is commonly observed as periods with relatively high frequency spiking, separated by periods with relative low frequency spiking or an absence of spikes. Bursts are known as the most important property in synaptic plasticity and information processing in the central nervous system, which enhances and prolongs signal strength for communication between individual neurons and small networks.

Spikes being the main means of communication in a neuronal network, voltage spike counting (spike rate) and sorting is often used in research to characterise network activity [2,3]. General burst related parameters that describe the behaviour of bursts are the mean spike frequency, number of spikes and percentage of spikes in bursts. In a cultured neuronal network the pattern of spontaneous spiking activity is characterised by short periods of synchronous firing at many of the recording channels. In a well matured culture (> 21 days *in vitro*) rich spontaneous dynamical behavior is detected, marked by the formation of temporal sequences of synchronised bursting events (SBEs). However, these SBEs have a very diverse spatiotemporal relationship and this diversity of bursts within a cultured neuronal network has been reported in a number of studies [4–6]. Further some studies on investigating inter and intra variability of the SBEs [5,6] also reported that the SBEs of a spontaneous spiking culture can be partitioned into statistically distinguishable subgroups, each with its own characteristic spatiotemporal pattern of activity.

Studying the existence of a causal relation between one or more recorded electrophysiological signals is a question of great interest in neurophysiology [8]. To this end the concept of Granger causality (GC) has been widely used as an effective connectivity measure in neurophysiology during the last four and a half decades. The idea behind GC states that two simultaneously measured time series can said to be causal, if by incorporating past knowledge of one series, provides a better prediction of the second series. Clive Granger [7] then formalised this notion using bivariate auto regression modelling.

In practice neural data is often recorded from multiple spatial positions (channels). However the main limitation of GC is that it is only precisely applicable for bivariate time series. Thus, genuine multi channel analysis was proposed in a number of studies through modelling the time series by using multivariate autoregressive (MVAR) modelling process, containing all available information from different recording sites [8–10]. Partial directed coherence (PDC) [10,11] is a widely used frequency domain measure of causality derived from MVAR modelling of the time series, the mathematical formulation will be discussed in details in Sect. 2.

Information carried by the spikes lies in their temporal pattern, not in their shape [12]. Therefore, effort has been put into investigating temporal patterns in terms of the information content of time series of spikes. GC-based metric estimates has been recently applied to spike train data to infer the causal strength and directional information within functional pathways *in vivo* [13,14] and *in vitro* [15,16]. MVAR models and PDC measures have been previously used in *in vivo* studies involving spike train data and have proven useful in characterising changing interaction patterns between neurons [9,17]. Apart from the aforementioned *in vivo* studies, MVAR modelling and PDC has been used for *in vitro* studies to infer the causality in the interdependency of the recorded multi channel data. In [18] the authors present a MVAR and a PDC technique to analyse the evolution of a cultured hippocampal neuronal network. Using complex network properties they show that by using the aforementioned methods the evolution of the culture can be described adequately.

In light of the studies discussed in [4–6], it shows that the functional connectivity within SBEs are playing a major role in identifying the communication pattern in a cultured neuronal network. However, the aforementioned studies only investigate the interdependency of the bursts without any specific information on the direction of their dependency. Further the studies in [16,18] present causal studies in cultured neuronal networks, however these studies are limited to investigating how the network structure evolves with the number of days in cuture. Our aim in this paper is to present proof−of−concept to infer the causality within a burst. In this paper we propose a MVAR-based connectivity analysis to identify the causal communication pattern within a SBE. In contrast to the functional connectivity studies presented in [5,6,16,18], our proposed method will involve directional quantification (causality) of the connectivity within and among SBEs to study their variability in neuronal network communication.

The rest of the paper is organised as follows. In Sect. 2, the MVAR framework and the mathematical formulation of PDC is presented followed by Sect. 3 carrying the experimental data analysis which includes burst detection and extraction from spike trains. Section 4 presents and discusses the burst connectivity analysis results using the proposed methods, while Sect. 5 concludes the paper.

2   Methodology

2.1   Multivariate Autoregressive Modelling

In order to determine the causality within a burst, first the mathematical formulation of the problem is presented here. The continuous signals of the n activated channels within the burst are formulated as a multivariate autoregressive (MVAR) model given by,

$$Y(t) = \sum_{i=1}^{p} A_r(i)Y(t-i) + E(t) \tag{1}$$

Where $Y(t) = [y_1(t), y_2(t), \ldots, y_n(t)]^T$, y_j denotes the jth channel, t denotes a time point with the superscript T stands for the transpose of the matrix. $A_r(i)$ is the $(k \times k)$ model coefficient matrix at lag i and p is the model order. $E(t) = [e_1(t), \ldots, e_n(t)]^T$ is a vector of zero mean white noise input with covariance matrix Σ_E.

The frequency domain representation of (1) can be given by first transforming the time domain to z domain and then to frequency domain,

$$E(f) = A(f)Y(f)$$
$$Y(f) = H(f)E(f) \tag{2}$$

where $H(f) = A^{-1}(f)$, $A(f) = [I_k - A_r(f)]$ and $A_r(f) = \sum_{m=1}^{p}(e^{-i2\pi fm}A_r(m))$ is the fourier transform of the MVAR coefficients. I_k is the $(k \times k)$ Identity matrix.

2.2 Partial Directed Coherence

Above frequency domain representation in (2) can be viewed as a linear filter with $E(f)$ the system input (white noise), $Y(f)$ the system output and $H(f)$ the system transfer function. Based on this frequency domain representation many connectivity measures have been suggested. In our study we adopt partial directed coherence (PDC) suggested in [11] to quantify the causality within a burst. The PDC measure can be mathematically given as,

$$PDC_{ij}(f) = \frac{A_{ij}(f)}{\sqrt{\sum_{m=1}^{k}|A_{mj}(f)|^2}} = \frac{A_{ij}(f)}{\sqrt{a_j^*(f)a_j(f)}} \tag{3}$$

where A_{ij} is the (i, j) element of the inverse transfer matrix $A(f)$. $a_j(f)$ is the jth column of $A(f)$ and the superscript $*$ denotes the transpose and the complex conjugate. PDC measures the conditional granger causality from $j \rightarrow i$, normalised by the total amount of causal outflow from j.

3 Experimental Data Analysis

The neocortex neurons from 14–16 day gestation C57BL/6 mice were cultured on a multi-electrode array (MEA) dish. After leaving for 21 days for cell growth and maturation the spontaneous activity of the matured culture was recorded using 60 electrodes MEA. Raw neural data was sampled at 25 kHz and recorded for 3 min through the MC-Rack software (Multi Channel System). Spike detection and analysis was performed using Neural data analyses software, developed by the Center for Intelligent Systems Research (CISR), Deakin University. The raw recorded data was pre-processed to threshold at 15 mV's to generate the spike trains. The activity raster map of the spontaneously spiking cultured network is shown in Fig. 1(a).

3.1   Burst Identification, Extraction and Filtering

Spike data were processed to detect bursts as follows. The provided parameters included the burst window (bw) in milliseconds and minimum number of electrodes active ($Min\_electrode$) within the window. For the current analysis $bw = 100ms$ and $Min\_electrode = 15$ were selected. The bursts were detected by estimating the peaks of filtered summed signal showed on Fig. 1(b). In the rest of the paper we will analyse the first burst detected to provide proof-of-concept for the proposed approach. The spatiotemporal spiking pattern for this burst is extracted through the activated 26 channels and their respective recorded spiking times of these channels. In order to apply a MVAR-based analysis it is necessary to convert the spike trains into continuous-time waveforms. General techniques used for this purpose are to convolve the spike trains with a Gaussian kernel [9], low-pass filter the spike train data to remove high frequency components [20] or by converting the spike trains in to instantaneous firing rates [19]. In the current study we have utilised a low-pass filter with a filter order of 1 and cut-off frequency at 10 % of the normalised Nyquist frequency [20] to pre-process the spike trains to generate continuous-time signals.

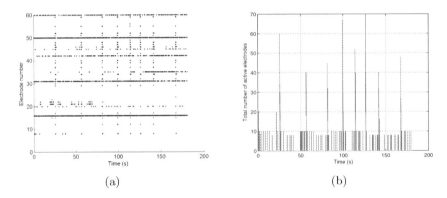

(a)   (b)

Fig. 1. (a) Raster plot of the spontaneous spiking activity of the cultured network (b) Burst detection-total activated number of electrodes is shown per time point for the whole length of the signal and 8 detected bursts are highlighted by overlaying in red to show the detected bursts.

4   Results and Discussion

In the current analysis we have assumed the signals are weakly stationary within a burst. In order to fit the MVAR model to the filtered data, the best model order should be selected. This is achieved by minimising the Schwarz-Bayes Criterion (SBC) defined by, $SBC(p) = N_T ln|\Sigma_E(p)| + ln(N_T)pM^2$. N_T is the total number of data points used to fit the model, M the number of variables and $\Sigma_E(p)$ is the estimated noise covariance matrix for the model order p. A widely used alternative model selection criterion is the Akaike Information Criterion (AIC). In the current study SBC is utilised as the preferred model selection criterion over AIC as it is a consistent estimator due to imposing heavier penalty at large model orders. The best model order was selected at $p = 20$ giving a clear minimum SBC value.

After fitting an order 20 MVAR model to the filtered (continuous) burst data the resulting directional connectivity patterns among the active 26 channels were inferred by using a statistical test based on surrogate data to evaluate the significance of the information flow between the channels. A total of 100 surrogate realisations were generated through random shuffling of the samples in time. The random shuffling saves the distributional properties of the data however, destroys the causal relations. Afterwards the 99 % confidence level was calculated as the threshold to determine statistically significant values of the PDC.

Figure 2 provides a visual representation for a section of the analysis results for seven channels. The full representation for the burst will require a 26×26 sub plots in the figure. Due to the high sampling rate of 25 kHz, the maximum frequency incorporated for the calculations is taken as 12.5 kHz (Nyquist frequency). Causal connections are clearly visible between channels $12 \rightarrow 17$, $12 \rightarrow 20$, $17 \rightarrow 12$, $17 \rightarrow 20$, $20 \rightarrow 12$ and $20 \rightarrow 17$. The complete network flow within the burst is given in Fig. 3.

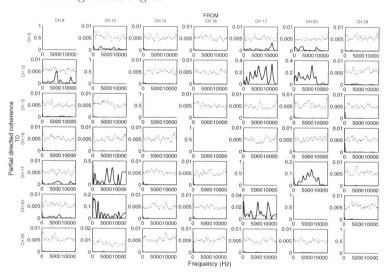

Fig. 2. Estimated directed connectivity patterns for seven MEA channels. Dotted lines represent the threshold and the solid lines represent the estimated PDC values. The diagonal is kept blank intentionally as they represent the self connectivity.

The preliminary results presented in the paper introduces a new dimension for analysing the characteristics of bursts through incorporating directional connectivity in network flow. Further the methods presented in the paper can be easily extended to obtain the time-varying causal communication between channels by using a short-time moving window approach or a Kalman filtering approach to determine temporal sequencing of the causality.

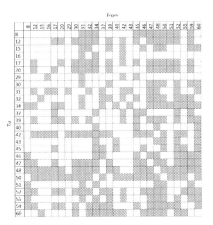

Fig. 3. Estimated directed connectivity patterns for the 26 MEA channels within the first burst. The boxes shaded in grey show a significant causal connectivity between the channels

5   Conclusion

The paper presented multivariate autoregressive (MVAR) modelling and Partial Directed Coherence (PDC) measure to characterise the directional network flow of the spiking activity within a burst. The proposed approach has a potential to explore intra-burst causal connectivity patterns in order to understand the spatiotemporal communication patterns in bursts. Due to the lack of availability of ground truth data for neural spiking activity in vitro studies, the significant directional connectivity is estimated employing surrogate statistics. Future work will involve incorporation of the proposed concept for multiple bursts to assess their spatiotemporal variability.

References

1. Johnstone, A.F.M., Gross, G.W., Weiss, D.G., Schroeder, O.-H., Gramowski, A., Shafer, T.J.: Microelectrode arrays: a physiologically based neurotoxicity testing platform for the 21st century. Neurotoxicology **31**(4), 331–350 (2010)
2. Zhou, H., Mohamed, S., Bhatti, A., Lim, C.P., Gu, N., Haggag, S., Nahavandi, S.: Spike sorting using hidden markov models. In: Lee, M., Hirose, A., Hou, Z.-G., Kil, R.M. (eds.) ICONIP 2013. LNCS, vol. 8226, pp. 553–560. Springer, Heidelberg (2013)
3. Haggag, S., Mohamed, S., Bhatti, A., Gu, N., Zhou, H., Nahavandi, S.: Cepstrum based unsupervised spike classification. In: Proceedings - 2013 IEEE International Conference on Systems, Man, and Cybernetics, SMC 2013, pp. 3716–3720 (2013)
4. Wagenaar, D.A., Madhavan, R., Pine, J., Potter, S.M.: Controlling bursting in cortical cultures with closed-loop multi-electrode stimulation. J. Neurosci. **25**(3), 680–688 (2005)
5. Segev, R., Baruchi, I., Hulata, E., Ben-Jacob, E.: Hidden neuronal correlations in cultured networks. Phys. Rev. Lett. 92(11), Article no. 118102 (2004)

6. Baruchi, I., Ben-Jacob, E.: Functional holography of recorded neuronal networks activity. Neuroinformatics **2**(3), 333–351 (2004)
7. Granger, C.W.J.: Investigating causal relations by econometric models and cross-spectral methods. Econometrica **37**(3), 424–438 (1969)
8. Pereda, E., Quiroga, R.Q., Bhattacharya, J.: Nonlinear multivariate analysis of neurophysiological signals. Prog. Neurobiol. **77**(1–2), 1–37 (2005)
9. Sameshima, K., Baccalá, L.A.: Using partial directed coherence to describe neuronal ensemble interactions. J. Neurosci. Methods **94**(1), 93–103 (1999)
10. Kamiński, M.J., Blinowska, K.J.: A new method of the description of the information flow in the brain structures. Biol. Cybern. **65**(3), 203–210 (1991)
11. Baccalá, L.A., Sameshima, K.: Partial directed coherence: a new concept in neural structure determination. Biol. Cybern. **84**(6), 463–474 (2001)
12. Cocatre-Zilgien, J.H., Delcomyn, F.: Identification of bursts in spike trains. J. Neurosci. Methods **41**(1), 19–30 (1992)
13. Kim, S., Putrino, D., Ghosh, S., Brown, E.N.: A granger causality measure for point process models of ensemble neural spiking activity. PLoS Comput. Biol. **7**(3), e1001110 (2011)
14. Nedungadi, A.G., Rangarajan, G., Jain, N., Ding, M.: Analyzing multiple spike trains with nonparametric granger causality. J. Comput. Neurosci. **27**(1), 55–64 (2009)
15. Cadotte, A.J., DeMarse, T.B., He, P., Ding, M.: Causal measures of structure and plasticity in simulated and living neural networks. PLoS One **3**(10), e3355 (2008)
16. Lamanna, J., Esposti, F., Signorini, M.G.: Study of neuronal networks development from in-vitro recordings: a granger causality based approach. In: Conference Proceedings of the Annual International Conference of the IEEE Engineering in Medicine and Biology Society, EMBS, pp. 4842–4845 (2010)
17. Fanselow, E.E., Sameshima, K., Baccalá, L.A., Nicolelis, M.A.L.: Thalamic bursting in rats during different awake behavioral states. In: Proceedings of the National Academy of Sciences of the United States of America, vol. 98(26), pp. 15330–15335 (2001)
18. Rodriguez, M.Z., Pedrino, E.C., Saito, J.H., Destro Filho, J.B.: Evolutionary dynamics of in vitro cultures of neurons in Multi Electrode Array - MEA. In: Proceedings of the IEEE International Conference on Systems, Man and Cybernetics, pp. 78–83 (2012)
19. Zhu, L., Lai, Y., Hoppensteadt, F.C., He, J.: Probing changes in neural interaction during adaptation. Neural Comput. **15**(10), 2359–2377 (2003)
20. Kamiński, M., Ding, M., Truccolo, W.A., Bressler, S.L.: Evaluating causal relations in neural systems: granger causality, directed transfer function and statistical assessment of significance. Biol. Cybern. **85**(2), 145–157 (2001)

Cognitive Load Driven Directed Information Flow in Functional Brain Networks

Md. Hedayetul Islam Shovon[1(✉)], D. (Nanda) Nandagopal[1],
Ramasamy Vijayalakshmi[2], Jia Tina Du[1], and Bernadine Cocks[1]

[1] Cognitive Neuroengineering and Computational Neuroscience Laboratory,
School of Information Technology and Mathematical Sciences,
University of South Australia, Adelaide, Australia
shomy004@mymail.unisa.edu.au, nanda.nandagopal@unisa.edu.au
[2] Department of Applied Mathematics and Computational Science,
PSG College of Technology, Coimbatore, Tamil Nadu, India

Abstract. The human brain connectome analysis describes the patterns of structural and functional brain networks and has become one of the most studied topics in computational neuroscience in recent years. Detailed investigation of functional brain networks based on the direction of information flow has subsequently gained significance. This study identifies changes in information flow direction between different brain regions during cognitive activity compared to baseline state using Normalized Transfer Entropy (NTE) estimated from electroencephalogram (EEG) signals. An algorithm is proposed for finding the cognitive state specific information flow direction patterns (IFDP) among various regions (lobes) of the brain. Results clearly demonstrate that IFDP based analysis is able to detect the changing information flow directional patterns during cognitive activity among four different brain regions: Frontal, Central, Parietal and Occipital. During cognitive activity, noticeable long range interconnections are established in the directed functional brain network from frontal to central, parietal and occipital lobes, and as well as from the central to occipital lobe. This suggests that the IFDP approach may have potential applications in the detection of cognitive impairments as well as in the clinical research e.g., for finding seizure foci in epilepsy.

Keywords: Transfer entropy · Information flow · Directed functional brain network · EEG · Cognitive activity

1 Introduction

The human brain is a large-scale adaptive complex network with billions of neurons generating nonlinear and non-stationary signals. A fundamental challenge to the research community therefore is to extract the information flow/exchange within such complex networks during different states of brain activity. Of the various techniques available, graph theoretical analysis of brain networks is perhaps the most powerful approach to understand such information exchange within brain networks [1–3].

© Springer International Publishing Switzerland 2015
S. Arik et al. (Eds.): ICONIP 2015, Part IV, LNCS 9492, pp. 332–340, 2015.
DOI: 10.1007/978-3-319-26561-2_40

Most previous graph theoretical analysis has, however, been applied only to undirected networks [2]; for a deeper understanding of information exchange in brain networks, directed network analysis is essential. Recent analyses have incorporated Granger causality to construct directed functional brain networks using resting state fMRI [4, 5], although Granger causality is limited to linear model of interaction. By comparison, the information theoretical measure of transfer entropy (TE) is a nonlinear measure which determines the direction and quantifies the information transfer between two processes [6, 7]. It has been demonstrated that TE sensitively detects cognitive load induced changes in directed functional brain networks (FBNs), but detailed investigation has not been performed based on the direction of information flow [3]. Various issues such as how information propagates from one brain region/lobe to another during cognitive activity and which regions act as drivers in a particular cognitive state remain unknown.

In an attempt to better identify the network bases of cognitive activity, the present study explores directed functional brain network architecture during cognitive activity. The objectives are twofold: (a) the construction and analysis of directed functional brain networks during both baseline and cognitive load states; and (b) the detection of information flow direction patterns during cognitive activity. In this regard, the directed brain networks were estimated by calculating normalized TE (NTE) between the EEG data of each pair of brain regions, with an information flow direction pattern (IFDP) algorithm proposed to analyze the directional information flow patterns in the functional brain networks. Given the EEG data during baseline and cognitive load states, the IFDP analysis is able to detect the cognitive load induced changes using directed information flow patterns across the scalp regions. The research literature on NTE and directed information flow in FBNs are reviewed and described in the following sections.

2 Normalized Transfer Entropy

Considering that two associated time series $X = x_t$ and $Y = y_t$ can be approximated by Markov process, the deviation from the following generalized Markov condition as shown in Eq. 1 [6, 8],

$$p\left(y_{t+1}|y_t^n, x_t^m\right) = p\left(y_{t+1}|y_t^n\right) \tag{1}$$

where $x_t^m = \left(x_t, \ldots, x_{t-m+1}\right)$, $y_t^n = \left(y_t, \ldots, y_{t-n+1}\right)$, while the subscript t denotes the considered state (or time step); m and n represent the orders (memory) of the Markov processes X and Y respectively; and $p(|)$ represents the conditional entropy which can be estimated using kernel based estimation or nearest-neighbor techniques [8, 9]. The transfer entropy from series X to series Y, written $TE_{X \to Y}$, can therefore be regarded as the information about future observations y_{t+1} gained from the past observations of y_t^n and x_t^m minus the information about future observations y_{t+1} gained from past observations of y_t^n only. $TE_{X \to Y}$ can be calculated using the summation as shown in Eq. 2 [6]:

$$TE_{X \to Y} = \sum_{y_{t+1}, y_t^n, x_t^m} p\left(y_{t+1}, y_t^n, x_t^m\right) \log\left(\frac{p(y_{t+1}|y_t^n, x_t^m)}{p(y_{t+1}|y_t^n)}\right). \tag{2}$$

TE is an asymmetric measure, $TE_{X\to Y} \neq TE_{Y\to X}$. TE can be in the range $0 \leq TE_{X\to Y} < \infty$. TE matrices usually contain some noise. Noise can be removed from the estimate of TE by subtracting the average TE from X to Y using a shuffled version of X denoted by $< TE_{X_{shuffle}\to Y} >$, over several shuffles. $X_{shuffle}$ contains the same symbol as in X but those symbols are rearranged in a randomly shuffled order [10]. Then, normalized TE is calculated from X to Y with respect to the total information in sequence Y itself. This will represent the relative amount of information transferred by X. The normalized transfer entropy (NTE) is shown in Eq. 3 as follows [11]:

$$NTE_{X\to Y} = \frac{TE_{X\to Y} - < TE_{X_{shuffle}\to Y} >}{H(Y_{t+1}|Y_t)} \tag{3}$$

In Eq. 3, $H(Y_{t+1}|Y_t)$ represents the conditional entropy of Y at time $t + 1$ given its value at time t as shown in Eq. 4.

$$H(Y_{t+1}|Y_t) = -\sum_{y_{t+1},y_t} p(y_{t+1}, y_t) \log \frac{p(y_{t+1}, y_t)}{p(y_t)} \tag{4}$$

NTE is in the range $0 \leq NTE_{X\to Y} \leq 1$. NTE is 0 when X transfers no information to Y, and is 1 when X transfers maximal information to Y.

3   Directed Information Flow in Functional Brain Networks

Many networks such as the World Wide Web, biological networks, food webs and even some social networks are inherently directed in nature and structure. The directions of edges within networks can therefore provide additional information which remains undetected in an undirected network. Several recent studies have demonstrated that edge direction can also contribute to identify certain crucial information about the network structure [12, 13]. A directed graph $G = (V, E)$ consists of two sets V and E. The elements of V are the nodes (vertices) of the graph G, while the elements of E are its links (edges). A directed graph has both an in-degree and an out-degree for each vertex, which are the numbers of incoming and outgoing edges respectively [14]. In the case of directed FBNs, scalp electrodes are considered as vertices which represent the activity of underlying neuronal populations. The directed connections/links from one vertex to another are measured using Granger causality or Transfer Entropy [3–5]. In this study, EEG has been used as it is an economical non-invasive tool. FBNs are constructed by computing the NTE between EEG channels.

4   Methods

4.1   Participants and EEG Data Acquisition

Six healthy, right-handed adults (4 males, 2 females) volunteered for EEG data collection (age range 19–59) at the Cognitive Neuroengineering Laboratory at the University of South Australia. The participants were recruited from staff and students of the university with all reporting normal hearing and normal or corrected-to-normal vision, and none reporting any psychological, neurological or psychiatric disorder. EEG data were acquired at a sampling rate of 1000 Hz through a 40 channel Neuroscan Nuamps amplifier using Curry 7 software [15]. The 30 electrode sites used were based on the international 10–20 convention: specifically, FP1, FP2, F7, F3, Fz, F4, F8, FT7, FC3, FCz, FC4, FT8, T3, C3, Cz, C4, T4, TP7, CP3, CPz, CP4, TP8, T5, P3, Pz, P4, T6, O1, Oz and O2. EEG data were collected in two different conditions. In the eyes open or baseline condition, participants were asked to simply stare at a blue colored star on a computer screen; they were not asked to perform any specific cognitive task. In this condition, continuous EEG data were collected for two minutes. To then stimulate complex cognitive activity, participants were asked to drive normally in a driving simulator maintaining a constant speed of 50–60 km/h on a virtual winding road for approximately four minutes. As the road was winding and a specific speed range was required, this ensured participants' attention to task.

4.2 EEG Signal Pre-processing

A band pass filter of 1–70 Hz and a notch filter of 50 Hz were applied to the EEG data. Eye blinks were removed using principal component analysis (PCA), with any residual bad blocks removed manually. From the continuous eyes open (EOP) and driving (Drive) data, 50 good epochs of 2 s length were chosen using the back-to-back epoching process of Neuroscan Curry 7 software [15], then averaged into a single two seconds data epoch.

4.3 Analysis Framework of Information Flow Direction Patterns

The pre-processed EEG data during EOP and Drive were then used for the construction of TE matrices, where each cell of the TE matrix represents the TE value from one electrode to another. For noise removal, an average shuffled TE matrix (noise matrix) was calculated and subtracted from the original TE matrix. The noise removed normalized TE matrix is called the NTE matrix. The computed NTE matrices were subsequently used for the construction of directed FBNs and further analysis. The data acquisition, preprocessing, information processing and associated computational steps are illustrated in Fig. 1.

This approach, involving information theoretical, graph theoretical and statistical analysis, detects the changes of direction of functional brain network during cognitive activity when compared to baseline. The applied algorithm for finding the information flow direction patterns is outlined as follows.

Algorithm: Information Flow Direction Patterns (IFDP)
Input: Pre-processed EEG data during Eyes Open (EOP) and Cognitive load (Drive) states
Output: Lobe-wise information flow network

(i)   For each participant,
    (a)   Compute NTE(EOP) for all pairs of electrodes for EOP.
    (b)   Compute NTE(Drive) for all pairs of electrodes for Drive.
    (c)   Extract sub-matrix SUB_NTE(EOP) from NTE(EOP) using electrodes: F3, Fz, F4, C3, Cz, C4, P3, Pz, P4, O1, Oz and O2.
    (d)   Extract sub-matrix SUB_NTE(Drive) from NTE(Drive) using electrodes: F3, Fz, F4, C3, Cz, C4, P3, Pz, P4, O1, Oz and O2.
    (e)   Compute binarized sub-matrix BSUB_NTE(EOP) and BSUB_NTE(Drive) from SUB_NTE(EOP) and SUB_NTE(Drive)
(ii) Construct graphs GROUP_BSUB_NTE(EOP) and GROUP_BSUB_ NTE(Drive) from BSUB_NTE(EOP) and BSUB_NTE(Drive) respectively by using those connections which are present in more than 80 % of the participants.
(iii) Compute the lobe wise information flow network using the graphs GROUP_BSUB_NTE(EOP) and GROUP_BSUB_NTE(Drive), where three electrodes of each lobe can be considered as a single node: F(F3, Fz, F4), C(C3, Cz, C4), P(P3, Pz, P4) and O(O1, Oz, O2) and link weight represents the total number of connection from one lobe's electrodes to that of another lobe.
(iv) Visualize the lobe-wise information flow network during EOP and Drive.

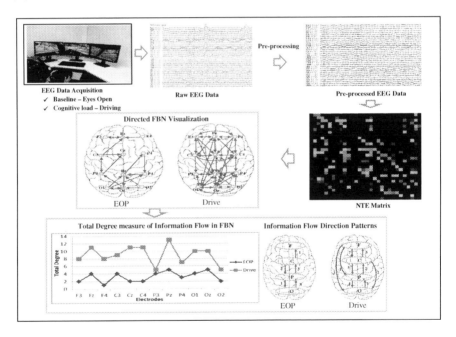

Fig. 1. Information flow direction patterns (IFDP) analysis framework

The advantage of the IFDP algorithm is that it constructs a lobe wise information flow network which identifies the subtle changes of information flow direction among different lobes of the brain during cognitive activity compared to baseline. This is essential for understanding brain function. The analysis results based on IFDP algorithm have been presented in the next section.

5 Results and Discussion

5.1 Construction and Visualization of Directed FBNs

From the NTE matrices (30×30 in size) of each participant, directed FBNs are constructed using 12 electrodes representing frontal, central, parietal and occipital lobes: F3, Fz, F4, C3, Cz, C4, P3, Pz, P4, O1, Oz and O2, where each of these electrodes can be considered as a node of the network. In this study, those 12 electrodes are chosen for the lobe wise information flow analysis purpose. The directed FBNs are then binarized using threshold = 0.002 to make binary directed FBNs. From the binary directed FBNs of each participant, group FBNs are constructed where each link of the group FBNs represents the directed connection which is present in five or more participant's FBNs (more than 80 % of participant's FBNs). For example, in Fig. 2a, there is a directed connection from F4 to Fz which means that this type of connection is present in at least five participants. The group FBNs during EOP and Drive have been shown in Fig. 2. The remaining analysis in this study has been performed on these group FBNs.

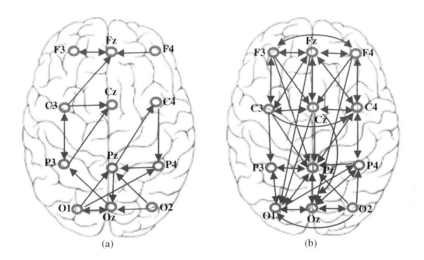

Fig. 2. Directed group FBN during (a) EOP (b) Drive (cognitive load)

5.2 Analysis of Directed Group FBNs Using Total Degree Measure

In a directed graph, the out degree of a node represents the number of outgoing edges; the in degree of a node represents the number of incoming edges; and node degree or total degree is the summation of in degree and out degree [1]. As indicated in Fig. 3, the total degree of directed group FBNs in the cognitive load state is significantly higher than the baseline condition in all of the nodes which suggest that each node sends and receives more information to facilitate cognitive function.

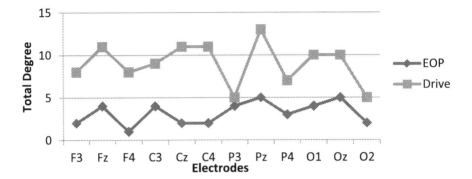

Fig. 3. Comparison of total degree during EOP and Drive

5.3 Information Flow Direction During Cognitive Activity

From the results of the total degree measure discussed in the previous subsection, it is clear that information flow in functional brain networks increases during cognitive activity, but the direction of this increased information flow during cognitive activity has not been explored. This subsection is dedicated to detect the information flow direction pattern during cognitive activity using the IFDP algorithm. To detect the information flow more efficiently, the electrodes in each lobe (frontal, central, parietal and occipital) are considered as single node. This results in four nodes (F, C, P and O) representing the respective lobes. The information transfer between these four nodes is shown in Fig. 4, where the edge weight represents the total number of connections from one lobe's electrodes to that of another lobe.

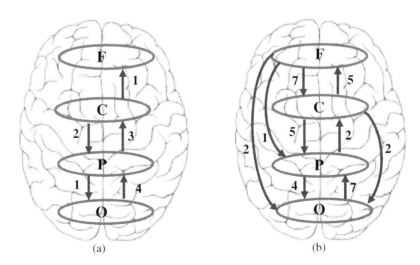

Fig. 4. Information flow among frontal, central, parietal and occipital lobes during (a) EOP and (b) drive

During cognitive load (Drive), the link weight increases compared to the baseline state. Moreover, information flows from the frontal lobe to the central, parietal and occipital lobes, and also from the central lobe to the occipital lobe during cognitive activity.

As is evident in Fig. 4, at least two distinct cognitive processes appear to be engaged to execute the requirements of the Drive condition. Firstly, the flow of information between the frontal and other lobes suggest higher-order planning and executive functions. Of note, as shown in Fig. 4(b), this appears to involve feedback loops whereby top-down (frontal lobe) activity interacts with bottom-up sensory information from the occipital, parietal and central lobes to adapt behavior in real time; that is, in response to changing virtual road conditions. Figure 4 also suggests a separate motor circuit from the central to occipital lobe which suggests the execution of the results of the planning/executive processing.

6   Conclusion

In the current study, directed functional brain networks were constructed and visualized using the information theoretic NTE measure. Cognitive load driven directed information flow among the various regions of the brain was also identified using the proposed IFDP algorithm. During cognitive load, information flow in the functional brain networks clearly increased when compared to baseline state, with the direct information flow from the frontal lobes to central, parietal and occipital lobes during cognitive load suggesting higher order planning and executive neural functions. The IFDP approach may therefore have potential applications in the clinical diagnosis of cognitive impairments. Future work will increase the sample size and apply various graph mining algorithms on the constructed directed functional brain networks to reveal more crucial information about brain function.

Acknowledgement. The authors wish to acknowledge partial support provided by the Defence Science and Technology Organisation (DSTO), Australia. The assistance and technical support provided by fellow researchers Mr Nabaraj Dahal and Mr Naga Dasari are greatly appreciated.

References

1. Rubinov, M., Sporns, O.: Complex network measures of brain connectivity: uses and interpretations. Neuroimage **52**, 1059–1069 (2010)
2. Bullmore, E., Sporns, O.: Complex brain networks: graph theoretical analysis of structural and functional systems. Nat. Rev. Neurosci. **10**, 186–198 (2009)
3. Shovon, M.I., Nandagopal, D., Vijayalakshmi, R., Du, J.T., Cocks, B.: Transfer entropy and information flow patterns in functional brain networks during cognitive activity. In: Loo, C.K., Yap, K.S., Wong, K.W., Teoh, A., Huang, K. (eds.) ICONIP 2014, Part I. LNCS, vol. 8834, pp. 1–10. Springer, Heidelberg (2014)
4. Liao, W., Ding, J., Marinazzo, D., Xu, Q., Wang, Z., Yuan, C., Zhang, Z., Lu, G., Chen, H.: Small-world directed networks in the human brain: multivariate Granger causality analysis of resting-state fMRI. Neuroimage **54**, 2683–2694 (2011)

5. Yan, C., He, Y.: Driving and driven architectures of directed small-world human brain functional networks. PLoS ONE **6**, e23460 (2011)
6. Schreiber, T.: Measuring information transfer. Phys. Rev. Lett. **85**, 461 (2000)
7. Vicente, R., Wibral, M., Lindner, M., Pipa, G.: Transfer entropy—a model-free measure of effective connectivity for the neurosciences. J. Comput. Neurosci. **30**, 45–67 (2011)
8. Lindner, M., Vicente, R., Priesemann, V., Wibral, M.: TRENTOOL: A Matlab open source toolbox to analyse information flow in time series data with transfer entropy. BMC Neurosci. **12**, 119 (2011)
9. Wibral, M., Vicente, R., Lindner, M.: Transfer entropy in neuroscience. In: Wibral, M., Vicente, R., Lizier, J.T. (eds.). UCS, vol. 93, pp. 3–36Springer, Heidelberg (2014)
10. Sabesan, S., Narayanan, K., Prasad, A., Iasemidis, L., Spanias, A., Tsakalis, K.: Information flow in coupled nonlinear systems: Application to the epileptic human brain. In: Pardalos, P.M., Boginski, V.L., Vazacopoulos, A. (eds.) Data Mining in Biomedicine, pp. 483–503. Springer, New York (2007)
11. Gourévitch, B., Eggermont, J.J.: Evaluating information transfer between auditory cortical neurons. J. Neurophysiol. **97**, 2533–2543 (2007)
12. Leicht, E.A., Newman, M.E.: Community structure in directed networks. Phys. Rev. Lett. **100**, 118703 (2008)
13. Fagiolo, G.: Clustering in complex directed networks. Phys. Rev. E **76**, 026107 (2007)
14. Newman, M.E.: The structure and function of complex networks. SIAM Rev. **45**, 167–256 (2003)
15. CURRY 7 EEG Acquisition and Analysis Software. Compumedics Neuroscan USA Ltd

Discrimination of Brain States Using Wavelet and Power Spectral Density

Raheel Zafar[1,2(✉)], Aamir Saeed Malik[1,2], Hafeez Ullah Amin[1,2],
Nidal Kamel[1,2], and Sart C. Dass[1,3]

[1] Centre for Intelligent Signal and Imaging Research (CISIR),
Universiti Teknologi PETRONAS, Perak, Malaysia
raheelsatti@gmail.com
[2] Department of Electrical and Electronic Engineering,
Universiti Teknologi PETRONAS, Perak, Malaysia
[3] Department of Fundamental and Applied Sciences,
Universiti Teknologi PETRONAS, Perak, Malaysia

Abstract. Cognitive task produces activation in the brain which are different from normal state. In order to study the brain behavior during cognitive state, different techniques are available. Wavelet energy and power spectral density (PSD) are well established methods for brain signal classification. In this paper, cognitive state of the brain is compared with the baseline using EEG. Data are taken from all lobes of the brain to see the effect of cognitive task in the whole brain and analyzed using wavelet energy and PSD. Graph of wavelet energy and power spectral density are plotted separately for each subject to see the effect individually. Individual results showed that the behavior of human brain change with the cognitive task and this change occurred in most of the human brain. This change is due to the neural activity which is increased during the cognitive task (IQ) and is better measured with wavelet compared to PSD.

Keywords: Electroencephalography (EEG) · Wavelet energy · Power spectral density

1 Introduction

The physiology of human cognition can be studied using different noninvasive techniques like Electroencephalography (EEG), Magnetoencephalography (MEG) and functional magnetic resonance imaging (fMRI).

From last few decades, EEG is commonly used technique to measure brain activities. fMRI is well established technique and gives good result but is expensive compared to EEG [1]. In EEG, electrical potentials on the scalp are measured which tells about the electrical activity of the brain tissue. EEG can differentiate the mental activities and cognitive states, which help in better understanding of brain functions. This is one of the reasons that EEG is particularly used for clinic diagnostics and Brain Computer Interface (BCI).

EEG is one of the main methods for brain studies but it is not easy to measure brain activities using EEG because EEG has some limitations like artifacts and quality of signals. These limitations have been overcome in many studies. It is widely used due to

© Springer International Publishing Switzerland 2015
S. Arik et al. (Eds.): ICONIP 2015, Part IV, LNCS 9492, pp. 341–347, 2015.
DOI: 10.1007/978-3-319-26561-2_41

its advantages like convenience and cost but the main cause of success is that the subjects are free to move around due to wireless EEG setup available now a days.

The human brain has billions of neurons which are responsible for different types of activities. EEG signal can be classified into different frequency bands, alpha (α), beta (β), gamma (γ), delta (δ) and theta (θ) bands. In brain, whenever there is a neural activity, there are activations in different brain regions. Different parts of brain are inter-connected and different brain regions are activated during single task. For example during a cognitive task, there is change in neural activity in the whole brain i.e. occipital, frontal, parietal and temporal.

The brain activity may varies for different participants due to some reasons like IQ level, age, attention and level of difficulty of the task. In different studies these factors are discussed. In a study [2] H.S. Locke and T.S. Braver defined that cognitive behavior during a task is dependent on the level of motivation. Similarly, some studies showed that cognitive behavior is also dependent on memory [3, 4].

In a study, L. J. Trejo et al. [5] measured EEG for mental arithmetic task. In this study, statistical model was created by using data from frontal and parietal regions.

P300 was also found for the cognitive tasks and showed that amplitude is high in high workload tasks [6]. It is also showed that due to long periods of wakefulness, the amplitude of P300 decreases [7]. Normally, participants who are strong in cognitive tasks are intelligent and smart [8].

Due to all these factors, we took young university students as subjects for this study and presented our results individually for every participant. EEG recording during eyes closed (EC) has been used for the baseline. There were differences in peak values among the subjects but the cognitive behavior of all participants is significantly different from the baseline (for details please see the result section).

In this paper, we discriminate between brain states during cognitive task and the baseline (EC) by using different techniques i.e. energy and power. We took eight subjects and the data was taken from 17 different channels which are from every part of the brain. These channels are from occipital, frontal, temporal and parietal. In previous studies, estimation of cognitive states have been done but with limited number of channels [9, 10]. The analysis is done by using two different techniques, i.e. Wavelet energy and power spectral density (PSD). The result shows that wavelet energy gives better results than PSD.

2 Energy and Power

The EEG signal has different spectral components. The magnitude of EEG signal taken from human brain is in the range of 10 to 100 μV and the important frequencies are in the range of 0 to 30 Hz. In this study, we have found the energy of EEG signal by using wavelet transform. Wavelets are best for non-stationary signal analysis as it decomposes the signal into bunch of signals. Wavelet transform gives us better information and localization about time and frequency. In this study, we used continuous wavelet transform (CWT).

The mathematical equation for CWT is described as

$$W(a,b) = \frac{1}{\sqrt{a}} \int_{-\infty}^{\infty} x(t) \cdot \psi^* \left(\frac{t-b}{a} \right) dt \qquad (1)$$

where W(a, b) is the CWT of function x(t) and ψ(t) is a continuous function in both the time domain and the frequency domain called the mother wavelet, while the * represents operation of complex conjugate. The main purpose of the mother wavelet is to provide a source function to generate the daughter wavelets which are simply the translated and scaled versions of the mother wavelet. We call it CWT because the parameter b and a are in continuous domain, where b is known as translation (shifting), a is known as dilation (scaling) and * represents operation of complex conjugate. The mother wavelet, also called the wavelet function, is the basic wavelet and all subsequent analysis is performed by using the shifted and scaled version of mother wavelet. In this study, we have used symlet as a mother wavelet.

The power spectral density (PSD) is a well-established method for EEG data analysis. It is used as a feature for signal classification to find power for EEG signals [11]. Different brain states can be distinguished by having different power in different frequency range. It shows the strength (power) of energy at different frequencies i.e. we can find that on which frequencies the energy is strong. It is also easily measurable and observable. In this study, the PSD is found using Welch method with 50 % overlapping. PSD is used for the continuous signal and the integral of PSD computes the average power within the signal over a given frequency band. The mathematical equation for PSD is as follows

$$P = \lim_{T \to \infty} \frac{1}{2T} \int_{-T}^{T} x^2(t) dt \qquad (2)$$

PSD is the frequency response of a random or periodic signal and the above equation tells where the average power is distributed as a function of frequency. P is the PSD of the signal x(t) which is integrated over a large time interval [−T, T].

3   Materials and Methods

3.1   Preprocessing

The raw EEG data had been filtered at 1–48 Hz frequency by using band pass filter and waveform tools is used to detect the artifacts. This waveform tool is available within net station software (EGI Inc). The detected artifacts were removed using regression based model [12] using electrooculography (EOG) channels (EOG channel more than 140 μV).

3.2   Participants

In this experiment there were 8 subjects. All subjects were volunteers and their ages were between 20–30 years. The subjects were healthy with normal vision. The age of subjects was in between 20 to 30. All subjects filled the consent form and the study was approved by the research ethics committee of Universiti Teknologi PETRONAS (UTP).

3.3   Tasks

This experiment involved 2 physiological conditions: eyes closed and cognitive task. In the task, 40 different multiple choice questions were presented to the subjects. The set of questions consists of both mathematical and figurative questions. The subject can choose only one option within given time. The given time of each question is 60 s. After the response, the subject was exposed to the next question immediately without waiting. In the beginning, questions were easy as compared to the later ones.

3.4   EEG Recording Procedure

Data were recorded using 128 channel Electrical Geodesics Incorporated (EGI) system. All recordings were performed at Centre of Intelligent Signal and Imaging Research, University Technology PETRONAS (UTP). Task details are given above and for the baseline readings; data was taken with eyes closed.

4   Results and Discussion

EEG data was measured from 8 subjects in two conditions i.e. cognitive (IQ) task and eyes closed (baseline). Data are taken from every part of the brain for analysis. We have taken channel 70, 75, 83, 90 and 65 from occipital region, 96, 58, 45 and 108 from temporal, 62, 92 and 52 from parietal, 11, 24, 33, 124 and 122 from frontal. A total of 17 channels were selected from the whole brain. All channels show significant difference during cognitive task which shows that neural activity is increased in every part of the brain. The peak is in frontal and temporal region which shows more neural activity during cognitive task in these regions.

The result of all eight participants is shown individually. In Fig. 1(a–h), the wavelet energy of every subject is shown individually. Sub1 means subject 1. In every figure, we can clearly see the difference between both states at each channel.

In x-axis we are representing the channels and y-axis describes the wavelet energy. The first channel is 70 from occipital and the last is 122 from frontal. Red line shows the wavelet energy during IQ task and the blue line shows the wavelet energy during baseline which is eyes closed. In all the graphs, the neural activity is high during the task compared to baseline. In some regions, it is higher than other like in frontal and temporal. The wavelet energy is increased with the cognitive task compared to baseline

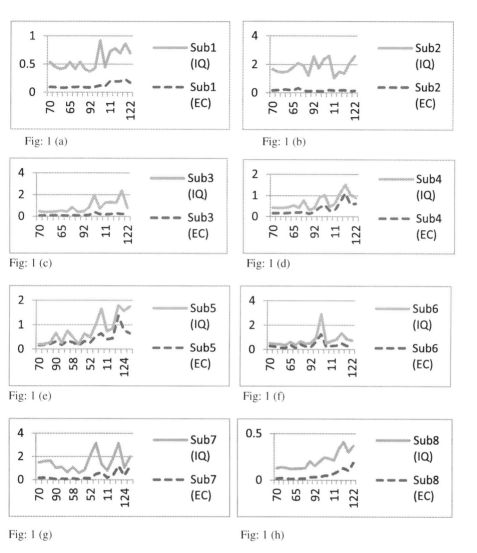

Fig: 1 (a)

Fig: 1 (b)

Fig: 1 (c)

Fig: 1 (d)

Fig: 1 (e)

Fig: 1 (f)

Fig: 1 (g)

Fig: 1 (h)

Fig. 1. (a–h) shows the wavelet energy of eight different subjects (1–8) respectively.

especially in channels 45 and 124. If we see first subject's data, from the first channel i.e. 70 till the end i.e. 122, a clear difference between both the states i.e. IQ and EC is seen.

We also found the difference between two states using power spectral density. Figure 2(a–h) shows the acceptable difference for all the eight subjects. Although, it is not as clear as in case of wavelets but it is enough to discriminate two states. During EC, the alpha frequency is dominant and alpha has a large magnitude so in PSD we did not have a clear difference between the two states.

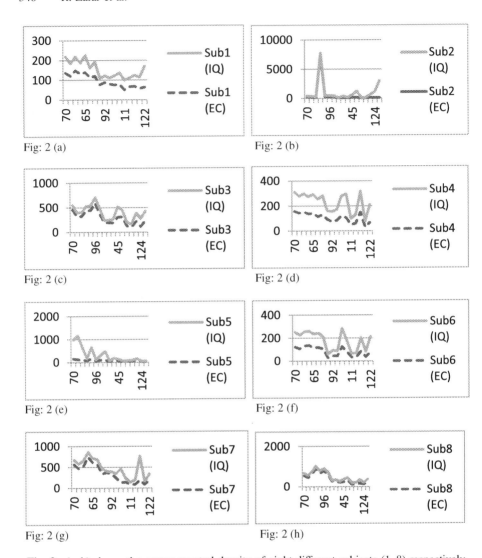

Fig. 2. (a–h) shows the power spectral density of eight different subjects (1–8) respectively.

If we summarize the results quantitatively then it can be said that the increase in wavelet energy during IQ is more than 100 % compared to baseline in most of the channels. So the difference between brain states can easily be measured. In case of PSD, the difference between IQ and EC is much lesser and in most of the cases it is less than 10 %.

5   Conclusion

Brain states can be measured using EEG but it requires good statistical analysis so better technique should be used to get significant information. Significant difference in brain states i.e. IQ task and relaxed state (EC) are found using wavelet and PSD. The result showed that wavelet energy gives better results compared to PSD. This is because in wavelet, the information is both in time and frequency domain. In results, higher wavelet energy was found during the cognitive task which showed that neural activity is increased during the cognitive task especially in frontal and temporal region. Power spectral density is also increased with cognitive task and gave acceptable results with the baseline.

Acknowledgment. This research has been funded by University Research Internal Funding (URIF: 0153AA-B26) and international grant (0153AB-E15), Universiti Teknologi PETRONAS, Malaysia.

References

1. Zafar, R., et al.: Decoding of visual information from human brain activity: a review of fMRI and EEG studies. J. Integrative Neurosci. **14**(2), 1–14 (2015)
2. Locke, H.S., Braver, T.S.: Motivational influences on cognitive control: behavior, brain activation, and individual differences. Cogn. Affect. Behav. Neurosci. **8**, 99–112 (2008)
3. Amin, H.U., et al.: Brain activation during cognitive tasks: an overview of EEG and fMRI studies. In: 2012 IEEE EMBS Conference on Biomedical Engineering and Sciences (IECBES), pp. 950–953 (2012)
4. Zafar, R., et al.: EEG Spectral Analysis during Complex Cognitive Task at Occipital (2014)
5. Trejo, L.J., et al.: Measures and models for predicting cognitive fatigue. In: Defense and Security, pp. 105–115 (2005)
6. Kramer, A.F., et al.: Psychophysiological measures of workload- potential applications to adaptively automated systems. Automation and Human Performance: Theory and Applications (A 98-12010 01-54), Mahwah, NJ, Lawrence Erlbaum Associates, Publishers, pp. 137–162 (1996)
7. Humphrey, D.G., et al.: Influence of extended wakefulness on automatic and nonautomatic processing. Hum. Fact. J. Hum. Fact. Ergonomics Soc. **36**, 652–669 (1994)
8. Sternberg, R.J.: The theory of successful intelligence. Interam. J. Psychol. **39**, 189–202 (2005)
9. Lan, T., et al.: Estimating cognitive state using EEG signals. J. Mach. Learn. **4**, 1261–1269 (2003)
10. Pan, J., et al.: Discrimination between control and idle states in asynchronous SSVEP-based brain switches: a pseudo-key-based approach. IEEE Trans. Neural Syst. Rehabil. Eng. **21**, 435–443 (2013)
11. Abdul-latif, A.A., et al.: Power changes of EEG signals associated with muscle fatigue: the root mean square analysis of EEG bands. In: Proceedings of the 2004 Intelligent Sensors, Sensor Networks and Information Processing Conference, pp. 531–534 (2004)
12. Hoffmann, S., Falkenstein, M.: The correction of eye blink artefacts in the EEG: a comparison of two prominent methods. PLoS ONE **3**, e3004 (2008)

A Cortically-Inspired Model
for Bioacoustics Recognition

Linda Main$^{(\boxtimes)}$ and John Thornton

Cognitive Computing Unit, Institute for Integrated and Intelligent Systems,
Griffith University, Gold Coast, Australia
{l.main,j.thornton}@griffith.edu.au

Abstract. Wavelet transforms have shown superior performance in
auditory recognition tasks compared to the more commonly used Mel-
Frequency Cepstral Coefficients, and offer the ability to more closely
model the frequency response behaviour of the cochlear basilar membrane. In this paper we evaluate a gammatone wavelet as a preprocessor
for the Hierarchical Temporal Memory (HTM) model of the neocortex
as part of the broader development of a biologically motivated approach
to sound recognition. Specifically, we apply for the first time, a gamma-
tone/equivalent rectangular bandwidth wavelet transform in conjunction
with the HTM's Spatial Pooler to recognise frog calls, bird songs and
insect sounds. Our audio feature detection results show that wavelets
perform considerably better than MFCCs on our selected datasets but
that combining wavelets with HTM does not produce further improvements. This outcome raises questions concerning the degree of match to
the biology required for an effective HTM-based model of audition.

Keywords: Signal processing · Wavelet transforms · Bioacoustics ·
Machine learning · Spatial pooling · Hierarchical temporal memory ·
k-NN classifier

1 Introduction

In order to apply machine learning to auditory detection and classification, a
suitable source signal preprocessing method is required. Such methods have the
goal of revealing salient features in the data that will best facilitate the learning
process. The choice of preprocessor is typically based on the nature of the signal
and the desired properties of the extracted features, without considering the
theoretical principles on which the learning algorithm is based.

Traditionally, Mel-Frequency Cepstral Coefficients (MFCCs) have been used
to preprocess signals for audio recognition. MFCCs are obtained via a short time
Fourier transform (STFT) to produce the power spectrum, which is thought to
model human vocal tract characteristics [1]. The spectrum is then warped on the
perceptual Mel-frequency scale in order to model the frequency response behaviour of the basilar membrane. The resulting features are therefore a combination
of modelling both speech production and auditory response mechanisms.

© Springer International Publishing Switzerland 2015
S. Arik et al. (Eds.): ICONIP 2015, Part IV, LNCS 9492, pp. 348–355, 2015.
DOI: 10.1007/978-3-319-26561-2_42

An alternative approach is the extraction of audio features by means of a wavelet function [2]. The wavelet transform (WT) uses basis functions with limited duration which are isolated in time and frequency, where each wavelet has a characteristic location and scale. Wavelets offer a number of benefits: improved time-frequency resolution compared to MFCCs; the wavelet envelope is mathematically tractable [3]; there are a variety of established basis functions available; and they can be used to model aspects of mammalian auditory perception.

In this paper we evaluate WTs as a preprocessor for the Hierarchical Temporal Memory (HTM) model of the neocortex [4], and as part of the broader development of a biologically motivated approach to sound recognition. HTM is a high level implementation of mammalian neocortical structure that aims to minimise unexpected interactions with the environment by learning to predict future input. As such it fits within the framework of Friston's free energy principle [5], but differs from the Hierarchical Predictive Coding (HPC) model by implementing neocortical regions as networks of artificial mini-columns that learn sequences of input received from lower regions with the assistance of feedback from higher regions.

To the best of our knowledge, HTM has not previously been applied to bioacoustic recognition, nor has the use of wavelets combined with HTM been explored. To address this we apply a gammatone/equivalent rectangular bandwidth (g/ERB) wavelet transform as a model of biological audition and evaluate its performance on three bioacoustic datasets, using a k-Nearest Neighbour classifier. Our results show the gammatone/ERB WT outperforms the previously reported best classification accuracies on two of our datasets. However, we found that classification accuracies fall significantly when the WTs are further processed through an HTM Spatial Pooler, indicating that raw WTs are not the best form of input for an HTM system. On this basis we conclude that further work, modelling processes already occurring in the brainstem, will be needed before an HTM can perform competitively in this domain.

2 Related Work

The primary challenges in bioacoustic recognition are the handling of a diverse range of animal sounds [6] and the difficulty of identifying them against a background of ambient noise. While MFCCs have been popular, it has been shown that other feature detection methods may offer better performance.

In 2006, Mitrovic et al. described a set of time-based low-level features which they compared to MFCCs. They found that a combination of their new features provided significant improvement over MFCCs in bioacoustic classification [7]. Gonzalez compared MFCCs against a selection of spectral features on a range of sound classes (e.g. music, frog calls, rain, etc.), and demonstrated that variations to the Principle Component Analysis (PCA) approach were competitive [8]. In the LifeCLEF 2014 Identification Challenge for birdsong classification, Stowell and Plumbley's winning audio-only submission used spherical k-means to learn features from PCA-whitened Mel spectral frames which significantly outperformed MFCC-based approaches [9]. Wavelets have also outperformed spectrogram template matching techniques for classifying Humpback whale song [10], and STFT and MFCC for classifying bat echolocation calls [11].

3 Wavelet Transforms

MFCCs, while efficient to compute and considered to perform well, are susceptible to noise and have poor time-frequency resolution. This means that while the transform is able to extract frequencies with a high degree of accuracy, the time at which they occur within the signal is lost. As sound has a fundamentally temporal nature, the loss of timing information is likely to impact on a system's ability to perform classification. The STFT attempts to address this shortcoming by applying the Fourier transform in sliding windows that move with time. The downside is that the length of the window limits the frequency resolution according to the Heisenberg-Gabor limit [3, pp. 43–45].

In order to achieve greater time resolution while maintaining good frequency detail, the WT may be used. The translation of the wavelet basis function across the signal allows identification of the temporal location of the obtained frequencies, and by varying the scale of the wavelet, high frequency resolution is maintained. This multi-scale approach makes wavelets ideal for extracting features from the non-stationary signals typical of bioacoustics.

An attractive aspect of WTs is the ability to modify the envelope (or window) of the wavelet and thereby optimise the quality of extracted features with respect to the target application. Various implementations using wavelet bases have been developed, with the discrete wavelet transform (DWT) being one of the most widely used. This is due to its non-redundant and invertible nature, which are key requirements for techniques aimed at signal compression and decompression (e.g. JPEG 2000) [12].

Morlet Wavelet: The Morlet (or Gabor) wavelet was one of the first basis functions developed [13]. The sinusoid of the Morlet wavelet is modified by a smooth Gaussian window, producing a waveform that is symmetrical about the peak amplitude. In a Morlet WT the translation and scaling factors are typically calculated as a linear progression.

Gammatone Wavelet: The mechanical frequency analysis of the cochlea is often modelled using a gammatone filter, which is considered to give a reasonable first-order approximation of basilar membrane impulse responses [14]. A gammatone filter is the product of a gamma distribution function and a sinusoidal tone centred at frequency f_c, calculated as:

$$g(t, B, f_c) = K \, t^{(n-1)} e^{-2\pi B t} e^{j2\pi f_c t} \qquad t > 0 \qquad (1)$$

where K is the amplitude factor; n is the filter order; f_c is the central frequency in Hertz; and B represents the duration of the impulse response [15].

Basilar membrane impulse responses are nearly linear for frequencies ranging between 20–1,000 Hz and approximately logarithmic between 1–20 kHz. Glasberg and Moore's Equivalent Rectangular Bandwidth (ERB) calculation may be used to model the basilar membrane's progressive bandwidth scaling [16]. By modifying the B term of the gammatone wavelet function according to the ERB scale, we obtain a filter bank considered to be a close match to the biology:

$$B = ERB(f) = 2.47 \times (4.37 \times f + 1) \tag{2}$$

Using the g/ERB wavelet transform, which models only biological audition, as a preprocessor for HTM, which models the neocortex, we can construct a biologically plausible pipeline which is coherently focussed on auditory processing.

4   Hierarchical Temporal Memory

The theoretical principles of HTM have been developed as a set of Cortical Learning Algorithms [17], which implement sparse coding, distributed representation, Hebbian learning, and inhibition techniques. HTM is distinguished from other related models (such as HPC) by the integration of *sequence prediction* as the *primary function* of the system. These properties are implemented in the interaction of artificial cortical mini-columns. Research on HTMs has steadily developed over the past ten years, with a focus on image processing [18].

HTMs are constructed by hierarchically arranging regions of cortical columns, where each column is a set of neurons with associated dendrites and synapses. Within each region, two functional processes cooperate to learn temporal sequences from their input, and then pass their learned patterns to the region above. A Spatial Pooler (SP) operates on the input first, with the objective of learning sparse, distributed representations. The spatial codes are then used by a Temporal Pooler (TP) to learn sequences within the data stream.

The spatial and temporal patterns learnt by HTM are represented by the activation levels of columns, rather than the responses of neurons. The role of neurons in HTM is to collectively determine the activity of the column. By adding columns as a feature of the model, a closer match to cortical structure [19] is achieved which permits more sophisticated processing.

Unlike other models, where synapses are associated with weights that modulate input signals, HTM dendrites are associated with *potential* synapses which become connected and *active* when their *permanence value* passes a certain threshold. Only dendrites with connected synapses relay their input to the column. All other synapses remain inactive, but potentially active if the column has not sufficiently participated in the learning process. The participation level of columns is controlled by inhibition, where strongly activated columns compete with and inhibit less active neighbours. The activation level of a column is determined by the sum of the inputs from its connected synapses.

Learning in SP is based on how well synapses of a column match the input to which they are connected. It is implemented by increasing permanence values of potential synapses connected to active input, and decreasing the same parameter for active synapses connected to inactive input. This method of altering synapse permanence values models the well established principle of Hebbian learning.

In order to focus on the relationship between data preprocessing and the initial operations of HTM, i.e. SP processing, we did not make use of the TP in this study. We refer the interested reader to [17].

5   Experimental Study

Datasets: Three bioacoustic datasets were used in this study. 'Frogs' are a set of 1,629 recordings of 73 different species of native Australian frog calls [20]. They are 250 milliseconds in duration, sampled at 22.05 kHz and 16 bits. The 'Insects' dataset consists of 381 insect species sounds, 5 seconds in duration, sampled at 44.1 kHz and 16 bits. The insects are categorised into four families: katydid, cricket, cicada, and others (i.e. bee, beetle, fruitfly, midges, mosquito, wasp).[1] The 'Birds' dataset was taken from the ICML 2013 Bird Challenge.[2] We used only the training files as the ground truths for the test set are not available. The training set comprises song recordings of 35 species of birds, 150 s in duration, sampled at 44.1 kHz and 16 bits.

Following the method of Gonzalez [8], we processed Insects using 1,024 samples per frame and no attempt was made to detect and remove ambient noise from the recordings. In order to accurately capture the lower frequencies typical of Frog calls, we increased the frame size to 5,120 samples. Because noise was not removed from the data, the Birds set produced a very large number of noise-only frames which distorted results. To counterbalance the extreme ratio of feature-to-noise frames, we increased the Bird frames to 32,768 samples.

Preprocessing: Wavelet features were obtained using Matlab R2012a and the UviWave.300 wavelet toolbox[3] running on a MacBook Pro with OS X Mavericks version 10.9.7. We used the UviWave.300 Morlet WT to produce individual scalograms for each sample frame. For g/ERB wavelet features, we extended the UviWave.300 toolbox by developing an ERB scaled gammatone wavelet function to replace the Morlet function when producing scalograms. To dimensionally reduce the scalograms we took the mean of each frequency band. For each dataset we produced two sizes of Morlet and g/ERB features, i.e. 36 or 100 coefficients. These feature set sizes were chosen as being the closest possible match to the feature set sizes used in [8], and which could be processed by the SP (which, for optimal performance, currently relies on input being a square matrix).

Spatial Pooling: Over the past few years the SP has been incrementally developed and used in a range of vision processing studies. The current version allows input of multiple channels per instance as separate matrices, but we disabled

---

[1] Compiled by Gonzalez [8] from various internet resources.
[2] Kaggle. https://www.kaggle.com/c/the-icml-2013-bird-challenge/
[3] Universida de Vigo, Spain. http://www.tsc.uvigo.es/~wavelets/uvi_wave.html

this feature and used only a single input matrix. The dimension of SP columns was set to match the dimension of input features, e.g. for 36 wavelet coefficients, 36 columns arranged in a six-by-six matrix were used. The SP was run for a maximum of 500 iterations at which time the 'best state' was used to obtain SP column codes. The best state was determined as the iteration during which the least number of synapses had their permanence values altered. SP column codes were output as the level of column activation, i.e. the sum of the input for all active synapses of the column.

Classification: Using a k-Nearest Neighbour (k-NN) classifier ($k = 1$), we performed ten-fold cross validation to evaluate all feature sets, which included features obtained after preprocessing by WTs, and those output by the SP. In [8], ten-fold cross validation using a k-NN classifier was also employed, so we are able to compare our Frog and Insect results against those achieved using spectral features.

6   Results and Discussion

The results of using WTs and the SP in this study are reported as the percentage of correcly classified instances. Table 1 summarises the results obtained using the Morlet and g/ERB wavelets. Due to our not being able to validate the test set of the ICML 2013 Bird Challenge, we cannot directly compare our results with those achieved in the competition. Nevertheless, we provide results on the Birds dataset as an extension to the range of bioacoustics investigated in this study.

Table 1. % of correctly classified instances using Morlet and gammatone/ERB WTs.

	36 Features		100 Features		36 Features		100 Features	
	Morlet	+SP	Morlet	+SP	G/ERB	+SP	G/ERB	+SP
Frogs	80.1 %	55.6 %	83.4 %	59.6 %	**95.1 %**	75.2 %	94.9 %	82.3 %
Insects	89.4 %	77.1 %	90.5 %	81.1 %	99.3 %	93.7 %	**99.5 %**	95.7 %
Birds	77.9 %	68.8 %	75.1 %	59.4 %	**92.9 %**	53.8 %	92.6 %	56.4 %

The gammatone/ERB wavelets outperformed Morlet wavelets on the Frogs dataset used in this study. A correct classification rate of 95.1 % was achieved using 36 g/ERB features, closely followed by 100 features which produced 94.9 %. These results represent a considerable improvement over those reported in [8]. For the Insects set, the g/ERB coefficients achieved 99.5 % with 100 features. Using 36 features, 92.9 % classification accuracy was achieved for Birds. The Insect result is an improvement on the previously reported classification rate of 99.2 %, and worth noting as any increase at these high levels of classification is difficult to achieve.

The use of the biologically-inspired gammatone/ERB wavelet consistently outperforms the linearly scaled Morlet wavelet on these datasets. We attribute this to the finer acuity achieved by the scaled wavelets, particularly in the higher frequency ranges typical of Insect and Bird sounds.

The inclusion of the SP reduced all classification accuracies (although not to the same degree for the Insects set[4]) suggesting that the SP encoding is degrading the salience of the g/ERB wavelet features, at least the features relevant for a k-NN classifier. This is an unexpected result, given that SP has performed well in other related domains [18] and suggests there is a missing element in our model of audition. One possibility is that the g/ERB wavelet is not a good model of the cochlear signal and so does not capture the salient features required for higher level neocortical processing. Another possibility is that the additional processing of the cochlea signal in the brainstem[5] changes the characteristics of the auditory signal so that it becomes suitable for neocortical processing. We consider this second option offers the most promising avenue for further research.

7   Conclusion

Details of preliminary work aimed at developing the HTM model for auditory recognition and classification of bioacoustics have been presented. A unique, biologically-inspired processing pipeline using WTs and the HTM SP was applied to three bioacoustic datasets and evaluated based on classification accuracy. These results showed that using gammatone/ERB WTs alone produced superior performance over previously published results for both frog call and insect sound classification. However, the inclusion of the SP caused classification rates to decline across all datasets. This suggests that the combination of gammatone/ERB WTs with an HTM Spatial Pooler does not accurately model the biological interaction between the cochlea and the neocortex. Further work is therefore required, particularly in studying and modelling the effects of brainstem activity on the auditory signals reaching the first region of the neocortex. We conjecture that such additional effects may be necessary for the neocortex (and an HTM system) to effectively classify bioacoustic data streams.

References

1. Murty, K.S.R., Yegnanarayana, B.: Combining evidence from residual phase and MFCC features for speaker recognition. IEEE Signal Process. Lett. **13**(1), 52–55 (2006)

---

[4] Unlike Frogs and Birds, the Insects samples are continuous and do not contain ambient noise. This suggests that the different context of the noise frames is impacting on the performance of the SP for Frogs and Birds.

[5] The cochlear nucleus of the brainstem provides considerable input to auditory processing due to a wide variety of neurons having distinct temporal and spectral response properties, e.g. cells of the posteroventral cochlear nucleus respond strongly to temporal features of complex tones. Higher within the brainstem, the superior olivary complex engages in binaural processing, while other regions of the brainstem handle reflexive and emotional responses to sound.

2. Strang, G.: Wavelet transforms versus Fourier transforms. Bull. Am. Math. Soc. **28**(2), 288–305 (1993)
3. Mallat, S.: A Wavelet Tour of Signal Processing: The Sparse Way. Academic Press, Burlington (2008)
4. Hawkins, J., Blakeslee, S.: On Intelligence. Henry Holt, New York (2004)
5. Friston, K.: The free-energy principle: a unified brain theory? Nat. Rev. Neurosci. **11**(2), 127–138 (2010)
6. Towsey, M.W., Planitz, B., Nantes, A., Wimmer, J., Roe, P.: A toolbox for animal call recognition. Bioacoustics Int. J. Anim. Sound Record. **21**(2), 107–125 (2012)
7. Mitrovic, D., Zeppelzauer, M. and Breiteneder, C.: Discrimination and retrieval of animal sounds. In: 12th International Multi-Media Modelling Conference Proceedings, pp. 339–343 (2006)
8. Gonzalez, R.: Better than MFCC audio classification features. In: Jin, J.S., Xu, C., Xu, M. (eds.) The Era of Interactive Media, pp. 291–301. Springer, Heidelberg (2013)
9. Stowell, D., Plumbley, M.D.: Audio-only bird classification using unsupervised feature learning. In: Working Notes of CLEF 2014 Conference (2014)
10. Seekings, P., Potter, J.R.: Classification of marine acoustic signals using wavelets and neural networks. In: Proceeding of 8th Western Pacific Acoustics Conference (Wespac8) (2003)
11. Mirzaei, G., Majid, M.W., Ross, J., Jamali, M.M., Gorsevski, P.V., Frizado, J.P., Bingman, V.P.: The bio-acoustic feature extraction and classification of bat echolocation calls. In: 2012 IEEE International Conference on Electro/Information Technology (EIT), pp. 1–4 (2012)
12. Usevitch, B.E.: A tutorial on modern lossy wavelet image compression: foundations of JPEG 2000. IEEE Signal Process. Mag. **18**(5), 22–35 (2001)
13. Daubechies, I.: Where do wavelets come from? a personal point of view. Proc. IEEE **84**(4), 510–513 (1996)
14. Schnupp, J., Nelken, I., King, A.: Auditory Neuroscience: Making Sense of Sound. MIT Press, Cambridge (2011)
15. Valero, X., Alías, F.: Gammatone wavelet features for sound classification in surveillance applications. In: 2012 Proceedings of the 20th European Signal Processing Conference (EUSIPCO), pp. 1658–1662 (2012)
16. Glasberg, B.R., Moore, B.C.J.: Derivation of auditory filter shapes from notched-noise data. Hear. Res. **47**(1), 103–138 (1990)
17. Hawkins, J., Ahmad, S., Dubinsky, D.: Hierarchical temporal memory including HTM cortical learning algorithms. Technical report, Numenta Inc, Palto Alto (2011)
18. Cowley, B., Kneller, A., Thornton, J.: Cortically-inspired overcomplete feature learning for colour images. In: Pham, D.-N., Park, S.-B. (eds.) PRICAI 2014. LNCS, vol. 8862, pp. 720–732. Springer, Heidelberg (2014)
19. Mountcastle, V.B.: Introduction to the special issue on computation in cortical columns. Cereb. Cortex **13**(1), 2–4 (2003)
20. Stewart, D.: Nature Sound. Australian Frog Calls: Subtropical East [Audio Recordings]. http://www.naturesound.com.au/cd_frogsSE.htm

Ontology-based Information Extraction for Residential Land Use Suitability: A Case Study of the City of Regina, Canada

Munira Al-Ageili and Malek Mouhoub[✉]

Department of Computer Science, University of Regina, Regina, Canada
{alageilim,mouhoubm}@uregina.ca

Abstract. In order to automate the extraction of the criteria and values applied in Land Use Suitability Analysis (LUSA), we developed an Ontology-Based Information Extraction (OBIE) system to extract the required information from bylaw and regulation documents related to the geographic area of interest. The results obtained by our proposed LUSA OBIE system (land use suitability criteria and their values) are presented as an ontology populated with instances of the extracted criteria and property values. This latter output ontology is incorporated into a Multi-Criteria Decision Making (MCDM) model applied for constructing suitability maps for different kinds of land uses. The resulting maps may be the final desired product or can be incorporated into the cellular automata urban modeling and simulation for predicting future urban growth. A case study has been conducted where the output from LUSA OBIE is applied to help produce a suitability map for the City of Regina, Saskatchewan, to assist in the identification of suitable areas for residential development. A set of Saskatchewan bylaw and regulation documents were downloaded and input to the LUSA OBIE system. We accessed the extracted information using both the populated LUSA ontology and the set of annotated documents.

Keywords: Ontologies · Geographic information system (GIS) · Multi-criteria decision making · Information extraction · Land use suitability

1 Introduction

Land use suitability analysis is used to assess the appropriateness of a specific area of land for a particular use. In the context of Geographic Information Systems (GIS), land use suitability is determined through a systematic multi-factor analysis of the different aspects of the landscape. Model input, therefore, can include factors related to physical and environmental sustainability, as well as factors pertaining to economic and cultural impacts. The results of the analysis are usually presented on maps that rate areas from high to low suitability. The maps may be the desired end result or they may be used as one of the inputs to a simulation model, such as cellular automata, for representing and predicting the spatial dynamics of land use such as urban growth. In this modeling

© Springer International Publishing Switzerland 2015
S. Arik et al. (Eds.): ICONIP 2015, Part IV, LNCS 9492, pp. 356–366, 2015.
DOI: 10.1007/978-3-319-26561-2_43

approach, the most important task is specifying the criteria and values that are applied to assess the suitability of the land for a particular kind of use in this study, for residential development. Determining the factors and their criteria and values helps in determining the data sets needed to create the GIS layers to be included in the evaluation. Each jurisdiction has its own regulations, bylaws, or policies that are applied to assess land use suitability for that jurisdictions geographic area. These provide the criteria for the factors to be included in the multi-criteria analysis. Bylaw and regulation documents available on the web are in natural language text and cannot be processed directly by machines to access and extract the information. Manually finding and extracting criteria and the specific values for the criteria can be a tedious and time-consuming task. In some cases, there may be no precise values for criteria and the actual or real values require expert judgment. For the purpose of automating the extraction of the criteria and values applied in land use suitability analysis (LUSA), we developed an ontology-based information extraction (OBIE) system to automatically extract the required information from bylaw and regulation documents related to the geographic area of interest. The output of the LUSA OBIE system can be presented as an ontology [1–3] populated with instances of the extracted criteria and property values, as a set of semantically (with ontology knowledge) annotated documents or the populated LUSA ontology can be exported to a database or a knowledge base or can be saved as an XML file. The user can retrieve the information from the ontology, view the annotated documents using an annotation editor, or query the database or knowledge base. The extracted criteria are then used to direct the process of obtaining the necessary data for the creation of the required land use suitability maps and also for determining the GIS operations that should be performed to create the GIS layers that represent the criteria. As shown in Fig. 2, the output from LUSA OBIE is applied in this paper to help produce a suitability map for the City of Regina, Saskatchewan to assist in the identification of suitable areas for residential development. A set of Saskatchewan bylaw and regulation documents were downloaded and input to the LUSA OBIE system. We accessed the extracted information (criteria and data property values) using both the populated LUSA ontology and the set of annotated documents.

The rest of the paper is organized as follows. The next section describes the spatial relations considered for this study. Our proposed method is then presented in Sect. 3. The case study for the city of Regina is described in Sect. 4. Finally concluding remarks are listed in Sect. 5.

2   Decision Criteria Based on Spatial Relations

Spatial relations between geographic objects are key elements of spatial modeling and spatial analysis. These relationships among objects in space result from their locations relative to each other. Geographic Information Systems (GIS) are often based upon spatial relations. A primary function of these systems is determining the relationships between objects in space. Spatial relations are classified into topological, orientation, and distance relations [4,5]. Topological relations describe the spatial relation between neighboring features such as adjacency, connectivity and containment. These relations are purely qualitative, and invariant under continuous transformations such as rotation and scaling. Direction relations help determine the orientation of a primary object relative to a reference object; they could be qualitative, such as Regina is east of Moose Jaw, or metric, such as a direction specified in degrees from a reference direction. Distance relations specify the distance between two objects; they could be metric distance relations, such as the distance separating a hazardous waste site and a residential subdivision should be at least 2 km, or qualitative distance relations, such as near, close to, far, and very far; for example, the house is close to the main road. Criteria (factors and constraints) involved in a multi-criteria analysis describe some of the spatial characteristics of the area under consideration. For example, in a land use suitability analysis for residential development, these criteria may include topographic properties of the land, such as slope of the terrain, and accessibility to amenities and services such as parks, roads, and fire, police and ambulance services. Spatial factors have a significant role in a land use suitability analysis process. Factors and constraints criteria are usually defined spatially in a way that depends on the GIS model (Raster or Vector) used to create them. Some criteria such as distance to roads are explicitly spatial and are usually created using GIS functionality. These criteria are represented as spatial data layers. Land use suitability evaluation criteria may include factors and constraints related to the physical attributes of the land (e.g., topography, soil characteristics, potential flooding, subsidence and erosion, and servicing). In [6], the authors distinguish between three classes of spatial relations applicable to multi-criteria decision-making: the location of the sites under consideration, their proximity to desirable or undesirable facilities or features, and the direction relative to certain facilities and the sites under consideration. The spatial relations primarily applicable to land use suitability analysis criteria are the location of the sites (or subdivisions) under consideration and their proximity to desirable or undesirable facilities. Direction relations are also important in some cases, such as directions of aircraft takeoff and landing relative to the location of the residential subdivision. However, direction relations are not dealt with in the selected domain documents. In addition, some criteria describe the spatial characteristics of the land, such as soil type, soil conditions, and slope or aspect of the land. Examples related to land use suitability analysis for residential development are shown in Table 1[1].

3   Proposed Method

In this work, A framework is proposed for integrating an OBIE [7] system and GIS-Based Multi-Criteria Decision Making approach (MCDM). The objective is to construct a land suitability map for residential development for the City of

---

[1] Saskatchewan subdivision guidelines. Retrieved from http://www.municipal.gov.sk. ca/Subdivision/Subdivision-Guide.

Table 1. Criteria examples extracted from selected domain documents

Spatial Relation/Characteristic	Criterion	Example
Location	Permitted land use	Subdivision zoned as residential
(derived from topology)		
		Building site on or near a drop off
	Topography	
	Servicing	Access to roads,
	(access to desirable facilities)	water or sewer connection
Proximity (distance)	Setbacks	Less than 457 meters from a landfill
	(close to undesirable facility)	
	Neighboring land use	sewage treatment plant; mining facility
	(close to undesirable area)	industrial development
	Soil type	Unsuitable soil type such as
		Loose or swampy soil
		Soils shifting, heaving or cracking
Land characteristics	soil condition	Steeply sloping land
		Polluted drainage onto the land
		from adjacent uses
	Slope	
	Surface and sub-surface drainage	

Regina. We developed an OBIE system for automating the extraction of criteria and their values that are applied in land-use suitability analysis, from bylaws and regulations documents, related to the geographic area of interest. These criteria represent the biophysical, social and economic factors that may be used in the construction of land-use suitability maps that support the process of evaluating the suitability of a particular area of land for a particular kind of use. The results obtained by LUSA OBIE (land use suitability criteria and their values) are then incorporated into the MCDM model applied for constructing a suitability map for residential development for the City of Regina.

3.1   LUSA OBIE

The LUSA OBIE system combines the use of ontology, domain-specific gazetteer lists, language processing tools and extraction rules based on regular expressions [8] to automatically add semantic annotation to domain documents (such as regulations and bylaws documents) and then extract the criteria to be applied in the process of land use suitability assessment. Figure 1 illustrates the overall architecture and the following components of LUSA OBIE.

Documents Selection and Analysis. A set of relevant bylaw documents in natural language, related to the geographic area of interest, is selected. The selected documents are carefully examined to help identify domain concepts to be included in the domain ontology, as well as for enumerating domain-specific gazetteer lists.

Linguistic Preprocessing. The selected domain documents are written in natural language, therefore it is difficult to directly process them and extract the information. The input text needs first to be structured to identify its essential lexical and syntactic constituents and make the knowledge accessible. We use a set of linguistic processing tools to process the text to obtain its various linguistic features. Text preprocessing includes tokenization (i.e., splitting the text into tokens), sentence splitting (segmenting the text into sentences), shallow lexico-syntactic analysis (such as Part of Speech (POS) tagging and morphological analysis) and concept lookup. The tools generate several linguistic annotations and features, which are used to build the extraction patterns to be matched in the text and extract the information.

LUSA Ontology Construction. The ontology is described as a formal specification of domain knowledge. In this context, the structure and type of the knowledge to be extracted from the selected documents is defined by the ontology. The criteria ontology encodes the categories of terms describing criteria (factors input to the suitability analysis model), their properties and the relationships that may exist among them, for which the selected documents can be searched. The ontology is provided as an input to the extraction system. The ontology guides the extraction process, providing the structure and the semantics of the knowledge to be extracted. The ontology is also populated with the extracted information.

Ontology-based Semantic Annotation. In this step the domain ontology is coupled with a set of domain-specific gazetteer lists and pattern matching rules. Gazetteer lists contain names of instances of domain concepts (instances of classes and property values). The documents are searched for instances of classes and property values defined in the ontology and instances of these concepts in the text are annotated with respect to classes and properties in the ontology. A set of grammar rules are used to annotate and extract more complex patterns and structures than can be done by gazetteer lists. The grammar rules check if instances found in the text belong to a class or a in the ontology and if so, they link the recognised instance to that same class and add ontology and class features to the annotations.

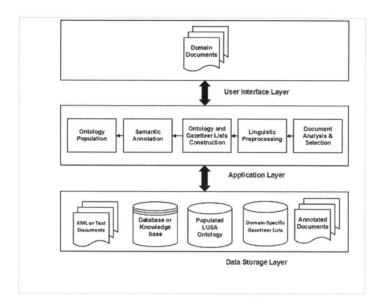

Fig. 1. LUSA OBIE architecture

Ontology Population. This component generates new instances in the ontology from the annotated text. Mentions of instances in the text are linked to instances of concepts in the ontology.

– Output Representation: The output of the system is presented as documents annotated with and linked to concepts in the ontology and the ontology populated with instances generated from the extracted information. The information in the populated ontology may be exported to a knowledge base, a database, an XML document or a text file for use or further analysis.

The extracted information (the criteria and their values) are then integrated into the MCDM model applied for constructing the residential development suitability maps for the City of Regina.

3.2 GIS-based Multi-criteria Decision Making (MCDM)

GIS-based Multi-Criteria Decision Making (MCDM) analysis (also known as Multi-Criteria Evaluation (MCE)) was applied in this work. Multi-criteria analysis combines various criteria into a single evaluation index that indicates the relative suitability of different locations for a specified use, such as residential development. Multi-criteria evaluation is determined through a method known as Weighted Linear Combination (WLC). Using WLC, the continuous criteria (factors) are standardized to a common numeric range, a weight is applied to each factor and then the weighted factors are combined to yield a weighted average. The result is a continuous map of suitability that can be masked by the

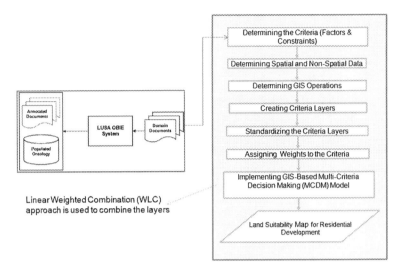

Fig. 2. Proposed LUSA OBIE and GIS-based multi-criteria evaluation for residential suitability

Boolean constraints to produce the final suitability map. Figure 2 summarizes the steps included in a GIS-based multi-criteria evaluation procedure for land use suitability analysis.

4 Case Study: The City of Regina

4.1 Study Area

Regina is the capital city of the Canadian Province of Saskatchewan. It is located at 50 27' 0"N / 104 37' 0" W. Regina is the second-largest city in Saskatchewan and represents a cultural and commercial centre for the southern part of the province. Figure 3 shows a land use map of the study area. Regina is experiencing both economic and population growth. According to the Statistics Canada, 2011 Census of population, the population grew to 193,100 in 2011 compared to 179,282 in 2006 (Statistics Canada).

4.2 Specification of the Criteria

The first step in a multi-criteria analysis is determining the criteria to be used. The criteria are of two types:

- Constraints: the regulations that limit the area available for development (those areas that are not suitable or not allowed for development under any circumstance; e.g., water bodies and already developed areas are restricted from development); constraints are thus Boolean.

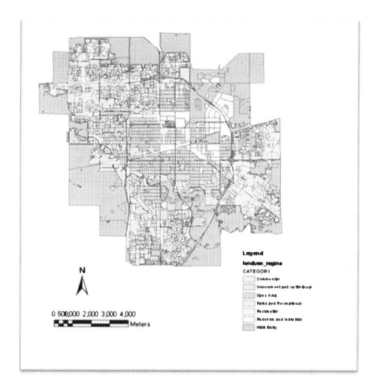

Fig. 3. Land use map for the City of Regina

– Factors: criteria that determine the relative suitability of the remaining areas for residential development; factors are continuous. For example, the type of existing land use is a factor that can increase or decrease the suitability of the land for development or, to a certain degree, areas close to major roads are preferable for development over areas that are distant from major roads.

Using our LUSA OBIE, we were able to identify a set of criteria pertaining to land use suitability analysis for residential development in Saskatchewan. For the purpose of this study several criteria were selected.

Constraints include the following: new development cannot occur within 100 meters of water bodies, areas that are already developed, and water bodies and roads are not considered suitable for new development under any circumstances.

Factors are as follows.

– current land use type: after eliminating areas that can be considered, remaining areas are rated according to their type; e.g., open areas are preferred to treed areas,
– distance to major roads, and
– distance to existing developed areas.

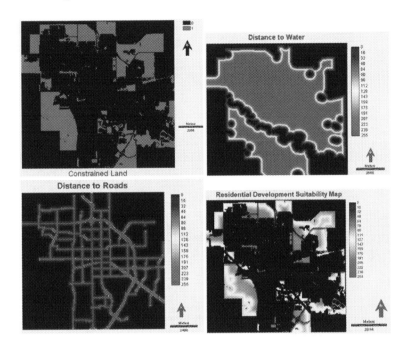

Fig. 4. Constrained land (top left), distance to water (top right), distance to roads (bottom left) and residential development suitability map (bottom right)

Other attractive, but non-essential, factors such as distance to schools and hospitals can be considered, but were not included in this analysis.

4.3 Selection of Data Layers

Following the selection of the criteria, the required image data necessary for the creation of factor and constraint layers (which will be combined to produce the final suitability map), were downloaded. The following datasets were used: land use map of Regina, road network, lakes and rivers, and wetlands. DMTI Spatial Inc. data layers were accessed via the Equinox website:
http://equinox.uwo.ca/EN/AdvancedSearch.asp.

4.4 Creation of Factor and Constraint Images

A raster image was created for each constraint and factor. The geospatial processes used to create the layers that represent the criteria applied in land use suitability analysis are reclassify, overlay, and distance. The raster images are shown in Fig. 4. The factor images created have different measurement units. To enable combining these factors, the images were standardized to a continuous scale from 0 to 255, where non-suitable areas are represented by 0 and the most suitable areas are represented by the value 255. Constraint images remain

Boolean, where non-suitable areas are assigned the value 0 and suitable areas are assigned the value 1. For the purpose of this work, the criteria were assigned equal weights. The Weighted Linear Combination approach was used to combine the layers and produce the final suitability map (see bottmo right of Fig. 4) for residential development.

5   Conclusion

Land use suitability maps can be useful to planners, developers, and environmentalists in their discussions and in making informed decisions on future, sustainable development. These maps can also be an effective means of presenting land use information to the public. The process of extracting the criteria and developing such maps also serves to identify information that should be important to the decision process but is not readily available: this may thus serve to initiate or support efforts to obtain such data. The extracted information can also be applied to assess the suitability of land for other types of development, such as agricultural, industrial or commercial, and to create the desired suitability maps. The resulting maps may then be integrated into a simulation model, such as cellular automata, to predict future growth in the City of Regina. The LUSA OBIE system assists in this process by automating the identification of the criteria and the data that must be obtained to carry out land use suitability analysis.

In the near future we are planning to expand the knowledge extraction rules in order to extract other forms of information such as tabular data, image data, and spatial and temporal data and relations. These latter will be expressed in the form of spatio-temporal constraints and preferences using the model we have developed in the past years [9,10].

References

1. Gruber, T.R.: A translation approach to portable ontology specifications. Knowl. Acquis. **5**, 199–220 (1993)
2. Gruber, T.R.: Toward principles for the design of ontologies used for knowledge sharing. Int. J. Hum. Comput. Stud. **43**, 907–928 (1995)
3. Smith, B.: Ontology. In: Floridi, L. (ed.) Blackwell Guide to the Philosophy of Computing and Information, pp. 155–166. Blackwell (2003)
4. Pullar, D., Egenhofer, M.: Toward formal definitions of topological relations among spatial objects. In: Third International Symposium on Spatial Data Handling (1988)
5. Sharma, J.: Integrated spatial reasoning in geographic information systems: combining topology and direction. Doctoral dissertation, University of Maine (1996)
6. Rinner, C., Heppleston, A.: The spatial dimensions of multi-criteria evaluation-case study of a home buyers spatial decision support system. In: Raubal, M., Miller, H.J., Frank, A.U., Goodchild, M.F. (eds.) Geographic Information Science. LNCS, vol. 4197, pp. 338–352. Springer, Heidelberg (2006)

7. Müller, H.M., Kenny, E.E., Sternberg, P.W.: Textpresso: an ontology-based information retrieval and extraction system for biological literature. PLoS Biol. **2**, e309 (2004)
8. Cunningham, H., Maynard, D., Bontcheva, K.: Text Processing with Gate. Gateway Press, Louisville (2011)
9. Mouhoub, M., Sukpan, A.: Managing temporal constraints with preferences. Spat. Cognit. Comput. **8**, 131–149 (2008)
10. Mouhoub, M., Liu, J.: Managing uncertain temporal relations using a probabilistic interval algebra. In: SMC 2008 IEEE International Conference on Systems, Man and Cybernetics, 2008, pp. 3399–3404. IEEE (2008)

Correlating Open Rating Systems and Event Extraction from Text

Ehab Hassan[✉], Davide Buscaldi, and Aldo Gangemi

LIPN, Université Paris XIII, CNRS UMR 7030, 93430 Villetaneuse, France
{ehab.hassan,davide.buscaldi,aldo.gangemi}@lipn.univ-paris13.fr

Abstract. Event extraction is a very important task for research textual information. This task can be applied to various types of written text, e.g. news messages, blogs, manuscripts, and user reviews for products or services. In this paper, we report results about an experiment in correlating event patterns obtained from machine reading, and ranking derived from open rating systems. The experiment is performed in the touristic domain, where there is some evidence of misalignment between the two sources of opinion.

Keywords: Event extraction · Machine reading · Semantic web · User reviews

1 Introduction

The web has significantly changed how people express themselves and interact with others. Now they can post reviews of products and services in merchant websites and express their opinions and interact with others through blogs and forums. It is now well agreed that user generated content contains valuable information that can be used for many applications. With the increasing amount of data on the Web, utilizing extracted information in the decision making process becomes urgent and difficult.

Open rating systems allow to synthetically grasp the opinion of the crowds with reference to specific entities: products, services, statements of ideas, etc. When opinion is given by both synthetic ranking and a review, one may wonder:

- if ranking corresponds to arguments given analytically, and vice-versa;
- what is, in summary, the analytical reason for the ranking;
- what is the lived experiences of users;
- what are the really cool or bad events one might expect for a certain class of rated entities.

In this paper, we describe our efforts to address the first task and the last task, i.e. if arguments given in reviews correlate with open rating, and if it is possible to extract relevant event dictionaries from user reviews. We formulate these tasks as a binary text classification task. We explored machine learning

© Springer International Publishing Switzerland 2015
S. Arik et al. (Eds.): ICONIP 2015, Part IV, LNCS 9492, pp. 367–375, 2015.
DOI: 10.1007/978-3-319-26561-2_44

techniques to build a classifier which allows to classify two types of reviews (Positive and Negatives) using event features. In order to extract events from user reviews, we use a *machine reader* to perform a deep semantic parsing of text which, allow us to obtain a RDF Linked-Data-ready graph representation of the text. The events are extracted as a sub-graph using SPARQL queries.

Our experiments show a F-Measure of 0.84 for the classification of either positive or negative reviews with regard to event types extracted from reviews, indicating a good correlation between the extracted event types and the rating that has been assigned by users.

2   Related Work

Previous works in event extraction focused largely on news articles; to our knowledge, this is the first study on event extraction from user reviews in opinion mining. Therefore, in this section we focus the existing representative approaches to event extraction.

Aone and Ramos-Santacrus [1] develop the system REES which extract 100 relation and event types, 61 of which are events, from a news source. They developed ontologies of the relation and events to be extracted for political, financial, business military, and life-related domains. The system achieved a 0.70 F-score. In this research, we aim to extract events from opinion reviews which differ from the work performed by [1]. The EVITA event recognition tool, developed by Sauri et al. [10], is used for event recognition and extraction in newsswire text. In this work, the authors perform the identification and tagging of event expression by combining linguistic and statistical methods to obtain a performance ratio of 0.80 F-measure. Bethard and Martin [2] show that the event identification task can be formulated as a classification task. They developed the system STEP that is capable to identify events in the purpose of question answering with a precision of 0.82 and recall of 0.71. Van Oorschot et al. [11] extract game events (e.g. goals, fouls) from tweets about football match to automatically generate match summaries. Events were detected and classified using a machine learning approach. Ritter et al. [9] present an open domain event extraction within Twitter. They propose an approach based on latent variable models to categorize events and classifying extracted events in an open-domain text with 0.64 of F-measure. Due to the difference in structure, this work is not suitable for extracting events from user reviews. In addition, tweets typically include a single event while the user reviews include sequence of events, which makes our task different from that discussed in [9]. Ploeger et al. [7] introduce an automatic activist events extraction method from various news sources using NLP tools. They extracted 1829 events with 0.71 precision 0.58 recall and 0.64 F-Measure. Unlike most approaches mentioned above, we do not use a predefined list of potentially interesting events, but we use a system of "machine reading", FRED [8], to automatically identify and extract events. FRED was compared with others machine reading systems able to extract events, such as, OpenCalais[1] and

---

[1] www.opencalais.com

CiceroLite[2], standing out clearly [4]. Finally, to our knowledge, no method has yet attempted to use event extraction for classifying polarity of reviews, or to correlate open rating with either stochastic classifiers.

3   Dataset

Ott et al. [6] have recently created the first publicly available[3] dataset for deceptive opinion spam research containing 800 positive reviews (400 truthful reviews and 400 fake reviews) which have been assigned with 5-stars in the open system ranking and 800 reviews for the negative polarity (400 truthful reviews and 400 fake reviews) which have 1-star in the open system ranking. In this work, we were only interested in the truthful reviews which are collected from the 20 most popular Chicago hotels on TripAdvisor[4]. We selected 600 reviews, 300 user reviews for the positive reviews (15 reviews for each hotel) and 300 user reviews for the negative ones. In our experimentation, 420 user reviews were used as a training set, 210 reviews for the positive class and 210 reviews for the negative one, and 180 user reviews were used to evaluate our classifier (90 for each class). We chose to not take into account any mixed review (that is, reviews having 3 stars in the system ranking) because they could contain mixed types of events (both positive and negative), while we had the objective to find discriminant events, or at least to have reviews in which most events show a correlation to the evaluation assigned by users.

4   Event Dictionary Construction

4.1   Event Extraction

Event extraction can be broadly defined as the creation of specific knowledge concerning facts and situations referred to in some content and/or data: texts, images, video, databases, sensors, etc. In this research, we focus on events expressed by verbs, propositions, common nouns, and named entities (typically proper nouns). In order to extract these events from user reviews, we employ a deep variety of machine reading [3], as implemented in the FRED tool[5] [8], which extracts knowledge (named entities, senses, taxonomies, relations, events) from text, resolves it onto the Web of Data, adds data from background knowledge, and represents all that in RDF and OWL. FRED is a tool to automatically transform knowledge extracted from text into RDF and OWL, i.e. it is a *machine reader* for the Semantic Web. It is event-centric, therefore it natively supports event extraction. FRED is available as a RESTful API and as a web application. In its current form, it relies upon several NLP components: Boxer[6] for

---

[2] www.languagecomputer.com

[3] Available by request at: http://www.cs.cornell.edu/~myleott/op_spam

[4] http://www.tripadvisor.com/

[5] http://wit.istc.cnr.it/stlab-tools/fred

[6] http://svn.ask.it.usyd.edu.au/trac/candc/wiki/boxer

the extraction of the basic logical form of text and for disambiguation of events to VerbNet, UKB[7] or IMS[8] or BabelNet API[9] for word sense disambiguation, and Apache Stanbol[10] for named entity resolution. More details on using FRED for event extraction are presented in [5]. In order to extract events, we used FRED to produce a semantic graph from the input review, and apply afterward a SPARQL query to this graph to finally select the events.

4.2   Dictionary Construction

We assume that events may be used to find whether a certain review is negative or positive: often, users describe events that led to an uncomfortable or enjoyable experience during their test of the products or visit to the hotels. For instance, if we look at the following sentence: *"When I asked for refund he was very rude, slamming things down on the counter and swearing. He finally refunded me after 5 min of arguing."*, FRED can identify, as it can be seen in Fig. 1, three events: *Ask*, *Refund* and *Slam*.

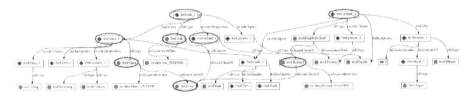

Fig. 1. The output of FRED for the example text: *"When I asked for refund he was very rude, slamming things down on the counter and swearing. He finally refunded me after 5 min of arguing."*.

Our approach for the construction of the dictionary is the following (Fig. 2):

1. Select a set R of positives and negatives reviews from TripAdvisor;
2. Extract all the events contained in R;
3. Consider the events that appear in the highly rated reviews as positive events (agreeable), and the events that appear in the worst rated reviews as negative events (disagreeable);
4. According to event frequencies, Select a limited number events to create the dictionary D_E;
5. Use the events in D_E as features for a multinominal Naïve Bayes classifier to check the correlation between events types and ranking.

Using our training set, we were able to recognize and extract 9160 events: 3645 for the positive reviews and 5524 for the negative ones. The number of aggregated events is 2267: 915 for the positive reviews and 1352 for the negative ones (Table 1).

---

[7] http://ixa2.si.ehu.es/ukb/
[8] http://www.comp.nus.edu.sg/~nlp/sw/
[9] http://lcl.uniroma1.it/babelnet/
[10] http://stanbol.apache.org

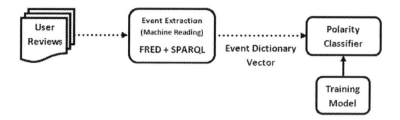

Fig. 2. Overview of the proposed approach

Table 1. The number of extracted events from the two classes of review.

Rev.class	Events	Aggregated events
Positive	3645	915
Negative	5524	1352
Overall	9160	2267

The resulting dictionary could not be used to discriminate between the two types of events since many reviews, positive and negative, since it may contain common events which are distributed in the two classes in a homogeneous manner. Therefore, we decided to look at the frequencies of events and remove these common events. We considered that an event is representative to a class if the probability $P(c|e) \geq \sigma$ where:

$c \in C = \{+, -\}$;
e: A generic event;
σ: A threshold that we determined empirically between 4 possibilities: $\{0.6, 0.7, 0.8, 0.9\}$. The best value of σ was 0.7.

For example, the event *Cancel* appears 11 times in the negative reviews and 3 times in the positive ones. It can be very useful to discriminate the negative reviews.

$$P(+|\text{``Cancel''}) = \frac{3}{3+11} \simeq 0.21.$$
$$P(-|\text{``Cancel''}) = \frac{11}{3+11} \simeq 0.79 > 0.7.$$

However, the event *Stay* occurs 321 times in the positive reviews and 262 times in the negative ones. This event appears almost identically in the two classes and should be deleted since it cannot be useful for our classification.

$$P(+|\text{``Stay''}) = \frac{321}{321+262} = 0.55 < 0.7.$$
$$P(-|\text{``Stay''}) = \frac{262}{321+262} = 0.45 < 0.7.$$

In the end, we removed 269×2 common events (269 from the positive list and 269 from the negative one) which do not help to discriminate the reviews types, and left 182 events among the events that exist in the two classes: the result was a dictionary containing 1547 characterizing events: 464 unique events for the positive reviews, 43 common events discriminating the positive reviews, 901 unique events for the negative reviews, and 139 common events but important to discriminate the negative reviews.

5   Evaluation

The 1547 events in the final dictionary were used as features for a multinomial Naïve Bayes classifier. Each review is represented by a feature vector where the i−th component value is the frequency of the i−th event in the review, and the last component is the type of the review (positive or negative). We chose the Weka[11] implementation for the classifier using cross-validation techniques to validate the model (in particular we used 10-fold cross-validation). Our approach is evaluated in terms of precision, recall, and F-measure. The obtained results for the training set can be observed in Table 2.

We assume that event qualities, i.e. the adverbial qualities which could be associated with events, can upgrade the classification task results. Therefore, we extract these qualities from reviews to use them with the extracted events as features in our classifiers. Table 2 shows the results of our classifiers using the events with their associated qualities.

To validate the obtained results, we used our event dictionary to classify the test set which contains 180 user reviews. The achieved results for this set can be shown in Table 2.

Table 2. Overall results for review classification using NB method with four configurations. *All_Events*: using all the events as features, $Event_\sigma = 0.7$: deleting the common events and using the rest as features, *All_Events_Quality*: using events with their qualities as features, and $(Event\text{-}Quality)\sigma = 0.7$: removing the common events and common qualities, and using the rest as features.

Features	Training set			Test set		
	Precision	Recall	F-Measure	Precision	Recall	F-Measure
All_Events	79.8 %	79.8 %	79.8 %	78.3 %	77.8 %	78 %
$Event_\sigma = 0.7$	82.6 %	82.6 %	82.6 %	75.5 %	73.3 %	74.4 %
All_Events_Quality	79.5 %	78.8 %	79.1 %	80.6 %	80.6 %	80.6 %
$(Event\text{-}Quality)\sigma = 0.7$	84%	84%	84%	80.6%	80.6%	80.6%

As shown in this table, $(Event\text{-}Quality)\sigma = 0.7$ achieves the best results for the training and test sets. In addition, the results using only events as features have been particularly good. Further adding quality features to events achieve even greater accuracy than event features due to the larger vocabulary size.

---

[11] http://www.cs.waikato.ac.nz/ml/weka/

5.1   Error Analysis

The classification errors are due to discriminants events for a type reviews, which exist in reviews of the other type. For example, from a positive review, according to a user, we extracted the following events: { *Get, Help, Put, Take, Experience* }. The classification results indicate that this review is negative. By looking in our dictionary, we only find the event { *Put* } among the other specified events. This event discriminates the negative reviews in our dictionary, but it exists in a positive review. For this reason, we are motivated to study and understand the impact of events. The other events of this review have been removed from the dictionary because they are common events and do not help to discriminate the two types of reviews.

5.2   Application to Data from the ESWC2014 Challenge

To be able to meaningfully evaluate our model, we compared our approach with the systems which participated in the Polarity Detection task, the elementary task in the ESWC-14 challenge on Concept Level Sentiment Analysis. The reviews for this task were extracted from the Blitzer dataset[12]. To build our classifier for this task, we extract events with their qualities from the training set which contains 8000 reviews (4000 positives and 4000 negatives). Then, we used them as features for a multinomial Naïve Bayes classifier.

Table 3 shows the results of our approach and the results of the top three participants in this challenge. The evaluation is carried out on the test, which is composed of 2429 sentences constructed in the same way and from the same sources as the Blitzer dataset. Our system achieved the best performance on Recall and the second best system in Precision and F-measure.

Table 3. Results of polarity detection task at ESWC2014

Participant	Precision	Recall	F-Measure	Final position
NCU	0.78	0.57	0.66	1
IBM	0.66	0.59	0.62	2
FBK	0.42	0.47	0.44	3
Our system	0.68	0.60	0.63	

6   Conclusions and Future Works

In this paper, we presented a completely event-based approach to detect and classify the polarity of customer reviews. We employed a machine reading system, the FRED tool, to extract event-based features and use them to predict the

---

[12] http://www.cs.jhu.edu/~mdredze/datasets/sentiments/

reviews ratings. We built an event dictionary, which can be used to discriminate the two types of reviews. We studied the correlation between the reviews and their ratings, finding that events can be effectively used to determine a correlation between the reviews and their ratings, confirming our initial hypothesis that some events have an influence on the rating scores given by users. We can conclude positively that accurate enough machine reading tools can be employed to investigate the correlation between synthetic and analytic judgments.

In the future, we plan to use events to extract "lived experiences". Lived experiences are related to user motivations and may be important to relate a potential customer profile to the profile of the customer that wrote the review. Lived experiences are usually centered around events that happened to customers.

Acknowledgements. This work is partially supported by a public grant overseen by the French National Research Agency (ANR) as part of the program "Investissements d'Avenir" (reference: ANR-10-LABX-0083).

References

1. Aone, C., Ramos-Santacruz, M.: REES: a large-scale relation and event extraction system. In: Proceedings of the Sixth Conference on Applied Natural Language Processing, pp. 76–83 (2000)
2. Bethard, S., Martin, J.H.: Identification of event mentions and their semantic class. In: Proceedings of the 2006 Conference on Empirical Methods in Natural Language Processing, pp. 146–154 (2006)
3. Etzioni, O., Banko, M., Cafarella, M.: Machine reading. In: Proceedings of the 21st National Conference on Artificial Intelligence (AAAI) (2006)
4. Gangemi, A.: A comparison of knowledge extraction tools for the semantic web. In: Cimiano, P., Corcho, O., Presutti, V., Hollink, L., Rudolph, S. (eds.) ESWC 2013. LNCS, vol. 7882, pp. 351–366. Springer, Heidelberg (2013)
5. Gangemi, A., Hassan, E., Presutti, V., Reforgiato, D.: Fred as an event extraction tool. In: Proceedings of the Workshop on Detection, Representation, and Exploitation of Events in the Semantic Web, p. 14 (2013)
6. Ott, M., Choi, Y., Cardie, C., Hancock, J.T.: Finding deceptive opinion spam by any stretch of the imagination. In: Proceedings of the 49th Annual Meeting of the Association for Computational Linguistics: Human Language Technologies, vol. 1, pp. 309–319 (2011)
7. Ploeger, T., Kruijt, M., Aroyo, L., de Bakker, F., Hellsten, I., Fokkens, A., Hoeksema, J., ter Braake, S.: Extracting activist events from news articles using existing nlp tools and services. In: Proceedings of the Workshop on Detection, Representation, and Exploitation of Events in the Semantic Web, p. 30 (2013)
8. Presutti, V., Draicchio, F., Gangemi, A.: Knowledge extraction based on discourse representation theory and linguistic frames. In: ten Teije, A., Völker, J., Handschuh, S., Stuckenschmidt, H., d'Acquin, M., Nikolov, A., Aussenac-Gilles, N., Hernandez, N. (eds.) EKAW 2012. LNCS, vol. 7603, pp. 114–129. Springer, Heidelberg (2012)

9. Ritter, A., Etzioni, O., Clark, S., et al.: Open domain event extraction from twitter. In: Proceedings of the 18th ACM SIGKDD international conference on Knowledge discovery and data mining, pp. 1104–1112 (2012)
10. Saurí, R., Knippen, R., Verhagen, M., Pustejovsky, J.: Evita: a robust event recognizer for qa systems. In: Proceedings of the Conference on Human Language Technology and Empirical Methods in Natural Language Processing, pp. 700–707 (2005)
11. Van Oorschot, G., Van Erp, M., Dijkshoorn, C.: Automatic extraction of soccer game events from twitter. In: Proceedings of the Workshop on Detection, Representation, and Exploitation of Events in the Semantic Web, p. 15 (2012)

Adaptive DDoS-Event Detection from Big Darknet Traffic Data

Nobuaki Furutani[1](✉), Jun Kitazono[1], Seiichi Ozawa[1], Tao Ban[2], Junji Nakazato[2], and Jumpei Shimamura[3]

[1] Graduate School of Engineering, Kobe University, Hyogo, Japan
146t255t@stu.kobe-u.ac.jp, kitazono@eedept.kobe-u.ac.jp,
ozawasei@kobe-u.ac.jp
[2] National Institute of Information and Communications Technologyi, Tokyo, Japan
{bantao,nakazato}@nict.go.jp
[3] Clwit Inc., Tokyo, Japan
shimamura@clwit.co.jp

Abstract. This paper presents an adaptive large-scale monitoring system to detect Distributed Denial of Service (DDoS) attacks whose backscatter packets are observed on the darknet (i.e., unused IP space). To classify DDoS backscatter, 17 features of darknet traffic are defined from IPs/ports information for source and destination hosts. To adapt to the change of DDoS attacks, we newly implement an online learning function in the proposed monitoring system, where an SVM classifier is continuously trained with darknet features transformed from packets during a certain period. In the performance evaluation, we use the MWS Dataset 2014 that consists of darknet packets collected from 1st January 2014 to 28th February 2014 (8 weeks). We demonstrate that the proposed system keeps good test performance in the detection of DDoS backscatter (0.98 in F-measure).

Keywords: Network traffic analysis · Network incident detection · DDoS attacks · Support vector machine

1   Introduction

More and more companies, regardless of their size or business, are encountering the threat from DDoS attacks – one of the most common and affordable cyber-weapons to date. According to the results of a survey conducted by Kaspersky Lab and B2B International, 1 out of 5 companies have experienced a DDoS attack within the year-long time frame, with damaging consequences including revenue and reputation losses.

In a popular botnet-assisted attack scenario, the attacker(s) commands a swarm of bot-infected computers to send out flooding packets to a target server, intending to reduce the services provide by the server, to a state where they cannot be accessed by legitimate users. It is essential to detect DDoS attacks

© Springer International Publishing Switzerland 2015
S. Arik et al. (Eds.): ICONIP 2015, Part IV, LNCS 9492, pp. 376–383, 2015.
DOI: 10.1007/978-3-319-26561-2_45

accurately in a timely fashion so that mitigation can be done before a server-down. In the field, there have been a good number of proposals to attain this goal from different network management perspectives. Early works on DDoS detection are featured by pattern matching and anomaly detection [1]. In [2], the authors introduced a method to detect TCP based DDoS attacks using protocol analysis. Recent works are motivated by the prominence of DDoS attacks that are making use of UDP protocols, generally known as the distributed reflection DoS (DRDoS) attacks [3,4].

Typical detection and mitigation of DDoS attacks can occur close to the victim, close to the attack source, or throughout the whole Internet infrastructure [4]. While detection at all the three access points may be technically achievable, approaches that address the victim networks or hosts have the unique advantage from a practical perspective: Operation is simpler and better motivated at the victim side. However, due to the high management cost for monitoring the network constantly, not all the service providers are DDoS ready.

When performing a DDoS attack, the attackers tend to send out flooding packets using spoofed source IP addresses to avoid filtering-out by simple methods such as black listing. When the spoofed IP is determined by a random number generator, the replying packets – referred as backscatters in the rest of the paper – from the victim server will be sent "uniformly" to the 32-bit IPV4 space. While the computation and communication costs for acquiring all these packets will be unaffordable in any sense, the monitoring of a subnet of unused IP-address space, a.k.a. a darknet, usually provides a good trade-off between the monitoring cost and global knowledge acquisition. A darknet, also known as network telescope, blackhole monitors, Sinkholes, or background radiation monitors, is a portion of routed, allocated IP space that contains no advertised services [5,6]. Because of the absence of legitimate hosts on the darknet, any traffic observed on a darknet is by its presence aberrant: it is either caused by malicious intent or mis-configuration. Assorted works have deployed darknets in existing networks to help identify the types and sources of malicious traffic present on the larger network of which they form a part, where darknets are used to host flow collectors, backscatter detectors, packet sniffers and so on [7,8].

In this paper, we present an effective DDoS-event detection system based on an analysis of backscatters collected from a large-scale darknet. The experiments show that our approach support fast and accurate detection of DDoS attacks in a world-wide range. Based on the discoveries, we can not only obtain the global trend of DDoS attacks but also discover new types of DDoS attacks as well. To the best of our knowledge, this is the first attempt to perform DDoS event detection based on darknet monitoring.

The remainder of this paper is organized as follows. Section 2 presents the system to detect DDoS backscatters from other types of darknet traffic. Section 3 details the experiment settings and performance evaluation. Discussions on the results is given in Sect. 4, where visualization of host activities is done using t-SNE [9]. Conclusions are drawn in Sect. 5.

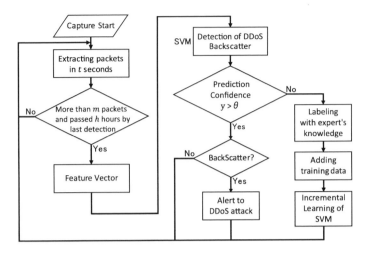

Fig. 1. Our proposed system to detect DDoS backscatter

2   Proposed System

2.1   System Constitution

We proposed a system to detect backscatter caused by DDoS attacks that generate feature vector from short darknet packet every host in feature extraction, and detect from learning with a classifier using generated feature vector [10].

Our proposed system is shown in Fig. 1. In the feature extraction shown in the left part of Fig. 1, we first sort packets observed in a darknet by source IP address. Then, we gather all packets in a fixed period of time from the time the first packet is observed and transform them into a feature vector. In the detection shown in the right part of Fig. 1, we give the input data to the classifier and distinguish backscatter from non-backscatter. Among detection results, if classifier output high prediction confidence and detect backscatter, it alert and inform a host receiving DDoS attack. On the other hand, if classifier output low prediction confidence, we think that the detection with classifier is difficult and perform evaluation by visualizing host activity. Then, we introduce incremental learning to conventional method. we perform incremental learning to distinguish similar data by next detection. In incremental learning, we give labels for the data with expert's knowledge by visualizing host activity if classifier output low prediction confidence. After labeling, we add it to training data and retrain classifier. We use support vector machine (SVM) [11] as the classifier because it has ability for high generalization in the classification of 2 classes.

2.2   Feature Extraction

In the feature extraction, we generate feature vectors every short activity per host from packets observed in darknet. We think that by using short data from the first observed darknet packet, we can detect a DDoS attack before a server is down. When packets from a host are observed less than m packets for t seconds, we think that the feature from too little number of packets is not enough for accurate distinction. Furthermore, when no packet from a host is observed for h minutes, we think that the attack or connection from the address is terminated. If new packets arrive, we generate a new feature vector.

As the goal of this study is the detection of the attack with various means not for specific port and protocols. We use features that are common to various communication forms, for example, statistics information of the source/destination port and destination IP, payload size. The deference by a protocol and the port disappears in feature vector that is generated by using features not to depend on port numbers and protocols. Then we normalize the values to $[0, 1]$ range.

2.3 Learning

Initial Learning and Parameter. We use SVM for detection of backscatters. We use radial basis function kernel (RBF kernel) as the kernel function in SVM classification. It is necessary to select the penalty parameter C and the kernel parameter γ of SVM beforehand to construct the hyperplane which separates two classes. However, the most suitable parameters are different for each data set. Therefore we perform grid search [12] and cross validation [13] to determine the appropriate values of C and γ. We use the pair (C, γ) that gives the lowest cross validation error for classification of test data.

In addition, because the number of the data is different in two classes, we apply cost-sensitive learning of SVM. In cost-sensitive learning, we assign a larger penalty value to the majority class than the minority class.

Incremental Learning. We think that activity of the hosts always changes, and distinguish new attacks or activity patterns only by initial learning is difficult. Therefore, if classifier output low prediction confidence, we consider that learning using only initial data is insufficient, and perform evaluation by visualizing the transmission of darknet packets from ports of a source host to the darknet IPs/ports to give class labels. After evaluation, we add it to training data and use it for retraining classifier. By this process, the classifier becomes able to learn the activity of hosts that have not appeared for conventional training data and detect it correctly. So, a detection performance gain would be expected.

3 Experiments and Performance Evaluation

3.1 Experimental Setup

The MWS Dataset 2014 [14] provided by NICT is used for performance evaluation. This dataset consist of darknet packets which were collected from 1st January, 2014 to 28th February, 2014 (8 weeks). In the performance evaluation,

Table 1. The number of generated feature vectors for performance evaluation.

	Week						
	1&2	3	4	5	6	7	8
DDoS backscatter	2,091	842	691	622	666	741	502
non-DDoS backscatter	378	201	189	232	160	175	264

we use TCP darknet packets except for part 80. The performance of the proposed system should be evaluated with the darknet packets that are not able to be judged by the existing rules. Since we have already proposed the detection rules for TCP/80 and UDP/53 darknet packets [10,15], these packets are removed from the evaluation datasets. In addition, SYN packets are hardly considered as DDoS backscatters because they are not reply packets.

To conduct the performance evaluation, the first thing to do is to give labels (DDoS or non-DDoS) for unlabeled darknet packets. The labeling is conducted with the expert's knowledge by visualizing the transmission of darknet packets from ports of a source host to the darknet IPs/ports. For the DDoS backscatter detection, we adopt the following 17 features of darknet packets that are sent from a single source host for 30 s [10]:

♯Packets sent
Average and standard deviation of time spans of packets
♯Source ports used
Average and standard deviation of ♯packets sent from source ports
♯Protocol types
♯Destination ports used
Average and standard deviation of ♯packets sent to destination ports
♯Destination IPs used
Average and standard deviation of ♯packets sent to destination IPs
Average and standard deviation of the difference of destination IPs
Average and standard deviation of payloads size

We generated a feature vector only for an active host which is defined as a host sending more than or equal to 20 packets for 30 s. After an hour, a new feature vector can be generated for the source host, because we consider such darknet packets are caused by a different cyberattack. That is, we conduct the DDoS backscatter detection at every 60 min.

Table 1 shows the number of generated feature vectors which are separated into the first 2 weeks and the remaining 6 weeks. Feature vectors of the first 2 weeks are used for initial training to learn SVM classifier, while those of the remaining 6 weeks are used for testing and retraining. After the initial training, we test the 1,043 feature vectors of week 3, and retrain SVM classifier with all feature vectors from the 1st week to the 3rd week. Then, the 880 features vector of week 4 are tested and retrained in the same way until the week 8.

Table 2. Experimental results

(a) Without incremental learning

Week	Precision	Recall	F-measure	Time (s)
3	0.966	1	0.982	120
4	0.969	0.998	0.983	—
5	0.987	1	0.992	—
6	0.963	1	0.981	—
7	0.983	1	0.991	—
8	0.967	0.998	0.982	—

(b) With incremental learning

Week	Precision	Recall	F-measure	Time (s)
3	0.966	1	0.982	120
4	0.974	0.998	0.986	237
5	0.987	1	0.992	368
6	0.964	1	0.982	531
7	0.983	1	0.992	676
8	0.967	0.998	0.983	880

We evaluate the performance with the following measures: precision, recall, F-measure, and learning time (sec.). We use LibSVM [16] for the L2-SVM classifier, where the parameters C and γ are optimized via the 5-fold cross-validation with the following parameter range: $C = 2^{-5}, 2^{-3}, \cdots, 2^{15}, \gamma = 2^{-7}, 2^{-5}, \cdots, 2^9$.

3.2 Experimental Results

The results without and with the incremental learning are summarized in Table 2(a) and (b), respectively. Table 2(a) shows that DDoS backscatter packets can be fairly-accurately detected. Especially, recalls reach almost one, that is, almost all the DDoS backscatter packets are detected. This indicates that the 17 features and the classifiers can capture the difference between DDoS backscatters and non-DDoS backscatters, and are effective for DDoS backscatter detection. Table 2(b) shows that, by the incremental learning, the detection performance is further improved in all weeks except for week 5. This implies that activity patterns become diverse over time and the incremental learning enables the system to respond to the diversification.

Computational time is not so high at this stage, as listed in Table 2(b). However, it will become too high to handle before long, because we retrained the SVM every week using all the data observed at that time. Online learning methods should be applied to our system in future work.

4 Discussion

As shown in Subsect. 3.2, the classification performance was improved by the incremental learning. This implies that new activity patterns appeared over time. To visualize such changes and diversification of activity patterns over time, we used a dimensionality reduction method, know as t-SNE [9]. By using t-SNE, the 17-D feature vectors are reduced to 2-D vectors, and are shown in scatter plots in Fig. 2. Figure 2(a)–(c) represent data observed during Jan. 1st to Jan. 7th (the first week), to Feb. 28th (about the first 8 weeks), and to June 31st, respectively. Red and blue indicate the DDoS backscatter and non-DDoS backscatter packets observed during the first 8 weeks, respectively. Green indicates unlabeled data

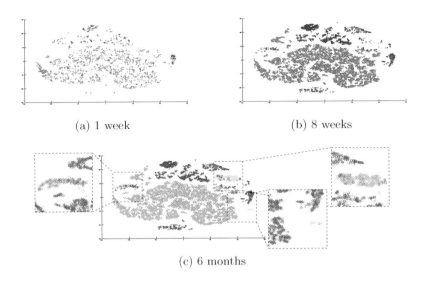

(a) 1 week

(b) 8 weeks

(c) 6 months

Fig. 2. Visualization of the host activity using t-SNE. Figure 2(a)–(c) represent data observed during Jan. 1st to Jan. 7th (the first week), to Feb. 28th (about the first 8 weeks), and to June 31st, respectively.

collected after the first 8 weeks, which were not used in the analysis in Subsect. 3.2. Compared with the distribution in Fig. 2(a), both the distributions of DDoS backscatter and non-DDoS backscatter packets in Fig. 2(b) spread more widely. This means that activity patterns became more diverse over time. Furthermore, clusters that are not in Fig. 2(a) emerge in Fig. 2(b). This means new types of activity patterns appeared after the 2nd week. Additionally, after the first 8 weeks, the distributions become wider and new clusters emerge, as shown in the insets in Fig. 2(c). These results reveal that the activity patterns of the hosts change over time and so that, to distinguish such new patterns, incremental learning is needed.

5   Conclusions

In this study, we proposed a system to detect DDoS-Event from packets observed in darknet. Although some of darknet packets are easily distinguishable by their source/destination port numbers, flags, etc., many of them are not. To detect such packets, our system extracts 17-dimensional feature vectors based on statistics of packets and classifies them using the SVM. Additionally, to deal with changes in activity patterns, the incremental learning is applied. In the experiments, our system was tested using the MWS dataset [14]. The F-measures were higher than 0.98 without incremental learning. Furthermore, they are improved by incremental learning. In addition, we visualized feature vectors using t-SNE [9]. We show that the activity patterns of the hosts become diverse and new types of patterns appear over time. This implies that without the incremental learning the performance falls steadily as time advances.

In future work, we will apply our system to packets other than TCP, such as UDP and ICMP, and incorporate online learning mechanism into SVM to reduce the computation time.

References

1. Mirkovic, J., Reiher, P.: A taxonomy of DDoS attack and DDoS defense mechanisms. SIGCOMM Comput. Commun. Rev. **34**(2), 39–53 (2004)
2. Wang, H., Zhang, D., Shin, K.: Detecting SYN floodingattacks. In: Proceedings of the 21st Annual Joint Conference of the IEEE Computer and Communications Societies, vol. 3, pp. 1530–1539 (2002)
3. Ryba, F.J., Orlinski, M., Wählisch, M., Rossow, C., Schmidt, T.C.: Amplification and DRDoS attack defense - a survey and new perspectives. CoRR, vol. abs/1505.07892 (2015)
4. Bardas, A.G., Zomlot, L., Sundaramurthy, S.C., Ou, X., Rajagopalan, S.R., Eisenbarth, M.R.: Classification of UDP traffic for DDoS detection. In: The 5th USENIX Workshop on Large-Scale Exploits and Emergent Threats (2012)
5. Bailey, M., Cooke, E., Jahanian, F., Nazario, J., Watson, D., et al.: The internet motion sensor - a distributed blackhole monitoring system. In: NDSS (2005)
6. Ban, T., Zhu, L., Shimamura, J., Pang, S., Inoue, D., Nakao, K.: Behavior analysis of long-term cyber attacks in the darknet. In: Huang, T., Zeng, Z., Li, C., Leung, C.S. (eds.) ICONIP 2012, Part V. LNCS, vol. 7667, pp. 620–628. Springer, Heidelberg (2012)
7. Harder, U., Johnson, M.W., Bradley, J.T., Knottenbelt, W.J.: Observing internet worm and virus attacks with a small network telescope. Electron. Notes Theor. Comput. Sci. **151**(3), 47–59 (2006)
8. Benson, K., Dainotti, A., Claffy, K., Aben, E.: Gaining insight into as-level outages through analysis of internet background radiation. In: IEEE Conference on Computer Communications Workshops (INFOCOM WKSHPS), pp. 447–452 (2013)
9. Van der Maaten, L., Hinton, G.: Visualizing data using t-SNE. J. Mach. Learn. Res. **9**, 2579–2605 (2008)
10. Furutani, N., Ban, T., Nakazato, J., Shimamura, J., Kitazono, J., Ozawa, S.: Detection of DDoS backscatter based on traffic features of darknet TCP packets. In: 2014 Ninth Asia Joint Conference on Information Security, pp. 39–43 (2014)
11. Vapnik, V.N.: Statistical Learning Theory, vol. 1. Wiley, New York (1998)
12. Hsu, C.W., Chang, C.C., Lin, C.J.: A practical guide to support vector classification. Technical report, Department of Computer Science, National Taiwan University (2003)
13. Kohavi, R.: A study of cross-validation and bootstrap for accuracy estimation and model selection. In: International Joint Conference on Artificial Intelligence, vol. 14, issue 2 (1995)
14. Kamizono, M.: Datasets for Anti-Malware Research (MWS Datasets 2014) (2014)
15. Nakazato, J., Shimamura, J., Eto, M., Inoue, D., Nakao, K.: Backscatter analysis toward clear categorization of DoS attacks. In: The 30th Symposium on Cryptography and Information Security (2013) (in Jananese)
16. Chang, C.C., Lin, C.J.: LIBSVM: a library for support vector machines. ACM Trans. Intell. Syst. Technol. **2**, 27:1–27:27 (2011)

Content Based Video Retrieval for Obscene Adult Content Detection

Hasini Yatawatte[1]([✉]) and Anuja Dharmaratne[2]

[1] University of Colombo School of Computing, Colombo, Sri Lanka
hyatawat@nd.edu
[2] School of IT, Monash University, Subang Jaya, Malaysia
anuja@monash.edu

Abstract. With the advancement in networking, content produced and distributed over the Internet is exponentially increasing. This imposes the threat of distribution of obscene content freely and largely, urging mechanisms to control access by minor aged users. Manual retrieval and indexing of material is impossible for large video repositories. This paper proposes a method to detect videos with obscene adult content using content based video retrieval techniques. We propose an algorithm to summarize the video by extracting keyframes that mark video shot boundaries and apply BoVW algorithm to classify keyframes indicating the presence of obscenity. Despite the ignorance of high-level features in temporal domain, a higher recognition rate of 85 % with spatial information alone is proved. Further, we show the irrelevance of color information to detect nudity in videos when using BoVW.

1 Introduction

The amount of public video content generated by private sources are incrementing, urging operating/ hosting companies to involve methodologies to classify their media repositories based on semantics. Content classification has a main role in access control. Specially, it is important for providing access control for focused user groups such as children, to avoid their access to improper content. Improperness of content may arise from security, social, religious and ethical concerns. Obscene adult content are one widely available improper information that causes unpleasant influence over social, religious and ethical realms. The possibility of children being exposed to obscene content which are freely and largely available over the Internet either intentionally or accidentally, is considerable [2]. Identifying the negative impact, some countries have established rules and regulations to ban and control the offensive content available online [1,3].

Filters and access control systems are occupied for remedying the aforementioned issues associated with obscene material including textual content, images and videos. Existing content blocking in the Internet mainly uses contextual keyword matching based on URLs, content in web pages or tags of videos. Service providers are also relying on their user community to detect videos with obscene

S. Arik et al. (Eds.): ICONIP 2015, Part IV, LNCS 9492, pp. 384–392, 2015.
DOI: 10.1007/978-3-319-26561-2_46

content, which is a slow and un-guaranteed process [7]. It is impossible to manually retrieve and index long video clips in large video repositories and human derivations of video descriptions suffer from subjectivity, biased results, incomplete and inaccurate results [5]. This urges tools for automatically classifying and retrieving videos using features associated with the video itself.

This paper addresses a solution to detect videos containing obscene adult scenes in a video repository for the aforementioned purpose.The proposed solution is based on spatial information in keyframes of the video. Video summarization is done by key frame extraction in order to reduce the time and processing overhead for real time usage. Video abstraction algorithm computes an adaptive threshold derived for a moving window of frames and uses that threshold to filter redundant video frames. Then the features are extracted from the selected key frames and classified as obscene or not using a decision system that consider Bag of Visual Words (BoVW).

2   Related Work

Previous attempts for detecting offensive adult content in videos can be categorized into image-based methods and video-based methods. Image based methods use only spatial information of single frames in the video while video-based approaches consider combined features such as spatial, temporal and motion information. Among the two, image based methods are considered as faster where as video based methods are giving higher accuracy at a high cost [4].

Most of the obscene adult visual media contain nudity, hence skin detection is used extensively in approaches that use spatial information. Skin detection is considered as a complicated process due to several reasons [13]; substantial non-Lambertian components available in skin's reflectance, often skin has bright areas or highlights that are de-saturated and illumination color vary from image to image resulting bluish or greenish off white skin regions. Hence, Wang et al. [13] claim that the model used to describe the skin must anticipate various luminance conditions and camera calibration problems.

Color based skin detection algorithms are based on parametric models, non-parametric models or explicit skin cluster definition models [7]. An online poll carried out by Liensberger et al. [7] show that even human are not capable of identifying skin color properly without knowing the context. Hence, their approach uses faces detected in the video frame as a basis for color model based skin detection, giving more recognition performance over pure color based approaches, working in real time and without requiring prior learning with training data sets. Kakumanu et al. [6] have concluded that using skin color analysis alone for detecting skin produces high true positives as well as high false positives. To reduce false positives other features such as shape, spatial and motion information are used together with the skin color information.

Detecting nudity in an image or a video is in fact a classification problem. The image or the video is labeled with a class category to indicate the presence or absence of nude people. Bag of Visual Features (BoVF), also referred as BoVW is a successful object recognition technique derived from the popular information retrieval algorithm called Bag of Words (BoW). BoVW is used to detect nudity considering the problem as a special case of object recognition task [8]. Based on the results obtained through the recognition, underlying material can be labeled with a class category to indicate the semantics. BoVW algorithm uses local descriptors fundamentally based on gray level values for recognition and is robust for occlusion. Most importantly, the results are robust against variations in lighting conditions, scale, shape and rotations [8].

3   Proposed Methodology

Proposed solution for detecting obscene content in videos is based on spatial information in individual frames. A long video clip is a sequence of small video shots and shots are distinguished by characteristics such as color distribution, texture information and available key points like corners and blobs of constituting frames. An extension of [14] is proposed for key frame extraction that compute an adaptive threshold within a sliding window of frames for video summarization. This reduces time and processing overhead for real time operation.

A recognition rate comparable with other methods can be achieved using BoVW classification alone and results are not affected by position and orientation of the objects in the video frame [11]. This ignores temporal information in videos that demand expensive computation and is independent from skin color or shape information, which significantly reduces false positive rates due to illumination differences and low resolutions. Hence, the proposed solution is based on BoVW classification on spatial information.

Our approach follows two phases; Training phase and Recognition phase. Training phase is an offline one-time process that builds elements of BoVW classifier and train a Support Vector Machine (SVM). Recognition phase detect the presence or absence of obscene content in a given video and obtain details such as timestamps of such scenes for filtering.

Extracting key frames requires comparing the similarity between video frames using HSV histograms. HSV color model is a non-linear transformation of RGB color space which is represented by Hue (H), Saturation (S) and Value (V) [10]. The computation of histograms with only H channel is easier and hue color model claims to be shift and scale invariant with respective to light intensities [12]. Since color information is sufficient to indicate the repetitiveness associated with video frames, only H channel is considered in histogram computation. Normalized H-histograms are compared using histogram intersection, which is given by Eq. 1[1].

$$d(H_1, H_2) = \sum_i \min(H_1(I), H_2(I)) \tag{1}$$

A threshold is determined to filter out redundant video frames and extract keyframes. The threshold is adaptive across a moving window of certain number of video frames. Proposed algorithm is as follows.

---

[1] H_1 and H_2 refer to normalized histograms of two frames.

1. Divide the video into N number of K size segments, where K is the number of video frames per segment.
2. Sequentially compare H-histograms of K frames within each segment and calculate mean (M) and standard deviation (SD) of differences per each segment.
3. Calculate the adaptive threshold (T_1) for the segment as M + C_1 * SD where C_1 is a constant.
4. Consider the first frame in the segment which is having a histogram difference greater than T_1 as a key frame.
5. Scan until the next frame that is having a histogram difference with the last selected key frame larger than the threshold T_2, where T_2 is calculated as C_2*M. C_2 is a constant.
6. Skip the next set of frames that are having a histogram difference less than T_2 with the current frame.
7. Repeat step 4 until all the segments are considered.

Once key frames are extracted, they are classified as obscene or not using the visual vocabulary and trained SVM. Advantages of BoVW algorithm mostly depend on the choice of feature detector and descriptor. Higher processing power and storage requirements are demanded by color descriptors than binary descriptors as they consider more spatial information.

If the video contains any frames that are detected as offensive adult content, then the video can be tagged as inappropriate for children. The degree of inappropriateness can also be expressed by the proportion of offensive frames to the total frames in the video. If the video is solely pornography, then the degree of unsuitability will be higher. Hence the inappropriateness is higher, where as a video with few shots of nudity may give less degree of inappropriateness. Such a measurement is important as the attitude about the unsuitability of content differs across cultures and also to distinguish attempts on using nudity in a tasteful manner to maintain aesthetic aspects in certain movies.

4   Implementation

A Proof of Concept (PoC) was implemented for testing and evaluating the proposed solution. The dataset published by Lopes et al. [8] was used for evaluating the proposed work due to the ability of comparing our results with their proposed methodology and the free availability of the dataset for research purpose. Dataset comprises images and videos containing obscene and non-obscene content and image dataset was used for the BoVW training process. Video dataset was used to evaluate keyframe extraction algorithm and the detection of obscene content in video segments. In addition, three long video clips that include first 5 min from the Avatar movie and first 15 and 30 min from the Sex and the City movie were used for overall testing.

The proposed solution requires several parameters to be optimized to achieve a better recognition rate (RR) and performance. Keyframe extraction algorithm demands adaptive thresholds T_1 and T_2, which are computed as per

Eqs. 2 and 3. This requires C_1 and C_2 to be determined through trial and error. This also requires a window size to be determined to compute adaptive thresholds. The optimum values were chosen by running the algorithm with different values for these three elements. Values that give minimum redundancy without losing important information was chosen as optimum values.

$$T_1 = M + C_1 * SD \qquad (2)$$

$$T_2 = M * C_2 \qquad (3)$$

BoVW classification requires a vocabulary size to define the number of clusters or BoVW. The optimum vocabulary size differs with the choice of feature detector and descriptor. Hence, a supervised learning approach for different vocabulary sizes and feature detectors/descriptors was carried out with the image dataset. BoVW classification uses a SVM. SVM parameters were tuned for two types of SVM kernels, RBF and Linear, using automatic training process offered by OpenCV machine learning library.

5   Results and Discussion

Accuracy of the keyframe extraction algorithm with HSV histograms and H-histogram was tested with number of video clips ranging from few seconds to 30 min (Table 1).

Table 1. Summary of keyframe extraction with window size of 50 frames

Histogram	Total length of video clips (sec)	Total number of frames	Total number of keyframes
HSV histogram	2764	30820	388
H histogram	2764	30820	307

HSV histogram reduced redundancy of video frames by 98.74 % while H histogram is reducing that by 99.00 %. The improvement given by h-histogram is due to the ignorance of S channel that depends on light intensities. HSV histogram resulted repetition of frames under low intensity, resolution and illumination conditions and we observed that the color information represented by hue channel is enough to extract frames with important information.

Despite the high-level features available in videos such as motion and audio information,the BoVW classification is based on rotation and scale invariant spatial features extracted from single frames of the video. Lopes et al. [9] have used BoVW algorithm for detecting nudity using two feature descriptors, SIFT and Hue SIFT. The original SIFT descriptor they have used uses grey level information. Hue SIFT is an extension that add hue component to SIFT which include color information to the analysis. Since Hue SIFT has given a RR higher than SIFT for all vocabulary sizes, they have emphasized the importance of color to recognize nudity when using BOVW classification.

As opposed to their claim, we argue that the color is not important for detecting nudity. Human can identify nudity even in a black and white image. This is because the interpretation of nudity is done based on features such as blobs, edges etc. available in the image rather than the color. Since BoVW representation of an image identifies such features, it should be able to provide this semantic even with the absence of color information.

We applied BoVW classifiers with different binary and color descriptors for different visual vocabulary sizes. Figures 1 and 2 show classification accuracy in terms of RR (Eq. 4) for two types of SVM kernels, RBF and Linear. It is visible that RR range for binary descriptors lie within the same range as color descriptors. Further, SURF has given the maximum RR[2]. Hence, these results confirm our argument that the color information is irrelevant for detecting nudity when using BoVW analysis.

$$RR = (TP + TN)/(TP + TN + FP + FN) \tag{4}$$

As shown in Figs. 1 and 2, a maximum RR of 85 % has been achieved with the SURF descriptor. This is with a small training dataset of 120 images. RR can be improved by using a better training dataset rich in number as well as quality of images. Further, SVM parameters other than the kernel function also can be fine-tuned to achieve an improved RR. Through these mechanisms RR can be improved beyond 85 %. Hence, we can conclude that the ignorance of temporal information can be tolerated by following a proper training mechanism.

Time taken to train the BoVW classifier with a RBF and Linear kernel was measured for different combinations of 120 frame sized datasets and Fig. 3 summarizes the results. SURF took significantly higher amount of time for training

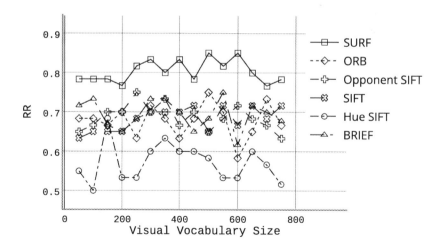

Fig. 1. Comparison of RR between binary and color descriptors with a RBF SVM

---

[2] TP- True Positive, TN- True Negative, FP- False Positive, FN- False Negative.

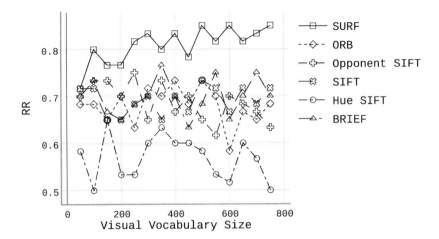

Fig. 2. Comparison of RR between binary and color descriptors with a linear SVM

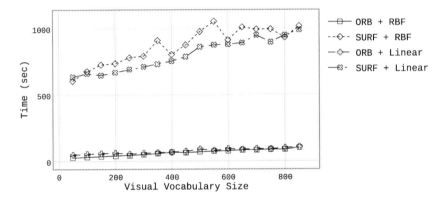

Fig. 3. Average time taken to train with 120 frames vs visual vocabulary size

compared with ORB and the processing time increased with the size of vocabulary under all cases. This is due to the complexity imposed by the process of encoding keypoints to visual words as the number of visual words are increased.

Figure 4 shows time taken to recognize an individual frame. It is visible that even the recognition is faster in ORB compared to SURF. This is due to differences in underlying feature detection and description mechanisms in both of these descriptors.

Some of the misclassifications resulted are shown in Fig. 5. Misclassifications can be due to inadequate training data used for initial vocabulary building and SVM training. Number as well as quality of training data affects the performance of the classifier. Hence a rich training dataset that contain a large amout of obscene and non-obscene material that cover a wide range of applications can increase the overall RR.

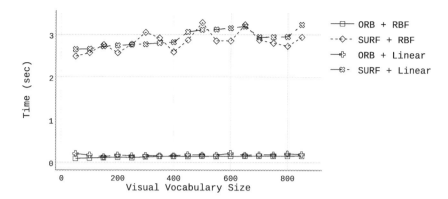

Fig. 4. Average time taken to recognize content in a frame vs visual vocabulary size

Fig. 5. Misclassifications. First row gives some of the false positive incidents where as bottom row gives some of the false negative incidents

6 Conclusion

The proposed methodology for recognizing obscene content in videos is based on spatial information in key frames of the underlying video that are classified using BoVW algorithm. Key frames of the video are extracted using an adaptive threshold calculated within a sliding window of video frames to filter out redundant video frames. This research emphasis the unimportance of color information for detecting nudity, resulting faster execution for real time processing. Results show that a RR of 85 % for detecting obscene content with nudity can be achieved with SURF.

References

1. Children's Internet Protection Act. https://en.wikipedia.org/wiki/Children%27s_Internet_Protection_Act/. Accessed 15 May 2013
2. Pornography Statistics. www.covenanteyes.com/. Accessed 15 May 2013
3. The Australian Communications and Media Authority. Online Regulation. http://www.acma.gov.au/theACMA/About/Corporate/Responsibilities/online-regulation-acma/. Accessed 15 May 2013
4. Behrad, A., Salehpour, M., Ghaderian, M., Saiedi, M., Barati, M.N.: Content-based obscene video recognition by combining 3D spatiotemporal and motion-based features. EURASIP J. Image Video Process. **2012**(1), 1–17 (2012)

5. Ghodeswar, S., Meshram, B.B.: Content based video retrieval. In: Proceedings of ISCET, p. 135 (2010)
6. Kakumanu, P., Makrogiannis, S., Bourbakis, N.: A survey of skin-color modeling and detection methods. Pattern Recogn. **40**(3), 1106–1122 (2007)
7. Liensberger, C., Stöttinger, J., Kampel, M.: Color-based skin detection and its application in video annotation
8. Lopes, A.P.B., de Avila, S.E.F., Peixoto, A.N.A., Oliveira, R.S., Coelho, M.D.M, Araujo, A.D.A.: Nude detection in video using bag-of-visual-features. In: 2009 XXII Brazilian Symposium on Computer Graphics and Image Processing (SIBGRAPI), pp. 224–231. IEEE (2009)
9. Lopes, A.P.B., de Avila, S.E.F., Peixoto, A.N.A., Oliveira, R.S., Araujo, A.D.A.: A bag-of-features approach based on hue-sift descriptor for nude detection. In: 2009 17th European Signal Processing Conference, pp. 1552–1556. IEEE (2009)
10. Satheesh, P., Srinivas, B., Sastry, R.V.L.S.N.: Pornographic image filtering using skin recognition methods (2012)
11. Truong, B.T., Venkatesh, S.: Video abstraction: a systematic review and classification. ACM Trans. Multimedia Comput. Commun. Appl. (TOMM) **3**(1), 3 (2007)
12. Van De Sande, K.E.A., Gevers, T., Snoek, C.G.M.: Evaluating color descriptors for object and scene recognition. IEEE Trans. Pattern Anal. Mach. Intell. **32**(9), 1582–1596 (2010)
13. Wang, D., Zhu, M., Yuan, X., Qian, H.: Identification and annotation of erotic film based on content analysis. In: Photonics Asia 2004, pp. 88–94. International Society for Optics and Photonics (2005)
14. Yusoff, Y., Christmas, W.J., Kittler, J.: Video shot cut detection using adaptive thresholding. In: BMVC, pp. 1–10 (2000)

Combined Document/Business Card Detector for Proactive Document-Based Services on the Smartphone

Yong-Joong Kim[1], Yonghyun Kim[1], Bong-Nam Kang[2], and Daijin Kim[1,2（✉）]

[1] Department of Computer Science & Engineering,
Pohang University of Science and Technology,
77 Cheongam-ro, Nam-gu, Pohang, Gyeongbuk 37673, Republic of Korea
{kimyj,gkyh0805,dkim}@postech.ac.kr
[2] Department of Creative IT Engineering,
Pohang University of Science and Technology,
77 Cheongam-ro, Nam-gu, Pohang, Gyeongbuk 37673, Republic of Korea
bnkang@postech.ac.kr

Abstract. In this paper, we present a novel combined detector of document and business card. To detect document or business card, our method firstly extracts a document object region from a given image, and then classifies it into positive or negative class. In the step of extracting the document object region, a block-based processing is exploited to efficiently find the line segment candidates of its boundary, and RANSAC-like method under three constraints is used to search its real boundary. In classification step, after performing image normalization on the extracted region, the Fisher vector is extracted to represent the document object, then it is classified by linear-SVM. For evaluating the proposed method, we carry out some experiments by using the collected images, and show that our method has achieved about 94 % accuracy.

Keywords: Block-based image segmentation · Document/business card classification · Document detector · Business card detector

1 Introduction

Thanks to the advance of hardware specification of smartphone, the computer vision applications working on the desktop environment have become executable on the smartphone. In particular, the document-based applications such as automatic business card scanning and optical character recognition (OCR) have been released and widely used by many smartphone users. For example, CidT Co., Ltd. has released a business card management application [2], which segments its region from an input image and then registers it into the application by performing OCR. Another example is the ABBYY Mobile OCR engine [1] released by ABBYY Co., which serves functions such as business card recognition, word search, and sentence translation using the result of OCR.

© Springer International Publishing Switzerland 2015
S. Arik et al. (Eds.): ICONIP 2015, Part IV, LNCS 9492, pp. 393–402, 2015.
DOI: 10.1007/978-3-319-26561-2_47

Even though many document-based applications are available, they however have a limitation in that their user has to give the positive image including a document object, namely no detection procedure. This is basically because the documents have various textures depending on document type, and consequently it makes building the document detector hard. If it is possible to build the document detector, it can be used as a service invocator of document-based applications on the smartphone.

In this paper, to build a service invocator of the two widely used applications, business card scanning and OCR, we propose a combined detector of document and business card, in which the detection problem is formulated as a three-class classification problem. Our detector consists of two steps:

- **Quadrilateral Region Extraction:** This procedure extracts a region candidate of document or business card from an input image via a block-based line fitting and the largest quadrilateral search.
- **Region Classification:** The extracted region from the previous step is normalized and classified into one of document object classes (document or business card) or negative class by using linear-SVM with the Fisher vector [18].

The remaining of this paper is organized as follows. In Sect. 2, the candidate region extraction method of the document object is described in detail, and Sect. 3 presents the description of region classification method. The experimental results are analyzed in Sect. 4, and finally Sect. 5 concludes this work.

2   Quadrilateral Region Extraction

The region extraction method of this paper is illustrated in Fig. 1. At first, an input image is partitioned into four blocks, and the probabilistic Hough transform (PHT) [14] is applied to the edge image of each block for obtaining line segment candidates of document object boundary. Finally, our method detects the boundary of the document object through searching the largest quadrilateral under three constraints.

2.1   Image Partitioning

When extracting region candidate from an input image, our key observation is as follows. One often takes a picture of some text-contained objects instead of note-

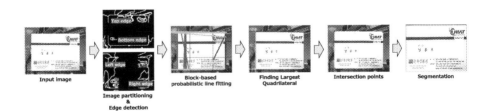

Fig. 1. Overview of our region extraction method.

taking. Thus the object captured by this purpose usually occupies a large space in the image and has little perspective distortion because one may think readability is important in this case. Following this observation, the input image is normalized to 320×240, and it is partitioned into four blocks (Top/Bottom/Left/Right) having some overlapped region (see Fig. 2). It means that each region of interest (ROI) to find the line segments of document object boundary is restricted to the specific size smaller than that of the whole image. In other words, the line segment fit on a particular block can be only the candidate corresponding to its side of document object boundary. We set the size of top and bottom block to 320×80 and that of left and right block to 100×240. It also means that the basic assumption about the size of document object region is that its size is greater than 120×80 (gray-colored region in Fig. 2).

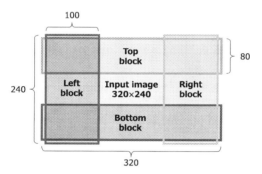

Fig. 2. The partitioning method of input image: top/bottom blocks and left/right blocks have the same size, respectively.

2.2 Block-based Line Fitting

To find line segment candidates of document object boundary, we apply the PHT [14] to the edge images of each partitioned block. Compared with the standard HT [8], the PHT reduces the computing time by considering the selected pixels instead of all pixels in voting procedure of the HT.

In line fitting step, the advantages which are able to be obtained when partitioning the input image are two folds. Firstly, we can reduce unnecessary line segments which can be fit on the inside region of document object. It leads to the search time reduction of finding the largest quadrilateral in a later step. Secondly, when finding the line segment candidates in each block through the PHT, we can use different parameter values such as minimum length of line segment. For example, we can set the minimum length of line segment in top and bottom blocks longer than that of left and right blocks.

The comparison results of several test images when using the PHT and the block-based PHT are shown in Fig. 3. We used different colors to distinguish the line segments of different blocks in Fig. 3(b). This figure also indicates our block-based processing is useful for finding true line segments of document object boundary well.

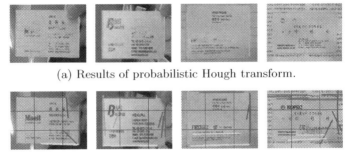

(a) Results of probabilistic Hough transform.

(b) Results of block-based probabilistic Hough transform.

Fig. 3. Comparison results of block-based probabilistic Hough transform with probabilistic Hough transform.

2.3 Searching Largest Quadrilateral

In this step, the RANSAC-like method [9] is used to search the largest quadrilateral, where one line segment is randomly selected from the detected candidates of each block, a quadrangle is formed by the selected four line segments, and the area of the quadrangle is calculated.

We denote N_i as the number of line segments extracted from i block, and denote $L_i = \{l_j^i | 1 \leq j \leq N_i\}$ as the set of line segments of i block, where $i \in \{Top, Bottom, Left, Right\}$. We also denote $\mathcal{Q} = \{Q_1, Q_2, ..., Q_k\}$ as the set of all possible quadrangles, where $Q_k = \{l_a^T, l_b^B, l_c^L, l_d^R\}, 1 \leq a \leq N_T, 1 \leq b \leq N_B, 1 \leq c \leq N_L, 1 \leq d \leq N_R$. Then, the total number of quadrangles is $|\mathcal{Q}| = \binom{N_T}{1} \times \binom{N_B}{1} \times \binom{N_L}{1} \times \binom{N_R}{1}$. To calculate the area of Q_k, its line segments are extended to straight lines, and their intersection points are calculated. Then, we can calculate its area by

$$Area(Q_k) = \frac{1}{2} pq \sin \theta, \tag{1}$$

where p and q are diagonal line segments, and θ is their included angle.

Using the above notations and Eq. (1), the largest quadrangle search problem can be formulated as

$$Q^* = \operatorname*{argmax}_{Q \in \mathcal{Q}} Area(Q). \tag{2}$$

When finding the maximum quadrangle Q^*, our method imposes three constraints on the quadrangle, following our observation (see Sect. 2.1). The first one is that the aspect value of business card must be in the range of lower to upper threshold value. In our experiments, we chose lower and upper threshold values as 0.55 and 0.75, respectively. The second one is that all vertical angles of quadrilateral must be greater than 75° and smaller than 105°. In other words, the pair of opposite sides of the quadrangle must be parallel as possible. The last one is the image boundary condition, which rejects the quadrangles touched on the image boundary.

To find the largest quadrilateral in Q under the constraints, our method iteratively does the largest quadrangle search procedure until it converges, or specific number is reached.

3 Region Classification

3.1 Previous Work on Document Image Classification

Before presenting our region classification method, we briefly review the related works with respect to the document classification. Various approaches have been proposed by researchers. Kang et al. [12] proposed to use convolutional neural network, in which the ReLU (Rectified Linear Units) [15] was used as activation function of neuron, and dropout [10] was employed to prevent overfitting. Kumar et al. [13] proposed a method to measure structural similarity for document image classification and retrieval. In their method, SURF-based codebook is constructed, horizontal-vertical pooling is applied by recursively partitioning the image in horizontal and vertical directions to compute features, and total feature is classified by random forest [6].

3.2 Proposed Method

Basically, object detection is a binary classification problem. However, in our case we formulate the detection problem as a three-class classification problem to build the combined detector of document and business card. Two classes are for document and business card, and the remaining one is for negative class.

In this step, the extracted region from the previous step is normalized by removing perspective distortion and reducing image size, and then the Fisher vector (FV) is extracted from the normalized image. Finally, it is classified via linear-SVM. The proposed method is depicted in Fig. 4.

Image Normalization. In the region extraction step, although input image is processed in the reduced size of 320×240, we can recover the original positions of the corner points of the extracted largest quadrangle by scale ratio,

$$S_w = \frac{W}{320}, \qquad S_h = \frac{H}{240}, \tag{3}$$

where S_w, S_h, W, and H are the scale ratios of width and height, original width, and height, respectively. After finding the real positions of four corner points, the document region is segmented and perspective distortion of the region is removed. Then, the size of the distortion removed image is normalized to 640×480.

Fisher Vector Extraction. We propose to use the FV [18] for representing a document object. Originally, its background theory was proposed by Jaakkola

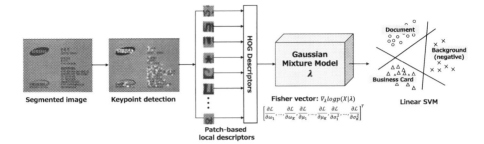

Fig. 4. Overview of our region classification method.

and Haussler [11] to combine the benefits of generative and discriminative methods, in which the Fisher kernel (FK) was derived from the generative model. The FK can be written as a dot-product between normalized vector $\mathscr{G}_\lambda$:

$$K_{FK}(X,Y) = \mathscr{G}_\lambda^{X\prime}\mathscr{G}_\lambda^Y, \tag{4}$$

where $\mathscr{G}_\lambda^X = L_\lambda G_\lambda^X = L_\lambda \nabla_\lambda \log u_\lambda(X)$. Here, $\mathscr{G}_\lambda^X$ is referred to as the FV of X.

For using the FV framework to encode the normalized document image, local descriptors are extracted from the patches segmented from keypoint locations on the image by using HOG [7], and then GMM is fit to the local descriptors calculated from training samples for representing generative model u_λ as in [16].

Classification. The FV representation can be seen as the extension of the BOV (Bag-of-Visual words) because it not only considers the number of occurrences of each visual word but it also encodes additional information about the distribution of the descriptors. Thus, we can regard the FV as a mapping result of an input to the FK space. For this reason, we use linear-SVM as a region classifier.

4   Experiments

In this section, the performance evaluation of our method is carried out. Firstly, we have built our combined detector using OpenCV [3] and VLFeat [5] libraries, and ported it to an Android-based smartphone. Throughout all experiments, we use the Samsung Galaxy S4 smartphone [4] as experimental device.

4.1   Data Collection

To evaluate the proposed detector, we have collected the total 2,839 images consisting of documents, business cards, and negative class images. The information of the dataset is shown in Table 1.

Table 1. Information of the collected dataset.

Item	Type	No. of images
Document	Magazine, paper, ppt slide, book	957
Business card	Various types	934
Background	Landscape, sign, photo, etc.	948
Total	–	**2,839**

4.2 Evaluation of Extracting Quadrilateral Region

For evaluating our region extraction method, we have sampled 300 business card images and 300 document images from the collected dataset, and made six test datasets in which each dataset consists of 100 images. This is because most of the images in our dataset have simple background. In constructing the six test datasets, we consider the background condition of images as shown in Table 2.

Table 2. Six test datasets for evaluating the quadrangle region extraction.

Dataset	Background description
Document1 (Do1)	Simple (100)
Document2 (Do2)	Simple (50), complex (50)
Document3 (Do3)	Complex (100)
Business Card1 (BC1)	Simple (100)
Business Card2 (BC2)	Simple (50), complex (50)
Business Card3 (BC3)	Complex (100)

In testing the datasets in Table 2, we use an evaluation criterion,

$$R_{overlap} = \frac{A(\text{ground-truth region}) \cap A(\text{detected region})}{A(\text{ground-truth region}) \cup A(\text{detected region})} \geq th, \qquad (5)$$

where $A(\cdot)$ and th are the area function and threshold value, respectively, and our method decides the success or failure of one test sample depending on whether the overlap ratio is over the threshold value or not. In our experiments, we used 0.8 and 0.9 as threshold values. Tables 3 and 4 show the evaluation results of the six datasets, and some examples of segmentation results are shown in Fig. 5. Average speed on the experimental device is about 1.2 fps.

4.3 Evaluation of Region Classification

To calculate local descriptors from a normalized image, we extract 300 key-points using FAST corner detector [17], segment 16×16 local patches from

Table 3. Segmentation rates with threshold value 0.8 on six datasets in Table 2.

Dataset	Line fitting	Block-based line fitting
Do1	90 %	98 %
Do2	83 %	94 %
Do3	80 %	90 %
BC1	85 %	91 %
BC2	91 %	96 %
BC3	93 %	100 %
Avg.	**87.0 %**	**94.8 %**

Table 4. Segmentation rates with threshold value 0.9 on six datasets in Table 2.

Dataset	Line fitting	Block-based line fitting
Do1	89 %	97 %
Do2	71 %	81 %
Do3	63 %	64 %
BC1	79 %	84 %
BC2	84 %	86 %
BC3	84 %	88 %
Avg.	**78.3 %**	**83.3 %**

Fig. 5. Examples of region extraction results.

each keypoint, and describe the patches using HOG. When fitting GMM to the HOG-based local descriptors calculated from training samples, we have experimentally set the number of Gaussian components as three, the minimum value that guarantees good performance.

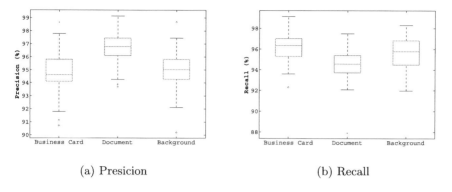

(a) Presicion   (b) Recall

Fig. 6. Experimental results of the twenty times of 4-fold cross validations on our dataset.

Table 5. Average precision, recall, and accuracy on our dataset.

Measure	Avg. precision	Avg. recall	Avg. accuracy
%	95.53 %	95.51 %	94.17 %

Using these experimental setting, we have carried out 4-fold cross validation twenty times by randomly dividing total dataset described in Table 1. The experimental results are shown in Fig. 6 and Table 5. Average speed on the experimental device is about 11.4 fps.

5   Conclusion

In this paper, we have presented a combined document/business card detector, which consists of two steps, quadrilateral region extraction and region classification. To extract a document object region, our method exploits the block-based line fitting and the largest quadrangle search. Then, the extracted region is classified into one of three categories: document, business card, and negative class. To do this, after normalizing the extracted region, the FV is extracted from the normalized image. Then, it is classified via linear-SVM.

In this work, we have only evaluated our method on the collected dataset, but, in future work we will evaluate several alternatives. Also, we will plan to improve the performance of quadrilateral region segmentation method when no line segments are fit on some blocks.

Acknowledgments. The research was supported by the Implementation of Technologies for Identification, Behavior, and Location of Human based on Sensor Network Fusion Program through the Ministry of Trade, Industry and Energy (Grant Number: 10041629), and also by the National Research Foundation of Korea (NRF) grant funded by the Korea government (MSIP) (No. 2015R1A2A2A01004282).

References

1. Abbyy mobile ocr engine. http://www.abbyy.com/mobile-ocr
2. camcard. http://www.camcard.com
3. Opencv. http://www.opencv.org
4. Samsung galaxy s4. http://www.samsung.com/global/microsite/galaxys4/
5. Vlfeat. http://www.vlfeat.org
6. Breiman, L.: Random forests. Mach. Learn. **45**(1), 5–32 (2001)
7. Dalal, N., Triggs, B.: Histograms of oriented gradients for human detection. In: Proceedings of IEEE International Conference on Computer Vision and Pattern Recognition, vol. 1, pp. 886–893 (2005)
8. Duda, R.O., Hart, P.E.: Use of the hough transformation to detect lines and curves in pictures. Commun. ACM **15**(1), 11–15 (1972)
9. Fischler, M.A., Bolles, R.C.: Random sample consensus: a paradigm for model fitting with applications to image analysis and automated cartography. Commun. ACM **24**(6), 381–395 (1981)
10. Hinton, G.E., Srivastava, N., Krizhevsky, A., Sutskever, I., Salakhutdinov, R.: Improving neural networks by preventing co-adaptation of feature detectors. CoRR abs/1207.0580 (2012)
11. Jaakkola, T.S., Haussler, D.: Exploiting generative models in discriminative classifiers. In: Proceedings of the 1998 Conference on Advances in Neural Information Processing Systems II, pp. 487–493 (1999)
12. Kang, L., Kumar, J., Ye, P., Li, Y., Doermann, D.: Convolutional neural networks for document image classification. In: Proceeding of 22nd International Conference on Pattern Recognition, pp. 3168–3172 (2014)
13. Kumar, J., Ye, P., Doermann, D.: Structural similarity for document image classification and retrieval. Pattern Recogn. Lett. **43**(1), 119–126 (2014)
14. Matas, J., Galambos, C., Kittler, J.: Robust detection of lines using the progressive probabilistic hough transform. Comput. Vis. Image Underst. **78**(1), 119–137 (2000)
15. Nair, V., Hinton, G.E.: Rectified linear units improve restricted boltzmann machines. In: Proceedings of the 27th International Conference on Machine Learning (ICML 2010), pp. 807–814 (2010)
16. Perronnin, F., Dance, C.: Fisher kernels on visual vocabularies for image categorization. In: Proceedings of IEEE International Conference on Computer Vision and Pattern Recognition, pp. 1–8 (2007)
17. Rosten, E., Drummond, T.W.: Machine learning for high-speed corner detection. In: Leonardis, A., Bischof, H., Pinz, A. (eds.) ECCV 2006, Part I. LNCS, vol. 3951, pp. 430–443. Springer, Heidelberg (2006)
18. Sánchez, J., Perronnin, F., Mensink, T., Verbeek, J.: Image classification with the fisher vector: theory and practice. Int. J. Comput. Vis. **105**(3), 222–245 (2013)

A Heuristic Data Mining Framework Towards Dynamic Data of Social Media

Estelle Xin Ying Kee and Jer Lang Hong[(✉)]

School of Computing and IT, Taylor's University, Subang Jaya, Malaysia
{estelle.kee,jerlang.hong}@taylors.edu.my

Abstract. The growth of social media over the past years has revolutionized the interaction between individuals as well as the business conduct of many industries. In contest for extraction of the valuable insights available, the dynamic nature of the social media data is seen to be a challenging factor. With many tools available in the market today, the ultimate solution is yet to suffice the needs of businesses today. Therefore, the concept and approach of a heuristic data mining framework is to be proposed in overcoming this task. By having a robust framework of which is proven to conduct the dynamic data mining, the essence sustained within social media of which industries are looking for will be extracted to achieve greater milestones.

Keywords: Social media · Data mining · Heuristic technique

1 Introduction

Having a glance throughout the past few years, many organizations despite public or private sectors have advanced further in their decision making and strategies of moving forward by leveraging on the tools and techniques used in data mining on various source of data banks. Being one of today's most eyed on data bank which has a dynamic nature, social media data mining has been the limelight of many organizations due to its potential of revealing much valuable yet often hidden and unknown data of high relevancy provided by their very own stakeholders' community. Social media data not only acts as a precursor to actual big data integration (Fig. 1), it also holds promising transformation of companies today to what we called "customer-centric businesses" [1].

Experiencing consequences of a change in realities within the context of economics and social views, the constantly expanding availability of a large-scale, real-world social media data has ushered in a new era of research and development in the field of data mining. Dubbing into the era of "big data" as everyone is most aware of, the amount of content-based data provided by social media has indeed open doors of great potential and unique challenges for researches today to dwell further into the extend of which frameworks and tools of data mining can achieve.

In this era of big data, the countless hours spent in social media is seen in various forms of human interaction and data is being generated at an unprecedented rate. This dynamic data though possesses great potential has also its limits such as being free-formatted.

© Springer International Publishing Switzerland 2015
S. Arik et al. (Eds.): ICONIP 2015, Part IV, LNCS 9492, pp. 403–409, 2015.
DOI: 10.1007/978-3-319-26561-2_48

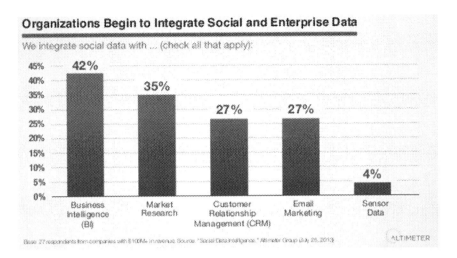

Fig. 1. Data integration

With the account of these limitations, this paper focuses on the presentation of a heuristic framework of which data mining can be carried out on the social media data of unstructured properties.

Out of the many trends and patterns available in the market which are used for data mining on various social media platforms, our publication aims on providing the insights and relevant support by highlighting a much improved yet heuristic data mining framework to researchers and professionals to advance further in their quest of extracting dynamic social media data.

2   Related Work

Rooting on deeper to today's contemporary Information Extraction (IE) systems, the concept of automatic rule generalization with machine language is said to over triumphs the written extraction rules by facilitated programmers. As we indulge into the current Wrapper Induction (WI) systems, the four (4) classes are namely the manually constructed IE systems, supervised IE systems, semi-supervised IE systems and the unsupervised/automated IE systems [4].

Using general programming languages or specially designed languages, the manually constructed IE systems depends heavily on user's vast knowledge on computing together with strong programming foundations. It is also required for the users to program a wrapper for each of the website by hand which leads to a high cost and utilization of resources such as time and budget. W4F [11], Minerva [5], XWRAP [9] and TSIMMIS are some of the early wrapper constructions which utilize the manual techniques.

To reduce cost of wrapper generation, the supervised WI systems were introduced as it enables general users to use the labeling GUI with minimal training. It is deemed workable as an initial set of labelled examples along with the GUI assist the process of labelling by suggesting additional pages to users. However, there are still seen shortcomings such as costly wrapper maintenance due to dynamic environment of the web and it is also not recommended for large number of sites due to the effort of manual labelling.

Moving on further on the scale of automation is the semi-supervised IE systems which comprises of IEPAD [3], OLERA [2] as well as Thresher [7]. Being in a completely opposite direction of the supervised approach as stated above, the OLERA and Thresher are able to accept rudimentary yet vague input from users to generate extraction rules. For IEPAD, though not needing labeled training pages, it is a must for user to choose a target pattern and indicate the data needed to be extracted [12]. With improve performance relative to supervised IE systems, the supervision of user still exists in the manner of which a GUI is still required for users to specify the extraction targets after the learning phase.

Due to limitations found in previous IE systems, the unsupervised/automated IE system was introduced. Not only extraction of big volume yet dynamic data is made convenient, data records often in tuple instances in sites are usually found encoded with small number of fixed templates which can be detected using the mining approach of repeated patterns. Without comprising to pre-extraction events such as programming wrappers, labelling and specifying extraction targets, the unsupervised/automated IE system is a much efficient extraction architecture to handle a large amount of dynamic data, which justify its suitability and relevancy in extracting social media data. Some of the examples of unsupervised/automated IE systems are categorized as the page-level solvers (RoadRunner & EXALG) and the record-level extractors (DeLa & DEPTA) [6, 10, 13].

3   Motivation and Problem Formulation

Spinning off huge chunks of data which organizations are utterly eager to attempt them along with the integration of their respective systems, social media sites are the data sets of which are deemed very useful in forming strategies and determining directions of which an organization should propel towards. Due to its massive size of data which changes in a real time manner, it is a challenge to keep track of the changes occurs especially mining the social media data from various social media platforms of different templates. Visible challenges such as social media data being of varying length, free-formatted, noisy as well as integrated multimedia elements within the important context of which is required to be extracted are all part and parcel of the issues that requires attention. Hence, this inspires us to further tap into the search of seeking a heuristic framework of obtaining social media data of dynamic nature.

With social media intercepting more and more of our time spent daily, it is undeni-ably true that this phenomenon has been eyed by people of many various fields to know the way of mining valuable data that is contained within the belly of the beast – the very core of social media itself. Data mining, the act of retrieving data of many forms (mostly in the form of unstructured, semi-structured and/or structured) from data sources has been deeply studied for the past years. For social media data which is contributed by billions of users all around the globe is seen growing in manner of exponential yet dynamic, it is rather challenging and fairly difficult to extract related data with similar methods used in search engines, e-commerce websites or even blog sites even though they are from the similar vibrant and lively web source. The root cause is seen as the unstructured layout of data in the social media platforms, each having their own different forms of presentation when displayed in uniquely designed user interfaces. Adding on to the complexity of social media data mining is the highly ambiguity of the human language [14] and complex social relations such as followers-followees. Henceforth, out of the 4 different architectures of Information Extraction (IE) systems, the most suitable architecture deemed fit to be utilize upon the social media sites is the unsupervised/automated information extraction system when compared to the other 3 (manual, supervised and semi-supervised). However, due to the inconsistency of information presentation by the social media sites, there is still no known data extraction tool which can be used in addressing this problem. This pressing need for new approaches and frameworks ushers the further exploration into social media data mining. Therefore, with this research done, we are determined to find a heuristic framework to extract valuable data from social media along with the entailed challenges.

4   Proposed Solution

Overview. To extract unstructured data from the social media, we need to gather a sample of social media datasets. Once the social media datasets are obtained, we develop a novel heuristic algorithm to extract this data. Before the data extraction is carried out, we first parse through the sample pages and construct DOM Tree accordingly. After that, the extraction will go through a series of steps where heuristic techniques are used. Details of these steps are described below (as shown in Fig. 2):

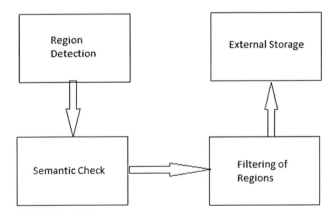

Fig. 2. Flow Chart of our algorithm

Step 1. Our technique parses through the DOM Tree in depth first manner. Once a block tag is detected (e.g. DIV Tag), we measure the size of the boundary of that block tag by calculating its width and height. If a block tag is of acceptable size (representing a region by a fairly big area), we tag it as potential region. Then our algorithm traverses through the sub tree of the block tag to find potential smaller regions in depth first manner. If smaller regions are found, we tag the new tag as potential region. The step is repeated until the smallest region is found. Otherwise, the next smallest region is chosen as the potential region by traversing upwards. We repeat the region detection process for the remaining nodes.

Step 2. Once the region detection step is carried out, we'll have a list of potential regions. From these regions, we perform semantic similarity check on the individual text nodes. We traverse through the region tree structure in depth first manner. Once a text node is found, we tokenize its content into individual words. We remove the stop words as well as stemming it to root word using WordNet stemmer. Once the root word is obtained, we use WordNet similarity checker (DISCO) to check the semantic similarity of the words. We use the algorithm of Jiang and Conrath to check the similarity of keywords as the study in shows that this algorithm works best among all the other algorithms. The word similarity checker will check the semantic similarity between two words (e.g. cat and dog is 70 % similar, cat and house is 10 % similar).

In some sentences, there may be a Word Disambiguation. For example, the word "interest" in the sentences "Interest in book" and "Interest in bank" have different meanings although they contain the same "interest" wording. To solve this problem, we implement Adapted Lesk algorithm which detects the neighboring words in conjunction to the word "Interest". To implement the word similarity check, we first compare the first word with that of the second word in the text node. If the two words are matched, then we consider them as matched and the text nodes are marked as containing similar words. Otherwise, we check the first word with that of third word, and repeat the step again for the remaining words. Once word matching is carried out, we then perform word disambiguation on the words in the text node. For every text node encountered, we consider the text node as containing semantically similar keywords if they have more than 3 words with similar meaning. The procedure is repeated again for the remaining text nodes in the region. We marked the region as semantically similar if the region contains more than 60 % text nodes with similar words.

Step 3. Step 3 will filter out irrelevant region from the list of potential regions. This step is to ensure that the region we extracted are the content of the social media, not other regions such as advertisement, menus, and so on. Our observations show that menus usually do not contain semantically similar content. To detect irrelevant regions from the list of potential regions, we parse through the region and divide them into smaller sub regions. For each of the sub regions, we match them using the same semantic rules in Step 2. If a region has more than 65 % similar sub regions, it is considered as relevant region. Otherwise, it is marked as irrelevant regions.

Step 4. Once the relevant regions are identified, we will then output them into an external file in a presentable form. We remove the styling properties where necessary, and even javascript codes.

5 Experimental Tests

We conduct our experimental tests on a wide range of datasets. We collect a random sample of 200 pages from the social media. These sample pages are written in different languages with complicated layout and structure. To measure the effectiveness of our algorithm, we use precision and recall which are formulated as follows:

Recall = Correct/Actual*100
Precision = Correct/Extracted*100

We compare our method against the state of the art data extraction algorithm by [8]. Correct depicts the number of pages where the relevant region is correctly identified. Actual is the actual number of pages containing relevant region. Extracted depicts the number of pages where relevant region is extracted.

Table 1 shows our experimental tests conducted on the sample pages. Our method outperforms the methods of [8] both in terms of recall and precision rates. This could be attributed to the fact that our method is able to detect the semantic properties of data in the web pages. Not only that, our method is multilingual, hence our method is able to work across web pages written in different languages. Due to the fact that our sample pages contain mostly social media data, our method is more efficient than the method of [8] for data extraction as the underlying layout and structure of social media is highly complicated.

Table 1. Experimental tests

	OntoExtract [8]	Our method
Actual	200	200
Extracted	184	195
Correct	146	192
Recall	73.00 %	96.00 %
Precision	79.35 %	98.46 %

6 Conclusions

Emphasizing on the nature of social media data which is highly ambiguous and dynamic, it is seen that a heuristic approach must be taken towards the extraction of the relevant data. Together with the analysis of the 4 classes of Wrapper Induction systems, the comparison of differences in the context of the characteristics and the strengths and weaknesses has been clearly addressed in this publication. As social media data is seen frequently under the unstructured category, the automated Information Extraction systems-based extraction framework is seen to be best fit. With the exponential increase of social media data today, we foresee that this framework proposed would be significantly value-adding to the community in broadening their perspective towards the possibilities of social media data utilization as well as creating open doors for researchers to further dwell into this area in the quest of seeking the ideal framework for dynamic data extraction.

References

1. Anthony, M.: Social Data Blooming into Social Intelligence. CMS Wire (2013)
2. Chang, C.-H., Kuo, S.-C.: OLERA: a semisupervised approach for web data extraction with visual support. IEEE Intell. Syst. **19**(6), 56–64 (2004)
3. Chang, C.-H., Lui, S.-C.: IEPAD: information extraction based on pattern discovery. In: Proceedings of the 10th International Conference on World Wide Web (WWW), pp. 223–231 (2001)
4. Chang, C.H., Mohammed, K., Moheb, R.G., Khaled, F.S.: A survey of web information extraction systems. IEEE Trans. Knowl. Data Eng. **18**(10), 1411–1428 (1999)
5. Crescenzi, V., Mecca, G.: Grammars have exceptions. Inf. Syst. **23**(8), 539–565 (1998)
6. Crescenzi, V., Mecca, G., Merialdo, P.: Roadrunner: towards-automatic data extraction from large web sites. In: Proceedings of the 26th International Conference on Very Large Database Systems (VLDB), pp. 109–118 (2001)
7. Hogue, A., Karger, D.: Thresher: automating the unwrapping of semantic content from the world wide. In: Proceedings of the 14th International Conference on World Wide Web (WWW), pp. 86–95 (2005)
8. Hong, J.L.: Ontoextract – automated extraction of records using multiple ontologies. In: IEEE International Conference on Fuzzy Systems, and Knowledge Discovery (2013)
9. Liu, L., Pu, C., Han, W.: XWRAP: an XML-enabled wrapper construction system for web information sources. In: Proceedings of the 16th IEEE International Conference on Data Engineering (ICDE), pp. 611–621 (2000)
10. Pinto, D., McCallum, A., Wei, X., Croft, B.C.: Table extraction using conditional random fields. In: Proceedings of the ACM SIGIR Conference, pp. 235–242 (2003)
11. Saiiuguet, A., Azavant, F.: Building intelligent web applications using lightweight wrappers. Data Knowl. Eng. **36**(3), 283–316 (2001)
12. Wang, J., Lochovsky, F.H.: Wrapper Induction Based on Nested Pattern Discovery, Technical Report HKUST-CS-27-02, Department of Computer Science, University of Science & Technology, Hong Kong, (2002)
13. Wang, J., Lochovsky, F.H.: Data extraction and label assignment for web databases. In: Proceedings of the 12th International Conference on World Wide Web (WWW), pp. 187–196 (2001)
14. Yang, G., Ramakrishnan, I.V., Kifer, M.: On the complexity of schema inference from web pages in the presence of nullable data attributes. In: Proceedings of the 12th ACM International Conference on Information and Knowledge Management (CIKM), pp. 224–231 (2003)

Abnormal Activity Detection Using Spatio-Temporal Feature and Laplacian Sparse Representation

Yu Zhao[1], Yu Qiao[1,2], Jie Yang[1,2(✉)], and Nikola Kasabov[3]

[1] Institute of Image Processing and Pattern Recognition,
Shanghai Jiao Tong University, Shanghai, China
{zhaoyunccq,qiaoyu,jieyang}@sjtu.edu.cn
[2] Key Laboratory of System Control and Information Processing,
Ministry of Education, Shanghai, China
[3] Knowledge Engineering and Discovery Research Institute,
Auckland University of Technology, Auckland, New Zealand
nkasabov@aut.ac.nz

Abstract. Abnormal activity detection in a video is a challenging and attractive task. In this paper, an approach using spatio-temporal feature and Laplacian sparse representation is proposed to tackle this problem. To detect the abnormal activity, we first detect interest points of a query video in the spatio-temporal domain. Then normalized combinational vectors, named HNF, are computed around the detected space-time interest points to characterize the video. After that, we utilize the Laplacian sparse representation framework and maximum pooling method to gain a more discriminative feature vector from the HNF set. Finally, the support vector machine (SVM) is adopted to classify the feature vector as normal or abnormal. Experiments on two datasets demonstrate the satisfactory performance of the proposed approach.

Keywords: Abnormal activity detection · Space-time interest points · Laplacian sparse representation · Maximum pooling · SVM

1 Introduction

In surveillance videos, the abnormal activities can be defined as aberrant events such as people fights, crowded escape activities. Thus, the abnormal activity detection is to identify these aberrant events from normal ones, which is a two-class classification problem. This problem is divided into the detection of global abnormal activity (GAA) and local abnormal activity (LAA). For GAA, there is only normal activity or abnormal activity at the same time in the entire scenario. For LAA, normal and abnormal activities emerge simultaneously in the scenario.

In some existing surveys, the global features are used to describe video clips for GAA. In [8], the streakline representation based on Lagrangian framework for fluid dynamics is utilized to detect abnormal activity. In [9], Mehran et al.

© Springer International Publishing Switzerland 2015
S. Arik et al. (Eds.): ICONIP 2015, Part IV, LNCS 9492, pp. 410–418, 2015.
DOI: 10.1007/978-3-319-26561-2_49

use the social force model [4] to analyze the human activity. The description of human activity in social force model is based on the intention of movement.

On the other hand, some surveys [10,11] utilize the Bag-of-Words (BoW) framework to tackle both GAA and LAA problem. The BoW model consists of three modules: (1) feature extraction; (2) codebook producing and feature quantization; (3) classification.

For the feature extraction, [10,11] use the local spatio-temporal feature to characterize video clips. They first detect the space-time interest points (STIP) [5] of video clips. Then spatio-temporal descriptors such as Histograms of Oriented Gradients (HOG) and Histograms of Optical Flow (HOF) are computed respectively around the detected STIP to describe video clips. However, HOG or HOF is not discriminative enough to encode the information of appearance and action. In [6], the HOG and HOF descriptors are combined into normalized vectors for the recognition of human actions. The combinational feature vectors of HOG and HOF, named HNF, are discriminative enough to describe the appearance and action in video clips. Note that detecting abnormal activity also needs to encode the appearance and action information. Therefore, the HNF is also a discriminative descriptor for the abnormal activity detection.

For the second module, BoW model first utilizes the K-means clustering to produce the codebook which contains several visual words. Then each feature is only assigned to its nearest visual word to generate a frequency histogram of visual words, which is called the feature quantization. However, the information loss [1] in feature quantization is severe because of the hard assignment approach used in BoW model. In order to reduce the information loss, [12] adopts the general sparse coding to generate the codebook and quantize features in abnormal activity detection, which achieves better performance than BoW model.

However, the general sparse coding disposes features separately, ignoring the similarity among features, which decreases the accuracy and robustness of sparse representation. In order to tackle this problem, Gao et al. [3] propose the Laplacian Sparse Representation (LSR) approach for image classification. In the LSR approach, the similarity matrix is used to preserve the similarity information among features, which can further reduce the feature quantization error and make the sparse representation more robust. Thus, we can also utilize the LSR

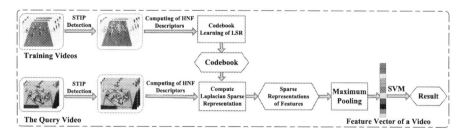

Fig. 1. The main framework of our approach

approach to learn a better codebook and produce a more descriptive video representation for abnormal activity detection.

As mentioned above, we propose an approach using the spatio-temporal feature and the Laplacian sparse representation for both GAA and LAA problem. The main framework is summarized in Fig. 1. Firstly, we detect the space-time interest points of a query video. Then the HNF descriptors are computed in the nearby 3D patches of the detected interest points. After that, the sparse representations of HNF descriptors are generated by applying the Laplacian sparse coding. Then we utilize the maximum pooling method among the entire Laplacian sparse representation set of the query video to obtain a more descriptive feature vector. Finally, we use the SVM to classify this feature vector and determine whether the query video includes the abnormal activity.

The rest of this paper is organized as follows: we present the details of our approach in Sect. 2. In Sect. 3, we report the experimental results and show the comparisons of different methods. We give the conclusion in Sect. 4.

2   Methodology

2.1   Spatio-Temporal Feature

In order to detect the abnormal activity, we first extract the spatio-temporal features from a query video, which contains two procedures. The Fig. 2 illustrates the extraction of spatio-temporal features.

Detection of Space-Time Interest Points. Detecting interest points in a video clip is to locate the local regions with intense variation of image intensity in the spatio-temporal domain. In this paper, we utilize the approach proposed by [5] to solve this problem.

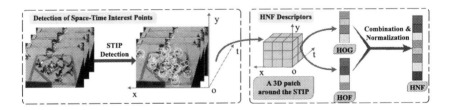

Fig. 2. The illustration of extracting spatio-temporal features

Firstly, let $f_{st} \colon \mathbb{R}^2 \times \mathbb{R} \mapsto \mathbb{R}$ represent a video clip. Then, we convert the f_{st} into linear scale space by applying the convolution of f_{st} with an anisotropic Gaussian kernel. After that, we compute the second moment matrix with respect to the linear scale-space representation of f_{st}. Then we integrate the second moment matrix with an Gaussian kernel to obtain a matrix Γ_{st}. Eigenvalues with large values of the matrix Γ_{st} indicate the local bricks which have intense variation of image intensity. Thus, following the work [5], the extended Harris corner function H_{st} with respect to the eigenvalues of Γ_{st} is introduced to tackle the problem.

$$H_{st} = \lambda_1 \lambda_2 \lambda_3 - \xi \cdot (\lambda_1 + \lambda_2 + \lambda_3)^3 = \lambda_1^3(\omega\eta - \xi \cdot (1 + \omega + \eta)^3) \qquad (1)$$

where $\lambda_1, \lambda_2, \lambda_3$ denote the eigenvalues of matrix Γ_{st}, $\omega = \lambda_2/\lambda_1$, $\eta = \lambda_3/\lambda_1$, $\xi \leq \omega\eta/(1 + \omega + \eta)^3$, and $H_{st} \geq 0$. We can detect space-time interest points by finding the local positive maximum values of H_{st}.

HNF Descriptors for STIP. After the detection of interest points, we use the HNF descriptors [6] to characterize the video. Firstly, Histograms of Oriented Gradient (HOG) and Histograms of Optical Flow (HOF) are computed respectively in the nearby 3D patches of detected interest points. Then, to obtain the HNF descriptors, we combine the HOG and HOF into vectors and normalize them. As shown in Fig. 2, the 3D video patch is divided into a grid with $3 \times 3 \times 2$ spatio-temporal bricks. In each brick, there are 4-bin HOG descriptors and 5-bin HOF descriptors. Therefore, one HNF descriptor vector has 162 dimensions, including 72 elements from HOG and 90 elements from HOF. With the combination of HOG and HOF, the HNF descriptor can provide more information of appearance and action.

2.2 Laplacian Sparse Representation for Spatio-Temporal Feature

In order to produce a more descriptive and precise representation for the appearance and action of a video clip, we utilize the Laplacian Sparse Representation (LSR) method [3] to encode spatio-temporal feature vectors. In the LSR method, the matrix S is used to preserve the similarity information among features and a regularization term with respect to the similarity matrix S is applied to improve the robustness and accuracy in feature quantization.

Let $X = [x_1, x_2, \ldots, x_N] \in \mathbb{R}^{d \times N}$ denote the spatio-temporal features set, and let $H = [h_1, h_2, \ldots, h_K] \in \mathbb{R}^{d \times K}$ be the codebook learned by LSR. Then the problem of LSR is formulated as follows:

$$\min_{H,M} \|X - HM\|_F^2 + \lambda \sum_i \|m_i\|_1 + \frac{\beta}{2} \sum_{ij} \|m_i - m_j\|^2 S_{ij}$$
$$= \min_{H,M} \|X - HM\|_F^2 + \lambda \sum_i \|m_i\|_1 + \beta \cdot trace(MLM^T) \qquad (2)$$
$$s.t. \ \|h_m\|^2 \leq 1$$

where the m_i denotes the Laplacian sparse representation of the i-th feature vector x_i and $M = [m_1, m_2, \ldots, m_N] \in \mathbb{R}^{K \times N}$. The λ is a regularization parameter with respect to the sparsity and β is the similarity constraint. The Laplacian matrix is $L = D - S$ and the $D_{ii} = \sum_j S_{ij}$ denotes the degree of the i-th node. We use the histogram intersection $I(x_i, x_j)$ to compute the similarity matrix S.

$$I(x_i, x_j) = \sum_{l=1}^{d} \min(x_{i_l}, x_{j_l}) \tag{3}$$

where x_{i_l} is the l-th entry of the d dimensional feature vector x_i. Let $S_{ij} = S_{ji} = I(x_i, x_j)$ if x_i is the k nearest neighbor of x_j, where $i \neq j$ and $k = 5$. Otherwise let $S_{ij} = 0$.

The problem defined by Eq. (2) is not convex if we optimize H and M concurrently. However, when one of H and M is fixed, the optimization for the other one is convex. Thus, H and M are optimized alternately following the work [3].

Firstly, we fix the codebook H, then each element in the sparse representation set M can be optimized individually. If we want to compute each m_i, the other sparse representations m_j, where $j \neq i$, should be fixed. Then the feature sign search algorithm [7] is used to solve the following optimization to produce m_i:

$$\min_{m_i} \|x_i - Hm_i\|^2 + \beta(m_i^T(ML_i) + (ML_i)^T m_i - m_i^T L_{ii} m_i) + \lambda \|m_i\|_1 \tag{4}$$

where the i-th column of the Laplacian matrix L is L_i, and the (i,i)-th element of L is L_{ii}. M is the initialized sparse representation set and should be updated after the optimization of m_i, where m_i denotes the i-th column of M.

When we fix the sparse representation set M, the learning of codebook H can be defined as:

$$\min_{H} \|X - HM\|_F^2 \qquad s.t. \ \|h_m\|^2 \leq 1 \tag{5}$$

The Lagrange dual proposed in [7] is utilized to solve the Eq. (5).

In our approach, we randomly extract some features to learn the codebook H by optimizing Eqs. (4) and (5) iteratively, which is an offline learning task. After we obtain the codebook H, we only need to solve Eq. (4) to learn the sparse representation for each feature in the spatio-temporal features set X.

2.3   Maximum Pooling

Through the Laplacian sparse representation method, we can obtain a set of sparse representations of a query video, denoted by $\{m_1, m_2, \ldots, m_n\}$. Then we utilize the maximum pooling over this sparse representation set to obtain a K dimensional feature vector ρ which is used to characterize the video. Let ρ_j denote the j-th element of ρ, then the maximum pooling can be defined as:

$$\rho_j = \max\{|m_{1_j}|, |m_{2_j}|, \ldots, |m_{n_j}|\}, j \in \{1, \ldots, K\} \tag{6}$$

where m_{i_j} is the j-th element of the vector m_i. Each column of the codebook H represents a basic pattern of the feature space. With the maximum pooling method, we reserve the strongest response to each basic pattern and produce a feature vector with K dimensions for a query video, which also reduces the influence of irrespective information and improves the robustness of our approach.

2.4   Abnormal Activity Detection via SVM

The detection of abnormal activity in a video is a two-class classification problem. In our approach, we utilize the support vector machine (SVM) [2] with a radial basis function (RBF) kernel to solve this problem. The kernel trick used in SVM can map the linearly inseparable features into the high-dimensional space, which can make features linearly separable in the new space. After we obtain the feature vector ρ, we use the SVM to classify ρ as normal or abnormal.

3   Experiments

In this section, we conduct experiments on the UMN dataset [9] and the Hockey dataset [10] to test our approach.

3.1   Experimental Results on the UMN Dataset

This dataset contains 7740 frames with the crowded escape activities in 3 scenes. And the resolution is 320×240 pixels. We cut the whole video into 258 clips, including 196 normal video clips and 62 abnormal video clips. All video clips are divided into 5 subsets to test our approach for the global abnormal activity detection with 5-fold cross validation. In addition, the size of codebook in this experiment is set to 1000.

We compare our approach with three different methods on the UMN dataset, including social force model [9], optical flow [9], and streakline representation method [8]. The ROC curves are shown in Fig. 3 and quantitative comparisons are reported with the area under ROC curve (AUC) in Table 1.

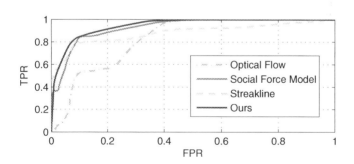

Fig. 3. The ROC curves on the UMN dataset

As reported in Table 1, the AUC of our approach is 0.971, which outperforms the optical flow method, streakline representation method and social force model. Experimental results on the UMN dataset have proved that our approach can perform effectively in the global abnormal activity detection.

Table 1. The performance comparisons on the UMN dataset with AUC

Method	Social Force Model [9]	Optical Flow [9]	Streakline [8]	Ours
AUC	0.96	0.84	0.90	0.971

We analyze the influence on our approach of parameters λ and β in Eq.(2). With the increase of λ, the sparse representation of feature is more sparse. We set λ among $0.3 \sim 0.4$ in the UMN dataset and find that $\lambda = 0.32$ performs best. The parameter β influences the similarity constraint in sparse representation generation. We set β among $0.1 \sim 0.3$ and find that our approach achieves good performance when $\beta = 0.1$.

3.2 Experimental Results on the Hockey Dataset

The Hockey dataset is composed of 1000 video clips with a resolution of 360×288 pixels, including 500 clips with fight events among athletes and 500 normal clips. Each video clip contains 50 frames. We test the proposed approach for the local abnormal activity detection on this dataset with 5-fold cross validation.

We present the comparisons among our approach, methods based on BoW model [10] and method using general sparse coding [12] in Table 2.

Table 2. The comparisons on the Hockey dataset with 5-fold cross validation

Visual Words	HOG with BoW [10]	HOF with BoW [10]	MoSIFT with BoW [10]	MoSIFT with sparse coding [12]		Ours	
	Accuracy	Accuracy	Accuracy	Accuracy	AUC	Accuracy	AUC
50	87.8%	83.5%	87.5%	90.9%	0.951	91.8%	0.955
100	89.1%	84.3%	89.4%	92.6%	0.958	93.0%	0.962
150	89.7%	85.9%	89.5%	93.4%	0.963	93.9%	0.966
200	89.4%	87.5%	90.4%	94.1%	0.971	94.7%	0.974
300	90.8%	87.2%	90.4%	94.1%	0.968	94.8%	0.976
500	91.4%	87.4%	90.5%	94.3%	0.971	95.1%	0.979
1000	91.7%	88.6%	90.9%	94.0%	0.967	95.3%	0.978

According to Table 2, the approach using general sparse coding [12] gains a higher prediction accuracy than approaches based on BoW model, which indicates that the sparse coding method achieves better performance than BoW model in the feature quantization. Our approach obtains a higher prediction accuracy than the other approaches, which indicates that HNF descriptor is discriminative. Furthermore, the AUC of our approach is larger than approach using general sparse coding, which indicates that Laplacian sparse representation method produces less feature quantization error because of the preserving of similarity among features. The experimental results indicate that our approach has achieved promising performance in the detection of local abnormal activity.

We also apply the same parameter analysis as the UMN dataset on this dataset. We find that our approach performs best when $\lambda = 0.3, \beta = 0.1$.

4   Conclusion

In this paper, we propose an approach based on spatio-temporal feature and Laplacian sparse representation to detect abnormal activity in a video. We use the HNF descriptor to characterize the appearance and action of a query video after the detection of space-time interest points. Then the Laplacian sparse representation and maximum pooling method are applied to obtain a more descriptive feature vector. With the introduction of similarity matrix in the LSR, we preserve the similarity among spatio-temporal features, which improves the accuracy and robustness in feature quantization. Experimental results on the UMN dataset and the Hockey dataset demonstrate that our approach can achieve satisfactory performance in the detection of both global abnormal activity and local abnormal activity.

Ackonwledgements. This research is partly supported by NSFC, China (No: 61273258) and 863 Plan, China (No. 2015AA042308).

References

1. Boiman, O., Shechtman, E., Irani, M.: In defense of nearest-neighbor based image classification. In: IEEE Conference on Computer Vision and Pattern Recognition, 2008. CVPR 2008, pp. 1–8. IEEE (2008)
2. Chang, C.C., Lin, C.J.: Libsvm: a library for support vector machines. ACM Trans. Intell. Syst. Technol. (TIST) **2**(3), 27 (2011)
3. Gao, S., Tsang, I.W., Chia, L.T., Zhao, P.: Local features are not lonely-laplacian sparse coding for image classification. In: 2010 IEEE Conference on Computer Vision and Pattern Recognition (CVPR), pp. 3555–3561. IEEE (2010)
4. Helbing, D., Molnar, P.: Social force model for pedestrian dynamics. Phys. Rev. E **51**(5), 4282 (1995)
5. Laptev, I.: On space-time interest points. Int. J. Comput. Vis. **64**(2–3), 107–123 (2005)
6. Laptev, I., Marszalek, M., Schmid, C., Rozenfeld, B.: Learning realistic human actions from movies. In: IEEE Conference on Computer Vision and Pattern Recognition, 2008. CVPR 2008, pp. 1–8. IEEE (2008)
7. Lee, H., Battle, A., Raina, R., Ng, A.Y.: Efficient sparse coding algorithms. In: Advances in neural information processing systems, pp. 801–808 (2006)
8. Mehran, R., Moore, B.E., Shah, M.: A streakline representation of flow in crowded scenes. In: Daniilidis, K., Maragos, P., Paragios, N. (eds.) ECCV 2010, Part III. LNCS, vol. 6313, pp. 439–452. Springer, Heidelberg (2010)

9. Mehran, R., Oyama, A., Shah, M.: Abnormal crowd behavior detection using social force model. In: IEEE Conference on Computer Vision and Pattern Recognition, 2009. CVPR 2009, pp. 935–942. IEEE (2009)
10. Bermejo Nievas, E., Deniz Suarez, O., Bueno García, G., Sukthankar, R.: Violence detection in video using computer vision techniques. In: Real, P., Diaz-Pernil, D., Molina-Abril, H., Berciano, A., Kropatsch, W. (eds.) CAIP 2011, Part II. LNCS, vol. 6855, pp. 332–339. Springer, Heidelberg (2011)
11. de Souza, F.D.M., Chávez, G.C., do Valle, E., Araujo, D.A., et al.: Violence detection in video using spatio-temporal features. In: 2010 23rd SIBGRAPI Conference on Graphics, Patterns and Images (SIBGRAPI), pp. 224–230. IEEE (2010)
12. Xu, L., Gong, C., Yang, J., Wu, Q., Yao, L.: Violent video detection based on mosift feature and sparse coding. In: 2014 IEEE International Conference on Acoustics, Speech and Signal Processing (ICASSP), pp. 3538–3542. IEEE (2014)

Noise-Robust Iris Authentication Using Local Higher-Order Moment Kernels

Keisuke Kameyama[1]([✉]), Trung Nguyen Bao Phan[2], and Miharu Aizawa[2]

[1] Faculty of Engineering, Information and Systems,
University of Tsukuba, Tsukuba, Japan
Keisuke.Kameyama@cs.tsukuba.ac.jp
[2] Graduate School of Systems and Information Engineering, University of Tsukuba,
1-1-1 Tennodai, Tsukuba, Ibaraki 305-8573, Japan

Abstract. A novel biometric authentication method using kernel functions of higher-order statistical feature of the iris texture is introduced. When the observed iris images include noise, direct estimation and use of Gabor and local higher-order moment (LHOM) features for iris code generation suffers from performance degradation. In order to solve this issue, we propose to use the LHOM kernel function of pairs of local textures on a single iris image. In the experiments, the proposed method using LHOM kernels of orders 2 to 6 proved to be significantly robust against noise when compared with the conventional method.

Keywords: Iris · Biometrics · Autocorrelation · Higher-order statistics · Kernel

1 Introduction

Due to an ever increasing need for personal identification for secure use of services, biometric authentication (BA) is becoming the means of choice. Among the various BA modalities, iris authentication system has attracted a lot of attention due to its distinctive features [9]. Since the 1990s, various feature extraction methods based on Gabor transform, orthogonal wavelets, biorthogonal wavelets, Hilbert transform and multi-resolution analysis of textural features have been sought [1,2] for further conversion to the iris code or the feature vector to represent the person. In this scheme, the stability and the robustness of the transformation from the iris image to the iris code, under various conditions of use, is the key to a successful implementation of the authentication system.

The Gabor wavelet-based characterization equivalent to utilizing the local autocorrelation feature was employed in the pioneering work of Daugman [2]. Direct use of local features will certainly be affected by the existence of noise, which will result in a different iris code. However, when the change in the observation is uniform (almost the same across the iris region) the relative relation between the local features extracted from two regions can remain stable.

© Springer International Publishing Switzerland 2015
S. Arik et al. (Eds.): ICONIP 2015, Part IV, LNCS 9492, pp. 419–427, 2015.
DOI: 10.1007/978-3-319-26561-2_50

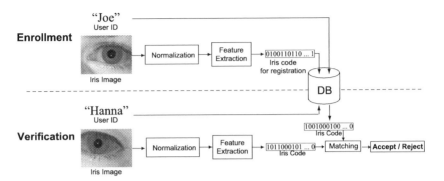

Fig. 1. The steps of enrollment and verification in the iris authentication system.

So far, the authors have been evaluating the utility of local higher-order moment (autocorrelation) features in texture classification, especially by way of using Local Higher-Order Moment (LHOM) kernels [3,4].

In this work, we propose a method for iris code generation based on LHOM kernel function of local textures at two different positions. The contribution is the confirmation of the fact that the kernel-based method is much robust against noise in the situation of authentication using noisy iris images. The low computational cost of iris code generation by the so-called kernel trick is also a merit.

This paper is organized as follows. In Sect. 2, the iris authentication scheme utilizing higher-order moment kernels is proposed. In Sect. 3, the authentication performances of the proposed method will be evaluated, and the paper will be concluded in Sect. 4.

2   Iris Authentication Using Local Higher-Order Moment Kernels

The iris authentication scheme used in this work follows the basic steps employed in the pioneering work by Daugman [2]. The procedures of enrolling and verifying the user is shown in Fig. 1. A user will be enrolled by submitting the iris image together with the user ID. The iris image will be normalized and the features of the texture will be extracted from the predetermined positions. Then, the features will be converted to a binary code which will be kept in the database paired with the user ID. Upon verification, the observed person's iris image will go through the same steps of normalization, feature extraction, and code generation. This code will be matched against the registered code of the user of the claimed user ID. Finally the decision to accept or reject this person will be made according to the similarity between the iris codes of registered and the observed iris images.

2.1 Iris Detection and Normalization

The iris within the observed image is detected by finding the two near-concentric-circular edges, namely the two borders between the pupil, the iris and the scelera. The search for the most likely center coordinate and the radius introduced in [2] is used in this work. After detecting the two circular edges, the iris region is nonlinearly converted to a rectangular image with radial (r) and angular (θ) coordinates. This conversion will normalize the iris region regardless of the pupilar constriction upon observation [2] as shown in Fig. 2.

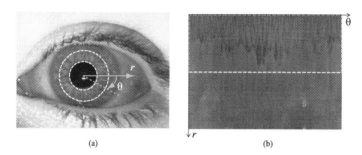

(a)   (b)

Fig. 2. Normalization of the iris region. (a) Observed image of the eye. (b) Normalized iris image. The texture in the region marked in red is used for feature extraction.

2.2 Local Higher-Order Moment Feature and the Kernel Function

After normalization, the features of the iris texture will be extracted from the inner regions (close to the pupil), marked in red in Fig. 2.

Let $s(t)$ be a real valued image signal defined on R^2. For $n \geq 2$, the n-th order local moment (or the $n - 1$ order local autocorrelation) [4,6] of $s(t)$ is defined as

$$m_{s,w,n}(\tau_1, \ldots, \tau_{n-1}, x) = \int_{R^2} w(t)s(t+x) \prod_{k=1}^{n-1} \{w(t+\tau_k)s(t+x+\tau_k)\}dt. \quad (1)$$

This is a higher-order generalization of the autocorrelation function with a window for characterizing the local signal. The Fourier transform of the n-th order local moment

$$M_{s,w,n}(\Omega, x) = \int_{R^{2(n-1)}} m_{s,w,n}(T, x) \exp(-j\Omega^T T)dT \quad (2)$$

with $\Omega = [\omega_1^T \omega_2^T \ldots \omega_{n-1}^T]^T$ and $T = [\tau_1^T \tau_2^T \ldots \tau_{n-1}^T]^T$ may also be used. The choice of a Gaussian window for $n = 2$ will give the local power spectrum of the signal obtained by an array of Gabor filters. Moments and moment spectra

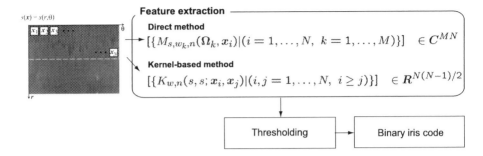

Fig. 3. Direct (conventional) and kernel-based (proposed) methods of iris code generation from normalized iris image.

of $n > 3$ will be referred to as local higher-order moments (LHOM) and local higher-order moment spectra (LHOMS).

In [2], the real and imaginary components of the complex response of the Gabor filter, at different positions on the normalized iris image for various central frequencies (ω) in the angular direction (θ) were used in the "direct" method as described in the following. However, direct estimation of LHOM or LHOMS feature can be computationally demanding.

In [5], McLaughlin and Raghu showed that inner product of moment functions of arbitrary order can be calculated without a drastic increase of computation. This achievement amounts to the finding of the kernel function of the higher-order moment (autocorrelation) function. The kernel function corresponding to the local moments (LHOM kernel [3]) is,

$$
\begin{aligned}
K_{w,n}&(s, v \; ; \; \boldsymbol{x}, \boldsymbol{y}) \\
&= \langle m_{s,w,n}(\boldsymbol{\omega}_1, \ldots, \boldsymbol{\omega}_{n-1}, \boldsymbol{x}), m_{v,w,n}(\boldsymbol{\omega}_1, \ldots, \boldsymbol{\omega}_{n-1}, \boldsymbol{y}) \rangle \\
&= (2\pi)^{-2(n-1)} \langle M_{s,w,n}(\boldsymbol{\omega}_1, \ldots, \boldsymbol{\omega}_{n-1}, \boldsymbol{x}), M_{v,w,n}(\boldsymbol{\omega}_1, \ldots, \boldsymbol{\omega}_{n-1}, \boldsymbol{y}) \rangle \\
&= \int \left[\int w(\boldsymbol{z}) s(\boldsymbol{z} + \boldsymbol{x}) w(\boldsymbol{z} + \boldsymbol{\tau}) v(\boldsymbol{z} + \boldsymbol{y} + \boldsymbol{\tau}) d\boldsymbol{z} \right]^n d\boldsymbol{\tau}.
\end{aligned}
\tag{3}
$$

This kernel function will give the relation between LHOM of $s(\boldsymbol{t})$ and $v(\boldsymbol{t})$ localized at $\boldsymbol{x}$ and $\boldsymbol{y}$, respectively.

2.3   Iris Code Generation and Matching

In Fig. 3, the conventional and proposed methods of iris code generation are depicted.

Direct Method. In the conventional method, LHOMS of predetermined N positions on the normalized iris image $\boldsymbol{x}_i$ $(i = 1, \ldots, N)$ are extracted. At each position, features for M different combinations of Gaussian window shapes and frequencies are extracted to make a MN-dimensional feature vector.

$$[\{M_{s,w_k,n}(\boldsymbol{\Omega}_k, \boldsymbol{x}_i)|(i = 1, \ldots, N, \ k = 1, \ldots, M)\}] \ \in \boldsymbol{C}^{MN}. \tag{4}$$

The vector components are further thresholded at θ_D to make a binary iris code. For $n > 3$ where LHOMS values are complex, thresholding is applied to real and imaginary components separately, thereby making the code length $2MN$.

Kernel-based Method. In the proposed method, values of the discrete version of the kernel $K_{w,n}$ [3] defined as

$$K'_{w,n}(s, v \ ; \ \boldsymbol{x}, \boldsymbol{y}) = \sum_i \left[\sum_j w(\boldsymbol{z}_j) s(\boldsymbol{z}_j + \boldsymbol{x}) w(\boldsymbol{z}_j + \boldsymbol{\tau}_i) v(\boldsymbol{z}_j + \boldsymbol{y} + \boldsymbol{\tau}_i) \right]^n \tag{5}$$

is evaluated for nonoverlapping pairs of positions $(\boldsymbol{x}_i, \boldsymbol{x}_j)$ from the selected N points as,

$$[\{K'_{w,n}(s, s; \boldsymbol{x}_i, \boldsymbol{x}_j)|(i, j = 1, \ldots, N, \ i \geq j)\}] \ \in \boldsymbol{R}^{N(N-1)/2}. \tag{6}$$

The real vector components are further thresholded at θ_K to make a binary code of $N(N-1)/2$ bits.

Matching. Upon verification of a user the registered and the newly observed iris codes will be compared to make an accept/reject decision. The normalized Hamming distance between two codes $\boldsymbol{a} = [a_1 \ldots a_L]$ and $\boldsymbol{b} = [b_1 \ldots b_L]$

$$d_H(\boldsymbol{a}, \boldsymbol{b}) = \frac{1}{L} \sum_{j=1}^{L} a_j(XOR)b_j \qquad (0 \leq d_H \leq 1) \tag{7}$$

is used as the measure of dissimilarity.

3   Experiments

Here, the performance of the direct and kernel-based iris authentication are evaluated. The following are the common experimental conditions.

Dataset: Iris images from the Bath Iris Image Database [8] were used. The database contains 1000 images from 50 eyes with 20 images taken from each eye. The dimension of all images is 1280×960 (pixel).

Preprocessing: Normalization was conducted following the Daugman's method as mentioned in Sect. 2. The database contained some iris images obtained from half-closed eyes in which the outer boundaries was indistinguishable. Thus, after removing those images that can cause improper localization, a set of properly normalized iris patterns were obtained. The dimension of the normalized iris image was 360×300 (pixel).

Table 1. The combinations of central frequencies used for LHOMS estimation.

Order n	Frequency component set	Code length
2	$\{(u_1,\, u_2,\, u_3)\}$	576
3	$\{(u_1,u_2),\, (u_2,u_3),\, (u_1,u_3)\}$	1152
4	$\{(u_1,u_2,u_3),\, (u_2,u_3,u_4),\, (u_1,u_3,u_4)\}$	1152
5	$\{(u_1,u_2,u_3,u_4),\, (u_2,u_3,u_4,u_5),\, (u_1,u_3,u_4,u_5)\}$	1152
6	$\{(u_1,u_2,u_3,u_4,u_5),\, (u_2,u_3,u_4,u_5,u_6),\, (u_1,u_3,u_4,u_5,u_6)\}$	1152

Feature Points: $N = 192$ regions centered at 4×48 points along the radial (r) and angular (θ) axes were selected to generate an iris code. The points were selected from the region near the pupil in order to avoid obscuration by the eyelid and eyelashes.

Performance Measure: For each condition of iris code generation, normalized Hamming distance d_H in Eq. (7) was evaluated for all possible pairs of normalized iris images. The verification performance under certain decision rule (thresholding on d_H axis) was evaluated from this dataset.

The tradeoff between False Acceptance Rate (FAR) and False Rejection Rate (FRR) was evaluated as the acceptance/rejection threshold used against d_H was varied in the $[0, 1]$ range. The Equal Error Rate (EER) which is the error rate of FAR=FRR on the FAR-FRR curve was also used.

3.1   Experiment 1: Direct Use of Local Higher-Order Spectral Features

The LHOMS $M_{s,w,n}$ in Eqs. (2) and (4) of orders $n = 2, 3, 4, 5, 6$ were used for feature extraction from selected $N = 192$ regions on the normalized iris image. A 2-dimensional Gaussian window with radial and angular width of 15 and 5 pixels respectively, were used. The central spatial frequencies in the angular (θ) direction were chosen from $(u_1, u_2, u_3, u_4, u_5, u_6) = (0.03, 0.06, 0.09, 0.12, 0.15, 0.18)$ $(pixel^{-1})$. Relation $\omega_i = 2\pi u_i$ $(radian/pixel)$ holds between the angular frequency mentioned in Eq. (2). The combinations of the frequency components and the iris code lengths are shown in Table 1.

Results: The FAR-FRR curves for LHOMS orders of $n = 2, \ldots, 6$ are shown in Fig. 4(a). It is found that lower orders perform better in direct LHOMS use. This result is compatible with the finding that higher-order features become more sensitive to small changes in [4].

3.2   Experiment 2: LHOM Kernel-based Method

The LHOM kernel $K_{w,n}$ in Eq. (5) for all possible pairs of local textures $(\boldsymbol{x}_i, \boldsymbol{x}_j)$ $(i, j = 1, \ldots, 192)$ of orders $n = 2, \ldots, 6$ respectively, were used for code generation (Eq. (6)).

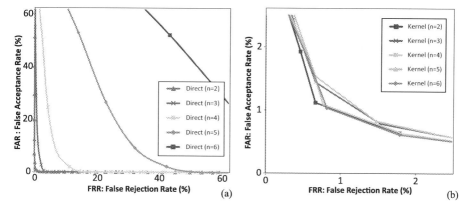

Fig. 4. FAR-FRR curves when using the direct method for orders $n = 2 \sim 6$. (a) Direct method. (b) Kernel-based method.

Here, the local texture used at each point is 31×31 (pixel) in size. The maximum shifting vector in Eq. (5) was set to $\tau = (15, 15)$. A Gaussian window having radial and angular width of 15 and 5 pixels were used. The iris code length was 18528 ($= 192 \times (192 + 1)/2$).

Results: The FAR-FRR curves for kernel orders of $n = 2, \ldots, 6$ are shown in Fig. 4(b). A novel finding here is that the performance is nearly constant for all orders, with a slight advantage when the even order ($n = 2, 4, 6$) kernels were used.

3.3 Experiment 3: Verification Using Noisy Data

Here, robustness of the direct and kernel-based code generation methods were evaluated by using noisy data in verification. Gaussian noise was added to the original image up to the specified signal-to-noise ratio (SNR). Noisy data having SNR values of 30, 10, 0 and -10 (dB) were used. On evaluation, normalized Hamming distances between iris codes of noisy and original data were calculated.

In addition to LHOM features, Local Binary Pattern (LBP)[7] feature was used for comparison. 256-bin LBP histograms of local texture of the same point set were concatenated to make a feature vector as the method corresponding to the direct scheme. Inner products of the two vectors were used as an equivalent method to the kernel-based scheme. The Euclidean distance was used as the dissimilarity measure between LBP codes.

Results: The Equal Error Rate (EER) for direct and kernel-based methods for orders $n = 2, 3$ are shown in Fig. 5. It is clear that the kernel-based method is much more robust against noise than the direct method as the EER is maintained low even in very noisy situations.

LBP feature turned out to be was far weaker against noise. However, when the images for registration and verification were both contaminated with noise, kernel-based method for LBP features also showed better EER to the direct method ($EER_{kernel-LBP} = 6\%$ vs $EER_{direct-LBP} = 10\%$ at SNR=0dB).

Discussion: This result shows that the LHOM kernel values remain largely unchanged under additive Gaussian noise. As the performance of the direct method degrades with the amount of noise, the local LHOM(S) feature values do change. However, this change occurs in the way that the inner product between the two features are largely maintained.

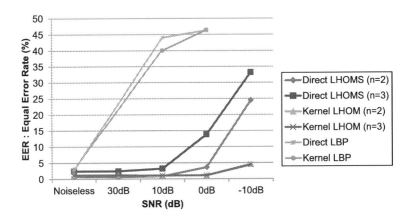

Fig. 5. Comparison of Equal Error Rate (EER) for LHOMS ($n = 2$),LHOMS ($n = 3$), kernel matrix ($n = 2$) and kernel matrix ($n = 3$) for different levels of noise

4   Conclusion

A novel scheme of iris code generation which is based on LHOM kernel was proposed. It proved to be robust against additive noise in the verification image, which enhances the usability of iris authentication systems. A deeper investigation on the mechanism of the kernel response robustness is underway.

References

1. Czajka, A., Strzelczyk, P.: Iris recognition with compact zero-crossing-based coding. In: Proceedings of SPIE 6347 (2006)
2. Daugman, J.G.: High confidence visual recognition of persons by a test of statistical independence. IEEE Trans. PAMI **15**(11), 1148–1161 (1993)
3. Kameyama, K.: Comparison of local higher-order moment kernel and conventional kernels in SVM for texture classification. In: Ishikawa, M., Doya, K., Miyamoto, H., Yamakawa, T. (eds.) ICONIP 2007, Part I. LNCS, vol. 4984, pp. 851–860. Springer, Heidelberg (2008)

4. Kameyama, K., Phan, T.N.B.: Image feature extraction and similarity evaluation using kernels for higher-order local autocorrelation. In: Lee, M., Hirose, A., Hou, Z.-G., Kil, R.M. (eds.) ICONIP 2013, Part III. LNCS, vol. 8228, pp. 442–449. Springer, Heidelberg (2013)
5. MacLaughlin, J.A., Raviv, J.: N-th autocorrelations in pattern recognition. Inf. Control **12**(2), 121–142 (1968)
6. Nikias, C.L., Petropulu, A.P.: Higher-Order Spectra Analysis - A Nonlinear Signal Processing Framework. Prentice Hall, Englewood Cliffs (1993)
7. Ojala, T., Pietikäinen, M., Harwood, D.: A comparative study of texture measures with classification based on feature distributions. Pattern Recognit. **19**(3), 51–59 (1996)
8. University of Bath, SmartSensors Inc.: Bath iris database. http://www.smart sensors.co.uk/products/iris-database/1000-samples/
9. Wildes, R.P.: Iris recognition : An emerging biometric technology. Proc. IEEE **85**(9), 1348–1363 (1999)

Myoelectric Control System of Lower Limb Exoskeleton for Re-training Motion Deficiencies

Vasily Mironov[✉], Sergey Lobov, Innokentiy Kastalskiy, and Victor Kazantsev

Laboratory of Intelligent Biomechatronic Technologies,
Lobachevsky State University of Nizhny Novgorod, Gagarin Avenue 23,
603950 Nizhny Novgorod, Russia
mironov@neuro.nnov.ru

Abstract. Technical progress in robotics has led to its expansion into different spheres of human life. Rehabilitation medicine has become one of those areas, actively adopting robotic principles and achievements. Over the last decade the progress in this area is inseparably linked with the development of robotic devices - exoskeletons designed to compensate lower limb disability arising as a result of brain or spinal cord injuries and diseases. In this paper, we propose a new control approach for robotic exoskeleton which integrates muscle bioelectric signals of the patient in the control loop. We believe that the proposed approach activates biofeedback mechanisms, which in turn intensify the rehabilitation process.

Keywords: Human-machine interface · Electromyography · Signal processing · Pattern recognition · Exoskeleton control

1 Introduction

The active implementation of human-machine interfaces, employing bioinformatic indicators for control purposes can promote the revolutionary changes in various technical fields. In particular, the application of such systems opens up new horizons in rehabilitation medicine. For example, such systems can be used in exoskeletons intend to restore lower limb motor function disturbed due to the brain injuries and diseases.

The most of modern exoskeletons are not focused on active human participation in the process of committing movements. The widely used approach involves the regular execution of a fixed motor pattern, adapted for the individual physiological characteristics of the patient. Such robotic exoskeletons (i.e., without the active participation of the patient) performing limbs passive movement are ReWalk[ref], eLegs[ref], Indego (exoskeleton of Vanderbilt University)[ref] and Rex[ref].

Another paradigm that is gaining popularity among researchers, involves the incorporation of the exoskeleton pilot in the device control loop. The strategy mentioned seems to be more promising in the context of the patient rehabilitation. The reason for this is the activation of so-called brain plasticity mechanisms.

S. Arik et al. (Eds.): ICONIP 2015, Part IV, LNCS 9492, pp. 428–435, 2015.
DOI: 10.1007/978-3-319-26561-2_51

This concept involves the registration of human bioelectric signals (e.g., brain or muscle activity), their interpretation in device control commands and making reinforcing movements. Thus, this leads the activation of afferent pathway and stimulation of the brain motor area performing reinforcement of structural changes.

One such device, demonstrating the benefit of using bioelectric signals in control scheme is the MindWalker exoskeleton[ref]. This device is intended for patients endured spinal cord injury, with consequent loss of the conduction of impulses travelling between brain and the peripheral nervous system. As a result, the person loses the ability of make limb movements. However, using EEG brain signals, MindWalker exoskeleton can realize some repertoire of movements controlled by user mental commands.

In our work a new exoskeleton control concept exploiting the signals recorded from lower limb muscles is proposed (Fig. 1). We believe that such a system will allow to realize a natural and an effective way to control the exoskeleton device, and help to mobilize patient resources to restore motor skills after stroke.

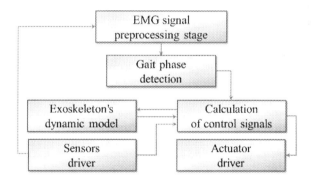

Fig. 1. The concept of exoskeleton control system exploiting the myoelectric signals recorded from the patient's lower limb muscles

2   Control Strategy for Guided Exoskeleton Motion

As already mentioned, a significant role in the proposed approach to building exoskeleton control system is given to interpretation of pilot myoelectric signals. Surface electrodes for patient's muscles signal acquisition are encouraged to use.

Whereas recorded data is always noisy the development of robust control system requires a preprocessing stage. Another one reason to separate out the preparatory treatment step is that each person has his own individual level of bioelectric potentials, and each electrode array set up contributes to the results variability. Also at preprocessing stage signal calibration and filtration happens together with fragmentation of dataflow. This is done by separating myoelectric activity record into frames for convenience of further work. Another purpose of the pretreating step is calculation of the statistical characteristics - both within each record frame, and for a certain period of time.

On the next stage the gait phase detection is occurring. This step is necessary due to the fact that bioelectric potentials by itself do not carry direct information about locomotion, the patient is trying to accomplish. At that step analysis of the muscle activity profile is performed and the decision to perform next gait phase or stop until appropriate myoelectric pattern is occurred.

At the further step, the calculation of control signals for robotic legs using the information from phase recognition stage and exoskeleton dynamic model is performed. The result of this stage is the values of the device actuator torques. Then computed values transferred to the actuator driver and motion happens.

The following sections provide a more detailed description of the principles of lower limb muscle activity registration and it interpretation.

3   Raw EMG Signal Acquisition

3.1   Sensor Placement

In order to obtain the most reliable EMG signals it is necessary to identify the optimal points of the lower limbs for the sensor placement. At first we are interested in measurement of the degree of driver's muscle contraction when piloting exoskeleton.

It is known that for the gait reconstruction with the virtual model is necessary to know the relative position of the thigh, shin and foot projected on the sagittal plane. Due to anatomical features, the most informative leg muscles for parallel surface electromyogram recording while walking are:

– Muscles of the thigh and upper leg:
  1. Quadriceps Femoris (rectus femoris);
  2. Biceps Femoris (long head and short head);
– Muscles of the thigh and upper leg
  1. Tibialis Anterior;
  2. Gastrocnemius Lateralis.

Statistically bioelectrical activity while walking has a high variability. It also has a short duration of individual motor patterns and indecipherable phases of steps.

3.2   Electrode Placement

In order to study the electrical activity of said muscle groups electrode placement was carried out by using a protocol similar to SENIAM (Table 1) http://www.seniam.org.

Two registration systems, running synchronously, was used for accurate detection of the step phase:

– Simi motion® platform for motion capture and movement analysis
– Trigno™ wireless EMG system (Delsys, Inc).

Simi motion® system was used in the primary stage of gait analysis to determine the time of the beginning and end of step.

Table 1. Electrode placement

Muscle name	Electrode placement
Quadriceps Femoris (rectus femoris)	At 1/2 on the line from the anterior spina iliaca superior to the superior part of the patella
Biceps Femoris (long head and short head)	At 1/2 on the line between the ischial tuberosity and the lateral epicondyle of the tibia
Tibialis Anterior	At 1/3 on the line between the tip of the fibula and the tip of the medial malleolus
Gastrocnemius Lateralis	At 1/3 of the line between the head of the fibula and the heel

Surface EMG signals measurement was carried by the base version of the Trigno™ complex. It allows you to record up to 8 channels of EMG in parallel. Each sensor is equipped with a battery, a combined indicator of the activity sensor / discharging the battery, electrodes, pre-board signal processing and wireless radio module. The sensor is composed of four electrodes made of 99 % silver, two of which are the reference. Data transmission to the base station is implemented on a proprietary Wi-Fi similar protocol with a frequency of about 2 kHz.

Thus, the obtained EMG signals are quite suitable for use as a biometric information for exoskeleton control, because of the optimal sensor placement on the lower extremities have achieved. A quite good sensor brand, the wireless nature of the data transmission and pre-treatment protocols also improve the quality of the sEMG signal.

4   Gait Phase Recognition

At the first stage of our investigation classification system is tuned by sEMG patterns recorded in the performance of healthy people of static postures, corresponding to different phases of gait:

1. single-phase support - standing on one leg with a small hand help to maintain support equilibrium - registration from support leg;
2. the end of the double support phase - standing on two legs, registration from leg on the back;

3. the transfer phase - as well as (1), but registration from raised legs;
4. the end of the transfer phase before heel contact - registration from leg on the front.

Performing static postures produces a large number of samples for machine learning and allow to real-time record of motor class. However, this approach is not very useful in the case of daily use of the exoskeleton and its application may not be possible in the case of sick people. Therefore, part of our investigation is dedicated to the development of automatic gait phases detecting and testing classification system during walking. Five subjects take part in this stage of our work. sEMG patterns are recorded from muscles of the right and left leg independently.

Pattern recognition system includes feature extraction and classification. Feature extraction begins from segmentation data stream on separate time windows. In the case of static postures 256 ms window is applied. Features are calculated from sEMG values of one window and then window shifted by 128 ms. This procedure is performed for all input channels. As features we use Mean Absolute Value (MAV) and 4 coefficients of autoregressive (AR) function. AR-coefficients are calculated recursively according to the Levinson-Durbin recursion.

As a classifier we use multi-layer perceptron with back-propagation learning algorithm. Network configuration is constructed in such a way that the number of inputs in neurons of the first layer is equal to the total size of the feature vector: 20 (4 channels by 5 features). The number of neurons in the last layer is chosen equal to the number of motor classes - 4 in the case of static postures. As a result of optimization, the classifier consists of two layers of neurons with 9 neurons in the first layer. The value of the learning rate is 0.01. The error rate of classification ranged from 0.5 % to 5.4 % for the different subjects

After setting up the classification system for static postures the analysis of sEMG gate patterns is obtained. This part of the study faces the following challenges: high variability of bioelectric activity during walking, short duration of motor patterns and the complexity of the phases step cycle logging.

Examples of variability of muscle activity during walking are shown in Fig. 2, where multiple records MAV of *Rectus Femoris* and *Gastrocnemius Lateralis* are combined. In Fig. 2a, we can mention two peaks of activity with a large variability in the amplitude of these peaks, but with stability over time with respect to the step cycle. The first peak is associated with *Rectus femoris* contraction in the support phase and, despite the great variability, you can choose the threshold value (25 %), the excess of which can serve as an indicator of this phase. Muscle *Gastrocnemius Lateralis* (Fig. 2b) is characterized by even greater variability of amplitude, which, unlike the *Rectus Femoris*, complicates any threshold indication.

Researchers suggest a sufficiently large number of classification schemes of cycle step phases. Thus, the total number of phases of the cycle reaches 8–10 (*Perry* 1992). It may be noted that typically these schemes are used for descriptive purposes, but not for automated recognition of the step cycle phase. Obviously, development of scheme is required in which phases would correspond to different sEMG patterns.

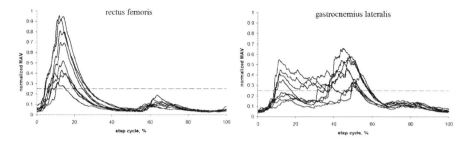

Fig. 2. Bioelectric activity of (a) *Rectus Femoris* (b) *Gastrocnemius Lateralis* in the step cycle. The activity obtained for 8 cycles. The dashed line shows threshold of 25 %.

The division cycle step by phases relates to task of phases logging, which consists in determining the exact time of the start and end of a phase. Moments of the initial heel contact in stance phase are recorded by video registration. Time between these moments in two steps is considered as step cycle and heel contact in the first step is the beginning of the cycle.

As an initial way of logging scheme the division cycle step into equal periods is tested. In particular, the division into 3, 4 and 5 periods has been implemented. We face with necessity of reduce the size of the window and window shifting in the segmentation data that has been associated with short duration of each phase. Thus, in experiments carried out, the total cycle time is 1.0 – 1.5 s, when divided by 5 periods, each period lasted 200–300 ms. In view of this, the window size is reduced to 50 ms, the shift - 25 ms. Among three variants of the uniform division step cycle the division by 4 period is characterized by the smallest classification error rate (15.3 % – 24.2 % errors). However, even in this case, the classification accuracy can not be considered acceptable. Division by 3 (19.8 % – 26.4 % errors) and 5 (23.0 % – 30.5 % errors) equal parts led to more recognition inaccurate.

Analysis of the classification accuracy shows that the first class is best recognized when division by 4 and 5 parts. This class includes pattern of high activity of all muscles in support phase when transferring body weight on one leg. The rest of the classes in these divisions, as well as all classes in the division into 3 parts were characterized by a high error rate that can be attributed of including in the same class very different motor patterns.

It has been observed that the detection of support phase can be carried out based on the sEMG signal using the channel registering *Rectus Femoris* activity. Such detection is combined with logging based on video registration. To eliminate false classification with non-standard steps (start, turn, stop), the detection is carried out only in the time slots marked with the help of video registration. Next we normalize step time based on activity maximum of support phases of two adjacent steps. After this time of all phases is set relative to normalized time step with following values: (Phase 1) 0.08 – 0.2 step cycle; (2) 0.4 – 0.52; (3) 0.7 – 0.8; (4) 0.85 – 0.95.

It should be noted that different values for the beginning and end of phases tested and the best are chosen in the context of recognizing problem. An example of the dynamics of MAV of four channels and the results of the classification in this method logging motor classes is shown in Fig. 3. Classification error rate of different subjects ranged from 12.3 % – 20.0 %.

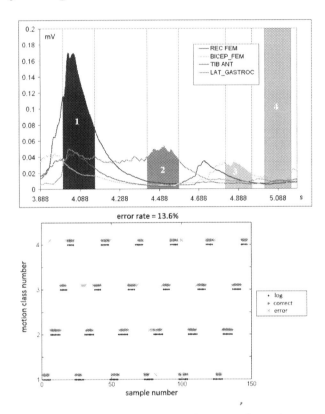

Fig. 3. Example of cycle step logging by dividing into four phases with the phase timing by detection activity *Rectus Femoris* (top) and the results of classification in this division (bottom).

Developed sEMG-based classification system allows us to identify motor patterns following four phases of gait:

1. Effort of all studied muscles, corresponding to the beginning of single support and transfer of body weight on the leg
2. Phase of *Gastrocnemius Lateralis* activity on the background of relative relaxation of the other three muscles. It corresponds to the maximum flexion in the ankle joint to in the final stage of the double support.
3. Activity *Biceps Femoris* on the background of small activity peaks of *Rectus Femoris* and *Tibialis Anterior*. *Gastrocnemius Lateralis* may also be active in this phase. It conforms to the initial phase of swing.
4. Phase of relative relaxation of muscles. It corresponds to the final stage of swing, when the leg moves substantially by inertia

We are going to use of simultaneous working of two classifier - for right and left legs. On the motor class definition in such a system it will be affected both the history of recognized phases and information about the detected phase of other leg. sEMG-based classification system is aimed at use in the exoskeleton for rehabilitation of patients with disorders of the lower limbs. If such patients have muscle activity, but it is not sufficient to walk or there are wrong pattern of muscle activity, the exoskeleton with such classification systems will act as a trainer with feedback as follows. During walking, the exoskeleton will fulfill the next phase of the step only when there is the correct pattern of muscle activity corresponding this moment. Such positive feedback will not only control the exoskeleton, but also will encourage patients to practice the lost functions.

5   Discussion

In this paper, we introduced the concept of an exoskeleton control system for patients with impaired motor function of the lower extremities. It should be noted that the presented approach is suitable for exoskeletons focused on rehabilitation of persons with preserved spinal conductivity and lost of movement ability due to diseases and injury of the brain (for example after stroke). Particular emphasis is placed on the use of biofeedback, i.e. mobilization of internal human mechanisms to improve the efficiency of traditional robotic therapy. Thus, this approach allows building patient-specific control system that takes into account the power and correctness of the pilot muscle activity profile. In addition, in our opinion, this system can be integrated into existing devices, but it requires some technical adaptation.

Acknowledgments. This study was supported by the Russian Ministry of Education and Science, the Federal Program(unique identification number RFMEFI58114X0011).

Reference

Perry, J.: Gait analysis Normal and pathological functions. SLACK Incorporated, p. 524 (1992)

A Medical Image Steganography Method Based on Integer Wavelet Transform and Overlapping Edge Detection

Hayat Al-Dmour$^{(\boxtimes)}$ and Ahmed Al-Ani

Faculty of Engineering and Information Technology, University of Technology,
Ultimo, Sydney 2007, Australia
{Hayat.Dmour,Ahmed.Al-Ani}@uts.edu.au

Abstract. Recently, there has been an increased interest in the transmission of digital medical images for e-health services. However, existing implementations of this service do not pay much attention to the confidentiality and protection of patients' information. In this paper, we present a new medical image steganography technique for protecting patients' confidential information through the embedding of this information in the image itself while maintaining high quality of the image as well as high embedding capacity. This technique divides the cover image into two areas, the Region of Interest (ROI) and the Region of Non-Interest (RONI), by performing Otsu's method and then encloses ROI pixels in a rectangular shape according to the binary pixel intensities. In order to improve the security, the Electronic Patient Records (EPR) is embedded in the high frequency sub-bands of the wavelet transform domain of the RONI pixels. An edge detection method is proposed using overlapping blocks to identify and classify the edge regions. Then, it embeds two secret bits into three coefficient bits by performing an XOR operation to minimize the difference between the cover and stego images. The experimental results indicate that the proposed method provides a good compromise between security, embedding capacity and visual quality of the stego images.

Keywords: Information hiding · Steganography · Edge detection · Human visual system (hvs) · Wavelet Transform · ROI · RONI

1 Introduction

Medical Information Systems (MIS) and Information Technology (IT) form the infrastructure of Telemedicine that incorporates computer to store, process and exchange medical information remotely [1]. During transmission, digital medical images and confidential patients' information should be protected from attackers. Therefore, different techniques have been presented to provide protection of medical images and electronic patient records (EPR) [2]. These methods are based either on cryptography or steganography techniques. Cryptography methods use encoding to hide the meaning of the message, but they cannot hide

© Springer International Publishing Switzerland 2015
S. Arik et al. (Eds.): ICONIP 2015, Part IV, LNCS 9492, pp. 436–444, 2015.
DOI: 10.1007/978-3-319-26561-2_52

its existence [3]. However, the transmission of encoded text certainly stimulate intruders' attention, whom may try to decode it. Steganography on the other hand is the art of concealing the existence of the secret message within other media, such as text, image, audio and video without drawing attention of unintended users [2,4]. Combining steganography and cryptography aim to make it difficult for attackers to extract the sensitive information [3]. This type of integration does not embed readable text, but embeds encrypted text to add another layer of security.

The Least Significant Bit (LSB) method is the most common steganography methodology [2,5]. It is capable of embedding large amount of secret data in a cover image, however it is vulnerable to statistical attacks [6].

In this paper, we propose a new medical image steganography for hiding the EPR into the coefficients of the Integer Wavelet Transform (IWT) high sub-bands to achieve high level of imperceptibility and security. The embedding process is accomplished in eight steps: (i) separate the cover image into two regions, ROI and RONI, (ii) reconstruct RONI pixels into a square matrix, (iii) histogram modification is performed on RONI matrix, (iv) the modified RONI is transformed using 2D IWT to get four sub-bands, (v) implement the detection of edge locations in high frequency sub-bands, (vi) apply an XOR operation, which minimizes the degradation caused by embedding the secret message, (vii) apply inverse IWT to recover RONI pixels, and (viii) combine the RONI to get the stego image. Our proposed algorithm utilizes the visual characteristic of overlapping blocks of pixels such that it achieves better capacity and maintains high visual quality.

The rest of this paper is organized as follows. The related work is described in Sect. 2. Details of the proposed method are presented in Sect. 3. Section 4 presents the experimental results, and the conclusion is given in Sect. 5.

2 Related Work

Exchange of medical information between hospitals and medical clinics placed in different areas is a habitual practice. Therefore, numerous methods have been introduced to provide secure storage and transmission using information hiding techniques.

Navas et al. [7] introduced a blind and reversible data hiding scheme for telemedicine applications that depends on IWT. The ROI is manually identified in the low coefficients sub-band as a rectangle shape. In order to obviate misdiagnosis, encrypted EPR is embedded in RONI and the ROI is stored without any noise. However, this method can hide at most 3400 characters into a cover image of size 512×512 and the computational cost is high. Zou et al. [8] proposed a reversible watermarking method to embed the secret data into the IWT high frequency sub-bands. The embedding procedure utilizes the high frequency sub-bands, where it divides each sub-band into non-overlapping blocks and performs a shifting operation to embed the data. This method is robust to JPEG2000 compression, however, it achieves low embedding rate. Prabakaran et al. proposed a method that utilizes IWT to secure an MRI medical image into a single holder. The method treats the patient's medical diagnosis image as a mystery image and applies the Arnold transform before a acquiring a scrambled secret image, which is hidden into a dummy holder, and the Inverse IWT is applied to get the secret image [9]. In [10], Solanki et al. proposed a watermarking scheme to hide the encrypted watermark in the CT or MRI images to ensure the security of EPR. This method begins by separating the original image into ROI and RONI. It identifies ROI by subtracting the cover image from the negative image, where the negative image is computed by subtracting the medical image from 255. To improve the security, the watermark is encoded by applying the RSA encryption method. Blockwise DWT is then performed over the image.

To overcome some of the drawbacks of the existing methods (i.e., the low embedding rates, low visual quality, underflow/overflow problem and protect ROI from any modifications), this paper proposes a new information hiding method in the IWT domain to embed the secret data in the edge regions of the high frequency sub-bands.

3   The Proposed Method

3.1   The Embedding Process

Almost every medical image contains a particularly important area that is known as the Region of Interest (ROI). This region should not be altered, and hence, when a steganography algorithm is applied the embedding of EPR data should be carried out in the Region of Non-Interest (RONI) only. The implementation of the embedding process is illustrated in Fig. 1 and explained in the following steps.

Algorithm 1: The Embedding Process.
Inputs: Cover image (C), Secret data (M), block size $(n \times n$, which is expected here to be $4 \times 4)$, threshold value (Th).
Output: Stego image (S).

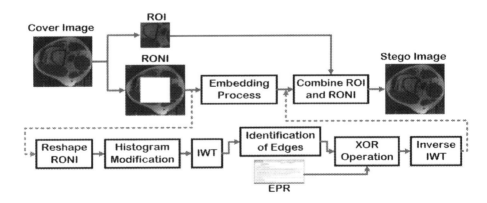

Fig. 1. The block diagram of embedding process.

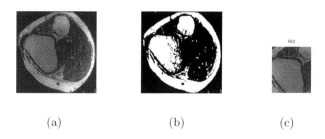

(a)   (b)   (c)

Fig. 2. (a) Cover Image 256×256, (b) Binary image , and (C) ROI.

1. *Identification of ROI and RONI:* The Otsu's method is utilized to identify the threshold value that reduces the difference of the black and white pixels. To separate the cover image into two regions, a threshold value is defined to differentiate between ROI and RONI i.e., the pixel is related to ROI, if the pixel value is greater than the threshold value. Otherwise, it is related to RONI. Figure 2(a) and (b) show the cover and binary images resulting from the Otsu's method. The ROI is selected in rectangular shape according to the white pixel intensity, as shown in Fig. 2(c).
2. *Reshape RONI:* RONI pixels are reshaped into a square matrix of size $r \times r$.
3. *Preprocessing:* The underflow/overflow problem occurs with the existence of pixels whose grey level values are near 0 or 255. This issue can be rectified by enforcing upper and lower limits (240 and 15 respectively) to the pixel's grey level value.
4. *Integer Wavelet Transform (IWT):* Apply the first level of IWT on the modified square of RONI to decompose the RONI into four sub-bands (LL1, HL1, LH1 and HH1). This will be followed by concatenating the HH1, LH1 and HL1 sub-bands in vertical order. In order to improve the embedding capacity and to detect the edge regions accurately, the absolute value is applied to the high sub-bands to generate positive coefficients. Also, the original coefficient sign (T_{sign}) is preserved in a temporary matrix to ensure a good quality of the stego image.
5. *Identification of Edges:* According to the fact that human eyes can bear more modifications in edge regions compared to uniform regions, we attempt to improve the stego image quality through embedding the message bits into edge regions. This will be achieved through an efficient edge detection scheme that uses overlapping blocks to identify and classify image blocks according to their edge strength, where more bits can be hidden into the strong edges and less bits into the smooth ones. Moreover, the aim of using overlapping blocks is to enhance the embedding capacity by reducing the number of unused pixels. The implementation of this process is explained in the following steps:

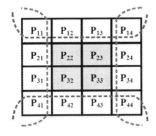

$$1^{st} D = avg\,||[P_{12}, P_{13}, P_{14}, P_{24}, P_{34}] - [P_{21}, P_{31}, P_{41}, P_{42}, P_{43}]||$$
$$2^{nd} D = avg\,||[P_{13}, P_{12}, P_{11}, P_{21}, P_{31}] - [P_{24}, P_{34}, P_{44}, P_{43}, P_{42}]||$$
$$H \quad = avg\,||[P_{11}, P_{12}, P_{13}, P_{14}] - [P_{41}, P_{42}, P_{43}, P_{44}]||$$
$$V \quad = avg\,||[P_{11}, P_{21}, P_{31}, P_{41}] - [P_{14}, P_{24}, P_{34}, P_{44}]|| \quad (1)$$

Fig. 3. 4×4 block

(a) Divide the image into overlapping blocks of size 4×4, as shown in Fig. 3, i.e., shifting of three columns/rows.

(b) Image edges are classified into four directions. Therefore for each block, the magnitude value for each direction is computed using Eq. 1.

(c) The final magnitude value (e) of each block is the maximum of the four values. The secret data (M) is embedded into the four shaded pixels $(P_{22}, P_{23}, P_{32}, \text{ and } P_{33})$.

To minimize the computational cost, edge detection method is applied only once using a threshold of value 4. Then, edge blocks are classified into five groups according to the magnitude value (e) as shown in Table 1, where sharper edges, such as G_5 and G_4, have a higher priority than the weaker edges, such as G_1 and G_2.

6. *The XOR operation*: n-bits are embedded in each RONI edge coefficient, where the value of n is determined by the magnitude value (e) of each block. The algorithm begins with G_5 coefficients by utilizing three bits from each coefficient to embed the secret message. If G_5 coefficients are not enough for embedding, then G_4 coefficients are utilized (three bits from each coefficient). If the message is not fully embedded, then move to the lower priority groups $(G_3$ then G_2 and finally $G_1)$. The XOR operations are applied to embed two secret bits m_1 and m_2 into three bits $p_1, p_2,$ and p_3 of the RONI coefficients as follows: $k_1 = p_1 \oplus p_2$, $k_2 = p_2 \oplus p_3$.

To complete the embedding process, the two secret bits m_1 and m_2 are compared to the calculated bits k_1 and k_2. There are four cases for this comparison:

Table 1. Numbers of bits that can be embedded in each of coefficients of an edge block based on the group it belongs to.

Group	G1	G2	G3	G4	G5
Range of Group	[4 - 7]	[8 - 15]	[16 - 63]	[64 - 127]	[128 - 255]
n (bpp)	1 bit	2 bits	3 bits	3 bits	3 bits

$$m_1 = k_1 \ and \ m_2 = k_2 \ \Rightarrow \ no \ change$$
$$m_1 \neq k_1 \ and \ m_2 = k_2 \ \Rightarrow \ p_1 = p_1\prime$$
$$m_1 = k_1 \ and \ m_2 \neq k_2 \ \Rightarrow \ p_2 = p_2\prime$$
$$m_1 \neq k_1 \ and \ m_2 \neq k_2 \ \Rightarrow \ p_3 = p_3\prime \tag{2}$$

Equation 2 shows that embedding two secret bits into three coefficients bits requires the modification of at most one bit. When the embedding process is complete, the original coefficient sign (T_{sign}) is assigned to the high frequency sub-bands.

7. *Inverse IWT*: Apply Inverse IWT on the four sub-bands to recover the RONI image of size $r \times r$.
8. Combine the ROI and RONI after the embedding process to construct the stego image.

3.2   The Extraction Process

Extraction of the secret message bits is implemented as follows:

1. Divide the stego image into ROI and RONI by applying the Otsu method.
2. RONI pixels are reshaped into a square matrix of size $r \times r$.
3. Apply the first level of IWT transform on RONI from the stego image to decompose the RONI into four sub-bands (LL1, HL1, LH1 and HH1). Concatenate HH1, LH1 and HL1 in vertical order and find the absolute value of the resulting matrix.
4. *Identification of Edges:* Identify the edge regions in the high frequency sub-bands by performing the edge detection procedure used in the embedding process. Then, classify the edge blocks into five groups. Finally, the secret bits m_1 and m_2 are extracted from the three stego bits $q_1, q_2,$ and q_3 using the following XOR extraction operations as follows: $m_1 = q_1 \oplus q_2, m_2 = q_2 \oplus q_3$.

4   Experimental Result and Discussion

The performance of the proposed method has been tested using 100 MRI images of size 256×256. The visual quality of the stego images are measured using the Peak Signal-to-Noise Ratio ($PSNR$) which is defined in Eq. 3.

$$PSNR = 10 \log_{10} \left[\frac{255^2 \times W \times H}{\sum_{i=1}^{W} \sum_{j=1}^{H} (c_{ij} - s_{ij})^2} \right] (dB) \tag{3}$$

where W and H are the width and height of the cover image respectively, and c_{ij} and s_{ij} are the gray values of pixel (i, j) of the cover and stego images respectively.

The weighted Peak Signal-to-Noise Ratio ($wPSNR$) is an alternative image quality measurement. It improves the classical $PSNR$ as shown in Eq. 4.

$$wPSNR = 10 \log_{10} \left[\frac{max(C)^2}{\|NVF(S-C)\|^2} \right] (dB) \tag{4}$$

where NVF is defined as:

$$NVF_{(i,j)} = \frac{1}{1 + \sigma^2_{L_{(i,j)}}} \tag{5}$$

where $\sigma^2_{L_{i,j}}$ denotes the local variance of an image in a window centered on the pixel with coordinates (i, j).

The length of the secret data is another measurement criterion, which is defined as $E = \frac{K}{WH}(bpp)$, where K is the number of the data message bits.

Figure 4(a) and (d) show one of the cover images used in the experiment and its histogram. Figure 4(b) and (c) are the stego images resulting from the proposed method with 5% and 30% embedding rates. The visual differences between the cover and stego images cannot be discovered by the human eye, and even the histograms of the stego images (illustrated in Fig. 4(e) and (f) are quite similar to that of the cover image.

In order to examine the performance of the presented scheme over other medical information hiding method, we compared our method with Navas's EPR hiding method [7] in term of visual quality and embedding rates. The obtained results that are listed in Table 2 indicate that the proposed method achieved better image quality for all metric measurements compared to Navas's method. Also, the proposed method provides a high embedding rate of up to 40%, while Navas's method could only achieve an embedding rate of up to 13%. The PSNR and wPSNR values of the proposed method are between (49.10 - 58.46) dB and (51.55 - 60.84) dB respectively. On the other hand, we have prevented any modifications to the ROI whereas Navas's method cannot guarantee this protection due to identifying the ROI from the low sub-band instead of identifying it in the spatial domain.

Table 2. Image quality evaluation between the proposed method and Navas [7] technique with various embedding rates over 100 stego images. The red values indicate the best result.

Embed Rate	Method	MSE	PSNR	wPSNR	Avg. Difference
5%	Navas [7]	1.247	47.54	58.89	0.1188
	Proposed	**0.096**	**58.46**	**60.84**	**0.0608**
10%	Navas [7]	1.774	45.64	56.75	0.2527
	Proposed	**0.225**	**54.68**	**57.04**	**0.1321**
25%	Navas [7]	-	-	-	-
	Proposed	**0.567**	**50.60**	**52.96**	**0.3032**
40%	Navas [7]	-	-	-	-
	Proposed	**0.801**	**49.10**	**51.55**	**0.3779**

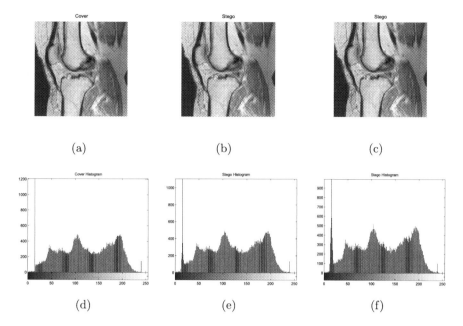

Fig. 4. (a) Cover Image 256×256, (b - c) Stego Images with 5 % and 30 % embedding rate, (d) Cover Image histogramand, and (e - f) Histograms of the corresponding stego images.

5   Conclusion

In this paper, a new and efficient medical image steganography is presented to hide the EPR data into the integer wavelet coefficients of RONI. It incorporates edge detection technique using overlapping blocks to identify and hide the secret data in the sharp regions of the image. The motivation behind using overlapping blocks is to improve the embedding payload by reducing the number of unused pixels. The difference between the cover and stego images is reduced by applying an XOR operation, which changes one bit at most to embed two secret bits into three pixel bits. The experimental results demonstrate improvements in the embedding capacity, and imperceptibility compared to one of the existing EPR hiding methods.

References

1. Singh Brar, A., Kaur, M.: High capacity, reversible data hiding using cdcs along with medical image authentication. Int. J. Sig. Proc. Image Proc. Pattern Recogn. **8**, 49–60 (2015)
2. Cheddad, A., Condell, J., Curran, K., Mc Kevitt, P.: Digital image steganography: Survey and analysis of current methods. Signal Proc. **90**, 727–752 (2010)

3. Ramaiya, M.K., Hemrajani, N., Saxena, A.K.: Security improvisation in image steganography using des. In: 2013 IEEE 3rd International Advance Computing Conference (IACC), pp. 1094–1099. IEEE (2013)
4. Al-Dmour, H., Al-Ani, A., Nguyen, H.: An efficient steganography method for hiding patient confidential information. In: 2014 36th Annual International Conference of the IEEE Engineering in Medicine and Biology Society (EMBC), pp. 222–225. IEEE (2014)
5. Hamid, N., Yahya, A., Ahmad, R.B., Al-Qershi, O.M.: Image steganography techniques: an overview. Int. J. Comput. Sci. Secur. (IJCSS) 6, 168–187 (2012)
6. Dumitrescu, S., Wu, X., Wang, Z.: Detection of lsb steganography via sample pair analysis. Signal Proc. IEEE Trans. 51, 1995–2007 (2003)
7. Navas, K., Thampy, S.A., Sasikumar, M.: Epr hiding in medical images for telemedicine. Int. J. Biomed. Sci. 3, 44–47 (2008)
8. Zou, D., Shi, Y.Q., Ni, Z., Su, W.: A semi-fragile lossless digital watermarking scheme based on integer wavelet transform. IEEE Trans. Circuits Syst. Video Technol. 16, 1294–1300 (2006)
9. Prabakaran, G., Bhavani, R., Rajeswari, P.: Multi secure and robustness for medical image based steganography scheme. In: 2013 International Conference on Circuits, Power and Computing Technologies (ICCPCT), pp. 1188–1193. IEEE (2013)
10. Solanki, N., Malik, S.K., Chhikara, S.: Roni medical image watermarking using dwt and rsa. Int. J. Comput. Appl. 96, 30–35 (2014)

Combining CNN and MIL to Assist Hotspot Segmentation in Bone Scintigraphy

Shijie Geng[1], Shaoyong Jia[1], Yu Qiao[1]([✉]), Jie Yang[1],
and Zhenhong Jia[2]

[1] Institute of Image Processing and Pattern Recognition,
Shanghai Jiao Tong University, Shanghai, China
{jeykigung,jiashaoyong,qiaoyu,jieyang}@sjtu.edu.cn
[2] School of Information Science and Engineering,
Xinjiang University, Urumqi, China
jzhh@xju.edu.cn

Abstract. Bone scintigraphy is widely used to diagnose tumor metastases. It is of great importance to accurately locate and segment hotspots from bone scintigraphy. Previous computer-aided diagnosis methods mainly focus on locating abnormalities instead of accurately segmenting them. In this paper, we propose a new framework that accomplish the two tasks at the same time. We first use sparse autoencoder and convolution neural network (CNN) to train an image-level classifier that label input image as normal or suspected. For suspected images, multiple instance learning (MIL) is applied to train a patch-level classifier. Then we use this classifier to produce a probability map of hotspots. Finally, level set segmentation is performed with the probability map as initial condition. The experimental results demonstrate that our method is more accurate and robust than other methods.

Keywords: Hotspot segmentation · Bone scintigraphy · Multiple instance learning · CNN · Level set method

1 Introduction

Bone scintigraphy is very effective in diagnosing cancer and tumor metastases [14]. The abnormalities in bone scintigraphy are called "hotspot", which generally appear to be brighter than its surroundings. It is of great clinical importance to accurate detect and segment hotspots from bone scintigraphy. Many computer-aided diagnosis (CAD) systems have been developed to detect and segment hotspots. May Sadik et al. [10] used adaptive threshold of a specific region for hotspot segmentation. Huang et al. [5] uses linear regression model to find regional threshold to extract hotspot. Chang et al. [3] proposed an algorithm utilizes Gaussian function to approximate intensity probability distribution and perform hotspot segmentation via adaptive region growing [4]. However, most of proposed methods focus on detection of hotspots rather than accurate segmentation. Adaptive threshold and region growing [4] are two most commonly

© Springer International Publishing Switzerland 2015
S. Arik et al. (Eds.): ICONIP 2015, Part IV, LNCS 9492, pp. 445–452, 2015.
DOI: 10.1007/978-3-319-26561-2_53

used approaches. Obviously, due to weak boundary contrast and low signal noise ratio of bone scintigraphy, these two methods will not achieve satisfactory segmentation results.

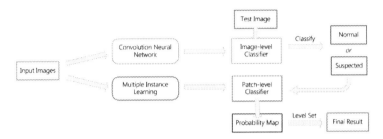

Fig. 1. Overview of our framework for accurate hotspot segmentation. We use convolution neural network to train an image-level classifier and multiple instance learning to train a patch-level classifier respectively. For test image, we first label it as normal or suspected. If the image is suspected, we will produce a probability map and perform segmentation via level set algorithm.

Due to advantages in sub-pixel accuracy and topology variability [11], level set methods [8,11] are widely used in image segmentation. They evolve a user-specified initial contour in order to minimize a given energy function. Level set methods can be divided into two types: edge-based and region-based. In terms of weak object boundaries, region-based methods have better performance compared with edge-based ones. In this paper, we choose region-based level set with local signed difference energy (LSD) [11] to perform accurate hotspots segmentation. In order to perform a level set segmentation, we need to set an initial contour for the algorithm. Generally, if the initial contour gets closer to the target object, the final segmentation result will be better. A probability map of hotspots is helpful to initialization of level set algorithm. In this paper, we apply convolution neural network (CNN) and multiple instance learning (MIL) to get the probability map.

Deep learning methods have achieve great success in these years. It has been applied to object recognition in ImageNet [7] and feature learning from unlabeled data [6,9]. Deep learning shows its robustness and discriminative power in feature learning and classification tasks. In this paper,we train an image-level classifier using sparse coding and CNN to classify input image as normal or suspected. If the input image is labeled as normal, it will not be segmented in following steps. MIL technique has been widely used in several scenarios, such as MIL-Boost [2] in object tracking, MCIL [13] in medical diagnosis, MILCUT [12] in natural image segmentation. It has many advantages, such as reducing the efforts in human annotations and automatically exploiting information from data [13]. In this paper, we apply MCIL algorithm to train a patch-level classifier. Given an input image, the classifier can produce a probability map of hotspots.

Figure 1 gives an illustration of our framework. Experiments in Sect. 3 demonstrate the effectiveness of our framework in hotspot detection and segmentation over previous methods.

2 Methodology

2.1 Using CNN to Train Image-Level Classifier

In this section, we use sparse autoencoder [9] to learn bases from image patches which extracted from bone scintigraphy. To extract feature representation from the hole image, we utilize CNN which consists of convolution and mean pooling layers. Finally, we input the pooled features to SVM and train an image-level classifier. Figure 2 is a diagram of the overall training process. To increasing discrimination between pixels with similar intensities, all images used in this section are first mapped to RGB color space by applying density slicing. The processed 3-channel images are showed in Fig. 2. Detailed explanations can be found in experiments section.

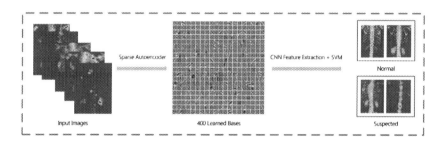

Fig. 2. Diagram of the overall training process: we first use sparse autoencoder to learn 400 bases from image patches. Then we extract feature representation using CNN. Finally we train image-level classifier via SVM.

Sparse autoencoder is an unsupervised learning algorithm that tries to learn higher-level representation of input images. It finds basic elements (bases) of image patches. We first randomly extract a large number of small patches from thoracic bone scintigraphy. Then we convert these $d \times d$ patches to unlabeled vectors $\{x_1, \ldots, x_m\}$, where $x_i \in \mathbb{R}^s$, $s = d \times d$. We denote the activation of basis ϕ_j for input x_i as $a_{i,j}$. Our task is to solve the following optimization problem:

$$\min_{a,\phi} \sum_{i=1}^{m} \left\| x_i - \sum_{j=1}^{k} a_{i,j}\phi_j \right\|_2^2 + \lambda \sum_{i=1}^{m} \sum_{j=1}^{k} \|a_{i,j}\|_1 \tag{1}$$

After solving the above problem, we get all bases vector $\Phi = \{\phi_1, \ldots, \phi_k\}$, where $\phi_j \in \mathbb{R}^s$. In this paper, we set k to 400, the middle of Fig. 2 shows the 400 learned bases via sparse coding. Since the size of input images are not uniform, as a preprocessing step, we resize them to uniform $n * n$ size. Then we apply convolution and mean pooling for these resized input images.

To get the convolved features, for every $d \times d$ region of the $n \times n$ image, we convolve it with the learned bases to get the feature activations. As a result, we will get $k \times (n-d+1) \times (n-d+1)$ array of convolved features. For mean pooling, we divide the convolved image into $l \times l$ equal parts, and calculate the mean for each part, then produce $k \times l \times l$ pooled features. We construct a training set T consists of pooled features and human labels, and train an image-level classifier with SVM algorithm.

2.2   Using MIL to Train Patch-Level Classifier

In this section, we use MIL method to train patch-level classifier which can produce probability map of hotspots. To learn detail information from input data, we need to set patches to a relatively small size. However, when the size of patches get smaller, it will be harder for human labeling. In multiple instance learning, instances are not directly labeled, it only needs bag labels. This property will largely reduce the efforts in human annotations. For this reason, we choose MIL to train patch-level classifier.

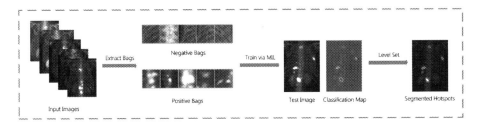

Fig. 3. The flow chart of our hotspot detection and segmentation process.

MCIL Algorithm. As shows in Fig. 3, we extract positive and negative bags consist of 4×4 instances from input images to construct training datasets. We denote a bag as $X_i = \{x_{i1}, \ldots, x_{in}\}$, in which i is the bag index and j is the instance index. The label of bag is denoted as $y_i \in \{-1, 1\}$, the label of instance $y_{ij} \in \{-1, 1\}$ is latent during training. During the training process, MCIL [13] tries to minimize the following negative log-likelihood loss function:

$$\mathcal{L}(H) = -\sum_{i=1}^{n} \left(\mathbf{1}(y_i = 1) \log H(X_i) + \mathbf{1}(y_i = -1) \log(1 - H(X_i)) \right) \qquad (2)$$

Here, $H(X_i)$ is the probability of bag i, $h(x_{ij})$ is the probability of instance j in bag i. The relationship between $H(X_i)$ and $h(x_{ij})$ is $H(X_i) = \max_{j}(h(x_{ij}))$. MCIL compute new weights as follows: $w_{ij} = -\frac{\partial \mathcal{L}(H)}{\partial h(x_{ij})}$, then use these weights to select best weak classifier h_t. Line search is then performed to find the coefficient α_t of the selected weak classifier. After training given number of weak classifiers, we get the final strong classifier : $h(x_{ij}) = \sum_{t=1}^{T} \alpha_t h_t(x_{ij})$.

Feature Extraction for Bags and Instances. In our work, we select relatively large regions from bone scan images and treat them as bags. For each bag, we and densely sample 4×4 patches as instances from it. The overlap step size is 2 pixels. We then extract a 29-dimension feature vector for each instance, including 11-dimension intensity features and 18-dimension texture features. Intensity features consist of statistical histogram, 4-neighborhood contrast and symmetric contrast. We use weighted difference to compute 4-neighborhood contrast: $Neighbor\ Contrast(M, N) = \frac{1}{10} \sum_{i=1}^{n} 2^i (M_i - N_i)$, where M and N are histograms of center patch and 4-neighborhood patch respectively. For symmetric contrast, we first find symmetric patch by calculating body central line as [3] does, then we compute symmetric contrast using modified chi-square distance: $Symmetry\ Contrast(M, S) = \sum_{i=1}^{n} \frac{(M_i - S_i)^2}{M_i + S_i + 1}$, where M and S are histograms of center patch and symmetric patch respectively. For texture feature, we use a variant of local binary patterns (LBP) [1], in which the threshold used is the mean intensity of the whole image. We collect 18947 instances to form 39 positive bags and 33 negative bags. Finally we use MCIL to train the patch-level classifier.

2.3 Using Level Set to Segment Hotspot

After getting the probability map of hotspot, we use local signed difference (LSD) [11] level set algorithm to segment hotspots from input image. LSD is able to deal with intensity inhomogeneity and weak object boundaries. It also consider global information – the order of local clusters, thus leads to robust segmentation performance.

Let C denote the contour evolving in image domain Ω, $f_1(x)$ and $f_2(x)$ are local clusters inside and outside contour C. The LSD energy is expressed as follows:

$$\mathcal{E}(C) = \int_\Omega \mathrm{sgn}(f_2(x) - f_1(x)) \, |f_2(x) - f_1(x)| \, dx + \mu Length(C) \qquad (3)$$

According to level set formulation, we use a Lipschitz function $\phi : \Omega \to R$ to represent LSD energy. To initialize the level set function ϕ_0, we utilize the probability map of hotspot $\mathcal{P}(x)$:

$$\phi_0 = \phi(x, t = 0) = \begin{cases} -c, & x \in \{x | \mathcal{P}(x) > \rho\} \\ 0, & x \in \{x | \mathcal{P}(x) = \rho\} \\ c, & x \in \{x | \mathcal{P}(x) < \rho\} \end{cases} \qquad (4)$$

We set ρ to 0.5 in this paper. Then we apply gradient descent method to minimize the LSD energy according to the equation: $\frac{\partial \phi}{\partial t} = -\frac{\mathcal{E}(\phi)}{\partial \phi}$.

3   Experiments

In this section, we conduct three experiments to demonstrate the effectiveness of our framework. We first give the reason why we use mapped color images to train image-level classifiers. In the second experiment, we make a comparison between multiple instance learning and supervised learning methods in terms of classification accuracy. At last, we show that our framework gets better segmentation results compared with previous methods. The bone scintigraphy images used in training are collected from Department of Nuclear Medicine, Shanghai Renji Hospital. Image labels used in training image-level classifier come from the gold standard of radiologist diagnosis.

Experiment I. In this experiment, we select 1030 images and resize them to 100×100. We extract 100000 patches with size of 11×11 to train sparse autoencoder. We respectively use mapped color images, original grayscale images as input images, and test recognition accuracy with 5-fold cross validation. Table 1 gives the tesing results. It shows that features learned from mapped color images have better discriminative ability compared with grayscale images.

Table 1. Cross validation accuracy of different input dataset

Input Dataset	Color Images	Grayscale Images
Cross Validation Accuracy	0.947	0.895

Experiment II. In this experiment, we respectively train patch-level classifier with MCIL [13] and SVM. The training dataset consists of 18947 instances from 39 positive bags and 33 negative bags. For SVM, bag labels are directly regarded as instance labels. We also construct a testing dataset consists of 1050 labeled instances, each instance forms a bag. Table 2 test the classification accuracy of two algorithms. It shows that multiple instance learning can exploit information from given data, and the classification accuracy is enough to get an approximate probability map of hotspot locations.

Table 2. Accuracy comparison between the two algorithms

Algorithm	Semi-supervised (MCIL)	Supervised(SVM)
Classification Accuracy	0.905	0.783

Experiment III. In this experiment, we compare our hotspot segmentation framework with two previous methods – adaptive region growing [3] and regional threshold [5]. We select 68 suspected thoracic images consists of 572 hotspots as testing dataset and invited expert radiologist to annotate the ground truths for these images. In this paper, Jaccard (J) and Dice (D) index are used to evaluate the performance of hotspot segmentation. They are defined as: $J = \frac{|S \cap G|}{|S \cup G|}$, $D = \frac{2|S \cap G|}{|S| + |G|}$. Table 3 shows the performance of the three methods in hotspot segmentation:

Table 3. The performance comparison on suspected thoracic images

Method	Our method	Region growing [3]	Threshold [5]
Jaccard (J)	0.8051	0.6505	0.5183
Dice (D)	0.8887	0.7613	0.6574

From Table 3, we can conclude that our framework gets more accurate segmentation results compared to the other two methods. In Fig. 4, we give a illustration of different segmentation results and human labeled ground truths. We can see that our method also get better segmentation results in terms of appearance. Our results are with smooth boundaries and very close to ground truths.

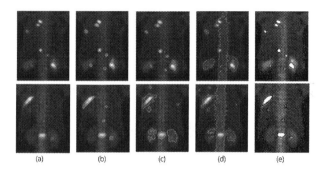

Fig. 4. Segmentation result comparison: (a) input image (b) our method (c) adaptive region growing [3] (d) regional threshold [5] (e) human annotations.

4   Conclusion

In this paper, we propose a novel framework which combine CNN and MIL to assist accurate hotspot segmentation. Experiments show that our method has advantage over existed methods and gets more accurate segmentation results. In future work, we will extend our framework from thoracic region to other regions in bone scintigraphy.

Acknowledgments. This research is partly supported by NSFC, China (No: 61375048).

References

1. Ahonen, T., Hadid, A., Pietikainen, M.: Face description with local binary patterns: Application to face recognition. IEEE Trans. Pattern Anal. Mach. Intell. **28**(12), 2037–2041 (2006)
2. Babenko, B., Dollár, P., Tu, Z., Belongie, S., et al.: Simultaneous learning and alignment: Multi-instance and multi-pose learning. In: Workshop on Faces in 'Real-Life' Images: Detection, Alignment, and Recognition (2008)

3. Chang, Q., Wang, Q., Qiao, Y., Zhu, Y., Huang, G., Yang, J.: Adaptive detection of hotspots in thoracic spine from bone scintigraphy. In: Lu, B.-L., Zhang, L., Kwok, J. (eds.) ICONIP 2011, Part I. LNCS, vol. 7062, pp. 257–264. Springer, Heidelberg (2011)
4. Hojjatoleslami, S., Kittler, J.: Region growing: a new approach. IEEE Trans. Image Procs. **7**(7), 1079–1084 (1998)
5. Huang, J.Y., Kao, P.F., Chen, Y.S.: A set of image processing algorithms for computer-aided diagnosis in nuclear medicine whole body bone scan images. IEEE Trans. Nucl. Sci. **54**(3), 514–522 (2007)
6. Kim, M., Wu, G., Shen, D.: Unsupervised deep learning for hippocampus segmentation in 7.0 tesla MR images. In: Wu, G., Zhang, D., Shen, D., Yan, P., Suzuki, K., Wang, F. (eds.) MLMI 2013. LNCS, vol. 8184, pp. 1–8. Springer, Heidelberg (2013)
7. Krizhevsky, A., Sutskever, I., Hinton, G.E.: Imagenet classification with deep convolutional neural networks. In: Advances in Neural Information Processing Systems, pp. 1097–1105 (2012)
8. Li, C., Xu, C., Gui, C., Fox, M.D.: Distance regularized level set evolution and its application to image segmentation. IEEE Trans. Image Proces. **19**(12), 3243–3254 (2010)
9. Raina, R., Battle, A., Lee, H., Packer, B., Ng, A.Y.: Self-taught learning: transfer learning from unlabeled data. In: Proceedings of the 24th International Conference on Machine Learning, pp. 759–766. ACM (2007)
10. Sadik, M., Hamadeh, I., Nordblom, P., Suurkula, M., Höglund, P., Ohlsson, M., Edenbrandt, L.: Computer-assisted interpretation of planar whole-body bone scans. J. Nucl. Med. **49**(12), 1958–1965 (2008)
11. Wang, L., Wu, H., Pan, C.: Region-based image segmentation with local signed difference energy. Pattern Recogn. Lett. **34**(6), 637–645 (2013)
12. Wu, J., Zhao, Y., Zhu, J.Y., Luo, S., Tu, Z.: Milcut: A sweeping line multiple instance learning paradigm for interactive image segmentation. In: 2014 IEEE Conference on Computer Vision and Pattern Recognition (CVPR), IEEE (2014)
13. Xu, Y., Zhu, J.Y., Eric, I., Chang, C., Lai, M., Tu, Z.: Weakly supervised histopathology cancer image segmentation and classification. Med. Image Anal. **18**(3), 591–604 (2014)
14. Yin, T.K., Chiu, N.T.: A computer-aided diagnosis for locating abnormalities in bone scintigraphy by a fuzzy system with a three-step minimization approach. IEEE Trans. Med. Imaging **23**(5), 639–654 (2004)

Topic and Thematic Description for Movies Documents

Manel Fourati[✉], Anis Jedidi, and Faiez Gargouri

Laboratory MIR@CL, University of Sfax, Sfax, Tunisia
Manel.Fourati@fsegs.rnu.tn, {anis.jedidi,
faiez.gargouri}@isimsf.rnu.tn

Abstract. The descriptions of audiovisual documents used in the interrogation process should not be limited to the identification of some keywords selected from signal, or from the forms presented in the image. They should be however, extracted basically from the content whilst exploiting the knowledge conveyed in the document. In this context, the topic and thematic description represents important information from the content. This importance result from the effective presence of the documents' content. Consequently, we concentrate efforts to propose a method that describe the theme and the topic of movies document based on the adaptation of the Latent Dirichlet Allocation (LDA) model by combining the textual and visual modalities from the pre-production movie document (Script) and from the superposed text in the image. The experiments results confirmed the interesting performance through two databases, namely, "Choi's dataset" and our own created database from the Internet Movie Database Imdb (http://www.imdb.com/years2012/2013).

Keywords: Topic · Theme · Movie · LDA · Textual modalities · Visual modalities

1 Introduction

The amount of information available of audiovisual documents requires some techniques for retrieving information through content. This need has an impact on different disciplines such as movies documents. In this context, the purpose is to extract descriptions of movie audiovisual from the content. In fact, the topic and thematic description can be classified among the most natural, practical descriptions inspired by the user. Such a primitive solution is already used in this respect which consists in extracting features from the audiovisual document. These features mainly came either through the low level descriptions (color, texture, movement …) or through the high-level descriptions (ancropersonne, event). However, these same high-level descriptions are not sufficient to provide semantically significant information. To face these multiple conditions, we must therefore concentrate efforts to achieve cinematic audiovisual documents description based on the topic and thematic description which has taken place in the last years. The remainder of the paper is organized as follows: Sect. 2 gives an overview of the state of the extracting thematic and topic description using textual analysis. In Sect. 3, a proposed approach is defined and described using the previous mentioned modalities. Thereafter, its evaluation is presented in Sect. 4. Finally, the last section concludes the paper and includes some future work.

© Springer International Publishing Switzerland 2015
S. Arik et al. (Eds.): ICONIP 2015, Part IV, LNCS 9492, pp. 453–462, 2015.
DOI: 10.1007/978-3-319-26561-2_54

2   Related Works

We used the textual pre-production document such as scenario and the text superposed in the image to extract description. So, we focus, in our state of the art, on the textual topic and thematic segmentation. There are many research works proposed to extract topic or theme documents. These existing methods can be classified in three main classes' approaches: Linguistic analysis, Statistic analysis and Semantic analysis. A few studies focus on the identification of the linguistic cues in the text to detect the thematic changes [1]. [2] proposes a system to analyze the discourse that allows segmenting the text into Elementary Discourse Unit (EDU) essentially based on linguistic cues. However, these approaches are not always valuable because they require the existence of markers in the text. Some works proposed methods for thematic segmentation of textual documents based on the statistical analysis such as TextTiling. In task-based information retrieval and Natural Language processing, [3, 4] used TextTiling as a pretreated phase in order to extract key terms to identify key concepts. Other research works considered the document as a set of latent topics. These methods can be classified into three sub-categories: semantic, probabilistic and probabilistic semantic method. LSA (Latent Semantic Analysis) is a semantic analysis technique that provides relationships between the terms related to the documents [5, 6]. Despite the performance proved by several studies that LSA is effective for information retrieval tasks, LSA provides very restricted set of semantic relationships mainly based on the statistical nature of the extracted lexico-semantic knowledge [7]. Consequently, PLSA (probabilistic latent semantic analysis) [8] is a probabilistic semantic analysis inspired from LSA by including a statistical model which models each word in a document as a sample from a mixture model with multinomial component models. Despite the effective results, the PLSA model is not suitable with the new documents not seen in the training data. The problem of the PLSA model must be considered as a starting point to propose the probabilistic model for the topic detection called Latent Dirichlet Allocation (LDA) [9]. It is a probabilistic generative model that allows organizing text documents based on the exploration of topics. It generates a set of hidden topics that is defined as a set of keyword. Indeed, several works have proven the importance of the LDA model to extract the hidden topics in different types of textual documents through several fields. For instance, event characterization based on topic modeling, is presented and discussed in [10–12]. In spite of the efficiency demonstrated by several studies of the LDA model in different tasks, there are still some lacks. Indeed, this model handles the document as a bag of words which allows obtaining a large number of sample data on the one hand and on the other hand the word order is not taken into account. Accordingly, LDA does not explicitly model the relationship between topics. In this work, we focus on topic extraction by adapting the LDA model to overcome these problems handled poorly.

3   Proposed Approach

In this section, it is interesting to gather the insights and knowledge gained from the study of the state of the art. These elements will direct our topic and thematic process. In our work, we are interested in describing the cinematic audiovisual documents. The proposed approach finds its originality in the extraction of semantic description from different analytical modalities such as textual and visual, the automation of the description process and the fusion between audiovisual modalities to improve the semantic description of movie document. In [13], we proposed an approach to genre and keyword identification of films' content and in [14] we are interested in the localization and the recognition of superposed texts from the image. The results of these woks were taken as a starting point to our proposed approach to extract topics and themes of movie audiovisual documents. Not only did we obtain these two results, but also we propose analyzing the discourse in the pre-production document of the movie such as; the scenario to extract themes and topics. In this section the proposed multimodal approach is presented. It consists of two phases: (i) the text segmentation phase and (ii) the description extraction phase.

3.1 Text Segmentation Phase

Starting from the definition proposed by Peter Stockinger [15] that "the thematic description of a document allows to describe and explain the speech"; we decide to analyze the discourse from the audiovisual documents. This analysis is based on the pre-production and the post-production documents like extracting the scenario and the text from the audiovisual flow. Such a semantic description based on the information extracted from the content, becomes a necessary condition for a better interrogation. In this context is located our work. To do so, we aim in this phase to extract the text related to the movie document based on a textual segmentation of the scenario and also of the superposed text in the images. Our textual segmentation step is based on the textual analysis through a study of the different structures of writing scripts proposed by the scriptwriter of films. When writing the script, the scriptwriter uses a well-defined structure which ensures the presentation of each scene separately. Starting from this predefined structure, we find that we can segment the scenario in discourse unit. Based on the cues of phrase from the slug line, we segment the discourse in elementary discourse units (EDU). Indeed, each line essentially begins with the following indications either by ".EXT", ".INT", or by "scene1", "scene2", etc. Once the segmentation of the discourse is obtained, a scene categorization of audiovisual flow will take place in order to obtain the text superposed in each scene. The purpose of this process is to gather all shot that has the same superposed text. Thereafter, a fusion modalities process allows combining the textual and the visual modalities by performing a getting closer procedure. This approaching is due to measure the similarity between the superposed texts of the images with the speech segmented from the scenario. The measurement of similarity cosine has been used to this end. Based on the textual segmentation of EDU and the superposed text steps, we have elaborated a description extraction process.

3.2 Description Extraction Phase

Consequently, the previous phase provides the segments and their texts of the audiovisual documents. The purpose of the description extraction phase is to identify the themes and topics of each segment. In this phase two different techniques are used: a probabilistic technique and a semantic technique. Figure 1 shows the steps to follow this phase.

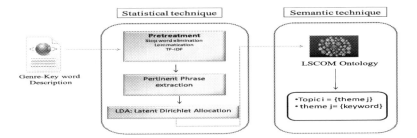

Fig. 1. Description extraction phase

(i) The first technique is the Latent Dirichlet Allocation (LDA). This probabilistic technique is used to extract the hidden topics contained in the textual document. In the literature, several works have proven the performance of this technique in the exploration of topics. However, several lacks are mentioned in the Sect. 2 of this paper. In our work, we propose adapting the LDA in order to address these problems. The general principle of such technique can be described mainly by the three following steps:

- Step 1: selection of representative words: To resolve the first problem that a large number of sample data, we propose a pretreatment steps for the LDA technique. This pretreatment step allows to: Remove all insignificant terms': stop words, lemmatize all terms through the Stanford lemmatizer and measure pertinence of each term. In order to measure the relevance of each term in the appropriate segments, we are interested in adapting the metric TF-IDF (Term Frequency-Inverse Document Frequency) in our result proposed in [13]. In fact, we measure the pertinence of the terms not only in the key words extracted from the synopses but also in all synonyms and hypenyms of all genres extracted from the content.
- Step 2: Selection of key phrases in each segment: Since LDA does not take into account the word order, instead of making all text as input of LDA, we find that it is important to select the key phrases in each segment. In fact, sentences that contain a sufficient number of representative terms are considered relevant sentences. In this regard, a pertinence measure of each phrase is defined. These measures provide relevant sentences related to each segment. In this context, we propose the following formula:

$$Pert\,(Phi) = \frac{\sum (Vpj)}{Nb\_mot\_p} \tag{1}$$

Where: *Pert (Phi)* represents the pertinence of the sentence Phi, *Vpj* represents the pertinence value of each term j considered as a relevant term in the document and *Nb_mot_p* represens the number of relevant terms of each sentence. To carry out the adaptation of the LDA model, this step allows reducing the number of data which are considered as input of the LDA model. Furthermore, the treated data represent the pertinent sentence in each segment. In fact, the more data are relevant, the more the result of topic modeling is approved.

- Step 3: Adaptation of the LDA model: As previously mentioned in the related works section, the LDA [9] is a modeling technique, to extract the hidden topics in the document. The benefits of the LDA model use are to generate not only a topic distribution associated with each document, but also a word distribution associated with each topic. The topic estimation is essentially based on the probabilistic process. More specifically, the generative process works is described as follows: (1) A number of topics, K, should be specified a priori representing the dimension of the topics set. (2) From a corpus of documents, two set of parameters to be estimated in a training process, namely: distribution of the topics Θ in a document and distribution of words in a topic, namely, $\Theta = \{\theta t, t = 1..K\}$, with t is a topic and $\Phi = \{\Phi tv\ t = 1..K, v = 1..V\}$, with V is the vocabulary size. In this context, two prior weight parameters must be defined such as, α is a prior weight of topic k in a document. This value must be between 0 and 1 and β is a prior weight of word w in a topic. This value must be between 0 and 1.

Once the distribution of the topic and the distribution of the words are estimated, a prediction of a novel document will take place in order to extract the topics and their related words. We note that the LDA model requires few empirical parameters to trained document, such as, Alpha, beta and K. It is worth mentioning that, in step1 of the description extraction phase of the proposed approach, the output of this step is a pertinent key words set and the value of pertinence of each key word. In the herein presented work, we propose automating the value of these parameters by using these pertinence values. For this purpose, an experimental study is performed, to define the ideal value of these parameters. This first technique (i) allows obtaining a set of topics with the necessary keywords, therefore the output does not define the "name" of all generic concepts but an unnamed concept represented by a word together. In our work, we need to segment the movie audiovisual document based upon topics, themes and key words. In this respect, we assume the output of the LDA model as a theme and we focus on using a second technique to identify the concept of each theme and key words sets.

(ii) In the second technique we are interest in extracting the concept of topics of the movie based on semantic analysis. This purpose requires a rich vocabulary of concepts in order to identify the concept of set themes. The principle of our method's concept identification is based on the exploitation of the Large-Scale Concept Ontology for Multimedia (LSCOM)[1]. LSCOM defines a set of 1,000 concepts which proposes a taxonomy "Generalization" relationships such as "IS-A" for a higher level concepts. It organizes the concepts into six categories such as: objects, activities events, locations, peoples, graphics, and program categories. This method allows identifying the concepts referred to key word sets of documents, after identifying the related concepts in the word; we note that two cases are then possible: (1) If the list of themes refers to a single concept c_i, then this latter is assumed as a pertinent topic of this document. We define 'C 'as a set of relevant concept with $C = \{c_i\}$. (2) If the list of themes refers more to a concept, then a disambiguation process is set up. Starting from the fact that an audiovisual segment cannot cover more than a topic, we can rely in this following reasoning, on

---

[1] http://www.ee.columbia.edu/ln/dvmm/lscom/.

deducing the pertinent concept of each segment. We then apply the following rule: if a set of themes refer more to a concept, we measure the similarity between these concepts and the set of relevant concepts 'C'. Indeed, the most similar concept (i.e. the concept that has a maximum similarity value) corresponds to a pertinent concept of this set of themes and exactly the segment. In order to measure the similarity between concepts, several works have proven the importance of the Wu and Palmer measure [4] which is based on measuring the number of edges between concepts.

4   Experiment Results

This section is reserved to presenting the different experiments'results conducted to evaluate our approach. In the current works, we are conserned about extracting topics and themes description. As these descriptions are based on the pre-production script and the post-production texts superposed in the image, we used two databases to experiment our proposed approach such as Choi's dataset[2] and our own created database from the online Internet Movie Database IMDb and from the script web[3] sites. The total number of English documents collected in our own created corpus is 120. Choi's [16] dataset is a benchmark that is very used in the textual segmentation [17–20]. The following table (Table 1) show the data distribution in this dataset.

Table 1. Data distribution of Choi's dataset

Subsets	"3-5"	"6–8"	"3–11"	"9–11"
Document	100	100	162	100

Such an approach is based on the adaptation of the Latent dirichlet allocation LDA model to detect topics and themes of movie audiovisual documents. In this context, two separate experiments have been performed to this end. The first series of experiments is designed to justify the selection's estimation parameters and their formula to automation. The second experimentation includes the study of comparison results of other works.

4.1   Evaluation of the Estimation Parameter

As previously mentionned in the second phase of our approach, the LDA model requires an estimation parameter to the training process such as, α, β and K. In our work, we are interested automating the value of these parameters. Recalling briefly that α is a prior weight of topic k in a document and it takes as default value "0.5" and β is a prior weight of word w in a topic and it takes as default value "0.1". Two formulas are proposed to

---

[2] http://web.archive.org/web/20040810103924/http://www.cs.man.ac.uk/~mary/choif/software.html.

[3] http://dailyscript.com/movie.html; http://www.moviescriptsandscreenplays.com/.

refer to the pertinence of key word values. Indeed, for each document in this base we propose calculating a score which is defined by:

$$Scorej = \frac{\sum_{i=0}^{m} perti}{m} \qquad (2)$$

That m represents the number of pertinent key words. From the score of each document in the dataset, we define the following alpha and beta values:

$$\alpha = \frac{\sum_{i=0}^{m} Scorej}{nb_{doc}}; \qquad (3)$$

$$\beta = max(perti) \qquad (4)$$

Taking the important value K that represents the number of topics generated, we start with the assumption that a movie segment cannot present more than a subject, to define the following value: $K = n/2$ That n represents the number of segment in the document. To validate and position these values, we have performed a successive testes and comparing the result obtained by this value with the other works using the same base. Table 2 shows the performance of the test results for the use of this value of α and β (LDAα and LDAβ).

Table 2. Comparison results

Method	3–5	6–8	9–11	3–11	All dataset
JTextTile [22]	0.473	0.513	0.533	0.524	0.510
C99 [16]	0.115	0.104	0.112	0.143	0.118
TextSeg [18]	0.074	0.052	**0.037**	0.106	0.067
MinCutSeg [19]	0.340	0.241	0.174	0.243	0.249
LDA [20]	0.230	0.158	0.144	–	0.177
LDA α	0.082	0.049	0.032	0.098	0.065
LDA β	0.069	0.059	0.058	0.102	0.072
LDA α β	**0.073**	**0.049**	0.042	**0.098**	**0.065**

4.2 Comparison and Discussion Results

For the last series of experiments, we present a comprehensive evaluation and discussion of the results compared to some similar work. As our work is based on the movie audio-visual of document and there is no benchmark for assessing filmic documents based on the description textual documents, two databases are used in our works. The first is the 'Choi's dataset' to compare our suggested approach with the works using the LDA models based on this same dataset. The second is our database previously defined which contains the pre-production movie document. To achieve our experiments, we used the probabilistic error metric Pk [22]. Table 2 highlights the results found in our work in comparison with four works using the same base. As is shown in Table 2, to evaluate our approach, we measure the Pk for LDAα (the LDA used in [21], replacing the α value by our proposed value), LDAβ (replacing the β value by our proposed value) and the LDA$\alpha\beta$ (replacing the α and β value by our proposed values). Our method yields a better result by comparing with other work. The probabilistic error metric is worth, in the all datasets, 0.065.

Furthermore, given that we are interested in our work by the movies, we used our collected from the online Internet Movie Database Imdb and from the script websites. As our work is based on the movie audiovisual documents and in order to ameliorate the training process, we have been inspired by the idea proposed by [23], to enrich our training dataset. In this context, a 'large Movie Review Dataset' dataset[4] is highlighted which consists of 2,000 movie reviews. The experiment demonstrates that the proposed method is effective.

5   Conclusion and Futures Works

In this paper, a new approach for multimodal description of movie audiovisual documents is presented. Such an approach achieves a description level that combines three types of analysis such as the linguistic analysis, the statistic analysis and essentially the semantic analysis. The framework includes two phases: the text segmentation phase and the description extraction phase. The first phase has used across two types of modalities such as: textual and visual modality. The description extraction process is based on the statistical analysis through the adaptation of the Latent Dirichlet Allocation (LDA). The objective behind this adaptation is, on the one hand, to address the three problems of LDA presented in the state of the art and on the other hand to extract the themes of movie documents. Thereafter, a semantic analysis is highlighted in order to identify concepts of each set of themes, in other words to extract topics of each set of themes. Indeed, the Large-Scale Concept Ontology for Multimedia (LSCOM) is used for this end. The performance of the proposed process has been evaluated by using two datasets. As a first perspective, we consider segmenting the audiovisual documents through the description extracted from the textual and visual modalities. As a long-term perspective, we will study the works related to retrieval information to choose the most appropriate approach to a semantic interrogation process.

References

1. Ferret, O., Grau, B.: A topic segmentation of texts based on semantic domains. In: ECAI, pp. 426–430 (2000)
2. Rahangdale, A., Agrawal, A.J.: Information extraction using discourse analysis from newswires. Int. J. Inf. Technol. Converg. Serv. **4**(3), 21 (2014)

---

[4] http://www.cs.cornell.edu/people/pabo/movie-review-data/.

3. Azaiez, I., Ben Ahmed, M.: An approach of a semantic annotation and thematisation of AV documents. In: 2010 10th International Conference on Intelligent Systems Design and Applications (ISDA), pp. 1406–1411. IEEE (2010)

4. Chabi, A.H., Kboubi, F., Ahmed, Ben, M.: Thematic analysis and visualization of textual corpus. arXiv Prepr. arXiv1112.2071 (2011)

5. Deerwester, S.C., Dumais, S.T., Landauer, T.K., Furnas, G.W., Harshman, R.A.: Indexing by latent semantic analysis. JAsIs **41**, 391–407 (1990)

6. Bellegarda, J.R.: A latent semantic analysis framework for large-Span language modeling. EUROSPEECH (1997)

7. Atkinson, J., Gonzalez, A., Munoz, M., Astudillo, H.: Web metadata extraction and semantic indexing for learning objects extraction. Appl. Intell. **41**, 649–664 (2014)

8. Hofmann, T.: Probabilistic latent semantic indexing. In: Proceedings of the 22nd Annual International ACM SIGIR Conference on Research and Development in Information Retrieval, pp. 50–57. ACM (1999)

9. Blei, D.M., Ng, A.Y., Jordan, M.I.: Latent dirichlet allocation. J. Mach. Learn. Res. **3**, 993–1022 (2003)

10. Dou, W., Wang, X., Skau, D., Ribarsky, W., Zhou, M.X.: Leadline: Interactive visual analysis of text data through event identification and exploration. In: 2012 IEEE Conference on Visual Analytics Science and Technology (VAST), pp. 93–102. IEEE (2012)

11. Vavliakis, K.N., Symeonidis, A.L., Mitkas, P.A.: Event identification in web social media through named entity recognition and topic modeling. Data Knowl. Eng. **88**, 1–24 (2013)

12. Pahal, N., Chaudhury, S., Gaur, V., Lall, B., Mallik, A.: Detecting and correlating video-based event patterns: an ontology driven approach. In: Proceedings of the 2014 IEEE/WIC/ACM International Joint Conferences on Web Intelligence (WI) and Intelligent Agent Technologies (IAT)-Volume 01, pp. 438–445. IEEE Computer Society (2014)

13. Fourati, M., Jedidi, A., Gargouri, F.: Automatic identification genre of audiovisual documents. In: 2014 IEEE/ACS 11th International Conference Computer Systems Applied (AICCSA) (2014)

14. Fourati, M., Jedidi, A., Ben Hassine, H., Gargouri, F.: Towards fusion of textual and visual modalities for describing audiovisual documents. Int. J. Multimedia Data Eng. Manag. (IJMDEM) **6**(2), 51–69 (2015)

15. Stockinger, P.: Audiovisual Archives: Digital Text and Discourse Analysis. John Wiley & Sons, New York (2013)

16. Choi, F.Y.Y., Wiemer-Hastings, P., Moore, J.: Latent semantic analysis for text segmentation. In: Proceedings of EMNLP. Citeseer (2001)

17. Ji, X., Zha, H.: Domain-independent text segmentation using anisotropic diffusion and dynamic programming. In: Proceedings of the 26th Annual International ACM SIGIR Conference on Research and Development in Informaion Retrieval, pp. 322–329. ACM (2003)

18. Utiyama, M., Isahara, H.: A statistical model for domain-independent text segmentation. In: Proceedings of the 39th Annual Meeting on Association for Computational Linguistics, pp. 499–506. Association for Computational Linguistics (2001)

19. Malioutov, I., Barzilay, R.: Minimum cut model for spoken lecture segmentation. In: Proceedings of the 21st International Conference on Computational Linguistics and the 44th annual meeting of the Association for Computational Linguistics, pp. 25–32. Association for Computational Linguistics (2006)

20. Misra, H., Yvon, F., Jose, J.M., Cappe, O.: Text segmentation via topic modeling: an analytical study. In: Proceedings of the 18th ACM Conference on Information and Knowledge Management, pp. 1553–1556. ACM (2009)

21. Beeferman, D., Berger, A., Lafferty, J.: Statistical models for text segmentation. Mach. Learn. **34**, 177–210 (1999)
22. Hearst, M.A.: Multi-paragraph segmentation of expository text. In: Proceedings of the 32nd Annual Meeting on Association for Computational Linguistics, pp. 9–16. Association for Computational Linguistics (1994)
23. Yu, K., Li, Z., Guan, G., Wang, Z., Feng, D.: Unsupervised text segmentation using LDA and MCMC. In: Proceedings of the Tenth Australasian Data Mining Conference vol. 134, pp. 21–26. Australian Computer Society, Inc. (2012)

Texture Based Method for Automated Detection, Localization and Evaluation of the Exudates in Retinal Images

Dan Popescu[(✉)], Loretta Ichim, and Traian Caramihale

Faculty of Automatic Control and Computers,
University Politehnica of Bucharest, Bucharest, Romania
dan.popescu@upb.ro, loretta.ichim@aii.pub.ro,
traian90@gmail.ro

Abstract. The detection, localization and evaluation of exudates can contribute to help to diagnosis and to automatic create a specific databases. In this paper, we propose a new method for detection of exudates using adaptive selection of features based on matching score. For image processing two algorithms were proposed. The first uses sliding box method for optic disc detection and the second uses non-overlapping boxes for exudates detection. The selected features combines spatial distribution, spectral properties and fractal characteristics of the region of interest and are based on sorted values. The algorithm has three steps: detection and removing of the optic disc, detection and localization of exudates and finally the evaluation of exudates size. In order to evaluate of algorithm efficiency we used 70 images with exudates. The experimental results prove the efficiency of the proposed method due to accurately detection of the exudates and the size evaluation.

Keywords: Retinal images · Image segmentation · Optic disc · Texture features · Exudates

1 Introduction

Exudates are lesions which appear like brightness regions with different shapes and locations in the retinal images and they indicate the presence of diabetic retinopathy, whose incidence is growing. After appearances, the exudates are classified in hard exudates and soft exudates [1]. The hard exudates are easier to detect because they are well delimited, having clear boundaries. The detection of the exudates in early stage, especially in the macula region (near fovea), is very important because this fact can be used to prevent blindness. Thus, in [2] an optimized algorithm based on AM-FM (multiscale amplitude-modulation frequency-modulation) is proposed to detect exudates in the macula.

Over the years, many papers have approached the detection of the exudates by mean of thresholding methods on the retinal images in the green channel [3–5]. Because the illumination conditions differ not only from one database to another, but also inside the same database, some algorithms for illumination and contrast correction are necessary. Also, combinations of different classification techniques are often used

© Springer International Publishing Switzerland 2015
S. Arik et al. (Eds.): ICONIP 2015, Part IV, LNCS 9492, pp. 463–472, 2015.
DOI: 10.1007/978-3-319-26561-2_55

[6]. Regions of interest like the optic disc (OD) and exudates have similar chromatic aspects and therefore, in general, they are difficult to distinguish. Papers like [7–9] proposed first the localization and elimination of the OD by masks and then the detection of the exudates. The retinal images are approximately at the same scale but at different resolutions, illumination and dominant color, depending on the concrete database. Therefore image preprocessing algorithms are necessary.

Just as in [8] the proposed method in this paper is a hierarchical one, but differs in the algorithm structure and also as content. The algorithm starts with detecting OD, and then determines the candidate areas that contain exudates. Next, the segmentation of the exudates and the evaluation of the size of the exudates are performed. Unlike the previously mentioned works, our method uses textural and colour features for local analysis in order to determine candidate patches for exudates.

In [10] it is presented an efficient method using sliding boxes for detection of the OD. This paper uses a different signature of features for determining the boxes with exudates. In addition, the method automatically provides the percentage of the exudates in the patches size.

The aim of this paper is to present an automatic and effective method for detection and evaluation of exudates in retinal image. The algorithm is hierarchical and is based on a matching score on different vector signatures.

2   Methodology and Algorithms

Since the OD has a similar signature with exudates, it need to be marked by corresponding box, according to the algorithm proposed in [10], and removing boxes that intersect with the OD box. Also, the boxes containing black pixels (background), which appear in the edge areas of the images, are needed to be removed. Initial retinal image is divided into square boxes with dimension of 128×128 pixels and the sliding box method is used to detect the box which completely contains the OD.

In the learning phase, we used 10 images containing exudates after eliminated the boxes which intersect OD and the boxes which intersect the background (black pixels on the image sides). Detection and localization of areas with exudates were done by dividing retinal image in non-overlapping boxes of size 64×64 pixels for an initial image resolution of 700×605 pixels (STARE database) [11]. Each box has a position (x-y) in the corresponding grid obtained by dividing the initial image. This position index also is the ID of sub-image. For these boxes it is necessary to calculate a vector of values for relevant features, which shows the presence of the exudates. To this end, the following features, on the color channels R, G and B, were tested: mean intensity (Im), variance (Var), entropy (Ent), energy (En), homogeneity (Hom), contrast (Con), correlation (Cor), mass fractal dimension (Dm) and lacunarity (L). To detect the sensitive interval to exudates, the features were ordered and then analyzed both in ascending direction and descending direction. In order to select the features, the results were compared with those obtained by direct marking by the ophthalmologist. For the above features, a statistical analysis was made and, as a result, a performance degree (PD) was assigned to each of them. To this end, we consider the rate of recognition of exudates in image i by the feature F as the probability $PD(F_i)$ of occurrence of a box with exudates in the first 10 positions in the corresponding vector of feature values. Thus, the *PD* for feature *F* is given by the average (1) of *PD*(F_i):

$$PD(F) = \frac{1}{10} \sum_{i=1}^{10} PD(F_i) \tag{1}$$

The results for the tested features are presented in Table 1, where the final letter attached to the feature acronym represents the color channel.

Table 1. The results obtained in learning phase for evaluating the *PD* for the tested features.

ImR	ImG	ImB	VarR	VarG	VarB	EntR	EntG	EntB
0.3	0.8	0.1	0.1	0.2	0.1	0.1	0.1	0.1
EnR	EnG	EnB	HomR	HomG	HomB	ConR	ConG	ConB
0.1	0.5	0.1	0.1	0.4	0.1	0.2	0.6	0.1
CorR	CorG	CorB	DmR	DmG	DmB	LR	LG	LB
0.1	0.3	0.1	0.1	0.9	0.1	0.1	0.2	0.1

From Table 1, we chose the five features with the best degree of performance that highlights some properties in exudates texture: *DmG*– descending order (*PD* = 0.9), *ImG* – descending order (*PD* = 0.8), *ConG* – descending order (*PD* = 0.6), *EnG* – descending order (*PD* = 0.5) and *HomG* – descending order (*PD* = 0.4). As it can be seen, the conclusive results were obtained on color channel G. The features *Con*, *En* and *Hom* are calculated from mean co-occurrence matrix [12], *Im* from histogram and *Dm* according to the algorithm presented in [13, 14]. It can be seen, that although the resolution of sub-images is very low, mass fractal dimension gives the best results.

The formulas used for calculating the proposed features are the following:

$$Im = \frac{1}{M \times N} \sum_{i=1}^{M} \sum_{j=1}^{N} I(i,j) \tag{2}$$

$$Dm = \frac{\log(\sum_{u} \sum_{v} n_r(u,v))}{\log r} \tag{3}$$

$$Con_d = \sum_{k=1}^{L} \sum_{l=1}^{L} (k-1)^2 N_d(k,l) \tag{4}$$

$$Hom_d = \sum_{k=1}^{L} \sum_{l=1}^{L} \frac{N_d(k,l)}{1 + |k-l|} \tag{5}$$

$$En_d = \sum_{k=1}^{L} \sum_{l=1}^{L} N_d^2(k,l) \tag{6}$$

In (2) $I(i,j)$ is the intensity value of the pixel in a $M \times N$ image. In the formulas (4)–(6) N_d (k,l) represents the element of the mean gray level co-occurrence matrix at a displacement d [12]. For Dm evaluation (3), r is the division factor and $n_r(u,v)$ is the difference between the maximum value and the minimum value of $I(i,j)$ in the square of dimension $r \times r$ with origin in $u \times v$ position from the grid [13, 14].

For each box (sub-image) of an image, an average score of rejection (SR) is established (7), taking into account the position in the table of the values of the selected features, in correct order. For each feature F, the box position $P_F(x,y)$ which this assimilated as a score associated with a particular feature:

$$SR(x - y) = \frac{1}{5}[P_{Im}(x - y) + P_{Dm}(x - y) + P_C(x - y) + P_{Hom}(x - y) + P_{En}(x - y)] \quad (7)$$

We considered the first 20 positions in the table. If a box is not in the top 20 positions, then this is penalized with 50 points.

Another important characteristic for identification of boxes with exudates from the feature table is the standard deviation (8):

$$\sigma(x - y) = \sqrt{\frac{1}{4}\sum_{i=1}^{5}(SR(x - y) - F_i(x - y))^2} \quad (8)$$

where $F_i(x-y)$ represents one of the selected features.

So, if SR and σ exceed some thresholds experimentally established, the interrogated boxes are considered without exudates and the boxes with lower value (under thresholds) are considered candidates.

After selection of candidate boxes, the next step is to separate the exudates inside from these boxes and to evaluate the local and global sizes of the exudates.

In the retained boxes (considered as candidates), the objects of interest (exudates) are segmented by Otsu method [15]. Thus, the exudates are considered as white on a black background. Then, the white pixels are counted and the result is divided by the total number of the pixels from the box. So, a local percentage of the exudates size is available. Similarly, by counting all the white pixels from the boxes and reporting them to the total number of the image pixels the total percentage of exudates can be obtained.

Based on above methodology we proposed the following algorithm for detection and evaluation of exudates:

1. Upload the image;
2. Image decomposition (R,G,B color channels);
3. Noise rejection by 3×3 median filter and by morphological operations (erosion and dilation, grey level case) on each channel;
4. Generation of the boxes with sliding box method for coarse OD detection [2];
5. Calculate the selected features and their extremes inside the boxes;
6. Calculate the signature of the corresponding boxes*;
7. Coarse OD detection, taken into account boxes of dimension 128×128 pixels and the extremes of the features from the signature;

8. Position determination of the box B containing OD;
9. Generation of the fixed non-overlapping boxes with dimension of 64×64 pixels for detection of the exudates;
10. Remove inadequate boxes**;
11. Calculate for each box the selected features and their extremes;
12. Calculate the matching score for each of five features and for 20 boxes;
13. Calculate the signature of the corresponding box;
14. Retain the candidate boxes;
15. Segmentation of exudates inside the selected boxes;
16. Calculate the percentage of the exudates of each box with exudates;
17. Calculate the percentage of the exudates in the analyzed retinal image.

Observations:
*The signature for coarse detection of OD contains the following elements [10]: $Dm_G = \min$, $Dm_R = \min$, $L_R = \max$, $L_G = \max$, $C_G = \max$. The OD is detected and localized if almost 4 from 5 elements of the signature belong to the same box (the box which contains OD).
**Position of the box B containing OD is (i_B, j_B), the left upper corner. Then the boxes which need to be removed as intersecting OD are the boxes (64×64 pixels) which contain rows between i_B and $i_B + 127$ or columns between j_B and $j_B + 127$. On the other hand, as it is previously noted, the boxes containing black pixels (background) are also removed.

The method used to compute the mass fractal dimension and lacunarity was implemented using FracLac [16], which is a plug-in of ImageJ. Several other programs from MATLAB were subsequently used for image processing and for computing the texture features. Especially, using parfor loops in MATLAB, the total time of execution was improved.

3   Experimental Results and Discussions

The proposed algorithm for exudates detection and localization was implemented taking into account a set of 70 retinal images with exudates from the STARE database. Among them 10 images were used in the learning phase and 60 images for testing. From these, 8 images were used for exemplification: 4 images for the learning phase and 4 for the testing phase (Fig. 1). Each image has assigned a unique ID.

The first phase of the algorithm consists on the OD localization and markup the boxes to be removed. Thus, in Fig. 2 we present an example of an image (ID 124) which contains OD, exudates and background. OD is framed in a 128×128 box (OD-ID 124). Sub-image (box) 1-3 contains black pixels (background) and sub-image 6-9 contains a portion of OD and they needed to be removed.

The second phase of algorithm consists in exudates detection and localization based on matching score. The boxes are indexed as elements of a grid $(i\text{-}j)$, $i = \overline{1, 9}$, $j = \overline{1, 10}$. For exemplification the images ID 058 and ID 001 from Fig. 1 were chosen.

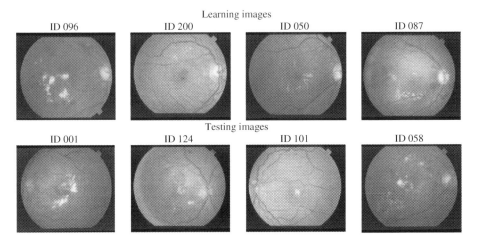

Fig. 1. Samples of learning and testing images.

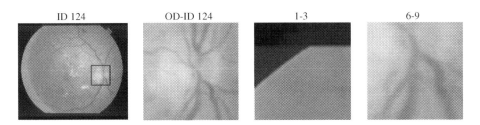

Fig. 2. Image (ID 124) which contains OD, exudates and background.

The results of feature values in order to obtain box signatures are presented in Tables 2 and 3 respectively. The boxes to be removed are the following: a) for OD: 5-8, 5-9, 6-8, 6-9 and b) for black area: 1-1, 1-2, 1-3, 1-4, 1-5, 1-6, 1-7, 1-8, 1-9, 1-10, 2-1, 2-2, 2-10, 3-1, 3-10, 4-1, 5-1, 6-1, 7-1, 7-10, 8-1, 8-2, 8-10, 9-1, 9-2, 9-3, 9-4, 9-5, 9-6, 9-7, 9-8, 9-9, 9-10. The boxes indicated with * will be considered in Tables 4 and 5 to calculate the SR and σ for identification of boxes with/without exudates. So for image ID 058 all the boxes in Table 4 contain exudates and for image ID 001 the last position (8-7) doesn't contain exudates.

The thresholds for SR and σ were chosen respectively as $SR = 20$ and $\sigma = 10$. Thus the boxes with exudates for images ID 058 and ID 001 are presented in Tables 4 and 5.

In Fig. 3 there are presented the first 5 candidates selected by the proposed algorithm for the images ID 058 and ID 001. For the last image, the process of segmentation is also presented in Fig. 3 and the local percentage is evaluated for each box (Table 6).

In the Table 6 there are presented 5 from the total number of 12 boxes with exudates from image ID 001. Noting that, the total percentage of the exudates in the image ID 001 is considered as the ratio between the total area of exudates, obtained as the sum of all local areas (8683 pixels) and the area of proper retinal image (in this case, 291886 pixels). It can be seen that the large exudates are firstly detected.

Table 2. The first 20 results obtained for image ID 058.

No.	Con	Box pos.	En	Box pos.	Hom	Box pos.	Im	Box pos.	Dm	Box pos.
1	0.531	*4-5	0.158	*5-5	0.789	*4-5	0.467	*5-5	2.452	*5-5
2	0.459	*5-5	0.211	*4-5	0.807	*5-5	0.422	*4-5	2.462	*4-5
3	0.388	*5-6	0.290	*5-6	0.835	*5-6	0.417	*2-6	2.474	7-4
4	0.323	2-7	0.295	3-5	0.839	2-7	0.402	3-7	2.486	6-4
5	0.310	*2-5	0.300	2-7	0.854	*2-5	0.401	*2-5	2.486	*2-5
6	0.290	3-5	0.313	5-7	0.855	3-5	0.382	3-6	2.495	*5-6
7	0.265	3-6	0.315	*2-5	0.867	7-10	0.382	2-7	2.500	4-4
8	0.257	8-6	0.318	7-9	0.867	3-6	0.380	*5-6	2.506	6-3
9	0.255	5-7	0.319	8-6	0.871	8-6	0.376	5-7	2.507	*2-6
10	0.252	7-9	0.341	8-4	0.873	5-7	0.373	2-4	2.508	7-5
11	0.231	7-4	0.351	*2-6	0.874	7-9	0.371	3-5	2.519	4-6
12	0.229	3-7	0.364	7-8	0.888	8-4	0.361	2-3	2.529	5-7
13	0.226	*2-6	0.373	8-8	0.888	*2-6	0.353	4-7	2.530	5-4
14	0.223	8-4	0.431	3-6	0.888	6-9	0.349	4-6	2.533	4-8
15	0.223	6-9	0.437	6-9	0.890	7-5	0.348	4-8	2.535	2-7
16	0.221	7-5	0.439	8-3	0.890	2-4	0.345	4-4	2.536	3-7
17	0.219	2-4	0.480	3-7	0.890	3-7	0.344	5-4	2.536	7-2
18	0.211	7-8	0.488	4-7	0.894	7-8	0.341	6-7	2.537	8-2
19	0.206	9-5	0.496	7-4	0.903	4-8	0.333	3-3	2.542	7-8
20	0.202	8-3	0.502	9-4	0.904	7-4	0.332	4-3	2.547	8-3

* marks the boxes with exudates.

Table 3. The first 20 results obtained for image ID 001.

No.	C	Box pos.	E	Box pos.	Hom	Box pos.	Im	Box pos.	Dm	Box pos.
1	0.918	*5-7	0.117	*5-7	0.756	*7-5	0.687	*5-7	2.454	*5-7
2	0.789	*6-6	0.132	*7-5	0.760	*5-7	0.593	*7-5	2.472	*7-5
3	0.785	*7-5	0.204	*5-6	0.787	*4-6	0.542	*6-6	2.472	7-6
4	0.700	*4-7	0.240	*4-6	0.796	*5-6	0.530	*5-6	2.479	6-5
5	0.553	*4-6	0.273	*6-6	0.802	*6-6	0.524	7-6	2.482	*6-6
6	0.486	*5-6	0.287	*4-5	0.824	*4-7	0.516	*4-6	2.482	*4-7
7	0.339	8-4	0.288	8-6	0.830	8-4	0.512	*4-7	2.494	*4-6
8	0.329	8-6	0.317	8-4	0.835	8-6	0.493	*4-5	2.499	6-7
9	0.323	7-6	0.324	2-8	0.843	4-9	0.478	4-4	2.517	5-2
10	0.315	*4-5	0.328	3-9	0.843	*4-5	0.475	5-4	2.521	7-4
11	0.311	*8-7	0.337	5-9	0.844	5-9	0.475	6-7	2.521	*5-6
12	0.285	5-2	0.343	7-8	0.861	3-9	0.463	6-5	2.527	3-2
13	0.276	3-9	0.370	*4-7	0.873	5-2	0.462	5-5	2.528	8-4
14	0.239	8-2	0.379	7-2	0.884	7-2	0.461	3-5	2.532	*4-5
15	0.231	7-2	0.380	8-3	0.886	8-5	0.457	6-4	2.533	4-2
16	0.277	8-5	0.400	2-6	0.888	7-6	0.457	3-6	2.541	2-6
17	0.217	6-7	0.458	7-6	0.891	7-8	0.450	7-4	2.546	8-5
18	0.216	7-8	0.449	3-9	0.892	2-6	0.435	5-2	2.547	3-7
19	0.211	3-3	0.445	7-8	0.894	3-3	0.433	6-3	2.549	8-6
20	0.206	3-7	0.441	8-5	0.901	8-3	0.432	5-3	2.550	2-3

* marks the boxes with exudates.

Table 4. Score of rejection and standard deviation for some boxes with exudates from ID 058.

Box Pos.	P_{Con}	P_{En}	P_{Hom}	P_{Im}	P_{Dm}	SR/σ
5-5	2	1	2	1	1	1.4/0.547
4-5	1	2	1	2	2	1.6/0.547
5-6	3	3	3	8	6	4.6/2.302
2-5	5	7	5	5	5	5.4/0.894
2-6	13	11	13	3	9	9.8/4.147

Table 5. Score of rejection and standard deviation for some boxes with/without exudates from ID 001.

Box Pos.	D(C)	D(E)	D(Hom)	D(Im)	D(Dm)	SR/σ
5-7	1	1	2	1	1	1.2/0.447
7-5	3	2	1	2	2	2/0.707
6-6	2	5	5	3	5	4/1.414
4-6	5	4	3	6	7	5/1.581
5-6	6	3	4	4	11	5.6/3.209
4-7	4	13	6	7	6	7.2/3.420
4-5	10	6	10	8	14	9.6/2.966
8-7	11	50	50	50	50	42.2/17.44

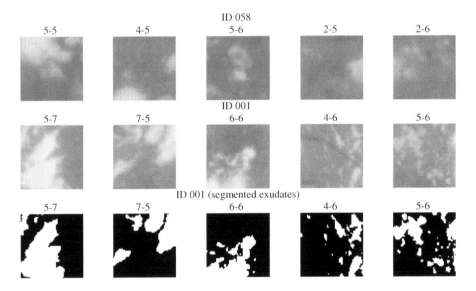

Fig. 3. Boxes with exudates from image ID 058 and ID 001.

In order to appreciate the sensitivity and specificity of our method we consider the positive rate and respectively the negative rate for the investigated boxes inside of the tested image. Then we consider the average of these rates over all images (70 with

Table 6. Area and percentage of exudates for different boxes and for whole image ID 001.

Box	ID 001/5-7	ID 001/7-5	ID 001/6-6	ID 001/4-6	ID 001/5-6	ID 001
Area (pixels)	1757	1127	983	622	961	8683
Percentage	42.9 %	27.5 %	24 %	15.18 %	23.46 %	3 %

exudates). The results are presented in Table 7 (as sensitivity, specificity and accuracy) and compared with other methods [17–19].

We tested 3710 boxes from 70 images (from STARE database) and the results are: false positive – 212 boxes and false negative – 257 boxes.

Table 7. Some performances of our method.

	Method in [17]	Method in [18]	Method in [19]	Our method
Sensitivity	93 %	77 %	88.45 %	93 %
Specificity	94.1 %	83 %	91 %	94.2 %
Accuracy	–	–	–	93.68 %

4 Conclusions

The proposed method differentiates exudates of non-exudates in retinal images after optic disc removal. The localization and size of the exudates are also evaluated. The algorithm is a hierarchical one, starting with OD detection and ended with the exudates size calculation. Both for OD detection and exudates detection we selected the features from retinal images, taking into account information about pixel distribution, color and fractal characteristics. Two important cues are considered: performance degree for feature selection and matching score for classification. For OD and exudates we considered two different local image processing: one based on overlapping boxes criteria and another based on non-overlapping box criteria. For OD and exudates the feature signatures are different. Due to the feature ordering and the voting scheme, the large lesions are firstly detected. Finally, statistical analysis validates the proposed method.

References

1. Akram, M.U., Khalid, S., Tariq, A., Shoab, K.A., Azam, F.: Detection and classification of retinal lesions for grading of diabetic retinopathy. Comput. Biol. Med. **45**, 161–171 (2014)
2. Agurto, C., Murray, V., Yu, H., Wigdahl, J., Pattichis, M., Sheila, N., Barriga, S., Soliz, P.: A multiscale optimization approach to detect exudates in the macula. IEEE J. Biomed. Health Inform. **18**, 1328–1336 (2014)
3. Phillips, R., Forrester, J., Sharp, P.: Automated detection and quantification of retinal exudates. Graefe's Arch. Clin. Exp. Ophthalmol. **231**, 90–94 (1993)
4. Ege, B., Hejlesen, O., Larsen, O., Moller, K., Jennings, B., Kerr, D., Cavan, D.: Screening for diabetic retinopathy using computer based image analysis and statistical classification. Comput. Meth. Programs Biomed. **62**, 165–175 (2000)

5. Usher, D., Dumskyj, M., Himaga, M., Williamson, T.H., Nussey, S., Boyce, J.: Automated detection of diabetic retinopathy in digital retinal images: A tool for diabetic retinopathy screening. Diabetic Med. **21**, 84–90 (2004)

6. Niemeijer, M., Russell, S.R., Suttorp, M.A., van Ginneken, B., Abràmoff, M.D.: Automated detection and differentiation of drusen, exudates, and cotton-wool spots in digital color fundus photographs for early diagnosis of diabetic retinopathy. Invest. Ophthalmol. Vis. Sci. **48**, 2260–2267 (2007)

7. Wisaeng, K.: Automatic detection of exudates in diabetic retinopathy images. J. Comput. Sci. **8**, 1304–1313 (2012)

8. Bu, W., Wu, X., Chen, X., Dai, B., Zheng, Y.: Hierarchical detection of hard exudates in color retinal images. J. Softw. **8**, 2723–2732 (2013)

9. Amel, F., Mohammed, M., Abdelhafid, B.: Improvement of the hard exudates detection method used for computer-aided diagnosis of diabetic retinopathy. I.J. Image, Graphics and Signal Processing 4, 19–27 (2012)

10. Popescu, D., Ichim, L., Dobrescu, R.: Sliding box method for automated detection of the optic disc and macula in retinal images. In: Ortuño, F., Rojas, I. (eds.) IWBBIO 2015, Part I. LNCS, vol. 9043, pp. 250–261. Springer, Heidelberg (2015)

11. Structured analysis of the retina. http://www.ces.clemson.edu/~ahoover/stare/

12. Popescu, D., Dobrescu, R., Angelescu, N.: Statistical texture analysis of road for moving objectives. U.P.B. Sci. Bull. Series C. **70**, 75–84 (2008)

13. Sarker, N., Chaudhuri, B.B.: An efficient differential box-counting approach to compute fractal dimension of image. IEEE Trans. Syst. Man Cybern. **24**, 115–120 (1994)

14. Chaudhuri, B.B., Sarker, N.: Texture segmentation using fractal dimension. IEEE Trans. Pattern Anal. Mach. Intell. **17**, 72–77 (1995)

15. Otsu, N.: A threshold selection method from gray-level histograms. IEEE Trans. Sys. Man. Cybern. **9**, 62–66 (1979)

16. Karperien, A.: FracLac for ImageJ. http://rsb.info.nih.gov/ij/plugins/fraclac/FLHelp/Introduction.htm

17. Osareh, A., Mirmehdi, M., Thomas, B., Markham, R.: Automated identification of diabetic retinal exudates in digital colour images. Br. J. Ophthalmol. **87**, 1220–1223 (2003)

18. Hatanaka, Y., Nakagawa, T., Hayash, Y., Fujita, A., Mizukusa, Y., Kakogawa, M., Kawase, K., Hara, T., Fujita, H.: CAD scheme for detection of hemorrhages and exudates in ocular fundus images. In: Proceedings of SPIE Medical Imaging 2007. Computer-aided Diagnosis, San Diego, vol. 6514, pp. 65142M-1-65142M-8 (2007)

19. Sharath Kumar, P.N., Rajesh Kumar, R., Sathar, A., Sahasranamam, V.: Automatic detection of exudates in retinal images using histogram analysis. In: 2013 IEEE Recent Advances in Intelligent Computational Systems (RAICS), pp. 277–281 (2013)

A Graph Community and Bag of Categorized Visual Words Based Image Retrieval

Wang Zhang, Haoyuan Lu, Shangdi Sun, and Xiaodong Gu[✉]

Department of Electronic Engineering, Fudan University, Shanghai 200433, China
{wangzhang12,sdsun14,xdgu}@fudan.edu.cn

Abstract. We propose a novel method for organizing image dataset with tags in which each image is regarded as a node of a complex network and the semantic information of the dataset can be grouped with community detection algorithm. The retrieval process are divided into two phases to improve accuracy by searching in a smaller sub-dataset. In the first phase, tags of a query image are taken as the priority matter to select target communities; in the second one, content based image retrieval is performed within the target communities. Bag of categorized visual words model is proposed as image content representation to improve description ability for objects compared with bag of visual words. Besides we also try to implement bag of visual words with SOM classifier.

Keywords: Image retrieval · Bag of categorized visual words · Tag graph · Community detection

1 Introduction

Recent years have witnessed the explosive growth of web images, especially in social networks like Flickr where users have access to upload their photos and corresponding textual descriptions.

Faced with retrieval challenges in such natural scene images, the combination of content and tag based methodology is a dominated approach. Li et al. [1] propose to learn tag relevance by accumulating votes from visual neighbors.

Tagging is a smart way to understand image content. Automatic annotation methods have been studied for missing and noisy tags [2–4]. Wu et al. [5] focus on noisy tag removal.

Kuo et al. [6] take each image as a node to construct two graphs: the visual graph for content similarity and the textual graph for tag similarity. Some clusters named visual clusters and textual clusters corresponding to these two graphs appear, and auxiliary features propagation is performed among the associated images in the extended visual or textual clusters. By doing so, content and tag information are combined to achieve better retrieval accuracy.

In this paper, we try to explore tag information from the aspect of complex networks for image retrieval, and take it as the primary concern rather than a supplement. Meanwhile, bag of categorized visual words model is developed as content representation to improve content retrieval accuracy compared with bag of visual words.

© Springer International Publishing Switzerland 2015
S. Arik et al. (Eds.): ICONIP 2015, Part IV, LNCS 9492, pp. 473–480, 2015.
DOI: 10.1007/978-3-319-26561-2_56

The rest of this paper is organized as follows. It starts in Sect. 2 with a brief review of bag of visual words model, and followed by a proposed version bag of categorized visual words model. The tag graph and community detection algorithm are introduced in Sect. 3. Section 4 shows experimental performance in image retrieval. Section 5 concludes the paper.

2   Bag of Categorized Visual Words

2.1   Bag of Visual Words

Bag of visual words (BOVW) originates from document retrieval technique. BOVW has become a popular image representation model for object retrieval, classification, and recognition since it was first proposed by Sivic and Zisserman [7] and modified in some related researches [8–11]. Here we briefly introduce BOVW representation for an image as follows:

1. Extract SIFT features for train images and get n descriptors $(\vec{x}_1, \vec{x}_2, \dots, \vec{x}_n)$;
2. Perform K-means clustering on descriptors $(\vec{x}_1, \vec{x}_2, \dots, \vec{x}_n)$ to obtain k $(k \ll n)$ centroids $\vec{m}_1, \vec{m}_2, \dots, \vec{m}_k$ which are called visual words, and k is the number of visual words;
3. Extract SIFT features for a given image, and find the nearest visual words for each descriptor in the image;
4. Count the frequency for each visual word found in Step 3 to list as a frequency histogram, and this frequency histogram is BOVW representation of an image.

2.2   Bag of Categorized Visual Words

Intuitively in BOVW, descriptors extracted from different image categories are taken as a whole sample set to cluster visual words. The description ability for particular object is weaken by such computation when observed from semantic level. We propose a modified model named bag of categorized visual words (BOCVW) to overcome this weakness.

BOCVW computes visual words for each semantic category in train set respectively and represents a given image in a histogram of all categories' words. We call the visual words computed from a particular train category category-description visual words, and the visual words for all categories categorized visual words. BOCVW algorithm is:

1. Extract SIFT features for train images in each category, and get N descriptors in each category $(\vec{x}_1, \vec{x}_2, \dots, \vec{x}_N)$, where N is not the same for different categories;
2. Perform K-Means clustering for each category samples to obtain n $(n \ll N)$ centroids $M_{Ci} = (\vec{m}_1, \vec{m}_2, \dots, \vec{m}_n)$, where M_{Ci} denotes category-description visual words for i^{th} category.
3. Concrete category-description visual words of all categories into categorized visual words $M_C = (M_{C1}, M_{C2}, \dots, M_{Ck})$, where k is the number of categories in train set;
4. Extract SIFT features for a given image, and find the nearest categorized visual words for each descriptor of the image;

5. Count the frequency for each categorized visual word generated in Step 3 to list as a frequency histogram, and this frequency histogram is BOCVW representation of an image.

   The experimental performance contrast of semantic description ability between BOVW and BOCVW is shown in Sect. 4.

3 Tag Graph Community Detection

3.1 Complex Networks and Community Detection

Studies on natural environment and human behaviors show that some phenomena such as spreading germs, scientific collaboration, web links can be investigated from network perspective when we observe the individuals as nodes and the relations between individuals as edges, so that graph theory can be employed in finding the rules hidden in these complex systems named complex networks.

There often appears some cliques in a complex network. Nodes inside a clique are generally connected densely, and these nodes have a relatively bigger probability of sharing similar properties; while nodes from different cliques have smaller chance to be connected and share less common properties. These cliques are communities in complex networks and we call clique partition community detection.

Many proven community detection algorithms have been developed in recent years. We employ a seed developing method [12] to find semantically similar images from tag information. This community detection algorithm comes from the idea that two neighboring nodes are likely to belong to the same community if they share the maximum number of common neighbors. The algorithm detects communities in two steps: finding the seeds and developing them. A detailed example experiment refers to [12].

1. Finding the seeds:
   (a) Obtain the adjacent list of a complex network, and count the number of common neighbors for each pair of nodes;
   (b) Find the maximum number of common neighbors for each node in adjacent list;
   (c) If the maximum number of common neighbors found in Step b for a node is shared with another node that has more common neighbors with others, this pair of nodes are ignored, otherwise taken as a seed.
2. Developing the seeds:
   (a) Define a parameter named absorption power to measure the connection density between the seed and its neighbor

$$f_s = \frac{w_b}{\sum k_s \cdot \sum k_n}, \tag{1}$$

where w_b is the number of links between the seed and its neighbor, $\sum k_s$ is the sum of the degrees of the members in seed s, and $\sum k_n$ is the sum of the degrees of the members in its neighbor n. Each neighbor of a seed may be a single node or another seed;

(b) Find the neighbor that has the maximum value of absorption power with a seed;
(c) Judge whether the seed can absorb its neighbor of maximum value of absorption power with the same rule as in finding the seeds;
(d) If a seed can absorb its neighbor, the seed grows by merging the neighbor to become a bigger community[1];
(e) Update the links and absorption power between each seed and its neighbor;
(f) Repeat from Step b to Step e until termination condition which can be set as the number of communities and absorption power threshold α.

3.2 Tag Graph

Two logical assumptions are made on intuition about the relationship between image semantics and its tags:

1. The closer semantic meaning between two images brings bigger chance of sharing common tags;
2. Two images sharing a common tag has a bigger chance to have similar semantic meaning than those sharing none.

Based on these two assumptions we construct a tag graph for an image dataset by denoting each image as a node and each edge as the relationship of sharing one common tag. For the circumstances of more than one common tags between two images, we still regard it as one edge, namely unweighted graph considered only in this paper.

After tag graph constructed, the community detection algorithm in Sect. 3.1 is conducted to roughly partition the dataset into some semantic groups according to tag information. By doing so, the content based image retrieval can be carried out in a smaller sub-dataset instead of the original whole dataset to achieve better accuracy.

4   Experimental Results

4.1   Dataset

We conduct image retrieval experiments on a Flickr image dataset[2]. The dataset has 7 categories 1282 ground truth photos taken and tagged by Flickr users. We preprocess the tags as follows and finally get a subset of 1215 images. It consists of 7 categories: Colosseum (88), Eiffel Tower (511), Golden Gate Bridge (162), Pisa (119), Starbucks (52), Tower Bridge (170), Are de Triomphe (113).

1. Only tags in English letters a ~ z (uppercase and lowercase treated equally) and character length between 3 and 19 investigated;
2. Count tag frequency in 1282 images and delete tags with frequency less than 3 times;
3. Delete frequently appeared semantic meaningless tags like *'travel'*, *'holiday'*, etc.;

---

[1] Note that the term seed has the same meaning as community in this algorithm.
[2] Available at: http://www.cmlab.csie.ntu.edu.tw/%7Ekuonini/Flickr11K.

4. Merge different-spelled tags in different languages, such as tags *'colosseo', 'coliseum', 'coloseum', 'coliseo'* are regarded as *'colosseum'* in colosseum images.

The number of query images for each category is chosen to be ten. Twenty images for each category are used for training the categorized visual words in BOCVW and visual words in BOVW.

4.2 BOCVW/BOVW Parameters

Regular grid SIFT features [8] with square grid width of 16 pixels and neighbor grids overlapping of 8 pixels are extracted for both models.

We adjust the number of visual words in BOVW from 400 to 2000 with spacing 200. The retrieval accuracy tends to grow with the increase of visual words number and it also tends to be saturated when the number reaches 1200. Therefore the number of visual words is set to be 1200 in BOVW. From the same criterion we get the best categorized visual words number (350×7), in which 350 category-description visual words are generated for each category.

4.3 Image Retrieval Framework

With the above mentioned ideas about tag graph and community detection, the flow chart for such a tag-first image retrieval method is shown in Fig. 1.

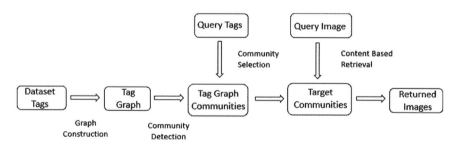

Fig. 1. Tag-first image retrieval framework flow chart

4.4 Tag Community Detection and Selection

The maximum absorption power threshold is set to be $\alpha = 5 \times 10^{-7}$ in the experiment. A case of 10 detected communities is shown in Table 1. The bold italic number "84" means there exists 84 images belonging to the first category in the first community while the total number of images in the first category is 88. The blanks in the table are filled with zeros.

The table demonstrates a majority of images in the 2nd and 7th categories fall into two different communities, while most of images in other categories only one community. The 1st, 2nd, 4th and 9th communities consist of two or three categories,

Table 1. Detected case of 10 communities

Community index	Community members' category belongs						
	1/88	2/511	3/162	4/119	5/52	6/170	7/113
1	84	3		104		3	
2		424		3	1	5	80
3		2					
4	2	73		6		6	33
5		2					
6		2					
7		1					
8		1				3	
9	2	3	162	6	51	153	
10						3	

while the rest only have a few images. The tag community detection method is of application value for image retrieval for its ability to roughly group images from tag perspective.

As community detection is carried out on tag information only, target communities for content based image retrieval are selected with the tags of a query image. The community selection rules are defined as follows:

1. Abandon the few-nodes communities (As in Table 1 case only the 1st, 2nd, 4th and 9th communities are investigated as the candidate communities);
2. List the tags of a query image $tag = \{tag_1, tag_2, \dots, tag_k\}$, where k is the number of tags the query image has;
3. Count the frequency each query tag appears in each candidate community $tagFreq = \left[\overrightarrow{tagFreq_1}, \dots, \overrightarrow{tagFreq_i}, \dots, \overrightarrow{tagFreq_k}\right]$, where $\overrightarrow{tagFreq_i}$ denotes the ith tag frequency in all candidate communities;
   $\overrightarrow{tagFreq_i} = \left[tagFreq_i^1, tagFreq_i^2, \dots, tagFreq_i^N\right]^T$, where N is the total number of candidate communities;
4. Compute mean frequency of query tags for the query image in all candidate communities. $\overrightarrow{tagFreq_{ave}} = \left[tagFreq^1, \dots, tagFreq^j, \dots, tagFreq^N\right]$, where $tagFreq^j$ denotes the mean value of all the tags' frequency of the query image in the jth community, and it's obtained by this equation:

$$tagFreq^j = mean\left(tagFreq_1^j, tagFreq_2^j, \dots, tagFreq_k^j\right);$$

5. Set the mean tag frequency threshold as $threshold = \max\left(tagFreq^j\right) * \beta$, where $j = 1, 2, \ldots, N$. In the experiment we set $\beta = 0.3$. The communities in which mean tag frequency of tags of a query image is no less than $threshold$ are selected as target communities.

4.5 Overall Experimental Performance

The retrieval results based on BOCVW/BOVW, and tag graph community detection are shown in Table 2. The table is a list of average retrieval accuracy at recall rates varying from 0.1 to 0.7 over 7 categories under BOCVW/BOVW, and the combination of tag graph community detection and BOCVW/BOVW. Retrieval performance with categorized visual words of 150×7, in which an image is represented by a 1050 dimensional vector, is also tested to compare storage cost between BOCVW and BOVW. The data in Table 2 demonstrate BOCVW increases retrieval accuracy by over 8 percentage points compared with BOVW at recall rates below 0.4, and some smaller improvement at higher recall rates. Besides BOCVW increases retrieval accuracy by over 5 % points without paying extra storage cost at recall rates below 0.4. The tag graph community detection method generally improves image retrieval accuracy by more than 20 % points (except with BOCVW at recall rate 0.1) compared at the same recall rate with the corresponding content based method.

Table 2. Image retrieval accuracy at different recall rates

	0.1	0.2	0.3	0.4	0.5	0.6	0.7
BOVW12DO	0.5912	0.4744	0.3806	0.3106	0.2482	0.2103	0.1858
BOCVW150 × 7	0.6530	0.5249	0.4479	0.3645	0.2856	0.2366	0.1975
BOCVW350 × 7	0.6955	0.5706	0.4735	0.3997	0.3235	0.2535	0.2121
Tag graph + BOVW1200	0.8105	0.7086	0.6343	0.5642	0.4916	0.4476	0.4135
Tag graph + BOCVW150 × 7	0.8453	0.7529	0.6890	0.6163	0.5351	0.4779	0.4270
Tag graph + BOCVW350 × 7	0.8859	0.8019	0.7259	0.6529	0.5689	0.4950	0.4410

5 Conclusions

Bag of categorized visual words (BOCVW) model and a framework of tag-first image retrieval approach are proposed and demonstrated to effectively help improve image retrieval performance. This tag graph and its community detection algorithm are also of applicable value for image dataset organization.

Besides the above BOCVW and BOVW representation of an image by clustering words with K-means and assigning each SIFT feature vector into the words, we are also trying to implement the models with SOM classifier for its potential of topological structure information over K-means clustering. Some partial results with visual words number below 400 have shown SOM classifier based BOCVW/BOVW are feasible though we cannot enlarge visual word size at present for the restriction by software and hardware performance. In future work we will try to realize it by optimizing the code and upgrading hardware. And we will also test the tag community detection method on large scale datasets.

Acknowledgements. This work was supported in part by National Natural Science Foundation of China under grant 61371148.

References

1. Li, X., Snoek, C.G.M., Worring, M.: Learning social tag relevance by neighbor voting. IEEE Trans. Multimedia **11**(7), 1310–1322 (2009)
2. Victor, L., Manmatha, R., Jiwoon, J.: A model for learning the semantics of pictures. In: Advances in Neural Information Processing Systems Conference, vol. 16 (2003)
3. Metzler, D., Manmatha, R.: An inference network approach to image retrieval. In: Enser, P.G.B., Kompatsiaris, Y., O'Connor, N.E., Smeaton, A.F., Smeulders, A.W.M. (eds.) CIVR 2004. LNCS, vol. 3115, pp. 42–50. Springer, Heidelberg (2004)
4. Carneiro, G., Chan, A.B., Moreno, P.J., Vasconcelos, N.: Supervised learning of semantic classes for image annotation and retrieval. IEEE Trans. Pattern Anal. Mach. Intell. **29**(3), 394–410 (2007)
5. Wu, L., Jin, R., Jain, A.K.: Tag completion for image retrieval. IEEE Trans. Pattern Anal. Mach. Intell. **35**(3), 716–727 (2013)
6. Kuo, Y.H., Cheng, W.H., HT, Lin, Hsu, W.H.: Unsupervised semantic feature discovery for image object retrieval and tag refinement. IEEE Trans. Multimedia **14**(4), 1079–1090 (2012)
7. Sivic, J., Zisserman, A.: Video Google: a text retrieval approach to object matching in videos. In: Ninth IEEE International Conference on Computer Vision, vol. 2, pp. 1470–1477. IEEE, Nice (2003)
8. Fei-Fei, L., Perona, P.: A Bayesian hierarchical model for learning natural scene categories. In: IEEE Computer Society Conference on Computer Vision and Pattern Recognition, vol. 2, pp. 524–531. IEEE, San Diego (2005)
9. Philbin, J., Chum, O., Isard, M., Sivic, J., Zisserman, A.: Lost in quantization: improving particular object retrieval in large scale image databases. In: IEEE Conference on Computer Vision and Pattern Recognition, pp. 1–8. IEEE, Anchorage (2008)
10. Nister, D., Stewenius, H.: Scalable recognition with a vocabulary tree. In: IEEE Computer Society Conference on Computer Vision and Pattern Recognition, vol. 2, pp. 2161–2168. IEEE, New York City (2006)
11. Lazebnik, S., Schmid, C., Ponce, J.: Beyond bags of features: spatial pyramid matching for recognizing natural scene categories. In: IEEE Computer Society Conference on Computer Vision and Pattern Recognition, vol. 2, pp. 2169–2178. IEEE, Los Alamitos (2006)
12. Ali, A.H.Z., Amir, H.D.: Finding communities in linear time by developing the seeds. Phys. Rev. E **84**(3), 1553–1563 (2011)

Nearest Neighbor with Multi-feature Metric for Image Annotation

Wei Wu$^{(\boxtimes)}$ and Guanglai Gao

Computer Science Department, Inner Mongolia University, Hohhot, China
{cswuwei,csggl}@imu.edu.cn

Abstract. Most of the Nearest Neighbor (NN) based image annotation (or classification) methods cannot achieve satisfactory performance. In this paper, we propose a novel Nearest Neighbor method based on a multi-feature distance metric, which takes full advantage of different and complementary features. We first establish a metric for each feature and assign a weight for every metric, and then linearly combine all of them together to form one distance metric, namely the multi-feature metric. After that, we construct an NN model based on "image-to-cluster" distances, which equals to the distances between an image and the clusters within an image category using our multi-feature based metric, and which is different from calculating Euclidean distances between two images. By introducing this multi-feature based distance metric, our NN based model can mitigate the semantic issues due to intra-class variations and inter-class similarities, and improve the image annotation performance. Experiments confirm the superiority of our model in comparison with both the traditional classifiers and the state of the art learning-based models.

Keywords: Nearest neighbor · Distance metric learning · Multi-feature metric · Image annotation

1 Introduction

Image annotation (or classification) and retrieval have drawn considerable attention in both research and industrial fields. Finding relevant images from web and other large-scale databases is not a trivial task because of semantic gaps between image content's semantic representations and user demands. The goal of image annotation is to automatically recognize visual concepts from image semantic concepts, including scenes (indoor, outdoor, landscape, etc.), objects (car, person, animal, etc.), events (work, travel, etc.), and even emotions (happy, unpleasant, etc.), and turns out to be greatly challenging due to large intra-class variations and inter-class similarities. There have been many research communities engaged into this work, such as ImageCLEF [1], Pascal VOC, ImageNet, and etc., which demonstrate the challenges in this field.

Recently, some works show promising results on the image annotation task, such as sparse coding dictionaries [2], kernel combination, and structured sparsity regularization [3]. But all of these methods often use learning-based classifiers, e.g. SVM [4], generative models [5] etc., but rarely use Nearest Neighbor (NN) based classifiers, because

© Springer International Publishing Switzerland 2015
S. Arik et al. (Eds.): ICONIP 2015, Part IV, LNCS 9492, pp. 473–480, 2015.
DOI: 10.1007/978-3-319-26561-2_57

it provides inferior performance relative to learning-based methods. But we may under-value the effectiveness of NN-based classifier. Boiman et al. [6] claim that the main reason resulting in the low performance of NN-based algorithms is the information loss when extracting image visual features, particularly when constructing the bag of visual words (BoVW) based visual features, which have generally shown promising perform-ance using learning-based classifiers.

The traditional NN methods generally use Eculidean distasnce, or other non-learning based metrics, such as Jensen-Shannon divergence (JSD), $\chi2$-Distance, Histogram Inter-section distance, Kullback-Leibler divergence (KLD), etc. Actually, learning a good metric is very important to many image applications. Recent years, many distance metric learning (DML) methods [7, 8] have been proposed and achieved remarkable successes on many image classification or annotation tasks. But how to learn a metric with multiple features have rarely been discussed in DML, except for simple feature concatenation. The drawback of feature concatenation is the risk of computational complexity growth in $O(n^2 d^2)$ scale, where n and d denote the number of features and their average number of dimension. To overcome this drawback, Wang et al. [7] propose a multi-feature metric which is based on the multi-task learning method. But Wang's method need division of tasks and learns a metric for each task, which increase the complexities of problem and computation.

In this paper, inspired by [7], but different from these works, we take full advantages of the NN model and metric learning method, and propose a novel NN-based method based on a multi-feature metric, which can greatly reduce the semantic information loss and computational complexity, thereby improving the performance of large scale image annotation. Firstly, we introduce the image class label information to learn a semantic metric for each visual feature, and then we combine these metrics to construct one metric, which is used in our NN model for computing distances. After that, we generate multiple clusters for each image category using the k-means algorithm based on this distance metric, and construct a novel NN-based classifier using these clusters.

Our method uses a different distance metric optimization strategy, and meanwhile, the computational complexity of our method is reduced from $O(n^2 d^2)$ to $O(Nd^2)$, where $N \ll n^2$, N and n represent the number of training samples, and d represents the dimension of features. Experiments on image dataset of ImageCLEF2012 [1] confirm the effec-tiveness of our method. Furthermore, our method outperforms the traditional classifiers and a new baseline of NN model [9], and is competitive compared with the state of the art models.

This paper is organized as follows: Sect. 2 describes our multi-feature based distance metric learning method, and Sect. 3 introduces our NN-based model. Section 4 describes the experiments and results. Finally, we conclude our work and shed light on the future work in Sect. 5.

2   Multi-feature Metric Learning Method

In order to reduce the semantic gap in calculating the distance between two images which have large intra-class variations or inter-class similarities, we introduce a distance metric learning (DML) scheme using image class information. In our method, we extract the pairwise constraints from training images for distance metric learning. We formalize the representation of the pairwise features constraints set $\{(I_{i1}^m, I_{i2}^m, y_i)\}_{i=1}^N$, where I_{i1}^m and I_{i2}^m denote the mth type of visual feature for image I_{i1} and I_{i2} respectively. If both I_{i1} and I_{i2} belong to the same image category, then $y_i = 1$, otherwise $y_i = -1$. It is worth noting that how to select pairwise constraints can greatly affect the annotation performance. For the image semantic annotation task, there are the large intra-class variations and inter-class similarities, so we comply with such selection criterion: one is that the features are of the same image category but with large variation, the other is that the features are of different image category but with large similarity.

When constructing pairwise constraints for training, we firstly extract features of all the training images and use the k-means algorithm in Euclidean distance space to cluster the image features for each image category, with the result that k feature centers are constructed for each image category. Then we regard these feature centers as visually different "images" in the same semantic category (namely, the images with large intra-class variation), and for each pair of these images, we construct pairwise constraints $(I_{i1}^m, I_{i2}^m, y_i = 1)$. Finally, for each feature center of an image category, we search for the closest image in Euclidean distance in any other image category (namely, the images with large inter-class similarity), and construct pairwise constraints $(I_{i1}^m, I_{i2}^m, y_i = -1)$. This strategy can greatly reduce the number of training samples from n^2 to N for distance metric learning algorithm, where n denotes the number of images and N is the number of pairwise constraints. Therefore, it can reduce the complexity of the algorithm.

Then, we learn a metric A_m for each visual feature m, and linearly combine them together to form a distance measure function. The following formula can represent this framework:

$$d(I_1, I_2) = \sum_{m=1}^M \beta_m d_{A_m}(I_1^m, I_2^m)$$

$$where \quad d_{A_m}(I_1^m, I_2^m) = \sqrt{(I_1^m - I_2^m)^T A_m (I_1^m - I_2^m)}$$

(1)

In formula (1), $d(I_1, I_2)$ means the distance between image I_1 and I_2, β_m denotes the weight for the mth visual feature, and $d_{A_m}(I_1^m, I_2^m)$ represents the Mahalanobis distance. Inspired by the multiple kernel method [10], we use the variance of the Parzen Gaussian kernel to determine the value of β_m. The Parzen density estimation is expressed as follows:

$$p(I^m) = \frac{1}{N} \sum_{j=1}^N K(I^m - I_j^m)$$

$$where \quad K(I^m - I_j^m) = \exp(-\frac{1}{2\sigma^2} d_{A_m}(I^m, I_j^m)^2)$$

(2)

Where $K(\cdot)$ is the Parzen Gaussian kernel, N is the number of training images.

To calculate $d_{A_m}(I_1^m, I_2^m)$, we need to find an optimal metric A_m, which minimizes the distances between the features of the same image category and meanwhile maximizes the distances between the features of different image category. We formulate this distance metric learning problem as the following optimization:

$$\min_{A_m, b} \sum_i y_i(||I_{i1}^m - I_{i2}^m||_{A_m}^2 - b) + \frac{\lambda}{2} tr(A_m^T A_m)$$

$$s.t. \quad y_i(||I_{i1}^m - I_{i2}^m||_{A_m} - b) \leq 1 \tag{3}$$

$$A_m \geq 0, ||A_m|| = 1/\sqrt{\lambda}$$

where $|| \cdot ||_{A_m}$ is the Mahalanobis distance between the two features under the metric A_m. With the first inequality constraints, minimizing this term will make the distance between two semantically identical image features closer. The second term of the objective function is the regularization term, which prevents the overfitting by minimizing this model. The second constraint is introduced to prevent the trivial solution by shrinking metric A_m into a zero matrix. Parameter λ is a constant, and b is a threshold variable that could be used to determine whether two features are similar or dissimilar. We can use a stochastic gradient search algorithm to solve this optimization problem. The stochastic gradient search algorithm is an iterative process, and for the M visual features, the computational complexity is $O(Md^2Nt)$, where N is the number of pairwise constraints (training samples), t is the number of iteration. Our algorithm is better than typical DML method on computational complexity (the typical computational complexity of DML is $O(M^2d^2n^2t)$, where n is the number of images).

We utilize formula (1) to combine all the metrics based on different visual features together, and to calculate the distance in our NN-based model.

3   Nearest Neighbor Model

Our NN based classifier is different from those traditional ones. We employ the "image-to-cluster" distance measurement strategy.

We first construct clusters for each training image category using the k-means method. Instead of using Euclidean distance, we use our multi-feature distance metric when running the k-means algorithm. Thus we get k cluster centre for each image category $C: I_1^C, I_2^C, \ldots, I_k^C$. Next, our work is to search out the image category which minimizes the following formula:

$$\hat{C} = \arg \min_C \sum_{i=1}^k d(I_{test}, I_i^C) \tag{4}$$

where the distance function $d(\cdot)$ is shown by formula (1), I_{test} is the test image, and C denotes the image category, I_i^C is the ith cluster of image category C, and k takes the same value for all the image category.

When applying our NN-based model (formula (4)) in the image annotation task, we only need compute the summation $\sum_{i=1}^{k} d(I_{test}, I_i^C)$ for each image annotation label C, and then sort these labels according to these summations in an ascending order.

4   Experiments and Results

Experimental images come from the image annotation and retrieval task of Image-CLEF2012 [1]. The objective of ImageCLEF2012 annotation task is to accurately detect a wide range of semantic concepts for the purpose of automatically image annotating. There are 94 concept labels in total including natural elements, environments, people, impression, transportation, etc. This task is a multi-label annotation problem in the sense that an image may have multiple concept labels. We need to allocate each test image with multiple concept labels, and then sort these labels according to the similarities between the image and labels. The evaluation measurement is the MiAP (Mean interpolated Average Precision) which is widely used in the field of image annotation and retrieval.

We select three diverse visual features for our model experiments: Color Histograms, FCTH (Fuzzy Color and Texture Histogram), and the SIFT (Scale-Invariant Feature Transform) based bag of words features [11].

4.1   Experiments with Our Model

Firstly, we test our NN model with a single feature based metric, then we test the multi-feature based metric (Formula (1)). Furthermore, we also verify the effectiveness of the weighting strategy adopted in our multi-feature metric (Formula (2)). The experimental results are plotted in Fig. 1.

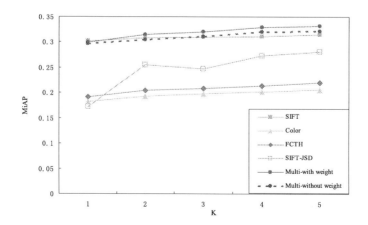

Fig. 1. Results of our NN model

In Fig. 1, the parameter k denotes the parameter of k-means clustering algorithm which means the number of clustering centers for each image category. The results of 3 features with our NN model are respectively denoted by SIFT, Color, and FCTH, and the best results are gained with the maximum value of parameter k. Compared with Fig. 1, we can see that the performance of our method is obviously better than the traditional k-NN method, regardless of using any single feature. As shown in Fig. 1, we can see that the SIFT local features get the best result among 3 features, MiAP reaches 0.3143, and color histogram's performance is not satisfactory, and the corresponding MiAP is just 0.2057 in the case of k equals to 5, but this result is greatly better than the traditional k-NN using the same feature.

In Fig. 1, Multi-with-weight and Multi-without-weight respectively represent our multi-feature metric based NN method with or without weights. We learn that our NN model with multi-feature metric attains the best result compared with any single feature metric, and the corresponding MiAP reaches 0.3211. We also learn that our weighting strategy improves the performance, the MiAP of which reaches 0.3321, 1 percent higher than that without the entropy weight, which confirms the effectiveness of our weighting strategy.

To test the effectiveness of the learned distance metric in our NN model, we use SIFT local features to test our NN model without using learned distance metric. As shown in Fig. 1, instead of using the learned distance metric, we use the traditional JSD distance for testing (denoted by SIFT-JSD), with the result that the usage of learned distance indeed contributes to performance. This shows that the introduction of the learned distance metric in our NN model is effective.

Furthermore, we also learn that the curve is relatively flat in Fig. 1 when using the multi-feature metric, which means that the parameter k (namely the number of clustering centers for each image category) has not much effect on performance, which means that we only use a small k can gain very good results.

In our experiments, we only test the value of parameter k up to 5, and it is shared by all the image annotation labels.

4.2   Experiments Compared with Other Methods

As shown in Table 1, the methods for comparison are k-NN, distance weighted k-NN (w-kNN) [12], Naive Bayesian (NB), NBNN method proposed by Boiman et al. [6], SVM, and a new Baseline of NN model [9], which is a combination model utilizing multiple low-level image features. The kernel function of SVM we used is Histogram Intersection kernel (HIK) [13].The results obtained by these methods use the same features as our method, but the combination strategy is different. These methods take a maximum voting combination strategy [14].

As illustrated by Table 1, we can see that the performance of our method outperforms all the other traditional methods and Baseline. The bottom row in Table 1 is the state of the art result published by ImageCLEF using multiple visual features, the MiAP gains 0.3481. This result is slightly higher than ours, which shows that our method is competitive.

Table 1. The results comparison with different methods

Methods	Features	MiAP
k-NN	Multiple	0.2702
w k-NN [12]	Multiple	0.2917
NB	Multiple	0.2732
SVM	Multiple	0.3015
NBNN [6]	Single	0.3004
Baseline [9]	Multiple	0.3025
Our method	Multiple	**0.3321**
ImageCLEF [2] (best result using visual features)	Multiple	**0.3481**

4.3   Experiments for Distance Metric Learning

Finally, we test the impact of the number of pairwise constraints for distance metric learning on annotation accuracy. We construct N visually different "images" by k-means algorithm (N is the number of clustering centers) for each image annotation label, and then we form pairwise constraint set $\{(I_{i1}^m, I_{i2}^m, y_i)\}_{i=1}^n$, where $n = N \times (N - 1)/2 + 94 \times (N \times (N - 1)/2 + 5 \times N)$. We take N values from 2 to 10 for testing, which corresponds that the number of pairwise constraints takes the value from 1034 to 8930. We use the same technique for each feature to perform experiment, and last we test the performance of multi-feature metric on different number of pairwise constraints. The experimental results are shown in Fig. 2.

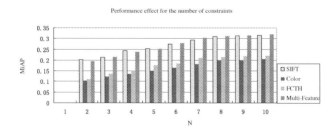

Fig. 2. Performance effects of the number of pairwise constraints

We can see from Fig. 2 that all the performance based on different features get better with the increasing of the number of pairwise constraints, and the multi-feature metric outperforms the other single feature metric, which confirms that our multi-feature metric strategy is significant and effective.

5   Conclusion

In this paper we described a novel NN-based classifier based on multi-feature distance metric. Our experiments on the ImageCLEF image annotation dataset achieved good results. Furthermore, the computational cost of our model is greatly reduced compared with other DML based models. In the future, we will combine more features to improve our distance metric learning algorithm, and to further improve the performance of our NN model.

References

1. Thomee, B., Popescu, A.: Overview of the ImageCLEF2012 flickr photo annotation and retrieval task. In: CLEF 2012 working notes, Rome, Italy (2012)
2. Yang, J., Yu, K., Gong, Y.: Linear spatial pyramid matching using sparse coding for image classification. In: Proceedings of CVPR, pp. 1794–1801. IEEE, Anchorage (2009)
3. Li, L.J., Su, H., Xing, E.P., et al.: Object bank: a high-level image representation for scene classification and semantic feature sparsification. Int. J. Comput. Vis. **107**(1), 20–39 (2014)
4. Wang, X., Du, J., Wu, S., et al.: High-level semantic image annotation based on hot Internet topics. Multimedia Tools Appl. **74**(6), 2055–2084 (2015)
5. Moran, S, Lavrenko, V.: Sparse kernel learning for image annotation. In: Proceedings of International Conference on Multimedia Retrieval. ACM, Glasgow (2014)
6. Boiman, O., Shechtman, E., Irani, M.: In defense of nearest-neighbor based image classification. In: Proceedings of CVPR, pp. 1–8. IEEE, Anchorage (2008)
7. Wang, S., Jiang, S., Huang, Q., Tian, Q.: Multi-feature metric learning with knowledge transfer among semantics and social tagging. In: Proceedings of CVPR, pp. 2240–2247. IEEE, Rhodes Island (2012)
8. Verma, Y., Jawahar, C.V.: Image annotation using metric learning in semantic neighbourhoods. In: Fitzgibbon, A., Lazebnik, S., Perona, P., Sato, Y., Schmid, C. (eds.) ECCV 2012, Part III. LNCS, vol. 7574, pp. 836–849. Springer, Heidelberg (2012)
9. Makadia, A., Pavlovic, V., Kumar, S.: A new baseline for image annotation. In: Forsyth, D., Torr, P., Zisserman, A. (eds.) ECCV 2008, Part III. LNCS, vol. 5304, pp. 316–329. Springer, Heidelberg (2008)
10. Gehler, P., Nowozin, S.: On feature combination for multiclass object classification. In: Proceedings of CVPR, pp. 221–228. IEEE, Anchorage (2009)
11. Jia, Y, Huang, C., Darrell, T.: Beyond spatial pyramids: receptive field learning for pooled image features. In: Proceedings of CVPR, pp. 3370–3377. IEEE, Rhodes Island (2012)
12. Zhang, L., Zhou, W.D.: Sparse ensembles using weighted combination methods based on linear programming. Pattern Recogn. **44**(1), 97–106 (2011)
13. Wu, J., Rehg, J.M.: Beyond the euclidean distance: creating effective visual codebooks using the histogram intersection kernel. In: Proceedings of ICCV, pp. 630–637. IEEE, Kyoto (2009)
14. Zeng, Z., et al.: A survey of affect recognition methods: audio, visual and spontaneous expressions. IEEE Trans. Pattern Anal. Mach. Intell. **31**(1), 39–58 (2009)

Number Recognition of Sudoku Grid Image
with Artificial Neural Networks

Selcuk Sevgen, Emel Arslan, and Ruya Samli[✉]

Department of Computer Engineering, Istanbul University, Istanbul, Turkey
{sevgens,earslan,ruyasamli}@istanbul.edu.tr

Abstract. In this study it is aimed to capture a SUDOKU grid image, to process this image, to recognize the numbers in the grid image with Artificial Neural Networks and finally to constitute a 9 × 9 number array with these numbers. The reason of choosing SUDOKU game as the input material is the thought of SUDOKU game as a prototype of real world fitting problems. After this number recognition is completed successfully, a robot software who finds the right solution of a SUDOKU game automatically will be developed. The next aim of this robot software is solving real world fitting problems.

Keywords: SUDOKU · Artificial neural networks · Image recognition

1 Introduction

The computer games is an important research area of computer science. The games can be thought as real world prototypes who have various mathematical, algorithmic and visual properties. Owing to these properties, they can be used for solving many real world problems like education problems [1–10] and mathematical problems [11–15]. Also it is known that some of the computer games like Tetris and Sokoban are used for solving fitting problems [16–18].

In this study, it is aimed to present an alternative solution for fitting problems by using SUDOKU game who is based on filling a 9 × 9 grid with digits so that each column, each row, and each of the nine 3 × 3 sub-grids that compose the grid contains all of the digits from 1 to 9. Although it has numerical values, SUDOKU game is not based on mathematical knowledge. A standart SUDOKU grid and a modified one are shown in Fig. 1. In this study a standart SUDOKU grid image is handled and investigated.

2 Method

In this section, the application steps are explained with details.

(1) Image Capturing: As all image processing studies, this study also starts with capturing clear images without any noise as much as possible. For this reason, some SUDOKU grid images are captured and some trials are made over them to obtain the most suitable image.

© Springer International Publishing Switzerland 2015
S. Arik et al. (Eds.): ICONIP 2015, Part IV, LNCS 9492, pp. 489–496, 2015.
DOI: 10.1007/978-3-319-26561-2_58

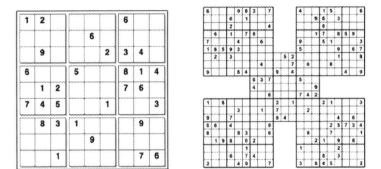

Fig. 1. SUDOKU grids

(2) Image Segmentation: For fitting the numbers on the image of SUDOKU grid in a 9 × 9 list, first of all, the image must be segmented properly. The aim of segmentation process is, dividing a one-piece grid image into 81 small square images who has the probability of containing 1–9 numbers. This segmentation process is implemented in MATLAB simulation environment. MATLAB implements every image as a matrix. In this study a 532 × 474 resolution image is obtained so it corresponds a 532 × 474 matrix in MATLAB. The 81 small squares mentioned above means 81 sub-matrices of this matrix.

(3) System Training: The 1–9 numbers captured from SUDOKU grid image are recognized by using a back-propagation Artificial Neural Networks (ANN). The training input number images and the SUDOKU grid images are chosen as differently. Thus, it is aimed to recognize images with different fonts. Every number image in the training set is constituted from 55 × 57 pixels. In the ANN model used, there are 3135 inputs (55 × 57 pixels), 9 outputs (1–9 number set) and a hidden layer with 10 neurons. For training process, first of all every number images are transformed to b/w images, so every pixel is represented by 0/1 values. Then, every matrix for a number are transformed a column vector form. These 9 column vectors are collected in an input matrix. sSo, every column in this matrix corresponds a number. As output values, a 9 × 9 identity matrix is constituted. Every column in this matrix matches 1 to 9 number respectively. 70 % of the input values are used for training, 15 % are used for validation and 15 % are used for test. The error value is chosen as 10^{-7}. After ANN training process, the 9 output neuron produces outputs. These outputs must be 0 or 1. The value set of the 9 output neurons corresponds a column vector in the output matrix. This column represents the number whose recognition is desired.

(4) Image Recognition: After the image is segmented, it is image recognition's turn. Respectively, sub-square images are transformed to binary images and some preprocesses are implemented. After them, edge detection process is applied to determine if there is an object or not in the square. If so, the numerical value of the square is recognized in MATLAB with the help of ANN.

(5) Numerical Array Constitution: In the following studies, we aim to develop a robot software who finds the right solution of a SUDOKU game automatically. This software will apply some mathematical algorithms, therefore it needs a numerical array. The sub-square images are transformed to a 9 × 9 with these principles:

> if a number determined in the sub-square, the numerical value in the array is its own numerical value,
> if any number cannot be determined in the other words if the image is empty, the numerical value in the array is 0,
> if there is an object determined but the system cannot decide which number is that, −1 value is returned.

If there is one or more −1 s in the array, it means that at least one number cannot be recognized. In this situation, because a solution cannot be found, this SUDOKU grid image must be trained again until it is determined as 0 or other number or the iteration number is equal to the decided iteration number. Unless both of the condition provide the system will be trained with different input number sets.

3 Results and Discussion

The results in the steps mentioned in the previous section are explained as follows:

(1) Image Capturing: The image captured from 9 × 9 SUDOKU grid has 532 × 474 resolution and it is shown in Fig. 2.

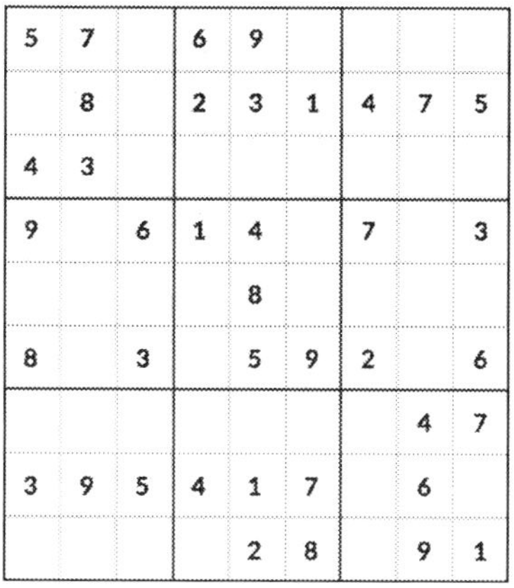

5	7		6	9				
	8		2	3	1	4	7	5
4	3							
9		6	1	4		7		3
				8				
8		3		5	9	2		6
							4	7
3	9	5	4	1	7		6	
				2	8		9	1

Fig. 2. 532 × 474 Resolution SUDOKU Grid Image

(2) Image Segmentation: With the segmentation of Fig. 2, 81 square images (with or without a number) are obtained. In the Fig. 3, an example of the empty square image and number images are given respectively.

(3) System Training: The number set given for training is given in Fig. 4.

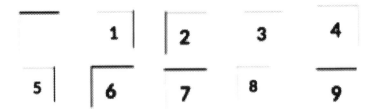

Fig. 3. Sample images obtained from segmentation process

1 2 3 4 5 6 7 8 9

Fig. 4. ANN training number set

As mentioned in the "Methods" section, the purpose of choosing these numbers in different fonts is to provide the system working without any dependency of fonts. The ANN structure used in this study is shown in Fig. 5.

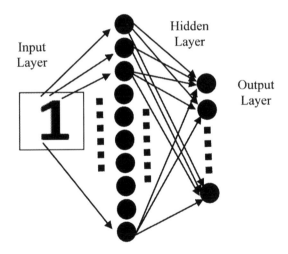

Fig. 5. ANN structure

A sample ANN training result of this number set is shown in Fig. 6. As shown in the figure, in 1613. iteration, the system reached the desired error value.

For example, the recognition results of 1, 2 and 5 numbers are shown in Fig. 7.

According to Fig. 7, the row whose numerical value is closest to 1 means the desired number got recognized.

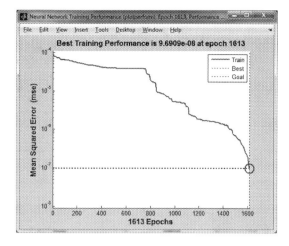

Fig. 6. A sample ANN training result

x =	x =	x =
0.9555	0.0004	0.2927
0.0000	0.9827	0.0000
0.0000	0.0026	0.0000
0.0031	0.0019	0.1344
0.0366	0.0000	0.7936
0.0000	0.0000	0.0000
0.0037	0.0007	0.0014
0.0000	0.0007	0.0000
0.0000	0.0000	0.0000

Fig. 7. The recognition results numbers 1, 2 and 5

(4) Image Recognition: The image recognition steps after system training can be thought as the transformations between RGB code of the image, binary code of the image, image after edge detection and image after recognition. For example, the recognition steps of number 5 is shown in Fig. 8.

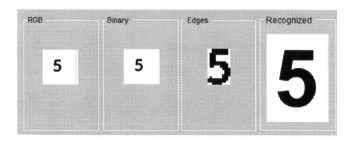

Fig. 8. Recognition of number 5

(5) Numerical Array Constitution: The numerical array that is constituted is given in Fig. 9. In this array, the numbers that can be determined from the SUDOKU grid image takes part with their own numerical values, while empty squares are represented with a 0. If the number in the square cannot be determined from the system, it is -1 in the array.

5	7	0	6	9	0	0	0	0
0	8	0	2	3	1	4	7	5
4	3	0	0	0	0	0	0	0
9	0	6	1	4	0	7	0	3
0	0	0	0	8	0	0	0	0
8	0	3	0	5	9	2	0	6
0	0	0	0	0	0	0	4	7
3	9	5	4	1	7	0	6	0
0	0	0	0	2	8	0	9	1

Fig. 9. The numerical array

4   Conclusion and Future Work

In this study, a SUDOKU grid image is captured, it is segmented in MATLAB, the numbers in the image are recognized by ANN and they constitute a 9×9 numerical array. Therefore 81 square images with or without any number are obtained by using MATLAB. From the segmented images, the numbers are recognized by using a back-propagation ANN. With this structure, it is not possible to recognize all the numbers in the system. Some of the possible reasons for this situation may be: the image does not have high-resolution enough, it has some noises and the input training set has different fonts from the numbers in the image.

This study aims to have different properties from similar studies via the reasons below:

it handles computer games subject which is an unpopular research study in computer science,
it helps SUDOKU to solve real world fitting problems and it establishes first step of this,
it does not only takes numbers as images, also the empty square is in the image set,
it does not only evaluate the images as an image, it also evaluates them with their numerical values,

it transforms both of the recognized images (number and empty square) to a numerical array,

this numerical arrays constitute an input for the mathematical algorithms which will be applied in the following studies.

In future work, our first aim is to constitute a robot software which can automatically solve SUDOKU game. Then, we will study for using this robot software to solve real world fitting problems. Also other computer games which can be thought as a prototype of real world fitting problems (2048, Tetris, Sokoban etc.) will be investigated and new solution alternatives will be presented.

References

1. Simkova, M.: Using of computer games in supporting education. Procedia – Soc. Behav. Sci. **141**, 1224–1227 (2014)
2. Ahmad, I., Jaafar, A.: Computer games: implementation into teaching and learning. Procedia – Soc. Behav. Sci. **59**, 515–519 (2012)
3. Kampf, R., Cuhadar, E.: Do computer games enhance learning about conflicts? A cross-national inquiry into proximate and distant scenarios in Global Conflicts. Computers in Human Behavior (2014, in press)
4. Mahmoudi, H., Koushafar, M., Saribagloo, J.A., Pashavi, G.: The effect of computer games on speed, attention and consistency of learning mathematics among students. Procedia – Soc. Behav. Sci. **176**, 419–424 (2015)
5. Bakker, M., Heuvel-Panhuizen, M., Robitzsch, A.: Effects of playing mathematics computer games on primary school students' multiplicative reasoning ability. Contemp. Educ. Psychol. **40**, 55–71 (2015)
6. Ke, F.: An implementation of design-based learning through creating educational computer games: a case study on mathematics learning during design and computing. Comput. Educ. **73**, 26–39 (2014)
7. Ke, F.: Computer-game-based tutoring of mathematics. Comput. Educ. **60**(1), 448–457 (2013)
8. Butler, Y.G.: The use of computer games as foreign language learning tasks for digital natives. System (2014, in press)
9. Smith, G.G., Li, M., Drobisz, J., Park, H.R., Kim, D., Smith, S.D.: Play games or study? Computer games in eBooks to learn English vocabulary. Comput. Educ. **69**, 274–286 (2013)
10. Alias, N., Rosman, F., Rahman, M.N.A., Dewitt, D.: The potential of video game in Malay language learning for foreign students in a public higher education institution. Procedia – Soc. Behav. Sci. **176**, 1020–1027 (2015)
11. Cooper, J., Kirkpatrick, A.: Critical sets for Sudoku and general graph colorings. Discrete Math. **315–316**, 112–119 (2014)
12. Jones, S.K., Perkins, S., Roach, P.A.: Properties, isomorphisms and enumeration of 2-Quasi-Magic Sudoku grids. Discrete Math. **311**(13), 1098–1110 (2011)
13. Mo, H.D., Xu, R.G.: Sudoku square - a new design in field. Acta Agron. Sin. **34**(9), 1489–1493 (2008)
14. Maji, A.K., Jana, S., Pal, R.K.: An algorithm for generating only desired permutations for solving Sudoku puzzle. Procedia Technol. **10**, 392–399 (2013)

15. Soto, R., Crawford, B., Galleguillos, C., Monfroy, E., Paredes, F.: A hybrid AC3-tabu search algorithm for solving Sudoku puzzles. Expert Syst. Appl. **40**(15), 5817–5821 (2013)
16. Dor, D., Zwick, U.: SOKOBAN and other motion planning problems. Comput. Geom. **13**(4), 215–228 (1999)
17. Hearn, R.A., Demaine, E.D.: PSPACE-completeness of sliding-block puzzles and other problems through the nondeterministic constraint logic model of computation. Theor. Comput. Sci. **343**(1–2), 72–96 (2005)
18. Anthony, T., Polani, D., Nehaniv, C.L.: General self-motivation and strategy identification: case studies based on Sokoban and Pac-Man. IEEE Trans. Comput. Intell. AI Games **6**(1), 1–17 (2014)

Counting Human Actions in Video During Physical Exercise

Burak Özeroğlu and Ediz Şaykol[✉]

Department of Computer Engineering,
Beykent University, Ayazağa, 34396 İstanbul, Turkey
burakozeroglu@windowslive.com, ediz.saykol@beykent.edu.tr

Abstract. We present a simple yet effective human action detection and counting scheme during physical exercise using video stream data. Counting human actions automatically is more meaningful for data analysis in sports centers, and for healthiness observations in rehabilitation centers. The identification of the action starts with the detection of crucial body regions, namely skeletal joints. We observed that hand-wrist, arm-elbow, and arm-shoulder points are crucial for human arm motion, whereas during leg motion ankle-knee, knee-waist, and waist-ankle points are critical. These body junctions get different angle values during physical exercise, which helps us track and count the action. We assumed a simple, cheap and effective solution for multi-tracking these joints, which are marked with a distinctive color. Color filtering and color-based tracking steps are then performed to detect and count the actions by tracing the angle variations between joints. The developed application and performance evaluation tests show that our technique provides a reasonable performance while providing a simple and cheap video setup.

Keywords: Human action detection · Color filtering · Color tracking

1 Introduction

The amount of physical exercise for human is very crucial to maintain skeletal health. Besides, the human action counts during physical exercise can be utilized for data analysis in sports centers, and for healthiness observations in rehabilitation centers. Hence, instead of manually analyzing the human actions during physical exercise, which is obviously a tedious and erroneous task, there exist various systems and devices for analyzing human motion automatically. Recent trends show reasonable results in detecting human actions from video streams.

When the human perception of motion is analyzed, it can be seen that the information for representing the motion is obtained from the changes in the speed and direction of the trajectory (e.g., [3,8]). In [7], a computational representation of human action to capture these dramatic changes using spatio-temporal curvature of 2-D trajectory is presented to analyze body pose estimations. Specialized sensors or RFID tags are generally used in the existing systems, however, there

© Springer International Publishing Switzerland 2015
S. Arik et al. (Eds.): ICONIP 2015, Part IV, LNCS 9492, pp. 497–504, 2015.
DOI: 10.1007/978-3-319-26561-2_59

are also techniques using depth images, such as using Microsoft Kinect data [2], for detecting physical exercise.

In this paper, we present a simple, cheaper yet effective video-based solution to analyze human actions for a later counting scheme during physical exercise. Counting human actions automatically from video streams is more meaningful in sports and rehabilitation centers for data analysis and healthiness observations. The identification of the action starts with the detection of crucial body regions, namely skeletal joints. We use blue bottle caps to identify the points to track, and employ a multi-object tracking algorithm to identify the human actions automatically. We observed that hand-wrist, arm-elbow, and arm-shoulder points are crucial for human arm motion, whereas during leg motion ankle-knee, knee-waist, and waist-ankle points are critical. These body junctions get different angle values during physical exercise, which helps us track and count the amount of action. We assumed a simple and cheap solution, when compared to existing depth sensor based techniques or RFID based techniques, for tracking these joints by manually marking them with a distinctive color. Color filtering and color tracking steps are then performed to detect and count the actions.

Some of the existing solutions for tracking human actions during physical exercise can be summarized as follows: In [5], an integrated system for body shape analysis (BSA) and physical fitness test (PFT)-HIMS is presented having 5 data acquisition modules for analyzing physiological status, an exercise load control module, a data processing module and a main control module. The parameters obtained from the tests include cardio-respiratory endurance, muscle strength, muscle endurance, agility, power, blood pressure at rest, blood pressure at recovery, body fat and body weight. This system has more functions, however is more complex and expensive to use in even very small gyms.

A motion capture system without markers is presented in [4] for exercise routines in a health club. This system uses image contour tracking and swarm intelligence methods to track the location of the spine and shoulders during treadmill, exercise bike, and overhead lateral pull-down. The system is not evaluated under various lighting conditions and as a consequence of the techniques used, it is very hard to track image contours under very low light. We used color region tracking to improve the performance with a cheaper setup.

The authors propose a two-stage identification and recognition process, based on velocity features and stochastic modeling of each motion in [6] from continuous time-series data of human movement. The approach is validated on 20 healthy subjects and four rehabilitation patients performing rehabilitation movements. There are also systems to define and evaluate the image quality and user satisfaction based on video captures during physical exercise in a public gym, especially for adults aged over 65 (e.g., [1, 10]). A set of human-computer interaction goals are generally tested such as effectiveness, efficiency, and user satisfaction. As expected, users want better image quality, when expectations are matched satisfaction increases, and unclear meaning of motion measures decreases satisfaction. These systems generally target for on line system, hence user-centered tests are performed.

The rest of the paper is organized as follows: Sect. 2 explains the preliminaries and the flow of execution to automatically identify and count a set of human actions based on color-based multi-object tracking. The proposed technique is evaluated in Sect. 3 via the developed application, and Sect. 4 concludes the paper and states future work.

2   Identification and Counting Human Actions

Color derives from the spectrum of light (distribution of light power versus wavelength) interacting the eye with devices with the spectral sensitivities of the light receptors. By defining a color space, color can be identified numerically by their coordinates. Red-Green-Blue (RGB) is one of the famous color space, which is known as its additive property to easy production and generation of compound colors. Color is one of the most important features that algorithms can utilize to detect the actions in the frame. Most image processing algorithms include a filtering step with respect to the color feature to ease its process and improve the results. Sample color blocks with RGB values are shown in Fig. 1.

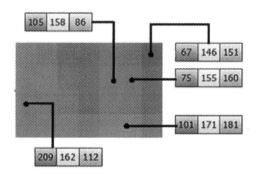

Fig. 1. Sample color blocks in RGB color space (Color figure online).

Image capture devices process frames in a certain speed and certain size. To identify and count human actions on-line in video, processing of a frame must be completed before grabbing next frame. To this end, we set the frame rate to 20 frames-per-second, and the resolution to 640×480 pixels. In many cases, working on gray-scale images yields better results, since it is in fact a quantization of the image intensity values with respect to predefined set of color (gray) levels [9]. A better conversion from RGB color space to gray level has to consider the perception of color in human vision system, which leads to using different coefficients for R-G-B values instead of a direct arithmetic mean. A popular formula for computing the gray-level intensity I as the weighted average is $I = 0.299R + 0.587G + 0.114B$, and the way that we utilized here is shown in Fig. 2.

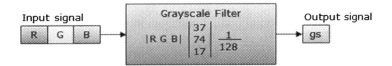

Fig. 2. Filtering-based conversion from RGB colors to gray level values (Color figure online).

The next step after the above image processing preliminary operations is detecting objects, that we marked with blue, before multi-object tracking. We utilized the following pseudo-code for detection multiple objects, i.e. blobs, in the frame:

```
Minimum_Object_Size = Pixel_Value_in_Color_Tracking;
Maximum_Object_Size = Pixel_Value_in_Color_Tracking;
Blobs = Find_Blob_Over_The_Image(Image);
Object_Coordinate_Array[] = Blobs(Objects);
while Object_Coordinate_Array[] > 0
    Mark Objects
    Draw a Rectangle Around Objects
```

Having provided the width and height information of the first object to detect the regions properly, all of the regions are detected and stored with their corresponding corner locations for later stages devoted to counting human actions. Once the minimum bounding rectangle (mbr) is drawn around a detected colored region, we can find the center of mass of the object as follows: Let $A(x_1, y_1)$ denote the top-left corner, $B(x_2, y_1)$ denote the top-right corner, $C(x_1, y_2)$ denote the bottom-left corner, and $D(x_2, y_2)$ denote the bottom-right corner of the mbr. Then, center-of-mass (x_c, y_c) can be found as $x_c = \frac{x_1 + x_2}{2}$ and $y_c = \frac{y_1 + y_2}{2}$.

The crucial points for tha salient human actions are skeletal joints, and a couple of them can be used for counting purposes. For example, the joints for human arm motion are hand-wrist, arm-elbow, and arm-shoulder points, whereas during leg motion ankle-knee, knee-waist, and waist-ankle points are critical. These body junction points get different angle values during exercise. By tracking the angle variations at these points and marking them with a distinctive color, the motion can be detected and statistical data for the detected exercise action can be calculated.

The angle variations are computed as follows: Two detected objects are processed as above and the center-of-mass values are computed. Let $C_1(x_1, y_1)$ and $C_2(x_2, y_2)$ denote these values for two skeletal joints, the vector between them can be found by $\overrightarrow{V} = C_2 - C_1$. The angle α between $\overrightarrow{V}$ and x-axis is traced to maintain a predefined value depending on the type of the motion.

This process is illustrated in Fig. 3 for a sample arm movement. In this case, α is checked against 120 degrees such that if $\alpha < 120$, the action count is incremented. By providing appropriate values, several set of actions can be identified and counted in a simple manner.

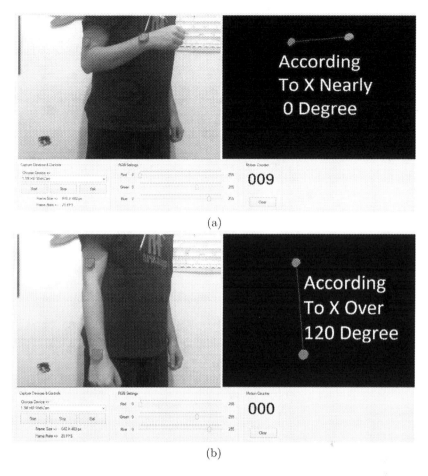

Fig. 3. Counting action for a sample arm movement, (a) start of the action, (b) end of the action. The angle variations are tracked for counting.

3   Performance Evaluation

In the evaluations, we developed an application in which the frame rate is set to 20 fps, and the frame resolution is set to 640 x 480 pixels. We used AForge library for frame processing algorithms. The main screen of the developed application is shown in Fig. 4. We employed $EuclideanColorFiltering()$ technique due to its high and fast adaptiveness to color variances. It uses a circular neighborhood to apply the filter with respect to color, and we set the required parameters as follows:

```
// RGB center color and the diameter of the
// color pallette is set experimentally
filter.CenterColor = Color.FromArgb(215,30,30);
filter.Radius = 100;
```

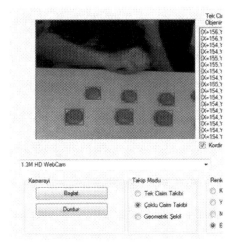

Fig. 4. The main application showing the results of the color filter for multi-object tracking.

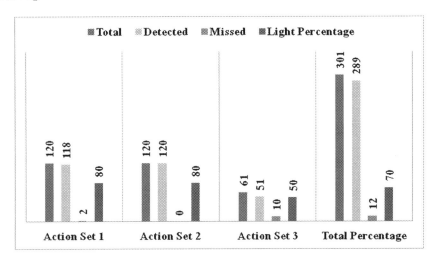

Fig. 5. The results of the performance evaluation tests.

We utilized color tracking based human action detection where critical joints need to be found in the scene. To provide a better evaluation, we created a dataset for various actions with one light source and two light sources. The main reason behind evaluating the performance with more than one light source is the fact that the color values on a surface might be different when we have more than one light source in the medium. Hence, our dataset includes 25 videos for different action sets and light sources, and the details are given in Table 1.

The results of the performance evaluation tests is given in Fig. 5. The performance of our technique is almost 90 % under single light source with adequate

Table 1. The dataset includes 25 videos for different action sets and light sources.

Action Set	Action Count	Light Source	Detected Action	Missed Action
Action 1, video 1	12	single	10	2
Action 1, video 2	12	single	12	0
Action 1, video 3	12	single	12	0
Action 1, video 4	12	single	12	0
Action 1, video 5	12	single	12	0
Action 1, video 6	12	single	12	0
Action 1, video 7	12	single	11	1
Action 1, video 8	12	single	11	1
Action 1, video 9	12	single	12	0
Action 1, video 10	12	single	12	0
Action 2, video 1	12	single	12	0
Action 2, video 2	12	single	12	0
Action 2, video 3	12	single	12	0
Action 2, video 4	12	single	12	0
Action 2, video 5	12	single	12	0
Action 2, video 6	12	single	12	0
Action 2, video 7	12	single	12	0
Action 2, video 8	12	single	12	0
Action 2, video 9	12	single	12	0
Action 2, video 10	12	single	12	0
Action 3, video 1	13	double	9	4
Action 3, video 2	12	double	10	2
Action 3, video 3	12	double	11	1
Action 3, video 4	12	double	11	1
Action 3, video 5	12	double	10	2

light percentage. We also altered the amount of light in the medium during tests to evaluate the performance under low light. Our technique, in fact, identifies the color blocks, however, based on the effect of the shadows, the performance lowers to 60 % under inadequate light.

4   Conclusion and Future Work

We present a human action detection and counting scheme during physical exercise using video stream data. Based on the observations, hand-wrist, arm-elbow, and arm-shoulder points are found crucial for human arm motion, and we provided a simple yet effective solution to count the amount of the action. The counting scheme assumes that the crucial body joints are marked with distinctive color apriori, and the angle variations between the salient joints are traced to count.

Counting human actions automatically is more meaningful for data analysis in sports centers, and for healthiness observations in rehabilitation centers. Our solution can be applied simple in almost all gyms and rehabilitation centers with cheap cameras. For personal use, a mobile camera is enough with at least two bottle caps for marking desired joints to count. This solution is very simple and cheap when compared to existing depth sensor based techniques or RFID based techniques. Besides, our tests show that using color filtering and tracking based on color provides reasonable performance under low amount of light.

Future directions include to devise simple counting schemes for other human actions to make the system complete. On the other side, using the video stream as time-series data to detect and count the actions is to be included to increase the performance under various conditions.

References

1. Alexander, G.L., Havens, T.C., Rantz, M., Keller, J., Abbott, C.C.: An analysis of human motion detection systems use during elder exercise routines. W. J. Nurs. Res. **32**(2), 233–249 (2010)
2. Ar, I., Akgul, Y.S.: A computerized recognition system for the home-based physiotherapy exercises using an RGBD camera. IEEE Trans. Neural Syst. Rehabil. Eng. **22**(6), 1160–1171 (2014). doi:10.1109/TNSRE.2014.2326254
3. Dönderler, M.E., Şaykol, E., Arslan, U., Ulusoy, Ö., Güdükbay, U.: Bilvideo: Design and implementation of a video database management system. Multimedia Tools Appl. **27**, 79–104 (2005)
4. Havens, T.C., Alexander, G.L., Abbott, C., Keller, J.M., Skubic, M., Rantz, M.: Contour tracking of human exercises. In: IEEE Workshop on Computational Intelligence for Visual Intelligence (CIVI 2009), pp. 22–28, Nashville, TN, USA (2009)
5. Kwak, H-M., Park, S-H., Yoon, Y-R.: An integrated system for body shape analysis and physical fitness test - HIMS. In: 27th Annual Conference of the IEEE Engineering in Medicine and Biology (EMBC), pp. 3742–3745, Shanghai, China (2005)
6. Lin, J.F.-S., Kulic, D.: Online segmentation of human motion for automated rehabilitation exercise analysis. IEEE Trans. Neural Syst. Rehabil. Eng. **22**(1), 168–180 (2014)
7. Rao, C., Yilmaz, A., Shah, M.: View-invariant representation and recognition of actions. Int. J. Comput. Vis. **50**(2), 203–226 (2002)
8. Şaykol, E., Güdükbay, U., Ulusoy, Ö.: Scenario-based query processing for video surveillance archives. Eng. Appl. Artif. Intell. **23**(3), 331–345 (2010)
9. Sezgin, M., Sankur, B.: Survey over image thresholding techniques and quantitative performance evaluation. J. Electron. Imaging **13**(1), 146–165 (2004)
10. Velloso, E., Bulling, A., Gellersen, H., Ugulino, W., Fuks, H.: Qualitative activity recognition of weight lifting exercises. In: Proceedings of the 4th Augmented Human International Conference (AH 2013), pp. 116–123. New York, NY, USA (2013)

Towards a Blind MAP-Based Traitor Tracing Scheme for Hierarchical Fingerprints

Faten Chaabane[1]([✉]), Maha Charfeddine[1], William Puech[2],
and Chokri Ben Amar[1]

[1] REGIM-Laboratory: REsearch Groups in Intelligent Machines,
University of Sfax, ENIS, BP 1173, 3038 Sfax, Tunisia
faten.chaabane@ieee.org
[2] LIRMM Laboratory, UMR 5506 CNRS,
University of Montpellier, 161, Rue Ada, 34392 Montpellier Cedex 05, France

Abstract. One of the key challenges of a traitor tracing scheme is to deal with the real scenarios. In this context, the tracing operation is usually constrained by the lack of information about the number of colluders and even the collusion channel. Indeed, the Tardos decoding is invariant regardless the type of collusion, which can be considered as a suboptimality of its accusation performance. In this paper, we propose to use a MAP-based estimation strategy which improves the Tardos decoding step and guarantees a good estimation results. Compared to the original version of Tardos code and the original MAP decoder operating in blind tracing scheme, the proposed technique takes the advantage of operating in hierarchical context to provide a more concise and accurate accusation decision, this in a short time.

Keywords: Tracing · Traitors · Blind · Estimation · Hierarchical

1 Introduction

The emergence of the multimedia distribution over Peer to Peer networks and web multimedia distribution platforms has encouraged several illegal manipulations tied to the copyright violation. One example of these illegal operations is the collusion attack made by a set of users to merge their releases in order to yield a copy close to the original. The tracing traitors, called also active fingerprinting was proposed to address this ever increasing phenomena. It consists in embedding, by the server distributor, a symbol sequence in each sold release of the same media content before its distribution. This sequence should allow to identify the release holder and mainly to trace back the dishonest actors of a collusion trial. Different approaches of sequence generating were proposed in tracing traitors' field which include probabilistic approaches [1,2]. The most powerful one is the Tardos code which was firstly proposed in [11]. Although this code provides a satisfying compromise between the bound of the code length and the colluders' detection rate, its decoding process suffers from heavy computational

© Springer International Publishing Switzerland 2015
S. Arik et al. (Eds.): ICONIP 2015, Part IV, LNCS 9492, pp. 505–512, 2015.
DOI: 10.1007/978-3-319-26561-2_60

costs, especially when it handles a great number of users. One other characteristic of the Tardos code is its invariance regardless the collusion channel. It is considered as an agnostic decoder since it does not need any information about the type of collusion [4]. To deal with such weaknesses, different techniques were striving to improve both the accusation Tardos functions and to decrease the complexity of its decoding process [9,10]. In this paper, we focus on two parts: the first one is the multi-level hierarchical fingerprint construction we proposed previously in [3] and the second is the migration of accusation Tardos functions in the blind scheme using the MAP Tardos functions proposed in [6]. According to the multi-level hierarchical fingerprint proposed in [3], we proceed by a two-level tracing strategy. In the first level, a group accusation based on the Boneh Shaw code is made to select at least one suspicious group, and in the second level, the Tardos accusation step is applied for these detected groups. Using the hierarchical structure and so the two-layered accusation strategy has the advantage to limit the space operating of the Tardos code. Instead of parsing the whole set of users, the Tardos detection is made only in retrieved groups. To show the efficiency of the proposed scheme, we add a crucial criteria in accusation performance which is the detection time, the time taken by the tracing algorithm to retrieve at least one colluder among the set of colluders. The paper is organized as follows: in Sect. 2, we remind the skeleton of the multi-level hierarchical fingerprint we proposed in [3]. In Sect. 3, we detail the structure of the proposed tracing scheme. In Sect. 4, we present the experimental evaluation, and finally we give some remarks in the conclusion.

2   The Multi-Level Hierarchical Fingerprint Construction

As a first step in this paper, we remind the different constraints we considered in the construction of our multi-level fingerprint. Indeed, as proposed by [12], in tracing traitor field, we can assume that users having common characteristics have more probability to cooperate together than with other users. Furthermore, a group-based accusation strategy takes the advantage of limiting the search space for Tardos code, which provides a less complex decoding process and a faster detection decision, compared to non-hierarchical tracing strategies. In [3], we proposed a five-level hierarchical fingerprint according to geographical, temporal and social constraints. Thus, for each user, the final identifier, as shown in Eq. 1 is the combination of its group identifier and its personal identifier within the group. The group identifier, ID_{group} is the concatenation of the five levels and is encoded with Boneh Shaw, whereas the personal identifier $personal_{id}$ is encoded with Tardos code.

$$[id_{level1} + id_{level2} + \cdots + id_{level5}] + personal_{id} \tag{1}$$

According to this hierarchical fingerprint, we apply a group-based accusation strategy which is more detailed in Sect. 3.

3   The Proposed Tracing Scheme

The main goal of the majority tracing traitor works is to provide an accurate decision about the colluders who cooperate in yielding the suspicious media release [7]. In this paper, the proposed technique consists in a migration from a non blind accusation scheme to a blind one. The aim is to simplify the MAP detector process and to have more accurate detection results. In the Section below, we detail the non blind scheme.

3.1   The Boneh Shaw - Tardos Accusation Strategy in a Non-Blind Scheme

To decrease the computational costs of the accusation step in a two-layered accusation process based on the Tardos code, [7] proposed a tracing algorithm based on a concatenation of two probabilistic codes, which are decoded successively in the accusation process. In a first step, a group selection process is performed using the Boneh Shaw code, and the second step consists in accusing users into these selected groups with Tardos code.

The Boneh Shaw with replication factor r, is a probabilistic tracing code which was proposed by [2], it depends only on r. It applies the Marking assumption and is represented by the couple (n,r) where n is the number of groups and r is the replication parameter value. The code length is defined by the equation:

$$m_{Boneh\ Shaw} = (n_g - 1) * r \tag{2}$$

The result of Boneh Shaw generation is a matrix of $(n-1)$ column types of the form $11\cdots00$ replicated r times and suiting the rule that the ith group has a 0 in the $(i-1)^{th}$ columns and 1 in the others. The set $\{g_1, g_2, \cdots, g_n\}$ represents the groups identifiers where g_i is a unique group identifier associated to one type transition. As a first step, the weight of each r-block is computed, W_{g_i}, the difference of 1's bits number between each bloc and the corresponding bloc in Y. Then, as described in [2], the accusation proceeds as follows:

1. If $W_{g_1} > 0$ then the first group is accused.
2. If $W_{g_{n-1}} < r$ then the last group is accused.
3. For $2 \leq s < n - 1$, let $k = W_s$,

    If $W_{s-1} < \frac{k}{2} - \sqrt{\frac{k}{2} \log \frac{2n}{\epsilon}}$, then the group s is accused, ε is the error probability value of Boneh Shaw code.

As shown in Fig. 1, once the Boneh Shaw accusation step is completed, at least one group is accused. We switch to the second level of the accusation by applying the Tardos code to the first accused group to retrieve at least one colluder.

We remind that users' codewords matrix X is a nxm matrix where X_{ji} is related to each user j. Computing a score related to each user is based on the Neyman-Pearson theorem [6], which states that to decide about his pertaining in the collusion trial, the optimal score σ_j^{NP} is as follows:

$$\sigma_j^{NP} = \frac{\mathbb{P}(y \mid x_j, s_j = 1, \mathbf{G}, \mathbf{p}, c)}{\mathbb{P}(y \mid x_j, s_j = 0, \mathbf{G}, \mathbf{p}, c)}. \tag{3}$$

To survive the constraint of missing information about the collusion size c and the applied strategy $\mathbf{G}$, [10] propose to compute the score in Eq. 3 using the symmetric accusation functions as below:

$$\sigma_j^T = \sum_{i=1}^{m} U(y_i, x_{ji}, p_i), \tag{4}$$

$$U(1, 1, p_i) = \sqrt{\left(\frac{(1 - p_i)}{p_i}\right)}, U(0, 0, p_i) = \sqrt{\left(\frac{p_i}{(1 - p_i)}\right)}, \tag{5}$$

$$U(1, 0, p_i) = U(1, 1, p_i), U(0, 1, p_i) = U(0, 0, p_i). \tag{6}$$

It is a threshold-based accusation step, where only users having scores above the threshold are accused. Although the accusation functions can be applied for all types of collusion strategies, providing an estimation about the applied strategy should bring more satisfying accusation rates.

In Sect. 3.2, we apply a MAP-based accusation functions proposed by [6] for the two-level accusation strategy to migrate from a non-blind tracing scheme to a blind one. This migration should optimize the performance of a simple decoder by reducing the time detection and probably improving the detection rates.

3.2　MAP Decoding Strategy

When considering the probabilistic model used by [6], it is important to remind some notations which will be mentioned throughout the paper as follows: X is a $\{0, 1\}^{m \times n}$ matrix representing users' codewords of length m where X_j is the embedded information in the sold release related to the user j. According to this, let s a $\{0, 1\}^n$ vector which assigns $s_j = 1$ to each user selected as a colluder, and otherwise $s_j = 0$. Typically, it is assumed that the ith symbol in the suspicious release depends only on the 1s' number in each position i in colluders' codewords. Obviously, a vector $t \in \{0, \ldots, c\}^m$ is constructed, representing the 1s' number encountered in all colluders' codewords for each position i, $i \in \{1, \ldots, m\}$. Thus, $t = Xs$. The probability of generating y from colluders' codewords consists essentially in a conditional probability as follows:

$$\mathbb{P}(\mathbf{y}|\mathbf{t}, \mathbf{G}) = \prod_i \mathbb{P}(y_i|t_i, \mathbf{G}), \tag{7}$$

where $\mathbf{G}$ denotes the collusion channel. It is assumed that $\mathbf{G}$, is a $m \times c$ matrix. The only requirement for g_{ik} elements is to respect the Marking Assumption [10], for $k \in \{0, c\}$ by following: $\forall i, g_{i0} = 1 - g_{ic} = 0$.

When addressing the problem of missing some information about the collusion channel, it is judicious to estimate the vector of accusation s from the codeword y. Furthermore, in practice, it was agreed upon that selecting at least one colluder is sufficient [5].

We turn our attention to the approach proposed in [6], the MAP-based decoder which was proposed to provide a tradeoff between two crucial requirements in a tracing scheme: the accusation accuracy and the computational costs. The MAP-based decoder is based on defining two non-informative priors, $\mathbb{P}(G)$ and $\mathbb{P}(c)$, which are respectively tied to the strategy $\mathbf{G}$ and the collusion size $\mathbf{c}$. To derive the decoder, and according to Bayesian statistics, authors exploit a joint probability in order to marginalize the unknown quantities as follows:

$$\mathbb{P}(\mathbf{y, t, X, p, s}) = \sum_c (\int \mathbb{P}(\mathbf{y, t, X, p, s}|c, \mathbf{G})\mathbb{P}(G)\,d\mathbf{G})\mathbb{P}(c). \tag{8}$$

From Eq. 8, the likelihood ratio σ_j^{MAP}, defined previously in Eq. 3 can be computed as follows:

$$\sigma_j^{MAP} = \frac{\mathbb{P}(s_j = 1|\mathbf{y}, \mathbf{x}_j, \mathbf{p})}{\mathbb{P}(s_j = 0|\mathbf{y}, \mathbf{x}_j, \mathbf{p})} = \frac{\mathbb{P}(\mathbf{y}, \mathbf{x}_j, \mathbf{p}, s_j = 1)}{\mathbb{P}(\mathbf{y}, \mathbf{x}_j, \mathbf{p}, s_j = 0)} \tag{9}$$

Let c, the discrete random variable describing the collusion size varying from 1 to $cmax$ and which distribution suits a uniform law so that:

$$\mathbb{P}(c) = \frac{1}{c_{max}} \tag{10}$$

In [6], the authors have performed several calculations which have lead to a calculable result of the likelihood ratio.

3.3   MAP Decoding in Two-Layered Tracing Scheme

In this paper, we focus on the MAP-based accusation functions which were detailed above, and we will apply them in our two-layered scheme. As depicted in Fig. 1, applying the two-layered accusation process should improve the tracing performance of a simple decoder. Moreover, by applying the Boneh Shaw selection in a first time, we guarantee a selection of at least one group. Switching to the second accusation level and computing scores using the MAP functions for users in the selected group should reduce the research space and eventually provides more accuracy to the accusation process.

4   Experimental Results

To show the efficiency of the proposed two-layered strategy accusation, we assess it in terms of detection rates for different collusion attacks and different code lengths. Another important criteria is the time estimation which computes the time required by the detector to estimate accurately the collusion channel. The aim of the proposed MAP-based detector for hierarchical fingerprints is to find a trade off between the detection runtime and the accusation rate, by taking less time for a good accusation rates. To estimate the decoder performance,

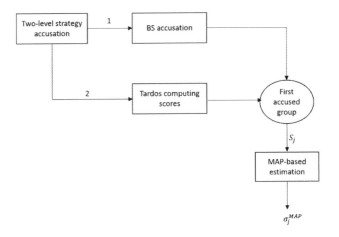

Fig. 1. The two-layered MAP-based tracing scheme

we used the modified Monte Carlo Estimation algorithm proposed by [8], and we compared the proposed detector, the MAP-based detector for hierarchical fingerprints, to the agnostic decoder, and also to the non-layered detector, called respectively MAP-H, T and MAP in this part. Each run of Monte Carlo estimation consists in 10^3 trials.

4.1 Results of Stability Against Different Collusion Strategies

In this part, we used ROC curves to compare the Tardos decoder, the MAP-based decoder in hierarchical context and the MAP decoder proposed by [6] in terms of different criteria. We evaluate the stability of the three decoders against different collusion attacks as follows. The deterministic attacks such as the Minority attack (the less frequent symbol in a colluders' codewords is selected), and the Uniform attack as a random attack (colluders select uniformly and randomly one symbol among their symbols). The 'min' and 'rand' state respectively for the Minority vote attack and the Uniform attack. The fingerprint code is of length m=300, the maximum number $cmax$ of colluders is equal to 10 and the number of colluders is c=6. In Fig. 2(a), compared to the two other detectors, the Tardos decoder varies a little bit for the two strategies. Although, it provides a good stability whatever is the collusion attack, it loses in detection accuracy especially for the minority attack, where we notice a large difference with the two other decoders. For the two strategies, we show that the proposed decoder improves the performance of the non-layered MAP-based decoder and provides a good performance and an accurate accusation rate especially for the minority attack.

4.2 Results When Varying the Code Lengths

In Fig. 2(b), we evaluate the performance of the MAP-based decoder for hierarchical fingerprints, the simple MAP decoder and the Tardos code when varying

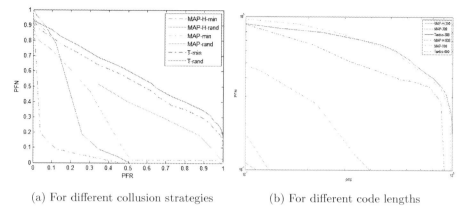

(a) For different collusion strategies   (b) For different code lengths

Fig. 2. ROC curves of the MAP hierarchical decoder, the MAP decoder and the Tardos decoder for c=6 and n=1000 users.

the code length (m=300 and m=800) and using the minority attack. We use the logarithmic scale for the probability of false negative and the probability of false alarm. We remark that the proposed decoder has the less detection errors compared to the other decoders, especially the Tardos code. In addition, when increasing the code length, the performance of our decoder is increasingly outstanding.

4.3 Time Estimation

The aim of the majority traitor tracing systems is to provide an accurate accusation of colluders despite the great number of users. In this part of the experimentation, we compared the time taken by the three decoders to estimate accurately the colluders. The experimental results shown in Table 1 are averaged values of 10^2 trials for different number of users. We show that the proposed technique reduces significantly the time estimation of the non-layered MAP decoder which can be explained by the group-based strategy. The MAP estimation is applied only in the selected group. Although, the Tardos code has a satisfying time accusation, the gap between the proposed detector and the other decoders increases as the code length is important. The proposed decoder provides a good tradeoff between the runtime estimation and the accusation rates.

Table 1. Time estimation for different number of users (in sec)

n	Tardos [10]	MAP [6]	MAP-H
10^3	8	830	5
10^4	600	1791	64
10^5	1800	31552	500

5   Conclusion

In this paper, we have proposed a blind MAP-based traitor tracing scheme for hierarchical fingerprints. The proposed system consists in a group-based accusation process. The MAP detector is applied only for the selected group. Experiments have shown that, compared to the non-layered MAP detector and to the original Tardos code, this proposed system provides not only an accurate and a fast decision about colluders but also has the less error probabilities in decoding process.

References

1. Boneh, D., Shaw, J.: Collusion-secure fingerprinting for digital data. In: Coppersmith, D. (ed.) CRYPTO 1995. LNCS, vol. 963, pp. 452–465. Springer, Heidelberg (1995)
2. Boneh, D., Shaw, J.: Collusion-secure fingerprinting for digital data. IEEE Trans. Inf. Theor. **44**(5), 1897–1905 (1998)
3. Chaabane, F., Charfeddine, M., Amar, C.B.: A multimedia tracing traitors scheme using multi-level hierarchical structure for tardos fingerprint based audio watermarking. In: SIGMAP 2014 Vienna, Austria, 28–30 August, 2014, pp. 289–296 (2014)
4. Charpentier, A., Xie, F., Fontaine, C., Furon, T.: Expectation maximization decoding of tardos probabilistic fingerprinting code. In: Media Forensics and Security, p. 72540 (2009)
5. Chor, B., Fiat, A., Naor, M.: Tracing traitors. In: Desmedt, Y.G. (ed.) CRYPTO 1994. LNCS, vol. 839, pp. 257–270. Springer, Heidelberg (1994)
6. Desoubeaux, M., Herzet, C., Puech, W., Guelvouit, G.L.: Enhanced blind decoding of tardos codes with new map-based functions. In: MMSP 2013, Pula, Sardinia, Italy, September 30 - October 2, 2013, pp. 283–288 (2013)
7. Desoubeaux, M., Guelvouit, G.L., Puech, W.: Fast detection of Tardos codes with boneh-shaw types. In: Proceedings of the SPIE 8303, Media Watermarking, Security, and Forensics (2012)
8. Furon, T., Pérez-Freire, L., Guyader, A., Cérou, F.: Estimating the minimal length of tardos code. In: Katzenbeisser, S., Sadeghi, A.-R. (eds.) IH 2009. LNCS, vol. 5806, pp. 176–190. Springer, Heidelberg (2009)
9. Laarhoven, T., de Weger, B.: Optimal symmetric tardos traitor tracing schemes. In: CoRR abs/1107.3441 (2011)
10. Skoric, B., Katzenbeisser, S., Celik, M.U.: Symmetric Tardos fingerprinting codes for arbitrary alphabet sizes. IACR Cryptology ePrint Arch. **2007**, 41 (2007)
11. Tardos, G.: Optimal probabilistic fingerprint codes. In: STOC, pp. 116–125 (2003)
12. Wang, Z.J., Wu, M., Trappe, W., Liu, K.J.R.: Group-oriented fingerprinting for multimedia forensics. EURASIP J. Adv. Sig. Proc. **2004**(14), 2153–2173 (2004)

Poisson Image Denoising
Based on BLS-GSM Method

Liangdong Li[1], Nikola Kasabov[2], Jie Yang[1(✉)], Lixiu Yao[1],
and Zhenghong Jia[3]

[1] Institution of Image Processing and Pattern Recognition,
Shanghai Jiao Tong University, 200240 Shanghai, China
{jamesleee,jieyang,lxyao}@sjtu.edu.cn
[2] Knowledge Engineering and Discovery Research Institute,
Auckland University of Technology, 92006 Auckland, New Zealand
nkasabov@aut.ac.nz
[3] School of Information Science and Engineering,
Xinjiang University, 830046 Urumqi, China
jzhh@xju.edu.cn

Abstract. Poisson noise removal is of significant importance for many applications such as spectral imaging, night vision and especially in medical imaging and astronomy. Gaussian scale mixture based methods have been widely used in image denoising. In this paper, we focus on the Poisson noise and propose a new strategy based on Bayesian least squares method for its removal. We begin with a method that removes Poisson noise by reducing it to an additive Gaussian noise with a Variance Stabilizing Transformation. Then we combine the localized version of BLS-GSM method to bring out a new denoising strategy for images corrupted by Poisson noise and experimentally show that it outperforms some of the best existing methods for Poisson noising removal both numerically and visually.

Keywords: Poisson noise removal · Variance stabilizing transformation · Gaussian noise · Localized BLS-GSM method

1 Introduction

In a wide range of imaging applications, observations are collected by counting photons hitting a detector array or vehicles passing a sensor. For example, in nuclear medicine, night vision, astronomy, traffic analysis and many other applications, there is a limited amount of available light. Even in well-lit environment, photon limitations can arise when using a spectral imager which heavily depends on the wavelength of each received photon. Thus Poisson noise removal is of significant importance in these applications above. As the noise variance equals to the expected value of the true value of observed images, Poisson noise depends on the true value. In this case, the premise of Poisson noise removal is very different from the scene of additive white Gaussian noise whose variance is usually assumed by signal processing designers and thus is known during preprocessing.

Up to now, many types of denoising methods specifically designed for Poisson noise removal have been proposed. Jin et al. [1] proposed an algorithm to restore the

© Springer International Publishing Switzerland 2015
S. Arik et al. (Eds.): ICONIP 2015, Part IV, LNCS 9492, pp. 513–522, 2015.
DOI: 10.1007/978-3-319-26561-2_61

Poisson noise by combing the special properties of the Poisson distribution and the idea of Optimal Weights Filter [2]. This algorithm can reconstruct the image contaminated by the Poisson noise efficiently and achieve good visual effect as well. Salmon et al. [3] proposed an algorithm which combines elements of dictionary learning and sparse patch-based representations of images. This method employs both an adaption of Principal Component Analysis (PCA) for Poisson noise and recently developed sparsity-regularized convex optimization algorithms for photon-limited images. Deledalle et al. [4] proposed an extension of the NL-means for images contaminated by the Poisson noise, based on probabilistic similarities to compare noisy patches and patches of a pre-estimated image, and on a minimization of the Mean Square Error (MSE) with respect to the denoising parameters. There are also many other methods (see [5, 6] for detail about the multiscale maximization a priori method and [7, 8] for regularization based on a total variation semi-norm).

Different from the algorithms mentioned above, the most popular methods of Poisson noise removal are usually performed by reducing it to nearly additive Gaussian noise through the following three steps. Firstly, the variance of the Poisson noise is stabilized by applying a Variance Stabilizing Transformation (VST) such as Anscombe root transformation [9], so that the transformed data are approximately homoscedastic and Gaussian. This step will produce a noisy image data in which the noise can be handled as additive Gaussian noise with unitary variance. Secondly, the noise is removed by a conventional denoising method for additive Gaussian white noise such as NL-means [10], BM3D [11] and BLS-GSM [12]. Thirdly, an inverse transformation is applied to the denoised data, obtaining the estimate of the signal of interest. Makitalo et al. [13, 14] focused on the last step, and introduce the Exact Unbiased Inverse (EUI) method. Authors [15–17] improved both the first and last step and achieved remarkable denoising results.

In this paper, we focus on a strategy for Poisson noise removal based on BLS-GSM [12, 18]. We begin with a Variance Stabilizing Transformation method that suppresses Poisson noise to an additive Gaussian noise with unitary variance. Then we combine the localized version of BLS-GSM method to bring out a new denoising strategy for images contaminated by Poisson noise. We will show that the resulting method is state-of-the-art in terms of PSNR.

The paper is organized as follows: in Sect. 2 we explain the Poisson noise removal problem with more details about VST, EUI, and BLS-GSM, and then propose our three-stage algorithm for Poisson denoising. In Sect. 3 we conduct some experiments and analysis to prove that EUI combined with localized BLS-GSM gives rise to a very efficient filtering solution that is competitive with some of the best existing methods for Poisson noise removal. In Sect. 4 we make some discussions to conclude this paper.

2   Method Description

2.1   Poisson Noise

Let $x_i, i = 1, \ldots, N$ be the observed pixel values obtained through an image sensor where x_i is an independent random Poisson variable whose mean $y_i \geq 0$ is the underlying intensity value to be estimated. Thus the discrete Poisson probability of each x_i is:

$$P(x_i|y_i) = \frac{y_i^{x_i} e^{-y_i}}{x_i!}.$$

(1)

As the mean of the Poisson variable is equal to its variance, so we have

$$E\{x_i|y_i\} = y_i = var\{x_i|y_i\}.$$

(2)

As we have mentioned above, the noise variance depends on the true intensity value, thus Poisson noise is signal dependent.

2.2 Our Algorithm

In this subsection we present our algorithm for removing Poisson noise and discuss in full detail in the following.

A standard way to denoise Poisson noise image is using a variance stabilizing transformation (VST). To remove the data-dependence of the noise variance, we apply VST to make it constant throughout the whole denoising procedure with a conventional Gaussian denoising method. In the recent years, many variance stabilization transforms have been developed [15–17], among which the most popular VST is the Anscombe transformation [9]:

$$f(x) = 2\sqrt{x + \frac{3}{8}}.$$

(3)

However, applying (3) directly to Poisson distributed data can only produce an approximately Gaussian distributed data with variance 1. This holds still when the mean of the Poisson data is greater than 4. The denoising of $f(x)$ produces a signal **D** that can be considered as an estimate of $E\{f(x)|y\}$. We will discuss later on.

After applying the Anscombe transformation, we need to apply conventional denoising technique for Gaussian noise such as Bayes Least Squares – Gaussian Mixture Model (BLS-GSM) method [12]. Instead of using it directly, we introduce an improved version (i.e. localized BLS-GSM). For rigorous reason we make some discussion and show more details than the original paper. The basic mind of the algorithm is to model a noiseless wavelet coefficient neighborhood **x**, by a Gaussian scale mixture which is defined as

$$\mathbf{x} = \sqrt{m}\mathbf{u}$$

(4)

where **u** is a zero-mean Gaussian random vector and m an independent positive scalar random variable. Without loss of generality, we can assume the expectation of $E\{m\} = 1$. Thus we have $\mathbf{C_x} = \mathbf{C_u}$, where **C** represents the covariance matrix of a vector. The idea of denoising is as follows: (1) decompose the noisy input image into a wavelet pyramid, (2) apply the whole denoising process on wavelet coefficients, (3) obtain the final denoised image by reconstruction. Besides, to avoid ringing artifacts in the reconstruction, a redundant version of wavelet transform steerable pyramid is used. For an $n_1 \times n_2$ image, the steerable pyramid is generated in $\log_2(\min(n_1, n_2)) - 4$ scales and eight orientations. Assume that the image is contaminated by independent additive Gaussian noise. Thus, a typical neighborhood of wavelet coefficients can be written as

$$\mathbf{y} = \mathbf{x} + \mathbf{w} = \sqrt{m}\mathbf{u} + \mathbf{w} \tag{5}$$

where $\mathbf{y}$ is the observed noisy neighborhood to be estimated, $\mathbf{x}$ the original neighborhood and w the independent additive white Gaussian noise signal with known variance σ. With the observed noisy vector $\mathbf{y}$, a Bayes Least Square (BLS) estimation of x, $E\{x|\mathbf{y}\}$ is calculated as follows

$$E\{x|\mathbf{y}\} = \int_0^\infty p(z|\mathbf{y})E\{x|\mathbf{y}, z\}dz. \tag{6}$$

According to the original work [12], discrete form of $E\{x|\mathbf{y}\}$ can be computed as

$$E\{x|\mathbf{y}\} = \sum_{i=1}^{13} p(\log_e(z_i)|\mathbf{y})E\{x|\mathbf{y}, \log_e(z_i)\}. \tag{7}$$

To get $E\{x|\mathbf{y}\}$, we just need to calculate $p(\log_e(z_i)|\mathbf{y})$ and $E\{x|\mathbf{y}, \log_e(z_i)\}$ respectively. For notational simplicity, we replace the $\log_e(z_i)$ term with z_i. The detailed derivation process of the two components can be found in [12]. Therefore, we omit the details in this paper and thus we have:

$$E\{x|\mathbf{y}, z_i\} = z\mathbf{M}\Lambda(z\Lambda + \mathbf{I})^{-1}\mathbf{v} \tag{8}$$

where $\mathbf{C}_w = \mathbf{S}\mathbf{S}^T$, $\mathbf{Q}\Lambda\mathbf{Q}^T = \mathbf{S}^{-1}\mathbf{C}_u(\mathbf{S}^{-1})^T$, $\mathbf{M} = \mathbf{S}\mathbf{Q}$, $\mathbf{v} = \mathbf{M}^{-1}\mathbf{y}$.

The discrete form of $p(z_i|\mathbf{y})$ is

$$p(z_i|\mathbf{y}) = \frac{p(\mathbf{y}|z_i)p_z(z_i)}{\sum_{j=1}^{13} p(\mathbf{y}|z_j)p_z(z_j)} \tag{9}$$

where $p_z(z_i) \propto \frac{1}{z_i}$.

The main point of this procedure is to use any wavelet coefficient, either extracted from a single orientation/scale or mixing orientations and scales. The idea is to denoise all these wavelet coefficients in the pyramid, and then a reconstruction step of the denoised image is performed by the inverse pyramid.

After discussing the BLS-GSM method, we realize the denoising procedure by firstly partition the input $n_1 \times n_2$ noisy image into $\sqrt{d} \times \sqrt{d}$ blocks, where $d = min(\lceil \sqrt{n_1} \rceil, \lceil \sqrt{n_2} \rceil)$. And then we apply BLS-GSM to each block.

Having finished the two successive steps above, the only work we need to do is applying an inverse transformation to $\mathbf{D}$ in order to obtain the denoised estimate of y. In [13] the authors proposed three inverse transformations: algebraic inverse, asymptotically unbiased inverse and exact unbiased inverse. In this paper we just focus on the EUI method which will be discussed in the following.

Given a perfect denoising, that is to say $\mathbf{D}$ can be treated as $E\{f(x)|y\}$, and the EUI of the Anscombe transformation f is an inverse transformation:

$$T : E\{f(x)|y\} \mapsto E\{x|y\}. \tag{10}$$

As mentioned in 2.1, for any given y, we have $E\{x|y\} = y$. Then the problem of finding the inverse transformation T reduces to computing the values $E\{f(x)|y\}$ defined as:

$$E\{f(x)|y\} = \int_{-\infty}^{+\infty} f(z)p(z|y)dz, \tag{11}$$

where $p(z|y)$ is the generalized probability density function of z conditioned on y. For discrete Poisson probabilities $P(x|y)$, we have the following form:

$$E\{f(x)|y\} = \sum_{x=0}^{+\infty} f(x)P(x|y). \tag{12}$$

Take (1) and (3) into consideration, we have the finally form:

$$E\{f(x)|y\} = 2 \sum_{x=0}^{+\infty} \left(\sqrt{x + \frac{3}{8}} \cdot \frac{y^x e^{-y}}{x!} \right). \tag{13}$$

To solve the problem of approximately transformation mentioned at the beginning of 2.2, for data in $\mathbf{D}$ that is greater than $2\sqrt{3/8}$, we set to 0.

3 Experiments and Analysis

Having introduced the main strategy in Sect. 2, in this section we implement our algorithm and make comparison with other Poisson denoising methods. As described above we will perform our method in three successive steps: (1) an Anscombe VST is used to stabilize the Poisson noise variance. (2) the localised BLS-GSM method is used for denoising the signal produced in the last step, which can be treated as additive white Gaussian noise. (3) the exact unbiased inverse is applied to the denoised signal, obtaining the estimate of the signal of interest. In the second step, we compare the performance of the localized BLS-GSM method with a few recent classic denoising algorithms for Gaussian noise removal: NL-means [10], K-SVD [19], BM3D [11]. We also conduct comparisons against algorithms specifically designed for Poisson noise removal such as Weights Optimization Filter (WOF) [1], Non-local PCA [3] and Unsupervised Non-local Means (UNL-Means) [4].

To describe quantitatively performance of denoising results, two classic measures are introduced, the Root Mean Square Error (RMSE) and Peak Signal to Noise Ratio (PSNR). The RMSE is computed as:

$$RMSE = \sqrt{\frac{\sum_i(\hat{y}_i - y_i)^2}{N}},$$

where $\hat{y}_i$ represents the estimated intensities, y_i the respective true value, N the total number of pixels in the image. The smaller the RMSE, the better the denoising. The PSNR is evaluated in decibels (dB):

$$PSNR = 20\log_{10}\left(\frac{max(y_i)}{RMSE}\right).$$

The larger the PSNR, the better the denoising.

To make objective and balanced comparisons, we conduct our experiments with the set of test images all of which are 256 × 256 in size provided by the authors of [17] showing in Fig. 1. They are Spots [0.03, 5.02], Galaxy [0, 5], Ridges [0.05, 0.85], Barbara [0.93, 15.73] and Cells [0.53, 16.93]. These test images can be downloaded from http://www.cs.tut.fi/foi/invansc/. Table 1 provides the numerical denoising results of these noise removal methods in terms of PSNR(dB)/RMSE. The experiments results show that the Localized BLS-GSM method is competitive with the more recent denoisng algorithms specifically designed for Poisson noise removal [1, 3] which achieve excellent denoising results. Figures 2 and 3 show the visual quality performance of these methods and the comparison of our strategy with other noise removal methods.

As shown in the Figs. 2 and 3, the Localized BLS-GSM algorithm preserves the sharpness of edges. The denoised images preserve more detail of the original ones than NL-Means(e), K-SVD(f) and UNL-Means(i) although a little inferior to BM3D ones. Besides, it also introduces fewer artifacts than most of the other methods. Overall, the visual performance is superior to most of the other recently proposed Poisson noise removal algorithms, while a little inferior to the BM3D's.

All in all, the performance of the Localized BLS-GSM algorithm is state-of-the-art in terms of both visual quality and numerical results of PSNR/RMSE.

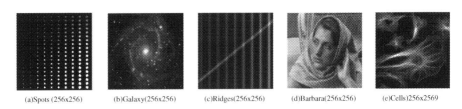

(a)Spots (256x256)   (b)Galaxy(256x256)   (c)Ridges(256x256)   (d)Barbara(256x256)   (e)Cells)256x2569

Fig. 1. Original test images used in the experiments.

Table 1. The numerical comparison in terms of PSNR/RMSE between LBLS-GSM combined with VST for Poisson noise removal method and other algorithms including Conventional Gaussian denoising methods and specifically designed for Poisson noise removal. The intensity range of each image is indicated in brackets.

	Conventional Gaussian denoising methods combined with VST				Specifically designed for Poisson noise removal methods		
	LBLS-GSM	BM3D [11]	NL-Means [10]	K-SVD [19]	WOF [1]	NL-PCA [3]	UNL-Means [4]
Spots [0.03,5.02]	31.47/0.1472	**31.96/0.1267**	29.98/0.2237	30.25/0.1672	31.52/0.1410	31.08/0.1522	30.58/0.1568
Galaxy [0.5]	**29.37/0.1485**	28.05/0.1980	26.69/0.3321	27.11/0.2120	**29.37/0.1485**	28.62/0.1913	27.24/0.2098
Ridges [0.05,0.85]	24.01/0.0568	**24.54/0.0507**	22.68/0.1030	23.87/0.0763	24.13/0.0520	**24.54/0.0507**	23.67/0.0800
Barbara [0.93,15.73]	**26.44/0.6493**	25.92/0.7956	24.97/0.9602	25.11/0.8972	25.78/0.8239	26.03/0.6233	25.32/0.8879
Cells [0.53,16.93]	**30.67/0.4936**	30.19/0.5240	28.71/0.8783	29.33/0.5567	29.62/0.5015	30.10/0.5103	28.99/0.8023

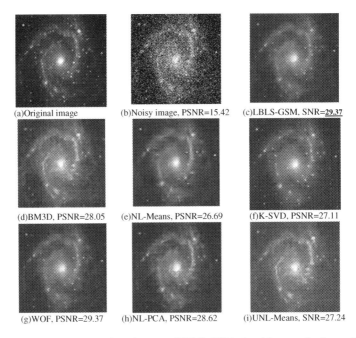

Fig. 2. Visual quality comparison between LBLS-GSM algorithms and others of Galaxy.

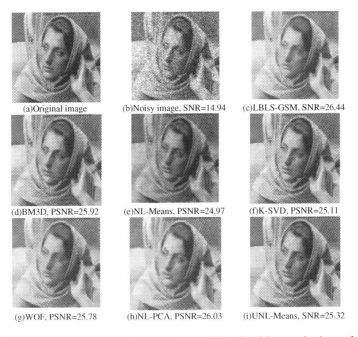

Fig. 3. Visual quality comparison between LBLS-GSM algorithms and others of Galaxy.

4 Discussion and Conclusion

In this part we present the experiments results and make some conclusions.

A new strategy for denoising images corrupted by Poisson noise is implemented. By combining the Localized BLS-GSM method with VST we implement Poisson noise removal effectively. Compared with other conventional Gaussian noise removal method and algorithms specifically designed for Poisson noise removal, the LBLS-GSM can achieve excellent effect in terms of both visual quality and numerical results of PSNR/RMSE.

Acknowledgement. This research is partly supported by NSFC, China (No: 61273258) and 863 Plan, China (No. 2015AA042308).

References

1. Jin, Q., Grama, I., Liu, Q.: A new poisson noise filter based on weights optimization. J. Sci. Comput. **58**(3), 548–573 (2014)
2. Jin, Q., Grama, I., Liu, Q.: Removing gaussian noise by optimization of weights in non-local means. arXiv, preprint arXiv:1109.5640 (2011)
3. Salmon, J., et al.: Poisson noise reduction with non-local PCA. J. Math. Imaging Vis. **48**(2), 279–294 (2014)
4. Deledalle, C.-A., Tupin, F., Denis, L.: Poisson NL means: unsupervised non local means for poisson noise. In: 17th IEEE International Conference on Image Processing (ICIP), pp. 801–804. IEEE (2010)
5. Nowak, R., Kolaczyk, E.D.: A multiscale MAP estimation method for poisson inverse problems. In: Signals, Systems & Computers, Conference Record of the Thirty-Second Asilomar Conference on, vol. 2, pp. 1682–1686. IEEE (1998)
6. Nowak, R.D., Kolaczyk, E.D.: A statistical multiscale framework for Poisson inverse problems. IEEE Trans. Inf. Theory **46**(5), 1811–1825 (2000)
7. Bardsley, J.M., Luttman, A.: Total variation-penalized Poisson likelihood estimation for ill-posed problems. Adv. Comput. Math. **31**(1-3), 35–59 (2009)
8. Beck, A., Teboulle, M.: Fast gradient-based algorithms for constrained total variation image denoising and deblurring problems. IEEE Trans. Image Process. **18**(11), 2419–2434 (2009)
9. Anscombe, F.J.: The transformation of poisson, binomial and negative-binomial data. Biometrika **35**, 246–254 (1948)
10. Buades, Antoni, Coll, Bartomeu, Morel, Jean-Michel: A review of image denoising algorithms, with a new one. Multiscale Model. Simul. **4**(2), 490–530 (2005)
11. Dabov, K., Foi, A., Katkovnik, V., Egiazarian, K.: Image denoising by sparse 3D transform-domain collaborative filtering. IEEE Trans. Image Process. **16**(8), 2080–2095 (2007)
12. Portilla, J., Strela, V., Wainwright, M.J., Simoncelli, E.P.: Image denoising using scale mixtures of Gaussians in the wavelet domain. IEEE Trans. Image Process. **12**, 1338–1351 (2003)
13. Makitalo, M., Foi, A.: On the inversion of the Anscombe transformation in low-count Poisson image denoising. In: International Workshop on Local and Non-Local Approximation in Image Processing, LNLA 2009. IEEE (2009)

14. Makitalo, M., Foi, A.: Optimal inversion of the Anscombe transformation in low-count Poisson image denoising. IEEE Trans. Image Process. **20**(1), 99–109 (2011)
15. Lefkimmiatis, S., Maragos, P., Papandreou, G.: Bayesian inference on multiscale models for Poisson intensity estimation: applications to photon-limited image denoising. IEEE Trans. Image Process. **18**(8), 1724–1741 (2009)
16. Luisier, F., et al.: Fast interscale wavelet denoising of poisson-corrupted images. Signal Process. **90**(2), 415–427 (2010)
17. Zhang, B., Fadili, J.M., Starck, J.L.: Wavelets, ridgelets, and curvelets for Poisson noise removal. IEEE Trans. Image Process. **17**(7), 1093–1108 (2008)
18. Rajaei, B.: An analysis and improvement of the BLS-GSM denoising method. Image Process. Line 44–70 (2014)
19. Lebrun, M., Leclaire, A.: An mplementation and detailed analysis of the K-SVD image denoising algorithm. In: Image Processing on Line (2012)

Exploring Relative Motion Features for Gait Recognition with Kinect

Ke Yang[1,2(✉)], Yong Dou[1,2], Shaohe Lv[1,2], and Fei Zhang[1,2]

[1] National Laboratory for Parallel and Distributed Processing,
National University of Defense Technology, Changsha 410073, China
[2] College of Computer, National University of Defense Technology,
Changsha 410073, China

Abstract. Gait is a very important biometric technology for human recognition. Gait feature can be divided into two categories: static and dynamic. Many previous works argue that, although motion reflects the essential nature of gait, the recognition performance based purely on the motion feature is limited. The root cause of the limited performance is however not yet to understand. In this paper, we study the gait recognition with motion feature by Kinect and show that, with a novel representation, the motion feature is still effective to distinguish the gaits from different human beings. In particular, relative distance-based motion features are proposed, which are extracted without calculating the gait cycle. Experimental results show that the accuracy of recognition with relative motion features is up to 85 %, which is comparable to that of static features. By combining the relative motion features and the static ones, the accuracy is above 95 %.

Keywords: Relative motion feature · Dynamic gait features · Gait recognition

1 Introduction

Accurate and effective human recognition is one of the major research areas of computer vision, pattern recognition, biometrics, and intelligent surveillance. Face recognition [1], fingerprint recognition [2] technologies are developed earlier and mature. But this biometrics need to be detected with subjects' cooperate, and require a high image resolution. Gait recognition technology can be detected in a long distance, and need not subjects' cooperate. And there are already studies using gait biometrics as a forensic tool [18].

Existing gait recognition methods can be roughly divided into two main groups: model-free and model-based. Model-free gait recognition technologies include a variety of silhouettes-based methods. Phillips et al. [17] proposed a baseline algorithm using the correlation of silhouettes. Gait Energy Image (GEI) [3] used the average image of the silhouettes in a gait period to characterizing gait features. Gait Entropy Image (GEnI) [4] and Chrono-Gait Image (CGI) [5] are similar to GEI [3]. Model-free method is simple and of low computational complexity, but performs poorly for the viewpoint and the occlusion problem. Model-based gait recognition methods model the structure of the

© Springer International Publishing Switzerland 2015
S. Arik et al. (Eds.): ICONIP 2015, Part IV, LNCS 9492, pp. 523–530, 2015.
DOI: 10.1007/978-3-319-26561-2_62

human body using body structure parameters. Early model-based methods conclude the pendulum model [6], stick model [7] and so on.

Gait features can be divided into two categories: dynamic and static. Silhouette-based methods employ both dynamic and static features for gait recognition except Phillips' [17] method. GEI [3] encodes static characteristic more, while GEnI [4] and CGI [5] encoding dynamic features more. Many previous works argue that, although motion reflects the essential nature of gait, the recognition performance based purely on the motion feature is limited [9, 15, 16]. However, is the distinguishing ability of the motion itself limited, or existing features doesn't adequately capture the characteristics of motion? This paper will explore this problem in the Kinect-based gait recognition.

Kinect's skeleton tracking function can provide real-time 3D coordinates of 20 human skeleton points, eliminating the need for complex extraction procedures of human model. Similarly, Kinect-based gait feature can also be divided into static and dynamic features. Static features mainly refer to the skeletons length, height and so on. For example, Araujo et al. [13] extracted a total of 11 features includes skeletons length and height as static features.

Dynamic features can be subdivided into four categories: (i) the intuitive dynamic features: step length, speed, and gait cycle. Preis et al. [8] extracted the step length and speed as dynamic gait features; (ii) angle-based dynamic features: statistical character-istics of bone angles. [9, 10] extracted statistical characteristics of the lower body angles as dynamic features; (iii) absolute motion features: statistical characteristics of skeleton points' absolute coordinates' changes. Vertical Distance Features (VDF) [11] described the statistical characteristics of the absolute coordinate changes of some joints when walking; (iv) relative motion features: relative motion features extract statistical char-acteristics from relative distances between joints, avoiding the problem of absolute motion features.

Chattopadhyay et al. [12] used the average relative distance between some joints as dynamic gait features and got a poor performance. But only using the mean value is not enough, standard deviation also implies sufficient dynamic characteristics of walking; the combination of the two will lead to a better result.

This paper extracted the relative motion features as dynamic gait features. Proposed robust relative motion features' recognition accuracy is up to 85 %, and don't need calculate the gait cycle. When combined with static features, recognition accuracy is above 95 %.

The rest of the paper is organized as follows. The proposed method is presented in Sect. 2, and the experiment and results are shown in Sect. 3. Finally, conclusion is drawn in Sect. 4.

2   Proposed Method

This section describes the proposed method and is divided into four parts. Firstly, Kinect skeleton data stream is introduced, followed by the extraction of dynamic and static features of gait, and the last parts are feature selection and classification.

2.1   Kinect Skeleton Data

This paper only uses skeleton data stream. Kinect v1.0 can provide 3D coordinates of the 20 joints of two human bodies at 30fps, as shown in Fig. 1(a). The coordinate system is shown in Fig. 1(b), the origin of the Cartesian coordinate system is the depth sensor's center; x-axis is parallel with Kinect; y-axis is vertical to bottom surface of Kinect; z-axis is the direction parallel to the sensor's normal direction. The units are in meters. This paper will extract dynamic and static gait features on Kinect skeleton data stream to provide for identifying people.

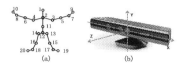

Fig. 1.   (a) 20 joints of skeleton; (b) Kinect depth sensor coordinate system.

2.2   Distance-Based Relative Motion Features Extraction

In practical application scenarios, persons will not walk in the preset direction as in the experimental environment; Kinect Sensor will move or rotate to get a larger view. Therefore, the absolute coordinates of human joints is not suitable for characterizing human gait features in the real scene. All dynamic features are extracted from the relative distances of the particular joints. The relative distances are divided into three groups according to directions: x direction, y direction and z direction. The directions are the same to the coordinate system's axis direction. The 11 relative distances are shown as formula (1),

$$
\begin{aligned}
Dx1 &= abs(x(17) - x(18)); \\
Dx2 &= abs(x(5) - x(6)); \\
Dx3 &= abs(x(9) - x(10)); \\
Dx4 &= abs(x(1) - (x(17) + x(18))/2); \\
Dx5 &= abs(x(11) - (x(17) + x(18))/2); \\
Dx6 &= abs(x(7) - x(8)); \\
Dx7 &= abs(x(3) - x(4)); \\
Dy1 &= abs(y(1) - (y(19) + y(20))/2); \\
Dy2 &= abs(y(1) - (y(15) + y(16))/2); \\
Dy3 &= abs(y(19) - y(20)); \\
Dz1 &= abs(z(9) - z(10));
\end{aligned}
\tag{1}
$$

Where Dxi (or Dyj, Dzk) stands for distances in x-axis (or y-axis, z-axis) direction, $x(i)$ (or$y(j)$, $z(k)$) stands for the ith(or jth, kth) joint's x (or y, z) coordinate. $abs(\cdot)$ stands for absolute value. For example, $Dx1$stands for distance between two ankles in x-direction.

Relative motion features of x (or y, z) direction mainly refers to statistical characteristic (i.e., mean and standard deviations) of relative distances of particular joints in the x (or y, z) direction, as shown in formula (2),

$$STD = std\{Dx1, Dx2, Dx3, Dx4, Dx5, Dx6, Dx7, Dy1, Dy2, Dy3\};$$
$$MEAN = mean\{Dx1, Dx2, Dx3, Dx4, Dx5, Dx6, Dy1, Dy2, Dy3, Dz1\}; \qquad (2)$$
$$Dynamic\ GF = \{STD, MEAN\};$$

Where $std(\cdot)$ stands for standard deviation function, $mean(\cdot)$ stands for mean value function. STD is the vector of standard deviation, $MEAN$ is the vector of mean value. The two sets together are combined into relative motion feature vector $Dynamic\ GF$, whose length is 20.

2.3   Static Gait Feature Extraction

Similar to [9, 13], proposed static gait features mainly refer to the length of the different body parts. All the 19 segments (bones in Fig. 1(a)) lengths set and Height together make up static gait features, totally 20. The height is defined as the sum of neck length, upper and lower spine length, and average leg length.

For each frame static gait features are calculated. The mean and standard deviation are calculated over the all frames, and means of each component is recalculated after remove outliers beyond two standard deviations from mean. And the new means make up the final static gait features vector, namely $Static\ GF$.

Proposed $Dynamic\ GF$ and $Static\ GF$ together make up the combined gait features vector $Combined\ GF$, whose length is 40.

2.4   Feature Selection and Classification

There are a lot noise when capture Kinect skeleton data stream. And what features are relevant are uncertain due to the unknown walking conditions, so a fixed feature selection is difficult. To track this problem, a classifier ensemble method based on Random Subspace Method (RSM) and Majority Voting (MV) is employed as feature selection and classification method [14]. And K-Nearest Neighbor (KNN) with Manhattan distance as the distance metric is employed for classification. As shown in Fig. 2(a), the classification results are obtained from the voting results of L weak classifiers, GF stands for gait features space, R stands for random subspace of GF. In order to achieve higher recognition rate, 10-fold cross-validation will be used to select the best K value for KNN. And the employment of RSM is to validate its effectiveness in this scenario, so parameter adjustment is not the focus. Parameters are set according to experience, therefore the parameters maybe not optimal.

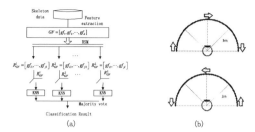

Fig. 2. (a) Feature selection and classification; (b) Semi-circular path used for subjects' walks, showing the Kinect sensor at the center equipped with dish to allow for tracking [9].

3 Experiment and Results

3.1 Skeleton Gait Dataset

Skeleton gait dataset used is public dataset Andersson et al. [9]. The dataset includes 140 subjects, each contains five sequences. In the data capturing procedure, each volunteer was asked to walk in front of the Kinect in a semi-circular. And a spinning dish helped the Kinect to keep the subject always in the center of its view. As shown in Fig. 2(b), each subject executed five round trip free cadence walk, starting from left, walking clockwise to the right and then back. Each sequence includes about 500–600 frames data.

3.2 Classification Results

Like most of the gait recognition algorithms, KNN is chosen as classifier. In order to achieve the highest recognition rate, 10-fold cross-validation was used for selecting K values i.e. the dataset was randomly partitioned in 10 subsets and training was performed 10 times, each time leaving one partition out of training process for testing; the accuracies are the average accuracy of these 10 executions. The search range of K is 1 to 70. Results showed that when K = 1, the three features sets' accuracy all reached highest value, so all subsequent experiments used the parameter K = 1 (i.e. 1-NN).

Table 1 shows the average recognition accuracies when using different set of features. The relative motion features' recognition accuracy on 140 subjects is 84.6 %, which is comparable to that of static features. When combined, recognition accuracy can reach 95.4 %, with a 9.3 % increase compared to static features, suggesting there is a good complementary between proposed relative motion features and static ones.

Table 1. Average recognition accuracy under different feature sets.

Feature set	Static GF	Dynamic GF	Combined GF
Accuracy	86.1 %	84.6 %	95.4 %

Taking into account the size of the gallery, proposed method was experimented under different gallery sizes. For a size P (P = 10, 20... 130), 10 subsets of this size are randomly drawn for gallery. The accuracies are the average over 10 subsets'. The results are shown in Fig. 3(a), proposed three gait features sets' recognition accuracies are not very sensitive to changes in gallery size, especially when using combined gait features. Experimental results on various gallery sizes further validated that motion has sufficient recognition capability as the essential nature of gait.

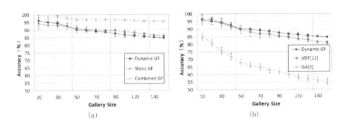

Fig. 3. (a) Average recognition accuracy under different gallery sizes and different feature sets. Error bars represent one standard deviation; (b) VDF[11] and GA[9] and proposed relative motion features' average recognition accuracy under different gallery sizes and different feature sets. Error bars represent one standard deviation.

When introducing the relative motion features of skeleton-based gait, it was mentioned that using only the mean value of relative distances is not enough; the standard deviation also represents the motion well. In order to prove this proposed viewpoint, *Dynamic GF*'s subsets *STD* and *MEAN* are used for recognition separately. The results are reported in Table 2. Though separately using *STD*'s performance is not very good, when combined with *MEAN*, the performance improves greatly than separately using *MEAN*, confirming the previous analysis.

Table 2. Average recognition accuracy under relative motion features and its subsets.

Feature set	Mean	Std.	Dynamic GF
Accuracy	72.6 %	41.6 %	84.6 %

Proposed method used RSM to do feature selection on the proposed gait feature space. To find how much improvement does RSM brings, this paper also experimented without RSM. The results showed that RSM brings 2.2 %, 0.8 % and 1.0 %'s accuracy improvement for static, dynamic, and combined gait features, respectively.

3.3   Comparison to Other Methods

Except method proposed in [11], proposed relative motion features outperforms other methods with a large margin (i.e., over 20 %). The recognition accuracies of intuitive feature-based [8] and angle-based motion features [9, 10] are very low. The intuitive feature-based features' poor performance is explained well in Sect. 1. Figure 3(b) shows how recognition accuracies of [9, 11] and proposed method vary for different gallery sizes. According to [9], the performance of Gait attributes (GA) decreases rapidly with the increase of gallery size. VDF developed in [11] belongs to absolute motion features, and achieve accuracy at 83.5 % on their 20 subjects' dataset. In this paper, VDF was implemented and experimented on the 140 subjects' dataset [9]. As shown in Fig. 3(b), when the gallery size is very small, such as 20, the implemented VDF's recognition accuracy is slightly higher than the proposed method, but the difference is not significant; the implemented VDF's recognition accuracy decrease more rapidly than proposed relative motion features as gallery size increase. To conclude, the discrimination of the relative motion features is higher than angle-based and absolute motion features.

In addition, when combined with the proposed static features, the recognition accuracy is up to 95.4 %.

4   Conclusion

This paper explored the relative motion features for gait recognition with Kinect. In this paper, relative distance-based motion features are proposed. Experimental results showed that the relative motion feature recognition accuracy is up 85 %, which is comparable to the static features. When we use the motion and static feature together, recognition accuracy is above 95 %.

These results suggest that motion is of significant recognition ability as the essential characteristics of gait; and relative motion gait features are an effective representation of gait, worthy of further study in a non-Kinect scene.

Acknowledgments.   This work was supported in part by National High Technology Research and Development Program of China under No. 2012AA012706.

References

1. He, X., Yan, S., Hu, Y., Niyogi, P., Zhang, H.J.: Face recognition using Laplacianfaces. IEEE Trans. Pattern Anal. Mach. Intell. **27**(3), 328–340 (2005)
2. Jain, A.K., Hong, L., Pankanti, S., Bolle, R.: An identity-authentication system using fingerprints. Proc. IEEE **85**(9), 1365–1388 (1997)
3. Han, J., Bhanu, B.: Individual recognition using gait energy image. IEEE Trans. Pattern Anal. Mach. Intell. **28**(2), 316–322 (2006)
4. Bashir, K., Xiang, T., Gong, S.: Gait recognition using gait entropy image. In: IET Conference Proceedings, p. 2. (2009)
5. Wang, C., Zhang, J., Pu, J., Yuan, X., Wang, L.: Chrono-gait image: a novel temporal template for gait recognition. In: Daniilidis, K., Maragos, P., Paragios, N. (eds.) ECCV 2010, Part I. LNCS, vol. 6311, pp. 257–270. Springer, Heidelberg (2010)
6. Cunado, D., Nixon, M.S., Carter, J.N.: Using gait as a biometric, via phase-weighted magnitude spectra. In: Bigün, J., Chollet, G., Borgefors, G. (eds.) AVBPA 1997. LNCS, vol. 1206, pp. 93–102. Springer, Heidelberg (1997)

7. Yoo, J.H., Nixon, M.S., Harris, C.J.: Extracting gait signatures based on anatomical knowledge. In: Proceedings of BMVA Symposium on Advancing Biometric Technologies, London, UK, pp. 596–606 (2002)

8. Preis, J., Kessel, M., Werner, M., Linnhoff-Popien, C.: Gait recognition with kinect. In: 1st International Workshop on Kinect in Pervasive Computing (2012)

9. Andersson, V., Araujo, R.: Person identification using anthropometric and gait data from kinect sensor. In: Proceedings of the Twenty-Ninth Association for the Advancement of Artificial Intelligence Conference. AAAI (2015)

10. Ball, A., Rye, D., Ramos, F., Velonaki, M., March.: Unsupervised clustering of people from 'skeleton' data. In: Proceedings of the Seventh Annual ACM/IEEE International Conference on Human-Robot Interaction, pp. 225–226. ACM (2012)

11. Ahmed, M., Al-Jawad, N., Sabir, A.: Gait recognition based on Kinect sensor. In: SPIE Photonics Europe, p. 91390B. International Society for Optics and Photonics (2014)

12. Chattopadhyay, P., Sural, S., Mukherjee, J.: Frontal gait recognition from incomplete sequences using RGB-D camera. IEEE Trans. Inf. Forensics Secur. **9**(11), 1843–1856 (2014). IEEE Biometrics Compendium

13. Araujo, R.M., Graña, G., Andersson, V.: Towards skeleton biometric identification using the Microsoft kinect sensor. In: Proceedings of the 28th Annual ACM Symposium on Applied Computing, pp. 21–26. ACM (2013)

14. Guan, Y., Li, C.-T., Roli, F.: On reducing the effect of covariate factors in gait recognition: a classifier ensemble method. IEEE Trans. Pattern Anal. Mach. Intell. **37**(7), 1521–1528 (2015)

15. Veeraraghavan, A., Roy-Chowdhury, A.K., Chellappa, R.: Matching shape sequences in video with applications in human movement analysis. IEEE Trans. Pattern Anal. Mach. Intell. **27**(12), 1896–1909 (2005)

16. Lombardi, S., Nishino, K., Makihara, Y., Yagi, Y.: Two-point gait: decoupling gait from body shape. In: 2013 IEEE International Conference on Computer Vision (ICCV), pp. 1041–1048. IEEE, December 2013

17. Phillips, P.J., Sarkar, S., Robledo, I., Grother, P., Bowyer, K.: The gait identification challenge problem: data sets and baseline algorithm. In: Proceedings of the 16th International Conference on Pattern Recognition, pp. 385–388. IEEE (2002)

18. Bouchrika, I., Goffredo, M., Carter, J., Nixon, M.: On using gait in forensic biometrics. J. Forensic Sci. **56**(4), 882–889 (2011)

Characterisation of Cognitive Activity Using Minimum Connected Component

Ramasamy Vijayalakshmi[1(✉)], D. (Nanda) Nandagopal[2], M. Thilaga[1], and Bernie Cocks[2]

[1] Computational Neuroscience Laboratory,
Department of Applied Mathematics and Computational Sciences, PSG College of Technology,
Coimbatore, Tamil Nadu, India
rvpsgtech@gmail.com
[2] Cognitive Neuroengineering Laboratory,
School of Information Technology and Mathematical Sciences, University of South Australia,
Adelaide, Australia
nanda.nandagopal@unisa.edu.au

Abstract. The concept of functional brain networks offers new and interesting avenues for studying human brain function. One such avenue, as described in the current paper, involves spanning subgraphs called Minimum Connected Components (*MCC*) that contain only the influential connections of such networks. This paper investigates cognitive load driven changes across different brain regions using these MCC sub-graphs constructed for different states of brain functioning under different degrees of cognitive load using the graph theoretic concept of clique. The presence of cliques signifies cohesive interconnections among the subsets of nodes in *MCC* that are tightly knit together. To further characterise the cognitive load state from that of the baseline state, the hemisphere wise interactions among the electrode sites are measured. The empirical analysis presented in this paper demonstrates the efficiency of the *MCC* based clique analysis in detecting and measuring cognitive activity with the technique presented potentially having application in the clinical diagnosis of cognitive impairments.

Keywords: Spanning subgraph · Functional brain network · EEG · Cognition · Clique

1 Introduction

The application of graph theoretic analysis and advances in neurophysiological recording of brain activity have significantly enhanced the understanding, modelling and characterisation of complex functional brain networks (*FBN*s) [1, 2]. In this paper, a graph theoretic spanning subgraph based concept known as the Minimum Connected Component (*MCC*) is used as a representative weighted sub-network of the functional brain network corresponding to a particular cognitive state [3]. The *MCC* subnetwork, G_{MCC} in general, is an undirected weighted subgraph extracted from a complete graph G which contains only the set of top high valued edges. It is a special spanning subgraph portraying prominent features of the functional brain network (*FBN*) of a particular brain state such as baseline or

S. Arik et al. (Eds.): ICONIP 2015, Part IV, LNCS 9492, pp. 531–539, 2015.
DOI: 10.1007/978-3-319-26561-2_63

cognitive load. The sub-networks thus constructed are then used to estimate cognition driven specific changes between different brain regions as a measure of the distance or difference between the two sub-networks representing the corresponding brain states. The graph theoretic concept of *clique*, which is a subgraph in which each node is connected with every other node, is used to analyse the *MCC* subnetworks relating to different brain states. The presence of cliques signifies cohesive interconnections among the subsets of nodes that are tightly knit together. The current study by using the *MCC* and clique analysis clearly differentiates changing neuronal patterns before and during cognitive activity.

2   Functional Brain Network Analysis Using Graph Theoretic Concepts

Among the many non-invasive neuroimaging (e.g., *MRI, fMRI, DTI*) and neurophysiological (e.g., *EEG, MEG*) techniques available to measure brain structure and activity, *electroencephalography* (*EEG*) provides the most accessible large spatio-temporal datasets for analysis [4]. Complex network metrics and graph theoretic analysis have been applied to EEG data at both whole brain and regional levels during a number of brain states suggesting that functional brain networks exhibit task-induced changes in connectivity and community structure at the regional level [5, 16].

A FBN is constructed by considering the *EEG* electrodes as *nodes* (*vertices*), and the pair-wise functional associations/connectivity between electrodes (brain regions) estimated using various linear/non-linear statistical measures as *edges* (*links*). The weighted symmetric adjacency matrices (pair-wise associations) thus computed from the *EEG* data are complete graphs where the weights on the edges between the nodes depict the varying levels of correlations between the signals.

These weighted undirected graphs (FBNs) are then used to study the dynamics of the functional connectivity of the underlying neuronal assemblies [6]. Removal of false positive connections that might exist in the resulting network is, however, a crucial task and it is thus essential to analyse only the influential connectivity patterns of the network that contribute to the study of cognitive activities of the brain. To construct such an influential sub-network for further analysis, the *MCC* concept is used in this paper.

3   *FBN* Analysis Using *MCC* and Cliques

The Minimum Connected Component, as proposed here, extracts a unique spanning subgraph from the fully connected functional brain network. The novelty of the proposed algorithm is that it extracts the significant component of a graph without using any user specified threshold. The concept of *MCC* is briefly described below.

Let $G = (V, E)$ be an undirected weighted graph with a set of V nodes, a set of E edges where each edge (u, v) is an unordered pair of nodes, and $w(u, v)$ is the weight of the edge connecting nodes u and v. A Minimum Connected Component, $MCC = G_{MCC} = G_{MCC}(V_{MCC}, E_{MCC})$ of graph G is defined by $G_{MCC} = \phi[G(V, E)]$, where ϕ is a unary graph operator that extracts only the minimum number of maximum weighted edges just needed to connect all the nodes of G forming a special spanning subgraph that satisfies the following properties:

(i) G_{MCC} is a spanning subgraph of graph G such that $E_{MCC} \subseteq E$, and $V_{MCC} = V$.
(ii)   A weighted undirected graph G has exactly one G_{MCC}.
(iii) G_{MCC} of a graph G containing n nodes has at least $n- 1$ edges forming a spanning tree and at most $n(n - 1)/2$ edges and $G = G_{MCC}$.

The function of the operator ϕ is described as follows. The graph operator ϕ is applied on any non-empty weighted connected subgraph G that picks only the highest valued edges from G and includes them into a forest containing all the nodes of G. Finally, when one (or more) edge (edges with the same weight) is (are) appended into the subgraph such that, it is transformed into a connected spanning subgraph, the algorithm terminates resulting in a connected spanning subgraph, the Minimum Connected Component, G_{MCC}. The *MCC* contains the set of all maximum weighted edges constituting a thresholded subgraph of the graph G based on the unique constraint enforced by the graph operator ϕ. The *MCC* is considered as a special functional brain network that is responsive to the cognitive load induced changes in *EGG* data. The *MCC* thus provides an approach that effectively accentuates only the cognitive load driven changes in the functional brain networks. The *MCC*s obtained for different states of brain are then subjected to various analyses using graph mining techniques namely clique analysis and distance computation to identify and quantify the subtle changes in neuronal patterns during the cognitive function.

Given an undirected graph represented by $G = (V,E)$, where V is the set of vertices and E is the set of edges such that $E \subseteq V x V$, a clique in G is defined as a subset $S \subseteq V$ if every pair of vertices $u,w \in S$ and the edge (u,w) exists in E. The cliques are fully connected subgraphs (complete subgraphs) with every node in the clique connected to every other node. In network analysis applications, clique detection is primarily used to find tightly knitted communities (cohesive subgroups) and overlapping communities [7]. A clique M is maximal in G if no vertex $u \in V - M$ can be added to M to form a larger clique; that is, a maximal clique is a complete subgraph that is not contained in any other complete subgraph [8]. Various techniques have been proposed to address the problem of finding maximal cliques within a graph in areas such as social networks, behavioural and cognitive networks, and terrorist networks [9–12].

4   Data Collection and Pre-processing

EEG data was acquired at 1000 Hz via a Nuamps amplifier using *Curry V7* software [13] from 10 healthy adult volunteers (9 male and 1 female) each wearing an appropriately sized 40 channel Quikcap (*Ag/AgCl* electrodes). Following experimental briefing, participants were allowed to practice virtual driving (US Standards) in a Simuride simulator. Impedances were then checked via *Curry (V7)*, with no recording undertaken until all were < 50 KΩ. The average of the two earlobes (A1, A2) was used as the reference.

Baseline continuous *EEG* was then recorded for 4 min while the participant drove a winding road without distraction (*Drive* state). Continuous *EEG* was then recorded as participants undertook an audio distraction task for about 4 min (*DriveAdo* state), whereby participants were asked to repeat the *Drive* task (primary task) while concurrently responding to audio stimuli. These stimuli consisted of a combination of letters/digits (e.g., A5, B9) read by either a male or female voice, presented by *STIM2* software [17]. Each participant underwent 80 trials in this task with equal numbers of male/female voice stimuli. The *EEG* data was band passed limited from 1 – 70 Hz with a 50 Hz notch filter applied to avoid line noise. Eye blink artifacts were removed post-recording using Principal Component Analysis (*PCA*) [14]. Due to the large variations in reaction times (RTs), only those trials with RTs from 1500–3000 ms were selected, with 2 s epochs (0.5 – 2.5 s) subsequently extracted and stored in the database for further analysis (approximately 50 % of the stimuli). Forty chunks of *EEG* data of 2 s duration were then selected randomly from the continuous *EEG* data of the participant while driving without distraction. The *EEG* data were then down sampled to 128 Hz before further analysis.

5   Measuring Cognitive Activity in *FBN* Using *MCC* and Clique Analyses

A three phase approach to measuring and distinguishing cognitive load induced changes in functional brain networks using *MCC* and clique analysis is described below:

(i)   *Construction of Graph Databases (GD):* The association between all pairs of multi-channel *EEG* data is computed in the time domain using Normalised Mutual Information (*NMI*) for different brain states thereby constructing graph databases [15].

(ii) *Construction of MCC:* The minimum connected component is constructed from the graph database for each state using the special graph operator ϕ that contains only the influential connections having high correlation values.

(iii) *MCC Sub-network Analysis:* The *MCC* subnetworks obtained for baseline and cognitive load states are analysed using an algorithm for counting the cliques. In the first instance, the *MCC* derived *FBN* dynamics are identified by detecting newly developed neural connections during cognitive load, the missing neural connections, and those connections which are strengthened among many of the associations. Further analyses are performed on the ability of nodes to form cliques within and across brain hemispheres.

The fully connected functional brain networks are constructed using *NMI* that delineates the non-linear dependencies between all pairs of electrodes constituting edges between them. The edge weights computed from the *EEG* data of all the pairs of electrodes using *NMI* represent the amount of information exchange between them in the range [0, 1]. The resulting symmetric *NMI* graphs of the baseline and cognitive load events are then stored in a graph database (*GD*). The special operator ϕ described earlier efficiently extracts the influential subnetwork with maximum weighted edges that are sufficient to connect all the electrode sites. This spanning subnetwork is considered as an effective representative of the original functional brain network [3]. Various Analyses on the *MCC* subnetworks MCC_{Drive} and $MCC_{DriveAdo}$ obtained for the driving (baseline) and driving with audio distraction (cognitive load) states respectively are discussed below.

The connection patterns between these subnetworks which are common in both the states, the connections newly developed and those which are missing during cognitive state and strengthening/weakening of connections during cognition among the common connections are computed as follows.

$Common\ Edges\ (CE) = Edges\ in\ MCC_{DriveAdo} \cap Edges\ in\ MCC_{Drive}$

$New\ Edges\ (NE) = Edges\ in\ MCC_{DriveAdo} - Edges\ in\ MCC_{Drive}$

$Missing\ Edges\ (ME) = Edges\ in\ MCC_{Drive} - Edges\ in\ MCC_{DriveAdo}$

$Strong\ Edges\ (SE) = CE \cap (Edges\ in\ MCC_{DriveAdo} > Edges\ in\ MCC_{Drive})$

$Weak\ Edges\ (WE) = CE \cap (Edges\ in\ MCC_{DriveAdo} < Edges\ in\ MCC_{Drive})$

The cliques present in the MCC_{Drive} and $MCC_{DriveAdo}$ are identified using the algorithm proposed by Bron et al. [9]. To quantify the influential electrode sites by comparing the number of cliques in which each electrode is a member in the baseline and cognitive load states, the *CliqueCount* algorithm is proposed and is presented below.

Algorithm CliqueCount(MCC $_{Drive}$, MCC$_{DriveAdo}$)

1: For each state of MCC, S_i

2:   For each Electrode E_j

3:   Compute number of cliques C_k in which E_j is a member

4:   End For

5: End For

6: For each Electrode E_j

7:   If $C_k(S_{DriveAdo}) > C_k(S_{Drive})$

8:   Cliquish($S_{DriveAdo}$)=1

9:   Else

10:   Cliquish(S_{Drive})=1

11:   End If

12: End For

The number of cliques formed in baseline and cognitive load states are compared and the position of the electrode in an array of the respective state is set as 1 if it has more number of cliques than that of the other state, otherwise as 0. The results are presented in the following section using customized topoplot for visualization.

6   Results and Discussion

The MCC subnetworks MCC_{Drive}(baseline) and $MCC_{DriveAdo}$(cognitive load state) for all the participants are computed as discussed earlier. The results and plots include only those nodes which exhibited a common behavior across all the participants.

6.1   Clusters of Connectivity Patterns

Firstly, the common, new, missing, strong and weak connections are computed by comparing the baseline and cognitive load states. The functional connections of brain regions (connections between electrode sites) represented on the brain images for better visualization shown in Fig. 1 illustrate the development of various neural clusters during cognitive activity (for only two participants due to space constraint).

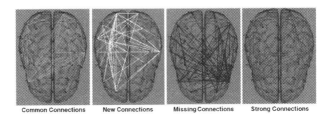

Common Connections   New Connections   Missing Connections   Strong Connections

Fig. 1. Neuronal Clusters - MCC_{Drive} Vs. $MCC_{DriveAdo}$ for Subject DP2

It is interesting to note that all the common connections are not only common during cognitive load and baseline states but they grew comparatively stronger during cognitive load condition in all participants. Since the common connections are all stronger in cognitive load states, there are no weak connections found in $MCC_{DriveAdo}$ when compared to MCC_{Drive} network.

6.2   Clique Analysis

To identify the cohesively working brain regions clique analysis is performed. All the cliques present in MCC_{Drive} and $MCC_{DriveAdo}$ subnetworks are computed and the analysis on these cliques depicts the following interesting results. The number of cliques in which the various electrodes are members is then computed for all participants and summarised in Table 1.

Table 1. Electrode wise Comparison of Cliques in MCC_{Drive} and $MCC_{DriveAdo}$

Electrodes	FP1	FP2	F7	F3	Fz	F4	F8	FT7	FC3	FCz	FC4	FT8	T3	T4	TP7
Drive	0	0	0	0	0	1	1	0	0	1	1	0	0	0	0
DriveAdo	1	1	1	1	1	0	0	1	1	1	0	1	1	1	1
Electrodes	TP8	C3	Cz	C4	CP3	CPz	CP4	T5	P3	Pz	P4	T6	O1	Oz	O2
Drive	1	1	1	0	1	1	1	0	1	1	1	0	1	1	1
DriveAdo	0	0	0	1	1	0	0	1	0	0	0	1	0	0	0

The empirical results show that most of the frontal electrodes form a greater number of cliques during the cognitive load state when compared to the baseline state. Also, central and parietal electrodes form more cliques in the baseline state. The results are presented as customized topoplots for better visualization in Fig. 2 which clearly demonstrates the ability of clique analysis of the MCC subnetworks in detecting cognitive load induced changes in EEG specifically, enables the identification of the predominant features during cognitive activity in healthy human subjects.

The number of cliques formed by the electrodes of Left Hemisphere, LH (FP1, F7, F3, FT7, FC3, T3, C3, TP7, CP3, T5, P3, O1), and Right Hemisphere, RH (FP2, F4, F8, FC4, FT8, C4, T4, CP4, TP8, P4, T6, O2) among themselves (within the hemisphere) and with each other (across the hemisphere) are computed for the baseline and Cognitive Load states with the results illustrated in Fig. 3.

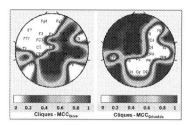

Fig. 2. Electrode wise Cliques Comparison - MCC_{Drive} Vs. $MCC_{DriveAdo}$

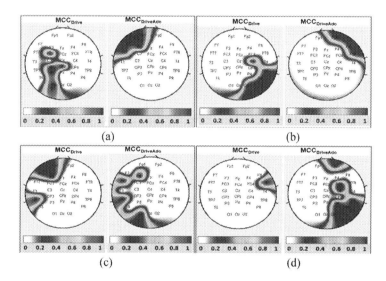

Fig. 3. Topoplots of Electrodes forming many Cliques (a) with LH (b) with RH (c) LH Electrodes forming many Cliques with RH and (d) RH Electrodes forming many Cliques LH

Of particular note, during cognitive load and baseline states, the frontal electrodes FP1, F7, F3, and FT7, and the electrodes FC3, CP3, P3 and O1 form more cliques within the LH respectively Fig. 3(a). Similarly during cognitive load and baseline states, the electrodes FP2, F8, and FT8, and the electrodes C4, P8, P6 and O2 form more cliques within the RH respectively Fig. 3(b). Also of note is the number of cliques formed across hemispheres as shown in Fig. 3. During cognitive load, the LH electrodes F3, FT7, TP3 and O1 form more cliques with the RH electrodes while during the baseline state, the LH electrodes FP1, F3, F7 and TP7 form many cliques with that of RH electrodes Fig. 3(c). As far as the RH electrodes are concerned, the electrode FT8 of baseline state and FP2, FC4, C4, P4, P8, TP8 and O2 of cognitive load electrodes form many cliques across the hemispheres Fig. 3(d).

7 Conclusion

The efficacy of MCC in detecting cognitive load induced changes has been clearly demonstrated. The MCC extracted from the complete FBN represents the most influential subgraph for that particular brain state. The concept of *MCC* has been extended to finding predominant patterns using cliques to understand the underlying neuronal connectivity and behavioural structures. It serves as an estimation of the subtle changes occurring in functional brain networks during cognitive load states as well. However, further studies on a larger data set are required to ensure consistency in results and develop a MCC based quantitative metric to measure cognitive activity in the brain.

Acknowledgement. This work is being supported by Cognitive NeuroEngineering Laboratory (CNeL), University of South Australia, Adelaide, Australia.

References

1. Sporns, O., Tononi, G., Kotter, R.: The human connectome: A structural description of the human brain. PLOS ONE. 1 (4) (2005)
2. Power, J.D., Cohen, A.L., Nelsom, S.M., Wig, G.S., Barnes, K.A., Church, J.A.: Functional network organization of the human brain. Neuron **72**, 665–678 (2011)
3. Vijayalakshmi, R., Nandagopal, D., Dasari, N., Cocks, B., Thilaga, M.: Minimum connected component - a novel approach to detection of cognitive load induced changes in functional brain networks. Neurocomputing (2015). doi:10.1016/j.neucom.2015.03.092
4. Nunez, P.L.: Electroencephalography. Encyclopedia of Human. Brain **2**(2), 1348 (2002)
5. Tomita, E., Tanaka, A., Takahashi, H.: The worst-case time complexity for generating all maximal cliques and computational experiments. Theor. Comput. Sci. **363**(1), 28–42 (2006)
6. Palla, G., Dernyi, I., Farkas, I., Vicsek, T.: Uncovering the overlapping community structure of complex networks in nature and society. Nature **435**(7043), 814–818 (2005)
7. Eppstein, D., Strash, D.: Listing all maximal cliques in large sparse real-world graphs. In: Pardalos, P.M., Rebennack, S. (eds.) SEA 2011. LNCS, vol. 6630, pp. 364–375. Springer, Heidelberg (2011)
8. Bron, C., Kerbosch, J.: Algorithm 457: Finding all cliques of an undirected graph. Commun. ACM **16**(9), 575–577 (1973)
9. Bron, C., Kerbosch, J.: Algorithm 457: Finding all cliques of an undirected graph. Commun. ACM **16**(9), 575–577 (1973)
10. Tomita, E., Tanaka, A., Takahashi, H.: The worst-case time complexity for generating all maximal cliques and computational experiments. Theor. Comput. Sci. **363**(1), 28–42 (2006)
11. Faust, K., Wasserman, S.: Social network analysis: Methods and applications. Cambridge University Press, Cambridge (1995)
12. Boginski, V., Butenko, S., Pardalos, P.M.: Statistical analysis of financial networks. Comput. Stat. Data Anal. **48**(2), 431–443 (2005)
13. Dumermuth, G.: Quantification and analysis of the EEG. Schweiz. Arch. Neurol. Neurochir. Psychiatr. **115**(2), 175–192 (1974)
14. Delorme, A., Makeig, S.: EEGLAB: An open source toolbox for the analysis of single-trial EEG dynamics including independent component analysis. NCBI **134**(1), 9–21 (2004)
15. Nandagopal, D., Vijayalakshmi, R., Cocks, B., Dahal, N., Dasari, N., Thilaga, M.: Computational neuroengineering Approaches to characterising cognitive activity in EEG data. Knowl. Based Inf. Syst. Pract. Smart Innovation Knowl. Syst. Technol. **30**, 115–137 (2015)

16. He, Y., Evans, A.: Graph theoretical modeling of brain connectivity. Curr. Opin. Neurol. **23**, 341–350 (2010)
17. CURRY 7 EEG Acquisition and Analysis Software. Compumedics Neuroscan USA Ltd

Real-Time People Counting Application by Using GPU Programming

Yasemin Poyraz Kocak[1] and Selcuk Sevgen[2(✉)]

[1] Vocational School of Technical Sciences, Istanbul University, Istanbul, Turkey
yasemin.poyraz@istanbul.edu.tr
[2] Department of Computer Engineering, Istanbul University, Istanbul, Turkey
sevgens@istanbul.edu.tr

Abstract. This study focuses on people counting in a video stream captured from a fixed camera. Aforementioned counting process is implemented by graphical processing unit (GPU) programming real-timely. For this reason, two video streams with different resolution and two different NVIDIA graphic cards are used. For all combinations of these video streams and graphic cards, the number of people are obtained in the video streams and they are compared with regard to performances. Consequently, it is examined that real-time people counting process can be successfully implemented by compute unified device architecture (CUDA) programming on NVIDIA graphic cards.

Keywords: Graphics processing unit · Compute unified device architecture · Image processing · People counting

1 Introduction

Recently, image processing is a popular subject used in several study fields. Because it needs enormous computational power, it is a challenging problem for classical single-thread computer architectures and they remain incapable for processing huge data. At this point, parallel computing concept is offered as a novel method. GPU programming which provides a flexible to design future multi-core processor architecture is a sample of this method [1, 2].

When compared to central processing unit (CPU), GPU has more cores and transistors in kernel. These properties provide more parallel processing than CPU.

People counting is an important element for security systems. There are many researches about this subject in the literature.

People counting can be classified as two main approaches. The first one is line of interest (LOI), in which a camera is fixed on a virtual line and people passing on this line are counted [4, 5]; the second one is region of interest (ROI) [6, 7], in which people in a certain area are counted. In this study, LOI approach is chosen because of using video streams captured from a high-position camera.

In this study, the current people counting algorithm is obtained from existing study in the literature [8] and implemented in MATLAB. In this scope, the people counting algorithm which takes a long time when implemented in MATLAB is speed up and the effect of graphic cards to the performance is observed.

© Springer International Publishing Switzerland 2015
S. Arik et al. (Eds.): ICONIP 2015, Part IV, LNCS 9492, pp. 540–547, 2015.
DOI: 10.1007/978-3-319-26561-2_64

In one of the few researches in the literature about counting with CUDA [3], Bjerge counted the objects in a series of still images. Different from present study, object counting is implemented in discrete images instead of video streams.

The study mentioned about how people counting algorithm by GPU programming is implemented and how different GPUs affect the performance of implementation. The application is experienced on two different video streams and graphic cards. In addition the effects of the distance between the people and the quantity of light in the environment are measured.

The components used in the paper can be summarized as follows: GeForce GT 630 2 GB and GeForce GTX 550Ti as graphic cards, NVIDIA CUDA Toolkit v6.0 as software development environment and OpenCV 2.4.5 for video capturing.

The purpose of the present study is to investigate, experimentally, the counts of people in a video stream captured from a fixed camera. For this, steps of the implementation and experimental study are carried out.

2   People Counting Application

2.1   Counting Algorithm

The first step of the algorithm is background prediction by subtracting two successive frames from each other. With this background image, the moving objects are segmented. Tracking and counting are performed in the case of people existence.

To implement an algorithm by CUDA, it is necessary to perform the steps of the algorithm parallelly. This implementation of the algorithm given in Fig. 1 and used in this study is constituted by parallel versions of the processes like gray-scale transform, background prediction, morphological operations, tracking and counting.

2.2   Capturing Images

The function that provides real-time execution is run on host CPU. The video stream is acquired by cvCreateFileCapture() function in OpenCV. With this function, properties of video stream like the number of frames and height-width of each frame were obtained. All of the steps in Fig. 1 were executed iteratively during video stream.

2.3   Gray-Scale Transform

The image was transformed to gray-scale by using the R,G,B (0-255) values which are acquired from each pixel using (1).

$$Gray\ Scale\ =\ R*0.3+\ G*0.6+\ B*0.1 \tag{1}$$

All of the pixels are transformed to gray-scale simultaneously with our parallel running rgba_to_grayscale kernel.

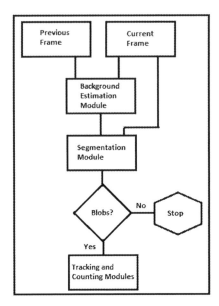

Fig. 1. Counting algorithm flowchart

2.4   Background Subtraction

One of the most important steps of people counting algorithm is background subtraction which is based on people detection in foreground. As seen in Fig. 2, Difference Algorithm is used for this step [13, 14].

In this algorithm, mean filter method was applied to minimize the noise arised from the camera and changing of light.

The average of the pixel values in the same position on all frame images is determined and using the obtained values the new values is provided (Fig. 3).

A new image is established by the absolute differences of successive images pixel to pixel. The non-zero value in this new image means motion. However it is possible to obtain deceptive images because of the camera quality and changing of light although there is no motion. To avoid this, a threshold value is determined. If the difference value is greater than the threshold, it is accepted as foreground and pixel value is set to 1, otherwise it is accepted as background and pixel value is set to 0 [13, 14].

2.5   Morphological Operations

The threshold values determined to constitute binary images cannot eliminate all noises. For this reason, opening operation is used in the algorithm as in Fig. 4.

This morphological operation has two operator called as erosion and dilation. The purpose of erosion is to shrink borders of foreground objects. So, foreground objects become smaller and the holes in the objects become larger. The dilation operation aims to expand the borders of the foreground objects. So, foreground objects become larger and the holes in the objects become smaller or disappear completely.

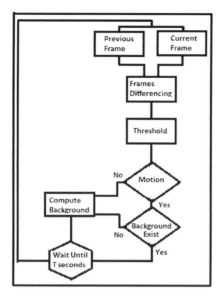

Fig. 2. Frame difference algorithm

Fig. 3. A sample of background estimation frame

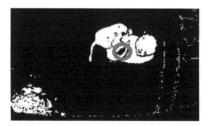

Fig. 4. A sample of noisy frame

Because small objects are eliminated by erosion, they cannot appear in dilation operation again [13].

The equations of erosion and dilation operations are labeled below.

$$\text{Erosion equation :} \quad \in_B (X) = XB = \{x \backslash B_X \subset X\} \tag{2}$$

$$\text{Dilation equation:} \quad \sigma_B(X) = X \oplus B = \{x + b \backslash b \in B, x \in X\} \tag{3}$$

where X is a subset of $\in$ and B is a structural element [13].

2.6   Blob Analysis and Counting

Due to object detecting algorithms evaluate all pixels, they present a low performance when they are applied with sequential programming techniques.

To track objects, it is necessary to define all objects and their properties. This operation is called as blob analysis. In this study, blob analysis is used for object tracking. To determine if the object is a person or not, an average size is decided according to the camera position. There are two different way of connection in blob analysis algorithm. One of them is vertical and horizontal adjacent pixels and the other one is diagonal adjacent pixels (Fig. 5) [13].

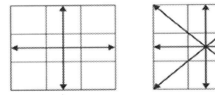

Fig. 5. Two rules of adjacent pixels.

The performance of blob analysis is depended on segmentation process. A weak segmentation algorithm can detect noises as blobs. The shortest way of blob analysis is to determine the central position, convex-hull, height and width of the blob. The algorithm must detect the new blobs. Also another factor affecting the result of blob analysis is adjacent objects. The counting process is performed owing to a virtual line separating entrance and exit area exactly. Two successive images are sufficient to count people. For example, if a blob in an image is below than the virtual line and it is above in the following image then the counter is incremented by one [9–12].

3   Experimental Results

In the experimental part, a real-time people counting algorithm is implemented by using two different graphic cards (GeForce GT 630 2 GB, GeForce GTX 550 Ti) and two different resolution (288 × 384 and 320 × 448) video streams.

The properties of graphic cards and the running times of every graphic card-video stream combinations are presented in tables. Table 1 presents the properties of graphic cards while Tables 2 and 3 show running times of first video stream and second video stream on different graphic cards respectively. The running times in Tables 2 and 3 are constituted by measuring running time of every step for current frame.

Table 1. Properties of two graphic cards

	GeForce GT 630 2 GB	GeForce GTX 550 Ti
Total Memory	2048 MB	1024 MB
GPU Clock rate	1.62 GHz	1.82 GHz
Total Texture Size	65536	65536
Total Constant Memory	65536	65536
Total Shared Memory per block	49152	49152
Warp Size	32	32
Number of thread	1024	1024
Dimension of a thread block	1024, 1024, 64	1024, 1024, 64
Dimension of a grid	65535,65535,65535	65535,65535,65535
Number of CUDA core	96	192

Table 2. Running times for 288×384 resolution video stream

	GeForce GT 630 2 GB	GeForce GTX 550 Ti
Gray-Scale transformation	0.067 ms	0.030 ms
Background Estimation	1.955 ms	0.880 ms
Erosion	0.099 ms	0.045 ms
Dilation	0.211 ms	0.095 ms
Blob Analysis and Counting	32.087 ms	21.901 ms
Total Running Time	34.421 ms	22.954 ms

Table 3. Running times for 320×448 resolution video stream

	GeForce GT 630 2 GB	GeForce GTX 550 Ti
Gray-Scale transformation	0.099 ms	0.041 ms
Background Estimation	2.280 ms	1.571 ms
Erosion	0.140 ms	0.063 ms
Dilation	0.298 ms	0.135 ms
Blob Analysis and Counting	164.287 ms	70.010 ms
Total Running Time	167.104 ms	71.82 ms

When the results in Tables 2 and 3 are compared, it is realized that more resolution in the video stream means more running time because the number of processed pixels in the video streams increased.

There are 2456 and 3181 frames in first and second video streams respectively, according to these values the total running times are calculated as follows: the total running times of the video streams for GeForce GT 630 2 GB graphic card are: 84,537 s and 531,557 s while they are 56,375 s and 228,459 s for GeForce GTX 550 Ti graphic card.

The total times (running time + duration time) of the video streams for GeForce GT 630 2 GB graphic card are: 182,537 s and 637,557 s while they are 154,375 s and 334,459 s for GeForce GTX 550 Ti for graphic card.

The processing times were measured as 31 min for the first video stream and 93 min for second video stream in MATLAB. The corresponding values in the study are approximately 1,4 min. and 0,9 min. for the first video stream and 10,6 min. and 5,57 min. for the second video stream. As seen from the results, GPU programming is more successful than classical sequential programming.

Respectively, the algorithm counted 22/31 people and 29/36 people passing through the camera fields in the first and second video streams. These ratios correspond to 70 % and 80 % counting success. The reason of the higher success rate in the second video stream depends on the distance between the people passing through. It is known that when there is more distance between the people, the success of counting algorithm increases. The adjacent people cause morphological operations merge the gaps between them and detect them as only one person. By improving morphological operations, the algorithm can detect more people and have higher success rates.

This people counting process which was implemented real-time by using GPU programming with 70 % and 80 % success rate, 31 and 93 min for two video streams respectively. These results show that GPU programming have up to 34,4 times faster than classical sequential programming for the first video stream and up to 16,7 times faster for the second video stream.

4   Conclusion

In this study a people counting algorithm implemented on GPU programming shows this approach gives faster results than classical sequential programming. For this implementation, a graphical card which can be found in any computer systems is used. This means speeding up the image processing applications can be possible with common and cheap tools.

To determine the relationship between the hardware and implementation, two different graphic cards with different properties are used. It is realized that the properties of graphic cards affect the performance of the implementation directly proportionally.

Another criteria that affects the performance is the resolution of video streams. The higher resolution means the more running time.

In future work, the people counting algorithm with different combinations of graphic cards and video streams can be examined to improve the performance. Also it is observed the blob analysis and background subtraction processes take the most time in the people counting algorithm. So, to improve these steps or to use different methodologies can also gain better counting performance.

References

1. NVIDIA CUDA, C Programming Guide Version 6.0 (2014)
2. NVIDIA CUDA, CUDA C Best Practices Guide (2014)
3. Bjerge, K.: Dynamic counting objects in images optimized for data-parallel computing, Aarhus University, Department of Computer Science
4. Barandiaran, J., Murguia, B., Boto, F.: Real-time people counting using multiple lines. In: 9th International Workshop on Image Analysis for Multi. Interactive Services, pp. 159–162 (2008)
5. Yu, S., Chen, X., Sun, W., Xie, D.: A robust method for detecting and counting people. In: International Conference on Audio, Language and Image Processing, pp. 1545–1549 (2008)
6. Ryan, D., Denman, S., Fookes, C., Sridharan, S.: Crowd counting using multiple local features. In: Digital Image Computing: Techniques and Applications, pp. 81–88 (2009)
7. Vicente, A.G., Munoz, I.B., Molina, P.J., Galilea, J.L.L.: Embedded vision modules for tracking and counting people. I IEEE T Inform. Theor. **58**(9), 3004–3011 (2009)
8. Guler, E., Gecer, B.: People Counting: People Counting. Technical Report (2015). http://www.ebubekirguler.com/goruntu-isleme-yontemleri-ile-insan-sayma
9. El-Azim, S.A., Ismail, I., El-Latiff, H.A.: An efficient object tracking technique using block-matching algorithm. In: 19th National, Radio Science Conference, pp. 427–433 (2002)
10. Chen, T.H., Chen, T.Y., Chen, Z.X.: An intelligent people-flow counting method for passing through a gate. In: IEEE International Conference on Cybernetics and Intelligent Systems, pp. 573–578 (2006)
11. Chen, T.H., Lin, Y.F., Chen, T.Y.: Intelligent vehicle counting based on blob analysis in traffic surveillance. In: IEEE International Conference on Innovative Computing, Information and Control, p. 238 (2007)
12. Karaulova, I., Hall, P., Marshall, A.: A hierarchical model of dynamics for tracking people with a single video camera. In: British Machine Vision Conference, pp. 352–361 (2000)
13. Lefloch, D.: Real-Time People Counting system using Video Camera, Gjøvik University College, Master Thesis (2007)
14. Gutchess, D., Trajkonic, M., Cohen-Solal, E., Lyons, D., Jain, A.K.: A background model initialization algorithm for video surveillance. In: 8th IEEE International Conference on Computer Vision, pp. 733–740 (2001)

Activity and Flight Trajectory Monitoring of Mosquito Colonies for Automated Behaviour Analysis

Burhan Khan[1(✉)], Julie Gaburro[1,2], Samer Hanoun[1],
Jean-Bernard Duchemin[2], Saeid Nahavandi[1], and Asim Bhatti[1]

[1] Centre for Intelligent Systems Research (CISR),
Deakin University, Geelong, Australia
{burhan.khan,jgaburro,samer.hanoun,saeid.nahavandi,
asim.bhatti}@deakin.edu.au
[2] Australian Animal Health Laboratory (AAHL), CSIRO, Geelong, Australia
jean-bernard.duchemin@csiro.au

Abstract. Monitoring and tracking of mosquitoes using image processing is important to facilitate the mosquitos' behaviour analysis automatically over longer period of times. In this paper, we propose a simple methodology to monitor mosquitos' activity using multiple cameras optimally placed. In order to ensure optimal camera coverage for the area of observation and desired image quality; we propose to simulate the experimental setup in a 3D virtual environment to obtain one-off optimum camera placement parameters. Our proposed methodology is demonstrated to have improved the activity monitoring process using two cameras for accurate count of occluded mosquitoes and 3D trajectory path reconstruction. This framework will enable working out more challenging tasks of constructing 3D trajectories using information received from multiple low quality cameras, which provide inconsistent and discontinuous trajectories.

Keywords: 3D tracking · Behaviour analysis · Camera placement · Mosquito activity

1 Introduction

Mosquitoes are known for their capacity to transmit pathogens worldwide to the humans. The diseases induced by these pathogens are called 'mosquito-borne diseases' meaning that bacteria, viruses or parasites are transmitted by mosquitoes. Meanwhile, the mosquitoes are themselves infected but mostly not affected. It has been estimated that about three billion people are threatened by blood-feeding arthropod diseases, making the mosquitoes the most deadly animal for mankind [1]. The consequences are not concerning only human health but also socio-economy and politics.

The most important vector-borne disease is malaria caused by five Plasmodium species [2]. But arboviruses (meaning virus transmitted by arthropods to vertebrates) have a worldwide distribution and a dramatic global emergence or resurgence in the twentieth century [3]. About one hundred arboviruses could infect humans and about forty could infect livestock [4]. The most historically famous and important of them,

© Springer International Publishing Switzerland 2015
S. Arik et al. (Eds.): ICONIP 2015, Part IV, LNCS 9492, pp. 548–555, 2015.
DOI: 10.1007/978-3-319-26561-2_65

are Yellow Fever Virus [5] or Dengue virus [6]. The last one is transmitted by few species of Aedes genus. This genus is medically very important as Chikungunya virus also could be transmitted by Aedes albopictus and Ae. aegypti [7]. Closely related to this virus, Ross River Virus could be transmitted by a wide range of vectors including Aedes species and Culex species. Moreover, Culex mosquitoes are important vectors for West Nile virus [8] with disease affecting some birds, horses and humans.

Since the discovery of the mode of transmission of these mosquito-borne diseases, scientists are trying to find solutions to prevent, control or treat them. Important work has been done on mosquito's biology, including the insect immune system, but still there is a gap of understanding how viruses affect mosquito's behaviour. Host seeking behaviour is important as it is at the core of the processes involved in the contact between vectors and to vertebrates and so humans. In laboratory conditions, it is easy to have thousands of mosquitoes under different experimental conditions, but very difficult to extract data or have a whole overview of behaviours with simple observations. This is where computing, automatic monitoring, and tracking are important to develop.

Thanks to the evolution of technologies during this last decades, many key automated image-based tracking systems have been developed and been very useful to ecologists [9]. Video recording enables the analysis of a single or multiple insects performing complex behaviours. However, most available literature concerning mosquito behaviours use visual observations, or single-camera recording of the flight activity of individual mosquitoes. Indeed, monitoring and tracking mosquitoes is a challenging work. Most of them are only active during dusk, dawn or at night; they fly in a very unpredictable pattern with variable speed and are relatively small (in average 5-6 mm in length in a cage 200 times bigger. Manual analysis of mosquito behaviour is possible. For example, Yee and Foster counted by eye the 40 to 70 mosquitoes attacking the probing screen in the same cage by trial during their study on host seeking behaviour with three different mosquitoes species [10]. But this is a very tedious work and time consuming, which leads easily to errors (lack of attention, tiredness). Recently, Wilkinson et al. improved the single camera video imaging method to study male mosquitoes over a long period of time, and observed two different algorithms for their effectiveness in flight tracking [11]. However, till today the only automatic video tracking system designed for mosquitoes is Track3D plug-in, developed by the commercial software Ethovision and Noldus for flight behaviour analysis. They are relatively expensive and are not easily affordable for developing countries, where people pay a heavy burden to vector-borne diseases. Few studies have been done on flight behaviour of the malaria vector Anopheles mosquitoes [12–14]. And such methods have been used only once for Culex mosquitoes for trap bait efficiency [15], despite their important vector role. Nevertheless, all these studies are limited to the flight behaviour and mainly target only a few or one individuals.

Several attempts are reported in the literature which makes use of image processing methods to aid the process of behaviour analysis. Vein, Butail S et al. [16] proposed a methodology to track 3D flight paths of mating mosquitoes in open air swarms. Stereo vision techniques are employed to identify and reconstruct the paths of mating mosquitoes even at 25 frames per second (fps). Straw et al. [17] proposed to simulate 3D tracking of flying animals in relatively larger space. They propose to use a network of large number of cameras with high spatial resolution and high frame rate, such as 100 fps.

Multiple cameras allow them to tackle the problem of occlusion and refine the information obtained for constructing 3D trajectories. A single video camera with limited number of mosquitoes was used by Wilkinson et al. [11] in the experimental setup and discussed the limitations of a single camera and challenges faced in using multiple cameras for 3D flight path extraction. Most recent work relevant to tracking mosquitoes suggested by McMeniman et al. [18], use Track3D software at 210 FPS to capture the active mosquitoes in a swarm. They studied the effect of carbon dioxide on the behaviour of insect feeding behaviour.

Our goal is to achieve mosquito monitoring in terms of activity and flight trajectory employing a simple two camera setup. The focus of this work is twofold: (1) optimize the cameras placement in order to achieve the best possible coverage of the area of interest with minimum occlusions and desired minimum pixel resolution for the objects of interest, which are mosquitoes in this case; and (2) monitor the mosquitos' colony in terms of their temporal activity as well as flight behaviour.

2   Camera Placement

The operational success of any tracking system is highly dependent on the placement of the cameras used in the monitoring process to minimize occlusions. In this work, our focus is monitoring of mosquito colonies to characterise their activity over defined periods of time. Special attention has focused on ensuring the top, base, sides and corners of the insect cage are adequately visible with a minimum defined resolution to ensure robust tracking and monitoring.

In this paper, we employ a multi-camera setup using the camera placement algorithm presented in [19]. The algorithm is provided with regions of interest that are uniformly spread in the mosquitoes cage where these regions of interest are required to be fully covered at least with 2 cameras. This enables each of these regions to appear in at least 2 cameras; therefore the 3D tracking and monitoring objectives are satisfied, which is mandatory to track flying insects within an open space [12, 16]. A set of mounting locations are provided to the algorithm in order for it to inform the best among these locations to mount our cameras. Each mounting location is described in 3D Cartesian coordinate system in terms of (X, Y, Z) coordinates. The regions of interest are associated with a specific Resolution of Detection (ROD) which is required to satisfy adequate tracking and monitoring. Figure 1C, shows a schematic diagram of the placement of our proposed multi-camera system. Two cameras are required to provide full coverage of the regions of interest shown in Fig. 1A and B. Figure 1D, shows the actual setup with the mosquito's cage and the 2 used cameras. Figure 1A and B, show the simulated views of camera 1 and camera 2 respectively, rendered by the camera placement optimization algorithm. Red and Green markers highlight the regions of interest that require full coverage. The cage top, each side as well as the corners are marked for full coverage. Two PTZ cameras are employed for mosquito monitoring to capture the colony of 21 mosquitoes. Camera parameters that camera optimizer takes into account for optimum camera placement are presented in Table 1. This optimizer provides optimal mounting positions for the cameras to achieve maximum coverage and minimal occlusion, as well as ensures minimum pixel resolution for the mosquitoes. As illustrated in Fig. 2A and C, a mosquito activity highlighted by the red circle is overlooked by the camera view shown in Fig. 2B due to the occlusion. This failure of activity detection is also visible in Fig. 2D.

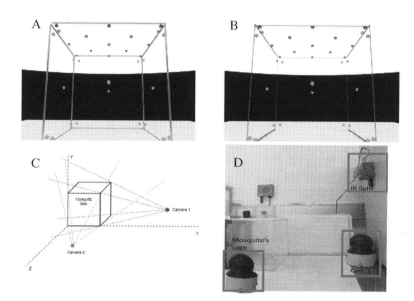

Fig. 1. (A & B) camera view perspectives simulated for optimal monitoring area coverage, (C) optimum location for cameras placement, and (D) actual camera setup

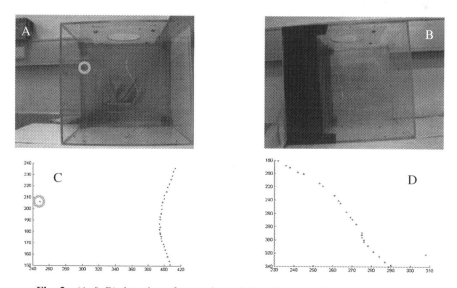

Fig. 2. (A & B) detection of mosquito activity, (C & D) 2D path reconstruction

Table 1. Multiple camera optimized configurations

Configuration	Pan	Tilt	Focal length	Location (x, y, z)
Camera 1	180	260	3.4	0.00, 0.55, 0.10
Camera 2	90	260	3.4	−0.55, 0.00, 0.10

3 Activity Monitoring

The experimental setup, as shown in Fig. 1D, employs two pan-tilt-zoom (PTZ) IR cameras mounted at locations recommended by the optimization algorithm in Sect. 2. Light background is used to create contrast in the IR image. In this work, the focus is to accurately estimate colony activity with respect to time. A two camera setup ensures continuous activity monitory as well as smooth 2D trajectory which will be extended to 3D in future work. A graphical user interface, shown in Fig. 3, adopting .Net framework 4.5 and open source "Emgu CV" wrapper available for using with "OpenCV" image processing library is developed.

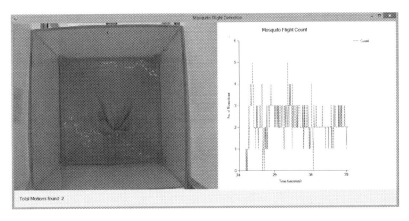

Fig. 3. Flight tracking application; left: video stream obtained from a specific camera, right: graph showing the activity of mosquitoes

Behaviour analysis of mosquitoes requires persistent observation of mosquito's activities over long periods of time. Usually, this procedure is conducted by taking a sequence of images with regular intervals for a desired duration of time and analysing those images to observe the behaviour of mosquitoes. Keeping track of more than one mosquito manually is challenging and therefore video analysis of intricate behaviour of multiple mosquitoes simplifies this process [11].

In order to facilitate the process of mosquito behavioural analysis, a simple image processing application to extract the information of flight activities in real time from live streaming videos is developed. In this image processing application, a mosquito flight activity is categorised into two distinct types such as active or non-active mosquito. This classification allows monitoring the chaotic behaviour of mosquitoes by responding to any internal or external stimuli.

To track the activity of mosquitoes, a simple image processing algorithm is proposed for the detection of active mosquitoes in real time. The proposed algorithm can be seen in Table 2.

Table 2. Proposed algorithm for mosquito activity monitoring

Algorithm

```
A ← Area Threshold
P ← List of Active Points
F ← List of Captured Frames
C ← List of 3D Coordinates
i ← Camera ID
for each fᵢ(x,y) in F do
    fᵢ(x,y) ← Apply Gaussian Blur to fᵢ(x,y)
    fᵢ(x,y) ← Apply Background Mask to fᵢ(x,y)
    B ← Get Components/Blobs from fᵢ(x,y)
    for each b in B do
      if b ≤ A then
      Pi ← Add b
    end if
end for
C ← Combine overlapping P to construct 3D path
end for
```

---

4 Flight Trajectory Estimation

In our approach, we use simple methodology to construct an approximated 3D trajectory of mosquito flight activity pattern for behaviour analysis, as illustrated in Fig. 4. This approach allows to construct the flight trajectory using synchronised frames obtained from multiple cameras, as:

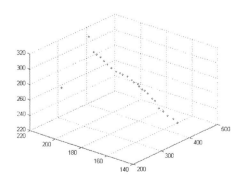

Fig. 4. 3D flight trajectory estimation of mosquitos

$$F(x, y, z) = (\cup_{i \in N} F_i(x, y)) \forall i = 1, 2, \ldots, N \qquad (1)$$

where 3D flight trajectory F is obtained through the union of overlapping 2D frame coordinates.

5 Conclusion

In this paper we presented a straightforward and simple methodology of approximated 3D reconstruction of flight trajectories of mosquitoes. We used two PTZ cameras to obtain multiple 2D images from different perspectives in order to tackle the occlusion problem. In the laboratory setup it is highly likely to have obstructions in front of camera viewing angles. Our experiments have been validated at several occasions where active mosquitoes were hidden from one viewing angle and captured in different viewing angle. Finally, approximated 3D paths were reconstructed to analyse the flight behaviour of normal and virus prone mosquitoes. This works leads path for further improvement on 3D tracking of individual mosquitoes.

References

1. Becker, N., Petrić, D., Boase, C., Lane, J., Zgomba, M., Dahl, C., et al.: Mosquitoes and Their Control, vol. 2. Springer, Heidelberg (2010)
2. Ghosh, A., Edwards, M., Jacobs-Lorena, M.: The journey of the malaria parasite in the mosquito: hopes for the new century. Parasitol. Today **16**, 196–201 (2000)
3. Gubler, D.J.: Human arbovirus infections worldwide. Ann. N. Y. Acad. Sci. **951**, 13–24 (2001)
4. Monath, T.P.: The Arboviruses: Epidemiology and Ecology. CRC Press, Boca Raton (1988)
5. Monath, T.P.: Yellow fever: an update. Lancet Infect. Dis **1**, 11–20 (2001)
6. Murrell, S., Wu, S.-C., Butler, M.: Review of dengue virus and the development of a vaccine. Biotechnol. Adv. **29**, 239–247 (2011)
7. Pialoux, G., Gaüzère, B.-A., Jauréguiberry, S., Strobel, M.: Chikungunya, an epidemic arbovirosis. Lancet Infect. Dis. **7**, 319–327 (2007)
8. Petersen, L.R., Brault, A.C., Nasci, R.S.: West Nile virus: review of the literature. JAMA **310**, 308–315 (2013)
9. Dell, A.I., Bender, J.A., Branson, K., Couzin, I.D., de Polavieja, G.G., Noldus, L.P., et al.: Automated image-based tracking and its application in ecology. Trends Ecol. Evol. **29**, 417–428 (2014)
10. Yee, W.L., Foster, W.A.: Diel sugar-feeding and host-seeking rhythms in mosquitoes (Diptera: Culicidae) under laboratory conditions. J. Med. Entomol. **29**, 784–791 (1992)
11. Wilkinson, D.A., Lebon, C., Wood, T., Rosser, G., Gouagna, L.C.: Straightforward multi-object video tracking for quantification of mosquito flight activity. J. Insect Physiol. **71**, 114–121 (2014)
12. Beeuwkes, J., Spitzen, J., Spoor, C., Van Leeuwen, J., Takken, W.: 3-D flight behaviour of the malaria mosquito Anopheles gambiae s.s. inside an odour plume. Proc. Neth. Entomol. Soc. Meet. **19**, 137–146 (2008)

13. Spitzen, J., Spoor, C., Kranenbarg, S., Beeuwkes, J., Grieco, F., Noldus, L., et al.: Track3D: visualization and flight track analysis of Anopheles gambiae ss mosquitoes. Proc. Measur. Beh. **2008**, 133–135 (2008)

14. Spitzen, J., Spoor, C.W., Grieco, F., ter Braak, C., Beeuwkes, J., van Brugge, S.P., et al.: A 3D analysis of flight behavior of Anopheles gambiae sensu stricto malaria mosquitoes in response to human odor and heat. PLoS One **8**, e62995 (2013)

15. Cooperband, M., Carde, R.: Orientation of culex mosquitoes to carbon dioxide-baited traps: flight manoeuvres and trapping efficiency. Med. Vet. Entomol. **20**, 11–26 (2006)

16. Butail, S., Manoukis, N., Diallo, M., Ribeiro, J.M., Lehmann, T., Paley, D.A.: Reconstructing the flight kinematics of swarming and mating in wild mosquitoes. J. R. Soc. Interface **9**, 2624–2638 (2012)

17. Straw, A.D., Branson, K., Neumann, T.R., Dickinson, M.H.: Multi-camera real-time three-dimensional tracking of multiple flying animals. J. R. Soc. Interface **8**, 395–409 (2011)

18. McMeniman, C.J., Corfas, R.A., Matthews, B.J., Ritchie, S.A., Vosshall, L.B.: Multimodal integration of carbon dioxide and other sensory cues drives mosquito attraction to humans. Cell **156**, 1060–1071 (2014)

19. Hanoun, S., Bhatti, A., Creighton, D., Nahavandi, S., Crothers, P., Esparza, C.G.: Target coverage in camera networks for manufacturing workplaces. J. Intell. Manuf. **26**, 1–15 (2014)

Facial Emotion Profiling Based on Emotion Specific Feature Model

Md. Nazrul Islam and Chu Kiong Loo$^{(\boxtimes)}$

Department of Artificial Intelligence,
Faculty of Computer Science & Information Technology,
University of Malaya, Lembah Pantai, 50603 Kuala Lumpur, Malaysia
nazrul_cse@siswa.um.edu.my, ckloo.um@um.edu.my

Abstract. Facial emotion profiling is rapidly becoming an area of intense interest in machine vision society for decade. In spite of major efforts, there are several open questions on how to embed the emotional intelligence in machine to respond immediately and precisely over facial expressions. In this sense, this paper presents an automatic facial emotion profiling from emotion specific feature model. A 17-point feature model on the frontal face region is proposed to track per frame facial emotion robustly. A measurement vector is formed based on a set of geometric distance displacements of a pair of feature points between neutral and expressive face frame. A two-stage fuzzy reasoning model is proposed to classify universal facial expressions. In the first stage measurements are fuzzified and mapped onto an Action Units (AUs) and later AUs are fuzzified and mapped onto an Emotion in the second-stage of fuzzy reasoning model. The overall performance of the proposed system is evaluated on two publicly available facial expression databases, average emotion recognition accuracy of 91 % was achieved for RaFD and 94 % for CK + database.

Keywords: Facial emotion recognition · Measurement vector · Fuzzy reasoning model · Recognition accuracy

1 Introduction

Facial emotions profiling is a significant step in our socio-emotional life aspects. They are visually observable, conversational, and interactive signals that indicate the emotional states, intentions, behaviours, and feelings of a person [1]. By synchronizing the dialogue by means of lip reading, brow raising, and nodding, facial emotions can be used to clarify the content and intent of what is said, identify focus of attention, and regulate interactions with environment, as well as individuals in our vicinity [2].

As of now, Facial Action Coding System (FACS) [3] is the standard and most commonly used in psychological research for automatic measurements of facial muscle movement (Action Unit). FACS defined six basic emotional facial emotions amongst forty six are defined as the persistent across cultures: Surprise, sadness, fear, anger, disgust and happiness, while contempt is newly defined as another basic emotions. Although humans can easily recognize the appearance of distinct facial emotions with naked eyes, automatic recognition by machine vision system is still a challenge.

© Springer International Publishing Switzerland 2015
S. Arik et al. (Eds.): ICONIP 2015, Part IV, LNCS 9492, pp. 556–565, 2015.
DOI: 10.1007/978-3-319-26561-2_66

However, it is believed that automated analysis of facial emotions and embedding emotional intelligence in computing system, future generations of human-computer interaction (HCI) will be more sophisticated and human-like [4].

In general, two major segments are distinguished for facial emotion recognition: feature extraction and emotion classification. Two different research trends are considered in the past exploration of facial feature extraction: Appearance-based and Model-based. Appearance-based models are linear-nonlinear and fully person dependent where model-based provides 2D-3D person-independent face fitting model. Some of the well-known approaches are based on Gabor Wavelets [5], Principal Components Analysis [6], Active Appearance and Geometric Models [7], Active Shape Model [8], Optical Flow and Deformable Models [9, 10], Discrete Cosine Transform in combining with Neural Networks [11], Online Clustering [12], and others. The most challenging issues related to these techniques are occlusion, precise region selection, robustness, manual initialization of feature points, tracking loss, and finally the illumination effect which can degrade the recognition system performance.

On the other hand classification techniques are well practiced by different authors for techniques such as: General Rule-base [8], Fuzzy Reasoning Model [13, 14], and Support Vector Machine [15], etc.

The main contribution of this paper is to profile universal facial emotions with a feature model and a two-stage fuzzy reasoning model.

The paper is organized as follows. The facial emotion profiling is described in Sect. 2. A 17-point feature model with two-stage fuzzy reasoning model is also presented in this section. Section 3 evaluates the performance of the proposed system. A graphical user interface is also presented in this section for clear view of our proposed system. Finally, Sect. 4 concludes the proposed facial emotion recognition model.

2 Facial Emotion Profiling

Our approach uses a Constrained Local Model (CLM) [16] based feature representation of facial data for Fuzzy classifier. Firstly, we define a feature model and a measurement vector. Later, a two-stage fuzzy rule-based approach is applied to classify prototypical facial expression.

A. Emotion Specific Feature Model. In automatic facial emotion recognition, the most crucial step is to track the prominent facial features surrounding different face regions such as mouth, eyes, eyebrows, etc. The PDM used in this paper is organized in a way so it can address the basic facial expressions on frontal face regions. The feature points are mapped (red dots) as 4 on eyebrows, 4 on eyes, 3 on nose, and 6 on mouth regions with their significance to address specific emotion as shown in Fig. 1. Each feature point is located in the frontal face region to address specific facial expression is also shown in this figure.

For predicting the amount of muscle movement during tracking, a measurement vector $(M_1,.....,M_{15})$ representing a set of distance displacement between neutral

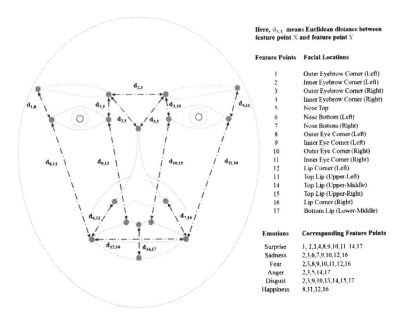

Fig. 1. Visualization of feature points with specific emotions and distance vector.

(N) and expressive (E) face frame is built, as shown in Table 1. Then, based on the measurement vector, related AUs and emotions are defined for the proposed classifier according to FACS standard and proposed system requirements.

Table 1. Definition of measurement vector.

Code	Measurement name	Measurements
M_1	Left inner eyebrow displacement	$E(d_{2,9}) - N(d_{2,9})$
M_2	Right inner eyebrow displacement	$E(d_{3,10}) - N(d_{3,10})$
M_3	Left outer eyebrow displacement	$E(d_{1,8}) - N(d_{1,8})$
M_4	Right outer eyebrow displacement	$E(d_{4,11}) - N(d_{4,11})$
M_5	Mouth width vertical displacement	$E(d_{14,17}) - N(d_{14,17})$
M_6	Mouth corners horizontal displacement	$E(d_{12,16}) - N(d_{12,16})$
M_7	Left mouth corner-eyelid displacement	$E(d_{8,12}) - N(d_{8,12})$
M_8	Right mouth corner-eyelid displacement	$E(d_{11,16}) - N(d_{11,16})$
M_9	Inner eyebrow corners displacement	$E(d_{2,3}) - N(d_{2,3})$
M_{10}	Left upper lip vertical displacement	$E(d_{9,13}) - N(d_{9,13})$
M_{11}	Right upper lip vertical displacement	$E(d_{10,15}) - N(d_{10,15})$
M_{12}	Left inner eyebrow-nose root displacement	$E(d_{2,5}) - N(d_{2,5})$
M_{13}	Right inner eyebrow-nose root displacement	$E(d_{3,5}) - N(d_{3,5})$
M_{14}	Left mouth to nose displacement	$E(d_{6,12}) - N(d_{6,12})$
M_{15}	Right mouth to nose displacement	$E(d_{7,16}) - N(d_{7,16})$

Figure 2 shows an example of facial feature points detection and precise localization on publicly available databases (RaFD [18] and CK + [19]) for distinct facial emotion.

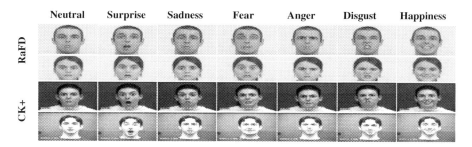

Fig. 2. Facial feature points detection and localization.

B. Two-Stage Fuzzy Rule-Based Classification. After extraction of necessary information from facial expression, a Two-Stage Fuzzy Rule-based approach (as shown in Fig. 3) is applied, where first-stage will predict AU intensity and second stage will predict emotion intensity [13]. For each input, three linguistic phrases, small, medium, and large is defined, which represents to the degree of output (as shown in Table 2). In order to give a shape of the membership functions, triangular and trapezoid form is selected to represent small, medium and large range.

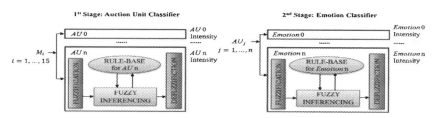

Fig. 3. Overview of two-stage fuzzy rule-based approach.

3 Results and Discussion

A. Dataset Description. For the evaluation of our work, we have used two publicly available databases, i.e., the extended Cohn-Kanade (CK +) database [19] and the Radboud faces database (RaFD) [18]. The CK + database contains 305 sequences of 123 subjects corresponding to the universal facial expressions. The RaFD contains a generic facial expression database of 67 subjects that includes gender, ethnic, age variations with diverse facial positioning.

Table 2. Two-stage fuzzy rules for AUs and emotions (e.g., surprise case) recognition.

First-stage rules														Second-stage rules											
Ms		Ψ	Ms		Ψ	Ms		Ψ	Ms			Ψ		Ψ			Es	Ψ			Es	Ψ			Es
M1	M2	ρ	M1	M2	ρ	M3	M4	Φ	M3	M4	M5	Φ	Ω	ρ	Φ	Ω	S	ρ	Φ	Ω	S	ρ	Φ	Ω	S
l	l	H	l	l	L	s	l	H	s	l	A	l	l	s	s	s	H	m	m	m	A	l	m	s	A
l	m	s	l	m	s	s	m	s	s	m	M5	m	m	s	s	l		l	m	m		m	l	s	
m	l	m	m	l	m	m	l	m	m	l	Q	l	l	s	s	m	s	m	m	m	s	s	s	m	
m	m	A	m	m	A	m	m	A	m	m	H	m	l	l	s		m	l	l	l	l				
l	s		l	s	s	l	s	l	l	s	A	s		l	l	l	l	m	s	l	s	m	m		
s	l		s	l	l	s	l	L	s	l	L	l		s	s	s	s	s	l	l	l	m	L		

l:large; *m*:medium; *s*:small; *H*:High; *A*:Average; *L*:Low; *Ms*:Measurements; *ρ*:AU1; *Φ*:AU2; *Ω*:AU2; *Ψ*: Action Units (AUs); *Es*: Emotions; *S*:Surprise

B. Emotion Profiling Interface. The prototype of the proposed system with CLM and fuzzy model has been developed using Visual Studio 2010 Professional with the support of OpenCV Toolbox. Figure 4 shows the Graphical User Interface (GUI) on facial emoion profiling of a sample image sequence of CK + database. The GUI contains six components as described below:

- **Preview**: The static or dynamic image sequences is shown here.
- **Expression Analysis:** The proposed 17-point feature model on frontal face region is tracked here.
- **Expression Intensity:** The percentage of each basic expression per frame is presented here.
- **Expression Timeline:** Final expression per frame is shown here for an image sequence.
- **Expression Plot:** The intensity of each expression is ploted here for an image sequence.
- **Expression Summary:** The overall percentage of each basic expression for an image sequence.

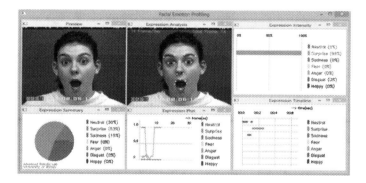

Fig. 4. Facial emotion profiling, image sequence from CK + database.

C. Emotion Recognition Accuracy. The results of emotion recognition using proposed classifier is shown in Figs. 5 and 6 for CK + and RaFD database. An unlabeled emotion class is reported by the proposed system, which means incoming emotion does not belongs to any of the existing classes including neutral emotion. The mouth corner tracking error, subject's over exposing expression, and incorrect neutral face frame is responsible for this unlabeled class.

Due to insufficient Lip Tightener (AU23), and Nose Wrikler (AU9) was counted for RaFD, the unlabelled percentage was also observed for anger and disgust emotions. However, the tracking robustness boost up the overall recognition accuracy to a high degree. Figure 7 shows an overall emotion recognition accuracy for CK + and RaFD facial database respectively.

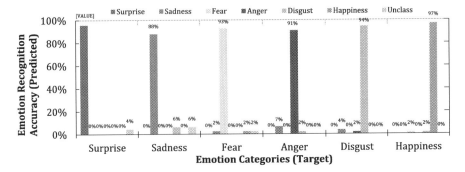

Fig. 5. Emotion recognition accuracy, target vs. predicted of CK + database.

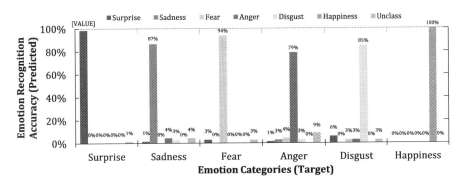

Fig. 6. Emotion recognition accuracy, target vs. predicted of RaFD database.

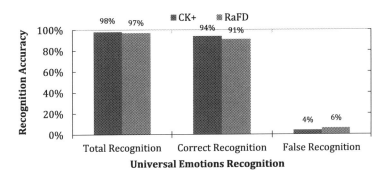

Fig. 7. Overall emotion recognition accuracy for five basic emotion categories.

D. State-of-the-art Comparisons. The state-of-the-art accuracy comparison is presented as shown in Table 3. It is worth mentioning that the performance of existing methods for emotion recognition are not exactly comparable [20]. This is due to each method reports its accuracy for emotion recognition with a specific face database. In addition, the recognition accuracy varies for extracted features, and classifiers.

Table 3. Performance comparison of the proposed and similar state-of-the-art prototype.

References	Tracking approach/#Landmarks	Extracted feature	Emotion classification approach	Classifier type	No. of emotions	Recognition accuracy
Tsalakanidou [8]	ASM/81	Distance displacement	General rule-based classifier	Knowledge-based	4	89.50 %
Besinger et al. [21]	Optical flow/26	Feature point displacement	Feature point displacement calculation	Knowledge-based	3	85.00 %
Contreras et al. [13]	No tracking/21	Distance difference	Fuzzy classifier	Knowledge-based	5	81.40 %
Rao and Koolagudi [22]	No tracking/Appearance-based analysis	Temporal variation of pixel	Autoassociative neural network	Trained	4	88.14 %
Proposed system	**CLM/17**	**Distance displacement**	**Two-stage fuzzy reasoning model**	**Knowledge-based**	**6**	**CK + 94.00 %, RaFD 91.00 %**

With person-independent tracking, automatic initialization of feature points, small number of landmarks, and trained-free classifier, the proposed system provides significant tracking and recognition accuracy as well as performs better than existing approaches of automatic facial emotion recognition.

4   Conclusion

In this paper, a fully automated system is proposed for measurement of facial muscle movement and emotion profiling. For the smooth continuation of tracking task a novel real-time CLM-based face tracker has been employed and a two-stage fuzzy reasoning model is developed for the task of classification. The proposed system supports the recognition of 6 universal emotions surprise, sadness, fear, anger, disgust and happiness. In addition, the proposed model is an evidence for real-time, person-independent, robust facial emotion recognition and classification demonstrating an increased accuracy of 91 % and 94 % for RaFD and CK + database.

Future work will exploit investigations of spontaneous emotion profiling without having a reference frame, i.e., neutral face frame in this paper. In this regard, unlabelled classification may occur if improper neutral face frame is considered as reference face frame. In real time, subjects are not always conscious about the first frame. After passing some frames, then subject can pay attention. In this regard, the proposed system may experience with improper neutral frame which can lead inaccurate emotion profiling. One possible way of avoiding this situation is firstly and identifying all the 46 AUs according to FACS standard of individual facial expression and finally to classify the universal facial emotions based on the inherent relationship of AUs among distinct facial expressions

Acknowledgments. This work was supported by University of Malaya HIR Grant UM. C/625/1/HIR/MOHE/FCSIT/10 of the University of Malaya.

References

1. Pantic, M., Rothkrantz, L.J.: Facial action recognition for facial expression analysis from static face images. IEEE Trans. Syst. Man Cybern. Part B Cybern. **34**(3), 1449–1461 (2004)
2. Russell, J.A., Fernández-Dols, J.M.: The Psychology of Facial Expression. Cambridge University Press, New York (1997)
3. Friesen, E., Ekman, P.: Facial action coding system: a technique for the measurement of facial movement, Palo Alto (1978)
4. Tie, Y., Guan, L.: Automatic landmark point detection and tracking for human facial expressions. EURASIP J. Image Video Process. **2013**(1), 1–15 (2013)
5. Bashyal, S., Venayagamoorthy, G.K.: Recognition of facial expressions using Gabor wavelets and learning vector quantization. Eng. Appl. Artif. Intell. **21**(7), 1056–1064 (2008)
6. Cho, K.S., Kim, Y.G., Lee, Y.B.: Real-time expression recognition system using active appearance model and EFM. In: 2006 International Conference on Computational Intelligence and Security, vol. 1, pp. 747–750. IEEE, November 2006

7. Sénéchal, T., Rapp, V., Salam, H., Seguier, R., Bailly, K., Prevost, L.: Facial action recognition combining heterogeneous features via multikernel learning. IEEE Trans. Syst. Man Cybern. Part B Cybern. **42**(4), 993–1005 (2012)
8. Tsalakanidou, F., Malassiotis, S.: Real-time 2D + 3D facial action and expression recognition. Pattern Recogn. **43**(5), 1763–1775 (2010)
9. Lin, D.T.: Facial expression classification using PCA and hierarchical radial basis function network. J. Inf. Sci. Eng. **22**(5), 1033–1046 (2006)
10. Saragih, J.M., Lucey, S., Cohn, J.F.: Face alignment through subspace constrained mean-shifts. In: 2009 IEEE 12th International Conference on Computer Vision, pp. 1034–1041. IEEE, September 2009
11. Kim, S.P., Simeral, J.D., Hochberg, L.R., Donoghue, J.P., Black, M.J.: Neural control of computer cursor velocity by decoding motor cortical spiking activity in humans with tetraplegia. J. Neural Eng. **5**(4), 455 (2008)
12. Nuevo, J., Bergasa, L.M., Jiménez, P.: RSMAT: robust simultaneous modeling and tracking. Pattern Recogn. Lett. **31**(16), 2455–2463 (2010)
13. Contreras, R., Starostenko, O., Alarcon-Aquino, V., Flores-Pulido, L.: Facial feature model for emotion recognition using fuzzy reasoning. In: Martínez-Trinidad, J.F., Carrasco-Ochoa, J.A., Kittler, J. (eds.) MCPR 2010. LNCS, vol. 6256, pp. 11–21. Springer, Heidelberg (2010)
14. Islam, M., Loo, C.K.: Geometric feature-based facial emotion recognition using two-stage fuzzy reasoning model. In: Loo, C.K., Yap, K.S., Wong, K.W., Teoh, A., Huang, K. (eds.) ICONIP 2014, Part II. LNCS, vol. 8835, pp. 344–351. Springer, Heidelberg (2014)
15. Kharat, G.U., Dudul, S.V.: Human emotion recognition system using optimally designed SVM with different facial feature extraction techniques. WSEAS Trans. Comput. **7**(6), 650–659 (2008)
16. Cristinacce, D., Cootes, T.: Automatic feature localisation with constrained local models. Pattern Recogn. **41**(10), 3054–3067 (2008)
17. Gross, R., Matthews, I., Cohn, J., Kanade, T., Baker, S.: Multi-pie. Image Vis. Comput. **28**(5), 807–813 (2010)
18. Langner, O., Dotsch, R., Bijlstra, G., Wigboldus, D.H., Hawk, S.T., van Knippenberg, A.: Presentation and validation of the radboud faces database. Cogn. Emot. **24**(8), 1377–1388 (2010)
19. Lucey, P., Cohn, J.F., Kanade, T., Saragih, J., Ambadar, Z., Matthews, I.: The extended cohn-kanade dataset (CK +): a complete dataset for action unit and emotion-specified expression. In: 2010 IEEE Computer Society Conference on Computer Vision and Pattern Recognition Workshops (CVPRW), pp. 94–101. IEEE, June 2010
20. Ilbeygi, M., Shah-Hosseini, H.: A novel fuzzy facial expression recognition system based on facial feature extraction from color face images. Eng. Appl. Artif. Intell. **25**(1), 130–146 (2012)
21. Besinger, A., Sztynda, T., Lal, S., Duthoit, C., Agbinya, J., Jap, B., Dissanayake, G.: Optical flow based analyses to detect emotion from human facial image data. Expert Syst. Appl. **37**(12), 8897–8902 (2010)
22. Rao, K.S., Koolagudi, S.G.: Recognition of emotions from video using acoustic and facial features. Signal Image Video Process., 1–17 (2013)

Heterogeneous Discriminant Analysis for Cross-View Action Recognition

Wanchen Sui[✉], Xinxiao Wu, Yang Feng, Wei Liang, and Yunde Jia

Beijing Laboratory of Intelligent Information Technology,
School of Computer Science, Beijing Institute of Technology,
Beijing 100081, People's Republic of China
{suiwanchen,wuxinxiao,fengyangbit,liangwei,jiayunde}@bit.edu.cn

Abstract. We propose an approach of cross-view action recognition, in which the samples from different views are represented by heterogeneous features with different dimensions. Inspired by linear discriminant analysis (LDA), we introduce a discriminative common feature space to bridge the source and target views. Two different projection matrices are learned to respectively map the data from two different views into the common space by simultaneously maximizing the similarity of intra-class samples, minimizing the similarity of inter-class samples, and reducing the mismatch between data distributions of two views. Our method is neither restricted to the corresponding action instances in the two views nor restricted to a specific type of feature. We evaluate our approach on the IXMAS multi-view dataset and the experimental results demonstrate its effectiveness.

Keywords: Cross-view action recognition · Transfer learning · Discriminant analysis · Heterogeneous domain adaption

1 Introduction

Human action recognition in videos plays an important role in computer vision due to its wide applications in human-computer interaction, smart surveillance, and video retrieval. However, the problem of viewpoint changes has posed a real challenge to it for the fact that the same action looks quite different when observed from different views. Both the data distribution and the feature space can vary drastically from one view to another. Hence, action models learned in one view tend to be incapable of the recognition in another different view.

Recently, lots of efforts have been made towards the problem of cross-view action recognition. A number of geometry-based approaches are motivated to perform by using the geometry measurement of body joints [12,16] or inferring 3D models of human subjects [13,17], usually requiring robust joint estimation which is still a challenging task. Another group of approaches tries to compute view-invariant human action representations that are stable across different viewpoints, such as temporal self-similarity matrix descriptors [3], view

© Springer International Publishing Switzerland 2015
S. Arik et al. (Eds.): ICONIP 2015, Part IV, LNCS 9492, pp. 566–573, 2015.
DOI: 10.1007/978-3-319-26561-2_67

and style-independent action manifold representation [5]. Wu et al. [14] proposed a latent kernelized structural SVM for view-invariant action recognition where the view is modeled as a latent variable. Several methods [1,6,8,9,15,18–20] have resorted to transfer learning, constructing the connections to bridge the gap between different views. Methods [1,8,18,20] rely on either feature-to-feature correspondence or video-to-video correspondence to transfer knowledge across views. Methods [6,9,19] require action features of the same type in different views. However, the corresponding data and homogeneous features in both views are not always available easily. In [15], Wu et al. proposed an iterative optimization algorithm to learn a common subspace for cross-view action recognition over heterogeneous feature spaces. Their method has less restrictions except that each action sample must be represented by a sequence of image features.

In this paper, we present a new transfer learning approach for cross-view action recognition with the heterogeneous features in source and target views. Our method is neither restricted to the corresponding action instances in the two views nor restricted to action features of the same type. Moreover, in this work, each action sample is represented by a commonly used feature vector. All these make our method more general than the existing ones. Specifically, in order to effectively utilize these heterogeneous features, we are encouraged to align the features from the two views via a common feature space with discriminant ability, where the action samples captured from different viewpoints can be compared directly. Our paper focuses on the construction of the common space. We aim to learn two different projection matrices to respectively map the data from different views to the common feature space. By introducing LDA to transfer learning, the two projections are learned by simultaneously maximizing the variance of intra-class samples, minimizing the variance of inter-class samples. We also add an effective nonparametric criterion into the objective function to reduce the mismatch between data distributions of different views. In addition, we try to extend the heterogeneous transfer discriminant analysis from linear space to kernel space, which leads to better performance.

2   Heterogeneous Transfer Linear Discriminant Analysis

In this paper, we focus on the problem of cross-view action recognition that the data from different views are represented by heterogeneous features. Given a large number of labeled training samples from the source view $\{(x_i^s, y_i^s)|_{i=1}^{n_s}\}$ with $x_i^s \in \mathbb{R}^{d_s}$, a few labeled samples from the target view $\{(x_i^l, y_i^l)|_{i=1}^{n_l}\}$ with $x_i^l \in \mathbb{R}^{d_t}$ and some unlabeled samples from the target view $\{(x_i^u, y_i^u)|_{i=1}^{n_u}\}$ with $x_i^u \in \mathbb{R}^{d_t}$, where n_s, n_l are the numbers of labeled samples from source and target views, n_u is the number of unlabeled samples from target view, y_i^s, y_i^l and y_i^u represent the labels of the samples x_i^s, x_i^l and x_i^u, and $y_i^s, y_i^l, y_i^u \in \{1, ..., c\}$ with c the number of action classes. In general, the feature dimensions from source and target views are not equal, i.e., $d_s \neq d_t$.

We aim to find two projection matrices w_s and w_t for respectively mapping data from source and target views into a common subspace via simultaneously maximizing the variance of inter-class samples, minimizing the variance of intra-class samples and reducing the mismatch between data distributions of different views.

There are equivalent variants of Fisher criterion to get the projection matrix w:

$$w = \arg\max_w \frac{|w^T S_B w|}{|w^T S_W w|} = \arg\max_w \frac{|w^T S_B w|}{|w^T S_T w|}, \tag{1}$$

where S_B is the between-class scatter matrix, S_W is the within-class scatter matrix, and $S_T = S_B + S_W$ is the total scatter matrix. As for this work, denoting $w = \begin{bmatrix} w_s \\ w_t \end{bmatrix}$ as the total projection matrix, the total scatter matrix $w^T S_T w$ and the between-class scatter matrix $w^T S_B w$ in the reduced common subspace can be defined as follows.

Total Scatter Matrix. The total scatter matrix is formulated as

$$w^T S_T w = \sum_{i=1}^{n_s} (w_s^T x_i^s - \mu)(w_s^T x_i^s - \mu)^T + \sum_{i=1}^{n_l} (w_t^T x_i^t - \mu)(w_t^T x_i^t - \mu)^T, \tag{2}$$

where μ indicates the global mean of all the projected labeled data from source and target view, defined by

$$\mu = \frac{n_s w_s^T \mu_s + n_l w_t^T \mu_l}{n_s + n_l}, \tag{3}$$

with μ_s and μ_l the mean of the labeled source and target samples.

Substituting (3) into (2), $w^T S_T w$ can be reformulated as

$$w^T S_T w = w^T \begin{bmatrix} S_{T,s} & 0 \\ 0 & S_{T,t} \end{bmatrix} w + \frac{n_s n_l}{n_s + n_l} w^T \begin{bmatrix} \mu_s \\ -\mu_l \end{bmatrix} \begin{bmatrix} \mu_s \\ -\mu_l \end{bmatrix}^T w, \tag{4}$$

where $S_{T,s}$ and $S_{T,t}$ are defined as $S_{T,s} = \sum_{i=1}^{n_s} (x_i^s - \mu_s)(x_i^s - \mu_s)^T$, $S_{T,t} = \sum_{i=1}^{n_l} (x_i^l - \mu_l)(x_i^l - \mu_l)^T$. Therefore, S_T can be written as

$$S_T = \begin{bmatrix} S_{T,s} & 0 \\ 0 & S_{T,t} \end{bmatrix} + \frac{n_s n_l}{n_s + n_l} \begin{bmatrix} \mu_s \\ -\mu_l \end{bmatrix} \begin{bmatrix} \mu_s \\ -\mu_l \end{bmatrix}^T. \tag{5}$$

Between-class Scatter Matrix. The between-class scatter matrix is

$$w^T S_B w = \sum_{j=1}^{c} n_j (\frac{w_s^T m_{sj} + w_t^T m_{tj}}{n_j} - \mu)(\frac{w_s^T m_{sj} + w_t^T m_{tj}}{n_j} - \mu)^T, \tag{6}$$

where n_j is the total number of the j-th class training samples from both source and target views, m_{sj} and m_{tj} represent the sum of the j-th class samples from source and target views, respectively. Combining (3) and (6), we can rewrite $w^T S_B w$ as

$$w^T S_B w = \sum_{j=1}^{c} w^T n_j (\begin{bmatrix} \frac{m_{sj}}{n_j} \\ \frac{m_{lj}}{n_j} \end{bmatrix} - \begin{bmatrix} \frac{n_s \mu_s}{n_s + n_l} \\ \frac{n_l \mu_l}{n_s + n_l} \end{bmatrix})(\begin{bmatrix} \frac{m_{sj}}{n_j} \\ \frac{m_{lj}}{n_j} \end{bmatrix} - \begin{bmatrix} \frac{n_s \mu_s}{n_s + n_l} \\ \frac{n_l \mu_l}{n_s + n_l} \end{bmatrix})^T w, \tag{7}$$

and S_B can be formulated as

$$S_B = \sum_{j=1}^{c} n_j \left(\begin{bmatrix} m_{sj} \\ n_j \\ m_{tj} \\ n_j \end{bmatrix} - \begin{bmatrix} \frac{n_s \mu_s}{n_s + n_l} \\ \frac{n_l \mu_l}{n_s + n_l} \end{bmatrix} \right) \left(\begin{bmatrix} m_{sj} \\ n_j \\ m_{tj} \\ n_j \end{bmatrix} - \begin{bmatrix} \frac{n_s \mu_s}{n_s + n_l} \\ \frac{n_l \mu_l}{n_s + n_l} \end{bmatrix} \right)^T. \tag{8}$$

In addition, as for heterogeneous discriminant analysis, we denote the matrix of projected data distribution difference between source and target views as $w^T S_D w$, defined by

$$w^T S_D w = w^T \begin{bmatrix} \mu_s \\ -\mu_t \end{bmatrix} \begin{bmatrix} \mu_s \\ -\mu_t \end{bmatrix} w, \tag{9}$$

where μ_t indicates the mean of all target samples, and S_D can be written as

$$S_D = \begin{bmatrix} \mu_s \\ -\mu_t \end{bmatrix} \begin{bmatrix} \mu_s \\ -\mu_t \end{bmatrix}. \tag{10}$$

In order to reduce the mismatch between data distributions of different views, we add a nonparametric criterion about $w^T S_D w$ into the discriminative function in (1). Hence, our objective function can be formulated as

$$w = \arg\max_w \frac{|w^T S_B w|}{|w^T S_T w| + \gamma |w^T S_D w|}, \tag{11}$$

where $\gamma > 0$ is the tradeoff parameter.

By solving the generalized eigen-decomposition problem $S_B w = \lambda(S_T + \gamma S_D)w$ with its leading eigenvalues, we can obtain the optimal projection matrix w and split it into w_s and w_t as $w_s = w(1 : d_s, :)$, $w_t = w(d_s + 1 : d_s + d_t, :)$. Then we use the two projection matrices to map the heterogeneous data into the common space and apply SVM to train the classifier by using the projected labeled training data from both source and target views.

3   Heterogeneous Transfer Kernel Discriminant Analysis

Considering linear discriminant is not complex enough for most real-world data, we extend the aforesaid heterogeneous transfer discriminant analysis from linear space to kernel space, in order to increase the expressiveness of the discriminant. In order to yield the projection matrices in the kernel space, we first map the data non-linearly into a high-dimensional kernel space and compute heterogenous linear discriminant analysis there. Let Φ be the non-linear mapping. $\Phi(x_i^s)$, $\Phi(x_i^l)$ and $\Phi(x_i^u)$ denote the transformed representation of x_i^s, x_i^l and x_i^u in the kernel space. According to the theory of reproducing kernels, the projections learned from the training samples lie in the span of all training samples. Therefore, the projection matrices w_s and w_t can be linearly represented as

$$w_s = \sum_{i=1}^{n_s} \alpha_i \Phi(x_i^s)$$

$$w_t = \sum_{i=1}^{n_l} \beta_i \Phi(x_i^l) + \sum_{i=1}^{n_u} \beta_{n_l + i} \Phi(x_i^u). \tag{12}$$

Denote $A = [\alpha_1, \cdots, \alpha_{n_s}]^T$, $B = [\beta_1, \cdots, \beta_{n_l+n_u}]^T$, $W = \begin{bmatrix} A \\ B \end{bmatrix}$. Substituting (12) into (4), we have

$$w^T S_T w = W^T \begin{bmatrix} S^\Phi_{T,s} & 0 \\ 0 & S^\Phi_{T,t} \end{bmatrix} W + \frac{n_s n_l}{n_s + n_l} W^T \begin{bmatrix} \mu^\Phi_s \\ -\mu^\Phi_l \end{bmatrix} \begin{bmatrix} \mu^\Phi_s \\ -\mu^\Phi_l \end{bmatrix}^T W, \qquad (13)$$

where $S^\Phi_{T,s} = \sum_{i=1}^{n_s} (\zeta^s_i - \mu^\Phi_s)(\zeta^s_i - \mu^\Phi_s)^T$, $S^\Phi_{T,t} = \sum_{i=1}^{n_l} (\zeta^l_i - \mu^\Phi_l)(\zeta^l_i - \mu^\Phi_l)^T$, with $\zeta^s_i = [k(x^s_i, x^s_1), \cdots, k(x^s_i, x^s_{n_s})]^T$, $\zeta^l_i = [k(x^l_i, x^l_1), \cdots, k(x^l_i, x^l_{n_l}), k(x^l_i, x^u_1), \cdots, k(x^l_i, x^u_{n_u})]^T$, $k(x_1, x_2)$ the inner product of $\Phi(x_1)$ and $\Phi(x_2)$; μ_s and μ_l are the mean vectors of all ζ^s and ζ^l, respectively. Therefore, denoting

$$S^\Phi_T = \begin{bmatrix} S^\Phi_{T,s} & 0 \\ 0 & S^\Phi_{T,t} \end{bmatrix} + \frac{n_s n_l}{n_s + n_l} \begin{bmatrix} \mu^\Phi_s \\ -\mu^\Phi_l \end{bmatrix} \begin{bmatrix} \mu^\Phi_s \\ -\mu^\Phi_l \end{bmatrix}^T, \qquad (14)$$

we have $w^T S_T w = W^T S^\Phi_T W$.

Similarly, we can obtain $w^T S_B w = W^T S^\Phi_B W$ and $w^T S_D w = W^T S^\Phi_D W$, where

$$S^\Phi_B = \sum_{j=1}^{c} n_j \left(\begin{bmatrix} \frac{m^\Phi_{sj}}{n_j} \\ \frac{m^\Phi_{tj}}{n_j} \end{bmatrix} - \begin{bmatrix} \frac{n_s \mu^\Phi_s}{n_s + n_l} \\ \frac{n_l \mu^\Phi_l}{n_s + n_l} \end{bmatrix} \right) \left(\begin{bmatrix} \frac{m^\Phi_{sj}}{n_j} \\ \frac{m^\Phi_{tj}}{n_j} \end{bmatrix} - \begin{bmatrix} \frac{n_s \mu^\Phi_s}{n_s + n_l} \\ \frac{n_l \mu^\Phi_l}{n_s + n_l} \end{bmatrix} \right)^T. \qquad (15)$$

with m^Φ_{sj} and m^Φ_{tj} the sum vectors of ζ^s and ζ^l from the j-th class;

$$S^\Phi_D = \begin{bmatrix} \mu^\Phi_s \\ -\mu^\Phi_t \end{bmatrix} \begin{bmatrix} \mu^\Phi_s \\ -\mu^\Phi_t \end{bmatrix}^T \qquad (16)$$

with μ^Φ_t the mean vectors of all ζ^l and ζ^u, $\zeta^u_i = [k(x^u_i, x^l_1), \cdots, k(x^u_i, x^l_{n_l}), k(x^u_i, x^u_1), \cdots, k(x^u_i, x^u_{n_u})]^T$.

Hence, by solving

$$W = \arg\max_W \frac{|W^T S^\Phi_B W|}{|W^T S^\Phi_T W| + \gamma |W^T S^\Phi_D W|}, \qquad (17)$$

A and B can be obtained, then the optimal projection matrices w_s and w_t can be finally gotten by different modalities of (12).

4 Experiments

4.1 Dataset

We evaluate our approach on the IXMAS multi-view action dataset [13], which consists of eleven daily-life acions, such as check watch, sit down, wave. Each action is performed three times by twelve subjects, taken from five different views. Figure 1 shows some exemplar actions. We extract the three dense trajectory features (i.e., HOG, HOF, MBH) proposed by Wang et al. [11]. For each descriptor, we use the bag-of-words approach and uniformly set the number of visual words 400. The 800-dimensional HOG/HOF feature (the concatenation of 400-dimensional HOG and 400-dimensional HOF) is adopted for source view, and 400-dimensional MBH feature for target view.

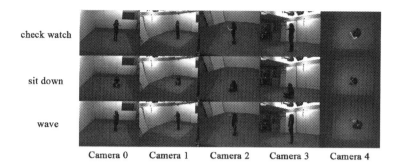

Fig. 1. Exemplar frames from IXMAS multi-view action dataset. Each row shows one action captured from different viewpoints.

4.2   Setup

In the experiments, we take one view as source view and another different view as target view. To enable appropriate verification, we look into all possible pairwise view combinations (20 in total for 5 views). The leave-one-subject-out cross validation strategy (i.e., 12-fold cross validation) is employed. For each time, we use videos of one subject from the target view for testing, the remaining videos for training, in which only two target subjects and all source subjects labeled.

As the source and target data are represented by heterogeneous feature vectors with different dimensions and there is no data-to-data correspondence, we compare our method with the state-of-the-art methods [2,4,7,10] of transfer learning on heterogeneous feature spaces. For DAMA [10], after finding the projection matrices, SVM is applied to train their final classifiers using the projected training data of pairwise views. For ARC-t [4], we construct the kernel matrix based on the learned asymmetric transformation metric, and then also apply SVM to train the final classifier. For MMDT [2], we find the transformation matrix using the max-margin constraints and simultaneously learn the final SVM classifier. For HFA [7], we obtain the two projection matrices using the standard SVM with the hinge loss. Moreover, we also set a baseline method, called SVMt, implementing SVM using the labeled target training data. For our method, the parameter γ is empirically set as $\gamma = 0.1$. For other methods, we validate the parameters chosen from $\{0.01, 0.1, 1, 10, 100\}$ according to their best performance. For all these methods, the parameter C in SVM is fixed as the default value (i.e., $C = 1$) and the RBF kernel is selected for fair comparison.

4.3   Results

As is shown in Table 1, the heterogeneous transfer discriminant analysis in kernel space generally achieves better results than that in linear space for all possible source-target view combinations. From Table 2, we can observe that our method outperforms other methods on the mean recognition accuracy for each target view. Compared with MMDT and ARC-t, our better performance may arise

Table 1. Accuracies of our method on the IXMAS dataset. The accuracy numbers in the bracket are the average recognition accuracies of our heterogeneous transfer discriminant analysis method in linear space and kernel space, respectively.

%	Target view1	Target view2	Target view3	Target view4	Target view5
Source View1		(71.3,77.5)	(71.3,79.9)	(67.4,75.0)	(61.1,72.2)
Source View2	(69.7,77.8)		(71.3,78.2)	(67.4,75.7)	(60.6,72.0)
Source View3	(69.7,76.9)	(71.3,77.5)		(67.4,74.1)	(60.9,72.0)
Source View4	(69.7,77.1)	(71.3,77.5)	(71.3,79.2)		(60.6,71.1)
Source View5	(69.7,76.2)	(71.1,77.1)	(71.3,78.5)	(67.4,74.3)	
Average	(69.7,77.0)	(71.2,77.4)	(71.3,78.9)	(67.4,74.8)	(60.8,71.8)

Table 2. Accuracies of different heterogeneous transfer learning approaches on the IXMAS dataset. Each column corresponds to one target view. The numbers are the mean recognition accuracies of each target view.

%	View 1	View 2	View 3	View 4	View 5	Average
SVMt	68.5	65.7	72.0	66.7	60.9	66.8
DAMA [10]	68.7	67.0	72.3	67.9	63.9	68.0
ARC-t [4]	68.8	67.6	72.5	68.2	63.7	68.1
MMDT [2]	71.1	71.9	73.0	73.1	64.1	70.7
HFA [7]	74.5	75.8	76.6	73.0	69.2	73.8
ours	**77.0**	**77.4**	**78.9**	**74.8**	**71.8**	**76.0**

from the utilization of unlabeled target training data. These data contribute to cope with the data distribution mismatch between the source and target views. DAMA performs only slightly better than SVMt, possibly due to the lack of strong manifold structure on this dataset.

5   Conclusions

We have presented a method for cross-view action recognition over heterogeneous feature spaces, and respectively accomplished the heterogeneous transfer discriminant analysis in linear and kernel space. We propose to seek a discriminant common subspace, where the samples from different views are comparable. In this way, two different matrices are learned to map the data from source and target views into the common subspace, by simultaneously maximizing the similarity of intra-class samples, minimizing the similarity of inter-class samples and reducing the mismatch between data distributions of different views. Our proposed method is capable of exploring the effective information contained in the labeled samples and unlabeled target samples, without any instance correspondence between two views. The promising results of our approach have been achieved on the IXMAS dataset for cross-view action recognition.

Acknowledgments. This work was supported in part by the Natural Science Foundation of China under Grant 61203274.

References

1. Farhadi, A., Tabrizi, M.K.: Learning to recognize activities from the wrong view point. In: Forsyth, D., Torr, P., Zisserman, A. (eds.) ECCV 2008, Part I. LNCS, vol. 5302, pp. 154–166. Springer, Heidelberg (2008)
2. Hoffman, J., Rodner, E., Donahue, J., Darrell, T., Saenko, K.: Efficient learning of domain-invariant image representations. In: ICLR 2013 (2013)
3. Junejo, I.N., Dexter, E., Laptev, I., Perez, P.: View-independent action recognition from temporal self-similarities. IEEE T-PAMI **33**(1), 172–185 (2011)
4. Kulis, B., Saenko, K., Darrell, T.: What you saw is not what you get: Domain adaptation using asymmetric kernel transforms. In: CVPR 2011, pp. 1785–1792 (2011)
5. Lewandowski, M., Makris, D., Nebel, J.-C.: View and style-independent action manifolds for human activity recognition. In: Daniilidis, K., Maragos, P., Paragios, N. (eds.) ECCV 2010, Part VI. LNCS, vol. 6316, pp. 547–560. Springer, Heidelberg (2010)
6. Li, R., Zickler, T.: Discriminative virtual views for cross-view action recognition. In: CVPR 2012, pp. 2855–2862 (2012)
7. Li, W., Duan, L., Xu, D., Tsang, I.W.: Learning with augmented features for supervised and semi-supervised heterogeneous domain adaptation. IEEE T-PAMI **36**(6), 1134–1148 (2014)
8. Liu, J., Shah, M., Kuipers, B., Savarese, S.: Cross-view action recognition via view knowledge transfer. In: CVPR 2011, pp. 3209–3216 (2011)
9. Rahmani, H., Mian, A.: Learning a non-linear knowledge transfer model for cross-view action recognition. In: CVPR 2015, pp. 2458–2466 (2015)
10. Wang, C., Mahadevan, S.: Heterogeneous domain adaptation using manifold alignment. In: IJCAI, vol. 22, p. 1541 (2011)
11. Wang, H., Klaser, A., Schmid, C., Liu, C.L.: Action recognition by dense trajectories. In: CVPR 2011, pp. 3169–3176 (2011)
12. Wang, J., Liu, Z., Wu, Y., Yuan, J.: Learning actionlet ensemble for 3D human action recognition. IEEE T-PAMI **36**(5), 914–927 (2014)
13. Weinland, D., Boyer, E., Ronfard, R.: Action recognition from arbitrary views using 3D exemplars. In: ICCV 2007, pp. 1–7 (2007)
14. Wu, X., Jia, Y.: View-invariant action recognition using latent kernelized structural SVM. In: Fitzgibbon, A., Lazebnik, S., Perona, P., Sato, Y., Schmid, C. (eds.) ECCV 2012, Part V. LNCS, vol. 7576, pp. 411–424. Springer, Heidelberg (2012)
15. Wu, X., Wang, H., Liu, C., Jia, Y.: Cross-view action recognition over heterogeneous feature spaces. In: ICCV 2013, pp. 609–616 (2013)
16. Xia, L., Chen, C.C., Aggarwal, J.: View invariant human action recognition using histograms of 3D joints. In: CVPRW 2012, 20–27 (2012)
17. Yan, P., Khan, S.M., Shah, M.: Learning 4D action feature models for arbitrary view action recognition. In: CVPR 2008, pp. 1–7 (2008)
18. Zhang, Z., Wang, C., Xiao, B., Zhou, W., Liu, S.: Cross-view action recognition using contextual maximum margin clustering. IEEE T-CSVT **24**, 1663–1668 (2014)
19. Zhang, Z., Wang, C., Xiao, B., Zhou, W., Liu, S., Shi, C.: Cross-view action recognition via a continuous virtual path. In: CVPR 2013, pp. 2690–2697 (2013)
20. Zheng, J., Jiang, Z.: Learning view-invariant sparse representations for cross-view action recognition. In: ICCV 2013, pp. 3176–3183 (2013)

A Comparison of Facial Features and Fusion Methods for Emotion Recognition

Demiyan V. Smirnov[1], Rajani Muraleedharan[2],
and Ravi P. Ramachandran[1(✉)]

[1] Rowan University, Glassboro, NJ, USA
smirno59@students.rowan.edu, ravi@rowan.edu
[2] Saginaw Valley State University, University Center, MI, USA
rmuralee@svsu.edu

Abstract. Emotion recognition is an important part of human behavior analysis. It finds many applications including human-computer interaction, driver safety, health care, stress detection, psychological analysis, forensics, law enforcement and customer care. The focus of this paper is to use a pattern recognition framework based on facial expression features and two classifiers (linear discriminant analysis and k-nearest neighbor) for emotion recognition. The extended Cohn-Kanade database is used to classify 5 emotions, namely, 'neutral, angry, disgust, happy, and surprise'. The Discrete Cosine Transform (DCT), Discrete Sine Transform (DST), the Walsh-Hadamard Transform (FWHT) and a new 7-dimensional feature based on condensing the Facial Action Coding System (FACS) are compared. Ensemble systems using decision level, score fusion and Borda count are also studied. Fusion of the four features leads to slightly more than a 90 % accuracy.

Keywords: Emotion recognition · Facial expression · Feature extraction · Linear discriminant analysis · K-nearest neighbor · Fusion

1 Introduction and Motivation

Emotion recognition is an important part of human behavior analysis, where applications including human-computer interaction (particularly the brain-computer interface) [1], human-robot interaction [2], computer games [3]. Driver safety [3,4], health care [5], stress detection [6] can be benefited. In addition, emotion recognition when combined with biometrics [7] is crucial for security purposes and improves law enforcement and forensic applications [8]. Humans express emotion using vocal, facial, gesture, body language, handwriting, and sign language. Identifying emotions through contents in an electronic conversation (like email) can maximize customer satisfaction, and is currently an important topic of research in marketing and customer service [9].

Face [2,3,10], speech [11,12] and physiological modalities are often used in emotion recognition. The use of physiological signals include the electrocardiogram (ECG) [13], electroencephalogram (EEG) [1], skin temperature and resistance, blood pressure and respiration [1]. Physiological modalities are highly

© Springer International Publishing Switzerland 2015
S. Arik et al. (Eds.): ICONIP 2015, Part IV, LNCS 9492, pp. 574–582, 2015.
DOI: 10.1007/978-3-319-26561-2_68

intrusive and cannot be measured remotely. However, speech or face signals can be acquired through remote audio and video surveillance, which is non-invasive and requires less cooperation from users.

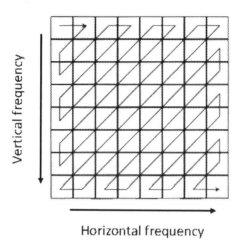

Fig. 1. Zig-zag scanning method for extracting a one-dimensional feature vector from a 2-D frequency transform

This paper compares the performance of various facial expression feature extraction and machine learning approaches for emotion recognition. The extended Cohn-Kanade database [4] is used to classify 5 emotions, namely, 'neutral, angry, disgust, happy, and surprise'. The main assumptions are (1) that the subject can be captured by at least one camera sensor to yield a digital image, (2) the subject is cooperative and the quality of the image is improved to remove illumination and noise effects and (3) biometric recognition of the face image is useful but not required for emotion identification (any biometric analysis would be in parallel with the system proposed in this paper). A pattern recognition [14,15] framework is used involving feature extraction, classification and a jackknife (or m-fold) strategy [14,16] with multiple trials for performance evaluation. The classifiers considered are linear discriminant analysis (LDA), and the k-nearest neighbor (kNN). Three of the facial expression features are the Discrete Cosine Transform (DCT), Discrete Sine Transform (DST) and the Walsh-Hadamard Transform (FWHT) [17]. A new 7-dimensional feature (recently proposed in [18]) based on condensing the Facial Action Coding System (FACS) [4] using 14 points is also analyzed. Ensemble systems using decision level, score fusion and Borda count are studied.

2 Facial Expression Feature Extraction

Each facial image in the database is first processed by the Viola-Jones face detector [7,19] and any non-face background portions of the image are removed.

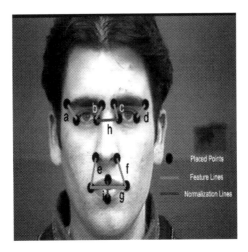

Fig. 2. The 14 point feature extraction method

Given a background removed face image, the 2-D DCT, 2-D DST and 2-D FWHT are calculated and scanned in a zig-zag fashion [17] as shown in Fig. 1 to get a one-dimensional feature vector.

Figure 2 shows a set of fourteen points, placed on a subject's face, based on the Facial Action Coding System (FACS) [4]. The FACS is used to quantify emotions by labeling muscle movements and facial feature changes by an Action Unit number [4]. The fourteen points and nine distances between the points are labeled as 'a' through 'i' in Fig. 2. The feature vector consists of 7 dimensions. The first six are the distances 'a' through 'f' each divided by the distance 'h' (accounts for image scaling and rotation). The seventh dimension is the horizontal distance at the mouth (labeled as 'g') divided by the vertical distance at the mouth (labeled as 'i'). This 7 dimensional feature vector is also is referred to as the 14 point feature extraction method.

The motivation of using this 14 point feature extraction method is that (1) the need to determine which Action Units a subject is/are displaying is avoided and (2) a vector of low dimension is configured by using normalized distances between key points corresponding to important facial attributes that indicate emotion. The specific points and distances were chosen as they represent movements of important muscles. The inner and outer eyebrows are accounted for, both when they are raised or lowered. In the same vein, the corners of the mouth produce longer lines when they are pulled down for a frown and shorter lines when pulled up for a smile. The status of the mouth (open or closed) can also be detected by the lines crossing on the lips.

3   Emotion Recognition Classifiers

Classifiers like neural networks and Gaussian mixture models require much data to obtain a good performance. In the extended Cohn-Kanade database, only the

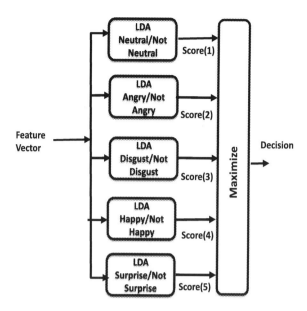

Fig. 3. LDAPA: Parallel arrangement of LDA classifiers

'neutral' emotion has much data (593 images). This is followed by 'happy' (69 images). The 'anger' emotion is only depicted by 45 images. In lieu of the limited amount of data, two simple classifiers, namely, linear discriminant analysis (LDA) and the k-nearest neighbor (kNN) are considered.

The LDA method is performed without dimensionality reduction. It is implemented in two ways. First, the LDA is trained using feature vectors from all 5 class labels (emotions) so that the feature space is partitioned into 5 distinct regions. It is assumed that the feature vectors for each of the five classes has a multivariate Gaussian density. For a linear discriminant function, each class has a different mean vector but the same covariance matrix. Each test image results in one test feature vector. The output of the LDA consists of 5 scores, each indicating the posterior probability that the test feature vector belongs to a particular 'emotion' class. The maximum score identifies the emotion.

The second LDA implementation is depicted in Fig. 3. The test feature vector is passed through a parallel arrangement of 5 LDA classifiers with each making a decision between two classes (emotion and not the emotion). This is referred to as LDA with a parallel arrangement (LDAPA). Each LDA classifier is trained using feature vectors representing the 'emotion' (like neutral) and feature vectors representing 'not the emotion' (like all emotions except neutral). Again, it is assumed that the feature vectors for each of the two classes has a multivariate Gaussian density. As before, each class has a different mean vector but the same covariance matrix. For a two class problem, a hyperplane divides the feature space into two regions that distinguish each class. The only score considered for each LDA classifier is the posterior probability that the test feature vector belongs to the 'emotion' class. The maximum score among the 5 classifiers identifies the emotion.

The kNN classifier stores all the training data and the associated class labels for the 5 emotions. The class labels of the k nearest neighbors to the test feature vector (in terms of squared Euclidean distance) are recorded. A majority vote among the class labels identifies the emotion. In the event of a tie, only the class labels involved in the tie are considered with the class label of the training vector closest to the test feature vector identifying the emotion.

4   Experimental Protocol

A jackknife (or m-fold strategy with $m = 7$) is used to randomly partition the image data into 7 non-overlapping subsets such that the amount of data for each emotion is about the same in each subset. Six of the subsets are used for training the classifier and one subset is used for testing. The subsets are revolving with each test so that each subset is used as the testing subset exactly once. Hence, seven test runs are performed for this specific partition. The Average Class Identification Rate (ACIR) is the number of times a test feature vector from a particular class is identified correctly with an average taken over the 7 test runs. There will be 5 ACIR values, one for each class or emotion. This process depicts one trial.

There are five trials performed, each with a different random partition of the image data into 7 subsets. The Average Trial Class Identification Rate (ATCIR) is the average of the ACIR values taken over the five trials. Again, there will be 5 ATCIR values, one for each class or emotion. The identification success rate (ISR) is the average of the ATCIR values taken over the 5 emotions.

Table 1. Identification success rate for individual features

Feature	Classifier	No SMOTE		With SMOTE	
		Best dimension	ISR (%)	Best dimension	ISR (%)
DCT	LDA	115	88.6	120	84.6
DCT	LDAPA	145	85.1	115	81.9
DCT	kNN	130	28.4	85	35.1
DST	LDA	120	88.1	95	85.2
DST	LDAPA	150	85.1	115	83.7
DST	kNN	85	28.2	130	35.8
FWHT	LDA	120	88.9	115	85.4
FWHT	LDAPA	150	86.0	115	82.5
FWHT	kNN	130	31.5	145	34.2
14 Point	LDA	7	49.9	7	63.3
14 Point	LDAPA	7	40.6	7	60.7
14 Point	kNN	7	58.1	7	61.6

The four features were each used with LDA, LDAPA and kNN to get an ISR value. The ISR values were obtained in two ways. The first was by using the unbalanced data in which there were many more samples of the 'neutral' emotion (593 feature vectors). The second was by using the Synthetic Minority Over-sampling Technique (SMOTE) [20] to increase the number of feature vectors from each class (except 'neutral') such that the data is balanced.

5 Performance of the Individual Features

Table 1 shows the ISR for the various feature/classifier combinations with and without SMOTE. The best dimension given in Table 1 is that which results in the maximum ISR for the cases when SMOTE is used and not used. For the DCT, DST and FWHT (with and without SMOTE), the performance was evaluated for dimensions ranging from 5 to 150. Figure 4 depicts the results for the case when the LDA classifier is used without SMOTE. For the kNN classifier, values of k equal to 1, 3, 5 and 7 were attempted and $k = 1$ gave the best performance.

The DCT, DST and FWHT show the best results using LDA or LDAPA. A one-tailed t-test with unequal variances [21] based on five trials and a 95 % confidence interval confirms the following for the DCT, DST and FWHT:

1. There is no statistical significant difference in performance a, omg the three features.
2. The LDA shows the best performance with statistical significance.
3. Not applying SMOTE shows a better performance with statistical significance.

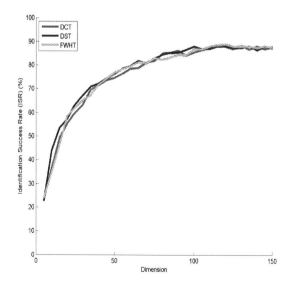

Fig. 4. ISR of DCT, DST and FWHT as a function of vector dimension for the LDA classifier (SMOTE not used)

The dimension of the vector resulting from the 14 point feature extraction method is fixed at 7. The performance of this feature is better when SMOTE is applied (with statistical significance). However, the performance is not as good as the DCT, DST or FWHT implemented with LDA or LDAPA. However, to achieve this better performance, the DCT, DST and FWHT require a much higher dimension. Future work is aimed at using more than fourteen points (based on the (FACS)), acquiring feature vectors of different dimensions (by labeling different points and taking distances between points as in Fig. 2) and investigating the ISR versus dimension. This will give a more clear comparison with the DCT, DST and FWHT. The aim is to get the high performance that DCT, DST and FWHT achieve but at a lower dimension that an FACS based method can potentially achieve.

6   Fusion

Since different feature/classifier combinations are used, an ensemble system [16] results, which naturally leads to the investigation of fusion. Decision level fusion is the simplest technique and involves taking a majority vote of the different features to get a final decision. For score fusion, the scores (or posterior probabilities) for each emotion of a single feature/classifier pair are converted to normalized scores such that their sum equals 1. For a particular emotion, the normalized scores generated by the different feature/classifier pairs considered are added to get a combined score. The maximum combined score identifies the emotion. The third fusion method is to use Borda count based on the normalized scores.

Table 2. Fusion results (Numbers expressed as a %)

Features and Classifiers	Fusion type	ISR (%) No SMOTE	ISR (%) SMOTE	ISR (%) Partial SMOTE
DCT/LDA and DST/LDA	Score	88.3	85.4	88.9
DCT/LDA and DST/LDA	Borda	89.1	82.8	89.8
DCT/LDA, DST/LDA and FWHT/LDA	Score	89.3	86.1	89.7
DCT/LDA, DST/LDA and FWHT/LDA	Borda	89.2	82.8	89.9
DCT/LDA, DST/LDA and FHWT/LDA	Decision	89.2	86.3	89.9
DCT/LDA, DST/LDA and 14 Point/kNN	Decision	88.8	85.5	90.4
DCT/LDA, DST/LDA, FWHT/LDA and 14 Point/kNN	Decision	89.7	86.6	90.9

Fusion experiments were performed using DCT/LDA, DST/LDA, FWHT/LDA and the 14 point feature extraction method with kNN. In the event of a tie due to fusion, the decision of the DCT/LDA is taken. Table 2 gives the results of the best approaches for the following cases:

1. No SMOTE: SMOTE not used for any of the features
2. SMOTE: SMOTE used for all of the features
3. Partial SMOTE: SMOTE not used for the DCT, DST and FWHT but used for the 14 point feature extraction method.

The best method is the decision level fusion of DCT/LDA, DST/LDA, FWHT/LDA and the 7 dimensional feature resulting from the 14 point feature extraction method (denoted as 14 Point/kNN). For this fusion method, partial SMOTE is the best (with statistical significance). This illustrates that the 7 dimensional feature is useful and should be explored further.

7   Summary and Conclusions

Various feature/classifier combinations have been compared for emotion recognition using facial features. Five emotions, namely, 'neutral, angry, disgust, happy, and surprise', from the extended Cohn-Kanade database are used in the experiments. Fusion of the features results in slightly more than a 90 % accuracy. Future work will aim to achieve a high performance with a low feature dimension.

Acknowledgement. This work was supported by the National Science Foundation through Grant DUE-1122296.

References

1. Huang, D., Zhang, H., Ang, K., Guan, C., Pan, Y., Wang, C., Yu, J.: Fast emotion detection from EEG using asymmetric spatial filtering. In: IEEE International Conference on Acoustics, Speech and Signal Processing, Kyoto, Japan, pp. 589–592 (2012)
2. Tariq, U., Huang, T.S.: Features and fusion for expression recognition - A comparative analysis. In: IEEE Computer Vision and Pattern Recognition Workshop, pp. 146–152 (2012)
3. Tsai, H.-H., Lai, Y.-S., Zhang, Y.-C.: Using SVM to design facial expression recognition for shape and texture features. In: International Conference on Machine Learning and Cybernetics, Qingdao, China, pp. 2697–2704 (2010)
4. Lucey, P., Cohn, J. F., Kanade, T., Saragih, J., Ambadar, Z., Matthews, I.: The Extended Cohn-Kanade Dataset (CK+): A complete dataset for action unit and emotion-specified expression. In: IEEE Computer Vision and Pattern Recognition Workshop, pp. 94–101 (2010)
5. Lucey, P., Cohn, J. F., Lucey, S., Matthews, I., Sridharan, S., Prkachin, K.: Automatically detecting pain using facial actions. In: International Conference on Affective Computing and Intelligent Interaction, pp. 1–8 (2009)

6. Nehra, D.K., Sharma, V., Mushtaq, H., Sharma, N., Sharma, M., Nehra, S.: Emotional intelligence and self esteem in cannabis abusers. J. Indian Acad. Appl. Psychol. **38**(2), 385–393 (2012)
7. Jain, A.K., Ross A., Nandakumar K.: Introduction to Biometrics. Springer, New York (2011)
8. Campbell, J.P., Shen, W., Campbell, W.M., Schwartz, R., Bonastre, J.-F., Matrouf, D.: Forensic speaker recognition. IEEE Signal Process. Mag. **26**, 95–103 (2009)
9. Gupta, N., Gilbert, M., Di Fabrizio, G.: Emotion detection in email customer care. Comput. Intell. 59 (2010)
10. Gupta, S.K., Agrwal, S., Meena Y.K., Nain, N.: A hybrid method of feature extraction for facial expression recognition. In: International Conference on Signal Image Technology and Internet-Based Systems, pp. 422–425 (2011)
11. Koolagudi, S.G., Kumar, N., Rao, K.S.: Speech emotion recognition using segmental level prosodic analysis. In: International Conference on Devices and Communications, pp. 1–5 (2011)
12. Busso, C., Deng, Z., Yildirim, S., Bulut, M., Lee, C.M., Kazemzadeh, A., Lee, S., Neumann, U., Narayanan, S.: Analysis of emotion recognition using facial expressions, speech and multimodal Information. In: ACM International Conference on Multimodal Interfaces, State College, Pennsylvania (2004)
13. Cai, J., Liu, G., Hao, M.: The research on emotion recognition from ECG Signal. In: International Conference on Information Technology and Computer Science, pp. 497–500 (2009)
14. Duda, R.O., Hart, P.E., Stork, D.G.: Pattern Classification. Wiley, New York (2001)
15. Bishop, C.M.: Pattern Recognition and Machine Learning. Springer, Heidelberg (2006)
16. Polikar, R.: Ensemble based systems in decision making. IEEE Circuits Syst. Mag. **6**, 21–45 (2006)
17. Bose, T.: Digital Signal and Image Processing. Wiley, New York (2004)
18. Smirnov, D.V., Banger, S., Davis, S.H., Muraleedharan, R., Ramachandran, R.P.: Automated human behavioral analysis framework using facial feature extraction and machine learning. In: 47th Annual Asilomar Conference on Signals, Systems, and Computers, Pacific Grove, California (2013)
19. Viola, P.A., Jones, M.J.: Robust real-time face detection. Int. J. Comput. Vision **57**, 137–154 (2004)
20. Chawla, N.V., Bowyer, K.W., Hall, L.O., Kegelmeyer, W.P.: SMOTE: synthetic minority over-sampling technique. J. Artif. Intell. **16**, 321–357 (2002)
21. Yates, R.D., Goodman, D.J.: Probability and Stochastic Processes. Wiley, New York (1999)

A Review and Comparison of Service E-Contract Architecture Metamodels

Ali Braytee[(✉)], Asif Qumer Gill, Paul J. Kennedy,
and Farookh Khadeer Hussain

School of Software, University of Technology Sydney,
Sydney, NSW 2007, Australia
{ali.braytee,asif.gill,paul.kennedy,farookh.hussain}@uts.edu.au

Abstract. An adaptive service e-contract is an electronic agreement which is required to enable adaptive or agile service sourcing and provisioning. There are a number of e-contract metamodels that can be used to create a context specific adaptive service e-contract. The challenge is which one to choose and adopt for adaptive services. This paper presents a review and comparison of well-known e-contract metamodels using the architecture theory. The architecture theory allows the analysis of the e-contract metamodels using a three-dimension analytical lens: structure, behavior and technology. The results of this paper highlight the metamodels structural, behavioral and technological differences and similarities. This paper will help researchers and practitioners to observe whether the existing e-contract metamodels are appropriate to the adaptive services or if there is a need to merge and integrate the concepts of these metamodels to propose a new unifying adaptive service e-contract metamodel. This paper is limited to the number of compared metamodels.

Keywords: Adaptive e-contract · Adaptive service · Contract metamodel · Foundational ontology

1 Introduction

In the modern service economy, focus shifts from the traditional upfront fixed service contract to the adaptive or agile service e-contract. This is because, the context of the modern enterprise encourages the use of agile services and e-contracts to deal with complex dynamic business requirements [1]. In service dominant logic (S-D logic), service is the application of competence (knowledge and skills) to benefit other parties [2]. The interaction between the parties in the service ecosystem may lead to value co-creation which means that all the engaged models earn value. The value in service science is the refinement of the system which co-creates the benefits or value to all involved parties. Traditional service provisioning is based on an upfront fixed contract, which is not appropriate and aligned to the modern business requirements where a dynamic response is required to deal with changing service demands [3]. The emerging service science

S. Arik et al. (Eds.): ICONIP 2015, Part IV, LNCS 9492, pp. 583–595, 2015.
DOI: 10.1007/978-3-319-26561-2_69

body of knowledge [4] suggests shifting focus from the traditional service delivery to a service offering through voluntary and dynamic interactions between service systems. A service system could be an organisational function or capability that offers services to other service systems. The adaptive enterprise service system theory states that a modern enterprise is an ecosystem of adaptive or agile service systems that voluntarily interact with each other and exchange adaptive or agile services for value co-creation through a dynamic or adaptive contract [3]. The type of dynamic interactions between service systems could be either bilateral relations which consist of service provider and service consumer, or multilateral relations between many service systems or parties [5]. The dynamic interactions between service systems could be enabled through a dynamic adaptive service contract (e-contract) as opposed to an upfront fixed service contract.

A contract is an agreement between two or more entities to create a business relation and to define a set of rules such as obligations, permissions and prohibitions for the business process [6]. A traditional contract is a static in nature which includes a static description of the business expectations of both parties. Furthermore, a traditional contract is not able to govern the relation between dynamic, ever-changing and adaptive services and it will not be able to monitor the clauses of such dynamic contracts. An e-contract is an electronic agreement between the service parties. It is composed of a set of entities and relationships in order to satisfy the service governance and compliance. In addition to the structural depiction, an e-contract has the ability to work with dynamic events and actions to handle the exceptions once the contract clauses are violated by either party. An e-contract can be established through a metamodel. There are a number of e-contract metamodels such as ER^{EC}, contract enforcement among others. Organisations are often unsure about the choice of a metamodel for establishing a context-specific e-contract. There is a need to understand the similarities and differences between the different metamodels. This study reviews the most well-known metamodels are the following: "ER^{EC} Metamodel", "Three-layered e-contract enforcement metamodel" and "Secure e-contract metamodel" by using the architecture theory (ISO/IEC 42010) in order to understand the scope and usability of the available metamodels. This work is significant because it helps practitioners and researchers in finding the shortcomings of the existing e-contract metamodels and serves as a knowledge base for developing a new unifying adaptive service e-contract metamodel. Additionally, It clarifies whether the existing e-contract metamodels are appropriate for adaptive service system or there is a need to propose a new adaptive service e-contract. The contribution of this work to the existing literature body is that, it is the first work of that systematically reviews and compares the most well-known contract metamodels. This paper is organised as follows. Section 2 presents the research method. Section 3 discusses the analysis. Section 4 presents the discussion before concluding in Sect. 5.

2   Research Method

This review paper applies the qualitative review approach [7] which is appropriate to address the research question in hand. The research process is two-fold. Firstly, we set the review criteria based on the architecture theory (ISO/IEC 42010). The review criteria provided us with a three-dimension analytical lens: structure, behavior, and technology. Secondly, we applied the review criteria to the most well-known e-contract metamodels identified from the literature which are the following: "ER^{EC} Metamodel", "Three-layered e-contract enforcement metamodel" and "Secure e-contract metamodel". The review criteria helped us to systematically analyse and compare each contract metamodel from the perspectives of the three dimensions. The structure perspective allowed us to analyse and compare the properties and relationships of the elements available in the three metamodels. The study of structure in its own right is not sufficient; we also need to analyse the behavior of the structure. The behavioral perspective allowed us to analyse and compare the semantics of the elements available in the reviewed metamodels. Finally, the technological perspective helped us to analyse and compare the supporting technology in the metamodels. This research approach provided a foundation for further research in the possible integration or merger of the well-known contract metamodels to create a more comprehensive unifying contract metamodel for agile or adaptive services. In the remainder of the paper, we will use terms electronic contract, e-contract and contract interchangeably.

3   Analysis: Metamodels

In this section, we will analysis the reviewed contract metamodels.

3.1 ER^{EC} metamodel

The ER^{EC} metamodel has evolved over a period of time since the first publication of the foundation in 2001. It has been continuously updated, the most recent updated version of the metamodel being published in 2013 [8]. The ER^{EC} model is shown in Fig. 1. It is based on the entity relationship model that captures the main concepts/entities of the contract and transforms the conceptualizations into workflows. This definition was used in the early versions of this metamodel, however its architecture has evolved over a period of time. Despite changes and evolution in the architecture of the ER^{EC}, the fundamental structure of the e-contract model remains the same [9]. In the most recent version of the ER^{EC} model [8], the authors proposed the meta-modeling approach which is able to define template for contracts.

This metamodel presents the main concepts of the e-contract. The evolved ER^{EC} metamodel organises the most abstract entities as core entities and defines a contract template. The template or model instantiates the ER^{EC} meta-model that includes the domain constraints and relationships for the e-contract application [8]. The use of the template allows the e-contract model to evolve. The e-contract template approach will be further discussed in this paper in the next section (Three-layered e-contract enforcement metamodel section). Furthermore, the template has several features such as adding new concepts into the model and reflecting the most real entities by modeling these concepts in the domain application. In addition, the latest versions of the ER^{EC} have proposed a meta-modeling approach to respond to the events that trigger run-time changes to the e-contract or exceptions due to violation in an existing contract clause [8]. Now, the ER^{EC} metamodel architecture will be analysed from its structure, behavior and technology perspectives. The structure of the ER^{EC} e-contract composed of the following core entities: parties, activities and clauses. The ER^{EC} e-contract must have two or more parties to initiate a contract and can have nested subcontracts. A subcontract is a contract which is created by one of the involved parties and may contain different parties, activities and clauses. A contract has a set of activities which are divided into tasks. The activities control the behavior of the e-contract. Each activity is used by the parties. The last core entity is the contract clause. An e-contract has a list of clauses and every clause can refer to other clauses. A clause is fulfilled when a relevant activity is successfully executed. Furthermore, clauses are considered as constraints or rules which may or may not be linked to the activities [9]. The other entities of the e-contract are financial entities such as payment and budget. The former captures the amount of service usage and the way of paying the other party, and the latter describes the maximum amount of money that the customer is obliged to pay. A contract has a specific period of time. More entities are declared in ER^{EC} such as role and exception [9]. Role depicts the position of the party which is linked to the parties via many-to-many relationships, and exception is executed once a clause is violated. Finally, an event can trigger action(s) related to an e-contract. An event can be predictable or unpredictable, and can be initiated at any time during the contract's execution. There are a number of events that may occur during contract execution, such as a contract database event, a temporal event or an external event. A contract event may refer to a situation such as add role, add party, or start payment. Furthermore, a temporal event can be specified as a time-based workflow, such as a payment frequency [10]. In summary, contract structure describes the contract elements and their relationships (see in Fig. 1).

Further, we need to analyse the behavior of the ER^{EC} e-contract structure, which is often linked to the contract run-time environment. The behavior of the e-contract model can be described and analysed in terms of contract activities or workflows. The ER^{EC} entities are connected to the workflow activities that execute the e-contract. The parties are related to roles and clauses to events that may trigger workflow activities or actions. The ER^{EC} workflow system uses the activity commit diagram (ACD), which is a list of atomic activity transactions. These transactions are executed sequentially. Krishna et al. (2004) states that "ACD ensures the consistent, atomic and durable execution of the activities". They specify a monitoring process to check the activity status and log messages. A contract activity can be rolled back in cases of failure and a message can be logged to report failures. Roll-back is not the only way to handle contract activity failure situations. There are a number of other options which can be applied to handle a failure situation, such as failure compensation, alternative activity, time-based retry or re-execution. ER^{EC} specifies the ECA (event-condition-action) concept in order to monitor the events in the e-contract model, however this procedure is based on a manual analysis process. It searches for a specific activity or behavioral-related words to create the conditions and actions, such as if-else, contract violations, among others [10]. In summary, the review of the ER^{EC} metamodel indicates that the ER^{EC} offers the ability to adapt to run-time changes. These changes may refer to contract updates, exceptions or failures. The following four behavioral operations related to the ER^{EC} e-contract model are identified in this review:

  i Adapt: This operation is used if the change is small, such as adding or modifying clauses in the e-contract and it doesn't modify the structure of the e-contract.
 ii Migrate: This is used to address the new requirements related to the existing e-contract by instantiating a new model, such as a subcontract.
iii Merge: This merges the current model with the new instantiated model.
 iv Build: This is used when there is a need to change the structure of the e-contract model by adding new concepts [8].

The analysis of the ER^{EC} structure and behavior provides important insights. However, the conceptual ER^{EC} structure and behavior implementation requires technological support. This analysis is further extended to include the technology perspective. The ER^{EC} specifies software components that implement the e-contract metamodel. A relational database is one such component that is used to store the structure of the e-contract, rules and workflows. Enterprise JavaBeans (EJB) components are used to support the presentation and business logic layers - similar to the model-view-controller architecture design patter (a.k.a. MVC). The web service server is implemented to allow interaction among inter-organizational parties. Finally, the E-ADOME Workflow Engine is used to support workflow management [10]. It is clear from the analysis that the ER^{EC} is not simply a conceptual metamodel, rather it specifies technological support, which highlights its practical applicability and usability.

3.2   Three-Layered E-Contract Enforcement Metamodel

E-contract enforcement architecture is composed of three layers: document layer, business layer and implementation layer. The document layer describes the clauses of the e-contract. The business layer or e-contract enforcement layer presents the business rules in the *event-condition-action* form. Finally, the implementation layer refers to the supported technologies that are used to implement the web services for e-contract enforcement. Furthermore, document and business layers are represented by the e-contract template and actual metamodel, respectively. The analysis of e-contract enforcement indicates that the dynamic

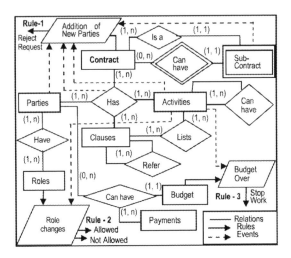

Fig. 1. ER^{EC} metamodel for the e-contract (Sourced from [8])

business interactions are not addressed in the e-contract clauses and there is a need to implement e-contract enforcement in order to perform monitoring process for the e-contract clauses [6]. In this section we analyse the architecture structure, behavior and technology of the e-contract enforcement metamodel.

E-contract enforcement architecture is organised into three layers. The document layer is the e-contract metamodel which comprises the e-contract template entity. The e-contract template is composed of a list of fixed entities and linked to the e-contract entity to refine the proposed template variables in the contract clauses. The e-contract entity also involves two or more parties which is related to two more entities: template variable and accepted value. A template variable is a set of variables that need to be refined during negotiation between the parties. Negotiation is an important concept of adaptive services. The result of the negotiation process will define the mutually agreed values, which can be transformed into the accepted values. The e-contract template, shown in Fig. 2, is composed of a set of contract clauses. Similar to the ER^{EC}, contract clauses can refer to each other and are nested clauses. Every contract clause has zero to many

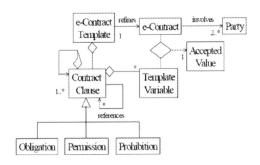

Fig. 2. A metamodel of an E-Contract Template in UML (Sourced from [6])

template variables. A contract clause is divided into three different types: obligation, prohibition and permission. Obligation describes the clauses that must be fulfilled (must do clauses), and prohibition presents the action that must not be performed. Finally, permission clauses depict permissible actions or behaviors [6]. The e-contract enforcement metamodel is shown in Fig. 3. It comprises the contract clause, which is related to the rules that enforce the clauses. A rule entity is related to event, condition and action. An event has three sub-types, which are business action invocation, exception and temporal event. There are also two types of events: external and internal events. An event triggers the rule. The rule evaluates the condition, which is a logical expression based on a business entity. If the condition is satisfied, then the enforcement action is executed. Typically, an action is structurally recursive; therefore, it has sub-actions and tasks. The remaining entity in a meta-model is a party which owns the business entity and communicates with the event to publish and subscribe to the events. The behavior of the contract enforcement depends on the ECA rules. The model presents the business action invocation object, which can be related to the temporal events (e.g. deadline or contract expiry date). For example, in the obligation enforcement clauses, it is obligatory to fulfil a certain clause in a specific time frame. If the condition is not satisfied by that time, it will raise an exception. The UML modeling tool can be used to design the e-contract metamodel and models. The e-contract metamodel is then implemented by using web services technology.

3.3 Secure E-Contract Metamodel

This section analyses the secure e-contract metamodel, which implements the security concerns of e-contract. Due to the nature of e-contract communication across organizations over the Internet, there is a need to address the contract security concerns of the involved contract parties such as authentication, integrity and non-repudiation. The appropriate handling of these concerns will help prevent the e-contract from being tampered or altered. The objective of the secure e-contract metamodel is to enhance the security and governance of the

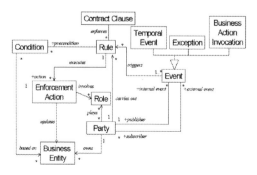

Fig. 3. A metamodel of contract enforcement in UML (Sourced from [6])

e-contract [11]. In this section, we analyse the secure e-contract metamodel from structure, behavior and technology perspectives. The secure e-contract meta-model structure is organised into seven entities. The core entity of the secure e-contract metamodel is the contract. Figure 4 shows contract attributes, such as contract duration and signature status. It also shows several relations to other entities. A secure contract has four sub-types: clause, party, exchange value and digital signature. A clause entity can refer to different clauses. A contract must have at least one clause. A party is one of the essential entities in the contract. A contract must have at least two parties. A party is linked to the digital signatures [11]. A party can have many digital signatures related to one or more contracts; however every digital signature is unique. Furthermore, a digital signature can refer to only one contract; however the contract may contain many signatures. The last sub-type of contract is the exchange value which represents the total cost and currency of the contract. The exchange value has a sub-type called service. A contract must have one and only one exchange value, and the exchange value can contain one or more services. The last entity in the secure e-contract metamodel is the activity, which is a sub-type of clause where every clause must have one or more activities [11]. The secure e-contract metamodel authors propose the ECRC web server to manage the e-contract formation. The ECRC system acts as a third party organization to offer services such as storing e-contract templates, downloading e-contract templates, querying existing contract templates and uploading new valid contracts. The process of creating an e-contract between the parties is as follows: party A (initiator) downloads an e-contract template that suits their requirements; the party updates the content, signs and sends the draft contract to the other party B. Party B checks the content and confirms the signature, then signs the contract. The two parties send a

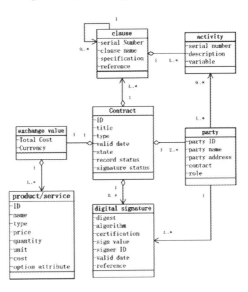

Fig. 4. UML model for secure e-contract metamodel (Sourced from [11])

Table 1. Review and comparison of e-contract metamodels

	ER^{EC} metamodel	Three-layered e-contract	Secure e-contract
Structure	Core entities: Party, activities and clause	Core entities: Clause, party and accepted value	Core entities: Clause, party, activity, exchange value and digital signature
	Secondary entities: Financial entities, roles and subcontract	Secondary entities: Event, rule, condition and enforcement action	
Behavior	Activities are mapped into contract workflows	Using e-contract template to create customized contracts	ECRC is a management system that shows the steps to create a secure e-contract from templates
	Tasks in activity can be committed or rolled back. Monitoring through log files	Event-Condition-Action rules are used for monitoring	
Technology	Enterprise JavaBeans	Web services	ECRC system
	Web services E-ADOME for workflows	UML	UML

copy of the valid contract to ECRC to save it [11]. The secure e-contract can be modeled using the UML modeling tool as shown in Fig. 4 below.

4   Discussion

In the previous section, we analysed three well-known e-contract metamodels. In this section, we analyse, compare and discuss the results, which are summarised in Table 1. The objective of this review and comparison is to understand and observe the similarities and differences between the e-contract metamodels from structure, behavior and technology perspectives and also to identify opportunities for their merger and interoperability for an adaptive service contract.

4.1   Structure

The review of the metamodels structure indicates that all three metamodels have similar entities or concepts. Additionally, these entities have similar semantics.

However, sometimes they have different names or spellings. We observe that the similar entities between all the metamodels are *'parties'* or *'party'* and *'clauses'* or *'clause'*. Furthermore, the *'activities'* or *'activity'* entity exist in the ER^{EC} metamodel and secure e-contract model, however it is not explicitly mentioned in the three-layered e-contract model. The three-layered e-contract model is divided into two metamodels which are: *'e-contract template'* and *'actual metamodel'* of contract enforcement. The event entity in the metamodel of contract enforcement has three subtypes which are: business action invocation, exception and temporal event. By applying the ontological analysis technique on these subtypes, we observed that business action invocation can be captured as an activity entity. Despite the similarities, Table 1 presents the differences between the contract metamodels. The ER^{EC} metamodel proposed financial entities such as payment and budget. Payment is the price of the service and budget is the maximum amount of money reserved for the service. Financial entities considered as an important concepts in the contract. However, surprisingly, they are not presented in the other metamodels of this study. Furthermore, the ER^{EC} metamodel presents the roles and subcontract entities; the former captures the roles of the parties and the latter describes the nested subcontracts as extensions of the main contract. A subcontract can be used by the service provider to delegate some tasks to the third party. On the other hand, the metamodel of contract enforcement in the three-layered e-contract presents several entities such as Event-Condition-Action, which are not modeled in the other metamodels reviewed in this study (see Table 1). Finally, the secured e-contract metamodel proposed a digital signature to ensure the integrity of the e-contract, which seems to be overlooked by the other two metamodels reviewed in this study. The structural review and comparison of the e-contract metamodels highlights the similar and different entities of the metamodels at a structural level. This review and comparison can be used as a baseline to further create a unified and richer structure of an e-contract for adaptive or agile services.

4.2   Behavior

This section discusses the behavioral similarities and differences among the three metamodels. The most common "behavioral" aspect of the metamodels is the template. The reviewed metamodels agree that a new e-contract can be derived from a ready-made template. This template contains the core concepts. The new e-contract derived from the template customizes the available entities and adds new concepts related to the e-contract instance. Despite the agreement between these metamodels for using the template as a basis for generating new e-contracts, the implementation is different in each metamodel. The ER^{EC} metamodel maps the contract template activities to workflows. Each activity contains at least one task and each task is able to commit or roll-back the operation. Also, it uses the log messages for monitoring purposes. However, the three-layered metamodel uses the ECA rules to monitor the actions of the e-contract. The secure e-contract also discusses the behavioral aspect through contract templates.

4.3   Technology

Finally, this section discusses the technological similarities and differences between the three metamodels. Table 1 shows that ER^{EC} and three-layered metamodels have used web services to access the operations of the e-contract interfaces. Also, three-layered and secure e-contract metamodels have used the ECRC system and UML modeling tool to design the e-contracts. Furthermore, ER^{EC} have used the enterprise JavaBeans for applications and E-ADOME for workflow management. All three metamodels discuss technological support, which is important to put the conceptual structure and behavior of the metamodels in practice.

In summary, the analysis and comparison of these metamodels indicate that they differ in defining some core concepts in contract. Further, it is noticed that surprisingly, the reviewed metamodels lack a theoretical underpinning and are not based on any theoretical or upper ontological model [12]. The similarities, differences and lack of a theoretical underpinning mark the need to create more comprehensive metamodels by incorporating elements from different metamodels and by using a solid foundation ontological theory. In order to address this issue in our current research, firstly, the core concepts of the future e-contract model will be collected from the three existing e-contract metamodels. These core concepts can be extended to contain other features of the different metamodels, such as the financial entities, security and exceptions. In the second stage, we will perform further analysis on the proposed concepts using the foundation ontology such as the Unified Foundational or upper-level ontology [12], the Descriptive Ontology for Linguistic and Cognitive Engineering (DOLCE) [13], the Basic Formal Ontology (BFO) [14] and others. The term upper ontology is defined as "a high-level, domain-independent ontology, providing a framework by which disparate systems may utilize a common knowledge base and from which more domain-specific ontologies may be derived" [15]. Of course, deriving the specific e-contract domain ontology based on the universal concepts of the foundation ontology will provide a strong theoretical under-pinning and accuracy in the context of correct modeling. Furthermore, the e-contract concepts will inherit important features such as interoperability. Therefore, the proposed e-contract metamodel will be composed of well-analysed core concepts and templates. Also, it is important to clarify that template metamodel is an abstraction of the real concepts that specifies the abstract contract concepts. The fundamental concepts and their relationships can be derived from universal objects of the foundation ontology. Hence, based on the analysis of this study, it can be observed that the existing major e-contract concepts are slightly different from each other and they lack from a theoretical and empirical research foundation. A foundational ontology-driven approach, as proposed here, is one systematic way of researching and creating a metamodel and related template for adaptive services. It is important to mention that this review paper is limited to only three e-contract metamodels. These metamodels are considered to be the most popular e-contract models in service contract research. In future, we will include more e-contract metamodels to further extend our studies in this important area of research.

5   Conclusion

This paper presents a comprehensive analysis, revision and cross-comparison of three well-known e-contract metamodels: ER^{EC} metamodel, three-layered meta-model and the secure e-contract metamodel. The analysis highlighted the structural, behavioral and supported technological similarities and differences of these three metamodels. Furthermore, this paper highlighted the lack of theoretical and empirical research underpinning of these metamodels, which is important for creating a comprehensive research-based metamodel. Hence, this study identifies the need for developing a unifying e-contract metamodel using the foundation ontology theory and elements from different e-contract metamodels. Also, it provides a platform for further theoretical and practical research in this important area of e-contract meta-modeling for adaptive services in the overall context of adaptive service systems. In the future, we will develop a new adaptive e-contract metamodel which contains the most appropriate entities in the reviewed metamodels in order to be considered as a standard e-contract.

References

1. Pourdehnad, J., Bharathy, G.K.: Systems thinking and its implications in organizational transformation. In: 3rd International Conference on Systems Thinking in Management, Philadelphia, PA (2004)
2. Gruhl, D., Bailey, J., Spohrer, J., Maglio, P.: Steps toward a science of service systems. Computer **1**, 71–77 (2007)
3. Gill, A.Q.: Applying agility and living service systems thinking to enterprise architecture. Int. J. Intell. Inf. Technol. (IJIIT) **10**(1), 1–15 (2014)
4. Martin, R.L.: The design of business: why design thinking is the next competitive advantage. Harvard Business Press, Boston (2009)
5. Klenk, A., Beck-Greinwald, A., Angst, H., Carle, G.: Iterative multi-party agreement negotiation for establishing collaborations. Serv. Oriented Comput. Appl. **6**(4), 321–335 (2012)
6. Chiu, D.K., Cheung, S.-C., Till, S.: A three-layer architecture for e-contract enforcement in an e-service environment. In: Proceedings of the 36th Annual Hawaii International Conference on System Sciences, 2003, pp. 10. IEEE (2003)
7. Corbin, J., Strauss, A.: Basics of qualitative research: Techniques and procedures for developing grounded theory. Sage Publications, Thousand Oaks (2014)
8. Krishna, P.R., Karlapalem, K.: A methodology for evolving e-contracts using templates. IEEE Trans. Serv. Comput. **6**(4), 497–510 (2013)
9. Karlapalem, K., Dani, A.R., Krishna, P.R.: A frame work for modeling electronic contracts. In: Conceptual ModelingâĂŤER 2001, Springer, Heidelberg, pp. 193–207 (2001)
10. Krishna, P.R., Karlapalem, K., Chiu, D.K.: An erec framework for e-contract modeling, enactment and monitoring. Data Knowl. Eng. **51**(1), 31–58 (2004)
11. Yu, X., Chai, Y., Liu, Y.: A secure model for electronic contract enactment, monitoring and management. In: Second International Symposium on Electronic Commerce and Security, 2009 ISECS 2009, vol. 1, pp. 296–300. IEEE (2009)
12. Guizzardi, G., Wagner, G.: Towards ontological foundations for agent modelling concepts using the unified fundational ontology (UFO). Springer, Heidelberg (2005)

13. Gangemi, A., Guarino, N., Masolo, C., Oltramari, A., Schneider, L.: Sweetening ontologies with DOLCE. Knowledge Engineering and Knowledge Management: Ontologies and the Semantic Web. LNCS, vol. 2473. Springer, Heidelberg (2002)
14. Guarino, N.: Formal ontology in information systems. In: Proceedings of the first international conference (FOIS 1998), June 6–8, Trento, Italy, vol. 46, IOS press (1998)
15. Semy, S.K., Pulvermacher, M.K., Obrst, L.J.: Toward the use of an upper ontology for us government and us military domains: An evaluation. Technical Report, DTIC Document (2004)

A Comparative Analysis of Scalable and Context-Aware Trust Management Approaches for Internet of Things

Mohammad Dahman Alshehri[1,2(✉)] and Farookh Khadeer Hussain[2]

[1] College of Computers and Information Technology,
Taif University, Taif, Saudi Arabia
Mohammad.D.Alshehri@student.uts.edu.au
[2] School of Software, Centre for Quantum Computation and Intelligent Systems –
QCIS, Decision Systems and e-Service Intelligence Laboratory - DeSI, University
of Technology Sydney, Sydney, NSW, Australia
Farookh.Hussain@uts.edu.au

Abstract. The Internet of Things – IoT – is a new paradigm in technology that allows most physical 'things' to contact each other. Trust between IoT devices is a critical factor. Trust in the IoT environment can be modeled using various approaches, such as confidence level and reputation parameters. Furthermore, trust is an important element in engineering reliable and scalable networks. In this paper, we survey scalable and context-aware trust management for IoT from three perspectives. First, we present an overview of the IoT and the importance of trust in relation to it, and then we provide an in-depth trust/reliable management protocol for the IoT and evaluate comparable trust management protocols. We also investigate a scalable solution for trust management in the IoT and provide a comparative evaluation of existing trust solutions. We then present a context-aware assessment for the IoT and compare the different trust solutions. Lastly, we give a full comparative analysis of trust/reliability management in the IoT. Our results are drawn from this comparative analysis, and directions for future research are outlined.

Keywords: Internet of things · Trust management · Trust protocol · IoT scalable · Context-aware assessment

1 Introduction

The Internet of Things (IoT) refers to a system of physical components or 'things' integrated with hardware, software, programming, sensors and a network to enable it to offer effective and efficient value in service through information sharing with manufacturers, users and/or connected devices [19]. Each component in the IoT can be uniquely identified but also has the capacity to inter-operate within the underlying internet infrastructure [16]. Scholars such as [15] use the expression 'Web of Things' to refer to the IoT. According to [14], the term was initially recorded in 1999 by Kevin Ashton, a British visionary. Typically, the IoT is relied upon to offer better integration and connectivity of devices, network infrastructures, systems, and services that span

© Springer International Publishing Switzerland 2015
S. Arik et al. (Eds.): ICONIP 2015, Part IV, LNCS 9492, pp. 596–605, 2015.
DOI: 10.1007/978-3-319-26561-2_70

connectivity beyond machine-to-machine links, and it is spread across a mixed collection of conventions or protocols, applications, and domains. The interconnection between these embedded components, coupled with the growing addition of intelligence, or 'smartness' in devices, is expected to introduce automation in almost all areas, while at the same time facilitating better applications such as Smart Grid [7].

Since 2014, the development of the IoT has rapidly grown because of the convergence of a range of technological advances, including remote connectivity via fault-tolerant networks, wireless communication, embedded systems, and micro-electro-mechanical systems [20]. This implies that the conventional areas of automation, remote sensors, control systems, embedded systems, and augmented reality all contribute to empowering the IoT. The idea of a system of intelligent devices has been discussed since the 1980s, when a Coca-Cola (Coke) machine was developed at Carnegie Mellon University which reported on its stock and the state of coldness of recently stacked beverages [4]. Mark Weiser's fundamental 1991 paper about computing anywhere, anytime in 1991, titled, 'The Computer of the 21st Century', gave rise to the expression 'ubiquitous computing' and is another milestone in the IoT.

Scholarly venues, for example, UbiComp, PerCom and IEEE Spectrum, created the modern concept of the IoT [19]. This concept was further galvanized in 1994 with conceptualization of 'moving little data packets to a huge collection of hubs', in order to incorporate and computerize everything ranging from personal, home and business appliances to complete factory operations [13]. In the period 1993–1996, organizations like Novell proposed such solutions as the Novell Embedded Systems Technology (NEST). In 1999, the field started to gain momentum with MIT's Auto-ID Center and related corporate sector publications [6].

In the IoT, 'things' include, but are not limited to: wearable devices such as heart monitoring tools, biochip transponders implanted in animals, cars with in-built sensors, electric clams used in coastal water areas, field operation equipment for rescue purposes, radio-frequency identification (RFID) applications, and surveillance devices. [6] argues that these devices are used to gather valuable information with the assistance of different innovations. The devices stream the information across other devices in their individual autonomous capacity. Current commercial IoT applications include: intelligent indoor regulator systems, health-oriented wearable devices to screen body temperature, heart rate and other wellbeing issues, spying devices, and home appliances that use Wi-Fi for remote operation and monitoring.

As with the plethora of new applications areas for internet-based automation to venture into, the IoT is likewise expected to create huge chunks of information that is rapidly amassed from disparate areas. As such, there is an increasing need to advance indexing, storage and processing capacity to derive value from the massively growing body of information [10]. [9] stated that the IoT has and will continue to expose people to privacy issues, especially with the 'big data' concept. As such, the IoT may erode the control we have over our own lives as corporations and governments try to amass huge volumes of data and endeavor to maximize financial advantage and control [11]. The importance of trust management in IoT will enable an IoT node to make reliable and context-aware assessments about its interacting partner.

In this paper we will focus on the existing work on trust management in the Internet of Things (IoT), with a view to identifying key shortcomings in this field. This will identify gaps in current state-of-the-art practice to facilitate the realization of a trust-worthy IoT network. We use the following classification to categorize the existing approaches:

- Trust/Reliable management protocol for IoT
- Scalable Solution for trust management in IoT
- Context-aware trust assessment in IoT Networks

The paper is organized as follows: Sect. 2 outlines the trust/reliable management protocol for IoT, and Sect. 3 presents s scalable solution for trust management of the IoT. Context-aware trust assessment in IoT networks is discussed in Sects. 4 and 5 offers a comparative analysis of trust/reliability management approaches in IoT and discussions.

2   Trust/Reliable Management Protocol for IoT

An IoT network includes a huge number of day-to-day life devices operating in heterogeneous networks, which creates a serious problem with regard to reliability and security management, notwithstanding which, all the elements of an IoT system need to inter-operate agreeably [17]. Reliability can be compromised by the failure to uphold acceptable levels of security, which exposes the system to attacks. Devices in the IoT framework are regularly open to the public and communicate wirelessly, thus creating vulnerability to breaches of security. Conventional approaches to trust protocol, network, system and data security, information management, identity administration, and fault tolerance and governance cannot accommodate modern IoT constraints because of the scalability, data explosion and high diversity of identity types [5]. Therefore, the types of relationship between devices in IoT environments are more complicated than ever before.

A trust management protocol for IoT frameworks was proposed in [2] that has two main goals: to provide an exact and flexible trust evaluation of the trust levels of IoT components, and to use the proposed protocol in different IoT applications to optimize application performance. The trust management protocol models a community-oriented social IoT setting by working with many social relationships across device owners. [2] they claim that social trust is clearly expected in such an environment. The system does not have a specialized trusted authority, but instead spreads the role of trust evaluation to individual nodes.

The underlying principle of the protocol rests in managing nodes in the IoT system to prevent them from misbehaving and to prevent malicious nodes from breaking into its primary functionality to launch trust-related attacks such as bad-mouthing. It considers an IoT framework that is being implemented in an intelligent group where every node self-sufficiently performs trust assessment. The authors give a formal consideration of the convergence, versatility, and accuracy properties of their trust management protocol.

A fuzzy-oriented trust management protocol was proposed by [3] for use in the IoT system that consists of wireless sensors only. The protocol uses Quality of Service (QoS) trust parameters such as energy utilization and packet transfer to delivery ratio.

Sensors may create direct communication links between themselves using the IPv6 over the Low the Power Wireless Personal Area Networks (6LoWPAN) protocol, a protocol used for IPv6 networking in devices with low data rates and low power radio transmission. A reputation and trust framework is perceived to be a critical means of preventing malicious nodes from accessing vast sensor IoT networks, because trust creation instruments can empower a coordinated effort across distributed things, support the discovery of malicious components, and facilitate the decision making process.

An Energy-efficient Protocol of Reliable Trust-based Data Aggregation (ERTDA) protocol was proposed in [5]. The objective of this protocol is to reduce the nodes' energy consumption using an effective routing and recovery approach. Path selection is also used to realize security and reliability in data segregation. The protocol ensures that security is upheld in data capturing, processing and sharing, in addition to identifying mutual trust relationships between nodes and excluding compromised components from the IoT network. This is achieved in three steps as follows: In step 1, every group of aggregated nodes should have its security guaranteed, and have adequate energy to support aggregation and data sharing; In step 2, link availability is ascertained based on the energy in neighboring nodes; and in the final step the importance of the outcome of data aggregation to allow selection of multiple paths is highlighted (Table 1).

Table 1. Overview and comparative evaluation of trust/reliability management protocol for the IoT

Trust management protocol for IoT approach	Description of the approach	Features of the approach	Issues/lacking of the approach
Trust management approach - Dynamic trust management Protocol (DTMP)	A distributed protocol based on a social IoT environment to model trust evaluation between nodes.	Trust is a factor of honesty, community-consciousness and cooperativeness; covers both encounter-based and activity-based incidents; trust evaluation is based on personal experience and recommendations from other common nodes.	A node can only manage its trust assessment to a limited collection of nodes, and thus cannot support trust management for large-scale IoT networks.
Trust, reputation and scalable management approach –(TRM-IoT)	A fuzzy-based trust and reputation management protocol for use in the IoT system consisting of strictly wireless sensors.	Considers a balance between battery drain and security guarantee; trust and reputation elements are derived from direct observation and recommendations; is meant for wireless sensor networks; trust metrics include: successful packet delivery and energy utilization.	Supports wireless sensor IoT networks only; devices with low data rates and/or low power radio transmissions may constrain coordination of trust evaluation across nodes.
Trust and reliability management approach – (ERTDA)	A trust and reliability evaluation protocol that relies on the observations of the cooperation between IoT nodes to enhance understanding of their behavior and detect incidents of compromised nodes.	Optimized routing to reduce energy consumption.	Computation complexity may arise in the course of election of parent node and intense routing.

The next section, we will focus on the scalable solution for trust management in IoT.

3 Scalable Solution for Trust Management in the IoT

As an IoT network connects a huge number of devices and applications, there is an increased challenge with respect to meeting the demands of scalability, dynamic adaptability and compatibility. [1] notes that the IoT assists applications such as continuous e-health and smart product management by capturing, processing and sharing data, which necessitates the use of effective trust management protocols to manage trust between different IoT entities. However, [11] argues that trust management is constrained by the vast quantity of IoT entities, which challenge scalability with respect to accommodating the growing number of computational and storage entities. In addition, IoT networks should evolve to adapt to nodes that are joining and leaving, while building up trust rapidly and accurately. This implies that trust management protocols for IoT networks should be highly resilient to trust-based attacks to endure security issues in hostile environments. According to [1], scalability should be a key consideration in the design of trust/reliable management protocols for IoT. In other works, the trust management protocols proposed by [2, 5, 8, 12, 18] did not address scalability, undermining their applicability in large-scale IoT networks. Therefore, it is important to consider trust management protocols that have been designed to address the scalability challenge. We now outline and discuss the working of each of these methods.

Firstly, [1] proposed the Scalable, Adaptive and Survivable Trust Management for Community of Interest (CoI) based IoT, recognizing that nodes in IoT networks are owned by individuals and interconnected by social networks. To achieve scalability, they designed a protocol whereby each node can store the trust relationship data of a set of nodes within its CoI, thus enhancing convergence. Nodes can dynamically join or leave while rapidly building up trust towards others due to the increased convergence in the CoI framework and enhanced survivability. Storage is optimized to ensure there is effective utilization of the constrained storage space and make it suitable for large-scale application.

Secondly, [11] proposed an IoT protocol framework for RFID-based devices - the Scalable RFID Security Framework and Protocol Supporting IoT (SRSFPSI). They noted that RFID frameworks should be installed with a comprehensive security structure for a secure, yet scalable operation. The proposal entails an effective ID procedure founded on a hybrid framework (group-based and collaborative technique) and highly adaptive security monitoring handoff for RFID IoT networks. The protocol offers adaptability and scalability while upholding secure and adaptable RFID networks. Other than preventing the introduction of malicious nodes and facilitating scalability, the protocol is integrated with a malware recognition tool.

Thirdly, [17] argued that trust management is a vital step in securing WSN and IoT environments characterized by frequent encounters with unknown agents. They proposed a scalable protocol for an IoT framework that is founded on existing IoT principles of trust management and reputation at semantic and data management levels. To establish tangible levels of scalability, there is no central database, which promotes global knowledge sharing as a means of evaluating earlier interactions. The approach scales well to meet the trust management demands of large sets of nodes, a feat achieved due to the implementation of completely IoT decentralized IoT systems (Table 2).

Table 2. Overview and comparative evaluation of scalable solution for trust management in the IoT

Scalable solution for trust management in IoT	Description of the approach	Features of the approach	Issues/lacking of the approach
Trust and adaptive scalable management approach - Scalable, Adaptive and Survivable Trust Management for Community of Interest (CoI) Based IoT.	A distributed, dynamic and scalable trust management IoT protocol based on CoI and storage management approach to extend the functionality of DTMP.	Distributed IoT protocol; trust relationships are evaluated on nodes within a CoI subset; uses a storage management approach to enhance scalability.	Recommendations may be biased if they are from nodes residing in different CoIs, especially in instances where minimal interactions have previously existed.
Trust and scalable management approach - SRSFPSI	A scalable trust framework for a highly mobile RFID-based IoT network.	Applicable in RFID IoT networks; incorporates malware detection capacity; ensures scalable implementation of RFID nodes for a distributed IoT.	Designed for RFID IoT networks only.

In the next section, we will investigate the context-aware assessment for IoT.

4 Context-Aware Trust Assessment in IoT Networks

In IoT networks, context awareness is the capacity to use environmental and situational data to predict instantaneous needs and offer relevant proactive responses [8]. IoT consists of the following technologies: embedded sensors, smart mobile devices, cloud computing, and big data analytics, which work collaboratively to collect, model and guide users. Modern computers, networks and, in this respect, the Internet, are completely dependent on people for data. The greater percentage of the approximately 50 Terabytes of information accessible on the Internet is a result of human effort such as typing, recording, taking digital pictures, or scanning [6]. The challenge lies in the fact that humans are constrained by time, accuracy, memory, and attention, implying that they are relatively poor at capturing information about real world things [2]. With a fully-functioning IoT, we would leverage information about all things, tracking and checking everything and significantly reducing waste and cost. In addition, it would be possible to identify things that require replacement, repair, review, or that are obsolete [12].

Trust management protocols were proposed by [1, 2, 5, 17] that did not address the context awareness issue. [9] argues that stakeholders in the IoT area of mobile, wearable and ubiquitous computing have recognized the need to secede from the conventional desktop model as more and more devices become mobile. As such, all services should be extended and enhanced to adapt to constantly changing contexts, but this complicates the implementation of trust management protocols in the IoT. [12] claimed that developing context-aware enabling technologies requires a well-defined security framework for IoT networks, whereby nodes are secure despite cutting across different settings – transportation, home, office and others. According to [8], network reactions in relation to user mobility and settings should be adjusted to meet different needs though real-time learning and monitoring to bolster precision.

The Context Awareness for Internet of Things (CA4IOT) framework proposed by [8] is based on automated filtering, synthesis, saving and reasoning in the realm of sensor data collection and the creation of meaningful information from raw data. The framework understands and maintains context data about sensors (such as location, nearby sensor, battery life and sampling rate) using appropriate annotations for quick retrieval. Relationships within different domains are learned from knowledge bases that amass information. The CA4IOT framework follows a layered architecture consisting of: the user – the device owner, application or service, user management, processing, reasoning, context discovery, data acquisition, and sensing

In a work by [18], it is apparent that the future of wireless systems is expected to be highly context-aware, to boost user experiences through personalized services. However, the area of context awareness is constrained by trust and security issues. The Context Broker Architecture (CoBrA) is a framework that facilitates the identification, acquisition, reasoning and presentation of context information. Additionally, it consists of privacy protection mechanisms. The fundamental assumption in CoBrA is that all context-oriented information providers (sensors) have past knowledge (stored in the database) about the presence of context brokers.

A context-aware trust management system was proposed by [12] for the IoT (CTMS4IOT) which adds an element of adaptability to meet the needs of today's dynamic IoT networks. The proposed model entails the following phases: information gathering, entity selection, transaction, reward and punish, and learning. For trust management, the approach uses past behavior and allows for fine-tuning to overcome challenges brought about by malicious nodes. It uses centralized trust management servers and prioritizes the context where evaluations are captured; therefore, appropriate trust management servers return context information with trustworthy values for each node (Table 3).

Table 3. Overview and comparative evaluation of context-aware trust assessments

Context-aware trust assessment in IoT approaches	Description of the approach	Features of the approach	Issues/lacking of the approach
Trust context-based – (CA4IOT)	A framework based on automated filtering, synthesis, saving and reasoning in sensor data collection and reasoning to derive valuable information.	Supports learning by understanding and maintaining context data in knowledge bases; uses appropriate annotations for quick retrieval; follows a layered architecture.	Relies on a dedicated server to facilitate knowledge sharing, thus is subject to a single point of failure which may challenge trust management; poor in scaling.
Ontology – (CoBrA)	A context-aware framework that relies heavily on a context broker to capture contextual information from disparate sources and integrate it into a unified model for sharing across computing devices in the IoT network.	The context broker is the fundamental component that maintains a context information sharing model for devices, agents, and services in IoT; uses ontology to model contexts, and supports privacy protection	In a dynamic environment, the assumption that information about context brokers is well-known in advance can lead to poor implementations that are incapable of handling inconsistent contexts; poor in scaling.
Trust context-based – (CTMS4IOT)	A context-aware distributed trust management system designed to address trust issues based on contextual information and learning.	Its operation is divided into five phases. Allows for fine tuning to meet disparate contextual constraints; modeled on a centralized server setting; support for learning.	Use of centralized trust management servers constrains scalability.

The next section evaluates the comparative analysis of trust/reliability management for the Internet of Things.

5 Comparative Analysis of Trust/Reliability Management Approaches in the IoT and Discussions

Table 4 presents a comparative analysis of trust/reliability management protocols in the IoT to measure the extent to which each protocol meets scalability and context-aware needs. Validation for compliance with trust and reliability considers both scalability and context-awareness. In the table:

- ✕Implies that a trust/reliability management protocol is neither scalable nor context-aware.
- ✓Implies that a trust/reliability management protocol is both scalable and context-aware.
- ✔Implies that a trust/reliability management protocol meets either the threshold for scalability or context-awareness, but not both.

Table 4. Comparative analysis of trust/reliability management in the IoT

Approach	Protocol/ Mechanism	Scalable	Context-Aware	Validation	Research Paper
Trust-based	DTMP	✕	✕	✕	[2]
Trust scalable and context-aware	TRM-IoT	✕	✕	✓	[3]
Trust-based	ERTDA	✕	✕	✕	[5]
Trust and scalable	Scalable, Adaptive and Survivable Trust Management for Community of Interest (CoI) Based IoT	✓	✕	✔	[1]
Trust scalable and context-aware	SRSFPSI	✓	✕	✓	[11]
Trust and scalable	IoT trust framework	✓	✕	✔	[17]
Trust and context-aware	CA4IOT	✕	✓	✔	[8]
Ontology	CoBrA	✕	✓	✔	[18]
Trust and context-aware	CTMS4IOT	✕	✓	✔	[12]

It is clear from the above comparisons that none of the existing methods for trust modeling in IoT combine the features of scalability and context-aware trust assessment, and validate the working of the proposed approaches. Hence, we can argue that there is a need for research to develop trust management methods that can scale to accommodate billions of IoT nodes and enable trustworthy assessments of IoT nodes.

6 Conclusion

This paper evaluates the existing approaches to trust management in the Internet of Things based on three parameters. The first parameter focuses on trust management protocol in IoT, the second parameter concerns scalable solutions for trust management in IoT, and the third parameter addresses context-aware assessment in IoT. We have given a comparative evaluation of each existing approach for trust modeling in IoT, based on these parameters. Further research into trust management in IoT is required to develop scalable and context-aware trust solutions in IoT networks, actually in the future we plan to focus to tackle that in our works.

References

1. Bao, F., Chen, I.-R., Guo, J.: Scalable, adaptive and survivable trust management for community of interest based Internet of things systems. In: 2013 IEEE Eleventh International Symposium on Autonomous Decentralized Systems (ISADS), pp. 1–7 (2013)
2. Bao, F., Chen, I.-R.: Dynamic trust management for internet of things applications. In: Proceedings of the 2012 International Workshop on Self-aware Internet of Things, pp. 1–6 (2012)
3. Chen, D., Chang, G., Sun, D., Li, J., Jia, J., Wang, X.: TRM-IoT: a trust management model based on fuzzy reputation for internet of things. Comput. Sci. Inf. Syst. **8**(4), 1207–1228 (2011)
4. Hersent, O., Boswarthick, D., Elloumi, O.: Z-wave. In: The Internet of Things: Key Applications and Protocols. John Wiley & Sons Ltd, Chichester (2011)
5. Ma, T., Liu, Y., Zhang, Z.: An energy-efficient reliable trust-based data aggregation protocol for wireless sensor networks. Int. J. Control Autom. **8**(3), 305–318 (2015)
6. Ning, H.: Unit and Ubiquitous Internet of Things. CRC Press, Boca Raton (2013)
7. Nixon, P., Terzis, S.: Trust Management, vol. 2692. Springer-Verlag, Berlin-Heidelberg (2003)
8. Perera, C., Zaslavsky, A., Christen, P., Georgakopoulos, D.: CA4IOT: context awareness for internet of things. In: Proceedings of the 2012 IEEE International Conference on Green Computing and Communications, pp. 775–782 (2012)
9. Perera, C., Zaslavsky, A., Christen, P., Georgakopoulos, D.: Context aware computing for the internet of things: a survey. IEEE Commun. Surv. Tutor. **16**(1), 414–454 (2014)
10. Pfister, C.: Getting Started with the Internet of Things. O'Reilly Media, Sebastopol (2011)
11. Ray, B.R., Abawajy, J., Chowdhury, M.: Scalable RFID security framework and protocol supporting internet of things. Comput. Netw. **67**, 89–103 (2014)

12. Ben Saied, Y., Olivereau, A., Zeghlache, D., Laurent, M.: Trust management system design for the Internet of things: a context-aware and multi-service approach. Comput. Secur. **39**, 351–365 (2013)
13. Sicari, S., Rizzardi, A., Grieco, L.A., Coen-Porisini, A.: Security, privacy and trust in internet of things: the road ahead. Comput. Netw. **76**, 146–164 (2015)
14. Singh, M.P.: The Practical Handbook of Internet Computing. CRC Press, Boca Raton (2004)
15. Tselentis, G., Domingue, J., Galis, A.: Towards the Future Internet: A European Research Perspective. IOS Press, Amsterdam (2009)
16. Uckelmann, D., Harrison, M., Michahelles, F.: Architecting the Internet of Things. Springer, Heidelberg (2011)
17. Wang, J., Bin, S.: Distributed trust management mechanism for the internet of things. In: Proceedings of the 2nd International Conference on Computer Science and Electronics Engineering (ICCSEE 2013), pp. 2197–2200 (2013)
18. Wrona, K., Gomez, L.: Context-aware security and secure context-awareness in ubiquitous computing environments. In: Proceedings of the XXI Autumn Meeting of Polish Information Processing Society, pp. 255–265 (2005)
19. Yan, Z., Zhang, P., Vasilakos, A.V.: A survey on trust management for internet of things. J. Netw. Comput. Appl. **42**, 120–134 (2014)
20. Zanella, A., Bui, N., Castellani, A., Vangelista, L., Zorzi, M.: Internet of things for smart cities. IEEE Internet Things J. **1**(1), 21–32 (2014)

Societal Intelligence – A New Perspective for Highly Intelligent Systems

László Barna Iantovics[1]([⊠]), László Szilágyi[2,3], and Camelia-M. Pintea[4]

[1] Department of Informatics, Petru Maior University of Tîrgu-Mureş,
Tîrgu-Mureş, Romania
ibarna@science.upm.ro
[2] Budapest University of Technology and Economics, Budapest, Hungary
[3] Sapientia University of Transylvania, Tîrgu-Mureş, Romania
[4] Technical University of Cluj-Napoca, Cluj-Napoca, Romania

Abstract. A novel concept of intelligence called "societal intelligence" and its related architecture for solving complex problems are introduced. The idea is based on what we consider on the "intelligence of human society". For illustrative purposes, a case study is realized, which involves the solution of a difficult problem in a societal multi-agent system where the agents operate in an unknown environment. In the simulated robotic mobile multi-agent system, the agents adapt their movement control in the environment based on some global knowledge constructed by the system. Besides the proposed architecture, a novelty presented in the paper is the demonstration that even in a simplified knowledge-based multi-agent system, if the principles of societal intelligence are followed, a powerful global intelligence emerges at the system's level.

Keywords: Hybrid knowledge base · Collective intelligence · Neural network · Learning · Cognitive multi-agent system · Complex system

1 Introduction

In this paper we present some novel concepts of intelligence as a property of a knowledge-based multi-agent system (MAS) - that allow the solving of complex problems. Specifically, we introduce a type of intelligence namely "societal intelligence" inspired by the "emergent intelligence of the human society". Furthermore, based on this concept, we introduce a novel MAS architecture. The overall concept of intelligence is supplemented by other related concepts encompassing different types of intelligence: "individual intelligence", "individual influenced intelligence", "global intelligence base", "emergent global intelligence" and "environment intelligence". We would like to note here that the term of societal intelligence has already been used in some studies with other meanings.

The paper is organized as follows. Section 2 presents several considerations related to the knowledge-based intelligent systems (KBIS); The newly proposed societal intelligence and its architecture are described in Sect. 3; Sect. 4 shows the realized case study, its results, and a statistical analysis. In the last section, the conclusions of the research are outlined.

© Springer International Publishing Switzerland 2015
S. Arik et al. (Eds.): ICONIP 2015, Part IV, LNCS 9492, pp. 606–614, 2015.
DOI: 10.1007/978-3-319-26561-2_71

2   Intelligent Knowledge-Based Agents

One of the main purposes of most KBIS consists in attempting to obtain improvements in problems solving while comparing it against a system that is perceived not to have any intelligence, whatsoever. Increased intelligence is usually noted by being more efficient, flexible and accurate solution of difficult problems (such as the detection of clinically relevant inconsistencies [7], for example).

We would like to note here that an important aspect that should be treated, during the development of intelligent systems is: an analysis of the necessary intelligence. Sometimes, an increased intelligence can even have disadvantages. For example, if we were to consider an extremely intelligent agent as one that uses complex specializations for processing but it must solve very simple problems.

Some researches [9,10] focused on the study of decision making in the frame of cooperative coalitions, which many times outperform the decisions of individuals that operate in isolation. In many cooperative MAS, the intelligence could be considered at the level of the whole system. The intelligence in these systems is higher than the individual member agents intelligence. There are some developed systems composed of simple agents that as a whole could be considered intelligent [4,8]. The literature has not defined effective metrics that could give a quantitative evaluation to the collective intelligence.

Kun and Galis [5] present an intelligent mobile MAS composed from simple reactive agents (with knowledge retained as a set of rules) specialized in a computer network administration. The MAS simulates the behavior of a human network administrator, however, as a whole could be considered intelligent.

A novel self-adaptive MAS called ERMS that can solve problems using genetic algorithms has been proposed in [4]. The self-adaptability of ERMS consists in the capacity to autonomously reorganize its structure based on the pattern that respects the problems transmitted for solving. The reorganizations of the system are described in a rule base constructed using an evolutionary learning algorithm. The results prove that some MAS successfully can use evolutionary algorithms in order to discover emergent patterns of reorganization for efficient solving of the undertaken problems. In case of a complex network of agents, this reorganizing behavior could be associated with the intelligence.

3   A New Perspective for Highly Intelligent Multi-agent Systems

We consider the complexity handling of current scientific realizations of the human society and attaining the current stage of evolution, is the result of the so called societal intelligence of the human society.

3.1   Characterization of the Intelligence of Human Society

Each human has its own intelligence which allows problems solving, like: specialty problems (e.g. medical diagnosis problems) using specialty knowledge (e.g. medical diagnosis knowledge) and solution of everyday life problems using commonsense knowledge (e.g. how to keep a house clean).

A human is able to take decisions based on the: commonsense knowledge (impossible to attain by computing systems [1]); specialty knowledge, attainable by expert systems [2] and human intuition defined by the psychologist and psychiatrist C. Jung. The intuition is very specific to humans and is almost totally missing from even the systems perceived to be intelligent [3]. It allows decisions taking in difficult situations, in the context of missing or erroneous information.

The following main concepts are defined for the human society and extended to MAS in the next section.

Definition 1. The *societal intelligence* is the intelligence of society where there is a requirement of solving a particular set of complex problems.

Definition 2. The *individual intelligence* defines the capacity of an individual to solve different complex problems.

We consider the existence of distributed knowledge as the global intelligence base of a society.

Definition 3. The *global intelligence* base of the society encompasses the entire knowledge of a society used for problem-solving.

Definition 4. The *influenced intelligence* of individuals defines how they learn from the global intelligence base and to learn from other individuals.

The influenced intelligence has an increasing effect to the individual intelligence. A human have the possibility to contribute to the global intelligence of the society by adding new knowledge and modifying existent knowledge. Based on these reasons the knowledge of the society is changing continuously. However, the intelligence of a human could influence the intelligence of other humans.

Definition 5. The *environment intelligence* is the potential of the environment to give information to individuals in order to solve problems in a more intelligent way (accurate, efficient and/or flexible).

Definition 6. The *emergent global intelligence* of the society is the intelligence that emerges at the human society level.

The recent results in the scientific evolution of the society are usually obtained by the collective effort of many individuals. They are based on collective intelligence, expertise, effort of many humans using distributed resources for creation/development of very complex artifacts as space rockets, airplanes etc.

3.2   A Novel Societal Multi-agent System Architecture

In this section we introduce the general architecture of a societal MAS, denoted with $IMG = \{IAg_1, IAg_2, \ldots, IAg_n\}$ (see Fig. 1). The agents have the same goal to solve more efficiently the problems at the level of the whole system. They belong to the same society where each agent must undertake at least one

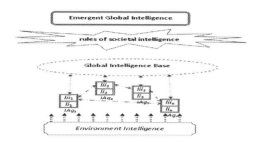

Fig. 1. The proposed societal multi-agent system architecture

role. The roles of an agent define the contribution of the agent to the problems solving in the frame of the MAS. To be able to undertake a role, an agent must have the necessary capacity (processing and memory) and capability (problem-solving specializations). If an agent undertakes a role it must fulfill all the commitments associated with that role (corresponding to the role it must be able to fulfill different functionalities in the frame of society). We denote with $Roles = \{rl_1, rl_2, \ldots, rl_m\}$ the existent roles in IMG. The notions society, role, commitment are already defined in [4].

Each agent have an "individual intelligence" (Definition 2) that defines its intelligence in problems-solving and contribution to the fulfilling of different functionalities of the MAS.

We denote with li_k the individual intelligence of the IAg_k agent, $li_k = locint(IAg_k)$. $li = [li_1, li_2, \ldots, li_n]$ denotes the individual intelligence of all agents from IMG.

The "global intelligence base" (Definition 3) consists in a specific collective intelligence formed at the level of the MAS where the agents belong. It contains some global data, information, knowledge and metaknowledge that govern the IMG system coherent, efficient and flexible operation. The existent knowledge in the global intelligence base may have different representations, like: rule base, neural network, semantic net etc. or combinations of these. The content of the global intelligence base is available to the agents, they must be able to read and sometimes to write data onto it. Parts of the global intelligence base are also detained by some of the agents from the system. There exists knowledge detained by agents that is not included in the global intelligence base.

Each agent has an "influenced intelligence" (Definition 4) defining how to use the environment intelligence, learning from other agents and from the global intelligence base. Using the influenced intelligence an agent may be able to increase its individual intelligence. Improvements given by the influenced intelligence may be given by learning, adaptation and/or evolution. Using its influenced intelligence an agent should be able to extend and sometimes modify the global intelligence base and also increase the individual intelligence of other agents. We denote with lii_k the influenced intelligence of the IAg_k agent, $lii_k = locinfint(IAg_k)$. lii_k may increase the individual intelligence li_k of IAg_k; $li_k = fincr(li_k, lii_k, inf_p)$; $fincr$ describes how the old individual intelligence and the influenced intelligence emerge in a new individual intelligence; inf_p is data/information and/or knowledge used for the new individual intelligence. The individual influenced intelligence of agents is denoted by $lii = [lii_1, lii_2, \ldots, lii_n]$.

We mention a difficult medical case, the diagnosis of a patient with co-morbidity when the necessity of the individual influenced intelligence is highly motivated. The physician in order to elaborate a diagnostic needs the assistance of an agent able to take decisions using its influenced intelligence by consulting some cognitive agents. The "intelligence of the environment" (Definition 5) sometimes chooses existent methods or combines more methods. An agent may be able or not to use emergent information/data offered by the environment.

To illustrate the concept of environment intelligence is considered the existence of ordering information related to a large set of numbers. If it is known that the numbers are ordered it can be applied the binary search method. The minimal "intelligence" - called intelligence for illustrative purposes only-in the environment (the information: the numbers are ordered), allows the improvement of the problem solving if the problems solver agent is able to understand it (understand: the numbers are ordered) and use this information (apply the most appropriate detained search technique).

The emergent global intelligence (Definition 6) is the problem-solving intelligence of the IMG system that emerges at the level of the whole system.

The societal intelligence in a societal MAS is defined by some specific rules that govern the organization and intelligence of the society of agents. Establishes how the global intelligence base is constructed and used by the individuals from the society. It defines the individuals personal intelligence and influenced intelligence (how they learn from each others, from the environment and from the global knowledge of the society). The application of the societal rules has as effect the emergence of a global intelligence at the society level.

4   Case Study of a Simple Societal Multi-agent System

Many times the intelligence of a system is considered based on capabilities like [4,6]: learning, self-adaptation and evolution. The existence of some of these capabilities does not implicitly allow a quantitative evaluation of the intelligence, they just prove its existence. We consider that the evaluation of a system's intelligence must be based on some metrics. Such metrics should be determined based on considerations that include problems solving point of view, like: difficult problems solving, capacity to treat different types of uncertainties etc.

Different studies related with robots are realized in [11]. We have implemented a simplified societal MAS. Notions related with different types of intelligence (and knowledge), like individual intelligence and individual influenced intelligence are used with the significance of intelligence just for illustrative purposes. We have simulated a two-dimensional virtual software environment where operates an n-element robotic mobile MAS denoted $SMG = \{RA_1 \ldots, RA_n\}$. Each of the agents has the same role to search for the same specified object. The agents using their effectors are able to move in the environment.

Using their sensors they are able to explore the environment during their movement and read-write information onto the global intelligence base. The environment map is not known to the agents. Each agent from the system uses the same method denoted $Eval$ for the objects evaluation: $Eval : Ob \rightarrow R+$. An object denoted Ob_k ($Ob_k \in Ob$; Ob denotes the objects' space) based on the similarities can be more or less appropriate with the value of the searched object. At each problem-solving cycle, SMG must solve a single problem denoted Pr of finding a specified object. SMG solve in a cooperative way the problem (find the object or fail). The individual intelligence of an RA_k agent consists in the followings:

- capacity to move in the environment;
- capacity to evaluate the objects in the environment;
- capacity to take into consideration the detained local history denoted $Best_k = (BestVal_k; Position_k)$; $BestVal_k$ denotes the best value found until that time; $Position_k$ denotes the position of the best value;
- Local history constructed during a problems solving.

The global intelligence base denoted GIB is dynamically constructed step-by-step by all the agents during a problem's solving. GIB contains the following data at each problem-solving cycle, $Global = (GlobalVal; GlobalPosition)$: $GlobalVal$ denotes the highest value found; $GlobalPosition$ denotes the global position of the best value found at system's level. The individual influenced intelligence of an agent RA_k consist in the capacity to take into consideration the best solution detained at a moment of time in the global intelligence base. Each agent RA_i is able to modify GIB when it finds a better solution. Each agent RA_i may take into consideration in a certain degree the personal best previously found during its history $PersBest_i = (BestVal_i; Position_i)$. Each agent may take into consideration to a certain degree the best location that any agent has ever found the global best $GlobalBest = (GlobalVal; GlobalPosition)$.

Cooperative Problem Solving presents the main steps of the SMG system operation at a problem's solving cycle. A problem-solving cycle begins when a problem is submitted to be solved and finishes when is obtained a solution or the running time expires. Gen denotes the admitted maximal time steps. It is not possible the theoretical determination of the necessary time, however,

```
Step 1.
@Creation of the environment and the distributed objects.
@Creation of the societal multi-agent system.
@Specification of the searched object.
Step 2.
while (Object not found) and (Gen not reached) do
 |  @Cooperative search for the object in the environment
end
Step 3.
@Obtained solution is reported.
```

Algorithm 1: Cooperative problem solving algorithm

Table 1. Results of 20 simulations

In	Table 1.1. $apb = 5$, $agb = 5$				Table 1.2. $apb = 8$, $agb = 8$			
	20	40	60	80	20	40	60	80
3	550.8 / 8	503.1 / 5	495.0 / 4	490.6 / 2	560.9 / 9	510.6 / 7	470.5 / 5	495.4 / 4
2	510.1 / 6	497.7 / 4	430.2 / 3	428.2 / 3	532.3 / 7	493.1 / 6	469.7 / 4	451.5 / 4
1	420.0 / 3	390.6 / 3	330.8 / 2	326.4 / 2	335.0 / 3	312.9 / 3	290.6 / 2	260.7 / 1

Table 2. Results of the two-factor ANOVA tests, $\alpha = 0.05$

	Fc	Table 2.1. Problem solving time			Table 2.2. Number of missed solutions		
		P-value	Fr/Fcr	Remark	P-value	Fr/Fcr	Remark
1-R	5.14	≈ 0.000036	87.64	$Fr > Fcr$	≈ 0.061	4.61	$Fr < Fcr$
2-R	5.14	≈ 0.0000009	303.27	$Fr > Fcr$	≈ 0.00068	31.2	$Fr > Fcr$
1-C	4.76	≈ 0.0023	17.43	$Fc > Fcr$	≈ 0.037	5.54	$Fc > Fcr$
2-C	4.76	≈ 0.0018	19.09	$Fc > Fcr$	≈ 0.0053	12.55	$Fc > Fcr$

is empirically established as $Gen = 600$. In different implementations depends on factors like the environment intelligence, the initial distribution of the agents in the environment and values of different parameters of the algorithm.

The parameters values of the simulation were $|SMG| \in [20, 80]$; *environment intelligence* (with significance "degree of unknown") denoted ei, $ei \in \{1, 2, 3\}$; consideration of the personal best denoted apb, with $apb \in [0, 10]$; consideration of global best denoted agb, with $agb \in [0, 10]$. Based on the simplified environment we have defined three types of environment intelligence: $ei = 3$-no intelligence (randomly generated and distributed objects); $ei = 2$-very small intelligence; $ei = 1$-small intelligence.

Emergent global intelligence measured by the elaborated metric considers two criteria, the average problem solving time of more simulations (until the solution has been found) and the number of missed solutions in that simulations (the solution is not obtained in the admitted time steps Gen). A result of 20 problem's solving cycle of the same problem is denoted with $Result = \{Ind_1; Ind_2\}$; Ind_1-shows the averaged number of time steps when the solution is obtained during 20 simulations; Ind_2- shows how many times the solution was missing during 20 simulations. $In = ei/|SMG|$. Table 1 presents some simulation results.

In order to prove that the number of agents and the environment intelligence influences the problem solving time, verification of statistical difference we consider a two-factor analysis of variance ANOVA without replication by taking into consideration ei, $|SMG|$ (independent) and the problem solving time (dependent). It is also considered a two-factor analysis of variance ANOVA without replication by taking into consideration ei, $|SMG|$ (independent) and the number of times when the solution have been missed (dependent).

For both ANOVA analysis it is considered the significance level $\alpha = 0.05$. The results of the ANOVA tests in the Table 2 are presented (conclusions that can be obtained are based on the details presented in the remark columns). Can be concluded a decrease of problem solving time (statistical difference at $\alpha = 0.05$) based on the increase of environment intelligence and the number of agents. The misses time (statistically analyzed at $\alpha = 0.05$), is influenced by the number of agents, the environment intelligence proved influence just in some conditions.

We have observed in our research that in a highly complex societal MAS an agent could not detain by itself the maximal individual intelligence for the most optimal operation and problem solving. The agent must have influenced individual intelligence with learning capabilities, learning from other agents and from the global intelligence base. In our case-study the use of information related with the global best improves the problem solving time.

5   Conclusion

We have proposed a novel type of intelligence for difficult problems solving called societal intelligence, inspired by the "intelligence of the human society". For proving the effectiveness of a societal MAS, we have realized a case study. The main conclusion of the considered case study was that using the principles of societal intelligence even if the intelligence of the considered agents is very limited, in the system emerges an increased global intelligence at the system's level. Based on a comprehensive study of the scientific literature we consider that our proposal is original and will represent the basis for many future researches, including elaboration of some metrics for the measurement of complex systems' intelligence able to solve extremely difficult problems. In our approach, there are considered different types of intelligence that can be measured and based on them can be established the intelligence of a highly complex system as a whole.

References

1. Zang, L.J., Cao, C., Cao, Y.N., Wu, Y.M., Cao, C.G.: A survey of commonsense knowledge acquisition. J. Comp. Sci. Technol. **28**, 689–719 (2013)
2. Duda, R.O., Shortliffe, E.H.: Expert systems research. Science **220**, 261–268 (1983)
3. Frantz, R.: Herbert Simon: Artificial intelligence as a framework for understanding intuition. J. Economic. Psychol. **24**, 265–277 (2003)
4. Iantovics, L.B., Zamfirescu, C.B.: ERMS: an evolutionary reorganizing multiagent system. Int. J. Innov. Comput. Inf. Control **9**, 1171–1188 (2013)
5. Kun, Y., Galis, A., Guo, X., Liu, D.: Rule-driven mobile intelligent agents for real-time configuration of IP networks. In: Palade, V., Howlett, R.J., Jain, L. (eds.) KES 2003. LNCS, vol. 2773, pp. 921–928. Springer, Heidelberg (2003)
6. Mamei, M., Menezes, R., Tolksdorf, R., Zambonelli, F.: Case studies for self-organization in computer science. J. Syst. Architect. **52**, 443–460 (2006)
7. McShane, M., Beale, S., Nirenburg, S., Jarrell, B., Fantry, G.: Inconsistency as a diagnostic tool in a society of intelligent agents. Artif. Intell. Med. **55**, 137–148 (2012)

8. Micacchi, C., Cohen, R.: A framework for simulating real-time multi-agent systems. Knowl. Inform. Syst. **17**, 135–166 (2008)
9. West, D., Dellana, S.: Diversity of ability and cognitive style for group decision processes. Inform. Sci. **179**, 542–555 (2009)
10. Zamfirescu, C.B., Duta, L., Iantovics, L.B.: On investigating the cognitive complexity of designing the group decision process. Stud. Inform. Contr. **19**, 263–270 (2010)
11. Moldovan, L.: Geometrical method for description of the 6-PGK parallel robot's workspace. In: CANS Conference, pp. 45–51. IEEE Computer Society Press (2008)

DOTS: Drift Oriented Tool System

Joana Costa[1,2(✉)], Catarina Silva[1,2], Mário Antunes[1,3],
and Bernardete Ribeiro[2]

[1] School of Technology and Management, Polytechnic Institute of Leiria,
Leiria, Portugal
{joana.costa,catarina,mario.antunes}@ipleiria.pt
[2] Department of Informatics Engineering, Center for Informatics and Systems
of the University of Coimbra (CISUC), Coimbra, Portugal
{joanamc,catarina,bribeiro}@dei.uc.pt
[3] Center for Research in Advanced Computing Systems, INESC-TEC,
University of Porto, Porto, Portugal
mantunes@dcc.fc.up.pt

Abstract. Drift is a given in most machine learning applications. The idea that models must accommodate for changes, and thus be dynamic, is ubiquitous. Current challenges include temporal data streams, drift and non-stationary scenarios, often with text data, whether in social networks or in business systems. There are multiple drift patterns types: concepts that appear and disappear suddenly, recurrently, or even gradually or incrementally. Researchers strive to propose and test algorithms and techniques to deal with drift in text classification, but it is difficult to find adequate benchmarks in such dynamic environments.

In this paper we present DOTS, Drift Oriented Tool System, a framework that allows for the definition and generation of text-based datasets where drift characteristics can be thoroughly defined, implemented and tested. The usefulness of DOTS is presented using a Twitter stream case study. DOTS is used to define datasets and test the effectiveness of using different document representation in a Twitter scenario. Results show the potential of DOTS in machine learning research.

Keywords: Drift · Text classification · Learning algorithms · Software tool

1 Introduction

The usual challenge for machine learning approaches is to build models that can perform well in classifying new data in production settings. Research efforts are usually put in proposing, implementing and testing innovative algorithms and techniques to deal with such challenges.

One major risk is that once a model is deployed, its performance can be significantly reduced when there is a drift in the distribution generating the new data when compared with the distribution that generated the data used to define the model. In machine learning this can be called as *model ageing*, in an analogy

© Springer International Publishing Switzerland 2015
S. Arik et al. (Eds.): ICONIP 2015, Part IV, LNCS 9492, pp. 615–623, 2015.
DOI: 10.1007/978-3-319-26561-2_72

with *software ageing* [1]. Not all models age well and there are several recent approaches [2–4] that try to deal with such challenges, namely by proposing dynamic techniques that try to detect, or deal, with drifts in different scenarios.

In the presence of drift, learning is not an easy task and requires special approaches, different from those commonly used, as the arriving instances can not be treated as equally important contributors to the final model [5]. In non-stationary environments, e.g. social networks as the Twitter stream, effective learning requires a learning algorithm with the ability to detect context changes without being explicitly informed about them, quickly recover from the context change and adjust its hypothesis to the new context [2]. It should also make use of previously experienced situations when old contexts and corresponding concepts reappear [6]. Additionally, drifts can have different patterns and thus must be treated differently. Four types of drift can be identified, as can be seen in Fig. 1: sudden, gradual, incremental and reoccurring [7].

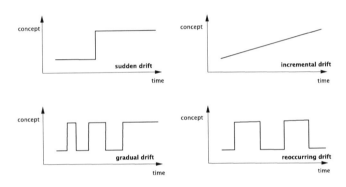

Fig. 1. Types of drift.

- **Sudden Drift** occurs when the speed of the drift is high and a concept appears or disappears in an abrupt way. Although it is usually stated as sudden or abrupt drift, it is also sometimes referred as concept change.
- **Gradual Drift** is another type of drift characterized by a low drift rate. Occurs when the probability of a given context to be associated with a concept decreases during a certain period of time. Moreover, the probability to be associated with another context increases proportionally.
- **Incremental Drift**, also considered a subgroup of gradual drift, can be considered differently because the change between the two concepts is very slow being just perceived when looking to what is occurring during a larger period of time.
- **Reoccurring Drift** occurs when previously active concept reappears after a certain period of time. It is noteworthy referring that the seasonality of the change must be previously unknown, otherwise the core assumption of the uncertainty about the future would be compromised.

To build drift aware datasets with blended temporal distributions is a challenging task. Therefore, it is not straightforward to find acceptable benchmarks in dynamic environments. To tackle this issue, in this paper we propose the Drift Oriented Tool System (DOTS), a framework that allows for the definition and generation of text-based datasets and can be used to simulate a set of different drift patterns with a temporal basis. The datasets obtained are then used to evaluate and validate learning strategies used in dynamic environments, including pre-processing strategies.

To validate DOTS we also present a case study using the Twitter stream that shows DOTS can be effectively used to define datasets and test learning algorithms and techniques.

The rest of the paper is organized as follows. We start in Sect. 2 by presenting the DOTS framework. We then proceed in Sect. 3 with a Twitter stream case study to show the effectiveness of using DOTS. Finally, in Sect. 4 we present the main conclusions of the work and delineate some directions for future work.

2   Drift Oriented Tool System

2.1   Introduction

DOTS is a drift oriented framework developed to dynamically create datasets with drift. It is a simple-to-use freeware application with a friendly interface as shown in Fig. 2. It can be download at http://dotspt.sourceforge.net/.

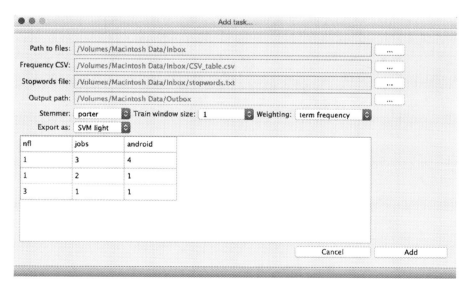

Fig. 2. DOTS interface.

The main purpose of DOTS framework is to represent drift patterns in a text-based dataset. Therefore, the framework input is a set of text document files,

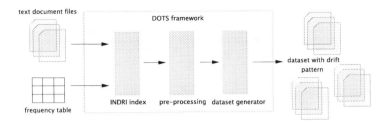

Fig. 3. The DOTS framework.

each representing a class, and a frequency table representing the drift patterns (see Fig. 3). The initial phase of DOTS processing includes building an INDRI index, provided by INDRI API, from the Lemur Project[1].

The DOTS framework represents each document in a vector space model, also known as *bag of words*, a commonly used text document representation.

Two problems arise when a vector with one element for each term occurring in the whole connection is used to represent a document: space high-dimensionality and overfitting. High dimensional space can cause computational problems and overfitting can easily prevent the classifier to generalize and thus the prediction ability can be endangered. To tackle both problems pre-processing methods were also integrated in the DOTS framework and are the second phase of the processing. Those methods are part of the INDRI API and aim at reducing the size of the document representation and prevent the mislead classification. Besides stopwords removal, DOTS also permits stemming. This pre-processing method consists in removing case and inflection information of a word, reducing it to the word stem. Stemming does not alter significantly the information included, but it does avoid feature expansion. Two important stemming algorithms were included: the Porter algorithm [8] and Krovetz algorithm [9].

It is also possible to define the weighting scheme used to represent each word of a document, that is the weight of each feature of a document. Two weighting schemes were defined, namely term frequency (*tf*) and term frequency-inverse document frequency, commonly known as *tf-idf*. Considering the defined input, DOTS will create a word index that allows users to use different strategies of filtering and analyzing data.

A major characteristic of DOTS is the possibility of defining the exact time, more precisely the exact time window, where each document appears, being thus possible to define time drifts. This is done by the frequency table that is also an input of the DOTS framework (see Fig. 3). DOTS takes no regard on the real counterpart dimension of the time window, since it constitutes an abstract realization of an amount of time (in Twitter streams it can represent seconds, but in other applications it can represent hours). The main idea is to use the frequency to reproduce artificial drifts.

---

[1] http://www.lemurproject.org/

DOTS output is thus a set of datasets, including the defined strategies related to vector space model, pre-processing strategies, feature representation and document division in time-windows.

Three output formats were implemented: Comma-Separated Values (CSV) file format, the Attribute-Relation File Format (ARFF) used in the widely used WEKA software, and the SVMLight file format. The possibility of using the concept of training windows is also possible, as users can define for each time-window the previous amount of data that should also be considered. By defining different training window sizes one can represent the memory properties of the training models, because it mimics a storage mechanism.

2.2   Specific Features of DOTS

The input of the DOTS framework is two fold, as seen in Fig. 3, and is composed by text documents and a frequency table. Each text document file represents the documents of the same class and the frequency table is use to define the drift patterns of the scenario. The frequency table must be in the CSV format and each row corresponds to a time instance. It is not important if a time instance represents a minute, an hour, or a day, but it is assumed that all of them correspond to the same amount of time. The first row contains the name of the class, and each cell of all the other rows contain the number of documents of a given class that occur in a given time instance. Consider Fig. 2 that represents a task to be added to the framework. By using as input a frequency table as the example given, we represent three classes: `nfl`, `jobs` and `android`, and 3 time-windows. As depicted in this example, in the first time-window there is 1 document of the class `nfl`, 3 from the class `jobs` and 4 of the class `jobs`.

Additional parameters can also be defined, like a stopword file and a stemmer algorithm. The stopwords file is a text file containing stopwords that will not be considered in the documents representation, and the stemmer algorithm will be used to reduce the document inflected words to their root form. Two stemmer algorithms were implemented: `porter` and `korvetz`.

It is also possible to define multiple training window sizes, multiple weighting schemes, and multiple export file formats. The training window size will define in each time-window how many previous time-windows will be considered, as this can be important for testing learning models with its memory capabilities. For instance, to perceive for how long it is relevant to keep previously gathered information and how that can affect the learning model capabilities. Weighting scheme will be used to define the document representation weighting scheme, like `term-frequency` or `tf-idf`. By exporting in multiple file formats, DOTS permits to create datasets that can be used in different classification frameworks, like SVM `Light` and `Weka`.

As it is often relevant to define various testing scenarios, DOTS will also permit adding tasks using INI files. INI files are structured files with "`key = value`" pairs, that will allow the definition of multiple tasks at once. A complete tutorial about DOTS can be download at http://dotspt.sourceforge.net/.

DOTS framework tries to fill an existing gap in machine learning research for text applications, by making possible to generate benchmark datasets with thoroughly controlled drift characteristics. We will now present an example of the potential use of the framework in a Twitter stream case study.

3   Case Study: Twitter Stream

3.1   Drift Example in Twitter

To validate the proposed framework and to show how DOTS can be used, we will present a Twitter stream case study. It constitutes a paradigmatic example of a text-based scenario where drift phenomena occur commonly. *Twitter* is a micro-blogging service where users post text-based messages up to 140 characters, also known as *tweets*. It is also considered one of the most relevant social network, along with *Facebook*, as millions of users are connected to each other by a following mechanism that allows them to read each others posts. *Twitter* is also responsible for the popularization of the concept of *hashtag*. An *hashtag* is a single word started by the symbol "#" that is used to classify the message content and to improve search capabilities. Besides improving search capabilities, *hashtags* have been identified as having multiple and relevant potentialities, like promoting the phenomenon described in [10] as *micro-meme*, i.e. an idea, behavior or style that spreads from person to person within a culture [11]. By tagging a message with a trending topic hashtag, a user expands the audience of the message, compelling more users to express themselves about the subject [12].

Considering the importance of the hashtag in Twitter, it is relevant to study the possibility of evaluating message contents in order to predict its hashtag. If we can classify a message based on a set of *hashtags*, we are able to suggest an hashtag for a given *tweet*, bringing a wider audience into discussion [13], spreading an idea [14], get affiliated with a community [15], or bringing together other Internet resources [16].

3.2   Twitter Classification

This case study aims to classify Twitter messages. A Twitter classification problem can be described as a multi-class problem that can be cast as a time series of tweets. It consists of a continuous sequence of instances, in this case, Twitter messages, represented as $\mathcal{X} = \{x_1, \ldots, x_t\}$, where x_1 is the first occurring instance and x_t the latest. Each instance occurs at a time, not necessarily in equally spaced time intervals, and is characterized by a set of features, usually words, $\mathcal{W} = \{w_1, w_2, \ldots, w_{|\mathcal{W}|}\}$. Consequently, instance x_i is denoted as the feature vector $\{w_{i1}, w_{i2}, \ldots, w_{i-W-}\}$.

When x_i is a labelled instance it is represented as the pair (x_i, y_i), being $y_i \in \mathcal{Y} = \{y_1, y_2, \ldots, y_{|\mathcal{Y}|}\}$ the class label for instance x_i.

We have used a classification strategy previously introduced in [17], where the Twitter message *hashtag* is used to label the content of the message, which means that y_i represents the *hashtag* that labels the Twitter message x_i.

Notwithstanding it is a multi-class problem in its essence, it can be decomposed in multiple binary tasks in a one-against-all binary classification strategy. In this case, a classifier h^t is composed by $|Y|$ binary classifiers.

3.3   Using DOTS to Twitter Classification

In this particular case, DOTS is used to create datasets able to test multiple learning scenarios. DOTS receives a document set for each class of tweets containing the same hashtag. A CSV table with different drift patterns was also defined, reproducing artificial drifts, like sudden, gradual, incremental, reoccuring and normal. As an example, a sudden drift might be represented by tweets from a given hashtag that in a given temporal moment start to appear with a significant frequency. Each tweet was represented by DOTS as a vector space model and pre-processing methods were applied, like stopword removal and stemming. We have exported, for our convenience, in SVMLight format, as is required to use Support Vector Machines (SVM) as the learning model of this case study. Two different weighting scheme in document representation were tested, term-frequency and *tf-idf*. Table 1 presents the obtained results of the performance obtained using both weighting schemes. In order to evaluate the possible outcomes of the classification, we used the van Rijsbergen F_β measure: $F_\beta = \frac{(\beta^2+1)P \times R}{\beta^2 P + R}$ with $\beta = 1$. F_1 is an harmonic average between precision and recall as is widely used in text classification problems. The results shed light on the use of term-frequency to classify Twitter messages in the presence of drift. By using term-frequency to represent document features one can improve the overall classification performance even in the presence of different drift patterns.

Table 1. Performance measure for the results obtained with Twitter stream.

Drift	Hashtag	F_1 using tf	F_1 using tf-idf
Sudden	#syrisa	79.37 %	75.32 %
Gradual	#airasia	57.88 %	56.08 %
Incremental	#isis	83.49 %	81.75 %
Reoccurring	#android	60.66 %	59.49 %
Normal	#sex	74.88 %	74.56 %

4   Conclusions and Future Work

In this paper we propose DOTS, a drift oriented framework developed to dynamically create text datasets with drift. This tool is specifically designed to text classification and its major goal is to create text labeled datasets that can be used to simulate different drift patterns. DOTS performs pre-processing methods like stopword removal and stemming and can be used to evaluate and validate learning strategies in dynamic environments.

We have also presented a case study based on a Twitter drift scenario. The aim of the study is to classify Twitter messages using their hashtags. We have generated multiple datasets and have tested different classification strategies, to define the best characteristic of a learning model in this particular scenario.

We have presented results that show DOTS capabilities with a Twitter scenario on testing different learning strategies in a dynamically created set of test based datasets, in which examples have a temporal meaning. DOTS is a very flexible tool as it can be used to validate learning strategies with text based datasets in different contexts rather than Twitter.

Regarding future work we will look at including new features in the DOTS framework, namely, how to automatically add noise to the datasets and new file exporting types.

Acknowledgment. We gratefully acknowledge iCIS project (CENTRO-07-ST24-FEDER - 107002003).

References

1. Wu, Q., Hu, W., Wang, B., Han, Z., Qi, Y.: Software aging mechanism analysis and rejuvenation. Int. J. Digit. Content Technol. Appl. **6**(22), 552 (2012)
2. Costa, J., Silva, C., Antunes, M., Ribeiro, B.: Concept drift awareness in Twitter streams. In: Proceedings of 13th International Conference on Machine Learning and Applications, pp. 294–299 (2014)
3. Mejri, D., Khanchel, R., Limam, M.: An ensemble method for concept drift in nonstationary environment. J. Stat. Comput. Simul. **83**(6), 1115–1128 (2013)
4. Ditzler, G., Polikar, R.: Incremental learning of concept drift from streaming imbalanced data. IEEE Trans. Knowl. Data Eng. **25**(10), 2283–2301 (2013)
5. Tsymbal, A., Pechenizkiy, M., Cunningham, P., Puuronen, S.: Dynamic integration of classifiers for handling concept drift. Inf. Fusion **9**(1), 56–68 (2008)
6. Widmer, G., Kubat, M.: Learning in the presence of concept drift and hidden contexts. Mach. Learn. **23**(1), 69–101 (1996)
7. Zliobaite, I.: Learning under concept drift: an overview. Vilnius University, Faculty of Mathematics and Informatic, Technical report (2010)
8. Willett, P.: The porter stemming algorithm: then and now. Program **40**(3), 219–223 (2006)
9. Krovetz, R.: Viewing morphology as an inference process. In: Proceedings of 16th Annual International ACM SIGIR Conference on Research and Development in Information Retrieval, pp. 191–202. ACM (1993)
10. Huang, J., Thornton, K.M., Efthimiadis, E.N.: Conversational tagging in Twitter. In: Proceedings of 21st ACM Conference on Hypertext and Hypermedia, pp. 173–178 (2010)
11. Merriam-webster's dictionary, October 2012
12. Zappavigna, M.: Ambient affiliation: a linguistic perspective on Twitter. New Media Soc. **13**(5), 788–806 (2011)
13. Johnson, S.: How Twitter will change the way we live. Time Mag. **173**, 23–32 (2009)

14. Tsur, O., Rappoport, A.: What's in a hashtag?: content based prediction of the spread of ideas in microblogging communities. In: Proceedings of 5th International Conference on Web Search and Data Mining, pp. 643–652 (2012)

15. Yang, L., Sun, T., Zhang, M., Mei, Q.: We know what @you #tag: does the dual role affect hashtag adoption? In: Proceedings of 21st International Conference on World Wide Web, pp. 261–270 (2012)

16. Chang, H.-C.: A new perspective on Twitter hashtag use: diffusion of innovation theory. In: Proceedings of 73rd Annual Meeting on Navigating Streams in an Information Ecosystem, pp. 85:1–85:4 (2010)

17. Costa, J., Silva, C., Antunes, M., Ribeiro, B.: Defining semantic meta-hashtags for twitter classification. In: Tomassini, M., Antonioni, A., Daolio, F., Buesser, P. (eds.) ICANNGA 2013. LNCS, vol. 7824, pp. 226–235. Springer, Heidelberg (2013)

Approximately Similarity Measurement of Web Sites

Doru Anastasiu Popescu[1]([⊠]) and Dan Radulescu[2]

[1] Faculty of Mathematics and Computer Science,
University of Pitesti, Piteşti, Romania
dopopan@gmail.com

[2] National College "Radu Greceanu", Slatina, Romania
dan_radulescu96@yahoo.com

Abstract. In this paper we will present a way to approximately measure the similarity of two web sites. The web sites considered will have only HTML web pages. We will present an algorithm which chooses a number of significant pages for each site and it will determine the similarity using the information from the selected web pages. We will use a genetic algorithm in order to select significant web pages. To implement the algorithm and show the results we used Java language.

Keywords: Tag · HTML · Web page · Genetic algorithm · Similarity

1 Introduction

Genetic algorithms are part of the Evolutionary Algorithms having a lot of applications. They solve optimization problems using operations inspired from reality such as mutation, selection or cross-over. Genetic algorithms are applied in a wide variety of fields like computer science, computational physics or statistics. They can be used at agricultural algorithms (in [6] we have an algorithm developed for irrigating a crop), at mathematical algorithms or at image processing algorithms. They can be introduced in algorithms which determine the similarity of two web sites as we are going to show in this paper (examples of other algorithms which determine the similarity of two web applications are presented in [7, 8]).

In this paper we will use the genetic algorithm to determine a number of significant web pages for each web site. We will use a fixed number of web pages considered representative for two web sites WA1 and WA2 for measuring their similarity. This genetic algorithm will improve the execution time (we will have less pages to verify) but it will return approximate results. In the second part of this paper we will present an example on which the algorithm can be applied. In the third part we will present the algorithm and in the fourth one we will show tables with results and charts obtained after testing.

To be more precisely we will define the genetic elements used in our algorithm. The chromosome will be a set of k pages (where k is fixed to reduce the execution time, in Sect. 4 we used k = [N/3], N is the total number of pages of the web site). The chromosome and the results returned by the fitness function will be memorized in a

© Springer International Publishing Switzerland 2015
S. Arik et al. (Eds.): ICONIP 2015, Part IV, LNCS 9492, pp. 624–630, 2015.
DOI: 10.1007/978-3-319-26561-2_73

one-dimensional array, the first k components are the chromosome's genes and the k + 1 component is the value of the fitness function. This function will evaluate the number of tags from each page and it will sum them. This is the fitness value. Other fitness functions can be used: we can consider a binary tree with tags from all the genes of a chromosome. The total number of tags can be the fitness value. The rest of the operations will be presented in the third part of the paper.

2   The Selection of the Significant Web Pages of a Web Site

Usually, web sites contain lots of web pages. For the existing algorithms (like in paper [7, 8, 10], similarity algorithms) this results into inefficient methods. By determining some significant pages we improve the execution time, so the efficiency. We determine significant pages for each web site separately. We consider the next three web pages and we will use them as examples in this chapter (Table 1):

Table 1. HTML code for three web pages

p1.html	p2.html	p3.html
<HTML>	<HTML>	<HTML>
<HEAD> </HEAD>	<HEAD> </HEAD>	<HEAD> </HEAD>
<BODY> E1	<BODY>	<BODY>
<I> e 1 </I>	<I> e 1 </I> 	<BODY> E1
<I> e 2	<I> e 2 </I> 	<I> e 1 </I>
</BODY>	</BODY>	
</HTML>	</HTML>	</BODY>
		</HTML>

Now let's consider a web site with this three web pages. We will have only one significant web page. It is going to be the one with the biggest number of tags, so it is p2.html.

Now, for calculating the similarity we consider the next definitions ([8, 9] have similar definitions):

Definition 1. Let TG be a set of tags, p_i and p_j two web pages from P. We say that T_i is in T_j as a sequence, if there exists an index in the sequence T_j from where there appears from left to right, one after the other, all the tags of T_i.

Definition 2. Let TG be a set of tags, p_i and p_j two web pages from P. We say that p_i is in relation R with p_j and we write p_i R p_j, if:

(i) T_i is in T_j as a sequence;
(ii) Any tag <Tg> from T_j which appears in T_i as well, if it has a closing tag

</Tg> in T_j, then </Tg> is also in T_i.

Definition 3. Let WS1 be a web site with pages WS1 = {p_1, p_2, ..., p_n} and WS2 a web site with pages WS2 = {q_1, q_2, ..., q_m}. We consider the similarity rank between them the next formula: A/N (where A is the total number of pages from WS2 (the total number of q_i) in relation with at least one page from WS1(a p_j page)).

For our example we have T_1 = (<I>, </I>, <I>); T_2 = (<I>, </I>, <I>, </I>); T_3 = (<I>, </I>) that p_1 R p_2, p_3 R p_2 and p_3 R p_1, if we consider TG = {<HTML>, </HTML>, <HEAD>, </HEAD>, <BODY>, </BODY>,
}.

We will calculate the similarity rank, after the genetic algorithm, using the next formula: **a/k** (where a is the number of significant pages from the second web site in relation with at least one significant page from the first web site). (The similarity rank has to be between 0 and 1).

So, let's consider WS1 a web site including pages p2 and p3 and WS2 a web site including only p1. The similarity rank between those web sites is 1/2.

Other ways for measuring the similarity of two web sites can be defined using the results obtained in [9, 10, 12].

3   Applied Genetic Algorithm

We will show the classic scheme of a genetic algorithm (Fig. 1), as it is described in [1].

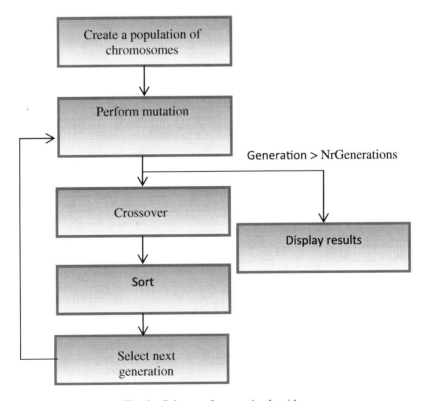

Fig. 1. Scheme of a genetic algorithm

Before the pseudocode we will present the most important variables and the data structures.

NrMutation = the number of mutations at each step
NrPop = the number of chromosomes for one generation
NrGenerations = the number of generations
pages[] = array with the number of tags for each page of the website (it has N elements)
population[][] = represents the population of chromosomes at each step.
k = the number of significant pages

We will start with the main sequence of the algorithm:

```
for i=1, NrGenerations do
        mutation();
        crossover();
        sort();

endfor
```

We will apply this sequence for each web site. Now we will present the mutation and the crossover sequences:

```
MUTATION():

for i=1,NrMutation do
        choose a chromosome;
        choose a page from the chromosome;
        swap the choosen page with one from the site which doesn't
        belong to the current chromosome;
endfor
```

```
CROSSOVER():

for i=NrPop+1,NrPop+100 do
        choose first chromosome;
        choose second chromosome;
        combine chromosome1-chromosome2;
        check new chromosomes;
endfor
```

SORT() is a simple sequence (it can be a QuickSort() type algorithm) which sorts descending the chromosomes after the the fitness value.

The execution time has a polynomial order. The algorithm can return more sets of significant pages. Using those sets we can calculate the similarity rank as an average between all the results obtained (we can choose the best two, three or more chromosomes). Using our fitness function it can be simpler to sort all the pages descending and choose the first k pages. But we will obtain only one result and we are interested in finding more solutions (these solutions can be found with the genetic algorithm).

4   The Implementation

The algorithm was implemented in NotePad, using the Java language. The computer which ran the tests has a 3.10 GHz Processssor and 8.00 GB RAM. We tested the algorithm for three values of NrGenerations. We will present the results in three tables and two charts. We are going to use websites [11] for our examples (Figs. 2 and 3) (Tables 2, 3 and 4).

In the first chart we observe that the execution times are proportional with the generations number, and in the second one we observe that we obtain the same results

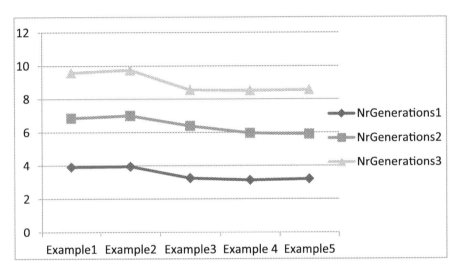

Fig. 2. Execution time chart

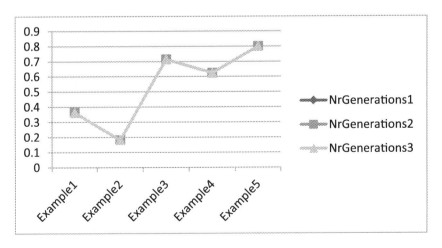

Fig. 3. Similarity rank chart

Table 2. NrGenerations = 300

Example	WA1	WA2	Similarity rank	Execution time
1	sofa2010	jatit	0.3636	3.91 s
2	jatit	civile.utcb.ro	0.1818	3.95 s
3	sofa2012	sofa2005	0.7142	3.25 s
4	sofa2007	sofa2010	0.625	3.13 s
5	civile.utcb.ro	sofa2009	0.8	3.20 s

Table 3. NrGenerations = 600

Example	WA1	WA2	Similarity rank	Execution time
1	sofa2010	jatit	0.3636	6.85 s
2	jatit	civile.utcb.ro	0.1818	7 s
3	sofa2012	sofa2005	0.7142	6.39 s
4	sofa2007	sofa2010	0.625	5.96 s
5	civile.utcb.ro	sofa2009	0.8	5.90 s

Table 4. NrGenerations = 900

Example	WA1	WA2	Similarity rank	Execution time
1	sofa2010	jatit	0.3636	9.58 s
2	jatit	civile.utcb.ro	0.1818	9.75 s
3	sofa2012	sofa2005	0.7142	8.55 s
4	sofa2007	sofa2010	0.625	8.52 s
5	civile.utcb.ro	sofa2009	0.8	8.56 s

for the similarity rank.(this is something normal because we compare the same pairs of web sites. It means that our algorithm returns a good value starting from lower values of NrGenerations).

5 Conclusions and Future Work

In this paper we presented a genetic algorithm which determines a number of significant web pages from a web site. We can use those pages to determine the similarity between two web sites. Using this algorithm we obtain a good efficiency and good execution times but we obtain approximated results. We decided to use a genetic algorithm as we wanted a good result for the similarity. If we only sort the pages after the fitness function we obtain a single set of k pages and this does not ensure a good result. This is the reason why we used a genetic algorithm.

Future work includes using a new fitness function (based on binary trees, [3, 4]). As in [2, 5, 13] there are other methods of calculating the similarity rank (as the edit distance for example).

References

1. https://en.wikipedia.org/wiki/Genetic_algorithm
2. Bollegata, D., Matsuo, Y., Ishizuka, M.: Measuring semantic similarity between words using web search engines. In: IW3C2 (2007)
3. Balcau, C.: Combinatorics and Graph Theory. University of Piteşti Publishing, Piteşti (2007)
4. Constantin, D., Samarescu, N.: Modern Techniques of Using the Computer. Tiparg Publishing, Piteşti (2009)
5. Lin, D.: An information-theoretic definition of simimarity. In: ICML 1998, pp. 296–304. ACM
6. Popescu, D.A., Radulescu, D.: Monitoring of irrigation systems using genetic algorithm. In: ICMSAO 2015. IEEE Xplore (2015)
7. Popescu, D.A., Danauta, C.M.: Similarity measurement of web sites using sink web pages. In: 34th International Conference on Telecommunications and Signal Processing, TSP 2011, 18–20 August 2011, pp. 24–26. IEEE Xplore, Budapest (2011)
8. Popescu, D.A., Nicolae, D.: Determining the similarity of two web applications using the edit distance. In: SOFA. LNCS (2014). http://trivent.hu/2014/sofa2014/documents/sofa2014_final_program.pdf
9. Popescu, D.A.: Sink web pages in web application. In: Schwenker, F., Trentin, E. (eds.) PSL 2011. LNCS, vol. 7081, pp. 154–158. Springer, Heidelberg (2012)
10. Torres, G.J., Basnet, R.B., Sung, A.H., Mukkamala, S., Ribeiro, B.M.: A similarity measure for clustering and its applications. In: ICASA 2008 (2008)
11. Jeh, G., Windom, J.: SimRank: a measure of structural-context similarity. In: KDD 2002, pp. 538–543. ACM (2002)
12. Pushpa, C.N., Thriveni, J., Venugopal, K.R., Patnaik, L.M.: Web search engine based semantic similarity measure between words using pattern retrieval algorithm. In: CS & IT-CSCP 2013 (2013)
13. Zhao, P., Han, J., Sun, Y.: P-Rank: a comprehensive structural similarity measure over information networks. In: CIKM 2009. ACM (2009)

Using NLP Specific Tools for Non-NLP Specific Tasks. A Web Security Application

Octavia-Maria Șulea[1,2,3]($\boxtimes$), Liviu P. Dinu[1,2], and Alexandra Pește[2,3]

[1] Center for Computational Linguistics,
University of Bucharest, Bucharest, Romania
mary.octavia@gmail.com
[2] Faculty of Mathematics and Computer Science,
University of Bucharest, 14 Academiei Street, sector 1, Bucharest, Romania
ldinu@fmi.unibuc.ro
[3] Bitdefender, 24 Delea Veche Street, sector 2, Bucharest, Romania
apeste@bitdefender.com

Abstract. In this paper we look at the task of detecting URLs corresponding to infected web pages using Machine Learning and Natural Language Processing specific features. We show that these features render better performance than the previously used hand-crafted *lexical* features and render similar results to the more expensive *host-based* features. We also introduce a new adjacent task, that of identifying URLs pointing to the download of portable executable files, and show that our models perform very well on this task too.

1 Introduction

With the web taking over more and more of our daily activities, the contexts for online crime have multiplied and this in turn has lead to a shift in the hacker's general goal from detecting and exploiting system vulnerabilities for fun to obtaining revenue out of their exploits [8]. Security online has become, thus, as vital as the everyday security we are accustomed to receive in our time offline. The first step in the majority of these criminal cases involves a user clicking on a link. Therefore, a lot of efforts have been put into automatically identifying malicious Uniform Resource Locators (URLs).

Beyond blacklisting heuristics, many of the attempts so far ([1,4,9,10]) have chosen to employ machine learning to classify webpages as either malicious or benign, prestigious or dubious, relevant or irrelevant. Since the least expensive information that can be gathered about a web page is its URL, the majority of works in this literature have attempted to find the best classification features which can be extracted only from the URL, rather than by visiting the potentially malicious site and extracting features from the crawled content. One common type of feature used in these studies has been the so called *lexical* type which takes advantage of the textual properties of the URL, such as length, number of dots, number of tokens in the URL, etc., the commonly accepted intuition behind

© Springer International Publishing Switzerland 2015
S. Arik et al. (Eds.): ICONIP 2015, Part IV, LNCS 9492, pp. 631–638, 2015.
DOI: 10.1007/978-3-319-26561-2_74

the choice of such features being that malicious URLs *look different* [10]. However, contrary to this intuition, all previous systems employed, beside the lexical, a suite of non-lexical (and more expensive) features not necessarily related to the URL, leading to the impression that the lexical were insufficient in achieving good performance.

In what follows, we confirm that these lexical features do poorly on their own. We propose a model that takes advantage of the general intuition that URLs of dubious sites look strange by extracting n-grams and tf-idf scores from them. We train and test using Logistic Regression, Linear SVC [6], open-source XGBoost[1], and Multilayer Perceptron and compare the results obtained using NLP features with those obtained using lexical and host-based features and show that the former perform similarly if not better. We also introduce a new adjacent task of classifying URLs based on whether they link to portable executables and report good results. The aim in both tasks is to create a machine learning based filter that would aid in the focused crawling of malicious sites.

2   Related Work

Although a seemingly evident first approach due to the URL's tendency of appearing at the threshold between natural and formal language, to our knowledge, only two studies attempted to extract NLP features from URLs: [9], which used word (token) level n-grams (while we used character n-grams) for a different web classification task (whether the page was relevant to a search query), reported 76 % accuracy without using the page's text and 81 % with text and [1], which extracted n-grams and tf-idf scores from the URL as well as the content, reported 87 % average precision (and no mention of recall, F1, or accuracy). While they shared our intuition of using text processing features on the URL, neither used *only* these features when reporting best scores. Also, a potentially serious problem with [1] is that the dataset was artificially generated using Genetic Algorithms to mutate a very small initial dataset (200 URLs) into a very large one, carrying out an oversampling of both classes. This can lead to the problem we have also encountered with our initial corpus, which contained URLs with the same domain, causing a bias in the NLP models and an ambiguity in the lexical model.

From the other machine learning approaches previously cited, [4] reported 98 % accuracy when using a plethora of lexical, host-based, link popularity, and web content features, while [10] reported 95–99% accuracy when using similar features, but ignoring the content of the webpage. Link popularity and host-based features are more expensive to retrieve than lexical features, n-grams or tf-idf scores, since they imply consulting third-party online tools, but not as costly as crawling the content of the potentially dangerous page itself. Therefore, these approaches differ from ours mainly in the cost of obtaining the features.

---

[1] https://github.com/dmlc/xgboost

3   Our Approach

For the datasets, we used two sources[2]: one containing URLs coming from exitnodes in the TOR network; another one containing URLs of pages accessed by anonymized users of Bitdefender security products. The data coming from these two sources was first pruned for uniqueness and then, for each URL, we consulted the URL Status tool[3] devised by Bitdefender and used its output to label the dataset: 0 for *benign*, 1 for *infected*. Infected here meant that the URL was found by the Bitdefender tool (through heuristics or checking its crawled content) to either be a phishing, spam or fraud site, or to contain and distribute malicious software (malware). Clean URLs on gray domains were considered benign.

After an initial set of experiments conducted on the TOR dataset with results (Table 2) comparable with the state-of-the-art host-based systems, we realized that, although this dataset contained unique URLs, many of them were URLs which shared the same domain and subdomain. Since we expected the path structure to be influenced by the domain (e.g. all youtube video URLs have the 5 gram *watch* before the unique ID) and since indirectly using the domain name as a predominant feature would be reminiscent of manual whitelisting, which we wanted to move away from, we pruned the dataset for duplicates relative to the domain and subdomain using urlparse' netloc[4] to see the extent to which removing the popular domain bias would affect our learners. Thus, from an initial TOR dataset containing 215, 299 unique entries, we were left with a new corpus (henceforth, TOR2) containing 23, 392 URLs. An *extended corpus* (henceforth, EC) was achieved by merging TOR2 URLs with those pruned similarly from the second source, leading to 59, 871 unique URLs.

Finally, for the second task, since the majority of the *infected* URLs were of the malware-containing sub-type, we wanted to see if we can detect, only by looking at the URL, links leading to the download of portable executables. To our knowledge, such a dataset does not exist and the creation process is not trivial, yet the URL Status tool can return the Content-Type header of the page. Thus, we gathered 139, 505 unique URLs from our second source and we labeled them with 1 if the Content-Type was set to *application/octet-stream* or 0, otherwise. The intent here was to see if our NLP features were appropriate for such a distinction so that, if they were, we would be one step closer to automatically finding malicious software distributed online. What must be noted here is that there is no biconditional relation (*if and only if*) between the presence of a file extension in the URL and its leading to the download of a PE, nor between the presence of *application/octet-stream* in the Content-Type header of a page and the download of a PE, although the left to right implication of the latter is usually true. Our dataset also lacks the former relation, but its labeling is based on the assumption that the latter is always true.

---

[2] All datasets used in this paper can be accessed from here: https://github.com/mary-octavia/URL-datasets

[3] http://www.bitdefender.com/oem/url-status.html

[4] https://docs.python.org/2/library/urlparse.html

Regarding class ratio and base distributions, what needs to be mentioned here is that, in a normal flow, only around 1–4% of URLs visited in a day are actually *infected*. Since it is known ([12,14]) that binary classifiers have a bias toward the majority class (here, benign URLs) and since [13] indicates that balanced training sets lead to better (test) performance even when the test sets are unbalanced, we undersampled the majority class until we reached a more balanced ratio. Since a slight imbalance remained (33 % were malicious in TOR, 34 % in EC, and 40 % in PE), we decided to use scikit-learn's StratifiedCross-Validation, which is able to maintain class ratio when folding, this way avoiding the risk of our learner never seeing examples from the minority class in a fold.

For each URL, we extracted, as in [4,10], the following lexical features: number of tokens, number of dots, length of the URL, length of the domain, average and maximum length of a domain token, average and maximum length of a path token. These were scaled using scikit-learn's [11] StandardScaler and used to obtain baseline scores. We proceeded to gather categorical features not related to any textual property of URLs like those used in previous studies: port number, protocol (i.e. http, https, ftp), top-level domain name, and top 500 most frequent query field names. Then we extracted character level (i.e. analyzer = 'char') n-grams and tf-idf scores using scikit-learn's CountVectorizer and TfidfVectorizer, respectively. These two vectorizers were applied to different versions of the URL leading to different results which will be discussed in the following section. The two NLP features were either used on their own, or combined with the scaled lexical features using scikit-learn's FeatureUnion.

For Logistic Regression and Linear SVC we used the scikit-learn implementations and for the MLP, pylearn2 [7]. The only parameter modified for LinearSVC was C which was set to 0.1 in all experiments, as our grid search indicated this to be the best. We used 10-fold cross-validation. The MLP with the best results had two layers containing 200 and 100 nodes, respectively, with pylearn2's default Sigmoid on the first layer and SoftMax on the second. It had 20 epochs, a batch size of 10, and 0.03 learning rate. It was run on character bigrams combined with lexical, resulting in a smaller feature set than the previous NLP models (3464 features in total). The best XGBoost classifier had eta = 0.07, max_depth = 8, gamma = 0.5, and num_rounds = 150 and also used only character bigrams plus the lexical. All other parameters were kept with their default values.

4   Results

In Tables 1, 2, 3 and 4, all measures are given for cross-validation test results, while precision, recall, and F1 strictly refer to the infected class. First, we looked at the results obtained when using different forms of the URL (Table 1). With the same n-gram range ([2,5]) and estimator (LogisticRegression), CountVectorizer led to best results when the delimiter characters [:/.?&–_.*+, %=—] were kept in the URL (form 1) and worst when only the delimiters were kept (form 3). On the other hand, TfidfVectorizer performed best when the delimiters, instead of being removed completely (form 2), were replaced with the same character

Table 1. TOR2 results using various URL forms

Tool	Form	Form example	Prec.%	Rec.%	F1%	Acc.%
ngram	1	http://www.google.com/search?q=hello	79	82	81	83
ngram	2	http://www.google.comsearchqhello	78	81	80	82
ngram	3	://../?=	71	70	71	74
tfidf	2	http://www.google.comsearchqhello	79	75	77	80
tfidf	4	http://www.google.comsearchqhello	80	77	79	81

Table 2. Results on the TOR and TOR2 datasets

Estimator	Features	Dataset	Prec.%	Rec.%	F1%	Acc.%	AUC%
SVC	lex	TOR	76	29	42	73	71
SVC	lex	TOR2	61	51	56	64	69
SVC	ngram	TOR	97	97	97	98	99
SVC	ngram	TOR2	78	82	80	82	89
LR	ngram	TOR2	79	82	81	83	90
LR	tfidf	TOR2	80	77	79	81	89
SVC	lex+ngram	TOR2	80	82	81	83	90
LR	lex+tfidf	TOR2	80	78	79	82	89

(form 4). We changed the estimator to LinearSVC, and the scores maintained this behavior. Thus, all further experiments were carried out on form 1 when using CountVectorizer, and on form 4 when using TfidfVectorizer.

As can be seen from Tables 2, 3 and 4, the lexical model is the weakest for all datasets. Conversely, for the initial TOR dataset, n-grams alone (where n = [2, 5]) were capable of reaching the 98 % test accuracy which was also reported by [4] and is within the accuracy score interval reported by [10]. For TOR2 and EC, which were pruned of near-duplicates relative to the subdomain, all models see a drop in scores. This is as expected, since the extent to which the learner could see almost the same training example during test time was decreased.

When applying LinearSVC and LogisticRegression to TOR2 and EC, the n-gram range had to be adjusted to [3, 6] in order to reach accuracy closer to the TOR one, however in doing so we started to see signs of overfitting, since from this range forward the test precision for the infected class starts to drop, while recall rises, meaning that, although more URLs are labeled correctly as infected, among the ones predicted to be infected lie more that are not actually infected. The results also show that there is little to no significant increase in scores when lexical features are combined with NLP features. The best scores, on TOR2 and EC, seem to be for the model using LinearSVC with n-grams and lexical features, although LogisticRegression on n-grams performs similarly. For the EC dataset, we also ran a few experiments using MLPs and XGBoost. Table 3 also shows the results obtained using all the features *except* the NLP

Table 3. Results on the EC dataset

Estimator	Features	Prec.%	Rec.%	F1%	Acc.%	AUC%
SVC	lex	67	56	61	75	81
XGBoost	nonNLP	79	69	74	83	91
SVC	ngram	82	85	83	88	95
LR	ngram	82	86	84	89	95
SVC	lex+ngram	83	85	84	89	95
LR	lex+tfidf	81	81	81	87	94
MLP	lex+2gram	79	83	81	84	92
XGboost	lex+2gram	82	85	83	86	94

Table 4. Classifier results on the PE dataset

Estimator	Features	Prec.%	Rec.%	F1%	Acc.%	AUC%
SVC	lex	79	65	72	79	89
LR	lex	80	66	72	80	90
SVC	tfidf	95	94	94	95	99
LR	tfidf	95	92	93	95	98
SVC	ngram	93	94	93	95	98
LR	ngram	96	94	95	96	99
XGBoost	lex+2gram	95	93	94	95	99

ones (1409 features). This model was slightly better than the purely lexical, but inferior to the NLP models.

When we applied the same classifiers to the PE corpus (Table 4) and varied the features as above, we obtained consistently higher results. For XGBoost, regardless of the parameter selection, all measurements varied with less than one percentage point (0.1 %), making it the most robust estimator. As with the previous corpora, LogisticRegression using frequency of n-grams as features seems to perform best (96 % accuracy). The n-gram range here was smaller ([2,3]). We investigated possible reasons, beside the more balanced class ratio, for the higher overall scores and found that the PE corpus had more URLs with the same domain but different subdomains which made our pruning method keep them in the dataset, this in turn introducing a bias in the n-gram models. We also noticed that a few of our URLs had the port specified which lead to urlparse thinking it was part of the domain and our pruning mechanism to find URLs with the same domain but different ports as needing no pruning.

To address these problems, we pruned the PE dataset further. We used Mozilla's effective top-level domain name list[5] to create a python script capable of identifying only the domain name. We created a dictionary with these

---

[5] https://publicsuffix.org/list/effective_tld_names.dat

domains (i.e. excluding from urlparse's netloc result the top-level domain, subdomain, and port) and used it to keep only the longest URL pertaining to each of these domains. On this new dataset, the NLP models see a drop in scores of at most 2 % for all measures except AUC, which is maintained at 98 %. Surprisingly, although the NLP models remain the best and the lexical models see a 3 % drop in precision, they also see an 11 % rise in recall, leading to a 5 % increase in F1 and a 2 % increase in accuracy. The AUC, however, is maintained at 90 %.

5 Conclusions and Further Work

In this paper, we've looked at identifying malicious URLs, a task lying on the outskirts of NLP, and shown that this can be solved using supervised machine learning with NLP specific features. We've also shown that the previously used *lexical* features, although intuitively appearing useful, perform poorly on their own and offer very little boost (around 1 %) in performance when combined with NLP features, like character n-gram counts or tf-idf scores. We've seen that even when a popularity bias is removed from the dataset, our models perform comparably to the host based models. Since the host-based features are more expensive to obtain (e.g. DNS fluxiness implying several DNS look-ups), and the lexical have proven to be weak features, our choice of NLP features, which are relatively easy to extract directly from the URL, appears to be a favorable one. We have also introduced a new related task (along with a dataset), that of classifying URLs based on whether they link to downloadable binaries or not, and shown the appropriateness of our models.

A final interesting conclusion relates to the results in Table 1. These results, beside showing the usefulness of including the delimiter characters in the feature space, echo multiple results in the field of authorship attribution ([2,3,5]), which show that adding punctuation mark frequencies to the feature space improves performance. It seems that URLs, although barely belonging to natural language (parts of the link representing filenames should intuitively be closest), display a similar characteristic of syntactic like structures betraying their identity, as choice of punctuation betrays syntactic structure preference and, ultimately, the author's stylistic identity (stylom).

As for future work, since class distribution may change when near-duplicates are removed (i.e. 33 % infected in TOR and 44 % infected in TOR2), methods of eliminating the models' bias for the popular domains while keeping all URLs in the dataset are worth further investigation. Also beneficial would be the development of other methods for gathering a corpus of URLs containing portable executables against which to test and validate current and future supervised machine learning models designed for this task.

Acknowledgements. The authors would like to thank the reviewers for their input and their Bitdefender collegues for the support given. All authors contributed equally. The work of Octavia-Maria Șulea was also supported by the strategic grant POS-DRU/187/1.5/S/155559. The work of Liviu P. Dinu was supported by UEFISCDI, PNII-IDPCE-2011-3-0959.

References

1. Aldwairi, M., Alsalman, R.: Malurls: a lightweight malicious website classification based on url features. J. Emerg. Technol. Web Intell. **4**, 128–133 (2012)
2. Baayen, R.H., van Halteren, H., Neijt, A., Tweedie, F.: An experiment in authorship attribution. In: IEEE Intelligent Systems and Their Applications (2002)
3. Chaski, C.E.: The computational-linguistic approach to forensic authorship attribution. In: Olsen, F., Lorz, A., Stein, D. (eds.) Law and Language: Theory and Practice. Düsseldorf University Press (2008)
4. Choi, H., Zhu, B.B., Lee, H.: Detecting malicious web links and identifying their attack types. In: Fox, A. (ed.) 2nd USENIX Conference on Web Application Development, WebApps 2011, 15–16 June 2011. USENIX Association, Portland (2011)
5. Dinu, L.P., Popescu, M., Dinu, A.: Authorship identification of romanian texts with controversial paternity. In: Proceedings of the International Conference on Language Resources and Evaluation, LREC 2008, 26 May – 1 June 2008. European Language Resources Association, Marrakech (2008)
6. Fan, R.E., Chang, K.W., Hsieh, C.J., Wang, X.R., Lin, C.J.: Liblinear: a library for large linear classification. J. Mach. Learn. Res. **9**, 1871–1874 (2008)
7. Goodfellow, I.J., Warde-Farley, D., Lamblin, P., Dumoulin, V., Mirza, M., Pascanu, R., Bergstra, J., Bastien, F., Bengio, Y.: Pylearn2: a machine learning research library (2013). arXiv preprint arXiv:1308.4214
8. Harper, A., Harris, S., Ness, J., Eagle, C., Lenkey, G., Williams, T.: Gray Hat Hacking: The Ethical Hacker's Handbook, 3rd edn. McGraw-Hill, New York (2011)
9. Kan, M., Thi, H.O.N.: Fast webpage classification using URL features. In: Herzog, O., Schek, H., Fuhr, N., Chowdhury, A., Teiken, W. (eds.) Proceedings of the 2005 ACM CIKM International Conference on Information and Knowledge Management, pp. 325–326. ACM, Bremen, October 31 - November 5 (2005)
10. Ma, J., Saul, L.K., Savage, S., Voelker, G.M.: Beyond blacklists: learning to detect malicious web sites from suspicious urls. In: IV, J.F.E., Fogelman-Soulié, F., Flach, P.A., Zaki, M.J. (eds.) Proceedings of the 15th ACM SIGKDD International Conference on Knowledge Discovery and Data Mining, pp. 1245–1254. ACM (2009)
11. Pedregosa, F., Varoquaux, G., Gramfort, A., Michel, V., Thirion, B., Grisel, O., Blondel, M., Prettenhofer, P., Weiss, R., Dubourg, V., Vanderplas, J., Passos, A., Cournapeau, D., Brucher, M., Perrot, M., Duchesnay, E.: Scikit-learn: machine learning in python. J. Mach. Learn. Res. **12**, 2825–2830 (2011)
12. Van Hulse, J., Khoshgoftaar, T.M., Napolitano, A.: Experimental perspectives on learning from imbalanced data. In: Proceedings of the 24th International Conference on Machine Learning, ICML 2007, pp. 935–942. ACM, New York (2007). http://doi.acm.org/10.1145/1273496.1273614
13. Wei, Q., Dunbrack Jr., R.L.: The role of balanced training and testing data sets for binary classifiers in bioinformatics. PLoS One **8**(7), e67863 (2013)
14. Weiss, G., Provost, F.: The effect of class distribution on classifier learning: an empirical study. Technical report (2001)

Towards Soft Computing Approaches for Formulating Viable Service Level Agreements in Cloud

Walayat Hussain[1(✉)], Farookh Khadeer Hussain[1], and Omar Khadeer Hussain[2]

[1] School of Software, Decision Support and e-Service Intelligence Lab,
Centre for Quantum Computation and Intelligent Systems,
University of Technology Sydney, Sydney, NSW 2007, Australia
walayat.hussain@student.uts.edu.au, farookh.hussain@uts.edu.au
[2] School of Business, University of New South Wales Canberra, Canberra, Australia
o.hussain@adfa.edu.au

Abstract. A service level agreement (SLA) is a legal document that binds consumers and providers together for the delivery of specific services for a certain period of time. Providers need a viable SLA to maintain successful relationships with consumers. A viable SLA, based on the previous profile of a consumer, will help a service provider determine whether to accept or reject a consumer's request and the amount of resources to offer them. In this paper we propose a soft-computing based approach to form a personalized and viable SLA. This process is carried out in the pre-interaction time phase. We build a Fuzzy Inference System (FIS) and consider a consumer's *reliability value* and *contract duration* as the input factors to determine the amount of resources to offer to the consumer. In addition to the Fuzzy Inference System, we tested various Neural Network-based methods for viable SLA formation and compared their prediction accuracy with the output of the FIS.

Keywords: Viable SLAs · SLA monitoring · SLA violation prediction · Fuzzy Inference System

1 Introduction

A service level agreement (SLA) is a legal document between a service provider and its consumers that describes the services that will be offered by the provider and the obligations of the both parties. An SLA is used by a consumer to quantify, compare and, in many cases, select services offered by a provider with that of other providers. An SLA is composed of one or many service level objectives (SLOs). The reliability, or trustworthiness, of a service provider depends on the successful completion or failure of all committed SLOs. Additionally, the reliability of the consumer is computed as a function of successfully completing their part of the specified obligations in the SLA. Subsequent to the completion of each transaction, the provider evaluates the promised behaviour and the actual behavior of a given consumer and this impacts the overall reputation of the consumer.

© Springer International Publishing Switzerland 2015
S. Arik et al. (Eds.): ICONIP 2015, Part IV, LNCS 9492, pp. 639–646, 2015.
DOI: 10.1007/978-3-319-26561-2_75

Given a cloud consumer's demand for resources can fluctuate, cloud service providers virtualize the underlying resources offered, so they can be varied on request It is vital for cloud providers, particularly small ones, to optimally allocate and commit their resources across their consumer base. Therefore, it is important for cloud providers to form SLAs that take into account both the specific requirements of the consumer and the provider's available resources – this makes a personalized and viable SLA.

In this paper, we propose a personalized and viable SLA lifecycle that assists cloud providers to form optimized SLAs with consumers based on their previous usage profile. We propose the use of a Fuzzy Inference System (FIS) and apply a Mamdani approach for determining resource requirements. The FIS uses two inputs, the *consumer's reliability* and the *contract duration*, to determine the amount of resources to be offered to the consumer. We subsequently make use of: four supervised neural network learning methods – feed-forward, cascade-forward, Elman back propagation, and nonlinear autoregressive exogenous model (NARX); one un-supervised neural network – radial basis function (RBF); and a generalized regression method for resource allocation determination. Finally, we compare the performance of the FIS with that of the proposed neural network-based approaches to determine the optimal method for forming a viable and personalized SLA. This is the first work that proposes a viable SLA lifecycle. Furthermore, this is the first paper that explores the formation of personalized and viable SLAs using soft-computing-based approaches such as FIS, neural-network and regression methods.

The rest of the paper is structured as follows. Section 2 outlines related work and identifies gaps in existing literature. Section 3 describes our proposed viable SLA lifecycle. Section 4 describes the FIS system for viable SLA formation and variable resource allocation. Section 5 validates the FIS and analyzes its accuracy by comparing it with different neural network approaches. Section 6 concludes the paper and outlines directions for future work.

2   Related Work

The process of cloud service management is divided into two time phases, the pre-interaction time phase and the post-interaction time phase[1]. The pre-interaction time phase includes all processes prior to finalizing an SLA, such as negotiation and prediction of likely resource usage. The post-interaction time phase includes all activities and processes subsequent to both parties forming an SLA, such as service monitoring, violation prediction and recommendations.

There is some related literature that describes different proactive and reactive SLA-monitoring approaches. Proactive monitoring approaches, predict SLA violation before it occurs and reactive monitoring approaches, monitor SLAs during runtime. In our previous work [2] we categorized existing SLA-monitoring approaches into four classes based on their attributes. A key shortcoming of existing literature is that none of the approaches are able to form a viable SLA based on predictions of likely resource usage in the pre-interaction time phase. Viable SLAs enable service providers to better manage their resources by optimally allocating and utilizing all of them. Additionally, they achieve effective SLA monitoring by initiating SLA monitoring in the pre-interaction time phase. Some approaches to SLA monitoring are described below.

The traffic anomaly detection system (TADE), proposed by Oliviera et al. [3], provides a decision support system that assists cloud providers and consumers in determining the various parameters of a transaction, like resource arrangement, calculation of cost and resource prediction for future use. Their proposed monitoring architecture is comprised of five modules and the architecture was validated in a federated cloud environment. The authors conclude that displaying all relevant information assists cloud providers and consumers in negotiating, forming and finalizing their SLA.

An automated service level agreement management system was first proposed by Lu et al. [4], in which the life cycle of an SLA is automatically managed by agents. To achieve a fault tolerant system, they developed a hierarchy of agents and established relationships between them. They proposed an SLA life cycle comprising of three phases – creation, operation and removal. These three phases were then further decomposed into six sub-steps – three in the creation phase, one in the operation phase, and two in the removal phase. The authors defined an SLA as the process during which the consumers negotiate with service providers for a required service, then both parties consent and establish their agreement. During SLA monitoring the service provider monitors the behavior of the consumer and detects any violations.

A prediction model, based on the Naive Bayesian Classifier, was proposed by Tang et al. [5]. The authors compared different prediction methods on an existing dataset of response times and found that the Bayes method gave optimal results compared to other prediction approaches.

In our previous work Hussain et al. [6], we proposed a profile-based SLA violation prediction approach to predict likely violations in post-interaction time phase. We predicted the consumer's future resource usage based on their previous usage profile. We used Pearson's Correlation Coefficient, by selecting its top-k nearest neighbors and considered their previous resource usage profile. We employed a user-based collaboration filtering method for resource usage prediction. Based on experimental results, we found a consumer's optimal likely resource usage by combining the consumer's resource usage profile and that of its nearest neighbors.

A workflow SLA violation detection control model (WSVDC) was proposed by Sun et al. [7], in which they considered a utility function and control charts for each SLA. Their proposed framework comprised three workflow modules – SLA aggregation and assessment, calculation of SLA, and detection of violation. Performance was measured against four QoS parameters: response time; cost; reputation; and reliability. They used a Western Electric rule to monitor service behavior and detect service violations.

3   Viable SLA lifecycle

The basic lifecycle of an SLA was proposed by Ron et al. [8]. It comprises three phases a) finding the best-matched service provider; b) accessing the services; and c) terminating the SLA when delivery is complete. Wustenhoff et al. [9] further divided these three phases into six steps – (i) find a suitable service provider; (ii) describe the SLA;

(iii) formalize the agreement; (iv) monitor the SLA; (v) terminate the SLA; and (vi) apply penalties to the violating party. Existing literature on SLA lifecycles takes into account most of the phases of interaction between the consumer and provider, however there are still many aspects that are missing. One of these aspects is the formation of a viable SLA between the consumer and provider. Viable SLAs predict likely violations of the SLA prior to its formation, and offer variable resources to consumers, based on their *reliability value* and *contract duration*.

Our proposed viable SLA lifecycle, from the provider's perspective, comprises six steps as shown in Figure 1.

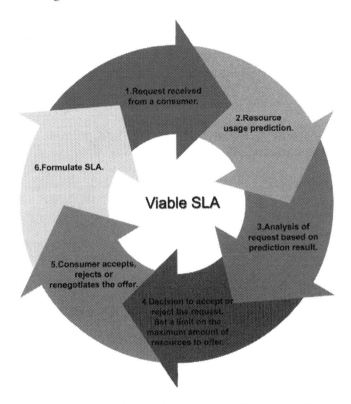

Fig. 1. Viable SLA lifecycle from the provider's perspective

A viable SLA lifecycle starts when the provider receives a formal service request from the consumer detailing: all required deliverables (perhaps articulated as service level objectives (SLOs)), QoS parameters for each SLO, a quantitative description of the requested resources, and the duration and importance of each resource. The likely resource usage of the consumer is then predicted based on their previous usage profile. The decision to accept or decline the consumer's request is based on these results. If the provider decides to accept the request, the amount of resources offered is determined by taking into account the *reliability of the consumer* and the *duration* for which those resources are needed. Offered resources may vary from one consumer to another. Priority is given to those consumers whose reliability value is high and contract duration is small, as this presents the lowest risk of violation and highest degree of flexibility in resource allocation. Subsequently, the consumer is informed that their request for resources has been accepted with a proposed resource allocation, or it has been rejected. The consumer may accept, reject or negotiate the offer of resources and this process continues until both parties mutually agree.

4   Fuzzy Inference System (FIS) for Dynamic Resource Allocation

In this section we outline and discuss the workings of a Fuzzy Inference System (FIS) to form viable SLAs. In this FIS system we make use of the Mamdani approach and use two parameters as input – the *reliability of the consumer* and the *contract duration* to determine the resources to be offered. The reliability of the consumer is determined by successfully committing to previously agreed objectives. At the end of each transaction, the provider assesses the consumer's commitment against the agreed objectives. These objectives are expressed in terms of resource usage. Consumers that have successfully complied with the transaction's commitments gain a one point increase to their reliability value and vice versa. Consumers are then categorized into three classes – bronze, silver and gold – based on their reliability value as shown in Table 1.

Table 1.   Input 1- Consumer class and their reliability value

Linguistic value	Numerical range for reliability values
Bronze	[01- 40]
Silver	[35- 75]
Gold	[70 -100]

Table 1 shows three classes of consumers. Bronze consumers are those consumers who are relatively new (compared to other customers) and hence their reliability value is not very good. The reliability value for a bronze consumers ranges from 1 to 40. The silver consumer's reliability ranges from 35 to 75. Gold consumers are the most reliable and their reliability value ranges from 70 to 100. To achieve gold status, a consumer needs to successfully complete at least 71 transactions.

The second input to the FIS is contract duration. Providers prefer requests for resources over short or medium lengths of time. We categorize contract duration into three categories – *short, medium* and *long* – as represented in Table 2.

The output of the FIS is the amount of resources, as a percentage of the provider's total resources, to be offered to the requesting consumer. We categorize the amount of resources to be allocated into three categories – low, medium and high. These ranges are represented in Table 3.

Table 2. Input 2 - Contract and their duration

Linguistic value	Numerical range for contract duration
Short	[01- 04] months
Medium	[03- 08] months
Long	[07 -12] months

Table 3. Output - Percentage of resources to offer

Linguistic value	Numerical range for resources to offer
Less resource	[01- 30] %
Medium resource	[20- 60] %
High resource	[50 -100] %

The FIS rules for resource allocation are represented in Table 4.

Table 4. FIS rules for resource allocation

Rule	If reliability is	and contract duration is	Then	Resources to be allocated
1	bronze	short		medium
2	silver	short		high
3	gold	short		high
4	bronze	medium		low
5	silver	medium		medium
6	gold	medium		medium
7	bronze	long		low
8	silver	long		low
9	gold	long		medium

5   Validate the FIS and Analyze Using Neural Nand Regression-Based Methods

To validate the performance of the fuzzy inference system we considered the reliability and contract duration from 10 prospective consumers and determined the amount of resources to be offered to those requests. The results are shown in Table 5.

Table 5. Amount of resource to offer using soft computing methods

Input 1 Reliability	Input 2 Contract duration (months)	FIS	Feed-forward back prop	Cascade-forward back prop	Elman backprop	Generalized regression	Radial basis (fewer neurons)	NARX
12	10	15%	15%	14%	15%	15%	15%	14%
42	3	75%	64%	70%	69%	61%	65%	60%
43	9	15%	15%	19%	14%	16%	15%	12%
14	1	40%	35%	37%	38%	40%	40%	39%
6	8	15%	15%	15%	16%	15%	15%	18%
64	10	27%	28%	27%	29%	28%	25%	27%
80	8	40%	40%	42%	39%	40%	43%	39%
10	5	15%	15%	13%	14%	16%	14%	16%
43	1	75%	71%	70%	72%	75%	57%	75%
19	3	40%	44%	44%	40%	33%	35%	43%

In the above table, we set the FIS output as the benchmark and compared evaluations using a regression-based method with various neural network methods to assess the optimal approach. The neural network methods we chose were: feed-forward back propagation; cascade-forward back propagation; Elman back propagation; Radial basis function and NARX. The regression-based method was generalized regression. We trained the neural network methods on an inputs and output dataset of 1000 consumers derived from the FIS. The inputs were *reliability of consumer* and *contract duration* and the output was the percentage of resources to allocate. We used the Root Mean Square Error (RMSE) and Mean Absolute Deviation (MAD), represented in Table 6, as the benchmark to evaluate the accuracy of each method.

Table 6. RMSE and MAD of neural network methods

	Feed-forward back prop	Cascade-forward back prop	Elman back prop	Generalized regression	Radial basis (fewer neurons)	NARX
RMSE	1.671552	1.234634	0.914864	1.995465	2.737062	1.952632
MAD	0.407625	0.411317	0.271173	0.367436	0.661833	0.441053

From the above table we can see that among all neural network methods, the Elman back propagation method gives most optimal results with the FIS output. The RMSE for the Elman back propagation method is 0.914864 and the MAD is 0.271173 which is less than all other neural network methods.

6 Conclusion

To avoid SLA violation and to increase trust values, service providers always want to predict service violations before they happen. Existing literature focuses on SLA violation in the post-interaction time phase when both parties have formed and finalized their SLA. However, there is a need to form a viable SLA in the pre-interaction time phase to determine likely the resource usage of requesting consumers and offer an amount of resources based on their reliability or trustworthiness value. We applied different soft computing approaches by giving reliability and contract duration as inputs, and a proportion of the provider's resources to offer as an output to formulate a viable SLA. In the future we will investigate the impact of a viable SLA in the post-interaction time phase.

References

1. Hussain, O.K., et al.: Risk assessment and management in the networked economy. Vol. 412. Springer, Heidelberg (2012)
2. Hussain, W., Hussain, F.K., Hussain, O.K.: Maintaining Trust in Cloud Computing through SLA Monitoring. In: Loo, C.K., Yap, K.S., Wong, K.W., Beng Jin, A.T., Huang, K. (eds.) ICONIP 2014, Part III. LNCS, vol. 8836, pp. 690–697. Springer, Heidelberg (2014)
3. Oliveira, A.C., et al.: Efficient network service level agreement monitoring for cloud computing systems. In: 2014 IEEE Symposium on Computers and Communication (ISCC). IEEE (2014)
4. Lu, K., et al.: Fault-tolerant Service Level Agreement lifecycle management in clouds using actor system. Future Generation Computer Systems (2015). https://scholar.google.com.au/scholar?hl=en&q=Fault-tolerant+Service+Level+Agreement+lifecycle+management+in+clouds+using+actor+system&btnG=&as_sdt=1%2C5&as_sdtp=
5. Tang, B., Tang, M.: Bayesian model-based prediction of service level agreement violations for cloud services. In: Theoretical Aspects of Software Engineering Conference (TASE). IEEE (2014)
6. Hussain, W., F.K. Hussain, and O.K. Hussain. Comparative Analysis of Consumer Profile-based Methods to Predict SLA Violation. in IEEE International Conference on Fuzzy Systems: Istanbul. IEEE, Turkey (2015)
7. Tan, W., Li, L., Sun, Y.: A Novel Performance Prediction Framework for Web Service Workflow Applications. In: Zu, Q., Hu, B., Gu, N., Seng, S. (eds.) HCC 2014. LNCS, vol. 8944, pp. 55–68. Springer, Heidelberg (2015)
8. Ron, S., Aliko, P.: Service level agreements. Internet NG. Internet NG project (1999-2001) http://ing.ctit.utwente.nl/WU2, 2001
9. Wustenhoff, E., BluePrints, S.: Service level agreement in the data center. Sun Microsystems Professional Series (2002)

ARPPS: Augmented Reality Pipeline Prospect System

Xiaolei Zhang[1], Yong Han[1(✉)], DongSheng Hao[1], and Zhihan Lv[2]

[1] Engineering Research Center of Marine Information Technology,
College of Information Science and Engineering, Ocean University of China, Qingdao,
People's Republic of China
chinahanyong@126.com
[2] Shenzhen Institutes of Advanced Technology (SIAT), Shenzhen, People's Republic of China

Abstract. Outdoor augmented reality geographic information system (ARGIS) is the hot application of augmented reality over recent years. This paper concludes the key solutions of ARGIS, designs the mobile augmented reality pipeline prospect system (ARPPS), and respectively realizes the machine vision based pipeline prospect system (MVBPPS) and the sensor based pipeline prospect system (SBPPS). With the MVBPPS's realization, this paper studies the neural network based 3D features matching method.

Keywords: Augmented reality · Pipeline prospect · ARGIS · Neural network

1 Introduction

Since the human activities are related with the geographic information all the time, it is very natural and extremely important to integrate the augmented reality technology with the GIS. Since its generation, the augmented reality has gone through several development phases: computer indoor augmented reality, helmet-type outdoor augmented reality and mobile based outdoor augmented reality. Its application is gradually converted from the traditional exhibition to the application combined with the industry. This paper studies the mobile based outdoor augmented reality geographic information system (ARGIS) [1–3] and analyzes it by combining the examples for underground pipeline prospect. In the process of the realization of the examples, this paper also studies the neural network based 3D features matching method to optimize the feature matching process.

The traditional GIS application is displayed on the map in the form of 2D symbols after analyzing the result based on the 2D maps. However, due to the lack of strong intuition, it is difficult for the user with weak geographic knowledge to distinguish it; even the field experts also have some difficulties in distinguishing in a complicated environment. For recent years, with the 3DGIS and VRGIS development, the 3D virtual model of real environment was established to replace the 2D map [4], which improves the intuition of marker distinguishing and promotes the visualization development of GIS. However, the 3D virtual model has high requirements on hardware while it still has difference with the real world scene. As a matter of fact, the real world scene is a complicated and perfect "3D space map" for itself. If the real world scene is taken as the "map" for GIS application operation and the inquired and analyzed result is displayed

© Springer International Publishing Switzerland 2015
S. Arik et al. (Eds.): ICONIP 2015, Part IV, LNCS 9492, pp. 647–656, 2015.
DOI: 10.1007/978-3-319-26561-2_76

on the real world scene via virtual information to achieve the interaction between the virtual space geographic information and the real world scene, so the GIS application experience can be enhanced.

Carry out scene augmentation [5, 6] for human's visual system by organically integrating the 2D or 3D pictures, text notes and other virtual information generated by the computer into the real world scene which can be seen by the user. The augmented reality has the effect of expressing the sense of reality but cannot store and control data. However, GIS has the functions of storage, management and analysis of space information, which precisely makes up for this deficiency. The integration of them can provide the user with the location-based service and can combine all kinds of virtual space information stored in the GIS space data with the scene actually observed [7]. Besides, such integration can not only enhance the GIS user's sense of real environment and the interactive experience but also supplies a new means for the research on GIS visualization.

Sun M [8] (2004) brought forward the concept of ARGIS(Augmented reality geographic information system), a geographic information system, digitally describes, stores and controls the objective geographic world, meanwhile, integrates such descriptions into the real world, offers the space information of a designated object and supplies the outdoor mobile information interaction. Guo Y and others (2008) pointed out that the significance of ARGIS is to apply the mobile computing and augmented reality technologies to the traditional space information service to change the traditional location-based service mechanism, which will provide the space-based industry system with a brand-new industrial pattern [9].

Gethin D's article [10] of 2002 adapted GPS/INS for 3D registration of outdoor geographic information, realizing the visual presentation of outdoor underground facility structure. The subsurface data visualization system of University of Nottingham, which uses the GPS/ INS integrated system in system registration, can carry out the 3D visualization for all kinds of the following subsurface characteristics such as geologic structure, underground pipe system, underground land pollution zone etc. Gerhard Schall has published 15 articles on outdoor augmented reality since 2008, emphasizing the application of hardware sensor. One of these articles written in 2009 [11] designs an underground facility visualization augmented reality system frame and an integrated-form handheld device and integrates GPS, camera, wireless network and other devices but is not the mobile-phone-based application. The 3D registration uses the pure hardware sensor technology (GPS + inertia measurement equipment). Gerhard Schall started to research the visual and sensor hybrid tracking registration method [12] in 2010 and obtained a series of research achievements [13–15]. The article [16] designs an industrial solution, using GPS positioning and CAD pipeline data in the real environment; it highly depends on the accuracy of GPS and there is no explanations about the orientation, tracking and other technical details [17–19].

2   Methods

Based on ARGIS framework, this paper achieves the mobile terminal augmented reality application system for urban underground pipeline prospecting (ARPPS). The system takes the underground pipeline data in Qingdao (500 square kilometers of construction area and 10 million population) as the experimental data to have achieved the machine vision based underground pipeline prospect system (MVBPPS) and sensor based underground pipeline prospect system (SBPPS).

2.1   Spatial Data Organization and Distribution

Shape-format vector pipeline data is totally 2.7 G. According to geometrical characteristics, data are divided into two types: pipeline and pipe point; according to the functional attributes, data are divided into 13 types including covered channel, power line carrier, power supply pipeline, monitoring signal pipeline, street lamp pipeline, hot water pipeline, drinking water (feed water) pipeline, natural gas pipeline, communication pipeline, sewage pipeline, rainwater pipeline, integrated pipeline, reclaimed water pipeline.

As for the Shape-format vector pipeline data, the spatial database is imported using ArcGIS; the PostGIS open source database is adopted as the spatial database; the spatial data network services interface satisfying OGC standards is researched and developed independently. The data services interface receives the parameters within the visional field of the camera at client side, and the visional field of camera is expressed with a rectangle, and the range of visional field is the coordinates of the lower left corner and upper right corner of the rectangle [Lon_Min, Lat_Min, Lon_Max, Lat_max]; parameters are transmitted in the form of URL parameters under http protocol (range of visional field obtained by calculation at client side). These parameters are used for pipeline data retrieval, and results are returned to client side in the GeoJSON format [38–42, 44, 45].

2.2   3D Registration and Tracking

MVBPPS: The SFM (Structure from Motion) technology is adopted to achieve the 3D registration and tracking. SFM is a kind of registration and tracking method based on natural scene reconstruction and is a kind of on-line reconstruction method. Systems mentioned in this paper utilize the PTAM (Parallel Tracking and Mapping) system, can process thousands of natural feature points in a real-time way under the PC environment, and have good performance in the aspects of precision and robustness. Klein [20] and other persons proposed the on-line reconstruction method to solve the problem of natural scene registration in mobile augmented reality, and they revised the PTAM system to achieve its operation in mobile equipment; experimental results indicate that the revised system can achieve the reconstruction and registration of small scenes, although its precision and robustness reduce slightly [56].

PTAMM(Parallel Tracking and Multiple Mapping) [21] extends PTAM system to allow it to use multiple independent cameras and multiple maps.

Recently, many scholars have used neural network to optimize the computing process of features matching. Transiently chaotic neural network (TCNN) [22] is one of the most representative. TCNN exploiting the rich behaviors of nonlinear dynamics have been developed as a new approach to extend the problem solving ability of standard HNN [23]. The TCNN model can be presented as follows:

$$x_i(t) = \frac{1}{1 + e^{-y_i(t)(1+\varepsilon)}} \qquad (1)$$

$$y_i(t+1) = ky_i(t) + \alpha \left(\sum_{j=i,j\neq ij}^{n} w_{ji}x_j(t) + I_i \right) - z_i(t) \left(x_i(t) - I_0 \right) \qquad (2)$$

$$z_i(t+1) = (1 - \beta)z_i(t) \qquad (3)$$

where $(i = 1,2,..., n)$, x_i = output of neuron i, y_i = internal state of neuron i, w_{ij} = connection weight from neuron j to neuron i, $w_{ij} = w_{ji}$, I_i = input bias of neuron i, α = positive scaling parameter for inputs, k = damping factor of nerve membrane ($0 \leq k \leq 1$), $z_i(t)$ = self-feedback connection weight (refractory strength) ≥ 0, β = damping factor of $z_i(t)$, $0 < \beta < 1$, I_0 = positive parameter, ε = steepness parameter of the output function ($\varepsilon > 0$).

When matching the image features, the neural network structure is a 2D array. If the reference image is represented by a set of feature points G, which size is M; Then an input scene image that consist of several overlapping objects, can be represented as another set of feature points S, which size is N. The number of neurons in the network will be MN. The output status of the neural v_{ij} represents the matching state of point No. i, and No. j. If i matched j, v_{ij} will be set to 1, else v_{ij} will be set to 0 [24].

The common method for SFM is feature based approach (see [25] for more details). The feature base SFM depends on robust feature detection and matching, geometric image transformation, image stitching and adjustment. For robust feature detection and matching in two corresponding images, SIFT feature [26] detector with FLANN [27] match is normally employed to extract the key points in a given video sequence [28–30].

SBPPS: Gyroscope, GPS, electronic compass, accelerometer and other sensor data are integrally used and fused to achieve 3D registration and tracking of virtual objects.

The difference between the 3D registration process and the traditional augmented reality registration process lies in determination of spatial range of visualized pipeline data. In the process of determining the spatial range of visualized pipeline data, the positioning parameters [x, y] of GPS and the loading radius r are utilized; spatial range parameters of rectangle are [x-r, y-r, x + r, y + r], in which x is longitude, y is latitude and r is the distance value of 10 m range in the World Coordinate System. After determination of spatial range for pipeline, data are obtained through service invocation at the server side, and are rendered to the screen or projector through coordinate transformation [43].

The process of tracking is achieved by adopting the angular acceleration of gyroscope and accelerometer as well as the direction parameters of compass to perform the real-time monitoring and to inversely calculate the position alteration of the virtual object. Currently, the problems of this method are the shaking and drifting phenomena due to the limited precision of mobile sensor. In this experiment, the weighed recursive average filtering algorithm [31] is adopted for data processing and shaking prevention. The weighed recursive average filtering algorithm is a kind of low-pass filtering algorithm; take the accelerometer as an example, and the corresponding parameters of the three coordinates x, y and z are Ax, Ay and Az respectively. Firstly, define the distance between new and old parameters as the standard for the degree of parameter shaking; new parameters are defined as A'_x, A'_y, A'_z.

$$d = \sqrt{(A'_x - A_x)^2 + (A'_y - A_y)^2 + (A'_z - A_z)^2} \tag{4}$$

Then, take the distance as the independent variable to obtain the weighting coefficient α of low-pass filter; the value of α is obtained based on the range segmentation of d. Finally, multiply the parameter variation value with the factor weighting coefficient in the weighed form and superpose it to the before-change data to obtain the new parameter value [32]. After the filtering processing of sensor signal, the shaking is reduced to some extent, and users' experience effect is improved; however, due to the limited precision of mobile sensor, simply adopting a low-pass filtering method cannot thoroughly solve the problem [51–55].

2.3 Mapping the Sense of Reality on Virtual-Real Fusion

In this paper, the virtual sectioning is adopted for both MVBPPS and SBPPS, to achieve the mapping of the sense of reality on fusion between underground pipeline and ground; MVBPPS adopts the all-sight rectangular tunnel sectioning, while SBPPS adopts the 180° front-sight circular tunnel sectioning. As for achieving the real fusion between grounds and sectioning tunnel, the OpenGL ES 2.0 mixed technology is utilized to achieve the transparent sheltering and to make the tunnel have a good immersion feeling on ground [33–35].

3   Operational Results

Operating environment of application is nexus 5 mobile, and operating system is android 4.4/5.0.1. Data server: i5-3470 CPU, 16G memory, window server 2013.

MVBPPS is subjected to multi-environment tests, and operational results are tested in different lighting environments and under different ground texture conditions; testing environments include: sufficient sunshine at noon, twilight in the afternoon, floor tile covered parking lot, bituminous street, lawn, and indoor marble. The results indicate that the system operation will be hardly affected by illumination, as long as the ambient light can guarantee the accurate focusing of the camera; but it has certain influence on the variation of environment textures, and the response time of initial 3D registration is mainly affected. The experimental results are shown as Fig. 1. **SBPPS** will not be affected by illumination or texture environment, so the multi-environment tests are not performed [46–50]. The operation effect is shown as Fig. 2.

Fig. 1. Results of tests on multiple pavements: (1) floor tile covered parking lot; (2) bituminous street; (3) lawn; (4) red floor tile; (5) indoor marble (Color figure online).

Fig. 2. SBPPS operation effect

4   Conclusion

Gerhard Schall's work [11] inspired our research. The 7-point Likert questionnaire method is adopted, and users include personnel in the public facilities GIS field and outdoor working staff. A series of problems about application in mobile terminal are put forward and need to be solved, including 3D modeling, registration and tracking, application interaction, etc. [36, 37].

Based on ARGIS framework, this paper utilizes the Qingdao World Horticultural Expo underground pipeline data to design and achieve the two systems: MVBPPS and SBPPS; MVBPPS adopts the SFM-based registration and tracking, and uses TCNN to optimize the process of 3D feature matching, while SBPPS adopts the multi-sensor processing based registration and tracking; expanded application is achieved based on Google Glass and Google Cardboard. It can be seen from research and analysis in this paper that machine vision and sensor show different advantages in the application research of outdoor augmented reality; with the improvement of intelligent mobile terminal processing capacity and sensor precision, it can be imaged that the application of mobile phone based outdoor augmented geologic information system will exist everywhere, changing people's working mode and lifestyle.

Acknowledgements. This research is supported by Innovation Fund for Technology Based Firms, China (No: 14c26211100180) and Qingdao science and technology project of China (No: 14-9-2-12-pt).

References

1. Wang, K., Zhou, X., Chen, H., Lang, M., Raicu, I.: Next generation job management systems for extreme-scale ensemble computing. In: Proceedings of the 23rd International Symposium on High-performance Parallel and Distributed Computing, pp. 111–114. ACM, Vancouver, BC, Canada (2014)
2. Li, T., Zhou, X., Brandstatter, K., Raicu, I.: Distributed key-value store on HPC and cloud systems. In: 2nd Greater Chicago Area System Research Workshop (GCASR). Citeseer (2013)
3. Lv, Z., Esteve, C., Chirivella, J., Gagliardo, P.: A game based assistive tool for rehabilitation of dysphonic patients. arXiv preprint arXiv:1504.01030 (2015)
4. Lv, Z., Réhman, S.U., Chen, G.: WebVRGIS: A P2P network engine for VR data and GIS analysis. In: Lee, M., Hirose, A., Hou, Z.-G., Kil, R.M. (eds.) ICONIP 2013. LNCS, vol. 8226, pp. 503–510. Springer, Heidelberg (2013)
5. Azuma, R.T.: A survey of augmented reality. Presence **6**, 355–385 (1997)
6. Zlatanova, S.: Augmented reality technology. Delft University of Technology, Faculty of Civil Engineering and Geosciences, Department of Geodesy, Section GIS Technology (2002)
7. King, G.R., Piekarski, W., Thomas, B.H.: ARVino-outdoor augmented reality visualisation of viticulture GIS data. In: 2005 Proceedings of Fourth IEEE and ACM International Symposium on Mixed and Augmented Reality, pp. 52–55. IEEE (2005)
8. Min, S., Mei, L., Feizhou, Z., Zhipeng, W., Daozheng, W.: Hybrid tracking for augmented reality GIS registration. In: Japan-China Joint Workshop on Frontier of Computer Science and Technology, FCST 2007, pp. 139–145. IEEE (2007)
9. Guo, Y., Qingyun, Y.L., Zhang, W., Xu, L.: Application of augmented reality GIS in architecture. Int. Arch. Photogrammetry Remote Sens. Spat. Inf. Sci. **37**, 331–336 (2008)
10. Roberts, G.W., Evans, A., Dodson, A., Denby, B., Cooper, S., Hollands, R.: The use of augmented reality, GPS and INS for subsurface data visualization. In: FIG XXII International Congress, pp. 1–12 (2002)
11. Schall, G., Mendez, E., Kruijff, E., Veas, E., Junghanns, S., Reitinger, B., Schmalstieg, D.: Handheld augmented reality for underground infrastructure visualization. Pers. Ubiquit. Comput. **13**, 281–291 (2009)
12. Schall, G., Mulloni, A., Reitmayr, G.: North-centred orientation tracking on mobile phones. In: 2010 9th IEEE International Symposium on Mixed and Augmented Reality (ISMAR), pp. 267–268. IEEE (2010)
13. Schall, G., Zollmann, S., Reitmayr, G.: Smart vidente: advances in mobile augmented reality for interactive visualization of underground infrastructure. Pers. Ubiquit. Comput. **17**, 1533–1549 (2013)
14. Langlotz, T., Degendorfer, C., Mulloni, A., Schall, G., Reitmayr, G., Schmalstieg, D.: Robust detection and tracking of annotations for outdoor augmented reality browsing. Comput. Graph. **35**, 831–840 (2011)
15. Schall, G., Mulloni, A., Reitmayr, G.: North-centred orientation tracking on mobile phones. In: 2010 9th IEEE International Symposium on Mixed and Augmented Reality (ISMAR), pp. 267–268. IEEE (2010)

16. Behzdan, A.H., Kamat, V.R.: Interactive augmented reality visualization for improved damage prevention and maintenance of underground infrastructure. In: Proceedings of the 2009 Construction Research Congress, Seattle, WA, pp. 1214–1222 (2014)

17. Yin, T., Han, Y., Chen, Y., Chen, G.: WebVR——web virtual reality engine based on P2P network. J. Netw. **6**, 990–998 (2011)

18. Lv, Z., Halawani, A., Feng, S., Li, H., Réhman, S.U.: Multimodal hand and foot gesture interaction for handheld devices. ACM Trans. Multimedia Comput. Commun. Appl. (TOMM) 11, 10 (2014)

19. Lv, Z., Halawani, A., Fen, S., Li, H.: Touch-less Interactive Augmented Reality Game on Vision Based Wearable Device. arXiv preprint arXiv:1504.06359 (2015)

20. Klein, G., Murray, D.: Parallel tracking and mapping on a camera phone. In: 2009 8th IEEE International Symposium on Mixed and Augmented Reality, ISMAR 2009, pp. 83–86. IEEE (2009)

21. Castle, R., Klein, G., Murray, D.W.: Video-rate localization in multiple maps for wearable augmented reality. In: 2008 12th IEEE International Symposium on Wearable Computers, ISWC 2008, pp. 15–22. IEEE (2008)

22. Chen, L., Aihara, K.: Chaotic simulated annealing by a neural network model with transient chaos. Neural Netw. **8**, 915–930 (1995)

23. Hopfield, J.J., Tank, D.W.: "Neural" computation of decisions in optimization problems. Biol. Cybern. **52**, 141–152 (1985)

24. Li, X., Chen, D.: Augmented reality in e-commerce with markerless tracking. In: 2010 The 2nd IEEE International Conference on Information Management and Engineering (ICIME), pp. 609–613. IEEE (2010)

25. Robertson, D., Cipolla, R.: Structure from motion. In: Practical image processing and computer vision. John Wiley, Hoboken, NJ, USA 49 (2009)

26. Lowe, D.G.: Distinctive image features from scale-invariant keypoints. Int. J. Comput. Vision **60**, 91–110 (2004)

27. Muja, M., Lowe, D.G.: Fast approximate nearest neighbors with automatic algorithm configuration. In: VISAPP, vol. 1, p. 2 (2009)

28. Richard, H., Andrew, Z.: Multiple View Geometry in Computer Vision. Cambridge University Press, New York (2003). ISBN 521, 8

29. Brown, M., Lowe, D.G.: Automatic panoramic image stitching using invariant features. Int. J. Comput. Vision **74**, 59–73 (2007)

30. Yousefi, S., Kondori, F.A., Li, H.: 3D gestural interaction for stereoscopic visualization on mobile devices. In: Real, P., Diaz-Pernil, D., Molina-Abril, H., Berciano, A., Kropatsch, W. (eds.) CAIP 2011, Part II. LNCS, vol. 6855, pp. 555–562. Springer, Heidelberg (2011)

31. Milette, G., Stroud, A.: Professional Android Sensor Programming. Wiley, Indianapolis (2012)

32. Ruiz-Ruiz, A.J., Lopez-de-Teruel, P.E., Canovas, O.: A multisensor LBS using SIFT-based 3D models. In: 2012 International Conference on Indoor Positioning and Indoor Navigation (IPIN), pp. 1–10. IEEE (2012)

33. Yang, Y., Pintus, R., Rushmeier, H., Ivrissimtzis, I.: A steganalytic algorithm for 3D polygonal meshes. In: 2014 IEEE International Conference on Image Processing (ICIP), pp. 4782–4786. IEEE (2014)

34. Zhao, D., Zhang, Z., Zhou, X., Li, T., Wang, K., Kimpe, D., Carns, P., Ross, R., Raicu, I.: FusionFS: Toward supporting data-intensive scientific applications on extreme-scale high-performance computing systems. In: 2014 IEEE International Conference on Big Data (Big Data), pp. 61–70. IEEE (2014)

35. Yang, Y., Pintus, R., Rushmeier, H., Ivrissimtzis, I.: A steganalytic algorithm for 3D polygonal meshes. In: 2014 IEEE International Conference on Image Processing (ICIP), pp. 4782–4786. IEEE (2014)
36. Yang, Y., Peyerimhoff, N., Ivrissimtzis, I.: Linear correlations between spatial and normal noise in triangle meshes. IEEE Trans. Visualization Comput. Graphics **19**, 45–55 (2013)
37. Yang, Y., Ivrissimtzis, I.: Polygonal mesh watermarking using Laplacian coordinates. In: Computer Graphics Forum, pp. 1585–1593. Wiley Online Library (2010)
38. Jiang, D., Ying, X., Han, Y., Lv, Z.: Collaborative multi-hop routing in cognitive wireless networks. Wireless Pers. Commun. (2015)
39. Wang, K., et al. Optimizing load balancing and data-locality with data-aware scheduling. In: 2014 IEEE International Conference on Big Data (Big Data). IEEE (2014)
40. Li, T., et al.: ZHT: A light-weight reliable persistent dynamic scalable zero-hop distributed hash table. In: 2013 IEEE 27th International Symposium on Parallel & Distributed Processing (IPDPS). IEEE (2013)
41. Zhou, X., et al.: Exploring Distributed Resource Allocation Techniques in the SLURM Job Management System. Illinois Institute of Technology, Department of Computer Science, Technical Report (2013)
42. Su, Y., et al.: In-situ bitmaps generation and efficient data analysis based on bitmaps. In: Proceedings of the 24th International Symposium on High-performance Parallel and Distributed Computing, pp. 61–72. ACM (2015)
43. Yang, J., He, S., Lin, Y., Lv, Z.: Multimedia cloud transmission and storage system based on internet of things. Multimedia Tools and Applications (2016)
44. Wang, Y., Su, Y., Agrawal, G.: A novel approach for approximate aggregations over arrays. In: Proceedings of the 27th International Conference on Scientific and Statistical Database Management. ACM (2015)
45. Wang, K., et al.: Using simulation to explore distributed key-value stores for exascale system services. In: 2nd Greater Chicago Area System Research Workshop (GCASR) (2013)
46. Ou, W., Lv, Z., Xie, Z.: Spatially regularized latent topic model for simultaneous object discovery and segmentation. In: The 2015 IEEE International Conference on Systems, Man, and Cybernetics (SMC 2015) (2015)
47. Lin, Y., Yang, J., Lv, Z., Wei, W., Song, H.: A self-assessment stereo capture model applicable to the internet of things. Sensors **15**(8), 20925–20944 (2015)
48. Chen, Z., Huang, W., Lv, Z.: Uncorrelated discriminant sparse preserving projection based face recognition method. Multimedia Tools and Applications (2016)
49. Gu, W., Lv, Z., Hao, M.: Change detection method for remote sensing images based on an improved Markov random field. Multimedia Tools and Applications (2016)
50. Zhang, S., Zhang, X., Ou, X.: After we knew it: empirical study and modeling of cost-effectiveness of exploiting prevalent known vulnerabilities across IaaS cloud. In: Proceedings of the 9th ACM Symposium on Information, Computer and Communications Security. ACM (2014)
51. Wang, K., et al.: Overcoming hadoop scaling limitations through distributed task execution. In: Proceedings of the IEEE International Conference on Cluster Computing (2015)
52. Dang, S., Ju, J., Matthews, D., Feng, X., Zuo, C.. Efficient solar power heating system based on lenticular condensation. In: 2014 International Conference on Information Science, Electronics and Electrical Engineering (ISEEE) (2014)
53. Yang, J., Chen, B., Zhou, J., Lv, Z.: A portable biomedical device for respiratory monitoring with a stable power source. Sensors 15(8), (2015)

54. Wang, K., et al.: Towards scalable distributed workload manager with monitoring-based weakly consistent resource stealing. In: Proceedings of the 24th International Symposium on High-Performance Parallel and Distributed Computing (2015)
55. Zhang, X., et al.: Spike-based indirect training of a spiking neural network-controlled virtual insect. In: 2013 IEEE 52nd Annual Conference on Decision and Control (CDC). IEEE (2013)
56. Su, T., Wang, W., Lv, Z., Wu, W., Li, X.: Rapid delaunay triangulation for random distributed point cloud data using adaptive hilbert curve. Comput. Graph. **54**, 65–74 (2015)

Identifying SQL Queries Similarity Using SVM

Jihad Zahir[1]([⊠]), Abderrahim El Qadi[2], and Ladjel Bellatreche[3]

[1] LRIT-CNRST, Mohammed V-Agdal University, Rabat, Morocco
jihad.zahir@gmail.com
[2] Team TIM, High School of Technology, Moulay Ismail University, Mekens, Morocco
elqadi_a@yahoo.com
[3] LISI/ENSMA, Poitiers University, Poitiers, France

Abstract. Recognizing that two SQL queries are similar is useful for many applications, such as query recommendation, plan selection and so on. However,questions such as which techniques are needed and which SQL query representation is best to produce accurate similarity estimation remain poorly addressed.

In this work we explore two SQL queries representations proposed in the literature, and study how SVM is accurate to predict SQL queries' similarity using these representations. We use RBF and polynomial kernels to build SVM models. As an additional contribution, we compute a personnalized kernel and compare it against kernels cited above. Results show that one of the studied representations gives better results than the other, and that our proposed kernel is comparable to RBF kernel in terms of accuracy.

Keywords: SVM · Similarity prediction · SQL queries

1 Introduction

Many application scenarios in different fields could benefit from the identification of similarities in the data. One of the intresting application using similarity identification is plan selection for SQL queries. Similarity identification in the context of plan selection means that two queries are said to be similar if the same access plan could be applied to both of them. In this paper, we are studying SQL queries similarity identification by comparing their execution plan and capturing similarity in the plan space not in the query space.To clarify more our definition of similarity, let us consider the following set of queries:

Q1: select ACTVARS.UNITSSOLD from ACTVARS,TIMELEVEL
where ACTVARS.TIME_LEVEL=TIMELEVEL.MONTH_LEVEL
and TIMELEVEL.QUARTER_LEVEL like '%Q2';

Q2: select * from ACTVARS,TIMELEVEL
where ACTVARS.TIME_LEVEL=TIMELEVEL.MONTH_LEVEL
and TIMELEVEL.MONTH_LEVEL='10';

S. Arik et al. (Eds.): ICONIP 2015, Part IV, LNCS 9492, pp. 657–666, 2015.
DOI: 10.1007/978-3-319-26561-2_77

Q3: select UNITSSOLD from ACTVARS,TIMELEVEL
where ACTVARS.TIME_LEVEL=TIMELEVEL.MONTH_LEVEL
and TIMELEVEL.MONTH_LEVEL='10';

Queries $Q2$ and $Q3$ appear very similar syntactically and would return exactly the same output but when given to the optimizer, they produce different execution plans. Oppositely, queries $Q1$ and $Q3$ are different in terms of output but have exactly the same execution plan. In the context of our approach,$Q1$ and $Q3$ are considered to be similar and $Q2$ and $Q3$ are not. Accordingly, we are studying queries' similarity from the execution plan perspective rather than queries' syntax or output. The contributions of this paper can be summarized as follows:

- We propose a summary structure, in the form of a hash table, that clusters queries before similarity prediction.
- We investigate SQL queries similarity prediction using SVM,
- We propose two kernels, use them to build SVM models and conduct an experimental comparison with the RBF and polynomial kernels.
- We explore two representations methods from the litterature and identify which one is better.

The rest of this paper is organized as follows. Section 2 gives an overview of some existing work in the field, Sect. 3 describes in details our approach to predicting similarity including SQL query representation, prediction process, summary structure and kernels building. We present the experimentations and discuss the results in the last section.

2   Related Work

Existent approaches in the literature perform SQL queries similarity identification based on two main phases: (i) Defining a pattern for an appropriate query representation, as well as (ii) developing a similarity function. Queries can be represented at the intentional level by considering the uninterpreted SQL sentence [3] or at the extensional level by using the set of tuples resulting from the query execution [4]. Other query representations range from vectors of features [1] to set of fragments [5]. Graphs [2,6] are sometimes used as a pattern for query representation as well.

The similarity function varies depending on the nature of the problem. [6] uses a simple equality test of queries patterns and the comparison in [7] is based on separate tests of query fragments. Other ways for establishing similarity are based on classical functions applied to the queries representation. For instance, authors of [4] use inner product, while similarity in [4,8] is identified based on cosine distance and Jaccard similarity, respectively. The work in [2] converts the queries similarity identification to a graph-isomorphism problem and compares the queries pattern using the VF graph matching algorithm [9].

3   Approach to Predicting Similarity

3.1   SQL Queries Representation

In this study, new queries are compared to old queries stored in the DBMS log. Since we are not interested in measuring similarity in the syntactical level of SQL statments, we need to represent these queries in a feature space in order to allow further processing. We have identified 2 methods from the litterature that have proposed 2 representation methods to allow comparison of SQL queries from the execution plan perspective.

A. Ghosh et al. [1] used structural and statistical features extracted directly from the inputs to the query optimizer while the work in [2] consists of representing queries in a labeled undirected graph format. The query graph is constructed such as each node N_i of the graph corresponds to a table in the query, and each edge $(N_i, N_i\prime)$ denotes the existence of a join relationship between the tables represented by the nodes N_i and $N_i\prime$. Then, nodes and edges are labeled by number of attributes and a cost model is applied.

We implement these two representations to prepare SQL statements, then we use them to build the SVM models and observe the accuracy of each one. In the rest of this paper representation of [1,2] will be referred to as $QtoF$ and $QtoG$ respectively.

3.2   Building a Summary Structure

We want to organize SQL queries from the DBMS log in a way that queries having the same execution plan are clusterd together. We know that each execution plan is identified by a plan hash value. When we execute a SQL statement in Oracle, a hash value is being assigned to its execution plan. Hence, every plan is identified by its hash value. Given two execution plans, comparing one plan hash value to another easily identifies whether or not the plans are the same. According to Oracle documentation, the plan hash value metric can be used as a shortcut to determine if two execution plans are the same. We benefit from the plan hash value metric to propose a summary structure that categorizes queries. Queries that are executed using the same access plan are mapped together. Figure 1 presents our structure in details. Building this structure will help us define the training set to use in order to generate SVM models. A set of SVM models $m_{0 \le i \le n}$ is stored in this summary structure. The number of models is equal to the number n of clusters that we obtain using the plan hash value metric. A Model $m_{0 \le i \le n}$ is built using a *training set* T which contains queries of cluster i and a set of queries from other clusters. These queries are respectively refered to as positive instances (PI) and negative instances (NI). For instance, in Fig. 1, we will use queries q_1, q_3 and $q_{(n-1)}$ as positive instances to train the model m_2. Models $m_{0 \le i \le n}$ are intended to predict the class value of new queries which is set to 1 if a similarity is detected, and 0 otherwise.

The proposed structure can be easily maintained. When a new query q_{new} arrives, it's tested sequentially by the models. If a similarity is detected by a

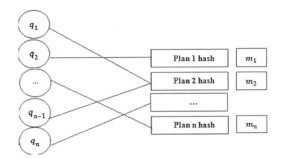

Fig. 1. Overview of the summary structure

model m_i, then q_{new} is executed using plan i, and is added to the set of queries that are associated to the hash value of plan i. If no similarity is detected after running all available models, q_{new} is executed by the DBMS optimizer, then, a new entry is created in the structure adding the hash value of plan $n + 1$ to the list of plans' hash values and associating q_{new} to it. A new model $m_{(n+1)}$ is generated as well.

3.3   Our Proposed Kernels

In the case of nonlinearly separable data, the basic idea of SVM [11–15] use a kernel function to make the data linearly separable. We could define a kernel $K(x, y)$ as a real-valued function $K : X \times X \to \mathbb{R}$ for which there exists a function $\varphi : X \to F$ with the property:

$$K(x, y) = \langle \varphi(x), \varphi(sy) \rangle \tag{1}$$

In order to characterize kernels we use Mercer's Theorem [12] which states: A symmetric function $k(x, y)$ can be expressed as an inner product $k(x, y) = \langle \varphi(x), \varphi(y) \rangle$ for some φ if and only if $k(x, y)$ is positive semidefinite (PSD), i.e.

Let's consider a training set T of n queries, a kernel K is simply a $n \times n$ matrix that contains the pairwise similarity between pairs of T. Thus, each element of K records a measure of similarity between two queries from T.

As a first step of the process of building kernels, we have to identify a similarity measure. In order to measure similarity between queries, we basically need a function that computes distance using queries representations in the feature space. Several distance metrics are available in the literature. Author of [16] gives a comprehensive survey on the most important distance/similarity metrics. Recall that a good distance measure should consider the nature of the input space. Hence, we use the distance function proposed in [1,2] in order to measure similarity between queries in $QtoF$ and $QtoG$ representation spaces respectively.

In the second step, we apply an exponential smoothing transformation to the distance function as presented in formula (2). Thus, we obtain kernels described in Table 1.

Table 1. Proposed kernels

Model	Distance measure	Kernel
$QtoF$	$dist_{ij}$ function	$k_1(x, y) = \exp(\frac{-dist_{ij}}{\sigma})$
$QtoG$	cost function	$k_2(x, y) = \exp(\frac{-cost_{ij}}{\sigma})$

$$k(x, y) = \exp(-\frac{distance\,function}{\sigma}) \tag{2}$$

4   Experimentations and Results

4.1   Experimental Methodology

As mentioned above, we need to store a set of SVM models $m_{0 \leq i \leq n}$ in the proposed summary structure. The number of models is equal to the number n of clusters that we obtain using the plan hash value metric. In our case, queries were categorized into 9 clusters. A Model $m_{0 \leq i \leq 9}$ is built using a *training set* T which contains queries of cluster i and a set of queries from other clusters. These queries are respectively refered to as positive instances (PI) and negative instances (NI). When a new query q_{new} arrives, model $m_{0 \leq i \leq 9}$ is intended to predict the class value of q_{new} which is set to 1 if a similarity is detected, and 0 otherwise.

Same characteristics (kernel, parameters and query representation) are used to build all models of the structure. The aim of our experiments is to identify which combinaison of characteristics is best to produce a good SVM model $m_{0 \leq i \leq 9}$.

Data: For our experiments, we use the dataset from APB1 benchmark [17]. The star schema of this benchmark has one fact table and four dimension tables, all described in Table 2. The dataset contains 55 queries' templates. At this stage we limit our study to the 31 templates containing 2 tables. We write a Java program in order to generate 1000 SQL queries using these templates.

Table 2. APB1 Benchmark Characteristics

Table Name	Cardinality	Size (MB)
ACTVARS	33323400	2085
CHANLEVEL	10	2.4×10^{-4}
CUSTLEVEL	990	2.4×10^{-2}
PRODLEVEL	9900	7.3×10^{-1}
TIMELEVEL	24	3.9×10^{-4}

Table 3. Experimental Datasets

Training Set	Positive Instances		Negative Instances	
	%	Number	%	Number
T_1	10	100	90	900
T_2	30	300	70	700
T_3	50	498	50	498

Training Set: Queries contained in the same cluster i will form the positive instances *(class value = 1)* in the training set of model m_i. However, we need to know how many negative instances we must add to the training set and whether it should be balanced or not. For this purpose, we suggest to conduct our experiments on 3 differents types of training sets as presented in Table 3, where: $T_1 = (10\% \text{ PI}, 90\% \text{ NI})$, $T_2 = (30\% \text{ PI}, 70\% \text{ NI})$ and $T_3 = (50\% \text{ PI}, 50\% \text{ NI})$.

Models Validation: We use k-fold cross validation method for model validation. The dataset is divided into k subsets, and the holdout method is repeated k times. Alternatively, each one of the k subsets is used as a test set and the other $k - 1$ subsets are put together to form a training set. Then the average error across all k trials is computed, and the variance of the resulting estimate is reduced as k increased. We take $k = 10$ in our experiments.

Evaluation Metrics: In this work, we use 3 metrics in order to measure the prediction accuracy.

1- Prediction Rate: Is the rate of correct predictions on the test set. Prediction rate represents the percentage of instances for which predicted class value equals the actual calss value.

2- RMSE: Root Mean Square Error (RMSE) is one of the most widely used statistics that measures the differences between values predicted by a model and the actual values.

3- Kappa Statistic: Kappa is used as a measure to assess model quality. It is always less than or equal to 1. A value of 1 implies perfect model and values less than 1 imply less than perfect models (0.01–0.2: Slight, 0.21–0.4: Fair, 0.41–0.60: Moderate, 0.61–0.80: Substantial 0.81–0.99: Almost perfect).

Tools: For preprocessing the data, generating and testing the models, we use WEKA [18]. We performed the SVM tests with the Sequential Minimal Optimization (SMO) algorithm, which is well-known for its simplicity and quickly solves the SVM Quadratic Problem [19]. We used Matlab to build the kernelts matrices described in the previous section in Table 1.

Table 4. Parameters' values

	Parameter	Description	Value
SMO	c	Complexity Parameter	10
RBF	γ	Free Paramater of the RBF function	0.03
Polynomial	d	Degree of Polynomial kernel	1
k_1, k_2	σ	Smoothing Parameter	0.8

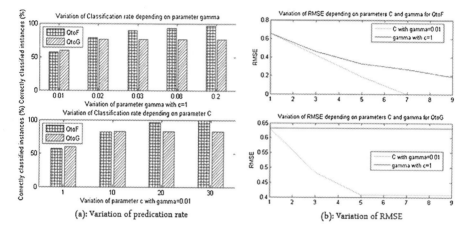

Fig. 2. Variation of prediction rate and RMSE depending on model parameters tuning and query representation

4.2 Results

Experiment 1: Similarity Prediction. Parameters highly affect the quality of prediction. For instance, Fig. 2 shows how changing values of parameters c and γ influences classification rate and RMSE score of a model obtained with RBF. In our work, we need to identify the best values for 4 parameters. In order to identify the best parameters to apply with RBF and SMO, we use the method proposed by Hsu et al. [19] for automatic selection of optimal parameters c and γ, while d and σ are identified empirically. The ranges for parameters c and γ are set within 1–30 and 0.1–1 respectively. Needed parameters for our study and their optimal values are all described in Table 4.

After tuning kernels' and SMO parameters, we start building the models. Recall that we have 3 training sets for each representation $QtoF$ and $QtoG$ and for each training we build 3 models using 3 different kernels. All models are validated using cross-validation with 10 folds. Table 5 summarizes our experiments for $QtoF$ and $QtoG$ respectively. We can see from the results that $QtoF$ generally gives better prediction quality than $QtoG$ representation, regardless of the used kernels. Regarding the training sets influence on the prediction, we observe that T_3 is producing much better prediction for both representations.

Table 5. comparison of prediction quality according to kernels and training sets for *QtoF* and *QtoG*

QtoF					QtoG			
Training sets	Kernel	Prediction rate(%)	RMSE	Kappa	Kernel	Prediction rate (%)	RMSE	Kappa
T_1	k1(x,y)	90	0.31	0	k2(x,y)	60	0.63	0
	RBF	90	0.31	0	RBF	85	0.38	0.65
	Polynomial	70	0.54	0	Polynomial	83.33	0.4	0.59
T_2	k1(x,y)	90.9	0.3	0.79	k2(x,y)	70	0.5	0
	RBF	95	0.22	0.64	RBF	91.2	0.30	0.79
	Polynomial	85	0.38	0.34	Polynomial	85	0.38	0.65
T_3	k1(x,y)	91	0.3	0.81	k2(x,y)	81.8	0.42	0
	RBF	100	0	1	RBF	91.66	0.28	0.83
	Polynomial	100	0	1	Polynomial	95.83	0.2	0.91

Table 6. Experimental Datasets

Dataset size	Time taken to build model (sec)
1000	0.65
21600	39.63

Accordingly, balanced training sets should be considered for building the models. Overall, the kernel that produces greatest prediction rate and kappa value, and lowest RMSE is RBF. Yet, it is worth noting that our proposed kernel $k1$ is doing better than both RBF and Polynomial kernels when used with T_2 and *QtoF*. Even if it shows a prediction rate of 90.9 % against 95 % for RBF but the kappa parameter values are 0.79 and 0.64 for k_1 and RBF respectively, which means that the model produced by k_1 is better than RBF in this case. From the above, we can conclude that paramaters presented in Table 5 with *QtoF* and RBF produce the best SVM model.

Experiment 2: Influence of the Data Set Size. After identifying the best model, we want to study the influence of the data size on the time taken to build the model. For this experiment, we use a balanced training set and the RBF kernel. Beside T_3 we create 2 more training sets with sizes 5000 and 22000 and note the time taken to build the model. Results of this experiment are reported in Table 6. They show that the model created with 22000 take 40 s approximatively to be built.

5   Conclusion

We present an approach to predict SQL queries' similarity using SVM. We learn similarity from already executed queries. We explore two representation techniques to model SQL queries as feature vectors, and identify the best one to assess similarity. Additionnaly, we propose a summary structure for organizing queries in the DBMS log. We compute two personnalized kernels and use them to build SVM models. Results show that our proposed kernels are comparable to RBF kernel and produce better results in some cases.

This approach of studying SQL queries' similarity can be particulary useful for recommending execution plans to a DBMS optimizer based on the prediction of similarity of new SQL queries, and could be of a good help to avoid generating new execution plan which is generally expensive. Moreover, the approach shows highly accurate results in terms of correctly predicting queries' similarity. However, the major constraint of this approach is the amount of time taken to build models for relatively big data sets. In future, we would like to propose a recommender system of execution plans,and use other machine learning algorithms for queries' similarity identification.

References

1. Antara, G., Jignashu, P., Vibhuti S., et al.: Plan selection based on query clustering. In: Proceedings of the 28th International Conference on Very Large Data Bases, VLDB Endowment, pp. 179–190 (2002)
2. Sadeg, L., Bellatreche, L.: Approach to query optimization led by the reuse of execution plans. Ph.D. thesis. ESI (2010)
3. Yao, Q., An, A., Huang, X.: Finding and analyzing database user sessions. In: Zhou, L., Ooi, B.-C., Meng, X. (eds.) DASFAA 2005. LNCS, vol. 3453, pp. 851–862. Springer, Heidelberg (2005)
4. Kostas, S., Marina, D., Pitoura, E.: You may also like results in relational databases. In: Proceedings of PersDB, vol. 9 (2009)
5. Aligon, J., Golfarelli, M., Marcel, P., Rizzi, S., Turricchia, E.: Mining preferences from OLAP query logs for proactive personalization. In: Eder, J., Bielikova, M., Tjoa, A.M. (eds.) ADBIS 2011. LNCS, vol. 6909, pp. 84–97. Springer, Heidelberg (2011)
6. Yang, X., Procopiuc, C.M., Srivastava, D.: Recommending join queries via query log analysis. In: IEEE 25th International Conference on Data Engineering, ICDE 2009. IEEE, pp. 964–975 (2009)
7. Gupta, A., Mumick, I.S., et al. (eds.): Materialized Views: Techniques, Implementations, and Applications. MIT press, Cambridge (1999)
8. Chatzopoulou, G., Eirinaki, M., Koshy, S., et al.: The QueRIE system for personalized query recommendations. IEEE Data Eng. Bull. $34(2)$, 55–60 (2011)
9. Cordella, L.P., Foggia, P., Sansone, C., et al.: Performance evaluation of the VF graph matching algorithm. In: Proceedings International Conference on IEEE of Image Analysis and Processing, pp. 1172–1177 (1999)
10. Vapnik, V.: The Nature of Statistical Learning Theory. Springer Science & Business Media, New York (2000)
11. Wu, X., Kumar, V., Quinlan, J.R., et al.: Top 10 algorithms in data mining. Knowl. Inf. Syst. $14(1)$, 1–37 (2008)
12. Minh, H.Q., Niyogi, P., Yao, Y.: Mercer's theorem, feature maps, and smoothing. In: Lugosi, G., Simon, H.U. (eds.) COLT 2006. LNCS (LNAI), vol. 4005, pp. 154–168. Springer, Heidelberg (2006)

13. Bennett, K.P., Campbell, C.: Support vector machines: hype or hallelujah? ACM SIGKDD Explor. Newslett. **2**(2), 1–13 (2000)
14. Herbrich, R.: Learning Kernel Classifiers. MIT Press, Cambridge (2002)
15. Hastie, T., Tibshirani, R., Friedman, J., et al.: The Elements of Statistical Learning. Springer, New York (2009)
16. Cha, S.-H.: Comprehensive survey on distance/similarity measures between probability density functions. City **1**(2), 1 (2007)
17. APB-1 olap benchmark. http://www.olapcouncil.org/research/spec1.htm
18. Hall, M., Frank, E., Holmes, G., et al.: The WEKA data mining software: an update. ACM SIGKDD Explor. Newslett. **11**(1), 10–18 (2009)
19. Platt, J., et al.: Fast training of support vector machines using sequential minimal optimization. In: Advances in kernel methodssupport vector learning, vol. 3 (1999)

Improving the Quality of Load Forecasts Using Smart Meter Data

Abbas Shahzadeh[(⊠)], Abbas Khosravi, and Saeid Nahavandi

Centre for Intelligent Systems Research, Deakin University, Geelong, Australia
{ashahzad,abbas.khosravi,saeid.nahavandi}@deakin.edu.au
http://www.deakin.edu.au/research/cisr

Abstract. For the operator of a power system, having an accurate forecast of the day-ahead load is imperative in order to guaranty the reliability of supply and also to minimize generation costs and pollution. Furthermore, in a restructured power system, other parties, like utility companies, large consumers and in some cases even ordinary consumers, can benefit from a higher quality demand forecast. In this paper, the application of smart meter data for producing more accurate load forecasts has been discussed. First an ordinary neural network model is used to generate a forecast for the total load of a number of consumers. The results of this step are used as a benchmark for comparison with the forecast results of a more sophisticated method. In this new method, using wavelet decomposition and a clustering technique called interactive k-means, the consumers are divided into a number of clusters. Then for each cluster an individual neural network is trained. Consequently, by adding the outputs of all of the neural networks, a forecast for the total load is generated. A comparison between the forecast using a single model and the forecast generated by the proposed method, proves that smart meter data can be used to significantly improve the quality of load forecast.

Keywords: Smart meters · Clustering · Neural networks · Load forecast · Wavelet transformation

1 Introduction

A study by Hobbs et al. in 1999, reports that in a 10000 MW system, when the forecast error is in the range of 3 % to 5 %, every 1 % reduction in forecast error can save between 0.1 % to 0.3 % in generation costs [1]. According to this reference, this saving would translate to $US1.6 million per year in 1999 for a 10 GW electrical system. Considering the fact that the electricity market has been restructured in many countries and that the price of electricity is directly affected by the size of demand, it is safe to assume that the savings due to an improvement in the accuracy of forecast, can be much bigger now than it used to be back in 1999. For instance, in many electricity markets, the trading is carried out in two phases namely day-ahead phase and real-time phase. Naturally the price per

© Springer International Publishing Switzerland 2015
S. Arik et al. (Eds.): ICONIP 2015, Part IV, LNCS 9492, pp. 667–674, 2015.
DOI: 10.1007/978-3-319-26561-2_78

MWh in the real-time phase is higher. Electricity market in California can be taken as an example. In California, the average day-ahead price is \$52/MWh while the average real-time price is \$70/Mwh [2]. Obviously in such a system, a utility company can make huge savings by optimizing its purchases in the day-ahead market which in turn requires an accurate demand forecast.

There has been a huge interest in load forecast in academia. Many forecast methods have been applied to this problem. Artificial Neural Networks [3], Regression Analysis [4], Fuzzy Modelling [5], Interval Type-2 Fuzzy Logic Systems [6] and Support Vector Machines [7] are some of the methods that have been used to provide solutions for the load forecast problem. However, despite years of study, there is still room for improvement in the performance of forecast.

At the same time, smart meters are about to completely replace traditional meters all over the world. The main difference between smart meters and traditional meters is the frequency of recording the electricity consumption of consumers. In the case of older meters, the consumed energy was usually recorded once per month. However smart meters can record energy consumption in every 15 min or at higher frequencies. Consequently a large volume of data is available now which was not accessible in the past.

In this paper it will be demonstrated how by using smart meter data, the forecast error can be significantly decreased. The rest of this paper is organized as follows. In Sect. 2, the dataset used for testing the proposed method in this paper is introduced. In Sect. 3, the problem is formulated and the concept of clustering is explained. In Sect. 4, a single model for forecasting the aggregated load is introduced and the results are reported to be used as a benchmark for comparison. In Sect. 5, wavelet transformation and a clustering method called I-k-means [8] are explained. In Sect. 6, the proposed method is applied to the dataset and the forecast results are compared with the single model method. Finally, Sect. 7 concludes the paper.

2   Smart Meter Dataset

A very important prerequisite for a research about smart meter data is having access to a dataset of smart meter records. Unfortunately, most utility companies are not willing to share their smart meter data with researchers mainly due to privacy concerns. However, there is a dataset provided by Irish Social Science Data Archive (ISSDA)[1], which has been made available for academic and research purposes. The data is gathered by Commission for Energy Regulation (CER) in Ireland and contains smart meter data of more than 6000 Irish consumers from 14 July 2009 to 31 December 2010 [9]. In this research only the smart meter records of 3176 residential consumers has been used.

3   Problem Statement

As explained in Sect. 2, the dataset used in this paper contains the information of $M = 3176$ consumers. The data is provided as M time series vectors. The vector

---

[1] www.ucd.ie/issda

corresponding to consumer number i is given in Eq. (1) in which, n shows the length of the time series.

$$\mathbf{l}_i = \{l_i^1, l_i^2, ..., l_i^n\} \tag{1}$$

The main goal of this paper is to generate two forecasts using two different methods and then compare the results. The first method uses only total load of the system consisted of all M consumers. The second method is explained by detail in Sect. 3.2.

3.1 Single Model Method Formulation

The total load of the system can be expressed as a time-series like Eq. (2). In this formula, n shows the number of time intervals in the period of interest.

$$\mathbf{L}_{total} = \{L_{total}^1, L_{total}^2, ..., L_{total}^n\} \tag{2}$$

Each element of $\mathbf{L}_{total}$ in (2) is calculated using (3). In other words, in each time interval like t the total load of the system can be easily calculated by adding the loads of all the consumers as expressed by Eq. (3). In (3), M shows the number of consumers in the network.

$$L_{total}^t = \sum_{i=1}^{M} l_i^t \tag{3}$$

A model like the one depicted in Fig. 1, can be used to generate a forecast for the total load. The model itself will be explained in detail in the next section. However the output of the model will be a time series given in Eq. (4).

$$\hat{\mathbf{L}}_{total} = \{\hat{L}_{total}^1, \hat{L}_{total}^2, ..., \hat{L}_{total}^n\} \tag{4}$$

To measure the difference between the actual load and the foretasted load, Mean Absolute Percentage Error (MAPE) calculated using Eq. (5) is used in this work.

$$MAPE = \frac{100}{n} \sum_{t=1}^{n} \frac{\hat{L}_{total}^t - L_{total}^t}{L_{total}^t} \tag{5}$$

3.2 Clustering-based Method Formulation

The main concept of the clustering method is portrayed in Fig. 2. In this method, the consumers are divided into C clusters denoted as S_1, S_2, ..., S_C. Number of clusters (C) can be any number between 1 and M (number of consumers). When $C = 1$, the clustering method basically reduces to the single model method explained in the previous subsection. $C = M$ means that for each consumer, an individual model will be trained. The generated forecast for cluster S_k is represented by $\hat{\mathbf{L}}_{S_k}$ as given in Eq. (6).

$$\hat{\mathbf{L}}_{S_k} = \{\hat{L}_{S_k}^1, \hat{L}_{S_k}^2, ..., \hat{L}_{S_k}^n\} \tag{6}$$

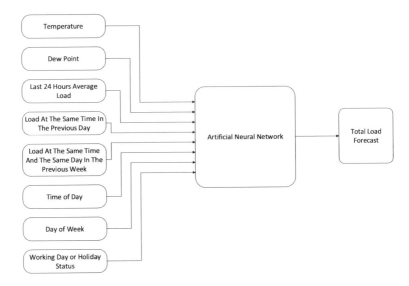

Fig. 1. The ANN model used for forecasting the total load of the system in the single model method and the total load of each cluster in the clustering-based method

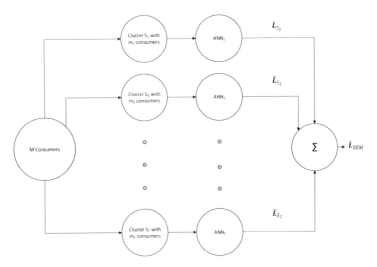

Fig. 2. In the clustering-based method, consumers are grouped into C clusters. For each cluster an individual ANN model generates the forecast. Adding all the forecasts, gives a forecast for the total load of the system.

The total load forecast using clustering method is calculated by adding all $\hat{\mathbf{L}}_{S_k}$ values together as given in Eq. (7) below.

$$\hat{\mathbf{L}}_{Sum} = \sum_{k=1}^{C} \hat{\mathbf{L}}_{S_k} \tag{7}$$

In (7), $\hat{\mathbf{L}}_{Sum}$ shows the total load forecast calculated using the clustering method. By applying formula Eq. (5) and using the elements of $\hat{\mathbf{L}}_{Sum}$ instead of $\hat{\mathbf{L}}_{total}$, the forecast error for this latter method can be calculated. A comparison between the two calculated MAPEs can reveal the improvement in the accuracy of forecast using the clustering method.

4   Single Model Forecast

The model portrayed in Fig. 1, was used to predict the total load of the network. This model has eight inputs with descriptive titles in the figure. The artificial neural network used here is a two-layer network, with 20 sigmoid neurons in the hidden layer and a linear neuron in the output layer. It was trained using the back-propagation algorithm called Levenberg-Marquardt in MATLAB.

Data from 1 February 2010 to 7 March 2010 was used for training the network. The period from 9 March 2010 to 16 March 2010 was used for testing the performance of this model. The experiment was repeated 150 times. The average MAPE using this method was 6.15 %.

5   Wavelet Transformation and Interactive k-means

In this section the Wavelet Transformation (WT) formulas will be presented and the concept of Interactive k-means method, which is used to group the consumers into different clusters will be explained. The formulas and notation used for explaining WT are mostly taken from [10].

Before the application of WT to a time series, the time axis of the time-series should be scaled to cover the range from 0 to 1. The signal should also be resampled and expressed as the summation of 2^n samples similar to Eq. (8) in which $u_i^{(n)}$ is defined by (9).

$$f(x) = \sum_{i=1}^{2^n} a_i \times u_i^{(n)}(x) \tag{8}$$

$$u_i^{(j)}(x) = \begin{cases} 1 & \text{if } \frac{i-1}{2^j} \le x < \frac{i}{2^j} \\ 0 & \text{otherwise} \end{cases} \tag{9}$$

The result of applying WT to the function described by (8) for l successive iterations, is given in (10). In this equation, l shows the level of WT and can be any number between 0 and n. In the case of $l = 0$, Eq. (10) will be the same as (8) and in the case of $l = n$, which corresponds to the deepest level of transformation, only one coefficient (the average of all samples) will remain. The step function term $u^{(n-l)}$ in (10) is again defined by (9).

$$f^{(l)}(x) = \sum_{i=1}^{2^{n-l}} a_i^{(l)} \times u_i^{(n-l)}(x) \tag{10}$$

Table 1. Interactive k-means algorithm for clustering consumers according to their daily average load profile

1	Calculate, re-sample and present each consumer's average daily load profile by a function similar to $f(x)$ in Eq. (8)
2	Set $l := n - 1$
3	Generate C centroids for clusters S_1 to S_C. If $l = n - 1$, generate centroids randomly, otherwise use the results of level $l + 1$
4	Apply the k-means algorithm. Use 2^{n-l} elements of $f^{(l)}$ from Eq. 10 as the feature set for each consumer
5	If $l = 0$ or no change in the clusters in the last run of k-means, stop. Otherwise set $l := l - 1$ and go to 3

It should be noted that WT converts a function to a sum of two sets of coefficients namely "detail coefficients" and "approximation coefficients". In this analysis only approximation coefficients are of interest. Eqs. (9) and (10) basically show how applying WT can give coarser approximations of the signal by increasing the number of iterations. The initial function has 2^n samples with each sample covering a width of $\frac{1}{2^n}$ of the time axis. After applying one iteration of WT, the converted function will have 2^{n-1} samples which is half as many as the initial function. The width of each sample will be doubled so that the new representation of the signal, covers the same area of the time axis as the original function. After the first iteration of WT, each sample will have a width of $\frac{1}{2^{n-1}}$.

Selecting the best number of WT iterations for optimum feature extraction often results in a dilemma. On the one hand, by opting for deeper levels of WT, the number of features in each time series will decrease which in turn can significantly lower the computational cost of data mining. On the other hand, selecting shallower levels of WT, i.e. a smaller number of WT iterations, preserves higher levels of detail in the signal which can improve the quality of the analysis.

Vlachos et al. in [8], have proposed a method to reconcile these two apparently opposing objectives. In their method, called "Interactive k-means", the clustering algorithm is first applied to a coarse representation of the data which is a result of a series of WT iterations. Then, the results of this initial clustering are used to calculate the centroids of the clusters for the next sweep of the algorithm which is applied to a finer representation of data. The same procedure is repeated until the finest level is reached (the original time series before the application of any transformation) or when using shallower levels of WT does not result in any changes in the clustering outcome. The algorithm is summarized in Table 1.

6   Case Study

For each consumer in the database, a mean daily load profile is calculated by averaging all 24-hour load profiles during the training period. Since the data is measured in half-hourly periods, each average load profile, initially has

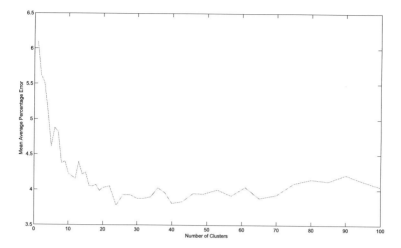

Fig. 3. The graph shows how MAPE changes when the number of clusters increases

48 elements. In order to use WT for feature extraction, first the time series signals are re-sampled. In this case, a re-sampling phase to convert 48 elements to 64 elements is used. The time axis is also scaled to cover the range between 0 and 1, so that each sample covers $\frac{1}{64}$ of the time axis. After this step, the average load profile of each consumer is represented by Eq. (8). Since there are $64 = 2^6$ elements in each time series, n is equal to 6 in this case. Consequently l can range between 0 and 6. The I-k-means algorithm as presented in Table 1, will start from $l = 5$ which corresponds to the approximation coefficients after the application of 5 iterations of wavelet transformations. At this level, the load profile of each consumer is condensed in only two coefficients. In the next step, the centroids of the calculated clusters are used to initialize the next k-means algorithm which works on 4 coefficients. The procedure continues according to the algorithm presented in Table 1.

After the clustering phase, a neural network is trained for each cluster. The predictions of these ANNs are added together to give a forecast for the total load. The procedure is repeated for different number of clusters ranging from 1 to 100 clusters. The MAPE for each number of clusters in this range is displayed in Fig. 3.

The results in Fig. 3, show that up to a certain point, as the number of clusters increases, the quality of forecast improves. By using 24 clusters, a MAPE as low as 3.78 % is achieved. However increasing the number of clusters beyond this point, does not result in any improvements in the accuracy of forecast. Therefore, in this case, using 24 clusters, gives the optimum results. By comparing the results attained by clustering (MAPE = 3.78 %) with the results from Sect. 4 (MAPE = 6.15 %), a decrease of almost 40 % in forecast error can be observed.

7 Conclusion

A method for using smart meter data to improve the accuracy of load forecast was presented. Wavelet Transformation was used in the feature extraction phase and then a clustering method called I-k-means [8] was utilized to group consumers according to their wavelet coefficients. Then an artificial neural network was used to generate a forecast for each group and eventually a forecast for the total load was generated. It was demonstrated that by using smart meter data and clustering, the forecast error can be decreased by almost 40 %.

References

1. Hobbs, B., Jitprapaikulsarn, S., Konda, S., Chankong, V., Loparo, K., Maratukulam, D.: Analysis of the value for unit commitment of improved load forecasts. IEEE Trans. Power Syst. **14**(4), 1342–1348 (1999)
2. Albert, A., Rajagopal, R.: Cost-of-service segmentation of energy consumers. IEEE Trans. Power Syst. **29**(6), 1–9 (2014)
3. Park, D., El-Sharkawi, M., Marks, R.J.I., Atlas, L., Damborg, M.: Electric load forecasting using an artificial neural network. IEEE Trans. Power Syst. **6**(2), 442–449 (1991)
4. Papalexopoulos, A., Hesterberg, T.: A regression-based approach to short-term system load forecasting. IEEE Trans. Power Syst. **5**(4), 1535–1547 (1990)
5. Mastorocostas, P., Theocharis, J., Bakirtzis, A.: Fuzzy modeling for short term load forecasting using the orthogonal least squares method. IEEE Trans. Power Syst. **14**(1), 29–36 (1999)
6. Khosravi, A., Nahavandi, S., Creighton, D., Srinivasan, D.: Interval type-2 fuzzy logic systems for load forecasting: a comparative study. IEEE Trans. Power Syst. **27**(3), 1274–1282 (2012)
7. Pai, P.-F., Hong, W.-C.: Support vector machines with simulated annealing algorithms in electricity load forecasting. Energy Convers. Manag. **46**(17), 2669–2688 (2005)
8. Vlachos, M., Lin, J., Keogh, E., Gunopulos, D.: A wavelet-based anytime algorithm for K-Means clustering of time series. In: Proceedings of Workshop on Clustering High Dimensionality Data and Its Applications, pp. 23–30 (2003)
9. Alzate, C., Sinn, M.: Improved electricity load forecasting via kernel spectral clustering of smart meters. In: Proceedings of IEEE International Conference on Data Mining, ICDM, pp. 943–948 (2013)
10. Nievergelt, Y.: Wavelets Made Easy, vol. 174. Springer, New York (1999)

Transport-Independent Protocols for Universal AER Communications

Alexander D. Rast[1]($\boxtimes$) , Alan B. Stokes[1], Sergio Davies[1],
Samantha V. Adams[2], Himanshu Akolkar[3], David R. Lester[1],
Chiara Bartolozzi[3], Angelo Cangelosi[2], and Steve Furber[1]

[1] School of Computer Science, University of Manchester, Manchester M13 9PL, UK
{rasta,daviess,dlester}@cs.man.ac.uk
{alan.stokes-2,steve.furber}@manchester.ac.uk
{samantha.adams,A.Cangelosi}@plymouth.ac.uk
Chiara.Bartolozzi@iit.it
http://www.cs.manchester.ac.uk/apt
http://www.plymouth.ac.uk
http://www.iit.it
[2] Plymouth University, Plymouth, UK
[3] Istituto Italiano da Technologia, Genoa, Italy

Abstract. The emergence of Address-Event Representation (AER) as a
general communications method across a large variety of neural devices
suggests that they might be made interoperable. If there were a standard
AER interface, systems could communicate using native AER signalling,
allowing the construction of large-scale, real-time, heterogeneous neural
systems. We propose a transport-agnostic AER protocol that permits
direct bidirectional event communications between systems over Ethernet, and demonstrate practical implementations that connect a neu-
romimetic chip: SpiNNaker, both to standard host PCs and to real-time
robotic systems. The protocol specifies a header and packet format that
supports a variety of different possible packet types while coping with
questions of data alignment, time sequencing, and packet compression.
Such a model creates a flexible solution either for real-time communications between neural devices or for live spike I/O and visualisation in a
host PC. With its standard physical layer and flexible protocol, the specification provides a prototype for AER protocol standardisation that is
at once compatible with legacy systems and expressive enough for future
very-large-scale neural systems.

Alexander Rast, Alan B. Stokes, Sergio Davies, David R. Lester and Steve Furber are
with the School of Computer Science, The University of Manchester, Manchester,
UK (email: rasta@cs.man.ac.uk). Samantha V. Adams and Angelo Cangelosi are
with Plymouth University, Plymouth, UK. Himanshu Akolkar and Chiara Bartolozzi
are with the Istituto Italiano da Tecnologia, Genoa, Italy. This work has been
partially supported by the European Union under grant nos. FP7-604102 (HBP),
FP7-287701 (BrainScales-Extension), and ERC-320689 (BIMPC), by EPSRC grant
EP/J004561/1 (BABEL) and by EPSRC grant EP/G015740/1 (BIMPA).

© Springer International Publishing Switzerland 2015
S. Arik et al. (Eds.): ICONIP 2015, Part IV, LNCS 9492, pp. 675–684, 2015.
DOI: 10.1007/978-3-319-26561-2_79

1   Neural AER Communications: The Case for Standardisation

Neuromorphic systems are entering the era of large-scale simulation. With an increasing diversity of available neuromorphic devices, it is desirable to link systems together into large, multisystem models having possibly millions of neurons. Many if not most devices use some form of Address-Event Representation (AER) [7] for communications, but there is as yet no *universal* AER standard. On the one hand it is useful to link event-generating sensors such as the various silicon retinas [8], to neuromorphic chips e.g. [6], whose event representation is a natural match, on the other to connect to very-large-scale systems in fixed locations that can provide the resource necessary for massive simulations [5,10]. A common standard for spike exchange would facilitate such goals.

Some progress has been made, particularly with respect to the low-level hardware details of connectors, cabling, signal voltages etc. By 2009 the CAVIAR project had developed an end-to-end neuromorphic system [11], linking sensors to processors to actuators using a common hardware standard. Beginning with early work [4] to develop a hardware interface, a group at the CapoCaccia neuromorphic workshop engaged in an ongoing effort at AER standardisation and by 2013 had solved most of the hardware-level issues [9] using a simple AER protocol over Ethernet using UDP. While this work clearly shows the benefits of leveraging industry-standard protocols and hardware up to the transport level, the AER protocol itself was preliminary and did not attempt to address many of the potential use cases. What is needed is a generic AER protocol that can work over a variety of transports and support a range of different packet types. Here we develop a specification for such a protocol and a reference implementation that permits communication over a common interface between multiple heterogeneous neuromorphic systems as well as ordinary industry-standard PCs.

2   A Proposed Protocol for AER Communications

Building upon the AEtheRnet interface introduced in [9], we have expanded the protocol into a data format specification for spikes to be sent over Ethernet between compatible AER-generating and receiving devices.

2.1   Comments on Data Handling

The protocol provides considerable flexibility in data handling at the receiver. The specification does not specify the transport. In principle, any Ethernet-compatible transport could be used. Packets are transported over the wire in big-endian format. No devices acknowledge receipt of packets; the protocol is fire-and-forget. There is no built-in error detection or correction although devices could use transport facilities (e.g. CRC) to recover from errors if desired. While the interface is bidirectional, it is not required that a given device must support both directions nor that it be able to service all packet types. Devices may limit packet size to less than the maximum of 256 events (e.g. to fit the Ethernet

MTU of 1500 bytes). Packets must be issued in strict event order and any spike buffering must meet any real-time spike timing constraints. Devices must drop late spikes, spikes with timestamps out of sequential order, and unsupported packet types.

2.2 Supported Formats

The protocol consists of a 16-bit header followed by data. Communication is stateless: all the information to interpret the data is contained in the header itself. The format of the header is:

Bit															
15	14	13	12	11	10	9	8	7	6	5	4	3	2	1	0
P	F	D	T	Type		Version		Count							

The decode of the header configuration bits is as follows:

Field name	Position	Value	Description
P & F	15 - 14	00	Basic data packet, no address prefix
		01	Command packet
		10	Data packet with lower halfword address prefix
		11	Data packet with upper halfword address prefix
D	13	0	No payload prefix
		1	With payload prefix
T	12	0	Payload are not timestamps
		1	Payload are timestamps
Type	11-10	00	16-bit address
		01	16-bit address and payload, alternating
		10	32-bit address
		11	32-bit address and payload, alternating
Version	9-8	00	Version 0 of the protocol

The count parameter indicates how many events there are in the packet. Any transmission needs to be limited to the smallest of the supported packet sizes between the sender and the receiver. Any prefix signalled in the packet header is ORed with the value in the data part of the packet to obtain the transmitted value. The header provides for a special command packet format and 8 data type formats.

For the command packet, the bits are organized as follows:

Bit			
15	14	13-0	Payload
0	1	Command ID	Command and device-specific

The command ID and packet payload are device-specific.

Data packet formats have one of the following general structures:

Bit						
15	14	13-0	Event	...	Event	
0	0	Header Info	Address (Payload)	...	Address (Payload)	

where fields in (round brackets) are optional. The number of events (address/pay-load pairs) is determined by the 'Count' parameter in the header field.

Bit						
15	14	13-0	Prefix	Event	...	Event
1	x	Hdr Info	Addr Prefix (Pay Prefix)	Address (Payload)	...	Address (Payload)

3   Implementation on a Real Platform

To implement the protocol, we chose UDP to provide the underlying transport since it is a natural fit to AER-type fire-and-forget systems, and is supported by a wide variety of AER-generating devices. As a test bed, we used two neuromorphic platforms and created a device simulator able to exercise the full feature set of the proposed protocol.

3.1   Platforms

The simulator is a host-based C++ application that includes options to configure it as a transmitter or receiver, or both, to generate patterns of events or pull them from a file, to dump received events to a file, reflect them back to the source, configure payloads and prefixes, and many other options.

One of the 'neuromorphic' platforms is a universal neural processor: SpiN-Naker (described exhaustively elsewhere, e.g. [5]). The chip is an homogeneous array of cores embedded in an AER network fabric. Because of this feature, it is possible to assign cores to special-purpose roles like I/O and have them be automatically mapped and routed via the SpiNNaker configuration toolchain from a PyNN-based [2] description.

The other is a robotic platform: the iCub, of which we used both the 'normal' version and the 'neuromorphic iCub', which replaces traditional frame-based cameras at the sensory periphery with AER-generating spiking retinas. An interface board located in the iCub head translates AER events into generic Yarp (Yet Another Robotics Platform) packets that can be distributed to other Yarp ports and devices.

3.2   SpiNNaker-Side: The EIEIO Interface

The SpiNNaker implementation of the protocol uses the External Internal Event Input Output (EIEIO) interface: a series of modules loaded on cores in the chip, taking advantage as noted in 3.1 of the ability to dedicate cores to application management tasks. 3 modules comprise the interface.

Root Monitor. This module resides on the core connected to the Ethernet and translates between external UDP and SDP, an internal communications layer able to transport messages around a SpiNNaker system using AER packets. The EIEIO interface uses the Root Monitor's 'SC&MP' system software to layer the external AER protocol onto SDP.

Live Packet Gatherer (LPG). This implements the spike output (from SpiN-Naker) side of the interface. It records spikes from tagged populations and immediately forwards them over SDP to the Root Monitor. One LPG can record and output from several populations (although not from an arbitrary subset of neurons). A Python module compatible with PyNN enables spikes to be directed to the LPG using the form `activate_live_output_for (<Population>)`.

Reverse IP Tag Multicast Source (RIPTMS). This implements the input (to SpiNNaker) side of the interface. It is an event-triggered listener which waits for incoming SDP packets from the Root Monitor, decodes the spikes contained therein, and then injects them directly onto the internal communications fabric. A spike translation decoder translates spikes represented in a user-specified external address space into the internal SpiNNaker address space. Like the LPG the RIPTMS has a Python module which instantiates it as a PyNN Population (from which projections can be made in normal PyNN style using the SpiNNaker software stack).

3.3   iCub-side: The AER Library

To provide users with an interface to the protocol without having to program low-level decoding details, we developed a general purpose C++ library and tested it on the iCub. Currently it supports only basic 32-bit data packets without prefixes, expecting timestamps in the input direction but issuing no payloads in the output direction. The library contains three main classes that are likely to be used in external programs:

SocketIF. Handles socket creation for send and receive and also handshaking with the SpiNNaker toolchain for synchronisation and retrieving neuron id – SpiNNaker key mappings.

EIEIOSender. Manages a FIFO queue of spikes for sending and bundles multiple spikes into a packet (max 63 spikes per packet). End users should call the addSpikeToSendQueue(int id) method in their program, passing in the neuron id as an argument.

EIEIOReceiver. Manages a FIFO queue of spikes decoded from incoming EIEIO packets. End users should call the getNextSpike() method in their program to pop the spikes from the queue.

Both EIEIOSender and EIEIOReceiver have independent sockets and can be used in the same program to enable concurrent send and receive.

4   Multi-system, Multi-modal Real-Time Operation

To test the AER interface we ran 3 different groups of experiments. The first group used the device simulator briefly described in sect. 3 to create a 'virtual device' on a laptop that could generate a variety of different I/O scenarios. The second group used the neuromorphic iCub to determine capacity limits and verify the ability to set up a connection. The third group used the regular iCub robot with the AER library to verify hardware connectivity with the new protocol.

4.1   Simulation with a Virtual Device

To verify the operation of the revised protocol using EIEIO, we tested 2 scenarios. The first scenario implements a use case where the protocol is being used exclusively for local I/O and visualisation. We configured a network with a chain of sequentially spiking neurons in 2 directions, whose input comes from a host-based spike source and whose output is redirected to a live visualisation tool. The host-based source was configured to inject a spike into, sequentially, neurons 0, 20, 40, 60, and 80 of 100 in the forward direction, and descending order (from 100) in the reverse direction, at a random interval, with weights from neuron to neuron in each population sufficient to trigger the next population in the sequence. Results (Fig. 1) show both the expected behaviour and the ability to receive spikes from an external host and display them in real-time using a live monitor, using the protocol.

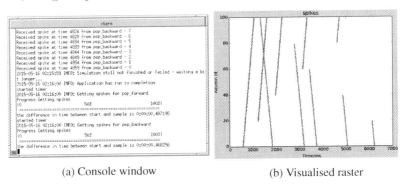

(a) Console window   (b) Visualised raster

Fig. 1. Test of local input/monitoring facilities using the protocol on SpiNNaker

The second scenario implements the use case of communication between 2 different neuromorphic devices, possibly in widely different geographic locations. We created a spike reflector: a virtual device that would accept input from SpiNNaker and reflect it back to the chip. We then configured a network in SpiNNaker with a source population of 10 neurons and a destination population of 10 neurons, a Live Packet Gatherer and a Reverse IP Tag Multicast Source, and specified the connectivity so that the source sent to the virtual device and the destination received from the virtual device, each neuron receiving one connection with weight high enough to cause it to spike. We tested this both with

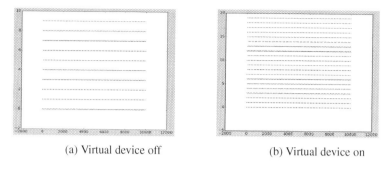

(a) Virtual device off   (b) Virtual device on

Fig. 2. External input/output using the device simulator

a direct network connection and through a routed network (by connecting the SpiNNaker chip to a public network). Figure 2 shows the results: with the virtual device off only the source population spikes; with it on, both populations spike and all spikes are received.

4.2 Hardware Operation with the Proposed Interface

As a preliminary to protocol deployment, we ran a series of tests with the neuromorphic iCub and SpiNNaker using an enhanced version of the AEtheRnet interface with the same packet size and arrangement as that of the new protocol using the Basic packet format with 32-bit addresses. We connected the systems through a host that converted spikes to Yarp bottles. We swept a bar-shaped object, (a power strip, in fact) long enough to occupy the entire visual field in one dimension, in horizontal and vertical directions of both sweep and bar orientation, and had the robot move its gaze in the direction of the bar when the bar was in a preferred orientation. Results (Fig. 3) demonstrate the basic feasibility of a real-time link using Ethernet and indicate a peak spike rate of about 8 K spikes/s.

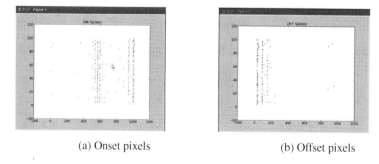

(a) Onset pixels   (b) Offset pixels

Fig. 3. Data from DVS sensor on iCub, as seen on SpiNNaker, horizontal object

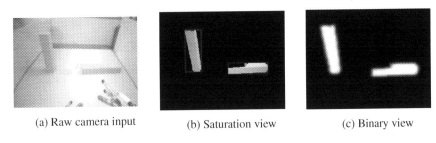

(a) Raw camera input   (b) Saturation view   (c) Binary view

Fig. 4. Camera input from the iCub robot

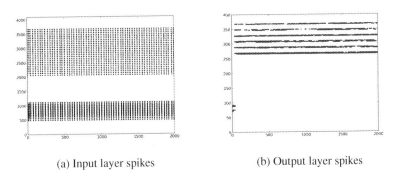

(a) Input layer spikes   (b) Output layer spikes

Fig. 5. Spiking activity for the largest network run with the iCub robot

To test the new EIEIO protocol (and the AER library described in Sect. 3.4) in a real-world scenario, scaled up versions of the attentional network used in [1] were run on a 48-chip SpiNNaker board interfacing to the regular iCub robot. We used two sizes: 32×32 input layer size (6,148 total neurons) and 64×64 input layer size (24,592 total neurons). The scenario was as used for previous work, with two objects, of different orientations, one 'preferred', one 'aversive', the task being to cause the iCub robot to look/point at the 'preferred' object. Examples of the raw and processed input camera images are given in Fig. 4. Examples of the input and output network spiking activity for the largest network tested are given in Fig. 5.

This generated spike rates of about 15 K spikes/s in the input layers and 2 K spikes/s in the output.

5 Discussion: Potential and Challenges

5.1 Current Work: Immediate Implications

The new EIEIO protocol yielded several improvements to the attentional network implementation in [1]. Throughput management was much easier using the AER library as spike packet construction is handled automatically with only minimal iCub host interfacing code required. The EIEIO protocol also handled SpiNNaker key generation and neuron id to key mapping automatically, reducing the possibility of errors.

One obvious use of the interface is to connect several AER-generating or receiving devices. But equally, since the interface provides a general mechanism for getting spikes in and out of a chip, with a general-purpose device such as SpiNNaker, it can be used as a means to control and monitor a simulation occurring on a single platform. Tests of this facility worked successfully and indicate that a universal AER interface not only permits direct inter-device connections but also provides a channel for real-time simulation management. In future such interfaces could be used for loading of configuration data for neuromorphic chips, possibly remotely.

The protocol specifies the ability to use timestamps but does not indicate a definitive reference for the meaning of timestamps to a given device. In general, asynchronous neuromorphic chips operating in possibly remote locations could have large differences in their local clocks, and possibly different time units. Although the interface cannot guarantee perfect synchronisation, command packets offer a solution that allows for a form of approximate synchronisation by exchanging timing messages.

5.2 Next Steps: Future Work

Experiments with the current protocol revealed that while functional, it could benefit from some additional features, and at the same time some of the more complex features could be streamlined in their definition. We are currently revising the proposal to allow packets containing both a timestamp and an additional payload, and to allow stateful transceiver interfaces supporting a user-defined link setup to describe things such as compression or time representation using a handshake with command packets. If reliable packet reception is required the protocol may also be implemented over TCP.

6 Conclusions

This work grew out of the earlier AEtheRnet protocol and a desire in the community to establish standards for remote neuromorphic communications. The earlier protocol established some basic properties of such a possible standard; this new work indicates some additional important features, which can be considered as the main contributions:

Transport Independence:
  The protocol is not bound to a particular underlying transport.
Payload Support:
  Packets can contain payloads and timestamps as well as addresses.
Partial Support Allowed:
  Device designers can choose what packet formats they support and drop others.
Idempotent Packets:
  Receivers are not required to have stateful interfaces.

With these features the protocol can support both simple and complex existing AER devices in a variety of use cases while suggesting the form of a built-in (on-chip) interface for future neuromorphic chips. Paired with an appropriate higher-level software interface (e.g. MUSIC [3]) which would allow ready exchange of spikes between heterogeneous simulators, this system would allow for one unified communications layer making interlinking and using spiking neural simulators as simple as, for example, using stream-based I/O in C/C++. While the high-level API still remains to be determined, we have developed an underlying base representation for these 'neural streams'.

Acknowledgements. This work has been partially supported by the European Union under grant nos. FP7-604102 (HBP), FP7-287701 (BrainScales-Extension), and ERC-320689 (BIMPC), by EPSRC grant EP/J004561/1 (BABEL) and by EPSRC grant EP/G015740/1 (BIMPA).

References

1. Adams, S.V., Rast, A.D., Patterson, C., Galluppi, F., Brohan, K., Pérez-Carrasco, J.-A., Wennekers, T., Furber, S., Cangelosi, A.: Towards real-world neurorobotics: integrated neuromorphic visual attention. In: Loo, C.K., Yap, K.S., Wong, K.W., Beng Jin, A.T., Huang, K. (eds.) ICONIP 2014, Part III. LNCS, vol. 8836, pp. 563–570. Springer, Heidelberg (2014)
2. Davison, A.P., et al.: PyNN: a common interface for neuronal network simulators. In: Frontiers in Neuroinformatics 2.11, January 2009
3. Djurfeldt, M., et al.: Run-time interoperability between neuronal network simulators based on the MUSIC framework. Neuroinformatics **8**(1), 43–60 (2010)
4. Fasnacht, D.B., Whatley, A.M., Indiveri, G.: A serial communication infrastructure for multi-chip address event systems. In: Proceedings of 2008 IEEE International Symposium Circuits and Systems (ISCAS2008), pp. 648–651 (2008)
5. Furber, S.B., et al.: Overview of the SpiNNaker system architecture. IEEE Trans. Comput. **62**(12), 2454–2467 (2013)
6. Indiveri, G., Chicca, E., Douglas, R.: A VLSI array of low-power spiking neurons and bistable synapses with spike-timing dependent plasticity. IEEE Trans. Neural Netw. **17**(1), 211–221 (2006)
7. Lazzaro, J., et al.: Silicon auditory processors as computer peripherals. IEEE Trans. Neural Netw. **4**(3), 523–528 (1993)
8. Lichsteiner, P., Posch, C., Delbrück, T.: A 128 × 128 120 dB 15 ms latency asynchronous contrast vision sensor. IEEE J. Solid-State Circ. **43**(2), 566–576 (2008)
9. Rast, A.D., et al.: A location-independent direct link neuromorphic interface. In: Proceedings of 2013 International Joint Conference Neural Networks (IJCNN2013), pp. 1967–1974 (2013)
10. Schemmel, J., et al.: A wafer-scale neuromorphic hardware system for large- scale neural modeling. In: Proceedings of 2010 IEEE International Symposium Circuits and Systems (ISCAS2010), pp. 1947–1950 (2010)
11. Serrano-Gotarredona, R., et al.: CAVIAR: A 45k Neuron, 5M Synapse, 12G Connects/s AER hardware sensory-processing-learning-actuating system for high-speed visual object recognition and tracking. IEEE Trans. Neural Netw. **20**(9), 1417–1438 (2009)

A Methodological Approach for Big Data Security: Application for NoSQL Data Stores

Houyem Heni[(✉)] and Faiez Gargouri

Miracl Laboratory, Higher Institute of Computer Science and Multimedia,
Sfax University, Sfax, Tunisia
houyem.heni@gmail.com, faiez.gargouri@isimsf.rnu.tn

Abstract. Securing big data is among the major challenges for information suppliers. Indeed, the lack of a robust methodological solution dedicated to the big data security makes the issues of privacy and personal data protection major research areas. In fact, many studies and works have dealt with the meeting between privacy and big information. Because of the huge volume of data that spread between social networks and clouds Application, we have to think about an approach that addresses enhancing data security in databases, specifically in the context of NoSQL environments. This paper introduces a new methodological approach for big data security based on data fragmentation.

Keywords: Big data · Security · Sensitive data · NoSQL · Mongo DB · Data fragmentation

1 Introduction

The quantity of heterogeneous data available on the web and in big databases is growing exponentially. According to research conducted by IDC, the volume of digital data will increase from 40 to 50 % every year, reaching a total volume of 40 bytes zeta-2020. As shown in [1], private and governmental organizations are increasingly gathering and maintaining vast amounts of data known as big data which often include sensitive personally identifiable information. Considering the three main characteristic of big data knowing as 3Vs: Volume Variety and velocity, we will focus on the security solution for big data. The privacy of the latter and the identity protection is, now, considered as a very important issue. Consequently, securing big data against some damages and information leakage is a critical goal, that's why the meeting between big data and security represents one of the biggest challenges for digital information suppliers. Because sensitive data is everywhere, the importance of data security has been growing during the last few years. In fact, some big business companies have to start investing in big data security such as IBM[1]. It could enlarge these perspectives on security while implementing 6,000 IBM Security experts worldwide, 3,000 IBM security patents, and 4,000 IBM managed security client services worldwide. Although other survey has also

---

[1] http://www.ibm.com/systems/data/flash/lv/pdf/venkateshSadayappan_IBMForum2013.pdf.

© Springer International Publishing Switzerland 2015
S. Arik et al. (Eds.): ICONIP 2015, Part IV, LNCS 9492, pp. 685–692, 2015.
DOI: 10.1007/978-3-319-26561-2_80

demonstrated that efficiently securing sensitive data has become an imperative concern. Following [2], a survey on enterprise data security conducted by Independent Oracle Users Group (IOUG) in 2012 shows that only 50 % of the inquired companies increased their investment in IT security. These factors include variables; such as large-scale cloud infrastructures, diversity of data sources and formats as well as streaming nature of data acquisition. Consequently, traditional security mechanisms are tailored to secure small-scale static [3]. This meeting between big data disturbed stores and security represents one of the biggest challenges for suppliers of digital information. The remainder of this paper is organized as follows. Section 2 depicts data security solutions. In Sect. 3, we discuss the related works that deal with different security techniques. Section 4 is devoted to a study that meets the big data and NoSQL. Section 5 represent an overview of our approach. Finally, we end up this paper with a conclusion.

2   A Survey on Big Data Security Techniques

When talking about information security, there are three things to keep in mind: Confidentiality, Integrity, and Availability, or CIA. Moreover As [4], there are other Concepts related to people who use that information. For instance, authentication is proving that a user is the person he or she claims to be. Besides, authorization is the act of determining whether a particular user or computer system has the right to carry out a certain activity [5]. In the following section, we present an overview of the most relevant works proposed in the literature related to the big sensitive information security.

Online big data applications are vulnerable to theft of sensitive information because adversaries can access to private data that can be captured or leaked by curious administrators. In fact, the problem of privacy and sensitive data protection have frequently been one of the major concerns for suppliers of vast and complex information databases during the last few years, especially in the context of social networking. As depicted in [1], the various security matters and upcoming challenges are reviewed in terms of standards; such as PCI-DSS, ITIL, and ISO- 27001/27002. In our approach, we will consider a survey on Big Data security. Security and privacy issues are magnified by the three V's of big data: Velocity, Volume, and Variety. These factors include attributes like large-scale cloud infrastructures, heterogeneous formats, streaming nature of data acquisition and the increasingly big volume of inter-cloud sharing[2]. As the Oracle big data appliance known as OBDA the Database, customers have benefited from a rich set of security features: encryption, redaction, data masking, database firewall and label-based access control. Oracle wants similar capabilities with their Hadoop cluster. In [7] the OBDA believe that an essential secret of an optimized appliance is protecting its data. Therefore, by default, the BDA delivers the "AAA of security": authentication, authorization and auditing for Hadoop cluster which represent the most important element in big data environment. Oracle, also, adds encryption of data-at-rest on Big Data Appliance. This encryption can be done in two modes. The first one leverages the

---

[2] https://cloudsecurityalliance.org/media/news/csareleases-the-expanded-top-ten-big-data-security-privacychallenges/.

Trusted Platform Module (TPM) on the motherboard to provide a key to encrypt the data on disk. This mode does not require a password or pass phrase but relies on the motherboard. The second mode leverages a passphrase, which in turn will be used to generate a private-public key pair generated with Open SSL.

3 Related Works

We notice that the harvesting of big data sets and the use of cloud applications imply security concerns. The tasks of ensuring security for sensitive and personal data and the privacy protecting at an age where big data has been transmitted and shared with high speed around the world, have become harder. Lots of research works have been done to treat these mass data security. In [8], the authors enumerate many benefits of Privacy in area of big data. They believe that Privacy and data protection laws are premised on individual control over information and on principles; such as data minimization and purpose limitation. However, they did not give a perfect solution for big data security. Given that the cloud is rich in big data thanks to its applications and distributed layers and it is important to know the security solution used in this dematerialized world as shown in [9], the authors proposed a combined approach which provides a way to protect the data and check the integrity and authentication by following the best possible industry mechanisms. This technique introduces the fragmentation of data into different sections: Index builder, 128-bit SSL encryption, message authenticate code and a double authentication of user; one by owner and the other by cloud and verification of the owner's digital signature. It, also, provides availability of data by surpassing many issues like data leakage, tampering of data and data encryption. In fact, data fragmentation was applied as a perfect solution to more ensure security in data bases. According to [1], fragmentation and encryption provide protection of data in storage. The authors address these issues by proposing a solution to enforce data collections privacy that combines data fragmentation with encryption. They model privacy requirements as Confidentiality constraints expressing the sensitivity of attributes and their associations. Then, they use encryption as an underlying measure for making data unintelligible, while exploiting fragmentation as a way to break sensitive associations among attributes. In [10], they introduced the problem of privacy-aware data partitioning; namely the problem of split-ting a sensitive dataset amongst untrusted parties. In their work, the authors present SPARSI, a theoretical framework that allows us to formally define the problem, as an optimization of the tradeoff between the utility derived by publishing the data and the maximum information disclosure incurred to any single adversary.

4 Big Data and NoSQL

The emergence of cloud computing and its distributed applications creates the need for huge bases to store the large amounts of structured, semi-structured and unstructured data. When we talk about big data and its storage we think about NoSQL data bases.

NoSQL[3] is not necessarily to be taken as "not", but "not only" SQL. It offers solutions that can complement conventional RDBMS solutions. Non-relational data stores have not yet reached security infrastructural maturity. NoSQL Databases were built to tackle different obstacles brought about by the analytics world; hence security was never part of the model at any point of its design stage. According to [6] Developers using NoSQL databases usually embed security in the middleware. Furthermore, NoSQL databases do not provide any support for enforcing it explicitly in the database. However, clustering aspect of NoSQL databases poses additional challenges to the robustness of such security practices. In fact, there are four types of NoSQL databases. In [11] the authors draw a comparison between these databases as follows:

- Key/value stores: store items as alpha-numeric identifiers (keys) and associate values in simple, standalone tables. Requests can only be performed against keys, not values.
- *Column stores*: do not store data in tables but in massively-distributed architectures. In column stores, each key is associated with one or more attributes (columns). The data stored is based on the sort order of the column family.
- *Document data bases*: are designed to manage and store documents encoded in a standard data exchange format; such as XML. Unlike the simple *key-value stores*, the value column contains semi-structured data, specifically attribute/value pairs. The number and type of recorded attributes can vary from row to row. Both keys and values are searchable. Document databases are good for storing and managing big data-size collections of literal documents like text documents, email messages, and XML documents.
- *Graph databases*: replace relational tables with structured relational graphs of interconnected key value pairings. The graphs are represented as an object-oriented network of conceptual objects (nodes), relationships (edges) and properties (object attributes expressed as key-value pairs).

We frame our work in the context of NoSQL databases. They are by far the most common solution for the management of the data privacy subject We are, also, convinced that document oriented databases are the suitable ones, because they are Scalable, schema-less and very flexible. Besides, these databases are intended to Store structured, semi-structured and unstructured data.

5   Overview of the Approach

As we mentioned above, one of the most important challenges of big data security is first the protection of sensitive data management and second how to determine appropriate methods and algorithms to ensure security. Big data has a disturbed technology. In fact, the data in big data environment has to go through different processing levels. Thus, the security mechanism should be efficient and provided at each step. Data should not succumb to the attackers trying to retrieve or tamper with it and even the big data provider should not be able to harm the data in any possible manner, because big data

---

[3] http://searchdatamanagement.techtarget.com/definition/NoSQL-Not-Only-SQL.

suppliers cannot be trusted with data of high sensitivity. For this reason, it is obvious that the proposed approach has been designed by keeping all these things in mind. Besides, in comparison to prior works, it provides all these required measures to protect data in a very efficient and organized manner. This approach will be applied to MongoDB. In order to explain this approach, Fig. 1 describes this method which consists of four phases:

1. Pre-processing phase: this phase is use to group automatically big data in NoSQL database.
2. Search and identify sensitive data phase: In this step, we can pass by a Learning Algorithms based on neural networks to identify sensitive data
3. Data Fragmentation phase: it is used to provide a very high security. We think about Data Fragmentation to better protect the data that has a sensitive and personal aspect. In this step, we must propose our algorithm based on data division followed by an encryption stage for the fragmented part.
4. Data Reconstruction phase

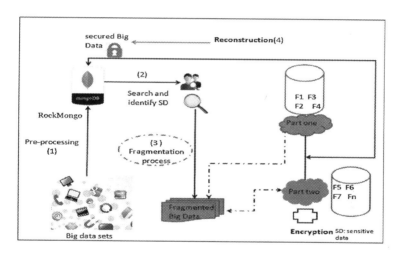

Fig. 1. Data security scenario based on data fragmentation

5.1   The Pre-treatment Phase

This step represents our preprocessing works. The main goal is to group automatically Big Data (structured, semi- structured and unstructured data) in a NoSQL database. Then, we feed MongoDB with heterogeneous data because it is known for its simplicity and high performance in processing big amounts of data. Since it is distributed in treatment, the MongoDB will facilitate the work of fragmentation. In this pre-processing step, we choose the Rock mongo[4] which is a MongoDB management tool, written in

---

[4] http://docs.mongodb.org/ecosystem/tools/administration-interfaces/

PHP 5 and dedicated for administering a mongo database to feed our data base and to insert heterogeneous data and save as json format in our document-oriented database.

5.2   Find and Identify Sensitive Attributes Phase

This phase deals with mechanisms and methods of classifying and identifying the sensitive attributes sets from collections stored in MongoDB. In many studies, the authors focus their work on attributes sets. For instance, in [1, 12], they worked on attributes sets and they split them into fragmentations sets f1...fn. In our approach, we will split the attribute itself after being fragmented.

During this step, we will categorize our data collections, i.e. group attributes of class from the collections. Then, we propose clustering algorithm. The used algorithm is a neural network which takes as input the MongoDB collection sets and is given as output many categories so each attributes will be classified in an appropriate category which allows two classes known as sensitive data and ordinary data. After this process, we will go through identification to highlight the difference between ordinary attributes and sensitive attributes. In the literature, there are many software used to discover sensitive data residing whatever in computers or in the massive parallel bases or even in the clouds. But, we believe that inventing a learning algorithm will better serve our request, especially when it comes to a huge volume of heterogeneous data. This learning algorithm aims at identifying sensitive data among ordinary information, i.e. if it comes from a succession of figures or address mail or secret code.

5.3   Sensitive Data Fragmentation Phase

This phase is summarizing the scenario of big data security. In many large-scale solutions, data is divided into separate partitions that can be managed and accessed separately. The partitioning strategy must be chosen carefully to maximize the benefits while minimizing adverse effects.This phase is further divided into two sub-sections.

Data Fragmentation

After identifying and searching sensitive data, each sensitive attribute must be split in two parts. As shown in [13], Data fragmentation is a process of division or mapping database where the database is broken down into a specific number of parts. Then, it is stored in the site or units of different computers in a data network. When applying fragmentation, data must meet several conditions to obtain optimal fragmentation. Below we illustrate some data fragmentation principles:

Completeness: a unit of data that is still in the main part of the relationship. Then, the data must be in one fragment. When there is a relation, the distribution of the data must be an integral part of the relationship.

Reconstruction: an original relation can be reused or combined return of a fragment. When it breaks down, data is still possible to be combined again with no change in the structure of data.

Disjointness: data within the fragment should not be included in the other fragments in order to avoid redundancy of data, except for primary key attributes of vertical fragmentation. Fragmenting means splitting sets of attributes so that they are not visible together; that is to say, the associations among their values are not available without access to the encryption key.

Data Encryption

As we mentioned above, to better ensure the security of sensitive attributes and after having split it in two parts, we will save the second part in somewhere else to prevent the hackers to find the full information. Thus, we think of a classic approach among the security algorithms. We choose encryption as a perfect solution. As [14, 15] demonstrate, the encryption is the best method to protect sensitive data at the database level while maintaining high database performance. Encryption is applied after the fragmentation sets level. Finally, the encryption key is given to the authorized users needing to access the information. Users that do not know the encryption key as well as the storing server are able neither to access sensitive information nor to reconstruct the sensitive associations. To better ensure security, we will apply an encryption algorithm after the Fragmentation process.

5.4   Data Reconstruction

Relation between sensitive attributes after fragmentation can be reused. When it breaks down, data is still possible to be combined again with no change in the data structure. It is worth-noting that in data reconstruction, there are many constraints; such as choosing the perfect fragment data to the adequate query.

6   Conclusion

In the early 21st century that witnessed the explosion of Big Data, the security of data with sensitive content is still considered as a very important and effective axis of research. This paper presents a big data protection scenario. The objective of this approach is to overcome the threads and leaks in big data environment, especially for NOSQL data bases. In this context, we use the data fragmentation combined with a data Encryption applied in MongoDB.

As future works, we envisage to automatically feed NoSQL database, to define Confidentiality constraint for NoSQL data bases, to implement learning algorithm, to identify sensitive attributes and to apply a fragmentation process appropriate to MongoDB collections.

References

1. Ciliani, V., De Capitani, S., Vimercati, D., Foresti, S.: Combining fragmentation and encryption to protect privacy in data storage. ACM **13**(3), 1–30 (2010)

2. McKendrick, J., IOUG Enterprise Data Security Survey: Closing the Security Gap, the Independent Oracle Users Group (IOUG) Security report, November 2012
3. Cloud security appliance, the-expanded-top ten-big-data-security-privacy challenges, CSA White Paper (2010)
4. dos Santos, R.J.R.: enhancing data security in data warehousing, Ph.D. thesis in Information Sciences and Technology, supervised by Professor Jorge Bernardino and Professor Marco Vieira, The University of Coimbra, February 2014
5. Pesante, L.: Introduction to Information Security. Carnegie Mellon University, Pittsburgh (2008)
6. Zvarevash, K., Mutandavari, M., Gotora, T.: A survey of the security use cases in big data. Int. J. Innov. Res. Comput. Commun. Eng. 2(5), 4259–4266 (2014)
7. Oracle Corporation: New Big Data Appliance Security Features. Oracle White Paper, November, 2013
8. Tene, O., Polonetsky, J.: Privacy in the age of big data: a time for big decisions, 64 Stan. L. Rev. Online 63, 2 February 2012
9. Sood, K.S.: A combined approach to ensure data security in cloud computing. J. Netw. Comput. Appl. 35(6), 1831–2012 (2012)
10. Rekatsinas, T., Deshpande, A., Machanavajjhala, A.: A SPARSI: partitioning sensitive data amongst multiple adversaries. PVLDB 6(13), 1594–1605 (2013)
11. Moniruzzaman, A.B.M., Hossain, S.A.: NOSQL database: new era of databases for big data analytics- classification, characteristics and comparison. Int. J. Database Theory Appl. 6(4) (2013)
12. Vinogradov, S., yak, A.P.: Evaluation of data anonymization tools. In: The Fourth International Conference on Advances in Databases, Knowledge, and Data Applications (2012)
13. Navaz, S.: A.S., Prabhadevi, C., Sangeetha, V.: Data grid concepts for data security in distributed computing (0975 – 8887). Int. J. Comput. Appl. 61(13), 6–11 (2013)
14. Oracle Corporation, "Security and the Data Warehouse", Oracle White Paper, April 2005
15. Oracle Corporation, Oracle Advanced Security Transparent Data Encryption Best Practices, Oracle White Paper, July 2010

Author Index

Printed in the United States
By Bookmasters